Supersymmetry

NATO ASI Series

Advanced Science Institutes Series

A series presenting the results of activities sponsored by the NATO Science Committee, which aims at the dissemination of advanced scientific and technological knowledge, with a view to strengthening links between scientific communities.

The series is published by an international board of publishers in conjunction with the NATO Scientific Affairs Division

A Life Sciences **B Physics**	Plenum Publishing Corporation New York and London
C Mathematical and Physical Sciences	D. Reidel Publishing Company Dordrecht, Boston, and Lancaster
D Behavioral and Social Sciences **E Engineering and Materials Sciences**	Martinus Nijhoff Publishers The Hague, Boston, and Lancaster
F Computer and Systems Sciences **G Ecological Sciences**	Springer-Verlag Berlin, Heidelberg, New York, and Tokyo

Recent Volumes in this Series

Volume 118—Regular and Chaotic Motions in Dynamic Systems
edited by G. Velo and A. S. Wightman

Volume 119—Analytical Laser Spectroscopy
edited by S. Martellucci and A. N. Chester

Volume 120—Chaotic Behavior in Quantum Systems: Theory and Applications
edited by Giulio Casati

Volume 121—Electronic Structure, Dynamics, and Quantum Structural Properties of Condensed Matter
edited by Jozef T. Devreese and Piet Van Camp

Volume 122—Quarks, Leptons, and Beyond
edited by H. Fritzsch, R. D. Peccei, H. Saller, and F. Wagner

Volume 123—Density Functional Methods in Physics
edited by Reiner M. Dreizler and João da Providência

Volume 124—Nonequilibrium Phonon Dynamics
edited by Walter E. Bron

Volume 125—Supersymmetry
edited by K. Dietz, R. Flume, G. v. Gehlen, and V. Rittenberg

Series B: Physics

Supersymmetry

Edited by

K. Dietz

R. Flume

G. v. Gehlen

and

V. Rittenberg

Institute of Physics
University of Bonn
Bonn, Federal Republic of Germany

Plenum Press
New York and London
Published in cooperation with NATO Scientific Affairs Division

Proceedings of a NATO Advanced Study Institute on
Supersymmetry,
held August 20-31, 1984,
in Bonn, Federal Republic of Germany

Library of Congress Cataloging in Publication Data

NATO Advanced Study Institute on Supersymmetry (1984: Bonn, Germany)
Supersymmetry.

(NATO ASI series. Series B, Physics; v. 125)
"Proceedings of a NATO Advanced Study Institute on Supersymmetry, held August 20–31, 1984, in Bonn, Federal Republic of Germany"—T.p. verso.
Includes bibliographies and index.
1. Supersymmetry—Congresses. 2. Supergravity—Congresses. I. Dietz, K. (Klaus), 1934- . II. North Atlantic Treaty Organization. Scientific Affairs Division. III. Title. IV. Series.
QC793.3.S9N36 1984 530.1 85-9395
ISBN 0-306-42012-0

A Division of Plenum Publishing Corporation
233 Spring Street, New York, N.Y. 10013

Printed in the United States of America

PREFACE

Ten years ago the Summer Institute on "Trends in Particle Theory" was held in Bonn. Among other subjects J. Wess spoke on his work with B. Zumino on "Fermi-Bose-Supersymmetry". In the meantime this subject has developed into one of the most exciting fields of research and the Bonn Summer Institute (August 20 - August 31 1984) was devoted exclusively to supersymmetry.

The main aim of the Summer Institute was to provide a comprehensive introduction for mainly young post-doctoral theoretical physicists to the currently most discussed subjects in supersymmetry and supergravity. The intention was to emphasize the pedagogical presentation. The seminars given in addition to the introductory lectures treat more specialized subjects and are closer to research reports. Since the topics, ranging from algebraic structures to experimental implications do not arrange easily in a one-dimensional array, the lectures are reproduced in alphabetic order of the lecturers.

The Summer Institute was made possible by the NATO Advanced Study Programme, additional support was granted by the Bundesministerium für Forschung und Technologie and the Ministerium für Wissenschaft und Forschung of Nordrhein-Westfalen. We gratefully acknowledge their cooperation.

It is a pleasure to thank the lecturers and seminar speakers for the effort invested in the presentation of the lectures and in their collaboration in preparing the written notes.

The smooth running of the School would not have been possible without the efficient and devoted effort of our Conference secretaries Mrs. D. Fassbender and Mrs. M. Künkel who took care of much of the technical work involved. We are also grateful to the members of the Theoretical Physics group of the Bonn University who helped in the organization and in resolving the various problems arising during the course of the School.

Bonn, January 1985

K. Dietz
R. Flume
G. v. Gehlen
V. Rittenberg

CONTENTS

LECTURES

SEMINARS

SUPERSYMMETRY AND INDEX THEORY

Luis Alvarez-Gaumé

Lyman Laboratory of Physics
Harvard University
Cambridge, Massachusetts 02138, U.S.A.

I. INTRODUCTION

It is a remarkable fact that the Atiyah-Singer index theorem[1] has provided theoretical physicists with a very powerful tool to obtain qualitative (and sometimes quantitative) information about a variety of physical systems. Among the best known examples in which the interplay between index theory and field theory has been useful, we have the ABJ[2] anomaly for gauge singlet currents, whose structure and normalization can be derived from the classical index theorem, the general restrictions obtained on fermion quantum numbers in Kaluza-Klein theories,[3] the understanding of the non-Abelian anomaly in terms of the index theorem for families of Dirac operators,[4] etc.

These lectures are mainly concerned with the fact that many of the ingredients and results in index theory have very simple analogs in supersymmetric quantum mechanical systems so that using elementary quantum mechanics one can obtain new proofs of the Atiyah-Singer index theorem and related results.

In order to exhibit such connections, in Section II we start by reviewing some simple ideas on index theorems and characteristic classes. Then in Section III we outline some simple results on fermion functional integrals which are useful later. In Section IV, we start our analysis of those supersymmetric quantum mechanical systems which appear naturally in dealing with index problems. Section V presents the proof of the classical index theorem. Section VI contains the derivation of the character valued index theorem. Section VII will present some applications of the methods developed in these lectures for the computation of anomalies.

Section VIII contains the conclusions and outlook. The presentation of these lectures is intended to be self-contained. Those readers already familiar with the basic results of differential geometry and characteristic polynomials and with the holomorphic representation of the functional integral formulation of quantum mechanics can skip Sections II and III and start with the study of these supersymmetric systems which play a central role in our discussion (Section IV). The references have been selected in order to provide further details about some of the topics to be presented, and do not represent a complete list of the contributions to the different subjects. I would therefore like to apologize to those authors whose work has not been included in the reference list.

II. FORMALISM AND GENERALITIES[5]

Throughout these lectures (M_n, g) will denote an n-dimensional Riemannian manifold with metric g_{ij}, $i,j = 1, \ldots, n$. M_n will always be compact and without boundary, and endowed with the usual Riemannian connection. $e^a = e^a{}_\mu dx^\mu$ represents the vielbein one forms, where the Latin indices a, b, c, ... refer to orthonormal frame indices. The basic geometrical objects which form the framework of Riemannian geometry are the spin connection ω, the curvature Ω and the torsion T. In terms of the vielbein form e^a, the line element ds^2 can be written as

$$ds^2 = g_{\mu\nu} dx^\mu \otimes dx^\nu = \eta_{ab}\, e^a{}_\mu\, e^b{}_\nu\, dx^\mu \otimes dx^\nu = \eta_{ab}\, e^a \otimes e^b \tag{2.1}$$

η_{ab} is a flat metric with either Euclidean (++...+) or Minkowskian (-++...+) signature depending on the signature of the manifold. The vielbein $e^a{}_\mu$ and its inverse $E^\mu{}_a$, $E^\mu{}_a = \eta_{ab}\, g^{\mu\nu} e^b{}_\nu$ are used to change tensor quantities referred to coordinate frames to tangent frames and vice-versa.

From (2.1) it is clear that the orthonormal tangent frames specified by $e^a{}_\mu$ are defined only up to local frame rotations. If e^a $(a = 1,n)$ is an orthonormal frame basis, then so is $(L^{-1})^a{}_b\, e^b$, with $L^a{}_b$ being an arbitrary space-time dependent element of SO(n) or SO(1,n-1) depending on the signature of M. The vielbein formalism, using orthonormal as well as coordinate frames, is only absolutely necessary for theories with fields of half-integer spin, since only these fields cannot be represented as ordinary tensor fields $T^{\alpha_1 \ldots \alpha_r}$ ($GL(n,\mathbb{R})$ is simply connected and has no spinor representations). Spinors must instead be introduced as spinor representations of the orthonormal frame rotation group.

To define the standard formulae of Riemannian geometry, we must introduce a connection which defines parallel transport of geometrical objects on the manifold. For instance, once we have erected a system of orthonormal frames on M, the spin connection defines the parallel transport of the orthonormal tangent frames along the manifold. Thus the spin connection is a one form taking values on the Lie algebra of the frame rotation group $\omega^a{}_b \equiv \omega^a{}_{b,\mu} dx^\mu$. Under a local frame rotation, $\omega^a{}_b$ transforms like an ordinary gauge connection:

$$\omega \to L^{-1}(\omega + d)L \tag{2.2}$$

with matrix indices suppressed. For notational simplicity we will assume that M_n has Euclidean signature, so that the tangent space group is SO(n), and $\eta_{ab} = \delta_{ab}$. The extension of the formalism to other signatures is straightforward. Given the vielbein and the connection, we can define the curvature and torsion two-forms. They simply follow if one tries to construct quantities from de^a and $d\omega^a{}_b$ which are covariant under local frame rotations. The torsion and curvature are respectively defined as

$$T^a = de^a + \omega^a{}_b \wedge e^b \equiv de + \omega e \tag{2.3}$$

$$\Omega^a{}_b = d\omega^a{}_b + \omega^a{}_c \wedge \omega^c{}_b \equiv d\omega + \omega\omega\ . \tag{2.4}$$

In (2.2)-(2.4), d represents the exterior derivative on forms, i.e. the operator $d = dx^\mu \wedge \partial/\partial x^\mu$ which produces a (p+1)-form out of a p-form, and $\wedge$ represents the exterior product of forms. Unless otherwise stated, the product of forms will always mean the wedge product, so that the wedge symbol will be suppressed whenever no confusion may arise in interpreting the formula. Thus ω^2 simply means the matrix $\omega^a{}_c \wedge \omega^c{}_b$, etc. The Bianchi identities are now a straightforward consequence of (2.3,2.4) and the fact that $d^2 = 0$.

$$dT = d^2e + d(\omega e) = d\omega e - \omega de = \Omega e - \omega T$$

$$dT + \omega T = \Omega e \tag{2.5}$$

and

$$d\Omega = d\omega\omega - \omega d\omega = \Omega\omega - \omega\Omega\ . \tag{2.6}$$

To any Riemannian manifold with metric there is associated a unique connection, called the Riemannian connection, for which the metric is covariantly constant, and the torsion vanishes. Using $\omega_{ab} = \eta_{ac}\omega^c{}_b$ and $de^a = \frac{1}{2}\xi^a{}_{bc}\, e^b e^c$, the condition $T^a = 0$ implies

$$\xi_{a,bc} = \omega_{ab,c} - \omega_{ac,b} \tag{2.7}$$

which can be solved by taking cycle permutations to give

$$\omega_{ab,c} = \tfrac{1}{2}(\xi_{a,bc} + \xi_{b,ca} - \xi_{c,ab})\ . \tag{2.8}$$

Finally, the definition of covariant derivatives is identical to the gauge theory case. For example, let Σ^a be a vector valued one-form, then the covariant derivative of Σ^a is defined by

$$D\Sigma^a = d\Sigma^a + \omega^a{}_b\, \Sigma^b \tag{2.9}$$

and similarly for other tensor valued forms.

Since we will be interested in index problems involving both gauge and gravitational fields, let us quickly explain the notation we will use: Let G be the gauge group, and $\hat{G}$ its Lie algebra; then the gauge connection is a one-form taking values on $\hat{G}$: $A = A_\mu dx^\mu$. We will always assume that A is antihermitian. Under a finite gauge transformation g(x), A transforms as

$$A \to g^{-1}(A + d)g \tag{2.10}$$

and the gauge field strength is a two-form defined by

$$F = dA + A^2 \ . \tag{2.11}$$

So far, we have been dealing exclusively with the local formulae of Reimannian geometry and gauge theories. Since we will eventually be interested in global topological properties of M_n and of the gauge fields defined on M_n, let us briefly review how to describe the global properties of M_n. Let M_n be the manifold we are interested in, and let B be a principal bundle with base space M_n and the group G as fibre. A curved M_n in general cannot be specified by a single coordinate patch, and requires instead a covering by several coordinate patches U_i, each diffeomorphic to $\mathbb{R}^n$, with well defined transition rules on their overlap. The gauge field (a connection on the principal bundle B) is defined separately on each patch with the requirement that gauge fields A_{U_i} and A_{U_j} on different patches U_i, U_j be related by a gauge transformation $A_{U_j} = g^{-1}_{ij}(A_{U_i} + d)g_{ij}$ on any non-empty overlap $U_i \cap U_j$. The topological information is thereby encoded in the transition functions g_{ij}. For example, let $M_n = S^n$, the n-dimensional sphere, and consider a gauge group G. The simplest covering consists of two coordinate patches, the upper and lower hemispheres D_+ and D_-, intersecting on an equatorial S^{n-1}. Letting $A_\pm$ denote the gauge field on $D_\pm$, we see that the bundle is defined by a single transition function g_{+-}, which in turn is a map from S^{n-1} into G. The bundle may thus be non-trivial if and only if $\pi_{n-1} G \neq 0$, i.e. if the (n-1)'st homotopy group of G is non-trivial. A familiar example of this construction is the instanton bundle on a compactified space-time manifold S^4. In this case the topological non-triviality is encoded by $\pi_3 G$ which is Z for all compact simple Lie groups.

From an elementary point of view, characteristic classes are local forms on M_n, constructed from the curvature two-forms, whose integrals over the manifold are sensitive to the existence of non-trivial topology, i.e., to the presence of homotopically non-trivial transition functions. The characteristic classes are closed forms (a p-form Ω_p is closed if $d\Omega_p = 0$), but not globally exact. By Poincaré's lemma, their closure property means that they can be written locally as exact forms $\alpha = d\phi_i$ on each patch U_i, so that

Stoke's theorem $\int_B d\phi = \int_{\partial B}\phi$ ensures that their integrals depend only on the transition functions. If the forms α_i piece together to form a global form, then Stoke's theorem ensures that their integrals over a manifold M_n without boundary vanish.

We now outline a general procedure which produces characteristic classes. (More details, and references to the original literature can be found in Ref. 5.) Let Ω be a matrix valued two-form taking values in $\hat{G}$, and let $P(\Omega)$ be a polynomial in Ω satisfying the invariance property

$$P(g^{-1}\Omega g) = P(\Omega) \tag{2.12}$$

for any $g \in G$. Ω will be taken to be the gauge field strength F or the curvature Ω so that G is respectively the gauge group or the frame rotation group. One can now show that any polynomial satisfying (2.12) is (i) closed: $dP = 0$, and (ii) has integrals which are topologically invariant (i.e. are invariant under deformations of the gauge field or connection and depend only on the transition functions). To make the discussion simpler, we note that it suffices to consider polynomials of the form $P_m = \mathrm{Tr}\,\Omega^m$, since all other invariant polynomials can be constructed as sums of products of the P_m's. The closure of P_m follows from the Bianchi identity:

$$dP_m = d\,\mathrm{Tr}\,\Omega^m = m\,\mathrm{Tr}\,d\Omega\,\Omega^{m-1} = m\,\mathrm{Tr}(d\Omega + \omega\Omega - \Omega\omega)\Omega^{m-1} = 0 \tag{2.13}$$

where we have used the cyclicity of the trace. This is legitimate because Ω is a two-form and thus behaves like an even element of a Grassmann algebra. To show that the integral of P_m over M_n is independent of the connection, consider two connections ω_0, ω_1 with the same transition functions, and define the interpolation

$$\begin{aligned} \omega_t &= \omega_0 + t(\omega_1 - \omega_0) \\ \Omega_t &= d\omega_t + \omega_t^2 , \qquad 0 \leqslant t \leqslant 1 . \end{aligned} \tag{2.14}$$

Using the relation $\partial\Omega_t/\partial t = d(\omega_1 - \omega_0) + \omega_t(\omega_1 - \omega_0) + (\omega_1 - \omega_0)\omega_t \equiv D_t(\omega_1 - \omega_0)$ gives

$$\begin{aligned} \frac{\partial}{\partial t} P_m &= \frac{\partial}{\partial t}\mathrm{Tr}\,\Omega_t^m = m\,\mathrm{Tr}\frac{\partial\Omega_t}{\partial t}\Omega_t^{m-1} \\ &= m\,\mathrm{Tr}\,D_t(\omega_1 - \omega_0)\Omega_t^{m-1} \\ &= m\,d\,\mathrm{Tr}(\omega_1 - \omega_0)\Omega_t^{m-1} , \end{aligned} \tag{2.15}$$

where in the last step we have used the chain rule and the Bianchi identity $D_t\Omega_t = d\Omega_t + \omega_t\Omega_t - \Omega_t\omega_t = 0$. Integrating in t from 0 to 1 gives

$$P_m(\Omega_1) - P_m(\Omega_0) = m\,d\int_0^1 dt\,\mathrm{Tr}\,(\omega_1 - \omega_0)\Omega_t^{m-1} , \tag{2.16}$$

which by Stoke's theorem implies $\int_M P_m(\Omega_1) = \int_M P_m(\Omega_0)$ when M has no boundary. We will in particular consider bundles with G = U(n) or SO(n). Let's therefore analyze some of the most useful characteristic classes for these cases. When G = U(n), the two-form Ω is an antihermitian matrix, and it can formally be diagonalized, with eigenvalues ix_j, j = 1, n. Then any polynomial satisfying (2.12) can be written in terms of symmetric functions of the x's. The symmetric and homegeneous function of the x's are known as the Chern classes, and they are generated by

$$c(\Omega) = \det\left(1 + \frac{1}{2\pi}\Omega\right) = 1 + c_1(\Omega) + c_2(\Omega) + \dots \tag{2.17}$$

Thus

$$(2\pi)c_1(\Omega) = \sum_k ix_k$$

$$(2\pi)^2 c_2(\Omega) = \sum_{i<j} (ix_i)(ix_j)$$

$$(2\pi)^j c_j(\Omega) = \sum_{i_1<\dots<i_j} (ix_{i_1})(ix_{i_2})\dots(ix_{i_j}), \text{ etc.} \tag{2.18}$$

the sum (2.17); $c(\Omega)$ is known as the total Chern class and it is multiplicative with respect to direct sums of bundles $c(E\oplus F) = C(E)c(F)$ where E and F are two U(n) vector bundles over some manifold M. Another useful set of characteristic classes are the Chern characters, which are generated by

$$\mathrm{ch}(\Omega) = \sum_i e^{ix_i/2\pi}\ . \tag{2.19}$$

(The Chern character is additive with respect to the direct sum of bundles: $\mathrm{ch}(E\oplus F) = \mathrm{ch}(E)+\mathrm{ch}(F)$, and plays a central role in the index theorem.)

When the bundle group is SO(n), Ω is an antisymmetric matrix, and can be skew diagonalized:

$$\Omega = \begin{pmatrix} 0 & x_1 & & & & & \\ -x_1 & 0 & & & & & \\ & & 0 & x_2 & & & \\ & & -x_2 & \cdot & & & \\ & & & & \cdot & & \\ & & & & & \cdot & x_n \\ & & & & & -x_n & 0 \end{pmatrix} \tag{2.20}$$

As before, the characteristic polynomials can be written in terms of the "eigenvalues" x_i. The Pontrjagin classes are defined by expanding $p(\Omega) = \det(1 + \Omega/2\pi)$. Since Ω is antisymmetric, only terms containing an even number of Ω's do not vanish:

$$p(\Omega) = \det(1 + \frac{\Omega}{2\pi}) = 1 + p_1 + p_2 + \ldots$$

$$(2\pi)^2 p_1 = \sum_i (x_i)^2$$

$$(2\pi)^4 p_2 = \sum_{i\,j} x_i^2 x_j^2$$

$$(2\pi)^6 p_3 = \sum_{i<j<k} x_i^2 x_j^2 x_k^2 \quad . \tag{2.21}$$

This concludes our short reviews of characteristic polynomials. Before explaining how an index problem is formulated, we first explain in more detail why the characteristic polynomials are intering when studying the topology of a manifold.

In a very loose sense, topology deals with continuous structures in a given space M and characterizes which properties of M do not change under continuous deformation. Suppose for example that one is concerned with the computation of the homotopy groups of a given manifold M_n. The m-th homotopy group of M_n (which we assume connected for simplicity) can be defined in terms of the following equivalence relation: Consider all the continuous maps f from S^m (m-dimensional sphere) into M_n such that for a given fixed point on S^m, y_0, $f(x_0) = y_0 \in M$. One says that two maps f_1, f_2 are homotopic if there is a continuous deformation from f_1 to f_2. More precisely, f_1 is homotopically equivalent to f_2 if there exists a function $g(x,t)$, $0 \le t \le 1$ such that $g(x,0) = f_1(x)$, $g(x,1) = f_2(x)$ and $g(x_0,t) = y_0$ for all t. Under this equivalence relation the maps $f = S^m \to M_n$ fall into equivalence classes, [f], and the m-th homotopy group of M_n is the set of these equivalence classes. Simple examples are provided by the maps from S^1 into S^1. In this case $\pi_1(S^1)$ has an infinite number of elements labeled by Z (integers) and each class is characterized by the number of times that the image of S^1 by the map winds around S^1. We then say that $\pi_1(S^1) = Z$. Similarly, it is not difficult to understand that for any sphere S^n, $\pi_n(S^n) = Z$ and $\pi_p(S^n) = 0$ for $p < n$. This last property is easily seen intuitively by recognizing that any "subsphere" of S^n can be obtained as the interaction of some hyperplane with S^n. To show that $\pi_p(S^n) = 0$, one can just slide the hyperplane up to the north pole of S^n where it would only have a common point with S^n. Since a constant map $f(x) = y_0$ for all x generates the trivial class, we see that $\pi_p(S^n) = 0$. For $\pi_n(S^n)$ we can for example consider the identity map $S^n \to S^n$ which clearly cannot be deformed to the trivial map. The reader is encouraged to show from analogy with the case of maps from $S^1 \to S^1$, that $\pi_n(S^n) = Z$. In general however, the computation of homotopy groups is rather difficult as evidenced by our ignorance even of the complete set of non-vanishing homotopy groups of spheres. Indeed one finds some interesting surprises. For example, $\pi_3(S^2) = Z$, $\pi_4(S^3) = \pi_5(S^3) = Z_2$, $\pi_{2n-1}(S^n) = Z$, $\pi_{n+2}(S^n) = Z_2$, etc. For Lie groups, there are many general results known about their homotopy groups, but once again the general results are not known. Useful results are that for any Lie group G, $\pi_2(G) = 0$, also for any semi-simple Lie group $\pi_3(G) = Z$. Furthermore, $\pi_5 SU(N) = Z$, $N \geq 3$, $\pi_7(SU(N)) = Z$, $N \geq 4$, etc. while π_5 of any other simple Lie group (with the exception of SU(2)) is zero; this turns out to be very useful in the topological interpretation of non-Abelian anomolies.[4]

There is another set of groups which are quite useful in understanding the topological properties of M_n. These are the homology groups, which roughly count the number of "holes" of a given dimension that the manifold may have. We can define the p-th homology group $H_p(M)$ by considering first all the closed submanifolds of dimension p. Out of those, some may be the boundaries of (p+1)-dimensional submanifolds of M_n. Let Z_p be the collection of closed p-manifolds on M_n, and B_p the collection of manifolds which are boundaries. Then $H_p(M)$ is equal to the equivalence classes of closed p-manifolds modulo boundaries: $H_p = Z_p/B_p$. Once these classes are identified, we can define "sums" of these classes with coefficients in different groups. Depending on the group we can distinguish between real homology $H_p(M,\mathbb{R})$, integer homology $H_p(M,Z)$, mod-2 homology $H_p(M,Z_2)$ according to whether different surfaces are summed with real, integer of Z_2 coefficients respectively. Let us consider some simple examples. First we can analyze S^2. The trivial homology group $H_0(S^2)$ has a single element; a zero manifold is just a point, and any two points represent the same class because S^2 is connected. Hence dim $H_0(S^2) = 1$. If we consider $H_1(S^2)$, any circle on S^2 is the boundary of a disc in S^2 (the upper or lower hemispheres in which the circle divides S^2). Thus $H_1(S^2) = 0$. Similarly $H_2(S^2)$ has one class, S^2 itself dim $H_2(S^2) = 1$. The dimension b_p of $H_p(M,\mathbb{R})$ is known as the p-th Betti number of M. The Poincaré polynomial of M, $P_t(M)$ is defined as $\sum_{i=0}^{n} b_i t^i$, and contains all the information concerning the Betti numbers. Since $b_i \geq 0$, $i = 1,n$, a rough measure of the topological complexity of M is given by $P_1(M) = \Sigma b_i$. The next simple example is provided by the 2-torus $T^2 = S^1 \times S^1$. As before, $b_0(T^2) = b_2(T^2) = 1$; now however $b_1(T^2) = 2$. Each of the classes of $H_1(T^2)$ are generated by one of the two circles generating the torus. Thus $P_t(T^2) = 1 + 2t + t^2 = (1+t)^2$. In general, $P_t(T^n) = (1+t)^2$. Similarly, it is not too difficult to show that $P_t(S^n) = 1 + t^n$.

A very remarkable result of the Hodge-DeRham theory[5] is that there is a dual representation of the homology groups in terms of differential forms. Let $\Lambda^p(M)$ be the space of p-forms on M and let g_{ij} be a Riemannian metric $\omega_p = 1/p!\ \omega_{i_1 \ldots i_p}\, dx^{i_1} \wedge \ldots \wedge dx^{i_p}$ where $i_1,\ldots,i_p = 1, n$, the x^i's are coordinates on M, and $\omega_{i_1 \ldots i_p}$ is a totally antisymmetric covariant tensor of rank p (for more details on differential forms see Professor Spindel's lectures in this volume). The exterior derivative d acting on $\Lambda^p(M)$ produces a (p+1) form

$$d\omega_p = \frac{1}{p!}\,\partial_{i_1}\,\omega_{i_2\ldots i_{p+1}}\,dx^{i_1}\wedge\ldots\wedge dx^{i_{p+1}} \ .$$

On the space of p-forms, we can define a scalar product by using the Hodge $*$-operation which transforms p-forms into n-p forms:

$$(*\omega_p)_{j_1\ldots j_{n-p}} = \frac{1}{p!}\,\varepsilon_{j_1\ldots j_{n-p}\,i_1\ldots i_p}\,\omega^{i_1\ldots i_p} \ .$$

Then for any two elements of $\Lambda^p(M)$ ω_p^1, ω_p^2, we can define the scalar product

$$(\omega^1,\omega^2) = \int_M \omega^1\wedge(*\omega^2) \ .$$

With respect to this scalar product, we can define the adjoint of d, d^*: $(\omega_p, d\,\Sigma_{p-1}) = (d^*\omega_p, \Sigma_{p-1})$. Since $d^2 = 0$, it also follows that $d^{*2} = 0$. One says that a p-form ω is closed, if $d\omega = 0$, and that it is co-closed if $d^*\omega = 0$. Finally, we can define the generalization of the usual Laplacian on scalar functions to p-forms by $\square = -(dd^* + d^*d) = -(d + d^*)^2$. The minus sign is included to ensure that $\square$ is a positive operator. Since $d: \Lambda^p \to \Lambda^{p+1}$ and $d^*: \Lambda^p \to \Lambda^{p-1}$, $\square$ transforms p-forms into p-forms $\square: \Lambda^p \to \Lambda^p$. It therefore makes sense to write down the eigenvalue problem $\square\omega_p = \lambda\omega_p$. If the manifold M_n is Riemannian (i.e. has positive signature) and compact without boundary, it is a consequence of the general theory of elliptic operators that $\square$ has a discrete spectrum, and that for each eigenvalue, the degeneracy is finite. In particular, we can look at the zero eigenvalues of $\square: \square\omega_p = 0$. Those p-forms such that $\square\omega_p = 0$ are called harmonic p-forms, and they generate a finite dimensional space $\mathrm{Harm}_p(M)$. The remarkable result mentioned of the Hodge-DeRham theory is that the homology groups $H_p(M)$ and the spaces of harmonic forms $\mathrm{Harm}_p(M)$ are isomorphic. Thus we can represent the classes of $H_p(M)$ in terms of harmonic forms. Furthermore, since $\square$ is positive semi-definite and d^* is the adjoint of d, $(\omega_p, \square\omega_p) = 0 = (d\omega_p, d\omega_p) + (d^*\omega_p, d^*\omega_p)$ which implies that a harmonic form is both closed and co-closed. In fact, any p-form ω_p can be represented as[5]

$$\omega_p = d\alpha_{p-1} + d^*\beta_{p+1} + \gamma_p$$

where γ_p is harmonic, and the decomposition is unique. A p-form like $d\alpha_{p-1}$ $(d^*\beta_{p+1})$ is called exact (co-exact), and another way of defining the space of harmonic forms is through the cohomology groups $H^p(M)$ whose elements are the set of closed p-forms modulo exact p-forms. What the Hodge-DeRham theory implies is that

$$H^p(M) \approx H_p(M) \approx \mathrm{Harm}_p(M) \ .$$

We will see in more detail later, that a bundle E over M is

characterized by a set of characteristic polynomials, some of which were introduced before. From the previous paragraphs, we can easily see that a non-trivial characteristic polynomial will generate an element of the appropriate cohomology group. In this respect, a characteristic polynomial is a map from a bundle over M to the cohomology groups of M_n. Thus knowing the non-trivial bundles which can be built over M gives a lot of information about the topological structure of M. This is one of the reasons why the characteristic polynomials play such an important role in modern topology and in particular, in index theory.

After all these preliminaries, we can formulate the index theorem. In a generic index problem, we have some elliptic operator D from some bundle E to some bundle F. In order to make things more concrete, we can think of D as the Dirac operator defined on M_{2n} in the presence of some gauge field A with group G. Then E is the space of positive chirality spinors (note that we are now taking M to be even dimensional so that chirality is defined, or, to put it more precisely, there are two independent spinor representations of the frame group SO(2n)), and F is the space of negative chirality spinors.

$$\not{D} = E_a{}^{\mu}\, \gamma^a(\partial_\mu + A_\mu + \tfrac{1}{2}\omega_{cd,\mu}\, \sigma^{cd})$$

$$\{\gamma^a,\gamma^b\} = 2\delta^{ab}\ , \qquad \sigma^{ab} = \tfrac{1}{4}[\gamma^a,\gamma^b]$$

$$\hat{\Gamma} = i^n \prod_{a=1}^{2n} \gamma^a\ , \qquad \hat{\Gamma}^2 = 1\ . \tag{2.22}$$

E and F in this case are characterized by having $\hat{\Gamma} = +1$ or $\hat{\Gamma} = -1$; and since we are also including a gauge field A, E and F carry some representation of the gauge group G. In this setting, the Weyl operator is $D_+ \equiv \not{D}(1+\hat{\Gamma}\,/\,2): E \to F$, and its adjoint $\not{D}_-: F \to E$. The collection $(\not{D}_+,\not{D}_-,E,F)$ forms an elliptic complex, and the index of the complex is defined by

$$\text{Ind}\ \not{D} = \dim\ker \not{D}_+ - \dim\ker \not{D}_-\ . \tag{2.23}$$

In general, if we have an elliptic operator $\hat{O}$ acting on some bundle E (spinor fields, vector fields, tensor fields, etc.) $\hat{O}: E \to F$, and $\hat{O}^\dagger: F \to E$, then the index of $\hat{O}$ is defined as in (2.23) Ind $\hat{O}$ = $\dim\ker\hat{O} - \dim\ker\hat{O}^\dagger$, where $\ker\hat{O}$ indicates the space of zero modes of the operator $\hat{O}$.

What the index theorem states is that Ind $\hat{O}$ is a topological invariant which depends only on E, F and M_n, and which moreover can be written as the integral of a characteristic class which measures the topological twisting of E, F and M. One of the aims of the index theorem is to provide explicitly the characteristic class which determines the index of a given elliptic complex. There are several approaches to computing the index of an operator in the mathematical literature.[5] The one which is the closest to methods used in physics is the heat kernel expansion.[6] This procedure is as follows:

Given the operators $\hat{O}$, $\hat{O}^\dagger$, we can construct two "Laplacians" $\Box_+ = \hat{O}^\dagger \hat{O}$, $\Box_- = \hat{O}\hat{O}^\dagger$. By construction, $\Box_\pm$ are elliptic, hermitian and self-adjoint, and since M_n is compact, the spectrum of $\Box_\pm$ is discrete, and the degeneracy of each eigenvalue is finite. Also $\Box_+ : E \to E$, $\Box_- : F \to F$ so that $\Box_\pm$ have well defined eigenvalue problems. Furthermore, let λ be an eigenvalue of $\Box_+$ with eigenfunction ϕ_λ: $\Box_+\phi_\lambda = \lambda\phi_\lambda$, then $\hat{O}\phi_\lambda \equiv \psi_\lambda$ is an eigenfunction of $\Box_-$ with the same eigenvalue $\Box_-\psi_\lambda = \hat{O}\hat{O}^\dagger\hat{O}\phi_\lambda = \lambda\hat{O}\phi_\lambda$, and vice-versa. We thus learn that the spectrum of non-zero eigenvalues of $\Box_+$ and $\Box_-$ are identical. This is not true however of the zero eigenvalues. The zeros of $\Box_+$ are the same as the zeros of $\hat{O}$, i.e. $\ker \Box_+ = \ker \hat{O}$ and $\ker \Box_- = \ker \hat{O}^\dagger$. These properties imply that:

$$\text{Ind}\, \hat{O} = \text{Tr}\, e^{-t\Box_+} - \text{Tr}\, e^{-t\Box_-} \,. \tag{2.24}$$

The first trace is taken over E, and the second over F. In the standard heat kernel expansion, one computes the asymptotic expansion in $1/t$ for $t \to \hat{O}^\dagger$ for both terms on the right-hand side of (2.24), and then picks out the piece which is independent of t. As we will show later on, supersymmetry provides a very powerful short cut for the computation of (2.24). In the supersymmetric form of the heat kernel expansion, the leading term is already the t-independent term in (2.24), thus providing a very quick identification of the characteristic class associated with a given elliptic complex.

In the following sections we will show how to use supersymmetric quantum mechanics to compute (2.24) efficiently.

III. QUANTUM MECHANICS AND FUNCTIONAL INTEGRALS

In this section we will review some useful formulae which will be the starting point of the supersymmetric heat kernel expansion. Our goal is to obtain a functional integral representation for the trace of any operator for a quantum mechanical system described by a Hamiltonian H. This is most easily accomplished by using the holomorphic representation of the functional integral.[7] Let us first consider a simple harmonic oscillator with unit mass and frequency. The Hamiltonian is $H = (p^2 + q^2)/2$, which in terms of creation and annihilation operators becomes $H = \hat{a}^\dagger\hat{a} + \frac{1}{2}$ $(a = (q+ip)/\sqrt{2})$ in terms of $\hat{a}$, $\hat{a}^\dagger$, the canonical commutation relation becomes $[\hat{a},\hat{a}^\dagger] = 1$, $[\hat{a},\hat{a}] = 0$. These commutators can be realized in the space of holomorphic functions of a single complex variable: $f = f(\bar{a})$. In terms of these functions $\hat{a},\hat{a}^\dagger$ are defined as follows:

$$\hat{a}^\dagger f(\bar{a}) = \bar{a} f(\bar{a})$$
$$\hat{a}\, f(\bar{a}) = \frac{d}{d\bar{a}} f(\bar{a}) \,. \tag{3.1}$$

In order to set up a well defined Hilbert space, we need to introduce a scalar product for the f's such that $\hat{a}$ and $\hat{a}^\dagger$ defined by (3.1) become each other's adjoints. The natural scalar product is

$$(f_1,f_2) = \int \frac{dz d\bar{z}}{2\pi i} \overline{f_1(\bar{z})}\, f_2(\bar{z})\, e^{-z\bar{z}} \tag{3.2}$$

$$dz d\bar{z}/2\pi i = dp dq/2\pi \qquad (\hbar = 1)$$

It is now a simple matter to check that $(f_1, \hat{a} f_2) = (\hat{a}^\dagger f_1, f_2)$ and that the functions $\psi_n = \bar{z}^n/\sqrt{n!}$ form an orthonormal basis. The Hilbert space so constructed is denoted by $\mathcal{H}$. The operators on $\mathcal{H}$ can be represented in two different ways: (i) in terms of their integral kernel, (ii) in terms of a polynomial of the $\hat{a}$'s and $\hat{a}^\dagger$'s.

(i) In integral form, an operator $\hat{A}$ is defined by

$$(\hat{A}f)(\bar{z}) = \int \frac{d\xi d\bar{\xi}}{2\pi i} e^{-\xi\bar{\xi}} A(\bar{z},\xi) f(\bar{\xi}) \;. \tag{3.3}$$

In the $\psi_n(\bar{z})$basis, $\hat{A} = \sum_{n,m} |\psi_n\rangle A_{nm} \langle\psi_m|$, $A_{nm} = \langle\psi_n|\hat{A}|\psi_m\rangle$, so that

$$A(\bar{z},z) = \sum_{n,m} A_{nm} \frac{\bar{z}^n z^m}{\sqrt{n!m!}} \tag{3.4}$$

In this form, the kernel of the product of two operators $\hat{A}_1,\hat{A}_2$ is given by a simple convolution formula

$$(\hat{A}_1,\hat{A}_2)(\bar{z},z) = \int \frac{d\xi d\bar{\xi}}{2\pi i} e^{-\xi\bar{\xi}} A_1(\bar{z},\xi) A_2(\bar{\xi},z) \;. \tag{3.5}$$

(ii) The operator $\hat{A}$ can also be defined by a normal order expansion in terms of $\hat{a}$ and $\hat{a}^\dagger$

$$\hat{A} = \sum_{n,m} K_{nm} \hat{a}^{\dagger n} \hat{a}^m \;. \tag{3.6}$$

From (3.6) we can define the normal order symbol of $\hat{A}$ by the polynomial $K(\bar{a},a) \equiv \sum_{n,m} K_{nm} \bar{a}^n a^m$. Both ways of defining the operator $\hat{A}$ are related by the simple formula

$$A(\bar{a},a) = e^{\bar{a}a} K(\bar{a},a) \;. \tag{3.7}$$

Since $\hat{A}$ is a linear operator, the relation between A and K must be linear; thus in order to prove (3.7) it is enough to consider an operator of the form $\hat{A} = \hat{a}^{\dagger k} \hat{a}^\ell$, i.e. $K(\bar{a},a) = \bar{a}^k a^\ell$. The matrix elements of $\hat{A}$ on the $\psi_n(\bar{z})$ basis are obtained by evaluating the integrals $\langle\psi_n|\hat{a}^{\dagger k} \hat{a}^\ell|\psi_m\rangle = A_{nm}$. A simple computation shows that

$$A_{nm} = \left|\frac{n!\,m!}{(n-k)!\,(m-\ell)!}\right|^{\frac{1}{2}} \theta(n-k)\theta(m-\ell)\delta_{n-k,m-\ell}$$

(where θ is the step function). Since $A(\bar{a},a) = \sum A_{nm} \frac{\bar{a}^n a^m}{\sqrt{n!m!}}$ we finally obtain

$$A(\bar{a},a) = e^{\bar{a}a} K(\bar{a},a) \ .$$

Exercise. Prove that the unit operator on $\mathscr{H}$ has $e^{\bar{a}z}$ as its kernel, i.e. show that

$$f(\bar{a}) = \int \frac{dzd\bar{z}}{2\pi i} e^{\bar{a}z} e^{-\bar{z}z} f(\bar{z}).$$

We are now ready to obtain general formulae for the trace of an operator $\hat{A}$:

$$\begin{aligned}
\mathrm{Tr}\,\hat{A} &= \sum_n \langle\psi_n|\hat{A}|\psi_n\rangle \\
&= \sum_n \int \frac{dzd\bar{z}}{2\pi i} e^{-z\bar{z}} \int \frac{d\xi d\bar{\xi}}{2\pi i} \frac{z^n}{\sqrt{n!}} A(\bar{z},\xi) \frac{\bar{\xi}^n}{\sqrt{n!}} \\
&= \int \frac{dzd\bar{z}}{2\pi i} \int \frac{d\xi d\bar{\xi}}{2\pi i} e^{-z\bar{z}-\xi\bar{\xi}} A(\bar{z},\xi) \sum_n \frac{z^n\bar{\xi}^n}{n!} \\
&= \int \frac{dzd\bar{z}}{2\pi i} \int \frac{d\xi d\bar{\xi}}{2\pi i} e^{-z\bar{z}-\xi\bar{\xi}} A(\bar{z},\xi) e^{\bar{\xi}z} \\
&= \int \frac{dzd\bar{z}}{2\pi i} e^{-\bar{z}z} A(\bar{z},z) \quad ,
\end{aligned}$$

i.e.

$$\begin{aligned}
\mathrm{Tr}\,\hat{A} &= \int \frac{dzd\bar{z}}{2\pi i} e^{-z\bar{z}} A(\bar{z},z) \\
&= \int \frac{dzd\bar{z}}{2\pi i} K(\bar{z},z) \ . \qquad (3.8)
\end{aligned}$$

Hence, for a product of operators

$$\mathrm{Tr}\,\hat{A}_1 \ldots \hat{A}_n = \int \frac{d\xi d\bar{\xi}}{2\pi i} \int \prod_{i=1}^{n-1} \frac{d\xi_i d\bar{\xi}_i}{2\pi i} e^{-\xi\bar{\xi} - \sum_{i=1}^{n-1} \bar{\xi}_i \xi_i} A_1(\bar{\xi},\xi_1) A_2(\bar{\xi}_1,\xi_2) \ldots A_n(\bar{\xi}_{n-1},\xi) \ . \qquad (3.9)$$

(3.8) and (3.9) are all we need to compute the traces needed in the supersymmetric heat kernel expansion. Consider for example the partition function $\mathrm{Tr}\, e^{-\beta H}$ for a system with one degree of freedom described by the Hamiltonian $H(\hat{a}^\dagger,\hat{a})$. We will assume that $H(\hat{a}^\dagger,\hat{a})$ has been normal order. Dividing up β in N intervals of length ε:

$$\mathrm{Tr}\, e^{-\beta H} = \mathrm{Tr}(e^{-\varepsilon H})^N$$

for ε small enough, $e^{-\varepsilon H} \approx 1-\varepsilon H(\hat{a}^\dagger,\hat{a})$, so that the normal symbol of $e^{-\varepsilon H}$ is $1-\varepsilon H(\bar{a},a)$, and its kernel $(e^{-\varepsilon H})(\bar{a},a) = e^{\bar{a}a-\varepsilon H(\bar{a},a)}$. Using (3.9) we get:

$$\mathrm{Tr}\, e^{-\beta H} = \int \frac{d\xi d\bar{\xi}}{2\pi i} \int \prod_{i=1}^{n-1} \frac{d\xi_i d\bar{\xi}_i}{2\pi i} \; e^{-\xi\bar{\xi} - \sum_{i=1}^{n-1} \xi_i \bar{\xi}_i} \cdot$$

$$\cdot \exp\Big\{\bar{\xi}\xi_1 + \bar{\xi}_1\xi_2 + \ldots + \bar{\xi}_{n-2}\,\xi_{n-1} + \bar{\xi}_{n-1}\,\xi -$$

$$- \varepsilon[H(\bar{\xi},\xi_1) + H(\bar{\xi}_1,\xi_2) + \ldots + H(\bar{\xi}_{n-1},\xi)]\Big\}$$

$$= \int (d\xi d\bar{\xi}) \exp\Big\{ - \bar{\xi}(\xi-\xi_1) - \bar{\xi}_1(\xi_1-\xi_2)\ldots - (\bar{\xi}_{n-1}-\xi) -$$

$$- \varepsilon\big(H(\bar{\xi},\xi_1) + \ldots + H(\bar{\xi}_{n-}\,,\xi)\big)\Big\} \qquad (3.10)$$

$$\mathrm{Tr}\, e^{-\beta H} = \int \prod_t \left(\frac{d\xi(t)\,d\bar{\xi}(t)}{2\pi i}\right) \exp\left(+\int_0^\beta \bar{\xi}(t)\,\frac{d\xi}{dt} - \int_0^\beta H(\bar{\xi},\xi)\,dt\right) \qquad (3.11)$$

which is the usual Euclidean functional integral representation of the partition function. Notice that this method automatically determines the boundary conditions: in the exponent of (3.10) we find the terms $- \bar{\xi}(\xi-\xi_1) - \ldots - (\bar{\xi}_{n-1}-\xi)$ which imply that $\xi_0 = \xi$, $\xi_n = \xi$. The extension of the above results to fermions is straightforward.[7] Again consider a fermionic system with a single degree of freedom. The canonical anticommutation relations are:

$$\hat{a}\,\hat{a}^{\dagger} + \hat{a}^{\dagger}\,\hat{a} = 1, \qquad \hat{a}^2 = 0, \qquad \hat{a}^{\dagger 2} = 0 \qquad (3.12)$$

which may also be represented in terms of holomorphic functions of a single complex Grassmann variable. Let $a,\bar{a}$ be two Grassmann numbers: $a\bar{a}+\bar{a}a = 0$, $a^2 = \bar{a}^2 = 0$. Then the holomorphic functions form a two-dimensional complex vector space:

$$f(\bar{a}) = f_0 + f_1\,\bar{a}$$

so that $\hat{a},\hat{a}^{\dagger}$ are represented by

$$\hat{a}^{\dagger} f(\bar{a}) = \bar{a}\, f(\bar{a}), \qquad \hat{a}\, f(\bar{a}) = \frac{d}{d\bar{a}}\, f(\bar{a}) \;. \qquad (3.13)$$

Using the standard rules of integration over Grassmann variables:

$$\int \bar{a} d\bar{a} = \int a da = 1, \qquad \int da = \int d\bar{a} = 0 \;.$$

It follows that $\psi_0 = 1$, $\psi_1 = \bar{a}$ form an orthonormal basis with respect to the scalar product

$$(f_1,f_2) = \int d\bar{a} da\; e^{-\bar{a}a}\; \overline{f_1(\bar{a})}\; f_2(\bar{a}). \qquad (3.14)$$

It is easy to show that $a, a^\dagger$ (3.13) are each other's adjoints for such a scalar product. With these definitions, the relation (3.7) between the kernel of an operator and its normal order symbol is still valid, and the kernel of the unit operator is also given by $e^{\bar{a}\xi}$. Putting all this together one can show that

$$(\hat{A}_1\hat{A}_2)(\bar{a},a) = \int d\bar{z}dz\ e^{-\bar{z}z}\ A_1(\bar{a},z)A_2(\bar{z},a) \tag{3.15}$$

$$\mathrm{Tr}\,\hat{A} = \int dz d\bar{z}\ e^{\bar{z}z}\ A(\bar{z},z)\ . \tag{3.16}$$

As a consequence, the partition function becomes:

$$\begin{aligned}
\mathrm{Tr}\,e^{-\beta H} = \mathrm{Tr}(e^{-\varepsilon H})^N &= \int d\xi d\bar{\xi} \int \prod_{i=1}^{N-1} d\bar{\xi}_i d\xi_i\ \exp\Big\{\bar{\xi}\xi - \sum_{i=1}^{N-1} \bar{\xi}_i\xi_i + \\
&\quad + \bar{\xi}\xi_1 + \bar{\xi}_1\xi_2 + \ldots + \bar{\xi}_{n-2}\,\xi_{n-1} + \bar{\xi}_{n-1}\,\xi - \\
&\quad - \varepsilon\big(h(\bar{\xi},\xi_1) + h(\bar{\xi}_1,\xi_2) + \ldots + h(\bar{\xi}_{n-1},\xi\,)\big)\Big\} \\
&= \int d\xi d\bar{\xi} \prod_{i=1}^{N-1} d\bar{\xi}_i d\xi_i\ \exp\Big\{\bar{\xi}_{n-i}(\xi-\xi_{n-1}) + \ldots + \\
&\quad + \bar{\xi}_2(\xi_3-\xi_2) + \bar{\xi}_1(\xi_2-\xi_1) + \bar{\xi}(\xi_1+\xi) - \\
&\quad - \varepsilon\big(h(\bar{\xi},\xi_1) + \ldots + h(\bar{\xi}_{n-1},\xi)\big)\Big\}\ .
\end{aligned} \tag{3.17}$$

From (3.17) we see that $\xi_n = \xi$ and that $\xi_0 = -\xi$, thus in cumputing $\mathrm{Tr}\ e^{-\beta H}$ for fermions we find:

$$\mathrm{Tr}\,e^{-\beta H} = \int_{\mathrm{A.B.C.}} d\bar{\xi}(t)d\xi(t)\ \exp\Big\{\int_0^\beta \Big(\bar{\xi}(t)\ \frac{d\xi}{dt} - H(\bar{\xi},\xi)\Big)dt\Big\} \tag{3.18}$$

(A.B.C. stands for antiperiodic boundary conditions $\xi(\beta) = -\xi(0)$.)

Finally, let us compute a functional integral representation for a trace which will appear often in subsequent sections:

$$Z_\theta = \mathrm{Tr}\ e^{i\theta\hat{a}^\dagger\hat{a}}\ e^{-\beta H}\ . \tag{3.19}$$

$\hat{a}^\dagger\hat{a}$ is the fermion number operator, and $\exp i\theta\hat{a}^\dagger\hat{a} \equiv Q_\theta$ is the operator which represents a U(1) rotation on the fermions

$$\begin{aligned}
Q_\theta\ \hat{a}\,Q_\theta^\dagger &= e^{-i\theta}\ \hat{a} \\
Q_\theta\hat{a}^\dagger Q_\theta^\dagger &= e^{i\theta}\ \hat{a}^\dagger\ .
\end{aligned} \tag{3.20}$$

In order to find a path integral representation for (3.19) we only need to compute the normal order symbol of Q_θ. Using the canonical anticommutation relations, we get

$$e^{i\theta\hat{a}^\dagger\hat{a}} = 1 + (e^{i\theta} - 1)\hat{a}^\dagger\hat{a}$$

i.e.
$$K(Q_\theta) = 1 + (e^{i\theta} - 1)\bar{a}a \tag{3.21}$$

so that the kernel of Q_θ becomes

$$\begin{aligned} A(Q_\theta) &= e^{\bar{a}a}\, K(Q_\theta) = (1+\bar{a}a)(1+(e^{i\theta}-1)\bar{a}a) \\ &= 1 + e^{i\theta}\,\bar{a}a = \exp(e^{i\theta}\,\bar{a}a)\ . \end{aligned} \tag{3.22}$$

Using (3.15)-(3.17) we get

$$\begin{aligned} \mathrm{Tr}\, Q_\theta\, e^{-\beta H} &= \int d\xi d\bar{\xi} \prod_{i=1}^{N-1} d\bar{\xi}_i d\xi_i\, \exp\Big\{\bar{\xi}\xi + e^{i\theta}\,\bar{\xi}\xi_1 - \bar{\xi}_1\xi_2 - \dots - \\ &\quad - \bar{\xi}_{N-1}\,\xi_{N-1} + \bar{\xi}_1\xi_1 + \bar{\xi}_2\xi_3 + \dots + \bar{\xi}_{N-1}\,\xi - \\ &\quad - \varepsilon\big(H(\bar{\xi}_1,\xi_2) + \dots\big)\Big\} \\ &= \int d\xi d\bar{\xi} \prod_{i=1}^{N-1} d\bar{\xi}_i d\xi_i\, \exp\Big\{\bar{\xi}_{N-1}(\xi - \xi_{N-1}) + \dots + \\ &\quad + \bar{\xi}_1(\xi_2 - \xi_1) + \bar{\xi}\, e^{i\theta}(\xi_1 + e^{-i\theta}\,\xi) - \varepsilon\big(H(\bar{\xi}_1,\xi_2) + \dots\big)\Big\} \end{aligned}$$

i.e.
$$\mathrm{Tr}\, Q_\theta\, e^{-\beta H} = \int \Pi\big(d\bar{\xi}(t)d\xi(t)\big)\; e^{\int_0^\beta (\bar{\xi}(t)\frac{d\xi}{dt} - H(\bar{\xi},\xi\ ) dt} \tag{3.23}$$
$$\xi(\beta) = -e^{-i\theta}\,\xi(0)\ .$$

Hence the only effect of Q_θ is to change the boundary conditions to $\xi(\beta) = -e^{-i\theta}\xi(0)$. Otherwise the integrand is as in (3.18). In particular, let $\theta = \pi$, then $Q_\theta = (-1)^F$, the operator which counts fermions modulo 2. In this case $\mathrm{Tr}(-1)^F\exp{-\beta H}$ implies that we have to integrate over the fermions with periodic boundary conditions (P.B.C.).

For the standard Atiyah-Singer index theorem one only needs to use $\mathrm{Tr}(-1)^F\, e^{-\beta H}$; for the character valued index theorem (the Atiyah-Hirzebruch index theorem) however, we will need (3.23) for arbitrary θ.

IV. SUPERSYMMETRIC σ-MODELS AND ELLIPTIC COMPLEXES

For our purposes we only need to deal with supersymmetric quantum mechanical systems, i.e. field theories in 0+1 dimensions. Given a quantum mechanical system whose dynamics is determined by a

Hamiltonian H, we say that H admits N supersymmetries if there exist N pairs of operators Q^i, Q^{*i}, $i,j = 1,N$ such that

$$\{Q^i, Q^{*j}\} = 2\,\delta^{ij}\,H$$

$$\{Q^i, Q^j\} = 0 \,. \tag{4.1}$$

For the time being we will not include the presence of central charges[8] in (4.1) because they are not needed to derive the classical index theorems. Apart from the fact that supersymmetry turns bosons into fermions, there is a property of supersymmetry which makes it very different from any other symmetry: namely the fact that the order parameter for supersymmetry breaking is the vacuum energy density. This follows from the fact that $H = \frac{1}{2}(Q+Q^\dagger)^2$, hence if the vacuum is not a supersymmetric invariant, $Q|0\rangle \neq 0$, and $H|0\rangle = E_0|0\rangle$ with $E_0 \neq 0$. However, computing E_0 in general is a very difficult problem. In a rather remarkable paper,[9] E. Witten bypassed the difficulties of computing the vacuum energy E_0 directly by introducing a topological index for the full quantum theory. Let us review some of the details of Witten's argument. Once we are given a supersymmetric Hamiltonian H and the supercharges $Q, Q^\dagger$, we can construct a Hermitian supercharge $S = (Q+Q^\dagger)/\sqrt{2}$ so that $H = S^2$. Let $|E\rangle$ be an eigenstate of H with energy $E \neq 0$. Then $S|E\rangle$ is another eigenstate of H with the same energy, but opposite fermion numbers because $(-1)^F S = -S(-1)^F$, and $[H,S] = 0$. Thus each energy level E furnishes a representation of supersymmetry with dimension larger or equal to two. If we instead look at zero energy states $|\Omega\rangle$, since $H|\Omega\rangle = 0$, we get $S|\Omega\rangle = 0$, and we conclude that the zero energy states furnish a one-dimensional representation of supersymmetry. The number of zero energy bosonic states is thus not necessarily balanced by the number of zero energy fermionic states. If we now compute the quantity $\Delta = \mathrm{Tr}(-1)^F$, or its regularized form $\Delta = \mathrm{Tr}(-1)^F \exp{-\beta H}$, we recognize that Δ is independent of β, and that the only contributions to Δ come from the zero energy sector:

$$\Delta = n_B(E=0) - n_F(E=0) \tag{4.2}$$

with $n_B(E=0)$ ($n_F(E=0)$) being the number of bosonic (fermionic) zero energy states. The most useful property of Δ is that it is invariant under continuous deformations of the supersymmetric Hamiltonian H. This can be seen in two different ways. Consider a bosonic state with energy $E \neq 0$. Due to the general structure of supersymmetry, there is a fermionic state with exactly the same energy E. As we change the Hamiltonian, these two states will retain the same energy, so that if the perturbation is smooth, the bosonic state will not lose its partner. In particular, the perturbation may be such that for some value of the parameters, E becomes zero, so that there is one more zero energy bosonic state. However, by continuity there will also be one more fermionic state of zero energy. This implies that Δ does not change. Similarly it could happen that under a perturbation a zero energy state acquires a non-vanishing energy. Again by continuity there must also be a zero energy fermionic state which under the same perturbation acquires a non-zero energy, thus leaving Δ unchanged.

Another way of obtaining the same conclusion is to go back to the description of S in terms of Q and $Q^\dagger$, and split the Hilbert space of the theory into its bosonic and fermionic parts $\mathcal{H} = \mathcal{H}_B \oplus \mathcal{H}_F$. With respect to this splitting, S is an off-diagonal operator: $S\mathcal{H}_B \subset \mathcal{H}_F$; $S\mathcal{H}_F \subset \mathcal{H}_B$. The zero energy bosonic (fermionic) states are given by the number of independent solutions of the equation $Q|\psi\rangle = 0$ ($Q^\dagger|\psi\rangle = 0$). Hence

$$\Delta = \ker Q - \ker Q^\dagger \tag{4.3}$$

and Δ is the standard mathematical index of the operator Q. It is invariant under smooth changes of the parameters of the theory which do not change the asymptotic behavior of the Hamiltonian for large values of the fields.

Using the fact that Δ is invariant under continuous deformation of the parameters, Witten was able to show that supersymmetry is unbroken for a wide class of interesting supersymmetric field theories. In particular, he was able to show that in supersymmetric non-linear σ-models in two dimensions, Δ is related to some topological invariants of the manifold M on which the σ-model is defined; namely, that Δ equals the Euler number of M $\chi(M)$.[9] Thus, by using the topological properties of M one can decide in many cases whether supersymmetry is unbroken dynamically. Take for example $M = S^{2n}$. The Euler number of S^{2n} is 2, thus $\Delta = 2$. Since Δ is a topological invariant, it is clear that there will be some states with zero energy for any values of the parameter of the theory (size of S^{2n}, volume of the space-time box on which the theory is defined, ...) which implies that supersymmetry is unbroken.

One can turn the argument around, and use suitable quantum mechanical supersymmetric systems to obtain a method to compute the index theorem for certain elliptic operators.[10] This procedure involves two steps. First one has to find a supersymmetric system whose supercharge is exactly the operator one is interested in, then one has to evaluate Δ in some reliable approximation scheme. It is quite remarkable that the index theorem for the classical complexes (which will be defined later on) can be obtained from simple variations of a simple supersymmetric system, namely the supersymmetric non-linear σ-model.

We now briefly describe some of the properties of σ-models which are relevant for our purposes. Further details and other applications of non-linear σ-models can be found in Dr. Bagger's lectures. The supersymmetric form of the σ-model can easily be obtained by finding the supersymmetric extension of the action for harmonic maps.[11] The bosonic σ-model in 0+1 dimensions is given by the Lagrangian:

$$L = \frac{1}{2} g_{ij}(\phi)\dot{\phi}^i \dot{\phi}^j \qquad \dot{\phi}^i = \frac{d\phi^i}{dt} \tag{4.4}$$

where (M_n, g_{ij}) is a compact Riemannian manifold without boundary and the bosonic field configurations $\phi^i(t)$ are maps from S^1 or $\mathbb{R}$ onto M_n. The supersymmetric extension of (4.4) is

$$L = \frac{1}{2} g_{ij}(\phi)\dot{\phi}^i \dot{\phi}^j + \frac{i}{2} g_{ij}(\phi)\bar{\psi}^i \gamma^0 \frac{D}{dt} \psi^j + \frac{1}{12} R_{ijk\ell}\, \bar{\psi}^i \psi^k \bar{\psi}^j \psi^\ell$$

$$\frac{D}{dt}\psi^i = \frac{d}{dt}\psi^i + \Gamma^i{}_{jk}\dot{\phi}^j \psi^k \tag{4.5}$$

$$\bar{\psi}^i{}_\alpha = \psi^i{}_\beta (\gamma^0)_{\beta\alpha}, \qquad \alpha,\beta = 1,2, \qquad \gamma^0 = \sigma_2 .$$

$\psi^i = \begin{pmatrix}\psi^i_1 \\ \psi^i_2\end{pmatrix}$ is a two-component real spinor. The action (4.5) is invariant under the following supersymmetry transformations

$$\delta\phi^i = \bar{\varepsilon}\psi^i$$

$$\delta\psi^i = -i\gamma^0 \dot{\phi}^i \varepsilon - \Gamma^i{}_{jk}(\bar{\varepsilon}\psi^j)\psi^k \tag{4.6}$$

ε being a real two-component constant anticommuting spinor. Although (4.5) looks a bit complicated, most of the arguments which follow can be made without using the explicit form of the Lagrangian. In order to understand the geometrical significance of (4.5), let us canonically quantize the theory. Shifting to a basis where γ^0 is diagonal, then with $\psi^i = \begin{pmatrix}\psi^i \\ \psi^{*i}\end{pmatrix}$ the canonical anticommutation relations become

$$\{\psi^i, \psi^{*j}\} = g^{ij}(\phi) \qquad \{\psi^i, \psi^j\} = 0 . \tag{4.7}$$

This implies that states without fermions are represented by functions on M. States with one fermion have the general form $\omega_i(\phi)\psi^{*i}|\Omega\rangle$, ($|\Omega\rangle$ is the fermionic vacuum), states with p-fermions have the general form $\omega_{i_1 \ldots i_p}(\phi)\psi^{*i_1}\ldots\psi^{*i_p}|\Omega\rangle$ and due to (4.7) $\omega_{i_1 \ldots i_p}(\phi)$ is a totally antisymmetric in its indices. Thus the Hilbert space of the theory is represented by the exterior algebra of $M = \Lambda^*(M)$, the collection of all 0-forms, 1-forms,..., m-forms. More precisely, $\Lambda^*(M)$ is an SO(n) vector bundle over M. The Hilbert space of wave functions of our system is the space of sections of $\Lambda^*(M)$. i.e. $\Gamma(\Lambda^*(M))$. Next we want to compute Q and $Q^\dagger$ for (4.5,6). We can do that in two different ways. The first one is to use Noether's theorem to calculate the charge associated to (4.6). The second and simplest, is to notice that Q and $Q^\dagger$ must satisfy a certain number of requirements which determine them up to a normalization factor. First of all, Q and $Q^\dagger$ are local square roots of a second-order elliptic operator: the Hamiltonian H. Thus Q and $Q^\dagger$ are first-order linear differential operators. Second, since the bosonic Hilbert space is made up of even forms, and since Q and $Q^\dagger$ change the fermionic character of the state on which they act, it follows that Q and $Q^\dagger$ transform even (odd) forms into odd (even) forms. Finally since Q and $Q^\dagger$ generate a supersymmetric algebra, they satisfy $Q^2 = Q^{\dagger 2} = 0$, $QQ^\dagger + Q^\dagger Q = H$. The only operators one can construct with these properties acting on forms are the exterior derivative d, and its adjoint d^* with respect to the Riemannian matrix g_{ij} defined on M_n. Thus the Hamiltonian associated with (4.5) is the standard Hodge-DeRham Laplacian on forms $H = dd^* + d^*d$. The zero energy states consist of the zero modes of the Laplacian. Such forms are known as harmonic, and play a central role in the Hodge-DeRham theory as explained in Section II. The Euler number of M, $\chi(M)$, is defined as the alternating sum of the Betti numbers. Thus[9]

$$\mathrm{Tr}(-1)^F e^{-\beta H} = \chi(M) = \sum_{k=0}^{n} (-1)^k b_k \ . \tag{4.8}$$

The elliptic complex whose index is given by the Euler number can be described as follows: the initial bundle E is the space of even forms, F is the space of odd forms, and the operator $\hat{O}$ is d so that $\hat{O}^\dagger = d^*$.

Another interesting topological invariant is the Hirzebruch signature of M, $\tau(M)$. To define $\tau(M)$ we use the Hodge *-operation, or Poincaré duality: to any p-form $\omega_{i_1 \ldots i_p}$, we can associate a n-p form as follows:

$$(*\omega_p)_{j_1 \ldots j_{n-p}} = \frac{1}{p!} \varepsilon_{j_1 \ldots j_{n-p}\, i_1 \ldots i_p}\, \omega^{i_1 \ldots i_p} \ .$$

ε is the totally antisymmetric Levi-Ciritta density. It is easy to show that the Laplacian $\Box = dd^* + d^*d$ and * commute. Thus we can simultaneously diagonalize $\Box$ and *. For an even dimensional manifold M_{2n}, $*^2 = (-1)^n$ so that for $n = 2k+1$, $*^2 = -1$, and the eigenvalues of * are $\pm i$. Since the operator $\Box$ is real, there are as many harmonic forms with $* = +i$ or $-i$. However, for $n = 2k$, $*^2 = +1$, so that the eigenvalues of * are ± 1 and the number of self-dual harmonic forms ($* = +1$) is not necessarily identical to the number of antiself-dual harmonic forms ($* = -1$). In order to identify where the assymmetry may come from, let ω_p be a harmonic p-form with $p \neq 2k$, i.e. $\omega_p \in H_p$. Then $*\omega_p \in H_{4k-p}$, so that $(\omega_p \pm *\omega_p) = \omega_\pm$ are such that $\omega_\pm = \pm *\omega_\pm$. Hence the asymmetry can only arise from H_{2k}, because $*(H_{2k}) \subset H_{2k}$. Thus, we can divide the space of harmonic 2k-forms into self-dual (H^+_{2k}) and antiself-dual (H^-_{2k}), and the Hirzebruch signature is defined by

$$\tau(M) = \dim H^+_{2k} - \dim H^-_{2k} \ . \tag{4.9}$$

$\tau(M)$ can be computed using the properties of (4.5). Notice that (4.5) admits a discrete symmetry $\mathcal{U}_5$. In the basis where γ^0 is diagonal $(\gamma^0 = \sigma_z)$, this discrete symmetry can be represented by $\psi \to \sigma_y \psi$. In other words, $\mathcal{U}_5$ is the symmetry which exchanges fermionic creation and annihilation operators $\psi^i \leftrightarrow \psi^{*i}$, $\psi^{*i} \to -\psi^i$, and at the same time exchanges the full and the empty Fermi sea. Geometrically, the action of $\mathcal{U}_5$ on $\Lambda^*(M)$ corresponds to transforming p-forms into (n-p)-forms. Thus $\mathcal{U}_5$ is essentially the Hodge *-operation. In order to compute $\tau(M)$ define $Q_\pm$ to be the linear combinations of $Q,Q^\dagger$ which diagonalize $\mathcal{U}_5$, i.e. $\mathcal{U}_5 Q_\pm = \pm Q_\pm \mathcal{U}_5$, and divide the spectrum into states of definite $\mathcal{U}_5$. If $|E,+\rangle$ is a state of energy $E \neq 0$ and $\mathcal{U}_5 = +1$, then $Q_-|E,+\rangle$ is a state of the same energy ($[H,Q_\pm] = 0$) but of opposote $\mathcal{U}_5$. Thus all non-zero energy states appear in pairs of opposite $\mathcal{U}_5$. Only the zero energy states $|0,\pm\rangle$ need not appear in $\mathcal{U}_5$-pairs. Therefore the Hirzebruch polynomial can be obtained from the following trace:[9]

$$\tau(M) = \mathrm{Tr}\,\mathcal{U}_5\, e^{-\beta H} = n^{E=0}(\mathcal{U}_5 = +1) - n^{E=0}(\mathcal{U}_5 = -1). \quad (4.10)$$

Let us now turn to a physically more interesting case: the Dirac operator. In order to obtain the Dirac operator from supersymmetry, let us impose on (4.5) the constraint $\psi_1^i = \psi_2^i = \psi^i/\sqrt{2}$ before quantizing the theory. The Lagrangian (4.5) is invariant under two supersymmetry transformations generated by two constant anticommuting real numbers $\varepsilon_1,\varepsilon_2$. After the constraint is implemented we obtain a Lagrangian

$$L = \frac{1}{2} g_{ij}(\phi)\dot{\phi}^i \dot{\phi}^j + \frac{i}{2} g_{ij}(\phi)\psi^i \frac{D}{dt} \psi^j \quad (4.11)$$

which is still invariant under a single supersymmetry transformation $\varepsilon_1 = -\varepsilon_2 = \varepsilon$. Notice that the quartic fermionic term in (4.5) drops out due to the first Bianchi identity. The canonical quantization of (4.11) is a bit trickier than the quantization of (4.5). First of all, the fermion number operator is well defined only for an even number of real fermionic variables. Thus we restrict our considerations to even dimensional manifolds. Now the canonical anticommutation relations become

$$\{\psi^i,\psi^j\} = g^{ij}(\phi) \ . \quad (4.12)$$

In order to make the arguments more clear, let us choose a different set of fermionic variables. Since the manifold is Riemannian, we can refer all geometrical objects to orthonormal frames. Let $e^a{}_i(\phi)$ be the vielbein field, and define $\psi^a = e^a{}_i(\phi)\psi^i$. The Lagrangian becomes

$$L = \frac{1}{2} g_{ij}(\phi)\dot{\phi}^i \dot{\phi}^j + \frac{i}{2} \delta_{ab}\psi^a \frac{D}{dt} \psi^b$$

$$\frac{D}{dt}\psi^a = \frac{d}{dt}\psi^a + \dot{\phi}^i \omega^a{}_{b,i}\psi^b \quad (4.13)$$

and the quantization rules for the ψ^a's become $\{\psi^a.\psi^b\} = \delta^{ab}$, i.e. the fermions generate a Clifford algebra on M_{2n} and $\omega^a{}_{b,i}$ is the spin connection introduced in Sec. II.

Since in going from (4.5) to (4.13) we have thrown away half of the fermions, only a particular Hermitian combination of Q and $Q^\dagger$ survives. Identifying it requires some effort. Using the equations of motion which follow from (4.11), one can show that

$$S = \psi^i g_{ij}(\phi)\dot{\phi}^j \tag{4.14}$$

is conserved, and that as usual in supersymmetry, the Hamiltonian is just $H = S^2$. After canonical quantization, the fermionic variables ψ^i become the generators of the Clifford algebra of M_{2n}, so that we can represent ψ^i by $\gamma^i/\sqrt{2}$, where the γ^i's are the Dirac matrices in M_{2n}. The canonically conjugate momentum to ϕ^i is:

$$p_i = g_{ij}(\phi)\dot{\phi}^j + \frac{i}{4}\omega_{ab,i}[\psi^a,\psi^b] \tag{4.15}$$

and it is replaced by $-i\partial/\partial\phi^i$ after quantization. Putting everything together, we obtain

$$S = -\frac{i}{\sqrt{2}}\gamma\left(\frac{\partial}{\partial\phi^i} + \frac{1}{8}\omega_{ab,i}[\gamma^a,\gamma^b]\right) = -\frac{i}{\sqrt{2}}\not{D} \tag{4.16}$$

$$\gamma^a = e^a{}_i(\phi)\gamma^i$$

which is the Dirac operator. As we pointed out at the end of Sec. II, for every even dimensional Clifford algebra, there is an operator $\hat{\Gamma}$ which anticommutes with all the generators of the Clifford algebra: $\{\hat{\Gamma},\gamma^a\} = 0$. Hence $\hat{\Gamma} = (-1)^F$ for the theory defined in (4.13). The Hamiltonian is $H = (i\not{D})^2/2$, and the Hilbert space of the theory is the space of spinors on the manifold M_{2n} (i.e. the Hilbert space of sections of the spin bundle over $M_{2n} = \Delta_+ \oplus \Delta_-$, $\psi \in \Delta_\pm$ iff $\hat{\Gamma}\psi = \pm\psi$). With these conventions, the space of zero energy bosonic (fermionic) states is the set of zero modes of $i\not{D}$ with positive (negative) chirality $\hat{\Gamma} = +1$ ($\hat{\Gamma} = -1$). In other words:

$$\text{Ind } i\not{D} = \text{Tr}(-1)^F e^{-\beta H} . \tag{4.17}$$

In this way, we can represent the Dirac complex in terms of supersymmetric quantum mechanics.

Before closing this section, we briefly sketch how to obtain the index of the Dolbeault complex from a supersymmetric system. In order to do that, we assume that M_{2n} admits a Kähler structure. This simply means that the manifold can be covered with complex coordinate charts, with the transition functions being holomorphic, and with the line element satisfying

$$dS^2 = 2\, g_{\alpha\bar{\beta}}\, dz^\alpha\, d\bar{z}^\beta$$

$$\partial_\alpha g_{\lambda\bar{\mu}} = \partial_\lambda g_{\alpha\bar{\mu}}$$

$$\partial_\alpha g_{\lambda\bar{\mu}} = \partial_{\bar{\mu}} g_{\lambda\bar{\alpha}} \tag{4.18}$$

$\alpha,\bar{\alpha} = 1, 2, \ldots, n$, where n is the complex dimension of M, and $\{z^\alpha\}$ is a set of complex coordinates of M. Under these conditions, it is possible to show that the σ-model has two supersymmetries,[12] and that the Kähler condition (4.18) is both necessary and sufficient to have N = 2 supersymmetry (for further details on these topics see Dr. Bagger's lectures). Given the Kähler structure, we can refine the exterior algebra $\Lambda^*(M)$ so that

$$\Lambda^*(M) = \bigoplus_{p,q=0}^{n} \Lambda^{p,q}(M)$$

$\Lambda^{p,q}(M)$ being the set of p-holomorphic, q-antiholomorphic forms. $\Lambda^{p,q}(M)$ is generated by $dz^{\alpha_1} \wedge \ldots \wedge dz^{\alpha_p} \wedge d\bar{z}^{\beta_1} \wedge \ldots \wedge d\bar{z}^{\beta_q}$, $\alpha_1 < \ldots < \alpha_p$; $\beta_1 < \ldots < \beta_q$. This also implies that after canonical quantization, the fermionic creation (annihilation) operators can be divided up between those creating (annihilating) antiholomorphic or holomorphic indices. Similarly, the space of harmonic forms can be refined to the different $H_{p,q}$; and the exterior derivative can be smoothly split into holomorphic and antiholomorphic pieces $d = \partial + \bar{\partial}$. If $\omega \in \Lambda^{p,q}$, $\partial\omega \in \Lambda^{p+1,q}$ and $\bar{\partial}\omega \in \Lambda^{p,q+1}$ the Dolbeault complex is defined by the index of the operator $\bar{\partial}$ restricted to the space of antiholomorphic forms:

$$\mathrm{ind}(\bar{\partial}) = \sum_{q=0}^{n} b^{0,q}(-1)^q \tag{4.19}$$

$b^{0,q}$ is the dimension of $H_{0,q}$. $\mathrm{Ind}(\bar{\partial})$ can be computed using $\mathrm{Tr}(-1)^F \exp{-\beta H}$, and restricting the trace to include only antiholomorphic forms. The index $\mathrm{ind}(\bar{\partial})$ is used in mathematics to analyze the number of Riemannian sheets of a complex analytic surface and plays a central role in the solution to the Riemann-Roch problem. Further properties of $\bar{\partial}$ can be found in Ref. 5.

The extension of the above to compute indices of operators including also gauge fields is relatively easy, and will be presented at the end of the next section, after we become familiar with the supersymmetric evaluation of the index theorem for the classical complexes: $\chi(M)$, $\tau(M)$, $\mathrm{ind}(i\not{D})$, $\mathrm{ind}(\bar{\partial})$.

V. THE INDEX THEOREM

We have so far accomplished the first part of our program. We have identified those supersymmetric systems which are closely related to the classical complexes. The next step in our program is to find an approximation scheme which permits a reliable computation of the traces representing the various indices.

Let us start with the simplest case, the Gauss-Bonnet-Chern-Avez formula for the Euler character.[5] Using the results of Section III, we know that $\mathrm{Tr}(-1)^F e^{-\beta H}$ has a functional integral representation

$$\mathrm{Tr}(-1)^F e^{-\beta H} = \int_{\text{P.B.C.}} (d\phi(t)\,d\psi(t)) \exp - S_E(\phi,\psi) \ . \tag{5.1}$$

(P.B.C. stands for periodic boundary conditions for bosons and fermions, see (3.23).) S_E is the Wick rotated version of the action associated to the Lagrangian (4.5). Let us make a change of fermionic variables to $\psi^i = (\psi_1^i + i\psi_2^i)/\sqrt{2}$, $\psi^{*i} = (\psi_1^i - i\psi_2^i)/\sqrt{2}$, so that (4.5) becomes

$$L = \frac{1}{2} g_{ij}(\phi)\dot{\phi}^i \dot{\phi}^j + i\,\psi^{*i} \frac{D}{dt} \psi^j g_{ij}(\phi) - \frac{1}{4} R_{ijk\ell}\, \psi^{*i} \psi^{*j} \psi^k \psi^\ell . \quad (5.2)$$

We only need to compute (5.1) in the limit when $\beta \to 0$ (recall that the result has to be β-independent). Thus we can expand (ϕ,ψ) in a Fourier series with frequencies $2\pi n/\beta$, $n \in Z$. By making β very small, the mass gap between the constant and non-constant modes can be made arbitrarily large, so that the only strongly coupled modes are the constant configurations, while the non-constant modes can be treated on a perturbative basis. Thus the functional integral splits into an integral over constant bosonic and fermionic configurations (a finite dimensional integral), and an integral over non-constant configurations. The latter can be evaluated in perturbation theory, and its first term is given by the ratio of the fermionic and bosonic determinants corresponding to the gaussian approximation to the action for non-constant modes. The ratio of these determinants is $1 + O(\beta)$ as a consequence of the supersymmetric Ward identities which imply that the zero point fluctuations of bosons and fermions cancel exactly. Thus in the small β-limit we get (after a rescaling of the constant fermionic modes by a factor of $\beta^{-\frac{1}{4}}$):

$$\chi(M) = \mathrm{Tr}(-1)^F e^{-\beta H}$$
$$= \frac{1}{(2\pi)^{n/2}} \int d(\mathrm{vol}) \int \prod_m d\psi_m^* \, d\psi_m \, \exp - \frac{1}{4} R_{ijk\ell}\, \psi^{*i} \psi^{*j} \psi^k \psi^\ell \quad (5.3)$$

where n is the dimension of the manifold. Notice that the factor $(2\pi)^{-n/2}$ comes from the standard Feynman normalization of the path integral over constant bosonic configurations, and that the first two terms in the Lagrangian (5.2) vanish for constant configurations, so that only the quartic fermionic interaction survives. Furthermore, if d is odd then $\mathrm{Tr}(-1)^F e^{-\beta H} = 0$, because we cannot saturate the Grassmann integrals with any term appearing in the expansion of the exponential. This shows that the Euler number of a compact odd dimensional manifold without boundary vanishes. This is in agreement with general arguments, because in general for compact M_{2n} (no boundary) the fact that the Laplacian on forms $\Box$ and the Hodge $*$-operator commute implies that $\dim H_p = \dim H_{n-p}$, i.e. $b_p = b_{n-p}$. If n is odd, then for p even (odd), n-p is odd (even) and results in a cancellation in the definition of $\chi(M) = \sum_{k=0}^{n} b_k(-1)^k$. For even n, we can easily compute (5.3) by simply expanding the exponent, and applying the rules of Grassmann integration. One gets

$$\chi(M_d) = \frac{(-1)^{d/2}}{2^d \left(\frac{d}{2}\right)! \, \pi^{d/2}} \int d(\mathrm{vol})\, \varepsilon^{i_1 j_1 \ldots i_n j_n} \varepsilon^{k_1 \ell_1 \ldots k_n \ell_n} R_{i_1 j_1 k_1 \ell_1} \ldots R_{i_n j_n k_n \ell_n} \quad (5.4)$$
$$d = 2n$$

which is the Gauss-Bonnet-Chern-Avez formula.[5] For a four-dimensional manifold, (5.4) implies the well known result:

$$\chi(M_4) = \frac{1}{32\pi^2}\int_M \varepsilon_{abcd}\, \Omega_{ab} \wedge \Omega_{cd} \ . \tag{5.5}$$

Ω_{ab} is the curvature 2-form referred to frames (2.4)

To get the index for other complexes requires some more work. Let us illustrate the basic ideas by computing the index density for the Hirzebrudh signature:

$$\tau(M) = \mathrm{Tr}\,\mathcal{U}_5\, e^{-\beta H} \ . \tag{5.6}$$

Using the methods of Section II, we can construct a functional integral representation for (5.6). In a Majorana basis $\psi^i = \begin{pmatrix}\psi_1^i \\ \psi_2^i\end{pmatrix}$ (ψ_α^i real), $\mathcal{U}_5$ acts as multiplication by $\gamma_5 = \sigma_3$. Thus ψ_1^i has eigenvalue +1 and ψ_2^i eigenvalue −1 with respect to $\mathcal{U}_5$. This implies that ψ_1^i must be integrated over with antiperiodic boundary conditions (A.B.C.), and ψ_2^i with periodic boundary conditions (P.B.C.) in β, $\psi_1^i(\beta) = -\psi_1^i(0)$, $\psi_2^i(\beta) = \psi_2^i(0)$, and

$$\tau(M) = \int_{\mathrm{P.B.C.}} (d\phi)(d\psi_2) \int_{\mathrm{A.B.C.}} (d\psi_1) \exp - S_E(\phi,\psi_1,\psi_2) . \tag{5.7}$$

In the small β limit, we can expand again in terms of the Fourier decomposition of ϕ^i, ψ_α^i. Note however that only ϕ^i and ψ_2^i can have zero modes because ψ_1^i is integrated with A.B.C. If we choose $(\phi^i(t),\psi_1^i(t),\psi_2^i(t)) = (\phi^i,0,\psi_0^i)$, it is easy to see that these constant configurations are solutions to the classical equations of motion with zero action. Thus, the first non-vanishing contribution comes from the small fluctuation determinants around (ϕ_0^i,ψ_{20}^i). These can be computed by expanding S_E around (ϕ_0^i,ψ_{20}^i) to second order in small fields. There are two methods to compute the action to second order. The first one requires writing (4.5) in superfield form, and then applying the background field expansion in terms of normal coordinates.[13] To illustrate the method, consider the purely bosonic σ-model

$$L = \tfrac{1}{2} g_{ij}(\phi)\dot{\phi}^i \dot{\phi}^j \tag{5.8}$$

and let $\phi_0^i(t)$ be a solution to the classical equations of motion

$$\frac{d^2\phi_0^i(t)}{dt^2} + T^i{}_{jk}\,\dot{\phi}_0^j\,\dot{\phi}_0^k = 0 \ . \tag{5.9}$$

Since the action and the S-matrix for this theory are invariant under coordinate reparameterization of M_n, it is useful to choose an expansion parameter around ϕ_0^i which transforms as vector rather than as a coordinate under coordinate changes. In order to do that, we can introduce Riemannian normal coordinates.[14] These are defined as follows: Let ϕ_0^i be some point on M_n, and let ϕ^i be some other point on a neighborhood of ϕ_0^i. Except for exceptional points, one can always find a geodesic which joins ϕ_0^i and ϕ^i. The geodesic is completely defined once we specify the tangent vector ξ^i at ϕ_0^i. Normalizing ξ^i so that ϕ^i is reached along the geodesic after a unit of proper time has elapsed, we can get ϕ^i as a function of ϕ_0^i and ξ^i by simply solving the geodesic equations via a Taylor series

$$\phi^i = \phi_0^i + \xi^i - \tfrac{1}{2}\Gamma^i{}_{jk}(\phi_0)\xi^j \xi^k + \ldots \tag{5.10}$$

If we choose $\xi^i(t)$ to be our expansion parameter, it follows that the expansion coefficients of $S[\phi] - S[\phi_0]$ will be covariant tensors constructed at ϕ_0. For instance, to second order one gets

$$g_{ij}(\phi) = g_{ij}(\phi_0) - \tfrac{1}{3} R_{ikj\ell}\,\xi^k \xi^\ell + O(\xi^3) \tag{5.11}$$

$$\begin{aligned}\dot\phi^i &= \dot\phi_0^i + \left(\frac{d\xi^i}{dt} + \Gamma^i{}_{jk}\,\dot\phi^j \xi^k\right) + \tfrac{1}{3} R^i{}_{k\ell j}\,\xi^k \xi^\ell + O(\xi^3)\\ &\equiv \dot\phi_0^i + \frac{D\xi^i}{dt} + \tfrac{1}{3} R^i{}_{k\ell j}\,\xi^k \xi^\ell .\end{aligned} \tag{5.12}$$

Substituting (5.11) and (5.12) in (5.8) gives

$$S[\phi] - S[\phi_0] = \int dt\, \tfrac{1}{2} g_{ij}(\phi_0)\, \frac{D\xi^i}{dt}\frac{D\xi^j}{dt} + O(\xi^3) \tag{5.13}$$

and the linear term in ξ^i drops out because ϕ_0^i satisfies the classical equations of motion.

We can readily extend this method to (4.5) written in terms of superfields by applying the normal coordinate expansion in superspace form.[13] Since this involves some technical details, we shall prefer our second method: a straightforward, naive background field expansion $\phi^i = \phi_0^i + u^i$, $\psi_2^i = \psi_0^i + \eta^i$, $\psi_1^i = \psi_1^i$. The quartic fermion term becomes

$$\begin{aligned}\tfrac{1}{12} R_{ijk\ell}\,\bar\psi^i \psi^k \bar\psi^j \psi^\ell &= \tfrac{1}{4} R_{ijk\ell}\,\psi_1^i \psi_1^j \psi_2^k \psi_2^\ell\\ &= \tfrac{1}{4} R_{ijk\ell}\,\psi_0^i \psi_0^j \psi_1^k \psi_1^\ell + \text{higher order.}\end{aligned} \tag{5.14}$$

The bosonic kinetic term becomes

$$\tfrac{1}{2} g_{ij}(\phi)\dot\phi^i \dot\phi^j = \tfrac{1}{2} g_{ij}(\phi_0)\dot u^i \dot u^j + O(u^3) . \tag{5.15}$$

The fermionic kinetic term for $\psi_1^i(t)$ is already second order in small fluctuations, and it contributes

$$\frac{i}{2}\psi_1^i \frac{d\psi_1^j}{dt} g_{ij}(\phi_0) . \tag{5.16}$$

The only term which requires some work is the kinetic term for $\psi_2^i = \psi_0^i + \eta^i$:

$$\frac{i}{2}\, g_{ij}(\phi_0+u)(\psi_0^i+\eta^i)\,\frac{D}{dt}(\psi_0^j+\eta^j) = \frac{i}{2}\, g_{ij}(\phi_0)\eta^i\,\frac{d\eta^j}{dt} - i\,\Gamma_{j,ki}\,\psi^i\,\eta^j\,\dot u^k + $$
$$+\frac{i}{2}\, g_{ij}(\phi_0)\psi_0^i\,\psi_0^\ell\,\partial_n\,\Gamma^j{}_{k\ell}\,u^n\,\dot u^k +$$
$$+\frac{i}{2}\,\psi_0^i\,\psi_0^\ell\,(\Gamma_{i,mj}+\Gamma_{j,im})\Gamma^j{}_{k\ell}\,u^n\,\dot u^k, \tag{5.17}$$

where we have used $\partial_k\, g_{ij} = \Gamma_{i,kj} + \Gamma_{j,ki}$, $\Gamma_{i,jk} = g_{i\ell}\,\Gamma^\ell{}_{jk}$. Redefining $\eta^i = \eta'^j - \Gamma^j{}_{k\ell}\,\psi_0^k\, u^\ell$ one obtains after a few lines of algebra

$$\frac{i}{2}\,\eta'^i\,\frac{d\eta'^j}{dt}\, g_{ij}(\phi_0) + \frac{i}{2}\,\psi_0^i\,\psi_0^j\, u^\ell\,\dot u^m\,(\partial_\ell\,\Gamma_{i,mj} - \Gamma_{r,i\ell}\,\Gamma^r{}_{mj}) \tag{5.18}$$

(all geometric quantities couputed at ϕ_0). Since

$$u^\ell\,\dot u^m = \frac{1}{2}(u^\ell\,\dot u^m + u^m\,\dot u^\ell) + \frac{i}{2}(u^\ell\,\dot u^m - u^m\,\dot u^\ell)$$
$$= \frac{1}{2}\,\frac{d}{dt}\,(u^\ell\, u^m) + \frac{1}{2}(u^\ell\,\dot u^m - u^m\,\dot u^\ell)\;. \tag{5.19a}$$

Using the periodicity of the boundary condition, the first term in (5.19) gives a vanishing contribution to the action. Thus (5.18) becomes

$$\frac{i}{2}\,\eta'^i\,\frac{d\eta^j}{dt}\, g_{ij}(\phi_0) + \frac{i}{4}\,\psi_0^i\,\psi_0^j\, u^\ell\,\dot u^m\,(\partial_\ell\,\Gamma_{i,mj} - \partial_m\,\Gamma_{i,j\ell} - \tag{5.19b}$$
$$- \Gamma_{r,i\ell}\,\Gamma^r{}_{mj} + \Gamma_{r,im}\,\Gamma^r{}_{\ell j})$$
$$= \frac{i}{2}\,\eta^i\,\frac{d}{dt}\,\eta^j\, g_{ij}(\phi_0) + \frac{i}{4} R_{ijk\ell}\,\psi_0^k\,\psi_0^\ell\, u^k\,\dot u^\ell\;,$$

where $R_{ijk\ell}$ is the Riemannian curvature tensor. Collecting (5.14, 5.15, 5.16, 5.19) we get

$$L^{(2)} = \frac{1}{2}\, g_{ij}(\phi_0)\dot u^i\,\dot u^j + \frac{i}{4} R_{ijk\ell}\,\psi_0^i\,\psi_0^j\, u^k\,\dot u^\ell +$$
$$+ \frac{i}{2}\,\psi_1^i\,\frac{d\psi_1^j}{dt}\, g_{ij}(\phi_0) + \frac{1}{4}\, R_{ijk\ell}\,\psi_0^i\,\psi_0^j\,\psi_1^k\,\psi_1^\ell +$$
$$+ \frac{i}{2}\,\eta^i\,\frac{d\eta^j}{dt}\, g_{ij}(\phi_0)\;. \tag{5.20}$$

At this point it is worth noticing that $u^i(t)$ and $\eta^i(t)$ must only include fluctuations without zero modes in order to avoid overcounting of the constant configurations with zero action. Rescaling u^i, ψ_1^i, η^i with the vielbein $u^a = e^a{}_i(\phi_0)u^i$, etc., we have

$$L^{(2)} = \frac{1}{2}\delta_{ab}\,\dot u^a\,\dot u^b + \frac{i}{4}R_{abcd}\,\psi_0^c\,\psi_0^d\, u^a\,\dot u^b + \frac{i}{2}\delta_{ab}\,\eta^a\,\frac{d}{dt}\,\eta^b + \frac{i}{2}\delta_{ab}\,\psi_1^a\,\frac{d}{dt}\,\psi_1^b +$$
$$+ \frac{1}{4}R_{abcd}\,\psi_0^c\,\psi_0^d\,\psi_1^a\,\psi_1^b\;. \tag{5.21}$$

Letting $M_{ab} \equiv \frac{1}{2} R_{abcd}(\phi_0)\psi_0^c\psi_0^d$, then, to leading order, (5.7) is given by a ratio of determinants

$$\left[\frac{\det' i \frac{d}{dt} \delta_{ab} \det(i \frac{d}{dt} \delta_{ab} + M_{ab})}{\det'(-\frac{d^2}{dt^2} \delta_{ab} + i M_{ab} \frac{d}{dt})}\right]^{\frac{1}{2}} \tag{5.22}$$

where the prime means that the zero mode has been removed and that the determinant is over configurations satisfying P.B.C. whereas $\det(i\, d/dt\, \delta_{ab} + M_{ab})$ is computed with A.B.C. Since M_{ab} is an anti-symmetric matrix, it can be formally skew-diagonalized

$$M_{ab} \sim \begin{pmatrix} 0 & x_1 & & & & \\ -x_1 & 0 & & & & \\ & & 0 & x_2 & & \\ & & -x_2 & 0 & & \\ & & & & \ddots & \\ & & & & 0 & x_n \\ & & & & -x_n & 0 \end{pmatrix} \tag{5.23}$$

In terms of the eigenvalues x_i, (5.21) becomes

$$\frac{\Pi\, 2 \prod_{n\geq 1}\left(1 + \frac{(x_i/2)^2}{\pi^2(n+\frac{1}{2})^2}\right)}{\prod_i \prod_{n\geq 1}\left(1 + \frac{(x_i/2)^2}{\pi^2 n^2}\right)} = \frac{\prod_i 2 \cosh x_i/2}{\prod_i - \frac{\sinh x_i/2}{x_i/2}} = \prod_i 2 \frac{x_i/2}{\tanh(x_i/2)} \tag{5.24}$$

and consequently

$$\tau(M) = \frac{1}{(2\pi)^{d/2}} \int d(\text{vol}) \int d\psi_0^1 \ldots d\psi_0^{2n} \prod_{i=1}^{n} 2 \frac{x_i/2}{\tanh(x_i/2)} \tag{5.25}$$

Formally, the constant anticommuting numbers ψ_0^a form a realization of the basis of 1-forms on the manifold, and the integral over the ψ_0's just projects out the term proportional to $\psi_0^1 \ldots \psi_0^{2n}$. More geometrically, let Ω_{ab} be the matrix of curvature 2-forms introduced in Sec. II, $\Omega_{ab} = R_{abcd}\, e^c \wedge e^d/2$ ($e^c = e^a{}_\mu\, dx^\mu$). What we have learned from (5.25) is that the formal characteristic polynomial (Hirzebruch L-polynomial)

$$L(M) = \prod_i \frac{(x_i/2\pi)}{\tanh(x_i/2\pi)} , \tag{5.26}$$

with the x_i interpreted as the formal skew-eigenvalue of Ω_{ab}, determines the signature $\tau(M)$ of the manifold. Since the x_i's are 2-forms, $L(M)$ contains only a finite number of terms, the highest order one (the "top form"), being proportional to the volume from of the manifold. Thus the top form of $L(M)$ for each dimension determines $\tau(M)$. In four dimensions,

$$\tau(M) = -\frac{1}{24\pi^2}\int \mathrm{Tr}\,\Omega \wedge \Omega = b_2^+ - b_2^- \ . \qquad (5.27)$$

In obtaining (5.24) one has to be careful about the normalization of the determinants. Since we are interested in the non-trivial dependence on the curvature eigenvalues, we normalize by dividing each determinant by the free determinant, i.e. with $R_{ab} = 0$. Another way of getting the same result is to notice that the integration over the ψ_1^a's, $a = 1,\ldots,2n$ is the partition function for n free fermion species moving on a line with masses x_i. Similarly, the bosonic integral is the partition function for a set of free oscillators of masses x_i^2 with the constraint that the zero momentum mode has been removed. Finally the factor of $(2\pi)^{-d/2}$ comes from the standard Feynman normalization of the functional integral. Note that (5.26) is symmetric under arbitrary exchange of x_i into $-x_i$. Thus L(M) can be written in terms of even symmetric polynomials in the x_i's, i.e. in terms of the Pontrjagin classes introduced in Sec. II.

The computation of the index of the Dirac operator is now straightforward. We have to compute $\mathrm{Tr}(-1)^F e^{-\beta H}$ for the Lagrangian (4.13). The procedure is identical to the computation we have just completed. First expand to second order in small fluctuations around the constant solutions to the classical equations of motion. The result is given by (5.19) with ψ_1^k set equal to zero. Thus we only have to compute the bosonic determinant. In this case the resulting invariant polynomial in the "eigenvalues" of the curvature 2-form Ω is known as the $\hat{A}$-genus or Dirac genus:

$$\hat{A}(M) = \prod_i \frac{x_i/2}{\sinh x_i/2} \ . \qquad (5.28)$$

In four dimensions we find

$$\mathrm{ind}\ i\not{D} = -\tau(M)/8 \ . \qquad (5.29)$$

For the Dolbeault complex, we recall that the manifold M_{2n} has to be chosen to be a Kähler manifold. If we restrict the computation of $\mathrm{Tr}(-1)^F \exp{-\beta H}$ to the antiholomorphic part of the Hilbert space, the computation is exactly the same as for the Dirac operator. We only have to notice that the background expansion around constant configurations has to be done using complex coordinates, and that the curvature 2-form Ω in these coordinates takes values on the Lie algebra of $U(n) \subset SO(2n)$.[5] It is left as an exercise to show that the invariant polynomial one obtains (the Todd genus) is

$$td(M) = \prod_\alpha e^{\omega_\alpha/2}\,\frac{\omega_\alpha/2}{\sinh \omega_\alpha/2} = \prod_\alpha \frac{\omega_\alpha}{1-e^{-\omega_\alpha}} \qquad (5.30)$$

where the ω_α's are now the eigenvalues of

$$\Omega_{\alpha\bar{\beta}} = \frac{i}{2\pi} R_{\alpha\bar{\beta}\gamma\bar{\delta}}\, dz^\gamma \wedge dz^{\bar{\delta}} . \qquad (5.31)$$

In the four-dimensional case,

$$\text{ind}(\bar{\delta}) = \tfrac{1}{4}\big(\chi(M) + \tau(M)\big) \ . \tag{5.32}$$

Before including gauge fields, it is useful to make contact with the more abstract nomenclature explained in Sec. II. This will give us a chance of writing down the general form of the Atiyah-Singer index theorem. In its general form, the index theorem can be stated as follows: Let M be a compact manifold, and let E,F be two vector bundles over M with some associated group G acting on them. Thus there is a principal G-bundle which can be constructed using the transition functions of E and F, and a connection on both bundles (which may or may not coincide with the one induced by the spin connection). If $\hat{O}$ is an elliptic operator interpolating between E and F, then[1]

$$\text{ind}\,\hat{O} = \int_{M_{2n}} \left[\big(\text{ch}(E) - \text{ch}(F)\big)\,\frac{\text{td}(TM \otimes C)}{e(TM)}\right]_{\text{vol}} \tag{5.33}$$

Here ch(F), ch(F) represent the Chern characters (2.19) of the two bundles E, F. td(TM ⊗ C) is the Todd genus for the complexified tangent space of M, and e(TM) is the invariant polynomial which determines the Euler number. If one rewrites equation (5.4) in terms of the skew eigenvalues x_α, $\alpha = 1,n$, of $\Omega/2\pi$, one finds that the characteristic polynomial is simply $\prod_{\alpha=1}^{n} x_\alpha$. Finally, the subscript (vol) in (5.33) means that one has to expand the characteristic polynomial in the integrand and consider only the term proportional to volume form of M. We now illustrate the use of (5.33) by directly obtaining some of the formulae we have derived for the index theorem for some of the classical complexes.

In order to obtain (5.28) from (5.33), we notice that the bundles E,F are the positive and negative chirality spinor bundles Δ_+,Δ_- are just those induced by the standard spin connection. In other words, ch(E) is obtained by computing Tr $\Omega_{ab}/4\pi\ \sigma_+^{ab}$ where σ_+^{ab} are the generators of SO(2n) in the spinor representation of positive chirality. Similarly, ch(F) is obtained by computing Tr exp $1/4\pi\ \Omega_{ab}\ \sigma_-^{ab}$, σ_-^{ab} being the generators of SO(2n) in the spinor representation of negative chirality. If for a moment we forget that $\Omega_{ab}/2\pi$ is a 2-form, and consider it just as an infinitesimal element of SO(2n), the traces to be computed are nothing but the characters of a generic element of SO(2n) in the two spinor representations. This can be done as follows. SO(2n) is a group of rank n, which means that the space of weights has n-generators e_i, $i = 1,n$. Then the weights of the spinor representations (i.e. the eigenvalues of the elements in the Cartan subalgebra of SO(2n)) are given by $\frac{1}{2}\sum_{a=1}^{n} \varepsilon_a e_a$, $\varepsilon_a = \pm 1$, and the two spinor representations are distinguished by the conditions $\prod_{a=1}^{n} \varepsilon_a = +1$ $(\hat{\Gamma} = +1)$ or $\prod_{a=1}^{n} \varepsilon_a = -1$ $(\hat{\Gamma} = -1)$.

In general, the weights Λ of a given representation of a group of rank r, G_r, are r-component vectors, and the character of an arbitrary group element g is

$$\mathrm{Tr}\, g = \sum_{\Lambda} \exp(i \sum_{i=1}^{r} \Lambda_i \phi_i) \quad .$$

The angles ϕ_i are determined by noticing that by a similarity transformation in G_r we can write $g = S(\exp i\, \phi_i H_i)S^{-1}$ where the H_i's are the generators of the Cartan subalgebra of G_r: the maximal set of commuting operators. Since the character involves the trace of the matrix representation of a group element, it is invariant under similarity transformations, and therefore only depends on the angles ϕ_i and on the weights of the representation (the eigenvalues Λ_i of the H_i's, $H_i|\Lambda\rangle = \Lambda_i|\Lambda\rangle$). After this small digression in group theory, we can now compute the Chern characters of the spinor bundles. If x_α, $\alpha = 1,n$, are the skew eigenvalues of $\Omega_{ab}/2\pi$, one can easily show that

$$\mathrm{Tr}_{\Delta+}(\exp \frac{1}{4\pi}\, \Omega_{ab}\, \sigma^{ab}) + \mathrm{Tr}_{\Delta-}(\exp \frac{1}{4\pi}\, \Omega_{ab}\sigma^{ab}) = \prod_{\alpha} 2 \cosh(\frac{x_\alpha}{2}) \tag{5.34}$$

$$\mathrm{Tr}_{\Delta+}(\exp \frac{1}{4\pi}\, \Omega_{ab}\, \sigma^{ab}) - \mathrm{Tr}_{\Delta-}(\exp \frac{1}{4\pi}\, \Omega_{ab}\sigma^{ab}) = \prod_{\alpha} 2 \sinh(\frac{x_\alpha}{2}) \quad . \tag{5.35}$$

(5.35) is the desired expression for $ch(\Delta_+) - ch(\Delta_-)$. The Todd genus of the complexified tangent bundle is obtained by taking $td(TM) \times td(TM)^*$ with $td(TM)$ the form already calculated in (5.30). Thus the characteristic polynomial which determines the index of the Dirac operator is:

$$\prod_{\alpha} 2 \sinh(\frac{x_\alpha}{2})\ \frac{1}{(\prod_\alpha x_\alpha)} \cdot \prod_{\alpha} \left(\frac{x_\alpha}{1-e^{-x_\alpha}}\right)\left(\frac{-x_\alpha}{1-e^{x_\alpha}}\right) = \prod_{\alpha} \frac{x_\alpha/2}{\sinh x_\alpha/2} \tag{5.36}$$

which is the same as (5.28). For the Hirzebruch polynomial, we only have to realize that the bundles E and F are respectively $\Delta_+ \otimes (\Delta_+ \oplus \Delta_-)$ and $\Delta_+ \otimes (\Delta_+ \oplus \Delta_-)$. This follows again from simple group theoretical considerations. If we consider the tensor product of two Dirac spinors of SO(2n) and compute the Clebsch-Gordon series, one obtains the direct sum of all the antisymmetric tensor representations of SO(2n). Thus, the whole De Rham complex (exterior forms) is equivalent to a Dirac bispinor. For this type of object, we can construct several projection operators corresponding to different eigenvalues of $\hat{\Gamma} \otimes 1$, $1 \otimes \hat{\Gamma}$, $\hat{\Gamma} \otimes \hat{\Gamma}$. It is left as an exercise to show that from this point of view, the duality operation is represented by $\hat{\Gamma} \otimes 1$. Hence the two bundles involved are $\Delta_+ \otimes (\Delta_+ \oplus \Delta_-) = E$, $\Delta_- \otimes (\Delta_+ \oplus \Delta_-) = F$ where $\Delta_+ \oplus \Delta_-$ represents the reducible bundle of Dirac spinors. Using

$$\mathrm{Tr}(A \otimes B) = \mathrm{Tr}\, A\ \mathrm{Tr}\, B$$

$$\mathrm{Tr}(A \oplus B) = \mathrm{Tr}\, A + \mathrm{Tr}\, B \tag{5.37}$$

we get

$$\mathrm{ch}(\Delta_{\pm}\otimes(\Delta_{+}\oplus\Delta_{-})) = \mathrm{ch}(\Delta_{\pm})(\mathrm{ch}\,\Delta_{+}+\mathrm{ch}\,\Delta_{-}) \tag{5.38}$$

and

$$\mathrm{ch}(E)-\mathrm{ch}(F) = (\mathrm{ch}(\Delta_{+})-\mathrm{ch}(\Delta_{-}))(\mathrm{ch}\,\Delta_{+}+\mathrm{ch}\,\Delta_{-}) = (\prod_{\alpha} 2\sinh\frac{x_{\alpha}}{2})(\prod_{\alpha} 2\cosh\frac{x_{\alpha}}{2}) \tag{5.39}$$

where we have used (5.34, 5.35). Substituting (5.39) in (5.33) and including the contribution from the Todd genus produces the Hirzebruch polynomial

$$L(M) = \prod_{\alpha} 2\,\frac{(x_{\alpha}/2)}{\tanh(x_{\alpha}/2)}\ . \tag{5.40}$$

Since one has to extract from (5.40) the volume form for each dimension, it follows that (5.40) gives the same result as

$$L(M) = \prod_{\alpha}\frac{x_{\alpha}}{\tanh x_{\alpha}}\ . \tag{5.41}$$

Simular arguments can be constructed for the Euler character and the Dolbeault complex. An important lesson which can be learned from the previous computations is that all the bundles considered, can be constructed in terms of spinor bundles tensored with some other vector bundles, i.e. any geometrical object can be considered from the index theorem point of view as a spinor taking values on some vector bundle. This implies that the general form of the index theorem is basically equivalent to the index theorem for the Dirac operator for spinors taking values on an arbitrary bundle. In order to finish the proof of the index theorem, we need to extend the arguments on the first part of this section to the Dirac operator on some M_{2n} with an arbitrary external gauge field A_{μ}, defining a connection on some bundle V over M_{2n} with bundle group (gauge group) G. The eigenvalue problem for the Dirac equation in this case is

$$i\gamma^{i}(\partial_{i}+\tfrac{1}{2}\,\omega_{iab}\,\sigma^{ab}+A^{\alpha}_{\ i}\,T^{\alpha})_{AB}\,\psi_{B} = \lambda\,\psi_{A}\ . \tag{5.42}$$

$(T^{\alpha})_{AB}$ ($A,B=1$, dim T, $\alpha=1$, dim G) are the antihermitian generators of G. In order to find the one-dimensional analog of (5.42), we introduce for each index $A=1$, dim T a pair of fermionic creation and annihilation operators C_A^*, C_A satisfying the usual canonical anticommutation relations

$$\{C_A, C_B^*\} = \delta_{AB} \qquad \{C_A, C_B\} = 0\ .$$

In the Hilbert space generated by the C's, we can consider states of the form

$$|\psi\rangle = \sum_{A}\psi_{A}(\phi)C_{A}^{*}|0\rangle \tag{5.43}$$

Then (5.42) can be written as

$$i\gamma^{i}(\partial_{i}+\tfrac{1}{2}\,\omega_{iab}\,\sigma^{ab}+A^{\alpha}_{i}\,C^{*}\,T^{\alpha}\,C)|\psi\rangle = \lambda|\psi\rangle\ . \tag{5.44}$$

A trivial feature of (5.42, 5.44) is that if $|\psi,\lambda>$ is an eigenfunction of the Dirac operator with eigenvalue $\lambda \neq 0$, then $\hat{\Gamma}|\psi,\lambda>$ is also an eigenfunction with opposite eigenvalue, in analogy with the situation where the gauge field is absent. This is simply a consequence of the fact that $i\not{D}$ contains an odd number of γ-matrices, and that $\{\hat{\Gamma},\gamma^i\} = 0$. Now it is easy to generalize (4.13) to a Lagrangian whose Hamiltonian is the square of the Dirac operator (5.42). This is given by

$$L = \frac{1}{2} g_{ij}(\phi)\dot{\phi}^i \dot{\phi}^j + \frac{i}{2} \delta_{ab} \psi^a \frac{D}{dt} \psi^b + i\, C^*_A \left(\frac{d}{dt} C_A - A^\alpha_i(\phi)\dot{\phi}^i\, T^\alpha_{AB} C_B\right) - \frac{i}{2} \psi^a \psi^b F^\alpha_{ab} C^*_A T^\alpha_{AB} C_B \; . \qquad (5.45)$$

Now the index theorem follows from

$$\mathrm{Tr}\, \hat{\Gamma}\, e^{-\beta(i\not{D})^2} = n^{E=0}(\hat{\Gamma} = +1) - n^{E=0}(\hat{\Gamma} = -1) \; . \qquad (5.46)$$

Next, we need to write a functional integral representation for (5.46) using the Lagrangian (5.45). If one looks at the space of states of c-fermions, one notices that only the set of one-particle states carries the representation T^α. All other particle states transform as tensor products of the representation T. Since we want to compute the index theorem for a spinor on the representation T, we have to restrict the functional integral representation of (5.46) to include only one-particle states of the c-fermions.

$$\mathrm{ind}\,\not{D} = \lim_{\beta\to 0} \int'_{\mathrm{A.B.C.}} (dC^* dC) \int_{\mathrm{P.B.C.}} (d\phi\, d\psi) \exp - \int_0^\beta L_E(t)dt \; . \qquad (5.47)$$

The prime indicates that one has to extract the result corresponding to one-particle states. In the small β-limit we again have to expand around the constant configurations ($\phi_0, \psi_0, c=0$) to second order. Note that the A.B.C. for the C's are required by the fact that the trace (5.46) contains $(-1)^F$ but not $(-1)^{F_c}$ where F_c is the fermion number operator for c-fermions. Since there are no constant configurations for the c's, the last two terms of (5.45) are already second order in the c's, and we can evaluate the coefficients of the C^*, C terms at the constant configurations (ϕ^i_0, ψ^a_0). This makes it particularly easy to compute the trace over c-fermions. The result is

$$\mathrm{ind}\, i\not{D} = (2\pi)^{-d/2} \int (d\phi_0 d\psi_0) \int (d\; d\nu)\, \mathrm{Tr}(e^{\frac{1}{2} i F^\alpha_{ab} T^\alpha \psi^a_0 \psi^b_0}) \cdot \cdot \exp - \int_0^\beta dt \left(\frac{1}{2}\delta_{ab} \dot{u}^a \dot{u}^b + \frac{i}{4} R_{abcd} \psi^c_0 \psi^d_0 u^a \dot{u}^b + \frac{i}{2} \eta^a \frac{d}{dt} \eta^a\right) \; . \qquad (5.48)$$

The rest of the computation is the same as in the absence of gauge fields. Finally note that the explicit trace in (5.48) is only over group indices. If we denote by ch(F) the Chern character of the vector bundle V, $ch(F) = \mathrm{Tr}\exp i F/2\pi$ $(F = \frac{1}{2} F^{\alpha}_{ab} T^{\alpha} e^a \wedge e^b)$, the index of $i\not{D}$ is given by

$$\mathrm{ind}(i\not{D}) = \int_M [ch(F)\hat{A}(M_{2n})]_{vol} . \qquad (5.49)$$

In particular, in four dimensions,

$$\mathrm{ind}(i\not{D}) = \frac{\dim T}{192\pi^2}\int \mathrm{Tr}\,\Omega \wedge \Omega + \frac{1}{8\pi^2}\int \mathrm{Tr}\, F \wedge F . \qquad (5.50)$$

(It is left as an exercise to show that starting from (5.49) one can derive the Hirzebruch polynomial and the Todd genus.) Notice that we could have started out by deriving (5.49) first, and then by restricting the bundle V to appropriate choice one could reobtain the results presented in the first half of this section. (This is essentially the approach followed by D. Friedan and P. Windey.[10])

Let us consider as a final application of (5.49) the computation of the index theorem for a spin 3/2 operator. In this case we only need to realize that a spin 3/2 object is described by a spinor taking values on the tangent bundle of the manifold M_{2n}. Hence we only have to compute $\mathrm{Tr}\exp 1/4\pi\, R_{ab}\, T^{ab}$ where T^{ab} are the generators of SO(2n) in the vector representation. In notation explained previously, the weights of the vector representation of SO(2n) are $\pm e_i$ so that the Chern character of the tangent bundle is $2\sum_\alpha \cosh(x_i/2)$, i.e.

$$\mathrm{ind}(D_{3/2}) = \int_M \left[(2\sum_\alpha \cosh\frac{x_i}{2}) \prod_\lambda \frac{x_\lambda/2}{\sinh x_\lambda/2}\right]_{vol} . \qquad (5.51)$$

Similar arguments can be applied to higher spin fields.

VI. CHARACTER VALUED INDEX THEOREM

In this section we present some other applications of supersymmetric quantum mechanics in the computation of character valued index theorems, or generalized fixed point theorems. These results have proved very useful in the general analysis of whether Kaluza-Klein theories can generate a realistic fermion spectrum at low energies.[3] To motivate the results which will be derived, let us quickly review one of the main problems encountered by purely gravitational theories in higher dimensions. In order to keep the discussion as simple as possible, we analyze the theory of a single chiral fermion in 4+N dimensions coupled to gravity. The extension of the arguments to cases involving gauge fields as well as spin 3/2 fields is not difficult, but the algebra is more complicated. In the Kaluza-Klein philosophy, one assumes that we are living in a 4+N dimensional space, and that the fundamental theory is invariant under reparametrizations plus local frame rotations of the 4+N dimensional space. The next assumption is that the ground state of the theory is of the form $M_4 \times B_N$, where M_4 is the usual Minkowski space, and B_N is a compact manifold whose "radius" is small enough to have thus far escaped detection. A consequence of this ansatz is that if B_N admits an isometry group, G, then from the purely four dimensional point of view G appears as the gauge group of the low energy world (i.e. concerned with energies far below the inverse of the compactification scale of B_N). There are many different schemes, and many different scenarios in Kaluza-Klein theories, and the reader is referred to Dr. Duff's lectures for more details.

In trying to reproduce the low energy phenomenology, of observed quarks and leptons, all the higher dimensional theories are faced with the fact that the ordinary quarks and leptons transform as a complex representation of the low energy gauge group SU(3) × SU(2) × U(1). For example, if we look at a single family of quarks and leptons, its group transformation properties in terms of purely left-handed fields are described by the representations $(3,2)_{1/3}$, $(\bar{3},1)_{-4/3}$, $(\bar{3},1)_{2/3}$, $(1,2)_{-1}$, $(1,1)_{+2}$ (where the first entry corresponds to the SU(3) representation, the second to the SU(2) representation, and the subscript is the hypercharge assignment, with the convention that the electric charge $Q = I_3 + Y/2$, and I_3 is the third component of weak isospin). If we embed SU(3) × SU(2) × U(1) in SU(5) in the standard way,[15] then each family of quarks and leptons follows from a $\bar{5}$ and a 10 representation of SU(5).

Thus one has to reproduce two basic facts of the low energy world: the gauge group is SU(3) × SU(2) × U(1), so that B_N must have an isometry group big enough to contain at least this group. The second problem is that the fermions belong to a complex representation of SU(3) × SU(2) × U(1), and therefore they remain massless until the scale of breaking of the weak group SU(2) × U(1). If we split the Clifford algebra of 4+N dimensional γ-matrices into $\gamma^0,\gamma^1,\gamma^2,\gamma^3$ referring to the usual four dimensional γ-matrices, and $\Gamma^1,\ldots,\Gamma^N$ referring to the γ-matrices of the internal space B_N, then the Dirac equation becomes $\not{D}^{(4)}\psi + \not{D}^{(N)}\psi = 0$. Thus the eigenvalues of $\not{D}^{(N)}$ on B_N appear as particle masses from the four dimensional point of view. In particular, we will have massless particles if $\not{D}^{(N)}$ has zero modes in B_N. The relevant question to ask is whether the zero modes of $\not{D}^{(N)}$ transform as complex representations of G. If they do, then we have a chance of explaining the low energy world in a Kaluza-Klein context. Since B_N is a compact manifold, it is possible to answer this question using general methods. Witten[3] used the Atiyah-Hirzebrudh theorem,[16] or the character valued index theorem,[1] to show that for spin 1/2 fields the zero modes always form real representations of G, and for spin 3/2 fields he showed that the same conclusion holds as long as the internal space B_N is a homogeneous space G_1/G_2. (G_1,G_2 are Lie groups, and G_2 is a subgroup of G_1.) We will show how to extend the arguments of previous sections to also produce a proof of the character valued index theorem.

In order to explain what the problem is, assume that a given even dimensional M admits a compact isometry group G. Let $K_i^{(a)}(\phi)$ $a = 1$, dim G, be the Killing vectors generating the isometries. The infinitesimal action of $K_i^{(a)}(\phi)$ on spinors is

$$\mathcal{L}(K^{(a)})\psi = (K^{(a)i}D_i + \tfrac{1}{4}[\Gamma^i,\Gamma^j]D_iK_j^{(a)})\psi . \tag{6.1}$$

Since the $K^{(a)}$'s generate isometries, it is easy to show that $\mathcal{L}(K^{(a)})$ commutes with the Dirac operator $[\not{D},\mathcal{L}(K^{(a)})] = 0$. Following Witten,[3] we may concentrate our attention on a single Killing vector K^i generating one of the U(1) subgroups of G. Since $[\not{D},\mathcal{L}(K)] = 0$, we can classify the eigenstates of $\not{D}$ according to how they transform under the group generated by K. We can write:

$$\mathcal{L}(K)\psi_\lambda = n\,\psi_\lambda$$
$$i\not{D}\,\psi_\lambda = \lambda\,\psi_\lambda . \tag{6.2}$$

Using once again that $\{\hat{\Gamma},i\not{D}\} = 0$, we learn that for each pair of eigenvalues (λ,n) $\lambda \neq 0$ of $i\not{D}$ and $\mathcal{L}(K)$ respectively, there is another eigenstate, $\hat{\Gamma}\psi_\lambda$, with eigenvalues $(-\lambda,n)$. If we consider the set of zero eigenvalues, then the eigenfunctions are not paired any more, and we may get that for a given n there are more eigenstates with $\hat{\Gamma} = +1$ than eigenstates with $\hat{\Gamma} = -1$. Let Q_θ be a finite U(1) transformation with U(1) parameter θ. What the character valued index theorem computes is the difference between the group characters of Q_θ for zero modes of $i\not{D}$ of positive and negative chirality, i.e.

$$\mathrm{ind}\, Q_\theta = \mathrm{Tr}\,\hat{\Gamma}\,Q_\theta . \tag{6.3}$$

Since we have shown that for each eigenstate $\psi(\lambda,n)$; $\hat{\Gamma}\psi_{(\lambda,n)}$ has eigenvalues $(-\lambda,n)$, it follows that only zero eigenstates contribute to the trace. Thus we can write (6.3) as

$$\mathrm{ind}\, Q_\theta = \mathrm{Tr}\,\hat{\Gamma}\,Q_\theta \exp - \beta(i\not{D})^2 . \tag{6.4}$$

Moreover, we can deform the operator $i\not{D}$ without changing (6.4) as long as the deformation preserves the symmetries $[\mathcal{L}(K),\not{D}+\delta\not{D}] = 0$ and $\{\hat{\Gamma},\not{D}+\delta\not{D}\} = 0$. One may for example change $i\not{D}$ into $i\not{D}_t = i\not{D} + t\Gamma^i K_i$,[3] for any real value of t. Hence

$$\mathrm{ind}\, Q_\theta = \mathrm{Tr}\,\hat{\Gamma}\,Q_\theta \exp - \beta(i\not{D}_t)^2 . \tag{6.5}$$

Equation (6.5) is exactly of the type that we can compute by using our supersymmetric representation of the Dirac operator. The only difference with respect to previous cases is that $i\not{D}_t$ is not the supercharge associated to the Lagrangian (4.13), and we have to find a suitable extension of (4.13) which has $(i\not{D}_t)^2$ as its Hamiltonian. It is not difficult to check that the following Lagrangian has all the desired properties[17]

$$L^{(t)} = \frac{1}{2} g_{ij}(\phi)\dot{\phi}^i\dot{\phi}^j + \frac{i}{2} g_{ij}(\phi)\psi^i \frac{D}{dt}\psi^j - \frac{t^2}{2} g_{ij}(\phi)K^i(\phi)K^j(\phi)$$
$$- i\,\frac{t}{2}(D_iK_j)\psi^i\psi^j . \tag{6.6}$$

For the Lagrangian (6.6) $\hat{\Gamma}$ still represents $(-1)^F$ in the quantized theory, so that (6.5) may be written as

$$\operatorname{ind} Q_\theta = \operatorname{Tr}(-1)^F \, Q_\theta \exp - \beta H_t \; . \tag{6.7}$$

Using the results derived at the end of Section III (see Eqs. 3.19-3.23) we can immediately write down a functional integral representation for (6.7). Since the fermions ψ^i transform like vectors under coordinate transformations, if we let θ be the parameter which describes the isometry group, then we define $\mu^i{}_\theta(\phi)$ to be the action of the isometry on the point ϕ, with parameter θ. Then the boundary conditions on ϕ^i are $\phi^i(\beta) = \mu^i{}_\theta(\phi^j(0))$, and similarly for the fermions, $\psi^i(\beta) = (\partial\mu^i{}_\theta(\phi)/\partial\phi^j(0)$. Thus

$$\operatorname{ind} Q_\theta = \int_{(\theta)} (d\phi d\psi) \exp - \int_0^\beta L_E^{(t)}(\phi,\psi)\,dt \tag{6.8}$$

where the subscript (θ) on the path integral means that we are integrating only over those configurations described in the previous paragraph. The easiest way to evaluate (6.8) is to consider β fixed but very small, and take the limit when t becomes very large. Since M_{2N} has a positive definite Riemannian metric, $g_{ij}(\phi)K^iK^j \geq 0$, and therefore in such a limit, the leading contributions to (6.8) come from the fixed points of the Killing vector K^i, i.e. those points where K^i vanishes. Let us first assume that $K(\phi)$ has only isolated fixed points $\phi_0^{(a)}$ $a = 1,r$. Since the fixed points are isolated, the matrix $D_aK_b(\phi_0^{(a)})$ (in vielbein components) is non-degenerate and antisymmetric. (A Killing vector K has $D_aK_b + D_bK_a = 0$.) Let $\omega_\alpha^{(a)}$, $\alpha = 1,n$, $a = 1,r$, be the skew eigenvalues of D_aK_b at $\phi_0^{(a)}$:

$$D_aK_b(\phi_0^{(a)}) \sim \begin{bmatrix} 0 & \omega_1^{(a)} & & & \\ -\omega_1^{(a)} & 0 . & & & \\ & & \cdot & & \\ & & & \cdot 0 & \omega_n^{(a)} \\ & & & -\omega_n^{(a)} & 0 \end{bmatrix} \equiv M(\phi_0) \; . \tag{6.9}$$

Then for each fixed point ϕ_0, we simply have to expand ϕ^i and ψ^i around ϕ_0 with the appropriate boundary conditions. In order to make the computation clear, expand ϕ^i around ϕ_0 using normal coordinates: $\phi^i = \phi_0^i + \xi^i - \frac{1}{2}\Gamma^i{}_{jk}\xi^j\xi^k + \ldots$ Since ξ^i is a vector, the boundary conditions for ξ^i in a basis which skew diagonalizes $D_aK_b(\phi_0)$ become $\xi(\beta) = \exp - (\theta M)\,\xi(0)$ using matrix notation, and similarly for the fermions. To second order in small fluctuations, the Lagrangian (6.7) becomes the Lagrangian of a collection of bosonic and fermionic oscillators with frequencies $t\omega_\alpha$. In each of the 2×2 subspaces of skew-eigenvalue ω_α, we can further diagonalize M by introducing complex combinations of the ξ's, so that we trade the 2n ξ^a variables for n complex variables Z^α and their complex conjugates $\bar{Z}^\alpha$.

In terms of these new variables, the boundary conditions simply become $Z^{\alpha}(\beta) = e^{i\theta\omega_\alpha} Z^{\alpha}(0)$, $\bar{Z}^{\alpha}(\beta) = e^{-i\theta\omega_\alpha} \bar{Z}^{\alpha}(0)$, and similarly for the fermions. Thus, if we Fourier expand the Z's in order to compute the bosonic and fermionic determinants which appear in the leading approximation to (6.8), the Fourier frequencies are $(2\pi n + \theta\omega_\alpha)/\beta$ for each α. Using now the same methods as in Section V, it follows that

$$\mathrm{ind}\, Q_\theta = \sum_{a=1}^{r} \prod_{\alpha=1}^{n} \frac{i}{2 \sin \theta \frac{\omega_\alpha}{2}} . \tag{6.10}$$

If the fixed points of $K^i(\phi)$ are not isolated, but rather form submanifolds of M_{2n}, it is straightforward to extend the result (6.10) to that more general case. Let m be a fixed submanifold of K. We can decompose the tangent space of M_{2n} close to m into two pieces. Those vectors which are tangent to m, and those which are orthogonal. For the latter, the associated ξ^a-fluctuations correspond to the non-vanishing eigenvalues of D_aK_b. For those directions, the result is as described in (6.10). For those directions tangent to m, the terms involving the Killing vector in (6.6) vanish, and the boundary conditions are ordinary periodicity. Thus the functional integral (6.8) along those directions tangent to m are identical to the computation of the index of the Dirac operator for m. Thus if $n_a \leq n$ labels the non-vanishing eigenvalues of D_aK_b, and $a = 1, r$ now indexes the fixed submanifolds of K, we get

$$\mathrm{ind}\, Q_\theta = \sum_{a=1}^{r} \left(\prod_{\alpha=1}^{n_a} \frac{i}{2 \sin \theta \frac{\omega_\alpha}{2}} \right) \mathrm{ind}_{m_a}(i\not{D}) \tag{6.11}$$

where $\mathrm{ind}_{m_a}(i\not{D})$ is the index of $i\not{D}$ on the manifold m_a. The equation (6.11) can also be generalized to deal with the Dirac operator with spinors taking values on some vector bundle V, and it is left as an exercise for the reader to find what is the appropriate extension of (6.11) in that case. These results were used by Witten in his analysis of the chiral fermion problem in Kaluza-Klein theories.[3] As pointed out at the beginning of this section, Witten showed that the zero modes of the Dirac operator on any compact Riemannian manifold form real representations with respect to the manifold's isometry group. This result represents a rather severe setback to the Kaluza-Klein program of explaining the low energy world from a purely gravitational theory in 4+N dimensions.

VII. ANOMALIES AND INDEX THEORY

As a simple application of the formalism we have developed, we can compute the anomalies in various currents[18] in terms of the index theorem and its supersymmetric representation. The idea behind the procedure to be presented is based on a method introduced by Fujikawa[19] which relates the anomaly to the fact that the measure for the fermionic functional integral is unexpectedly not invariant under certain changes of variables. For anomalies in global symmetries, like the axial baryon current anomaly in QCD, it has been known for a while[2] that the Atiyah-Singer index theorem and the anomaly are intimately connected. What Fujikawa's method has to its advantage is that it makes this connection very clear. The basic observation of Ref. 19 is to notice that the symmetries of the classical action are not necessarily symmetries of the measure in the functional integral which defines the quantum theory. Thus classical symmetries may cease to be conserved at the quantum level. Let us illustrate these issues by computing the gravitational contribution to the axial anomaly for a fermion coupled to gravity in an arbitrary number of dimensions. Let $\psi(x)$ be a Dirac fermion living on a Riemannian manifold M_{2n} and for simplicity again assume that the metric g_{ij} has Riemannian signature. The coupling of $\psi(x)$ to the gravitational field $g_{ij}(x)$ is described by the Lagrangian

$$\mathscr{L} = e\,\bar{\psi}(x) i \gamma^{\mu} D_{\mu} \psi$$

$$D_{\mu}\psi = \partial_{\mu}\psi + \tfrac{1}{2}\omega_{\mu ab}\sigma^{ab}\psi. \tag{7.1}$$

(7.1) is clearly invariant under the global U(1) axial transformation

$$\psi(x) \to e^{i\alpha\gamma_5}\psi(x) . \tag{7.2}$$

If we make an infinitesimal space-time dependent chiral rotation, the action (7.1) changes by

$$\delta S = \int dx \sqrt{g}\,\alpha(x) D_{\mu}(\bar{\psi}\gamma^{\mu}\gamma_5\psi) . \tag{7.3}$$

In order to check whether the axial current is still conserved at the quantum level, $\nabla_{\mu}\langle\bar{\psi}\gamma^{\mu}\gamma_5\psi\rangle = 0$, we consider the effective action

$$e^{-\Gamma[g]} = \int d\bar{\psi} d\psi \exp - S(e,\psi,\bar{\psi}) \tag{7.4}$$

and make a change of variables $\psi \to \psi = \psi' + i\alpha(x)\gamma_5\psi'(x)$. The only term which could lead to anomalous contributions to the Ward identities is the Jacobian induced by the measure under the change of variables. The easiest way to define the measure in general is to expand ψ and $\bar{\psi}$ in terms of the eigenfunctions of the Dirac equation

$$i \not{D} \psi_n = \lambda_n \psi_n \tag{7.5}$$

$$\psi = \sum a_n \psi_n \qquad \bar{\psi} = \sum \bar{b}_n \psi_n^{\dagger} \tag{7.6}$$

where the a_n's and $\bar{b}_n$'s are Grassmann numbers. The measure is now defined as $\prod_{n,m} d\bar{b}_n\, da_m$. Under the infinitesimal change $\psi \to \psi' + i\alpha(x)\gamma_5\psi'(x)$, the measure changes by a Jacobian factor

$$\prod_{n,m} d\bar{b}_n\, da_m \to \Big(\prod_{n,m} d\bar{b}_n\, da_m\Big) \exp - 2i \int dx \sum \psi_n^{\dagger}(x)\alpha(x)\gamma_5\psi_n(x) \tag{7.7}$$

(The minus sign on the exponent is due to Fermi statistics.) If we let $\alpha(x)$ be a constant for simplicity, we have to evaluate

$$\sum_n \int (dx) \psi_n^\dagger \gamma_5 \psi_n \, . \tag{7.8}$$

This trace is clearly ill-defined. A simple way to define (7.8) is to use a Gaussian cut-off[19]

$$\begin{aligned} \sum \int (dx) \psi_n^\dagger \gamma_5 \psi_n &\equiv \lim_{\beta \to 0} \int (dx) \psi_n^\dagger (x) \gamma_5 \psi_n(x) \, e^{-\beta \lambda_n^2} \\ &= \lim_{\beta \to 0} \mathrm{Tr}\, \gamma_5 \, e^{-\beta (i\not{D})^2} \end{aligned} \tag{7.9}$$

The anomaly, if any, is given by the β-independent term of the right-hand side of (7.9). The procedure used in Ref. 19 to evaluate (7.9) in terms of a plane wave expansion of the trace becomes very cumbersome when trying to obtain the gravitational contribution to the axial anomaly in an arbitrary number of dimensions. At this point, we can use the formalism developed in Sec. IV and Sec. V.[20] There, we showed that the N = 1/2 supersymmetric Lagrangian (4.13) is such that $Q = i\not{D}$ and $\gamma_5 = (-1)^F$. Thus we have already evaluated the anomaly (7.9). We know that it is given by the index theorem, and that (7.9) is given in terms of the top form of the $\hat{A}$-genus (5.28).

$$\hat{A} = 1 + \frac{1}{48} \mathrm{Tr}\, \Omega^2 + \frac{1}{5760} \left(\frac{5}{4} \mathrm{Tr}\, \Omega^2)^2 - \mathrm{Tr}\, \Omega^4 \right) + \ldots$$

A more complicated example for which the method of using the supersymmetric quantum mechanical system (4.13) is much more useful is the computation of the gravitational anomaly in covariant form.[20] In this case we start by considering a 2n-dimensional Weyl fermion coupled to an external gravitational field. The only difference between the classical action (7.1) for a Dirac fermion and that for a Weyl fermion is that for the latter $i\not{D}$ is replaced by $i\not{D}_L \equiv i\not{D}(1-\Gamma)/2$. We may now proceed to define the measure by expanding ψ in terms of the eigenfunctions of $(i\not{D}_L)(i\not{D}_L)^\dagger$.[19] Let us now choose the following coordinate transformation property for the spinor:

$$\begin{aligned} \delta_\xi \psi &= \xi^i D^i \psi \\ x^i &\to x^i + \xi^i(x) \, . \end{aligned} \tag{7.10}$$

This transformation differs from a pure coordinate transformation $\psi \to \psi + \xi^i \partial_i \psi$ by a local Lorentz transformation with infinitesimal parameter $\xi^i \omega_{iab}$. The Jacobian induced by (7.10) is given by

$$J = \exp\left(\sum_n \int (dx) \psi_n^\dagger(x) \xi^i(x) D_i \psi_n(x) - \sum_n \int (dx) \phi_n^\dagger(x) \xi^i D_i \phi_n(x) \right) \tag{7.11}$$

$$(i\not{D}_L)^\dagger (i\not{D}_L) \psi_n = \lambda_n \psi_n$$

$$(i\not{D}_L)(i\not{D}_L)^\dagger \phi_n = \lambda_n \phi_n \, .$$

Regularizing the trace as before gives

$$\sum_n \int (dx)\psi_n^\dagger \xi^i(x) D_i \psi_n - \sum_n \int (dx)\phi_n^\dagger \xi^i(x) D_i \psi_n \equiv$$

$$\equiv \lim_{\beta\to 0} \int (dx) \Big(\sum_n \psi_n^\dagger(x)\xi^i D_i e^{-\beta(i\not{D}_L)^\dagger(i\not{D}_L)} \psi_n -$$

$$- \sum_n \phi_n^\dagger(x) \xi^i D_i e^{-\beta(i\not{D}_L)(i\not{D}_L)^\dagger} \phi_n =$$

$$= \lim_{\beta\to 0} \mathrm{Tr}\, \xi^i D_i \gamma_5 \exp - \beta(i\not{D})^2 \qquad (7.12)$$

which is very similar to (7.9) except that now we have the linear operator $\xi^i D_i$ inside the trace. The nice thing about (7.12) is that it also admits a representation in terms of $N = 1/2$ supersymmetry. Using (4.13) it follows easily that $\xi^i D_i$ is represented in the $N = 1/2$ theory by $i\,\xi_i(\phi)d\phi^i/dt$. Thus (7.12) is identical to

$$\lim_{\beta\to 0} \int_{P.B.C.} \Big[d\phi(t)d\psi(t) \Big] \Big(-\xi_i(\phi) \frac{d\phi^i}{dt} \Big) \exp - \int_0^\beta L_E(\phi,\psi)dt \,, \qquad (7.13)$$

which can be computed by expanding around the constant configuration (ϕ_0,ψ_0) which dominate the path integral in the $\beta\to 0$ limit. Since we already know how to expand the exponent to second order, we only have to concentrate on expanding $\xi_i(\phi)\, d\phi^i/dt$ to second order. Since both $\xi_i(\phi)$ and $d\phi^i/dt$ transform like vectors under reparameterizations of the manifold, it is easy to obtain that the expansion to second order in small fluctuations. Instead of using the naive expansion of Sec. V, it is more suitable to use normal coordinates, so that $u^i = \zeta^i - \frac{1}{2}\Gamma^i{}_{jk}\zeta^j\zeta^k$, where the ζ^i's transform like vectors under coordinate reparameterizations. Then $\xi_i(\phi)\, d\phi^i/dt$ becomes to second order $D_i\,\xi_j(\phi_0)\zeta^i\dot{\zeta}^j$. The computation is further simplified if we exponentiate $D_i\,\xi_j(\phi_0)\zeta^i\dot{\zeta}^j$, and at the end expand to first order in ξ_i. Once we exponentiate, the only change with respect to the computation of the Dirac index, is that the "mass" term $\frac{1}{2}\Omega_{abcd}\psi_0^c\psi_0^d = \Omega_{ab}$ gets substituted by $\Omega_{ab} + D_a\eta_b - D_b\eta_a \equiv \Omega'_{ab}$. If we denote the skew eigenvalues of Ω'_{ab} by x'_α, $\alpha = 1,\ldots,n$, the result can be copied from (5.28):

$$\frac{1}{(2\pi)^n} \int d(\mathrm{vol}) \int (d\psi_0) \prod_\alpha \frac{x'_\alpha/2}{\sinh x'_\alpha/2} \,. \qquad (7.14)$$

Taylor expanding (7.14) in $\xi(\phi_0)$, the anomaly will be given by the term which is first order in η^i and contains 2n ψ_0's. Since the index of the Dirac operator is only different from zero in 4k dimensions, it is clear that from (7.14) one gets an anomaly only in 4k-2 dimensions.[20] Similar arguments can be carried out for spin 3/2 fields and also in the case where external gauge fields are included. The interested reader is referred to the literature for more details.[20] There is only one technical detail which should be emphasized. We have outlined a method of computing the anomaly with respect to coordinate transformations. What happens however is that the form of the anomaly so obtained does not satisfy the gravitational Wess-Zumino consistency conditions, so that one has to add contact terms to the current in order to obtain the form of the anomaly which satisfies the consistency conditions. Although the construction of the contact terms for the currents is a bit complicated,[21] one finds that the anomaly cancellation conditions between different representations remains the same.[20] Those interested should consult the literature for further details.

VIII. CONCLUSIONS AND OUTLOOK

Throughout these lectures we have tried to present a reasonably self-contained guide to the connections between supersymmetry and index theory. Although no new results were derived, we believe that the methods used are simple enough to deserve further attention. In particular, it is reasonable to expect a supersymmetric version of the index theorem for families of elliptic operators.[1] These types of generalized index theorems have recently been applied in the physics literature[4] in the context of gauge and gravitational anomalies. This preliminary use of the index for families of operators indicates that it may become a rather useful tool in obtaining qualitative understanding of non-perturbative aspects of the interplay between fermions and gauge fields.[22,23]

Acknowledgements

I would like to thank J. Bagger, P. Ginsparg and P. Windey for useful comments which helped improve the final form of these lectures. I would also like to thank the organizers of this school for giving me the opportunity to present this material in such an stimulating environment.

Research is supported in part by the National Science Foundation under Grant No. PHY82-15249

References

1. M. F. Atiyah and I. M. Singer, Ann. Math. 87 (1968) 485; M. F. Atiyah and G. B. Segal, Ann. Math. 87 (1968) 531; M. F. Atiyah and I. M. Singer, Ann. Math. 87 (1968) 546; *ibid.* 93 (1971) 119; 93 (1971) 139. A pedagogical exposition of the index theorem intended for physicists can be found in T. Eguchi, P. B. Gilkey and A. J. Hanson, Phys. Rep. 66 (1980) 213.
2. R. Jackiw, C. Nohl and C. Rebbi in *Particles and Fields*, D. Boal and A. Kamal, eds. (Plenum Press, N.Y., 1978); N. K. Nielsen, H. Römer and B. Schroer, Nucl. Phys. B136 (1978) 478. For a recent review see R. Jackiw in the Proceedings of the Les Houches

Summer School, August 1983, MIT-CTP-1089, 1149.
3. E. Witten in Proceedings of the Shelter Island Conference - II, June 1983. Princeton preprint.
4. M. F. Atiyah and I. M. Singer, Proc. Natl. Acad. Sci., USA 81 (1984) 2597; L. Alvarez-Gaumé and P. Ginsparg, Nucl. Phys. B243 (1984) 449; C. Gomez, Salamanca preprint (1983); O. Alvarez, I. M. Singer and B. Zumino, University of California at Berkeley preprint PTH-84/9 ; J. Lott, MIT-CTP-1146,1147; J. Mickels, HU-TFT-83-57; T. Sumitani, U. T. Komaba-84-7.
5. The material in this section is fairly standard and can be found in textbooks. See for instance, Lichnerowicz, *Global Theory of Connections and the Holonomy Group* (Noordhoff, Holland,1976); S. S. Chern, *Complex Manifolds without Potential Theory* (Springer-Verlag, N.Y., 1974); Shananan, *The Atiyah-Singer Index Theorem* Springer-Verlag, 1974); and the review of T. Eguchi, P. B. Gilkey and A. J. Hanson quoted in Ref. 1.
6. M. F. Atiyah, R. Bott, V. K. Patodi, Invent Math. 19 (1973) 279; see also P. B. Gilkey in *The Index Theorem and the Heat Equation* (Publish or Perish, Inc., Boston, 1974).
7. Most of the material contained in this lecture can be found in the following texts: Berezin, *The Method of Second Quantization;* L. Faddeev and A. Slavnov, *Gauge Fields: The Quantum Theory* (Benjamin, 1982).
8. R. Haag, J. Lopuszanski and M. Sohnius, Nucl. Phys. B88 (1975) 61.
9. E. Witten, Nucl. Phys. B202 (1982) 253.
10. L. Alvarez-Gaumé, Comm. Math. Phys. 90 (1983) 161, J. Phys. A16 (1983) 4177; P. Windey, Lectures delivered at the Zakopane School, Poland, June 1983; D. Friedan and P. Windey, Nucl. Phys. B235 (1984) 395. These papers use the functional integral formulation of quantum mechanics in order to derive the index theorem from supersymmetry. The Hamiltonian formulation has been presented in E. Getzler, Comm. Math. Phys. 92 (1983) 163, from a very mathematical point of view, and in B. Zumino, Proceedings of the Shelter Island Conference II, Preprint LBL-17972, and J. Mañes and B. Zumino (in preparation).
11. P. Di Vecchia and S. Ferrara, Nucl. Phys. B130 (1977) 93; E. Witten, Phys. Rev. D16 (1977) 2491; D. Z. Freedman and P. K. Townsend, Nucl. Phys. B177 (1981) 282.
12. B. Zumino, Phys. Lett. B87 (1979) 203; L. Alvarez-Gaumé and D. Z. Freedman, Comm. Math. Phys. 80 (1981) 443.
13. L. Alvarez-Gaumé, D. Z. Freedman and S. Mukhi, Ann. of Phys. 134 (1981) 85.
14. A very detailed exposition of normal coordinates can be found in Petrov, *Einstein Spaces* (Pergamon Press).
15. A thorough discussion of grand unified theories can be found in P. Langacker, Phys. Rep. 72 (1981) 185.
16. M. F. Atiyah and F. Hirzebruch, in *Essays on Topology*, ed. Haefliger and Narashiman (Springer-Verlag, 1980).
17. L. Alvarez-Gaumé and D. Z. Freedman, Comm. Math. Phys. 91 (1983) 87.

18. For reviews on anomalies see S. L. Adler in *Lectures on Elementary Particles and Quantum Field Theory*, eds. S. Deser *et al.* (M.I.T. Press, 1979), and R. Jackiw in *Lectures on Current Algebra and Its Applications* (Princeton University Press, 1970).
19. K. Fujikawa, Phys. Rev. Lett. 42 (1979) 1195; 44 (1980) 1733.
20. L. Alvarez-Gaumé and E. Witten, Nucl. Phys. B234 (1984) 269.
21. W. Bardeed and B. Zumino, Univ. of California at Berkeley preprint PTH-84/12; L. Alvarez-Gaumé and P. Ginsparg, Harvard preprint HUTP-84/A016, to appear in Annals of Physics.
22. C. Vafa and E. Witten, Princeton preprints, "Restrictions on Symmetry Breaking in Vector-Like Gauge Theories" and "Eigenvalue Inequalities for Fermions in Gauge Theories".
23. A fairly clear exposition of the index for families of elliptic operators, and K-theory has recently appeared. See G. Moore and P. Nelson, Harvard preprint HUTP-84/A076.

SUPERSYMMETRIC SIGMA MODELS*

JONATHAN A. BAGGER

Stanford Linear Accelerator Center
Stanford University
Stanford, California 94305

ABSTRACT

We begin to construct the most general supersymmetric Lagrangians in one, two and four dimensions. We find that the matter couplings have a natural interpretation in the language of the nonlinear sigma model.

1. INTRODUCTION

The past few years have witnessed a dramatic revival of interest in the phenomenological aspects of supersymmetry [1,2]. Many models have been proposed, and much work has been devoted to exploring their experimental implications. A common feature of all these models is that they predict a variety of new particles at energies near the weak scale. Since the next generation of accelerators will start to probe these energies, we have the exciting possibility that supersymmetry will soon be found.

While we are waiting for the new experiments, however, we must continue to gain a deeper understanding of supersymmetric theories themselves. One vital task is to learn how to construct the most general possible supersymmetric Lagrangians. These Lagrangians can then be used by model builders in their search for realistic theories. For $N = 1$ rigid supersymmetry, it is not hard to write down the most general possible supersymmetric Lagrangian. For higher N, however, and for all local supersymmetries, the story is more complicated.

In these lectures we will begin to discuss the most general matter couplings in $N = 1$ and $N = 2$ supersymmetric theories. We will start in $1 + 1$ dimensions, where we will construct the most general supersymmetric couplings of the massless spin $(0, \frac{1}{2})$ matter multiplet [3,4]. We will consider $N = 1$, $N = 2$ and $N = 4$ supersymmetric theories, and we shall find that as N increases, the matter couplings become more and more restricted.

We shall then drop to $0+1$ dimensions, where we will discuss supersymmetric quantum mechanics. Although supersymmetric quantum mechanics might seem irrelevant, it has important mathematical and physical consequences. For example, it has been used to demonstrate dynamical supersymmetry breaking [5], to prove the Atiyah-Singer index theorem [6], and even to invent a new branch of Morse theory [7].

After introducing supersymmetric quantum mechanics, we shall climb back to $3+1$ dimensions, where we will remain for the rest of the lectures. We will first consider $N = 1$ supersymmetry, both rigid and local. We will construct the most general Lagrangian containing the spin $(0, \frac{1}{2})$ and $(\frac{1}{2}, 1)$ supersymmetry multiplets [8–13]. The resulting Lagrangian will be long and complicated, but it will have a very simple geometrical interpretation. In the case of local supersymmetry, we shall see that global topology places important restrictions on the matter couplings. A surprising result is that in some cases, global consistency requires that Newton's constant be quantized in units of the scalar self-coupling [11].

For our final topic, we will begin to discuss the most general couplings of the massless spin $(0, \frac{1}{2})$ multiplet in $N = 2$ supersymmetry. We will not attempt to include masses, potentials or gauge fields, but we will still find a striking result: Matter couplings that are allowed in $N = 2$ rigid supersymmetry are forbidden in $N = 2$ local supersymmetry, and vice versa [14]. Furthermore, we shall see that the reduction from $N = 2$ to $N = 1$ is not trivial.

2. SUPERSYMMETRY IN ONE AND TWO DIMENSIONS

THE SUPERSYMMETRIC NONLINEAR SIGMA MODEL

As we shall see throughout these lectures, supersymmetric matter couplings generally induce complicated terms in the interaction Lagrangian. Fortunately, these terms have a relatively simple description in the language of the nonlinear sigma model. The most familiar sigma model is the famous $O(3)$ model introduced by Gell-Mann and Lévy [15]. In this model, three pion fields π^a are constrained to lie on the sphere S^3. The Lagrangian $\mathcal{L}$ is given by

$$\mathcal{L} \;=\; -\,\frac{1}{2}\, g_{ab}(\pi^c)\, \partial_\mu \pi^a \partial^\mu \pi^b \,, \tag{2.1}$$

where $g_{ab}(\pi^c)$ is the metric on S^3,

$$g_{ab}(\pi^c) \;=\; \left[(f^2 - \vec{\pi}\cdot\vec{\pi})\delta_{ab} + \pi_a \pi_b\right] \Big/ (f^2 - \vec{\pi}\cdot\vec{\pi}) \,. \tag{2.2}$$

From this example it is clear how to generalize the sigma model to other Riemannian manifolds $\mathcal{M}$. One simply views the scalar fields ϕ^a as maps from spacetime into $\mathcal{M}$,

$$\phi^a \;:\; \text{spacetime} \;\rightarrow\; \mathcal{M} \; . \tag{2.3}$$

The bosonic Lagrangian is given by

$$\mathcal{L} \;=\; -\,\frac{1}{2}\, g_{ab}(\phi^c)\, \partial_\mu\phi^a \partial^\mu\phi^b \; , \tag{2.4}$$

where $g_{ab}(\phi^c)$ is the metric on the manifold $\mathcal{M}$. Note that for each value of μ, $\partial_\mu\phi^a$ is a vector in the tangent space of $\mathcal{M}$. The Lagrangian $\mathcal{L}$ is constructed entirely from geometrical objects, and transforms as a scalar under coordinate transformations in $\mathcal{M}$. Furthermore, $\mathcal{L}$ is left invariant under the various isometries of $\mathcal{M}$.

In $1+1$ dimensions, it is easy to find a supersymmetric extension of (2.4) [3,4]. The superspace Lagrangian is given by[♮1]

$$\mathcal{L} \;=\; \int d^2\theta \; g_{ab}(\Phi^c)\, \overline{D}\Phi^a D\Phi^b \; . \tag{2.5}$$

Here Φ^a is a real scalar superfield in $1+1$ dimensions [17],

$$\Phi^a(x,\theta) \;=\; \phi^a(x) \;+\; \overline{\theta}\chi^a(x) \;+\; \frac{1}{2}\overline{\theta}\theta\, F^a(x) \; , \tag{2.6}$$

and $g_{ab}(\Phi^c)$ is again the metric. The component fields ϕ^a and F^a are real and bosonic, and χ^a is a two-component Majorana spinor. The spinor derivative

$$D \;=\; \frac{\partial}{\partial\overline{\theta}} \;+\; (\overline{\theta}\gamma^\mu)\partial_\mu \tag{2.7}$$

and the superfield Φ^a both contain the real Grassmann parameter $\overline{\theta} = (-\theta_2, \theta_1)$. The derivative D anticommutes with the spinor supercharge Q, so the Lagrangian (2.5) is manifestly supersymmetric.

♮1 Our metric and spinor conventions are those of Ref. [16].

To see that (2.5) corresponds to the sigma model (2.4), we must expand Φ^a in terms of component fields. Inserting (2.6) in (2.5), taking the highest component, and eliminating the auxiliary fields F^a, we find the following component Lagrangian [4],

$$\begin{aligned}\mathcal{L} = &-\frac{1}{2} g_{ab}(\phi^c)\,\partial_\mu\phi^a\partial^\mu\phi^b - \frac{1}{2} g_{ab}(\phi^c)\,\overline{\chi}^a\gamma^\mu \mathcal{D}_\mu\chi^b \\ &+ \frac{1}{12} R_{abcd}\,(\overline{\chi}^a\chi^c)(\overline{\chi}^b\chi^d)\,.\end{aligned} \tag{2.8}$$

The covariant derivative

$$\mathcal{D}_\mu\chi^b = \partial_\mu\chi^b + \Gamma^b{}_{cd}\partial_\mu\phi^c\chi^d \tag{2.9}$$

ensures that χ^b transforms in the tangent space of $\mathcal{M}$. The connection $\Gamma^b{}_{cd}$ is the Christoffel symbol on the (Riemannian) manifold $\mathcal{M}$, and R_{abcd} is the usual Riemann curvature.

The Lagrangian (2.8) is manifestly supersymmetric because it was derived from a superspace formalism. However, it is useful to check that it is invariant (up to a total derivative) under the following supersymmetry transformations:

$$\begin{aligned}\delta\phi^a &= \overline{\epsilon}\chi^a\,, \\ \delta\chi^a &= \gamma^\mu\partial_\mu\phi^a\epsilon - \Gamma^a{}_{bc}\delta\phi^b\chi^c\,.\end{aligned} \tag{2.10}$$

The term with the connection coefficient is required to ensure that supersymmetry transformations commute with the coordinate transformations, or diffeomorphisms, of the manifold $\mathcal{M}$.

Models with Extended Supersymmetry

The Lagrangian (2.8) demonstrates that $N = 1$ supersymmetric sigma models exist for all Riemannian manifolds $\mathcal{M}$. We would now like to know which manifolds give rise to sigma models with extended supersymmetry ($N > 1$). In this section we shall see that extra supersymmetries imply strong restrictions on the manifolds $\mathcal{M}$ [4,8].

To discover exactly what restrictions follow from additional supersymmetries, we take the most general ansatz for the transformation laws, consistent with dimensional arguments and Lorentz and parity invariance,

$$\begin{aligned}\delta\phi^a &= I^a{}_b\,\overline{\epsilon}\chi^b \\ \delta\chi^a &= H^a{}_b\gamma^\mu\partial_\mu\phi^b\epsilon - S^a{}_{bc}(\overline{\epsilon}\chi^b)\chi^c + V^a{}_{bc}(\overline{\epsilon}\gamma^\mu\chi^b)\gamma_\mu\chi^c \\ &\quad + P^a{}_{bc}(\overline{\epsilon}\gamma_5\chi^b)\gamma_5\chi^c\,.\end{aligned} \tag{2.11}$$

Here I, H, S, V and P are all functions of the dimensionless field ϕ^a. Since the transformations (2.11) must commute with diffeomorphisms, I, H, V and P must all be tensors.

We now demand that the Lagrangian (2.8) be invariant (up to a total derivative) under the transformations (2.11). Cancellation of the $\bar{\epsilon}\chi$ terms requires

$$g_{ac}I^c{}_b = g_{bc}I^c{}_a\,, \qquad \nabla_c I^a{}_b = \nabla_c H^a{}_b = 0\,, \tag{2.12}$$

while cancellation of the $(\bar{\epsilon}\chi)(\bar{\chi}\chi)$ terms implies

$$V^a{}_{bc} = P^a{}_{bc} = 0\,, \qquad S^a{}_{bc} = \Gamma^a{}_{cd}I^d{}_b\,, \qquad I^a{}_b H^b{}_c = \delta^a{}_c\,. \tag{2.13}$$

The conditions (2.12) and (2.13) lead automatically to the cancellation of the $(\bar{\epsilon}\chi)(\bar{\chi}\chi)(\bar{\chi}\chi)$ terms in the variation of $\mathcal{L}$.

The conditions (2.12) and (2.13) tell us that each additional supersymmetry requires the existence of a covariantly constant tensor $I^a{}_b$, such that

$$g_{ab}I^a{}_c I^b{}_d = g_{cd}\,. \tag{2.14}$$

For each such tensor, there is a fermionic transformation law given by

$$\begin{aligned} \delta\phi^a &= I^a{}_b\bar{\epsilon}\chi^b\,, \\ \delta\chi^a &= (I^{-1})^a{}_b\gamma^\mu\partial_\mu\chi^b\epsilon - \Gamma^a{}_{bc}\delta\phi^b\chi^c\,. \end{aligned} \tag{2.15}$$

Note, however, that the transformations (2.15) are not quite supersymmetries. To be supersymmetries, they must also obey the supersymmetry algebra

$$\{Q^{(A)}, \overline{Q}^{(B)}\} = 2\gamma^\mu P_\mu\,\delta^{AB}\,, \tag{2.16}$$

where $A, B = 1, \ldots, N$. If there are several supersymmetries, each with its own covariantly constant tensor $I^{(A)a}{}_b$, then (2.16) requires

$$I^{(A)}I^{(B)-1} + I^{(B)}I^{(A)-1} = 2\delta^{AB}\,, \tag{2.17}$$

where we have switched to matrix notation. Let us suppose that $A = 1$ or $B = 1$ denotes the first supersymmetry, so that $I^{(1)a}{}_b = \delta^a{}_b$. Then, when $A \neq 1$ and $B = 1$, (2.17) implies

$$I^{(A)} = -\,I^{(A)-1}\,, \tag{2.18}$$

or

$$I^{(A)a}{}_c I^{(A)c}{}_b = -\,\delta^a{}_b\,, \qquad I^{(A)}{}_{ab} = -\,I^{(A)}{}_{ba}\,, \tag{2.19}$$

(for $A \neq 1$). We see that each extra supersymmetry requires the existence of a covariantly constant tensor whose square is -1. Furthermore, equation

(2.17) implies a Clifford algebra structure for all supersymmetries beyond the first,

$$I^{(A)}I^{(B)} + I^{(B)}I^{(A)} = -2\delta^{AB} , \tag{2.20}$$

where $A, B \neq 1$.

The preceeding analysis tells us that any Riemannian manifold gives rise to a supersymmetric sigma model in $1+1$ dimensions. However, a second supersymmetry exists only if one can define a covariantly constant tensor field $I^a{}_b$, satisfying

$$I^a{}_c I^c{}_b = -\delta^a{}_b , \qquad g_{ab} I^a{}_c I^b{}_d = g_{cd} . \tag{2.21}$$

Such a tensor field is called a *complex structure*; it denotes "multiplication by i" in the tangent space. A manifold $\mathcal{M}$ that admits such a tensor field is called a *Kähler manifold*. In $1+1$ dimensions, $N = 2$ supersymmetry requires that $\mathcal{M}$ be Kähler [4,8].

Further supersymmetries require the existence of additional parallel complex structures satisfying (2.20) and (2.21). Note that if two such structures exist, their product automatically generates a third, so $N = 3$ supersymmetry automatically implies $N = 4$. A manifold with three parallel complex structures satisfying (2.20) and (2.21) is called a *hyperkähler manifold*. In $1+1$ dimensions, $N = 4$ supersymmetry requires that $\mathcal{M}$ be hyperkähler [4].

The various manifolds discussed above can be described very elegantly in terms of their holonomy groups G [18]. The *holonomy group* of a connected n-dimensional Riemannian manifold is the group of transformations generated by parallel transporting all vectors around all possible closed curves in $\mathcal{M}$. If the parallel transport is done with respect to the Riemannian connection, the transported vector V'^a will be related to the original vector V^a by a rotation in the tangent space. In infinitesimal form, we have

$$[\nabla_a, \nabla_b] V^c = R_{ab}{}^c{}_d V^d , \tag{2.22}$$

and we see that the holonomy group G is generated by the Riemann curvature tensor of the manifold.

In general, an n-dimensional Riemannian manifold has holonomy group $O(n)$, provided the parallel transport is done with respect to the Riemannian connection. If, however, the manifold admits a parallel complex structure, the holonomy group is not all of $O(n)$. Since the complex structure $I^{(A)a}{}_b$ is parallel, it commutes with the holonomy group,

$$[\nabla_a, \nabla_b]\, I^{(A)c}{}_d = 0 \qquad \Rightarrow \qquad I^{(A)c}{}_f\, R_{ab}{}^f{}_d - R_{ab}{}^c{}_f\, I^{(A)f}{}_d = 0 . \tag{2.23}$$

Manifolds $\mathcal{M}$ whose holonomy group G leaves invariant one complex structure are called Kähler. They necessarily have dimension $2n$, and their holonomy group is not all of $O(2n)$, but rather $G \subseteq U(n) \subseteq O(2n)$. Manifolds $\mathcal{M}$ whose holonomy group leaves invariant the quaternionic structure (2.20), (2.21) are called hyperkähler. They necessarily have dimension $4n$ and holonomy group $G \subseteq Sp(n) \subseteq O(4n)$. The relation between supersymmetry, sigma model manifolds and holonomy groups is summarized in Table 1.

Table 1
Rigid Supersymmetry in $1+1$ Dimensions

N	manifold $\mathcal{M}$	holonomy G
1	n-dimensional Riemannian	$\subseteq O(n)$
2	$2n$-dimensional Kähler	$\subseteq U(n)$
4	$4n$-dimensional hyperkähler	$\subseteq Sp(n)$

SUPERSYMMETRIC QUANTUM MECHANICS

With these results, it is easy to specialize to $0+1$ dimensions and discuss supersymmetric quantum mechanics [5]. One simply assumes that the fields ϕ^a and χ^a do not depend on the spatial coordinate x. Thus the action for supersymmetric quantum mechanics is given by

$$\begin{aligned} \mathcal{A} = \int dt \Big[-\frac{1}{2} g_{ab}(\phi^c)\, \dot{\phi}^a \dot{\phi}^b - \frac{1}{2} g_{ab}(\phi^c) \overline{\chi}^a \gamma^0 \frac{\mathcal{D}}{\mathcal{D}t} \chi^b \\ + \frac{1}{12} R_{abcd} (\overline{\chi}^a \chi^c)(\overline{\chi}^b \chi^d) \Big] , \end{aligned} \tag{2.24}$$

where, as before, $g_{ab}(\phi^c)$ is the metric on the manifold $\mathcal{M}$, R_{abcd} is the Riemann curvature tensor, $\mathcal{D}\chi^b/\mathcal{D}t = \dot{\chi}^b + \Gamma^b{}_{cd}\dot{\phi}^c\chi^d$ is the covariant derivative, and all fields depend only on time. One can think of this Lagrangian as describing an object with a tower of spins (up to spin $\frac{1}{2}n$) moving on the n-dimensional manifold $\mathcal{M}$. The action (2.24) can be used to derive the Atiyah-Singer index theorem for the de Rahm and signature complexes [6].

The Lagrangian (2.24) can be further restricted to have $N = \frac{1}{2}$ (but not $N = 1$) supersymmetry. To do this, one identifies the components of each Majorana spinor, ${\chi_1}^a = {\chi_2}^a = \chi^a$, $\epsilon_1 = \epsilon_2 = \epsilon$. Under this restriction, the curvature term in (2.24) vanishes, while the kinetic terms remain as before. The new Lagrangian has $N = \frac{1}{2}$ supersymmetry because the supersymmetry parameter ϵ contains half the number of degrees of freedom. The $N = \frac{1}{2}$ Lagrangian can be used to derived the Atiyah-Singer index theorem for the Dirac spin complex [6].

3. $N = 1$ SUPERSYMMETRY IN FOUR DIMENSIONS

$N = 1$ SUPERSYMMETRY AND KÄHLER GEOMETRY

Lest we be accused of spending too much time in too few dimensions, we shall now turn to the important question of supersymmetric matter couplings in $3+1$ dimensions. We shall start by considering the most general coupling of spin $(0, \frac{1}{2})$ multiplets in $N = 1$ rigid supersymmetry. We expect our $3+1$ dimensional results to be related to the $1+1$ dimensional results discussed earlier. This is because $(0, \frac{1}{2})$ multiplets exist in both $1+1$ and $3+1$ dimensions, and $N = 1$ supersymmetry in $3+1$ dimensions reduces to $N = 2$ supersymmetry in $1+1$ dimensions. Indeed, we will find that the most general $(0, \frac{1}{2})$ matter coupling in $3+1$ dimensions may be described by a sigma model. For $N = 1$ rigid supersymmetry, we will see that the scalar fields ϕ^i must be the coordinates of a Kähler manifold $\mathcal{M}$.

However, in $3+1$ dimensions this result arises in a different way than in $1+1$ dimensions. This is because in $3+1$ dimensions, spin $(0, \frac{1}{2})$ multiplets are described by *chiral* superfields Φ^i, $\overline{D}\Phi^i = 0$. The lowest component of a chiral superfield is a *complex* scalar field, so the natural way to describe a Kähler manifold in $3+1$ dimensions is in terms of complex coordinates.

Therefore, before actually constructing the sigma model, let us take a moment to discuss complex manifolds, in general, and Kähler manifolds, in particular [19,20]. An n-dimensional *complex manifold* is a $2n$-dimensional real Riemannian manifold whose $2n$ real coordinates can be regarded as n complex coordinates z^i together with their n conjugates z^{*j}. Consistency requires that neighboring coordinate patches be linked by holomorphic (analytic) transition functions, $z'^i = f^i(z^j)$.

In terms of the $2n$ real coordinates $x^a = \{\mathrm{Re}\, z^i, \mathrm{Im}\, z^j\}$, a complex manifold has a globally defined tensor field $I^a{}_b$, such that

$$I^a{}_b I^b{}_c \;=\; -\,\delta^a{}_c\,. \tag{3.1}$$

As before, the field $I^a{}_b$ defines multiplication by i in the tangent space, and is called an almost complex structure. All complex manifolds are endowed with an almost complex structure. However, not all manifolds with an almost complex structure are, in fact, complex. The necessary and sufficient condition for this to hold is that the Nijenhuis tensor must vanish:

$$I^d{}_a \partial_d I^b{}_c \;-\; I^d{}_c \partial_d I^b{}_a \;-\; I^b{}_d \partial_a I^d{}_c \;+\; I^b{}_d \partial_c I^d{}_a \;=\; 0\,. \tag{3.2}$$

Equation (3.2) is the integrability condition for the existence of complex coordinates. When (3.2) is satisfied, $I^a{}_b$ is called a *complex structure.*

Having defined complex manifolds, we shall now restrict our attention to hermitian manifolds. A *hermitian manifold* is a complex manifold on which the line element ds^2 takes the following form:

$$ds^2 = g_{ij^*}\, dz^i\, dz^{*j} . \tag{3.3}$$

The matrix g_{ij^*} is hermitian, so ds^2 is real. For $\mathcal{M}$ to be hermitian, the metric g_{ab} must be invariant under the complex structure,

$$g_{ab} = g_{cd}\, I^c{}_a I^d{}_b . \tag{3.4}$$

In this case $I_{ab} = g_{ac} I^c{}_b$ is an antisymmetric tensor.

On any hermitian manifold, the tensor I_{ab} defines a fundamental two-form Ω,

$$\begin{aligned} \Omega &= \frac{1}{2}\, I_{ab}\, dx^a \wedge dx^b \\ &= \frac{i}{2}\, g_{ij^*}\, dz^i \wedge dz^{*j} . \end{aligned} \tag{3.5}$$

The two-form Ω is known as the Kähler form; it is both real and nondegenerate. A *Kähler manifold* is a hermitian manifold on which the Kähler form is closed,

$$d\Omega = 0 . \tag{3.6}$$

This is equivalent to saying that the complex structure is parallel with respect to the Riemann connection

$$\nabla_a I^b{}_c = 0 . \tag{3.7}$$

Equations (3.1), (3.4), and (3.7) are the same as equations (2.12), (2.14), and (2.19) of Section 2.

The fact that the Kähler form is closed leads to important consequences. In terms of complex coordinates, equation (3.6) implies

$$\frac{\partial}{\partial z^k} g_{ij^*} - \frac{\partial}{\partial z^i} g_{kj^*} = 0 , \quad \frac{\partial}{\partial z^{*k}} g_{ij^*} - \frac{\partial}{\partial z^{*j}} g_{ik^*} = 0 . \tag{3.8}$$

This means that *locally* the metric g_{ij^*} is the second derivative of real scalar function $K(z^i, z^{*j})$:

$$g_{ij^*} = \frac{\partial}{\partial z^i} \frac{\partial}{\partial z^{*j}} K(z^k, z^{*\ell}) . \tag{3.9}$$

The function K is known as the Kähler potential. It is important to note that (3.9) does not determine K uniquely. The metric g_{ij^*} is invariant under shifts of the Kähler potential,

$$K(z^i, z^{*j}) \rightarrow K(z^i, z^{*j}) + F(z^i) + F^*(z^{*j}) , \tag{3.10}$$

where $F(z^i)$ is a holomorphic function of the coordinates. Although the metric g_{ij^*} is defined over the entire Kähler manifold, the potential K is in general defined only locally.

The Kähler condition (3.8) severely restricts the connection coefficients of the metric connection. In complex coordinates, the only nonvanishing components are

$$\Gamma^i{}_{jk} = g^{i\ell^*} \frac{\partial}{\partial z^k} g_{j\ell^*} , \qquad \Gamma^{i^*}{}_{j^*k^*} = g^{\ell i^*} \frac{\partial}{\partial z^{*k}} g_{\ell j^*} . \tag{3.11}$$

From these it is a simple exercise to compute the nonvanishing components of the curvature tensor,

$$R_{ij^*k\ell^*} = - R_{j^*ik\ell^*} = - R_{ij^*\ell^*k} = R_{j^*i\ell^*k} , \tag{3.12}$$

where

$$R_{ij^*k\ell^*} = - g_{mj^*} \frac{\partial}{\partial z^{*\ell}} \Gamma^m{}_{ik} . \tag{3.13}$$

We are now ready to discuss the $N = 1$ matter coupling. In $3+1$ dimensions, the most general coupling of chiral superfields is given by

$$\mathcal{L} = \int d^4\theta \, K(\Phi^i, \Phi^{*j}) + \int d^2\theta \, W(\Phi^i) . \tag{3.14}$$

The function K is real, and W is analytic. For the moment let us set W to zero.

To find the component Lagrangian, one must expand Φ^i in components,

$$\begin{aligned} \Phi^i(x,\theta,\overline{\theta}) = \; & \phi^i(x) + \sqrt{2}\, \overline{\theta}_R \chi_L{}^i + \overline{\theta}_R \theta_L F^i - \overline{\theta}_L \gamma^\mu \theta_L \partial_\mu \phi^i \\ & + \frac{1}{\sqrt{2}} \overline{\theta}_R \theta_L \, \overline{\theta}_L \gamma^\mu \partial_\mu \chi_L{}^i + \frac{1}{4} \overline{\theta}_R \theta_L \, \overline{\theta}_R \theta_L \, \partial^2 \phi^i , \end{aligned} \tag{3.15}$$

where $\chi_{L,R} = (1 \pm \gamma_5)\chi$. One must do the θ-integrals and eliminate the auxiliary fields F^i. It turns out that the component Lagrangian has a geometrical interpretation precisely when the function K is identified with the Kähler potential [8]. In this case the component Lagrangian takes the following form

$$\begin{aligned} \mathcal{L} \;=\; & - g_{ij^*}\, \partial_\mu \phi^i \partial^\mu \phi^{*j} \;-\; \frac{1}{2} \overline{\chi}^i\, \widehat{g_{ij^*}}\, \gamma^\mu \mathcal{D}_\mu \chi^j \\ & +\; \frac{1}{32} R_{ij^*k\ell^*}\, \overline{\chi}^i (1+\gamma_5)\gamma_\mu \chi^j\, \overline{\chi}^k (1+\gamma_5)\gamma^\mu \chi^\ell \,. \end{aligned} \qquad (3.16)$$

The hat notation is from Ref. [9]; it is basically a fancy way to write two-component spinors in four-component notation ($\widehat{A} = \operatorname{Re} A + i\gamma_5 \operatorname{Im} A$). The scalar fields ϕ^i parametrize a Kähler manifold $\mathcal{M}$. The spinors χ^i are four-component Majorana spinors; they lie in the complexified tangent bundle over $\mathcal{M}$. The covariant derivative is given by

$$\mathcal{D}_\mu \chi^i \;=\; \partial_\mu \chi^i \;+\; \Gamma^i{}_{jk} \partial_\mu \phi^j \chi^k \,, \qquad (3.17)$$

where the connection coefficients are given in (3.11). The Kähler curvature tensor $R_{ij^*k\ell^*}$ is exactly that given in (3.13). Note that $R_{ij^*k\ell^*}$ has precisely the right symmetries to appear in the four-fermion term.

The first thing to remark about (3.16) is that it is rigidly supersymmetric. This is guaranteed because (3.16) is derived from a superspace formalism. One can check, however, that $\mathcal{L}$ is invariant (up to a total derivative) under the following supersymmetry transformations:

$$\begin{aligned} \delta\phi^i \;&=\; \frac{1}{2}\sqrt{2}\,\overline{\epsilon}(1+\gamma_5)\chi^i \\ \delta\chi^i \;&=\; \sqrt{2}\,\gamma^\mu\, \widehat{\partial_\mu \phi^{*i}\epsilon} \;-\; \frac{1}{2}\sqrt{2}\,(\gamma_5 \chi^k)(\overline{\chi}^j\, \widehat{\Gamma^i{}_{jk}}\, \gamma_5 \epsilon) \\ & \quad -\; \frac{1}{2}\sqrt{2}\,\chi^k (\overline{\chi}^j \widehat{\Gamma^i{}_{jk}\, \epsilon}) \,. \end{aligned} \qquad (3.18)$$

As in (2.15), the term with the connection coefficients is required to ensure that supersymmetry transformations commute with diffeomorphisms.

The Lagrangian $\mathcal{L}$ also possesses the Kähler invariance (3.10). This is easy to verify directly. It may also be seen in superspace, provided one uses the fact that

$$\int d^4\theta\, F(\Phi^i) \;=\; \int d^4\theta\, F^*(\Phi^{*j}) \;=\; 0 \,. \qquad (3.19)$$

The Kähler invariance is necessary, but not quite sufficient, to prove that the Lagrangian $\mathcal{L}$ is well-defined over the whole Kähler manifold $\mathcal{M}$. One must still worry about problems that might arise globally. For the case of rigid supersymmetry, one may readily show that there are no such problems [8].

Having constructed the sigma model, we are now ready to couple it to gravity [21]. The first surprise is that in supergravity, the superspace sigma model Lagrangian is not

$$\int d^4\theta\, E\, K(\Phi^i, \Phi^{*j}) \,, \tag{3.20}$$

but rather

$$\mathcal{L} \;=\; -\,3 \int d^4\theta\, E \,\exp\left[-\frac{1}{3}\,K(\Phi^i, \Phi^{*j})\right] \,, \tag{3.21}$$

where E is the superdeterminant of the superspace vielbein [9–11].[♮2] Equation (3.21) is the choice that leads, after Weyl rescaling and eliminating the auxiliary fields, to a component Lagrangian with the correct normalizations for the Einstein action and for the matter kinetic energies. To make (3.21) more plausible, let us restore the factors of Newton's constant, $\kappa^2 = 8\pi G_N$, and expand the exponential. As $\kappa \to 0$, we find

$$\mathcal{L} \;\to\; -\,\frac{3}{\kappa^2} \int d^4\theta\, E \;+\; \int d^4\theta\, E\, K(\Phi^i, \Phi^{*j}) \;+\; \mathcal{O}(\kappa^2) \,. \tag{3.22}$$

The first term contains the Einstein and Rarita-Schwinger actions, which are frozen out as $\kappa \to 0$. The second term is simply (3.20), the flat sigma model Lagrangian. Higher order terms vanish as $\kappa \to 0$, so (3.21) has the correct flat-space limit. The final justification for (3.21) comes, however, from expanding in components and examining the normalizations of the various kinetic terms.

Precisely how to pass from the superspace Lagrangian (3.21) to the correct form of the component Lagrangian is beyond the scope of these lectures [22]. There are many different methods on the market [16,17,23], and each suffers from its own drawbacks. None is particularly easy to use. Suffice it

♮2 In these lectures, we consider the "$n = -1/3$" version of supergravity. Other versions are treated in Ref. [42].

to say that the component Lagrangian corresponding to (3.21) is given by

$$\begin{aligned}
\mathcal{L} \;=\; & -\frac{1}{2} eR \;-\; e\, g_{ij^*}\, \partial_\mu \phi^i \partial^\mu \phi^{*j} \\
& -\frac{1}{2} \epsilon^{\mu\nu\rho\sigma} \overline{\psi}_\mu \gamma_5 \gamma_\nu \mathcal{D}_\rho \psi_\sigma \;-\; \frac{1}{2} e\, \overline{\chi}^i\, \widehat{g_{ij^*}}\, \gamma^\mu \mathcal{D}_\mu \chi^j \\
& + \frac{1}{2} \sqrt{2}\, e\, \overline{\chi}^i g_{ij^*} \partial_\nu \phi^{*j} \gamma^\mu \gamma^\nu \psi_\mu \\
& + \frac{1}{16} e\, g_{ij^*} \overline{\chi}^i (1+\gamma_5) \gamma_\sigma \chi^j \left[\epsilon^{\mu\nu\rho\sigma} \overline{\psi}_\mu \gamma_\nu \psi_\rho \;-\; \overline{\psi}_\rho \gamma_5 \gamma^\sigma \psi^\rho \right] \\
& - \frac{1}{64} e \left[g_{ij^*} g_{k\ell^*} \;-\; 2 R_{ij^*k\ell^*} \right] \overline{\chi}^i (1+\gamma_5) \gamma_\mu \chi^j \overline{\chi}^k (1+\gamma_5) \gamma^\mu \chi^\ell \,.
\end{aligned} \tag{3.23}$$

Again g_{ij^*} is the Kähler metric and $R_{ij^*k\ell^*}$ is the Kähler curvature.

It is important to note that, in supergravity, the covariant derivatives contain several new pieces:

$$\begin{aligned}
\mathcal{D}_\mu \chi^i \;=\; & \partial_\mu \chi^i \;+\; \frac{1}{2} \omega_{\mu\alpha\beta}\, \sigma^{\alpha\beta} \chi^i \;+\; \Gamma^i{}_{jk} \partial_\mu \phi^j \chi^k \\
& - \frac{i}{2} \operatorname{Im} \left(\frac{\partial K}{\partial \phi^j} \partial_\mu \phi^j \right) \gamma_5\, \chi^i \\
\mathcal{D}_\mu \psi_\nu \;=\; & \partial_\mu \psi_\nu \;+\; \frac{1}{2} \omega_{\mu\alpha\beta}\, \sigma^{\alpha\beta} \psi_\nu \;+\; \frac{i}{2} \operatorname{Im} \left(\frac{\partial K}{\partial \phi^j} \partial_\mu \phi^j \right) \gamma_5\, \psi_\nu \,.
\end{aligned} \tag{3.24}$$

In the second-order formalism, the spin connection $\omega_{\mu\alpha\beta}$ is a function of the vierbein $e_{\mu\alpha}$ and the gravitino ψ_μ. It is present because of the fact that spacetime is curved. The Kähler connection is also present in the covariant derivative of χ^i. It just says that the χ^i transform in the (complexified) tangent bundle over $\mathcal{M}$. The striking new term is the $U(1)$ connection $\frac{i}{2} \operatorname{Im} [(\partial K / \partial \phi^j) \partial_\mu \phi^j]$. It occurs in the covariant derivatives of χ^i and ψ_ν. We shall see that this term gives rise to dramatic consequences for the matter coupling [11].

The Lagrangian (3.23) is manifestly supersymmetric because it is derived from a superspace formalism. It is not obvious, however, that (3.23) possesses the Kähler invariance (3.10). This must be verified if $\mathcal{L}$ is to be well-defined over the entire Kähler manifold. To do this, let us imagine that we cover our Kähler manifold by open sets $\mathcal{O}_A$. We assume that our covering is fine enough so that each intersection region $\mathcal{O}_A \cap \mathcal{O}_B$ is simply connected. On each open set we have a Kähler potential K_A. On the overlap regions $\mathcal{O}_A \cap \mathcal{O}_B$, K_A does not equal K_B, but rather

$$K_A \;-\; K_B \;=\; F_{AB} \;+\; F^*_{AB} \,. \tag{3.25}$$

Here F_{AB} is an analytic function, and $F_{AB} = -F_{BA}$. Note that (3.25) does not quite uniquely define the F_{AB}. If we let

$$F_{AB} \rightarrow F_{AB} + i\,C_{AB}\ , \tag{3.26}$$

where the C_{AB} are *real constants*, then F_{AB} is still analytic and (3.25) is still obeyed.

As we see from the covariant derivatives (3.24), the Lagrangian (3.23) in $\mathcal{O}_A \cap \mathcal{O}_B$ is invariant under Kähler transformations (3.25) only if the Kähler transformations are accompanied by chiral rotations of the Fermi fields:

$$\begin{aligned} \chi^i_{(A)} &= \exp\left[\frac{i}{2}\,\mathrm{Im}\,F_{AB}\,\gamma_5\right]\chi^i_{(B)}\ , \\ \psi_{\nu(A)} &= \exp\left[-\frac{i}{2}\,\mathrm{Im}\,F_{AB}\,\gamma_5\right]\psi_{\nu(B)}\ . \end{aligned} \tag{3.27}$$

The combined Kähler/chiral invariance is necessary, but not sufficient, to ensure that $\mathcal{L}$ is well-defined over the whole Kähler manifold $\mathcal{M}$. We must still worry about problems that might arise globally. In particular, we must worry about the consistency of (3.27). Inconsistency might arise in the triple intersection regions $\mathcal{O}_A \cap \mathcal{O}_B \cap \mathcal{O}_C$, since (3.27) relates $\chi^i_{(A)}$ to $\chi^i_{(B)}$, $\chi^i_{(B)}$ to $\chi^i_{(C)}$, and $\chi^i_{(C)}$ back to $\chi^i_{(A)}$. The consistency condition is

$$1 = \exp\left[\pm\frac{1}{4}\left(F_{AB} + F_{BC} + F_{CA} - F^*_{AB} - F^*_{BC} - F^*_{CA}\right)\right]\ . \tag{3.28}$$

It is this condition that leads to the quantization of Newton's constant in certain $N = 1$ supergravity theories [11].

To understand the consistency condition, note that the identity $(K_A - K_B) + (K_B - K_C) + (K_C - K_A) = 0$ implies

$$F_{AB} + F_{BC} + F_{CA} = -\left(F_{AB} + F_{BC} + F_{CA}\right)^*\ . \tag{3.29}$$

The left-hand side of (3.29) is an analytic function that equals minus its own complex conjugate. This implies that if we denote

$$2\pi\,i\,C_{ABC} = F_{AB} + F_{BC} + F_{CA}\ , \tag{3.30}$$

then the C_{ABC} are *real constants.* Note that the C_{ABC} are not quite uniquely defined. In view of (3.26), we are free to shift

$$C_{ABC} \rightarrow C_{ABC} + C_{AB} + C_{BC} + C_{CA}\ . \tag{3.31}$$

The consistency condition requires that we choose the C_{AB} such that the C_{ABC} are *even integers.*

On a general Kähler manifold, it is not always possible to choose the C_{ABC} to be integers. It is only possible to do this if $\mathcal{M}$ is a Kähler manifold of restricted type, or a *Hodge* manifold [11,20,24,25]. What, precisely, is a Hodge manifold? A mathematician would tell us that a Hodge manifold is a Kähler manifold on which it is possible to define a complex line bundle whose first Chern form is proportional to the Kähler form. How can we understand the consistency condition in this mathematical language?

First of all, the transformations (3.27) tell us that the χ^i are sections—not only of the tangent bundle—but also of a complex line bundle $\mathcal{E}$. The transition functions in this bundle are given by the $U(1)$ elements $\exp(\pm\frac{i}{2}\,\mathrm{Im}\,F)$. Furthermore, the covariant derivatives (3.24) imply that the connection ω on $\mathcal{E}$ is given by

$$\omega \;=\; \frac{i}{4}\left(\frac{\partial K}{\partial\phi^i}\,d\phi^i \;-\; \frac{\partial K}{\partial\phi^{*i}}\,d\phi^{*i}\right) . \tag{3.32}$$

The corresponding curvature tensor $\Omega = dw$ may be readily seen to be

$$\Omega \;=\; \frac{i}{2}\,\frac{\partial^2 K}{\partial\phi^i\partial\phi^{*j}}\,d\phi^i \wedge d\phi^{*j} . \tag{3.33}$$

This is just the Kähler form (3.5). Now, the first Chern form of any line bundle is proportional to the curvature Ω. Hence the first Chern form of our line bundle $\mathcal{E}$ is indeed the Kähler form, and our manifold $\mathcal{M}$ must be Hodge. The consistency condition (3.28) simply says that the first Chern form integrated over any closed two-cycle must give an (even) integer. When this is true, the complex line bundle $\mathcal{E}$ exists, and the Lagrangian $\mathcal{L}$ is well-defined over the whole Kähler manifold.

To see how the consistency condition (3.28) can lead to the quantization of Newton's constant, let us consider a simple example. For $\mathcal{M}$ we take CP^1, the ordinary two-dimensional sphere. This is a Kähler manifold, the Riemann sphere S^2. Deleting the point at infinity, we stereographically project S^2 onto the complex plane (see Fig. 1).

In terms of the complex coordinates $(z,\ z^*)$, an appropriate choice for the Kähler potential is $K(z,z^*) = n\log(1+z^*z)$. In this case the bosonic part of the Lagrangian (3.23) is given by

$$\mathcal{L} \;=\; \Lambda^2\sqrt{-g}\left[-\frac{1}{2}R \;-\; n\partial_\mu z\,\partial^\mu z^*/(1+z^*z)^2\right] , \tag{3.34}$$

where a constant, Λ^2, has been restored. The conventional form of the sigma

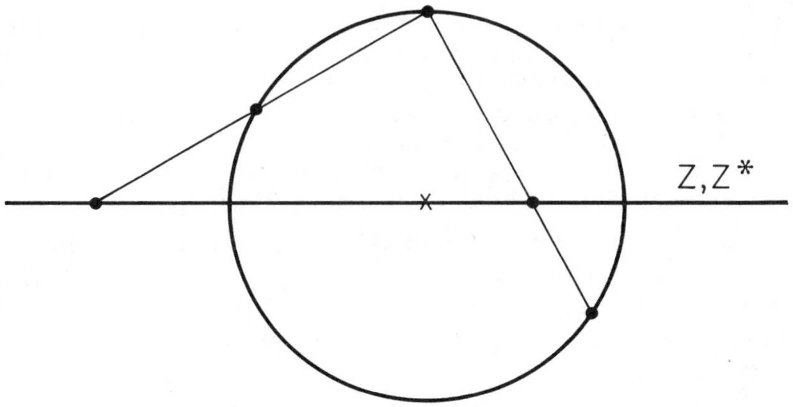

Figure 1: Complex projective coordinates for the sphere S^2

model would be

$$\mathcal{L} = \sqrt{-g}\left[-\frac{1}{16\pi G_N}R - f^2\partial_\mu z\partial^\mu z^*/(1+z^*z)^2\right] . \quad (3.35)$$

Comparing (3.34) and (3.35), we see that G_N and f are related by

$$G_N = n/8\pi f^2 . \quad (3.36)$$

As we shall see, the quantization condition for S^2 is that n must be an even integer. Equation (3.36) implies that Newton's constant must be quantized in units of the scalar self-coupling.

The sphere S^2 is so simple that we can work out the quantization directly, without resorting to the cocycle condition (3.28). The complex z-plane spans all of S^2 except the point at infinity. Letting $w = 1/z$, we see that the w-plane spans the entire sphere except for the point $z = 0$. In terms z, we take $K_z = n\log(1 + z^*z)$; for w, we take $K_w = n\log(1 + w^*w)$. The difference is

$$K_z - K_w = n\log z + n\log z^* . \quad (3.37)$$

Therefore we identify F as

$$F = n\log z . \quad (3.38)$$

The problem with this is that F is multi-valued—it cannot be defined consistently on the whole plane $z \neq 0$, $z \neq \infty$. However, in relating the description on the z-plane to the description of the w-plane, all we need is the single-valuedness of $\exp\frac{1}{4}(F - F^*)$. This is obeyed if and only if n is an (even) integer.

The quantization condition discussed here is exactly the same as the Dirac condition for a magnetic monopole. For the magnetic monopole, one must worry about consistently defining a $U(1)$ bundle over S^2. The integral of the first Chern form over S^2 measures the monopole charge, and consistency demands it be quantized with respect to the electric charge. The (even) integer n introduced above is the supersymmetric analogue of the "monopole charge" for the CP^1 sigma model.

Thus we have seen that in rigid supersymmetry, sigma models exist for all Kähler manifolds $\mathcal{M}$. Only a subclass of these models may be coupled to supergravity—those whose scalar fields lie on a Hodge manifold. For topologically nontrivial Hodge manifolds, this leads to the quantization of Newton's constant in terms of the scalar self-coupling.

4. SUPERPOTENTIAL AND GAUGE COUPLINGS

THE SUPERPOTENTIAL

in this section we shall finish our discussion of the $N = 1$ matter couplings. We will first include the superpotential $W(\Phi^i)$, and then we will gauge the (holomorphic) isometries of $\mathcal{M}$.

As mentioned in Section 3, the superpotential $W(\Phi^i)$ is an analytic function of the chiral superfield Φ^i. In flat superspace, it gives rise to an extra term in the superfield Lagrangian

$$\Delta\mathcal{L} \;=\; \int d^2\theta \, W(\Phi^i) \, . \tag{4.1}$$

This term can and should be included in the most general coupling of chiral superfields. In terms of components, equations (4.1) induces the following additional terms in the Lagrangian (3.16):

$$\Delta\mathcal{L} \;=\; -\,\frac{1}{2}\,\overline{\chi}^i \left[\nabla_i \frac{\partial W}{\partial \phi^j} \right] \chi^j \;-\; g^{ij^*} \left(\frac{\partial W}{\partial \phi^i} \right) \left(\frac{\partial W}{\partial \phi^j} \right)^* \,, \tag{4.2}$$

where ∇_i is the covariant derivative (3.11).

The first thing to note is that if W is quadratic, the terms in $\Delta\mathcal{L}$ give mass to the fields ϕ^i and χ^i. In supersymmetric theories, it is not usually so easy to give masses to fields, because massive and massless multiplets contain different numbers and types of fields. In $3+1$ dimensions, the $(0, \frac{1}{2})$ multiplet is unique in its ability to represent both the massless and massive supersymmetry algebra.

The second thing to note about (4.2) is that the scalar potential $\mathcal{V} = g^{ij*}(\partial W/\partial\phi^i)\;(\partial W/\partial\phi^j)^*$ is positive semidefinite. Supersymmetry is spontaneously broken if and only if $\langle \partial W/\partial\phi^i \rangle \neq 0$ for some value of i.

In local supersymmetry, it is also possible to include a superpotential. The superfield Lagrangian is given by

$$\Delta\mathcal{L} = \int d^2\theta\, \widetilde{\mathcal{E}}\, W(\Phi^i) + \text{h.c.}\,, \tag{4.3}$$

where $\widetilde{\mathcal{E}}$ is the chiral superspace density. Expanding Φ^i in components, doing the θ-integral, eliminating the auxiliary fields, and Weyl rescaling, one finds the following addition to the Lagrangian (3.23) [9–11,22]:

$$\begin{aligned} e^{-1}\Delta\mathcal{L} = & \exp(K/2)\Big\{\overline{\psi}_\mu \sigma^{\mu\nu}\widehat{W}^*\psi_\nu + \frac{1}{2}\sqrt{2}\,\overline{\psi}\cdot\gamma\,\widehat{D_iW}\chi^i \\ & - \frac{1}{2}\overline{\chi}^i\left[\mathcal{D}_j(D_iW)\right]\chi^j\Big\} \\ & + \exp(K)\left[3\,|W|^2 - (D_iW)(D_jW)^* g^{ij^*}\right]. \end{aligned} \tag{4.4}$$

Here $D_iW = \partial W/\partial\phi_i + (\partial K/\partial\phi^i)W$, and $\mathcal{D}_jD_iW = \partial\, D_iW/\partial\phi^j - \Gamma^k_{ji}D_kW + (\partial K/\partial\phi^j)\, D_iW$.

From the transformation law for χ^i,

$$\delta\chi^i = -\sqrt{2}\, e^{\frac{1}{2}K} g^{ij^*}(D_jW)^*\epsilon + \ldots\,, \tag{4.5}$$

we see that χ^i transforms by a shift whenever $\langle D_jW\rangle \neq 0$. This identifies $\langle D_jW\rangle \neq 0$ as the criterion for spontaneous supersymmetry breaking, and χ^i as the associated Goldstone fermion. Note that $\langle D_jW\rangle \neq 0$ reduces to $\langle \partial W/\partial\phi^j\rangle \neq 0$ as $\kappa \to 0$.

From the potential $\mathcal{V}$,

$$\mathcal{V} = e^K\left[(D_iW)(D_jW)^* g^{ij^*} - 3|W|^2\right] \tag{4.6}$$

we see that in local supersymmetry, it is possible to have spontaneously broken supersymmetry and *zero* cosmological constant. This requires that $\langle W\rangle \sim 1/\kappa^2$. Note that spontaneously broken supersymmetry gives no restrictions on the cosmological constant. Unbroken supersymmetry, however, requires that the cosmological constant be zero or negative [21,26]. In particular, this implies that unbroken supersymmetry cannot exist in de Sitter space.

The Lagrangian (4.4) possesses the Kähler invariance (3.10), but only if we also transform $W \to W \exp(-F)$. This means that if $\mathcal{M}$ is covered by open sets $\mathcal{O}_A$, on each of which we choose a superpotential W_A, then we must require

$$W_A = W_B \exp(-F_{AB}) \tag{4.7}$$

on $\mathcal{O}_A \cap \mathcal{O}_B$ for the two descriptions to match. The consistency condition is that the C_{ABC} must again be integers.

As before, all this has a standard mathematical interpretation. The W_A are sections of a holomorphic line bundle $\mathcal{F}$ over $\mathcal{M}$. On $\mathcal{F}$, there is a natural metric given by $\| W \|^2 = WW^* \exp(K)$. It is invariant under the Kähler transformations (3.10) and (4.7). On $\mathcal{F}$, there is also a holomorphic connection

$$\omega = i \frac{\partial K}{\partial \phi^i} d\phi^i . \tag{4.8}$$

As before, the first Chern form of this connection is proportional to the Kähler form. The consistency condition simply says that the first Chern form integrated over any closed two-cycle must give an even integer [11].

The potential (4.6) looks somewhat different than the potential usually used in supergravity model-building [9,10]. However, it is the form (4.6) that makes the geometrical structure manifest. To recover the formulae used by model-builders, one need only choose the gauge $W = 1$. In this gauge, the two formalisms are identical.

THE GAUGE INVARIANT SUPERSYMMETRIC NONLINEAR SIGMA MODEL

We would now like to gauge the bosonic symmetries of the Lagrangian $\mathcal{L}$. These symmetries are given by the *isometries* of the sigma model manifold. In this section, we shall gauge the (holomorphic) isometries of the manifold $\mathcal{M}$ [12].

To that end, let us assume that $\mathcal{M}$ admits a d-dimensional isometry group $\mathcal{G}$. The group $\mathcal{G}$ has d linearly independent generators,

$$\delta\phi^i = V^{(a)i}(\phi^j) , \qquad a = 1, \ldots, d . \tag{4.9}$$

Since we would like the symmetry rotations to preserve the complex structure, we require that the $V^{(a)i}$ be holomorphic,

$$\frac{\partial V^{(a)i}}{\partial \phi^{*j}} = \frac{\partial V^{(a)*i}}{\partial \phi^j} = 0 . \tag{4.10}$$

Furthermore, since the $V^{(a)i}$ must generate isometries, they must also obey

Killing's equation. On a complex manifold, Killing's equation has two parts,

$$\begin{aligned} \nabla_i V_j{}^{(a)} + \nabla_j V_i{}^{(a)} &= 0 , \\ \nabla_i V_{j^*}{}^{(a)} + \nabla_{j^*} V_i{}^{(a)} &= 0 . \end{aligned} \tag{4.11}$$

The first is satisfied automatically; it is a consequence of (4.10) and of the fact that $\nabla_i g_{jk^*} = 0$. The second implies that ***locally*** there exist d real scalar functions $D^{(a)}(\phi^i, \phi^{*j})$, such that

$$g_{ij^*} V^{(a)*j} = i \frac{\partial D^{(a)}}{\partial \phi^i} . \tag{4.12}$$

The functions $D^{(a)}$ are called *Killing potentials* because their gradients give the *Killing vectors* $V^{(a)i}$.

It is important to note that (4.12) defines the potentials $D^{(a)}$ only up to arbitrary integration constants $C^{(a)}$,

$$D^{(a)} \rightarrow D^{(a)} + C^{(a)} . \tag{4.13}$$

We will see that this freedom is related to the Fayet-Iliopoulos D-term in abelian gauge theories [27].

The Killing vectors $V^{(a)i}$ are well-defined over the entire Kähler manifold $\mathcal{M}$. The potentials $D^{(a)}$, however, are defined only locally. For example, if $\mathcal{M}$ is not simply connected and $\mathcal{G}$ contains a $U(1)$ factor, the $D^{(a)}$ may not be globally well-defined. We shall see that the global existence of the $D^{(a)}$'s is the necessary and sufficient condition to gauge the group $\mathcal{G}$.

Since the Killing vectors $V^{(a)i}$ generate a Lie group, they must obey the usual Lie bracket relations:

$$\left[V^{(a)}, V^{(b)}\right]^i = V^{(a)j} \frac{\partial}{\partial \phi^j} V^{(b)i} - V^{(b)j} \frac{\partial}{\partial \phi^j} V^{(a)i} = f^{abc} V^{(c)i} . \tag{4.14}$$

If the $D^{(a)}$ exist, they can be chosen to transform in the adjoint representation of the (compact) gauge group,

$$\left[V^{(a)i} \frac{\partial}{\partial \phi^i} + V^{(a)*i} \frac{\partial}{\partial \phi^{*i}}\right] D^{(b)} = f^{abc} D^{(c)} . \tag{4.15}$$

Note that (4.15) fixes the constants $C^{(a)}$ for nonabelian groups. For each $U(1)$ factor, however, there is an undetermined constant C. This will turn out to be the reason why Fayet-Iliopoulos terms only arise in abelian theories.

Locally, the Killing vector fields $V^{(a)i}$ generate the following motions:

$$\begin{aligned} \delta\phi^i &= \epsilon^{(a)} V^{(a)i} , \\ \delta\chi^i &= \epsilon^{(a)} \frac{\partial V^{(a)i}}{\partial\phi^j} \chi^j . \end{aligned} \tag{4.16}$$

These motions are isometries, so they leave the metric invariant. They do, however, shift the Kähler potential,

$$\delta K = \delta\phi^i \frac{\partial K}{\partial\phi^i} + \delta\phi^{*i} \frac{\partial K}{\partial\phi^{*i}} . \tag{4.17}$$

It is easy to check that (4.17) is a Kähler transformation, $\delta K = F + F^*$, provided

$$F = \epsilon^{(a)} \left[V^{(a)i} \frac{\partial K}{\partial\phi^i} + i D^{(a)} \right] \tag{4.18}$$

The function F is analytic because of (4.10) and (4.12). This shows explicitly that the isometries (4.9) leave invariant the Lagrangian (3.14).

We are now ready to gauge the group $\mathcal{G}$. This corresponds to setting $\epsilon^{(a)} \to \epsilon^{(a)}(x)$ in (4.16), exactly as in ordinary Yang-Mills theory. We would like to do this using superfields, in order to ensure that our resulting Lagrangian is supersymmetric. However, it is only easy to gauge that subgroup $\mathcal{H} \subseteq \mathcal{G}$ that leaves K invariant (a general transformation in $\mathcal{G}/\mathcal{H}$ shifts K by a Kähler transformation, $K \to K + F + F^*$) [28].

In flat superspace, the Lagrangian invariant under $\mathcal{H}$ is given by

$$\mathcal{L} = \int d^4\theta \, K \left[\Phi^i, (\Phi^+ e^{2gV})^{*j} \right] . \tag{4.19}$$

To this one must add the kinetic terms for the vector multiplet V,

$$\mathcal{L} = \int d^2\theta \, \mathrm{Tr}\, \overline{W}_R W_L + \text{h.c.} \tag{4.20}$$

From here it is straightforward to work out the component Lagrangian. One can then guess its extension to a new Lagrangian invariant under $\mathcal{G}$,

$$\begin{aligned}\mathcal{L} \;=\; &- g_{ij^*} \mathcal{D}_\mu\phi^i \mathcal{D}^\mu\phi^{*j} \;-\; \frac{1}{2} g^2 D^{(a)2} \\ &- \frac{1}{2}\overline{\chi}^i \,\widehat{g_{ij^*}}\, \gamma^\mu \mathcal{D}_\mu \chi^j \;-\; \frac{1}{4} F_{\mu\nu}{}^{(a)} F^{\mu\nu(a)} \\ &- \frac{1}{2}\overline{\lambda}^{(a)} \gamma^\mu \mathcal{D}_\mu \lambda^{(a)} \;-\; \sqrt{2}\, g\, \overline{\chi}^i g_{ij^*} V^{(a)*j} \lambda^{(a)} \\ &+ \frac{1}{32} R_{ij^*k\ell^*}\, \overline{\chi}^i (1+\gamma_5)\gamma_\mu \chi^j\, \overline{\chi}^k (1+\gamma_5)\gamma^\mu \chi^\ell \;,\end{aligned} \tag{4.21}$$

where

$$\begin{aligned}\mathcal{D}_\mu\phi^i &= \partial_\mu\phi^i \;-\; g A_\mu{}^{(a)} V^{(a)i} \\ \mathcal{D}_\mu\chi^i &= \partial_\mu\chi^i \;+\; \Gamma^i{}_{jk} \mathcal{D}_\mu\phi^j \chi^k \;-\; g A_\mu{}^{(a)} \frac{\partial V^{(a)i}}{\partial\phi^j}\, \chi^j \\ \mathcal{D}_\mu\lambda^{(a)} &= \partial_\mu\lambda^{(a)} \;-\; g\, f^{abc} A_\mu{}^{(b)} \lambda^{(c)} \;.\end{aligned} \tag{4.22}$$

The covariant derivatives (4.22) are coordinate and gauge covariant, as is evident from the transformations (4.16). The Lagrangian (4.21) is also gauge invariant under the isometry group $\mathcal{G}$. What is not evident, however, is that (4.21) is still supersymmetric. That must be checked by hand, using the following transformation laws:

$$\begin{aligned}\delta\phi^i &= \frac{1}{2}\sqrt{2}\,\overline{\epsilon}(1+\gamma_5)\chi^i \\ \delta\chi^i &= \sqrt{2}\,\gamma^\mu \widehat{\mathcal{D}_\mu\phi^{*i}}\epsilon \;-\; \frac{1}{2}\sqrt{2}\,(\gamma_5\chi^k)(\overline{\chi}^j\, \widehat{\Gamma^i_{jk}}\, \gamma_5\epsilon) \\ &\quad - \frac{1}{2}\sqrt{2}\,\chi^k\,(\overline{\chi}^j\, \widehat{\Gamma^i_{jk}}\, \epsilon) \\ \delta A_\mu{}^{(a)} &= -\, \overline{\epsilon}\gamma_\mu\lambda^{(a)} \\ \delta\lambda^{(a)} &= F_{\alpha\beta}{}^{(a)} \sigma^{\alpha\beta}\epsilon \;+\; i\, g\, D^{(a)}\gamma_5\epsilon \;.\end{aligned} \tag{4.23}$$

The Lagrangian (4.21) gives what might be called the gauge invariant supersymmetric nonlinear sigma model [12].[♮3]

♮3 Equations (4.21)–(4.23) are written as if $\mathcal{G}$ were a simple group. The generalization to an arbitrary group is trivial.

The Lagrangian (4.21) does not contain a superpotential W, which has been omitted for simplicity. Yet it still contains a scalar potential,

$$\mathcal{V} \;=\; \frac{1}{2} g^2 \, D^{(a)2} \, . \tag{4.24}$$

The potential (4.24) is a sigma model generalization of the so-called "D-term" familiar to model builders. Supersymmetry is spontaneously broken if $\langle D^{(a)} \rangle \neq 0$, for some value of $a = 1, \ldots, d$. The spinor $\lambda^{(a)}$ is the corresponding Goldstone fermion.

For $U(1)$ factors in the gauge group $\mathcal{G}$, the relations (4.12) and (4.15) do not completely determine the corresponding potentials D. They leave the D's undetermined up to additive constants C,

$$D \;\rightarrow\; D \;+\; C \, . \tag{4.25}$$

However, each D appears in the potential $\mathcal{V}$, and supersymmetry is spontaneously broken whenever any D develops a nonzero vacuum expectation value. By choosing the C's appropriately, it is possible to ensure that $\langle D \rangle \neq 0$. This is known as the Fayet-Iliopoulos mechanism for spontaneous supersymmetry breaking [27].

THREE EXAMPLES

To get a better feeling for the formalism developed above, let us consider three examples. For the first we take $\mathcal{M}$ to be the complex z-plane, and we choose to gauge the rotations about the origin. We set $K = z^* z + d$ and $D = z^* z + c$. Then the metric g_{zz^*} is 1, and $R_{zz^*zz^*} = 0$. The Killing vector V is simply $-iz$, so the covariant derivatives are

$$\begin{aligned} \mathcal{D}_\mu z \;&=\; \partial_\mu z \;+\; igA_\mu z \, , \\ \mathcal{D}_\mu \chi \;&=\; \partial_\mu \chi \;+\; igA_\mu \chi \, . \end{aligned} \tag{4.26}$$

We see that the first example corresponds to a renormalizable $U(1)$ gauge theory. The potential $\mathcal{V}$ is

$$\mathcal{V} \;=\; \frac{1}{2} g^2 (zz^* \;+\; c)^2 \, . \tag{4.27}$$

For $c > 0$, supersymmetry is spontaneously broken via the Fayet-Iliopoulos mechanism. This example can be generalized for any renormalizable gauge theory. One takes $\mathcal{M} = C^n$, and one gauges an appropriate subgroup of $U(n)$.

For our second example, we again take $\mathcal{M}$ to be C^1. This time, however, we choose to gauge translations in the y-direction on $\mathcal{M}$. (Note that we could have chosen to gauge translations in the x-direction, but because of (4.15), we cannot gauge both simultaneously.) As before, we take $K = z^*z + d$, so $g_{zz^*} = 1$ and $R_{zz^*zz^*} = 0$. For D we take $D = m(z + z^*)$, so $V = -im$. The covariant derivatives are

$$\begin{aligned} \mathcal{D}_\mu z &= \partial_\mu z + img A_\mu , \\ \mathcal{D}_\mu \chi &= \partial_\mu \chi . \end{aligned} \tag{4.28}$$

The field z gives a gauge invariant mass to the vector A_μ. It is a compensating field, analogous to the compensators introduced in conformal supergravity [21].

For our final example, we take $\mathcal{M} = CP^1 = S^2 = SU(2)/U(1)$. It is a Kähler manifold as well as a homogeneous space $\mathcal{G}/\mathcal{H}$. As in Section 3, we use projective coordinates z and z^*. In these coordinates, we take $K = \log(1+zz^*)$. We choose to gauge the entire isometry group $\mathcal{G} = SU(2)$. The isotropy group $\mathcal{H}$ is $U(1)$, and the functions D are as follows:

$$D^{(1)} = \frac{1}{2}\frac{z+z^*}{(1+z^*z)}, \quad D^{(2)} = -\frac{i}{2}\frac{z-z^*}{(1+z^*z)}, \quad D^{(3)} = -\frac{1}{2}\left(\frac{1-z^*z}{1+z^*z}\right). \tag{4.29}$$

From here one can work out the covariant derivatives $\mathcal{D}_\mu z$ and $\mathcal{D}_\mu \chi$. Because we have gauged the full $SU(2)$, we may go to the "unitary gauge" where $z = z^* = 0$. This gauge exhibits the particle content of the theory,

$$\begin{aligned} \mathcal{L} = &-\frac{1}{4}F^{(a)}_{\mu\nu}F^{\mu\nu(a)} - \frac{1}{2}\overline{\lambda}^{(a)}\gamma^\mu \mathcal{D}_\mu \lambda^{(a)} - \frac{1}{2}\overline{\chi}\gamma^\mu \mathcal{D}_\mu \chi \\ &- \frac{1}{2}g^2 W^+_\mu W^{-\mu} - \frac{1}{8}g^2 - ig\overline{\chi}\gamma_5\psi_2 + \frac{1}{4}(\overline{\chi}\chi)(\overline{\chi}\chi) , \end{aligned} \tag{4.30}$$

where

$$\begin{aligned} W^\pm_\mu &= \frac{1}{2}\sqrt{2}\left(A_\mu{}^{(1)} \pm i A_\mu{}^{(2)}\right) \\ \psi_{1,2} &= \frac{1}{2}\sqrt{2}\left(\lambda^{(1)} \pm i\gamma_5\lambda^{(2)}\right) \\ \mathcal{D}_\mu\chi &= \partial_\mu\chi + ig A_\mu{}^{(3)}\gamma_5\chi . \end{aligned} \tag{4.31}$$

The $SU(2)$ symmetry implies that $D^{(a)2}$ is a constant. The constant is positive, so supersymmetry is spontaneously broken. The mass spectrum is as follows. The charged vector mesons $W^\pm_\mu$ are massive; they have eaten the scalars z and z^*. The massless vector meson A_μ is the gauge field corresponding to the unbroken $U(1)$ symmetry. Its supersymmetry partner is the massless Goldstone spinor λ^3. The Majorana spinors χ and ψ_2 are massive; they have combined to form one massive Dirac spinor, of mass proportional to the inverse radius R^{-1}. Finally, ψ_1 is both massless and charged. Note that this is just what one wants for grand unified theories. The CP^1 model has spontaneously broken supersymmetry, no leftover Higgs, and massless Weyl spinors in complex representations of the unbroken gauge group.

Properties like these hold for other homogeneous spaces $\mathcal{G}/\mathcal{H}$ which form Kähler manifolds. If $\mathcal{G}$ is gauged, one finds $\mathcal{G}$ spontaneously broken to $\mathcal{H}$. One also finds that supersymmetry is spontaneously broken and that all scalars are eaten. Furthermore, at the tree level, there are charged massless spinors in complex representations of $\mathcal{H}$. For example, $SU(5)/SU(3) \times SU(2) \times U(1)$ yields a massless $(\underline{3}, \underline{2})$ of $SU(3) \times SU(2)$ Weyl fermions. Other examples are listed in Table 2 [29].

Table 2
Complex Fermions on Kähler Manifolds $\mathcal{G}/\mathcal{H}$

$\mathcal{G}$	$\mathcal{H}$	Complex Fermions
$SU(r+s)$	$SU(r) \times SU(s) \times U(1)$	$(\underline{r}, \underline{s})$ of $SU(r) \times SU(s)$
$SO(10)$	$SU(5) \times U(1)$	$\underline{10}$ of $SU(5)$
E_6	$SO(10) \times U(1)$	$\underline{16}$ of $SO(10)$
E_7	$E_6 \times U(1)$	$\underline{27}$ of E_6

Despite these miraculous features, one must not take these models too seriously. The models are nonrenormalizable, and there is no clear way to break $\mathcal{H}$ down to $SU(3) \times SU(2) \times U(1)$. In the full quantum theory, one often cannot gauge $\mathcal{H}$—much less $\mathcal{G}$—because of anomalies. And what is even worse, in many cases a coordinate anomaly on $\mathcal{M}$ implies that the quantum effective action does not respect the symmetries of the classical theory [30]. Still, Table 2 is intriguing, and these models are remarkable because the particle spins (as well as their masses) violate supersymmetry. No model with unbroken supersymmetry has the same spin spectrum.

COUPLING THE GAUGE INVARIANT SUPERSYMMETRIC SIGMA MODEL TO SUPERGRAVITY

Having constructed the gauge invariant supersymmetric nonlinear sigma model, we would now like to couple it to supergravity [10,13]. As before, we shall first use superspace techniques to gauge the linear subgroup $\mathcal{H}$ of the isometry group $\mathcal{G}$. We will then expand the Lagrangian in components, and guess its extension to the full group $\mathcal{G}$. Of course, we must then verify by hand that the resulting Lagrangian is still supersymmetric.

In Section 3 we found that supersymmetric sigma models may be coupled to supergravity if the scalar fields ϕ^i lie on a Hodge manifold $\mathcal{M}$. In the remainder of this section, we shall see that the gauge interactions lead to no new restrictions on $\mathcal{M}$. The final Lagrangian will give a geometrical interpretation to the most general gauge invariant coupling of chiral multiplets in supergravity. It clarifies the deep relation between chiral invariance, the Fayet-Iliopoulos D-term and supergravity. It also leads to formulae of interest to model builders.

In parallel with the previous case, we start our construction by gauging that subgroup $\mathcal{H}$ of $\mathcal{G}$ which leaves invariant the Kähler potential $K(\phi^i, \phi^{*j})$. Comparing (3.21) with (4.19), we see that

$$\mathcal{L} \;=\; -\,3 \int d^4\theta \; E \; \exp\left\{ -\frac{1}{3} K\left[\Phi^i, (\Phi^+ e^{2gV})^{*j}\right]\right\} \tag{4.32}$$

is the $\mathcal{H}$-invariant generalization of the sigma model Lagrangian in local superspace. To gauge the full group $\mathcal{G}$, we decompose (4.32) in component fields. We then add the kinetic terms for the gauge fields, eliminate the auxiliary fields, and guess the extension to the full group $\mathcal{G}$. After a long calculation, we find the coupling to supergravity of the full gauge invariant supersymmetric nonlinear sigma model [10,13]:[♮4]

♮4 The authors of Ref. [10] only gauge the linear subgroup $\mathcal{H}$ of $\mathcal{G}$. However, they consider a more general action for the gauge field multiplet.

$$
\begin{aligned}
\mathcal{L} \;=\; & -\frac{1}{2}eR \;-\; e\,g_{ij^*}\mathcal{D}_\mu\phi^i\mathcal{D}^\mu\phi^{*j} \\
& -\frac{1}{2}e\,g^2 D^{(a)2} \;-\; \frac{1}{2}\epsilon^{\mu\nu\rho\sigma}\overline{\psi}_\mu\gamma_5\gamma_\nu\mathcal{D}_\rho\psi_\sigma \\
& -\frac{1}{2}e\overline{\chi}^i\,\widehat{g_{ij^*}}\,\gamma^\mu\mathcal{D}_\mu\chi^j \;-\; \frac{1}{4}e\,F_{\mu\nu}{}^{(a)}F^{\mu\nu(a)} \\
& -\frac{1}{2}e\overline{\lambda}^{(a)}\gamma^\mu\mathcal{D}_\mu\lambda^{(a)} \;+\; \frac{1}{2}\sqrt{2}\,e\overline{\chi}^i g_{ij^*}\mathcal{D}_\nu\phi^{*j}\gamma^\mu\gamma^\nu\psi_\mu \\
& -\sqrt{2}\,eg\,\overline{\chi}^i\,g_{ij^*}V^{(a)*j}\,\lambda^{(a)} \;+\; \frac{i}{2}eg\,D^{(a)}\overline{\psi}_\mu\gamma^\mu\gamma_5\lambda^{(a)} \\
& +\frac{1}{4}e\overline{\psi}_\mu\sigma^{\alpha\beta}\gamma^\mu\lambda^{(a)}\left[2F_{\alpha\beta}{}^{(a)} \;+\; \overline{\psi}_\alpha\gamma_\beta\lambda^{(a)}\right] \\
& +\frac{1}{16}e\,g_{ij^*}\overline{\chi}^i(1+\gamma_5)\gamma_\sigma\chi^j\left[\epsilon^{\mu\nu\rho\sigma}\overline{\psi}_\mu\gamma_\nu\psi_\rho \;-\; \overline{\psi}_\rho\gamma_5\gamma^\sigma\psi^\rho\right] \\
& -\frac{1}{64}e\left[g_{ij^*}g_{k\ell^*} \;-\; 2R_{ij^*k\ell^*}\right]\overline{\chi}^i(1+\gamma_5)\gamma_\mu\chi^j\overline{\chi}^k(1+\gamma_5)\gamma^\mu\chi^\ell \\
& +\frac{3}{64}e\overline{\lambda}^{(a)}(1+\gamma_5)\gamma_\mu\lambda^{(a)}\overline{\lambda}^{(b)}(1+\gamma_5)\gamma^\mu\lambda^{(b)} \\
& -\frac{1}{32}e\,g_{ij^*}\overline{\lambda}^{(a)}(1+\gamma_5)\gamma_\mu\lambda^{(a)}\overline{\chi}^i(1+\gamma_5)\gamma^\mu\chi^j\;.
\end{aligned}
\tag{4.33}
$$

The covariant derivatives are as follows,

$$
\begin{aligned}
\mathcal{D}_\mu\phi^i \;&=\; \partial_\mu\phi^i \;-\; gA_\mu{}^{(a)}V^{(a)i} \\
\mathcal{D}_\mu\chi^i \;&=\; \partial_\mu\chi^i \;+\; \frac{1}{2}\omega_{\mu\alpha\beta}\sigma^{\alpha\beta}\chi^i \;+\; \Gamma^i{}_{jk}\mathcal{D}_\mu\phi^j\,\chi^k \\
&\quad -\; gA_\mu{}^{(a)}\frac{\partial V^{(a)i}}{\partial\phi^j}\chi^j \;-\; \frac{i}{2}gA_\mu{}^{(a)}\mathrm{Im}\,F^{(a)}\,\gamma_5\,\chi^i \\
\mathcal{D}_\mu\lambda^{(a)} \;&=\; \partial_\mu\lambda^{(a)} \;+\; \frac{1}{2}\omega_{\mu\alpha\beta}\,\sigma^{\alpha\beta}\lambda^{(a)} \;-\; g\,f^{abc}A_\mu{}^{(b)}\lambda^{(c)} \\
&\quad +\; \frac{i}{2}g\,A_\mu{}^{(b)}\mathrm{Im}\,F^{(b)}\,\gamma_5\,\lambda^{(a)} \\
\mathcal{D}_\mu\psi_\nu \;&=\; \partial_\mu\psi_\nu \;+\; \frac{1}{2}\omega_{\mu\alpha\beta}\,\sigma^{\alpha\beta}\psi_\nu \;+\; \frac{i}{2}gA_\mu{}^{(a)}\mathrm{Im}\,F^{(a)}\,\gamma_5\,\lambda^{(a)}\;.
\end{aligned}
\tag{4.34}
$$

The Lagrangian (4.33) together with the covariant derivatives (4.34) is invariant under the isometry group $\mathcal{G}$. It is also invariant under the following supergravity transformations,

$$
\begin{aligned}
\delta e_{\alpha\mu} &= \bar{\epsilon}\gamma_\alpha\psi_\mu \\
\delta\phi^i &= \frac{1}{2}\sqrt{2}\,\bar{\epsilon}(1+\gamma_5)\chi^i \\
\delta\chi^i &= \sqrt{2}\,\gamma^\mu\widehat{D_\mu\phi^{*i}}\epsilon \;+\; \frac{1}{4}\sqrt{2}\,(\gamma_5\chi^i)\left(\overline{\chi}^j\,\frac{\widehat{\partial K}}{\partial\phi^j}\,\gamma_5\epsilon\right) \\
&\quad - \frac{1}{2}(\gamma_\mu\epsilon)(\overline{\psi}^\mu\chi^i) \;-\; \frac{1}{2}(\gamma_5\gamma_\mu\epsilon)\,(\overline{\psi}^\mu\gamma_5\chi^i) \\
&\quad - \frac{1}{2}\sqrt{2}\,(\gamma_5\chi^k)(\overline{\chi}^j\,\widehat{\Gamma^i_{jk}}\,\gamma_5\epsilon) \;-\; \frac{1}{2}\sqrt{2}\,\chi^k\,(\overline{\chi}^j\,\widehat{\Gamma^i_{jk}}\,\epsilon) \\
\delta\psi_\mu &= 2D_\mu\epsilon \;+\; \frac{1}{4}g_{ij^*}(\sigma_{\mu\nu}\gamma_5\epsilon)\left[\overline{\chi}^i(1+\gamma_5)\gamma^\nu\chi^j\right] \\
&\quad - \frac{1}{4}\sqrt{2}\,(\gamma_5\psi_\mu)\left(\overline{\chi}^i\,\frac{\widehat{\partial K}}{\partial\phi^i}\gamma_5\epsilon\right) \\
&\quad - \frac{1}{4}\left[g_{\mu\nu} \;-\; \sigma_{\mu\nu}\right]\gamma_5\epsilon\,(\overline{\lambda}^{(a)}\gamma^\nu\gamma_5\lambda^{(a)}) \\
\delta A_\mu{}^{(a)} &= -\,\bar{\epsilon}\gamma_\mu\lambda^{(a)} \\
\delta\lambda^{(a)} &= \left[F_{\alpha\beta}{}^{(a)} \;+\; \overline{\psi}_\alpha\gamma_\beta\lambda^{(a)}\right]\sigma^{\alpha\beta}\epsilon \;+\; i\,g\,D^{(a)}\gamma_5\epsilon\;,
\end{aligned}
\tag{4.35}
$$

as may be checked by hand.

Form the superspace Lagrangian (4.32), we see that $\mathcal{L}$ is not invariant under Kähler transformations, $K \to K + F + F^*$. As before, $\mathcal{L}$ is invariant if and only if the Kähler transformations are accompanied by chiral rotations of the Fermi fields,

$$
\begin{aligned}
\chi^i &\to \exp\left[\frac{i}{2}\,\mathrm{Im}\,F\,\gamma_5\right]\chi^i \\
\lambda^{(a)} &\to \exp\left[-\frac{i}{2}\,\mathrm{Im}\,F\,\gamma_5\right]\lambda^{(a)} \\
\psi_\mu &\to \exp\left[-\frac{i}{2}\,\mathrm{Im}\,F\,\gamma_5\right]\psi_\mu\,.
\end{aligned}
\tag{4.36}
$$

These chiral rotations lead to the same quantization condition as in Section 2. They imply that only Kähler manifolds of restricted type may be coupled to supergravity.

The chiral rotations (4.36) give rise to terms proportional to $2i\,\mathrm{Im}\,F^{(a)} = V^{(a)i}(\partial K/\partial\phi^i) - V^{(a)*i}(\partial K/\partial\phi^{*i}) + 2iD^{(a)}$ in the covariant derivatives (4.34). This is as expected, for isometries in $\mathcal{G}/\mathcal{H}$ give rise to shifts in the Kähler potential, $\delta K = F + F^*$, where $F = \epsilon^{(a)}[V^{(a)i}(\partial K/\partial\phi^i) + iD^{(a)}]$. Invariance of the Lagrangian $\mathcal{L}$ requires that these shifts be compensated by the chiral transformations (4.36). It is no surprise that gauging the isometries of $\mathcal{M}$ requires the gauging of the associated chiral rotations.

It is important to note that the Lagrangian (4.33) contains explicitly the functions $D^{(a)}$. Their existence is both necessary and sufficient to gauge the group $\mathcal{G}$. If the group $\mathcal{G}$ contains a $U(1)$ factor, equations (4.12) and (4.15) do not determine the $D^{(a)}$ uniquely. There is an arbitrary integration constant associated with each $U(1)$ factor,

$$D \rightarrow D + C . \tag{4.37}$$

In the globally supersymmetric case, shifts of these constants were shown to give rise to Fayet-Iliopoulos D-terms. The same is true in supergravity. By shifting the functions D, it is easy to recover the gauge invariant supergravity version of $\delta\mathcal{L}_{\mathrm{FI}}$ [31]:

$$\delta\mathcal{L}_{\mathrm{FI}} = -\frac{1}{2}eg^2\,\xi^2 - eg^2\xi D + \frac{i}{2}eg\,\xi\,(\overline{\psi}_\mu\gamma^\mu\gamma_5\lambda) . \tag{4.38}$$

Note that all the spinor covariant derivatives now contain chiral pieces proportional to the Fayet-Iliopoulos parameter ξ.[♮5]

The Lagrangian (4.33) contains the kinetic pieces necessary to couple the gauge invariant nonlinear sigma model to supergravity. As written, however, it completely ignores the existence of the superpotential. As explained in Section 3, the superpotential is an additional interaction term which is of crucial importance to realistic models. In superspace, it is given by

$$\Delta\mathcal{L} = \int d^2\theta\,\tilde{\mathcal{E}}\,W(\Phi^i) + \text{h.c.} , \tag{4.39}$$

where W is an analytic function of the Φ^i, invariant under $\mathcal{H}$.

In the previous section, the superpotential was shown to be a section W of a holomorphic line bundle $\mathcal{F}$ over $\mathcal{M}$. The hermitian structure was shown to be

$$\| W \|^2 = e^K W W^* . \tag{4.40}$$

This is invariant under Kähler transformations provided

$$W \rightarrow W e^{-F} . \tag{4.41}$$

From this we see that a given superpotential W is gauge covariant if and

♮5 In the language of the trade, these are known as gauged R-transformations.

only if

$$W \;\rightarrow\; W \exp\left[\,-\,\epsilon^{(a)}\big(V^{(a)i}\,\frac{\partial K}{\partial\phi^{i}} \;+\; i\,D^{(a)}\big)\right] \tag{4.42}$$

under the isometry generated by $\epsilon^{(a)}V^{(a)i}$. Potentials not obeying this law explicitly break the gauge symmetry of $\mathcal{M}$.

The Lagrangian (4.39) may be readily decomposed in terms of component fields.

$$\begin{aligned} e^{-1}\Delta\mathcal{L} \;=\; & \exp(K/2)\Big\{\overline{\psi}_{\mu}\sigma^{\mu\nu}\widehat{W}^{*}\psi_{\nu} \;+\; \frac{1}{2}\sqrt{2}\,\overline{\psi}\cdot\gamma\,\widehat{D_{i}W}\,\chi^{i} \\ & -\;\frac{1}{2}\overline{\chi}^{i}\left[\mathcal{D}_{j}(D_{i}W)\right]\chi^{j}\Big\} \\ & +\;\exp(K)\left[3\,|W|^{2} \;-\; (D_{i}W)(D_{j}W)^{*}g^{ij^{*}}\right]\,. \end{aligned} \tag{4.43}$$

The covariant derivatives D_iW and $\mathcal{D}_j(D_iW)$ are given below (4.4). It is easy to verify that (4.43) is invariant under the transformations (3.10), (4.36) and (4.41). The combined Lagrangian (4.33) plus (4.43) is also invariant under the following supergravity transformations:

$$\begin{aligned} \delta e_{\alpha\mu} \;&=\; \overline{\epsilon}\gamma_{\alpha}\psi_{\mu} \\ \delta\phi^{i} \;&=\; \frac{1}{2}\sqrt{2}\,\overline{\epsilon}(1+\gamma_{5})\chi^{i} \\ \delta\chi^{i} \;&=\; \sqrt{2}\,\gamma^{\mu}\,\widehat{\mathcal{D}_{\mu}\phi^{*i}}\epsilon \;+\; \frac{1}{4}\sqrt{2}(\gamma_{5}\chi^{i})\left(\overline{\chi}^{j}\,\frac{\widehat{\partial K}}{\partial\phi^{j}}\gamma_{5}\epsilon\right) \\ & -\;\frac{1}{2}(\gamma_{\mu}\epsilon)(\overline{\psi}^{\mu}\chi^{i}) \;-\; \frac{1}{2}(\gamma_{5}\gamma_{\mu}\epsilon)(\overline{\psi}^{\mu}\gamma_{5}\chi^{i}) \\ & -\;\frac{1}{2}\sqrt{2}(\gamma_{5}\chi^{k})(\overline{\chi}^{j}\,\widehat{\Gamma^{i}{}_{jk}}\,\gamma_{5}\epsilon) \end{aligned}$$

$$
\begin{aligned}
& - \frac{1}{2}\sqrt{2}\,\chi^k(\overline{\chi}^j\,\widehat{\Gamma^i{}_{jk}\,\epsilon}) \; - \; \sqrt{2}\exp(K/2)g^{ij^*}(D_jW)^*\epsilon \\
\delta\psi_\mu \; &= \; 2\mathcal{D}_\mu\epsilon \; + \; \frac{1}{4}\,g_{ij^*}(\sigma_{\mu\nu}\gamma_5\epsilon)\left[\overline{\chi}^i(1+\gamma_5)\gamma^\nu\chi^j\right] \\
& - \frac{1}{4}\sqrt{2}(\gamma_5\psi_\mu)\left(\overline{\chi}^i\,\frac{\widehat{\partial K}}{\partial\phi^i}\,\gamma_5\epsilon\right) \\
& - \frac{1}{4}\left[g_{\mu\nu} \; - \; \sigma_{\mu\nu}\right]\gamma_5\epsilon\left(\overline{\lambda}^{(a)}\gamma^\nu\gamma_5\lambda^{(a)}\right) \; + \; \frac{1}{2}\exp(K/2)\widehat{W}\gamma_\mu\epsilon \\
\delta A_\mu{}^{(a)} \; &= \; - \; \overline{\epsilon}\gamma_\mu\lambda^{(a)} \\
\delta\lambda^{(a)} \; &= \; \left[F_{\alpha\beta}{}^{(a)} \; + \; \overline{\psi}_\alpha\gamma_\beta\lambda^{(a)}\right]\sigma^{\alpha\beta}\epsilon \; + \; i\,g\,D^{(a)}\gamma_5\epsilon\,.
\end{aligned}
\tag{4.44}
$$

The Lagrangian (4.33) plus (4.43) is rather complicated. The potential, however, takes a relatively simple form:

$$
\mathcal{V} \; = \; \frac{1}{2}g^2D^{(a)2} \; + \; e^K\left[(D_iW)(D_jW)^*g^{ij^*} \; - \; 3|W|^2\right]\,. \tag{4.45}
$$

In contrast to the globally supersymmetric case, it is no longer positive semidefinite. This gives us hope that we might construct realistic theories with spontaneously broken supersymmetry and no cosmological constant. Form the transformations (4.44), we see that χ^i and $\lambda^{(a)}$ transform nonlinearly whenever

$$
\left\langle e^{K/2}g^{ij^*}(D_jW)^*\right\rangle \neq 0\,, \qquad \left\langle D^{(a)}\right\rangle \neq 0\,. \tag{4.46}
$$

These conditions identify a linear combination of χ^i and $\lambda^{(a)}$ as the Goldstone spinor, and signal spontaneous supersymmetry breaking. From equation (4.45), we see that spontaneous supersymmetry breaking and no cosmological constant are consistent if and only if

$$
\langle W\rangle \neq 0\,. \tag{4.47}
$$

Because of (4.41), equation (4.47) breaks all the symmetries in $\mathcal{G}$ not in $\mathcal{H}$. Only symmetries for which $F = 0$ are left unbroken.

5. $N = 2$ SUPERSYMMETRY IN FOUR DIMENSIONS

$N = 2$ SUPERSYMMETRY AND HYPERKÄHLER GEOMETRY

In the previous sections, we derived the most general $N = 1$ supersymmetric Lagrangian in $3 + 1$ dimensions. We found that the matter couplings had a natural description in the language of the nonlinear sigma model. In this section we shall extend these results to $N = 2$. We will begin to construct the most general Lagrangian for the massless spin $(0, \frac{1}{2})$ multiplet [14,32]. As before, we shall see that our results have a geometrical interpretation in the language of the sigma model.

In what follows, we will not attempt to include mass terms or potentials, nor will we discuss the spin $(0, \frac{1}{2}, 1)$ gauge field multiplet. Including mass terms and potentials is more difficult in $N = 2$ supersymmetry than in $N = 1$. This is because the $N = 2$ multiplets contain different sets of component fields. As shown in Table 3, the $N = 1$ algebra admits both massless and massive spin $(0, \frac{1}{2})$ representations. The $N = 2$ algebra, on the other hand, has only massless spin $(0, \frac{1}{2})$ representations. The equivalent massive representations contain spins 0, $\frac{1}{2}$ and 1. Thus in $N = 2$ supersymmetry, one cannot pass from massless to massive representations simply by adding a superpotential to the Lagrangian $\mathcal{L}$. One must change the field content of the theory as well.

The question of gauge fields is also more subtle in $N = 2$ supersymmetry than in $N = 1$. Again this is because of the different field content of the $N = 2$ multiplets. In $N = 1$ supersymmetry, gauge field multiplets contain spins $\frac{1}{2}$ and 1. In contrast, $N = 2$ gauge multiplets contain spins 0, $\frac{1}{2}$ and 1. If the $N = 2$ Lagrangian is to be described in geometrical language, the extra scalar fields in the $N = 2$ gauge multiplets must have a sigma model interpretation. They too must lie on a sigma model manifold, whose dimension should be equal to the dimension of the isometry group $\mathcal{G}$.

Despite these apparent difficulties, much progress has been made on these issues. One elegant approach is to work in five or six dimensions, where the $N = 2$ supersymmetry becomes $N = 1$. In the higher dimensions, it is easier to add mass terms and gauge fields. One can then recover a $3 + 1$ dimensional model by dimensional reduction. Of course, there is no guarantee that this procedure gives the most general $3 + 1$ dimensional model, but it does give insight into the geometrical structure of the lower-dimensional theory. Recent developments along these and other lines are discussed in Ref. [33].

In the remainder of this section, we shall restrict our attention to massless spin $(0, \frac{1}{2})$ multiplets. We will work out the most general Lagrangian containing these fields, and we will describe the matter couplings in the language of the nonlinear sigma model. As before, we shall see that different $N = 2$ couplings correspond to different manifolds, and constraints on the matter couplings arise as restrictions on the manifolds $\mathcal{M}$.

Table 3
Supersymmetry Representations in $3+1$ Dimensions

N	Representation	Spin	Multiplicity	Type
1	massless matter	0	1	complex
		1/2	1	Majorana
1	massive matter	0	1	complex
		1/2	1	Majorana
1	massless gauge field	1/2	1	Majorana
		1	1	real
2	massless matter	0	2	real
		1/2	1	Majorana
2	massive matter	0	5	real
		1/2	4	Majorana
		1	1	real
2	massless gauge field	0	2	real
		1/2	2	Majorana
		1	1	real

In $N=2$ rigid supersymmetry, we shall find that sigma models exist for all hyperkähler manifolds $\mathcal{M}$ [4]. This is, of course, expected, since $N=4$ supersymmetry in $1+1$ dimensions is related to $N=2$ supersymmetry in $d=3+1$. However, we will see that hyperkähler manifolds arise in a different way than in the $1+1$ dimensional models of Section 2. After discussing rigid supersymmetry, we shall move on to consider $N=2$ local supersymmetry. We will see that $N=2$ local supersymmetry requires the scalar fields to be the coordinates—not of a hyperkähler manifold—but rather of a quaternionic manifold [14]. Quaternionic and hyperkähler manifolds are related to each other,

Thus one of our results is that matter couplings allowed in $N=2$ supersymmetry are forbidden in $N=2$ supergravity, and vice versa.[♯6] A second surprising result is that in $N=2$ supergravity, the matter couplings cannot be trivially reduced to $N=1$. The $N=1$ and $N=2$ rigid and local matter couplings are summarized in Table 4.

In $3+1$ dimensions, we will find that $N=2$ sigma models are best described according to their holonomy groups G. As discussed in Section 2, the holonomy group of a connected n-dimensional Riemannian manifold is

♯6 This result was partly anticipated in Ref. [35], in which an $N=2$ supersymmetric sigma model was found that could not be coupled to supergravity.

Table 4
Matter Couplings in 3 + 1 Dimensions

	rigid supersymmetry	local supersymmetry
$N = 1$	Kähler	Hodge
$N = 2$	hyperkähler	quaternionic

the group of transformations generated by parallel transporting all vectors around all possible closed curves in $\mathcal{M}$. If the parallel transport is done with respect to the Riemann connection, then the holonomy group G is contained in $O(n)$.

In Section 2, we defined a hyperkähler manifold as a $4n$-dimensional real Riemannian manifold endowed with three parallel complex structures, obeying the relations (2.20) and (2.21) [35]. In this section we take a slightly different point of view. We now *define* a hyperkähler manifold to be a $4n$-dimensional real Riemannian manifold whose holonomy group is contained in $Sp(n) \subseteq O(4n)$. It is easy to show that this definition is equivalent to that given in terms of the parallel complex structures.

As shown in Table 3, the $N = 2$ supersymmetry algebra has representations with 2 massless spin 0 states, and 2 massless spin $\pm\frac{1}{2}$ states. However, interacting field theories seem to exist only if the number of real scalar fields is divisible by four. This is related to the fact that $\mathcal{M}$ must be hyperkähler. Therefore, we consider theories with $4n$ real scalars ϕ^i, $2n$ Majorana spinors χ^Z, and 2 Majorana spinors ϵ^A. The spinors χ^Z are the supersymmetry partners of the scalars ϕ^i, and the spinors ϵ^A are the two supersymmetry parameters.

On dimensional grounds, the supersymmetry transformation of the scalar fields ϕ^i must take the following form

$$\delta\phi^i \;=\; \gamma^i_{AZ}(\bar{\epsilon}_R{}^A\chi_L{}^Z \;+\; \bar{\epsilon}_L{}^A\chi_R{}^Z)\;, \tag{5.1}$$

where γ^i_{AZ} is a nonsingular function of the ϕ^j. Since the variation of a coordinate is a vector, $\delta\phi^i$ takes its values in the $4n$-dimensional tangent bundle T of $\mathcal{M}$. The spinors ϵ^A and χ^Z take their values in 2 and $2n$-dimensional bundles H and P. The supersymmetry transformation (5.1) says that $\mathcal{M}$ must admit structures γ^i_{AZ} that split the tangent space in two, $T = H \otimes P$. This immediately gives a strong restriction on the allowed manifolds $\mathcal{M}$.

When we actually construct the sigma model, we shall see that supersymmetry demands an even stronger condition. Cancellation of the $\bar{\epsilon}\chi$ terms requires $\nabla_j \gamma^i_{AZ} = 0$. That is to say, there must exist suitable connections in T, H and P such that the γ^i_{AZ} are covariantly constant.

To make the γ^i_{AZ} more familiar, let us specialize for the moment to $n = 1$. Then $\mathcal{M}$ and T are four-dimensional, and H and P are both two-dimensional. In this case, H and P are bundles of left- and right-handed spinors. In four dimensions, T is indeed the product of two spinor bundles, and the γ^i_{AZ} are simply the Dirac γ-matrices. And in Riemannian geometry, the Dirac γ-matrices are indeed covariantly constant.

There is one more piece of information we have not yet used. That is that the bundle T is real. The fact that T is real implies that H and P are both real or both pseudoreal. As far as is known, H and P real does not lead to a sigma model. Therefore we take H and P to be pseudoreal. In down-to-earth terms, this just says that A is an $Sp(1)$ index, and Z is an $Sp(n)$ index.

In rigid supersymmetry, the supersymmetry parameters ϵ^A are constants, so the bundle H is trivial, and the $Sp(1)$ connection is flat. In local supersymmetry, we shall see that H cannot be trivial, so the holonomy group $G \subseteq Sp(1) \times K$, where $K \subseteq Sp(n)$.

To actually prove these assertions, one must do a little work. The fact that the γ^i_{AZ} are covariantly constant implies that they satisfy several relations similar to the Dirac algebra:[♯7]

$$
\begin{aligned}
\gamma^i_{AY}\,\gamma^j_{BZ}\,g_{ij} &= \epsilon_{AB}\,\epsilon_{YZ} \\
\gamma^i_{AZ}\,\gamma^{jBZ} + \gamma^j_{AZ}\,\gamma^{iBZ} &= g^{ij}\delta_A{}^B \\
\gamma^i_{AY}\,\gamma^{jAZ} + \gamma^j_{AY}\,\gamma^{iAZ} &= n^{-1}g^{ij}\delta_Y{}^Z\,.
\end{aligned}
\tag{5.2}
$$

Here ϵ_{AB} and ϵ_{YZ} are the totally antisymmetric $Sp(1)$ and $Sp(n)$ invariant tensors. Furthermore, the fact that $[\nabla_i, \nabla_j]\gamma^k_{AZ} = 0$ implies

$$
R_{ijk\ell}\,\gamma^\ell_{AY}\gamma^k_{BZ} = \epsilon_{AB}\,R_{ijYZ} + \epsilon_{YZ}\,R_{ijAB}\,,
\tag{5.3}
$$

where R_{ijAB} and R_{ijYZ} are the $Sp(1)$ and $Sp(n)$ curvatures, formed form the appropriate connections. In rigid supersymmetry, the ϵ^A are constants,

♯7 Our raising and lowering conventions are as follows: $\chi^A = \epsilon^{AB}\,\chi_B$, $\chi_A = \epsilon_{BA}\,\chi^B$ and $\epsilon_{AB}\,\epsilon^{BC} = -\delta_A{}^C$ (and likewise for ϵ_{XY}).

so $R_{ijAB} = 0$. In this case

$$R_{ijk\ell} = \gamma_k{}^{AZ} \gamma_{\ell A}{}^{Y} R_{ijYZ} \tag{5.4}$$

and we see that the holonomy group G is not all of $O(4n)$, but rather is contained in $Sp(n)$. That is why $N = 2$ rigid supersymmetry in $3+1$ dimensions demands that $\mathcal{M}$ be hyperkähler.

Because of the antisymmetry of ϵ_{AB}, it is easy to check that (5.4) implies

$$R_{ijk\ell} = - R_{ij\ell k} , \tag{5.5}$$

as required. Furthermore, the cyclic identity on the curvature requires

$$R_{ijYZ} = \gamma_i{}^{AW} \gamma_{jA}{}^{X} \Omega_{XYZW} . \tag{5.6}$$

The object Ω_{XYZW} is a totally symmetric $Sp(n)$ tensor; it is known as the hyperkähler curvature. Note that (5.4) and (5.6) are sufficient to prove that hyperkähler manifolds are Ricci flat.

The relations (5.2)-(5.6) are all that are needed to prove the invariance of the $N = 2$ supersymmetric nonlinear sigma model. The Lagrangian is as follows [4,14],[♯8]

$$\begin{aligned} \mathcal{L}_{SS} &= - g_{ij}\, \partial_\mu \phi^i \partial^\mu \phi^j - \frac{1}{2} \overline{\chi}_Z \gamma^\mu \mathcal{D}_\mu \chi^Z \\ &+ \frac{1}{16} \Omega_{XYZW} \left(\overline{\chi}_L{}^X \gamma_\mu \chi_L{}^Y \right) \left(\overline{\chi}_L{}^Z \gamma^\mu \chi_L{}^W \right) , \end{aligned} \tag{5.7}$$

where $\chi^Z_{L,R} = (1 \pm \gamma_5)\chi^Z$. The covariant derivative $\mathcal{D}_\mu \chi^Z$ contains the $Sp(n)$ connection coefficients $\Gamma^{iZ}{}_Y$,

$$\mathcal{D}_\mu \chi^Z = \partial_\mu \chi^Z + \Gamma_i{}^Z{}_Y\, \partial_\mu \phi^i \chi^Y , \tag{5.8}$$

and the transformation laws are given by

$$\begin{aligned} \delta \phi^i &= \gamma^i_{AZ} \left(\overline{\epsilon}_R{}^A \chi_L{}^Z + \overline{\epsilon}_L{}^A \chi_R{}^Z \right) \\ \delta \chi_L{}^Z &= 2\, \partial_\mu \phi^i\, \gamma_i{}^{AZ} \gamma^\mu\, \epsilon_{RA} - \Gamma_i{}^Z{}_Y\, \delta\phi^i\, \chi_L{}^Y . \end{aligned} \tag{5.9}$$

Cancellation of the $\overline{\epsilon}\chi$ terms implies that $R_{ijAB} = 0$.

♯8 When no helicity is specified, we adopt the following conventions: $\overline{\chi}_A \lambda^A = \overline{\chi}_{RA} \lambda_L{}^A - \overline{\chi}_L{}^A \lambda_{RA}$, $\overline{\chi}_A \gamma^\mu \lambda^A = \overline{\chi}_{LA} \gamma^\mu \lambda_L{}^A + \overline{\chi}_R{}^A \gamma^\mu \lambda_{RA}$ and $\overline{\chi}_A \sigma^{\mu\nu} \lambda^A = \overline{\chi}_{RA} \sigma^{\mu\nu} \lambda_L{}^A - \overline{\chi}_L{}^A \sigma^{\mu\nu} \lambda_{RA}$.

The Lagrangian (5.7) is invariant under rigid $N = 2$ supersymmetry transformations. To gauge them, we shall use the well-known Noether procedure, adding terms proportional to $\kappa^2 = 8\pi G_N$ to the Lagrangian and transformation laws. We begin by adding the pure $N = 2$ supergravity Lagrangian $\mathcal{L}_{SG}$ [36] to the globally supersymmetric matter Lagrangian $\mathcal{L}_{SS}$:

$$\begin{aligned}
\mathcal{L} &= \mathcal{L}_{SG} + \mathcal{L}_{SS} \\
\mathcal{L}_{SG} &= -\frac{1}{2\kappa^2}\, e R - \frac{1}{2}\,\epsilon^{\mu\nu\rho\sigma}\,\overline{\psi}_{\mu A}\gamma_5\gamma_\nu \mathcal{D}_\rho \psi_\sigma{}^A - \frac{1}{4}\, e\, F_{\mu\nu}F^{\mu\nu} \\
&\quad + \frac{1}{4}\sqrt{2}\, e\,\kappa\,\overline{\psi}_{\mu A}\left(F^{\mu\nu} + \frac{1}{2}\, e^{-1}\widetilde{F}^{\mu\nu}\gamma_5\right)\psi_\nu{}^A \\
&\quad - \frac{1}{8}\, e\,\kappa^2\,\overline{\psi}_{\mu A}\left(\overline{\psi}^\mu{}_B\psi^{\nu B} + \frac{1}{2}\, e^{-1}\,\epsilon^{\mu\nu\rho\sigma}\,\overline{\psi}_{\rho B}\psi_\sigma{}^B\gamma_5\right)\psi_\nu{}^A ,
\end{aligned} \tag{5.10}$$

where $\widetilde{F}_{\mu\nu} = \epsilon_{\mu\nu\rho\sigma}\,F^{\rho\sigma}$. The pure supergravity action contains two gravitinos $\psi_\mu{}^A$, and one $U(1)$ gauge boson A^μ. It is invariant under the following supergravity transformations:

$$\begin{aligned}
\delta e_{\mu\alpha} &= \kappa\,\overline{\epsilon}_A\gamma_\alpha\psi_\mu{}^A \\
\delta A_\mu &= \sqrt{2}\,\overline{\epsilon}_A\psi_\mu{}^A \\
\delta\psi_{\mu L}{}^A &= \frac{2}{\kappa}\,\mathcal{D}_\mu\epsilon_L{}^A + \frac{1}{2}\sqrt{2}\left(F_{\mu\nu} - \frac{1}{2}\, e\,\widetilde{F}_{\mu\nu}\gamma_5\right)\gamma^\nu\epsilon_R{}^A \\
&\quad - \frac{1}{2}\,\kappa\left(\overline{\psi}_{\mu A}\psi_\nu{}^A - \frac{1}{2}\, e\,\epsilon_{\mu\nu\rho\sigma}\,\overline{\psi}^\rho{}_A\psi^{\sigma A}\gamma_5\right)\gamma^\nu\epsilon_R{}^A \\
&\quad - \Gamma_i{}^A{}_B\,\delta\phi^i\,\psi_{\mu L}{}^B .
\end{aligned} \tag{5.11}$$

The covariant derivatives in $\mathcal{L}_{SS}$ and $\mathcal{L}_{SG}$ now contain the spin connection $\omega_{\mu\alpha\beta}$:

$$\begin{aligned}
\mathcal{D}_\mu\psi_\nu{}^A &= \partial_\mu\psi_\nu{}^A + \frac{1}{2}\,\omega_{\mu\alpha\beta}\,\sigma^{\alpha\beta}\psi_\nu{}^A + \Gamma_i{}^A{}_B\,\partial_\mu\phi^i\,\psi_\nu{}^B \\
\mathcal{D}_\mu\chi^Z &= \partial_\mu\chi^Z + \frac{1}{2}\,\omega_{\mu\alpha\beta}\,\sigma^{\alpha\beta}\chi^Z + \Gamma_i{}^Z{}_Y\,\partial_\mu\phi^i\,\chi^Y \\
\mathcal{D}_\mu\epsilon^A &= \partial_\mu\epsilon^A + \frac{1}{2}\,\omega_{\mu\alpha\beta}\,\sigma^{\alpha\beta}\epsilon^A + \Gamma_i{}^A{}_B\,\partial_\mu\phi^i\,\epsilon^B .
\end{aligned} \tag{5.12}$$

Note that the covariant derivatives for ϵ^A and $\psi_\mu{}^A$ also contain the $Sp(1)$ connection coefficients $\Gamma^{iA}{}_B$.

We shall use the 1.5-order formalism [16,37], which says that $\omega_{\mu\alpha\beta}$ obeys its own equation of motion and need not be explicitly varied under supergravity transformations. To lowest order, we couple $\mathcal{L}_{SG}$ and $\mathcal{L}_{SS}$ through the Neother current $J_\mu{}^A$:

$$\mathcal{L}_N = -\frac{1}{2}\overline{\psi}_{\mu A} J^{\mu A} = e\,\kappa\,\gamma_{iAZ}\left(\overline{\chi}_R{}^Z\gamma^\mu\gamma^\nu\psi_{\mu L}{}^A + \overline{\chi}_L{}^Z\gamma^\mu\gamma^\nu\psi_{\mu R}{}^A\right)\partial_\nu\phi^i\,. \tag{5.13}$$

These terms ensure that $\mathcal{L} = \mathcal{L}_{SG} + \mathcal{L}_{SS} + \mathcal{L}_N$ is invariant to order κ^0 under local supersymmetry transformations. The $\overline{\epsilon}\chi$ terms cancel provided $\nabla_i\gamma^j_{AZ} = 0$. The $\overline{\epsilon}\psi$ terms require

$$R_{ijAB} = \kappa^2\left(\gamma_{iAZ}\,\gamma_{jB}{}^Z - \gamma_{jAZ}\,\gamma_{iB}{}^Z\right)\,. \tag{5.14}$$

Substituting (5.14) into (5.3), and using the cyclic identity on the curvature, we find

$$R_{ijXY} = \kappa^2\left(\gamma_{iAX}\,\gamma_j{}^A{}_Y - \gamma_{jAX}\,\gamma_i{}^A{}_Y\right) + \gamma_i{}^{AW}\gamma_{jA}{}^Z\Omega_{XYZW}\,. \tag{5.15}$$

Equation (5.14) tells us, as mentioned before, that the $Sp(1)$ curvature is nonzero. A manifold with holonomy contained in $Sp(1)\times SU(n)$ and nonzero $Sp(1)$ curvature is called a *quaternionic* manifold [38,39]. When $R_{ijAB} \neq 0$, the supersymmetry parameter ϵ^A cannot be chosen to be covariantly constant.

Note that equations (5.3), (5.14) and (5.15) fix the scalar curvature in terms of Newton's constant:

$$R = -8\kappa^2\left(n^2 + 2n\right)\,. \tag{5.16}$$

This is the analogue of the quantization condition found previously for $N = 1$. Here, however, we find that only one value of the scalar self-coupling is consistent with supergravity. These results imply that the order κ^0 variations in $\mathcal{L}$ restrict $\mathcal{M}$ to be a quaternionic manifold of negative scalar curvature. They lead to the surprising conclusion that rigidly supersymmetric theories with $N = 2$ scalar multiplets may not be coupled to ($N = 2$) supergravity.[♮9]

♮9 Note that as $\kappa \to 0$, equation (5.15) approaches (5.6). This tells us that quaternionic manifolds of zero scalar curvature are precisely the hyperkähler manifolds discussed before.

It is remarkable that all this information can be gleaned from the order κ^0 variations of the supergravity Lagrangian. The higher order variations simply confirm the above results. It is a long and tedious calculation to compute the Lagrangian and transformation laws. The calculation is aided by using the 1.5-order formalism and by collecting various terms into the following "supercovariant" expressions:

$$\begin{aligned}\widehat{F}_{\mu\nu} &= F_{\mu\nu} - \frac{1}{2}\kappa\sqrt{2}\,\overline{\psi}_{\mu A}\psi_\nu{}^A \\ \widehat{D}_\mu\phi^i &= \partial_\mu\phi^i - \frac{1}{2}\kappa\,\gamma^i_{AZ}\left(\overline{\psi}_{\mu R}{}^A\chi_L{}^Z + \overline{\psi}_{\mu L}{}^A\chi_R{}^Z\right) .\end{aligned} \tag{5.17}$$

These expressions are supercovariant because their supersymmetry variations contain no $\partial_\mu\epsilon^A$ pieces.

Using (5.2), (5.3), (5.14), (5.15) and (5.16), one can verify that

$$\begin{aligned}\mathcal{L} = &-\frac{1}{2\kappa^2}\,e R - e\,g_{ij}\,\widehat{D}_\mu\phi^i\widehat{D}^\mu\phi^j - \frac{1}{4}\,e\,\widehat{F}_{\mu\nu}\widehat{F}^{\mu\nu} \\ &-\frac{1}{2}\,\epsilon^{\mu\nu\rho\sigma}\,\overline{\psi}_{\mu A}\gamma_5\gamma_\nu D_\rho\psi_\sigma{}^A - \frac{1}{2}\,e\,\overline{\chi}_Z\gamma^\mu D_\mu\chi^Z \\ &+\frac{1}{16}\kappa\sqrt{2}\,\overline{\psi}_{\mu A}\gamma_5\psi_\nu{}^A\left(e\widehat{F}^{\mu\nu} + \widehat{\widetilde{F}}^{\mu\nu}\right) \\ &+ e\,\kappa\,\gamma_{iAZ}\left(\overline{\chi}_R{}^Z\sigma^{\mu\nu}\psi_{\mu L}{}^A + \overline{\chi}_L{}^Z\sigma^{\mu\nu}\psi_{\mu R}{}^A\right)\left(\partial_\nu\phi^i + \widehat{D}_\nu\phi^i\right) \\ &+\frac{1}{4}\kappa\sqrt{2}\,e\,\overline{\chi}_Z\sigma^{\mu\nu}\chi^Z\widehat{F}_{\mu\nu} - \frac{1}{32}\,e\,\kappa^2\left(\overline{\chi}_Y\gamma_\mu\gamma_5\chi^Y\right)\left(\overline{\chi}_Z\gamma^\mu\gamma_5\chi^Z\right) \\ &+\frac{1}{16}\,e\,\kappa^2\left(\overline{\chi}_Y\sigma_{\mu\nu}\chi^Y\right)\left(\overline{\chi}_Z\sigma^{\mu\nu}\chi^Z\right) \\ &+\frac{1}{16}\,e\,\Omega_{XYZW}\left(\overline{\chi}_L{}^X\gamma_\mu\chi_L{}^Y\right)\left(\overline{\chi}_L{}^Z\gamma^\mu\chi_L{}^W\right)\end{aligned} \tag{5.18}$$

is invariant under the following supergravity transformations:

$$\begin{aligned}\delta e_{\mu\alpha} &= \kappa\,\overline{\epsilon}_A\gamma_\alpha\psi_\mu{}^A \\ \delta A_\mu &= \sqrt{2}\,\overline{\epsilon}_A\psi_\mu{}^A \\ \delta\phi^i &= \gamma^i_{AZ}\left(\overline{\epsilon}_R{}^A\chi_L{}^Z + \overline{\epsilon}_L{}^A\chi_R{}^Z\right) \\ \delta\chi_L{}^Z &= 2\,\partial_\mu\phi^i\,\gamma_i{}^{AZ}\gamma^\mu\,\epsilon_{RA} - \Gamma_i{}^Z{}_Y\,\delta\phi^i\,\chi_L{}^Y \\ \delta\psi_{\mu L}{}^A &= \frac{2}{\kappa}\,D_\mu\epsilon_L{}^A + \frac{1}{2}\sqrt{2}\left(\widehat{F}_{\mu\nu} - \frac{1}{2}\,e\,\widehat{\widetilde{F}}_{\mu\nu}\gamma_5\right)\gamma^\nu\epsilon_R{}^A \\ &\quad - \Gamma_i{}^A{}_B\,\delta\phi^i\,\psi_{\mu L}{}^B + \frac{1}{2}\gamma^\nu\gamma_5\epsilon_R{}^A\left(\overline{\chi}_Z\sigma_{\mu\nu}\gamma_5\chi^Z\right) .\end{aligned} \tag{5.19}$$

The spin connection ω contains both χ- and ψ-torsion, and the $4n$ real scalar fields are restricted to lie on a negatively curved quaternionic manifold with holonomy group G contained in $Sp(n) \times Sp(1)$.

Even if the holonomy group is contained in $Sp(n) \times Sp(1)$, the γ^i_{AZ} may not exist globally. [In the case $n = 1$, the condition that the γ^i_{AZ} exist globally is precisely the condition that the manifold admit a spin structure.] Because they appear explicitly in (5.18) and (5.19), the γ^i_{AZ} must exist globally for the Lagrangian and transformation laws to make sense. For quaternionic manifolds of positive scalar curvature, this imposes a severe restriction—the only allowed manifolds are the quaternionic projective spaces $HP(n)$ [39]. However equation (5.16) limits us to negatively curved (and typically noncompact) manifolds. It is not known if there are negatively curved quaternionic manifolds which do not admit globally defined γ's.

For $n > 1$, all quaternionic manifolds are Einstein space of constant (nonzero) scalar curvature. The only known compact cases are the symmetric spaces discussed by Wolf [40,41]. These consist of the three families

$$HP(n) = \frac{Sp(n+1)}{Sp(n) \times Sp(1)} \qquad X(n) = \frac{SU(n+2)}{SU(n) \times SU(2) \times U(1)}$$

$$Y(n) = \frac{SO(n+4)}{SO(n) \times SO(4)}, \tag{5.20}$$

where $n > 1$, as well as

$$\frac{G_2}{SO(4)} \qquad \frac{F_4}{Sp(3) \times Sp(1)} \qquad \frac{E_6}{SU(6) \times Sp(1)}$$

$$\frac{E_7}{SO(12) \times Sp(1)} \qquad \frac{E_8}{E_7 \times Sp(1)} . \tag{5.21}$$

Note that $X(2) \simeq Y(2)$. The only known noncompact examples are the noncompact analogues of (5.20) and (5.21), as well as the homogeneous but not symmetric spaces found by Alekseevskii [41].

By way of conclusion, let us now relate the Lagrangian (5.18) to the $N = 1$ results derived before. In rigid supersymmetry, the reduction from $N = 2$ to $N = 1$ is trivial. The reduction is trivial because any hyperkähler manifold is Kähler with respect to each of its three parallel complex structures. In local supersymmetry, however, the story is more complicated, and the reduction from $N = 2$ to $N = 1$ is *not* trivial. The simple truncation $\epsilon^1 = \psi^1_\mu = A_\mu = 0$ is not preserved by the supergravity transformations (5.19). This is easy to understand in mathematical terms. On a quaternionic manifold, three almost complex structures $I^{(A)i}{}_j$ are defined *locally*, and they locally obey the Clifford algebra relation (2.20). However, the three almost complex structure are *not* defined globally. As one moves from one coordinate patch to another, the almost complex structures rotate into each other via $Sp(1)$ transformations [38]. Since $N = 1$ supergravity requires that $\mathcal{M}$ be Hodge, and Hodge manifolds have globally defined complex structures, we see that the nonzero $Sp(1)$ curvature of a quaternionic manifold not only obstructs the global definition of the $I^{(A)i}{}_j$, but also prevents a trivial reduction from $N = 2$ to $N = 1$.

ACKNOWLEDGEMENTS

Much of the material discussed here was developed in collaboration with Edward Witten. My debt to him is substantial. I am also grateful to Luis Alvarez-Gaumé, Martin Roček, Stuart Samuel, and especially, Julius Wess, for many discussions on supersymmetry and sigma models. Finally, I would like to thank Vladimir Rittenberg and the Organizing Committee for a very pleasant and fruitful stay in Bonn.

REFERENCES

1] For an overview of supersymmetry model-building, see the lectures of S. Ferrara, this volume.

2] Other reviews include H. Haber and G. Kane, Michigan preprint UM-HE-TH-83-17 (1984), to appear in Phys. Rep.; H. Nilles, Geneva preprint UGVA-DPT 12-412 (1983), to appear in Phys. Rep. Early work is reviewed in P. Fayet, *Unification of the Fundamental Particle Interactions*, Erice, ed. S. Ferrara, J. Ellis and P. Van Nieuwenhuizen (Plenum Press, New York, 1980).

3] P. DiVecchia and S. Ferrara, *Nucl. Phys.* B130 (1977) 93; E. Witten, *Phys. Rev.* D16 (1977) 2991; D. Freedman and P. Townsend, *Nucl. Phys.* B177 (1981) 282. For a review of more recent developments, see M. Roček, Stony Brook preprint ITP-SB-84-30 (1984).

4] L. Alvarez-Gaumé and D. Freedman, *Comm. Math. Phys.* 80 (1981) 443.

5] E. Witten, *Nucl. Phys.* B188 (1981) 513; P. Salomonson and J. Van Holten, *Nucl. Phys.* B196 (1982) 509; D. Lancaster, *Nuovo Cim.* 79A (1984) 28.

6] See the lectures of L. Alvarez-Gaumé, this volume; as well as L. Alvarez-Gaumé, *Comm. Math. Phys.* 90 (1983) 161; D. Friedan and P. Windey, *Nucl. Phys.* B235 (1984) 395.

7] E. Witten, *J. Differential Geometry* 17 (1982) 661.

8] B. Zumino, *Phys. Lett.* 87B (1979) 203.

9] E. Cremmer and J. Scherk, *Phys. Lett.* 74B (1978) 341; E. Cremmer, B. Julia, J. Scherk, S. Ferrara, L. Girardello and P. Van Nieuwenhuizen, *Nucl. Phys.* B147 (1979) 105.

10] E. Cremmer, S. Ferrara, L. Girardello and A. Von Proeyen, *Phys. Lett.* 116B (1982) 231, *Nucl. Phys.* B212 (1983) 413.

11] E. Witten and J. Bagger, *Phys. Lett.* 115B (1982) 202.

12] J. Bagger and E. Witten, *Phys. Lett.* 118B (1982) 103.

13] J. Bagger, *Nucl. Phys.* B211 (1983) 302.

14] J. Bagger and E. Witten, *Nucl. Phys.* B222 (1983) 1.

15] M. Gell-Mann and M. Lévy, *Nuovo Cim.* 16 (1960) 705; see also Y. Nambu and G. Jona-Lasinio, *Phys. Rev.* 122 (1961) 345.

16] P. Van Nieuwenhuizen, *Phys. Rep.* 68 (1981) 189.

17] For an introduction to supersymmetry in superspace, see J. Wess and J. Bagger, *Supersymmetry and Supergravity* (Princeton University Press, Princeton, 1983); S. Gates, M. Grisaru, M. Roček and W. Siegel, *Superspace* (Benjamin/Cummings, Reading, 1983); P. Fayet and S. Ferrara, *Phys. Rep.* 32C (1977) 249.

18] A. Lichnerowicz, *Global Theory of Connections and Holonomy Groups*, (Noordhoff, Leyden, 1976).

19] A simple introduction to complex manifolds is contained in L. Alvarez-Gaumé and D. Freedman, *Unification of the Fundamental Particle Interactions*, Erice, ed. S. Ferrara, J. Ellis and P. Van Nieuwenhuizen (Plenum Press, New York, 1980).

20] S. Chern, *Complex Manifolds Without Potential Theory*, (Springer-Verlag, New York, 1979); R. Wells, *Differential Analysis on Complex Manifolds*, (Springer-Verlag, New York, 1980).

21] For an introduction to supergravity, see the lectures of D. Freedman and B. DeWit, this volume.

22] Many of the missing details are given in J. Bagger, *Matter Couplings in Supergravity Theories*, (University Microfilms, Ann Arbor, 1983); see also Ref. [9].

23] The tensor calculus approached is developed in K. Stelle and P. West, *Phys. Lett.* 74B (1978) 330; S. Ferrara and P. Van Nieuwenhuizen, *Phys. Lett.* 74B (1978) 333; S. Ferrara and P. Van Nieuwenhuizen, *Phys. Lett.* 76B (1978) 404; K. Stelle and P. West, *Phys. Lett.* 77B (1978) 376.

24] T. Eguchi, P. Gilkey and A. Hanson, *Phys. Rep.* 66 (1980) 213.

25] R. Bott and L. Tu, *Differential Forms in Algebraic Topology*, (Springer-Verlag, New York, 1982).

26] S. Deser and B. Zumino, *Phys. Rev. Lett.* 38 (1977) 1433; B. Zumino, *Nucl. Phys.* B127 (1977) 189; D. Freedman and A. Das, *Nucl. Phys.* B120 (1977) 221.

27] P. Fayet and J. Iliopoulos, *Phys. Lett.* 51B (1974) 461.

28] It is possible to gauge the full group $\mathcal{G}$ in superspace. See S. Samuel, Columbia preprint CU-TP-276 (1984); M. Roček, private communication (1983).

29] C. Ong, *Phys. Rev.* D27 (1983) 911, *Phys. Rev.* D27 (1983) 3044.

30] G. Moore and P. Nelson, Harvard preprint HUTP-84/A044 (1984); J. Bagger and C. Ong, in preparation.

31] D. Freedman, *Phys. Rev.* D15 (1977) 1173; A. Das, M. Fischler and M. Roček, *Phys. Rev.* D16 (1977) 3427; B. DeWit and P. Van Nieuwenhuizen, *Nucl. Phys.* B139 (1978) 216; K. Stelle and P. West, *Nucl. Phys.* B145 (1978) 175; R. Barbieri, S. Ferrara, D. Nanopoulos and K. Stelle, *Phys. Lett.* 113B (1982) 219.

32] A. Das and D. Freedman, *Nucl. Phys.* B120 (1977) 221; C. Zachos, *Phys. Lett.* 76B (1978) 329; Thesis, California Institute of Technology (1979). E. Fradkin and M. Vasiliev, Lebedev preprint N-197 (1976). P. Breitenlohner and M. Sohnius, *Nucl. Phys.* B165 (1980) 483.

33] B. DeWit, P. Lauwers, R. Philippe and S. Su, *Phys. Lett.* 134B (1984) 37; M. Günaydin, G. Sierra and P. Townsend, *Nucl. Phys.* B242 (1984) 244;, *Phys. Lett.* 133B (1983) 72; S. Samuel, Columbia preprint CU-TP-276 (1984); A. Galperin, E. Ivanov, S. Kalitzin, V. Ogievetsky and E. Sokatchev, Trieste preprint IC/84/43 (1984); H. Nishino and E. Sezgin, Trieste preprint IC/84/50 (1984); B. DeWit, P. Lauwers, A. Van Proeyen (to appear); E. Cremmer, J. Derendinger, B. DeWit, S. Ferrara, A. Van Proeyen (to appear).

34] T. Curtwright and D. Freedman, *Phys. Lett.* 90B (1980) 71, *(E* 91B (1980) 487).

35] E. Calabi, *Ann. Scient. Ec. Norm. Sup. (4)* 12 (1979) 269.

36] S. Ferrara and P. Van Nieuwenhuizen, *Phys. Rev. Lett.* 37 (1976) 1669.

37] P. Van Nieuwenhuizen and P. Townsend, *Phys. Lett.* 67B (1977) 439; P. West and A. Chamseddine, *Nucl. Phys.* B129 (1977) 39.

38] S. Ishihara, *J. Differential Geometry* 9 (1974) 483; F. Gursey and C. Tze, *Annals of Physics* 128 (1980) 29.

39] S. Salamon, *Invent. Math.* 67 (1982) 143.

40] J. Wolf, *J. Math. Mech.* 14 (1965) 1033.

41] D. Alekseevskii, *Math. USSR - Izv.* 9 (1975) 297.

42] S. Ferrara, L. Girardello, T. Kugo and A. Van Proeyen, *Nucl. Phys.* B223 (1983) 191; G. Girardi, R. Grimm, M. Müller and J. Wess, Karlsruhe preprint (1984); R. Grimm, M. Müller and J. Wess, CERN preprint CERN-TH-3881 (1984).

SUPERSTRINGS

Lars Brink

CERN
CH-1211 Geneva 23, Switzerland

1. INTRODUCTION

One of the most important discoveries in elementary particle physics in the 1960's was the existence of hadronic resonances lying on (nearly) linear Regge trajectories (spin vs. mass^2). To describe these states and their interactions one was lead to construct models with infinitely many states. This culminated in the Veneziano (dual) model[1], which was quite successful as a phenomenological model for mesons. The version of the model which best fitted the data, however, turned out to have theoretic shortcomings such as negative-norm states as intermediary states. It was soon realized that by choosing a specific non-physical (from a hadronic point of view) intercept for the leading Regge trajectory, the model possesses a huge symmetry, corresponding to an infinite algebra, the Virasoro algebra[2]. For this case it was finally proven by Brower[3], Goddard and Thorn[4] that there are no negative-norm states in the model (although the model has tachyonic states).

Shortly after the discovery of the Veneziano model it was realized by Nielsen, Nambu and Susskind[5] that the model describes the scattering of one-dimensional objects, strings. This is, in fact, a rather appealing picture of a meson; a string connecting a quark and an antiquark. Somehow it was believed by many that the missing parts in the Veneziano model were the quarks at the ends.

Another shortcoming of the model was the absence of fermions. In order to allow for fermions Ramond[6] generalized the Virasoro algebra to also include anticommuting generators, in fact laying the ground for supersymmetry! Soon after Neveu and Schwarz[7] found another bosonic model realizing such a symmetry. Eventually it was shown that the Ramond fermions fitted into the Neveu-Schwarz model building up one model with both bosons and fermions[8].

For both these models one found that the model could seemingly be consistent (apart from the tachyons) only in specific "critical dimensions of space-time", d=26 for the Veneziano model and d=10 for the Neveu-Schwarz-Ramond model. These facts found a natural explanation when Goddard, Goldstone, Rebbi and Thorn[9] considered the quantization of a relativistic string. The action for such an object, taken to be proportional to the area of the world-sheet swept out by the string had been proposed by Nambu, Hara and Goto[10]. When performing the quantization it was found that d=26 is in fact the natural choice. It took some ten years before Polyakov[11] showed how to quantize the theory in an arbitrary dimension, which in fact leads to the possibility of having string theories in 4 dimensions. This still seems quite likely to be a good model of hadrons as a limiting case of QCD in the multicolour limit.

The "spinning string" corresponding to the Neveu-Schwarz-Ramond model had to await the advent of supergravity before its complete action was constructed[12], but the basic structure became known long before that[13]. This formalism also showed clearly the uniqueness of the critical dimension d=10.

The spinning string theory is based on local supersymmetry in the world-sheet. It is natural to ask whether this symmetry can be extended, and in fact there exists an N=2 model[14]. However, the critical dimension for it turns out to be d=2 and hence it has no chance of being an interesting model in particle physics.

The Veneziano model was originally S-matrix Born terms. When loop corrections were attempted, it was found that certain loop graphs contain a new set of poles. In fact this is true only for d=26 and was the first hint of the critical dimension[15]. The new trajectory carries the quantum numbers of the vacuum and has double the intercept and half the slope of the leading "ρ-trajectory", which are the signs of the Pomeron. This trajectory had been missing in the model and its appearance aroused a lot of interest. In the string picture the new set of states was easy to understand. While the trajectories in the tree amplitudes correspond to strings with open ends, the new ones correspond to closed strings and in a string interaction it is natural that the two open ends of a string can join to make up a closed string.

The string theories were invented as candidates for interacting hadronic models. The spin-1 particle on the leading Regge trajectory was assumed to be the ρ-meson whose Regge trajectory has an intercept $\alpha_o \approx 1/2$. However, to enforce the Virasoro symmetry, the intercept has to be 1, demanding the spin-1 particle (on the leading trajectory) be massless. Neveu and Scherk[16)] then checked the scattering amplitudes for this state in the limit when the slope of the trajectory goes to zero (pushing all massive states to infinite mass) and found that the S-matrix is just the one of a Yang-Mills theory.

Still for some time the question seemed to be how to lower the ρ- and the Pomeron-trajectories, keeping the ratio of the intercepts, to the physical ones and still have a consistent model. This is so far an unsolved problem and the failure to solve it made the interest in the string models for hadronic physics fade away some 10 years ago. Many of the concepts developed in dual models have, however, influenced the modern thinking of gauge field theories and hadron physics in terms of QCD. A radical change in philosophy was instead proposed by Scherk and Schwarz[17)] who pointed out that the "Pomeron trajectory" with a massless spin-2 particle in the zero-slope limit interacts appropriately to be identified as a graviton, and that perhaps the string models should be regarded as unified models including gravity! Strings are characterized by a fundamental length scale L($\sim(\alpha')^{1/2} = (T)^{-1/2}$, where α' is the Regge slope and T the string tension). In the case of hadron physics L is taken to be of order 10^{-13} cm. If we instead regard the string quanta as fundamental quanta including gravity and Yang-Mills fields, the natural length scale is the Planck length, 10^{-33} cm.

The Neveu-Schwarz-Ramond model considered in this new light is still not a consistent model. It contains tachyons and a detailed study of some of the couplings, for example the ones of the closed string state of spin-3/2, would reveal inconsistencies. Gliozzi, Scherk and Olive[18)], however, pointed out that the tachyons could all be eliminated by considering only bosons with even "G-parity". We had been reluctant to impose this constraint since the "pion" in the model (which is tachyonic) has odd G-parity. However, once we give up the aspirations to construct hadronic amplitudes this step is most natural. Furthermore by making the fermions satisfy both the Weyl and the Majorana constraints, which is possible if d=10, the number of boson and fermion degrees of freedom are equal at the first few levels indicating a supersymmetric spectrum. This idea was finally completed in a series of papers by Green and Schwarz[19)] who have managed to construct the "Superstring Theories". These models are supersymmetric models and are in fact free from all the problems that plague the old models. There are two classes of models. The type I model which corresponds to the Neveu-Schwarz-Ramond model with the constraints above, has both open and closed strings and its zero slope limit (infinite tension limit) corresponds to N=1 supergravity coupled to N=1 Yang-Mills field theory (for d=10). By performing a trivial dimensional reduction to d=4, one obtains N=4 supergravity coupled to N=4 Yang-Mills theory. The type II models describe only closed strings and correspond to N=2 supergravity for d=10, and N=8 supergravity for d=4 in the limits taken above.

In retrospect it is almost inevitable to demand a global supersymmetry if one attempts to construct an interacting model of extended objects. Such objects can typically be decomposed into an infinite set of modes. When quantized such a set will provide infinite vacuum fluctuations. For a free string one can renormalize away such infinities. However, an interaction between two extended objects must be local in order to ensure causality and the infinite vacuum fluctuations will destroy such locality[20]. Supersymmetry, though, demands an equal number of bosonic and fermionic modes leading to a cancellation of all vacuum fluctuations. Explicit loop computations in the superstring theories in fact show remarkable properties[21]. The type II models are one-loop finite and the type I theories are divergent just in the right way to be renormalized by a renormalization of the string tension.

Einstein's gravity as well as supergravity extensions has one dimensionful coupling constant. Such theories are non-renormalizable. The only way they can be consistent theories, if they are quantized straightforwardly, is if all divergencies cancel. The supergravity theories are, in fact, shown to be two-loop finite at the S-matrix level but I find it most unlikely that the finiteness will persist to all orders. The proliferation of graphs at higher orders is so huge that no ordinary (perhaps so far hidden) symmetry would be capable of controlling all possible divergencies. To demand finiteness of a 4-dimensional theory with one dimensionful coupling constant would most probably constrain the theory to be trivial.

One way out of this dilemma is to introduce another dimensionful constant. The string tension is such a parameter. Thinking of it as the slope parameter relating an infinite set of states along a Regge trajectory one can see the extra states as regulators. In fact, the ordinary UV-divergencies are absent in a string theory since the trajectory amounts to Gaussian propagators. The infinite set of states can, however, give rise to other types of divergencies so it still remains to be checked if the superstring theories are finite or renormalizable to all orders. I find this much more likely than in the ordinary supergravity theories.

2. FREE BOSONIC STRINGS

Our goal will be to construct a full-fledged field theory of strings. Before getting there let us review the basic ingredients in a field theory of point particles. A spinless point particle falling freely will follow a one-parameter trajectory $x^\mu(\tau)$. The action must be independent of how the trajectory is parametrized and is taken to be proportional to the arc-length the particle travels.

$$S = m \int_{s_i}^{s_f} ds = m \int_{\tau_i}^{\tau_f} d\tau \sqrt{-\dot{x}^2(\tau)} , \tag{2.1}$$

m will be identified with the mass of the particle. The fact that the action should be reparametrization invariant has a ring of gravity to it. By regarding $x^\mu(\tau)$ as a set of scalar fields in one dimension we can couple them to a metric $g \equiv g_{\tau\tau}$ and write an action just the way scalar fields are coupled to gravity in 4 dimensions.

$$S = -\frac{1}{2} \int d\tau \sqrt{g} \, (g^{-1}\dot{x}^2 - m^2) \quad . \tag{2.2}$$

(Note that the mass term is like a cosmological constant.)

Eliminating the metric field g and inserting its solution back into (2.2) gives (2.1) so the actions (2.1) and (2.2) are equivalent, at least classically. In fact (2.2) is more general since it allows us to take the limit m=0.

From this action one can quantize and find the propagation of a free particle. This can so be elevated to a free field theory where we consider fields $\phi(x)$, which could be extended to interacting field theories. Let me not pursue that line further but instead ask that question if we restart the programme this time with a string.

We describe the simplest string by just its d-dimensional Minkowski coordinate $x^\mu(\sigma,\tau)$ (with space-like metric). The parameters σ and τ span the world-sheet traced out by the string when it propagates; σ is space-like and τ time-like. The action for the free string should be independent of the parametrization and it is easy to generalize (2.2) to obtain[22)]

$$S = -\frac{1}{2} T \int_0^\pi d\sigma \int_{\tau_i}^{\tau_f} d\tau \sqrt{-g} \, g^{\alpha\beta} \, \partial_\alpha x^\mu \, \partial_\beta x_\mu \quad , \tag{2.3}$$

$g^{\alpha\beta}$ is the inverse metric tensor and $g = \det g_{\alpha\beta}$. (Indices α and β take the values σ and τ.) T is a proportionality factor (which ensures that x^μ is a length) and will turn out to be the string tension.

The symmetries of Eq. (2.3) are global Poincaré invariance

$$\delta x^\mu = \ell^\mu{}_\nu \, x^\nu + a^\mu; \quad \delta g_{\alpha\beta} = 0 \tag{2.4}$$

and local reparametrization invariance:

$$\delta x^\mu = \xi^\alpha \, \partial_\alpha x^\mu$$

$$\delta g_{\alpha\beta} = \xi^\gamma \, \partial_\gamma g_{\alpha\beta} + \partial_\alpha \xi^\gamma \, g_{\gamma\beta} + \partial_\beta \xi^\gamma \, g_{\alpha\gamma} \tag{2.5}$$

together with Weyl invariance

$$\delta g_{\alpha\beta} = \Lambda(\sigma,\tau) \, g_{\alpha\beta} \; ; \; \delta x^\mu = 0 \tag{2.6}$$

Classically, one can eliminate $g_{\alpha\beta}$ from Eq. (2.3) by solving its field equations algebraically and substituting this solution back in the action. The resulting expression is the Nambu-Hara-Goto formula, namely the area of the world-sheet, (definitely a reparametrization invariant expression). Quantum mechanically the elimination of g involves performing a path integral. In general an extra "Liouville mode" is left over[11] (the Weyl-invariance is broken) except in the special case of d=26. Whether or not it is possible to make sense of the bosonic string theory for d<26 is still not completely settled. For the rest of these lectures we will only consider the string theories in their critical dimensions.

Note that we are not allowed to introduce a cosmological constant this time, if we insist on the Weyl invariance (2.6).

In order to quantize the system let us consider the equations of motion.

$$\partial_\alpha \, (\sqrt{-g} \, g^{\alpha\beta} \, \partial_\beta x^\mu) = 0 \tag{2.7a}$$

$$T_{\alpha\beta} = \partial_\alpha x^\mu \partial_\beta x_\mu - \tfrac{1}{2} \, g_{\alpha\beta} \, g^{\gamma\delta} \, \partial_\gamma x^\mu \, \partial_\delta x_\mu = 0 \tag{2.7b}$$

The invariances (2.5) and (2.6) allow us to choose the gauge

$$g_{\alpha\beta} = \eta_{\alpha\beta} \tag{2.8}$$

where $\eta = \begin{pmatrix}-1 & 0\\ 0 & 1\end{pmatrix}$ is the two-dimensional Minkowski metric. The equations (2.7) then simplifies into

$$\partial_\alpha \partial^\alpha x^\mu = 0 \qquad (2.9)$$

$$(\dot{x}^\mu \pm x'^\mu)^2 = 0 \quad , \qquad (2.10)$$

where we introduced the notation $\dot{x} \equiv \frac{\partial}{\partial\tau} x$ and $x' \equiv \frac{\partial}{\partial\sigma} x$.

At this point there are two alternatives approaches to studying the quantum theory. The first one is covariant quantization, which entails using an indefinite-metric Hilbert space containing unphysical (gauge) degrees of freedom (from x^0) as well as physical ones. The physical subspace is defined by the Gupta-Bleuler procedure, i.e. we impose as subsidiary conditions that the positive frequency part of the operators in (2.10) annihilate physical states. This leads, in fact, to the Virasoro operators.

The other approach, which is the one we will follow here, since it will turn out to be much more easy to extend to other string models, is to choose a physical gauge in which the Hilbert space consists entirely of physical states.

To see that such a gauge choice is possible, consider the solutions of (2.9) in the case of open strings. Then (2.9) should be supplemented by the boundary conditions

$$x'^\mu = 0 \quad \text{for } \sigma = 0, \ \pi \qquad (2.11)$$

in order that surface terms vanish. The general solution of Eqs. (2.9) and (2.11) is

$$x^\mu(\sigma,\tau) = f^\mu(\tau+\sigma) + f^\mu(\tau-\sigma) \ , \qquad (2.12)$$

where

$$\frac{\partial}{\partial\xi} f^\mu(\xi) = \frac{\partial}{\partial\xi} f^\mu(\xi+2\pi) \ . \qquad (2.13)$$

The system of equations (2.12), (2.13) and (2.10) still contains a conformal invariance. One can perform a transformation

$$\delta f^\mu(\xi) = \zeta(\xi) \frac{\partial}{\partial\xi} f^\mu(\xi) \quad . \qquad (2.14)$$

This symmetry is such that we can transform one component of x^μ to be essentially trivial. The natural choice is to introduce light-cone coordinates

$$x^{\pm} = \frac{1}{\sqrt{2}}\,(x^{0} \pm x^{d-1}) \tag{2.15}$$

and choose

$$x^{+}(\sigma,\tau) = x^{+} + p^{+}\tau \ , \tag{2.16}$$

where x^{+} and p^{+} are constants. This means that we identify the light-cone frame evolution coordinate x^{+} with τ, hence having the same time along the string. In principle this expression should involve T since we want x^{+} to be like a c.m.s. coordinate and p^{+} the total momentum in the + -direction. We will correct this later by choosing T to be the appropriate constant.

The momentum conjugate to x^{μ} in the gauge (2.8) is

$$p^{\mu}(\sigma,\tau) = \frac{\delta}{\delta\dot{x}_{\mu}(\sigma,\tau)} = T\,\dot{x}^{\mu}(\sigma,\tau) \quad . \tag{2.17}$$

The total momentum is

$$p^{\mu} = \int_{0}^{\pi} p^{\mu}(\sigma,\tau)\, d\sigma \quad . \tag{2.18}$$

It is easy to see that p^{μ} is a constant of motion using (2.9) and (2.11). Comparing (2.18) and (2.16) we find that we have made the choice $T = 1/\pi$. Note that the p^{+} distribution is constant along the string.

In the "light-cone gauge" (2.16) with x^{+} known, x^{-} can be solved for from the constraints (2.10)

$$\dot{x}^{-} = \frac{1}{2p^{+}}\,(\dot{x}^{i\,2} + x'^{\,2}) \tag{2.19a}$$

$$x'^{-} = \frac{1}{p^{+}}\,(\dot{x}^{i}x'^{i}) \quad , \tag{2.19b}$$

which can be integrated to give $x^{-}(\sigma,\tau)$ in terms of the transverse coordinates $x^{i}(\sigma,\tau)$, p^{+} and a single integration constant x^{-}.

The string in the light-cone gauge can be described by the action (with $T = 1/\pi$)

$$S = \frac{-1}{2\pi} \int_{0}^{\pi} d\sigma \int d\tau\ \eta^{\alpha\beta}\,\partial_{\alpha}x^{i}\,\partial_{\beta}x^{i} \quad . \tag{2.20}$$

This gives, in fact, <u>the complete description of the string</u>. The action is Lorentz-invariant, although the Lorentz transformations are <u>non-linearly</u> realized. We will return to these transformations later.

In the open-string case the complete solution to the eqs. of motion (2.9) is

$$x^i(\sigma,\tau) = x^i + p^i\tau + i \sum_{n=0} \frac{1}{n} \alpha_n{}^i \cos n\sigma\, e^{-in\tau} . \qquad (2.21)$$

The canonically conjugate momenta, which now follow directly from (2.20) are

$$p^i(\sigma,\tau) = \frac{1}{\pi} \dot{x}^i(\sigma,\tau) , \qquad (2.22)$$

and quantization amounts to

$$[p^i(\sigma,\tau), x^j(\sigma',\tau)] = -i\, \delta^{ij}\, \delta(\sigma-\sigma') . \qquad (2.23)$$

Inserting the solution (2.21) leads to

$$[\alpha_m{}^i, \alpha_n{}^j] = m\, \delta_{m+n,0}\, \delta^{ij} . \qquad (2.24)$$

Hence α's with negative indices are raising operators and the ones with positive indices are lowering operators.

Let us consider

$$p^- = \frac{1}{\pi} \int_0^\pi d\sigma\, \dot{x}^- = \frac{1}{2p^+\pi} \int_0^\pi d\sigma \left(\frac{1}{\pi^2} p^{i\,2}(\sigma) + x'^{i\,2}(\sigma)\right). \qquad (2.25)$$

Inserting the solution (2.21) we find

$$p^- = \frac{p^{i\,2}}{2p^+} + \frac{1}{p^+} \sum_{n=1}^{\infty} \alpha_{-n}{}^i \alpha_n{}^i \qquad (2.26)$$

which amounts to

$$p^2 = -2 \sum_{n=1}^{\infty} \alpha_{-n}{}^i \alpha_n{}^i = -m^2 . \qquad (2.27)$$

Here it is seen that the lowest mass for a classical string is zero. Reintroducing dimensions for a moment via the Regge slope parameter $\alpha' = 1/2\pi T$, we have

$$\alpha' m^2 = \sum_{n=1}^{\infty} \alpha_{-n}{}^i \alpha_n{}^i . \qquad (2.28)$$

From the relations (2.24) it is clear that we can regard the modes of the string as an infinite set of harmonic oscillators. When quantized each oscillator will generate a zero-point fluctuation and the mass of the lowest state will be[23)]

$$\alpha' m_o^2 = \frac{d-2}{2} \sum_{n=1}^{\infty} n \quad . \tag{2.29}$$

This is clearly a divergent sum which must be regularized. We do so by comparing the sum to the Riemann ζ-function[24)]

$$\zeta(s) = \sum_{n=1}^{\infty} n^{-s} \quad \text{Re } s > 1 \quad . \tag{2.30}$$

This is a function which can be analytically continued to $s = -1$

$$\zeta(-1) = -\frac{1}{12} \quad . \tag{2.31}$$

In this way the infinite series has been regularized into

$$\alpha' m_o^2 = -\frac{d-2}{24} \quad . \tag{2.32}$$

There is also a more standard way of obtaining this result by adding counter terms to the action (2.20), which amounts to renormalizing the speed of light[23)].

The really important consequence of Eq. (2.32) is that the lowest state must be a tachyon. It means that the quantum theory does not exist and it is wrong to try to base an interacting model on this string. It was mentioned in the introduction that the Veneziano model, which is based on this string, does have some inconsistencies and they are related to this fact.

We here see the first hint that a quantum theory of strings will face more constraints than a quantum theory of point particles. I do find this fact quite inspiring!

I have not mentioned the closed string sector of the model yet. It is obtained by using periodic boundary conditions in σ and essentially amounts to double the set of harmonic oscillators. Also in this sector one finds tachyons. The closed string sector will play an important rôle in the superstring and we will dwell on it there at length.

The action (2.20) must be Lorentz invariant. It follows from the way it was derived. However, the Lorentz transformations are non-linear. In order to derive them, one considers the linear transformations (2.4), makes compensating gauge transformations (2.5) to stay in the gauge (2.16) and substitutes (2.19). The resulting transformations will then be non-linear. If one checks the algebra in the quantum case, it turns out to work only when d=26. This is the origin of the critical dimension as seen in the light-cone gauge.

3. THE SPINNING STRING

We saw in the preceding section that it is not enough to consider strings described by only its coordinate. There must be another set of modes which can compensate the zero-point fluctuations coming from the x-modes. The most natural thing to do is to introduce a set of anticommuting harmonic oscillators. This is, in fact, what Ramond did[6)], although from a completely different point of view. If we introduce as many anticommuting modes as commuting we can get all zero-point fluctuations to cancel.

The first problem to solve is to determine in which representation of the Lorentz group the new oscillators should be chosen. The really relevant group to consider is the transverse subgroup SO(d-2). The x-coordinates belong to the vector representation. We can always try this representation also for the new set, which we shall do first, but we should keep in mind, that for certain values of d-2, there are other representations with the same dimension as the vector one.

Consider hence a set of anticommuting harmonic oscillators $d_n^{\,i}$ satisfying

$$\{d_m^{\,i},\ d_n^{\,j}\} = \delta_{n+m,0}\ \delta^{ij}\ . \tag{3.1}$$

If the relevant mass formula is

$$\alpha' p^2 = -\Big(\sum_{n=1}^{\infty} \alpha_{-n}^{\,i}\alpha_n^{\,i} + \sum_{n=1}^{\infty} n\ d_{-n}^{\,i} d_n^{\,i}\Big) \tag{3.2}$$

this will correspond to a massless quantum ground state. Our burden is hence to find a light-cone action which leads to (3.2) and which is Lorentz invariant.

Let me introduce two normal-mode expansions

$$\lambda^{1i} = \sum_{n=-\infty}^{\infty} d_n^{\,i}\ e^{-in(\tau-\sigma)} \tag{3.3a}$$

$$\lambda^{2i} = \sum_{n=-\infty}^{\infty} d_n^{\,i}\ e^{-in(\tau+\sigma)}\ . \tag{3.3b}$$

We first rewrite (3.2) as (again going back to $\alpha' = 1/2$)

$$p^- = \frac{{p^i}^2}{2p^+} + \frac{1}{p^+}\left(\sum_{n=1}^{\infty} \alpha_{-n}{}^i \alpha_n{}^i + \sum_{n=1}^{\infty} n\, d_{-n}{}^i d_n{}^i\right) . \qquad (3.4)$$

In the light-cone frame p^- is conjugate to x^+ which is the evolution parameter. Hence p^- is a Hamiltonian. However, the momenta is defined with τ as the evolution parameter, but

$$\frac{\partial}{\partial\tau} = p^+ \frac{\partial}{\partial x^+} \quad , \qquad (3.5)$$

i.e.

$$H = \frac{{p^i}^2}{2} + \left(\sum_{n=1}^{\infty} \alpha_{-n}{}^i \alpha_n{}^i + \sum_{n=1}^{\infty} n\, d_{-n}{}^i d_n{}^i\right)$$

$$= \frac{1}{2\pi}\int_0^{\pi} d\sigma\left(\frac{1}{\pi^2}\, {p^i}^2 + {x'^i}^2 - i\lambda^{1i}\lambda'^{1i} + i\lambda^{2i}\lambda'^{2i}\right) . \qquad (3.6)$$

From this Hamiltonian it is straightforward to write down the corresponding action. We combine the two λ's into a 2-dimensional 2-component Majorana spinor and use the Majorana representation of the 2x2 Dirac matrices, here called ρ^α. Then (with the tension T)

$$S = -\frac{T}{2}\int_0^{\pi} d\sigma \int d\tau \left[\eta^{\alpha\beta}\, \partial_\alpha x^i \partial_\beta x^i + i\bar{\lambda}^i \rho^\alpha \partial_\alpha \lambda^i\right] . \qquad (3.7)$$

It remains to be shown that the action is Lorentz invariant. Before doing that let us investigate closer the physical states. Consider open-string states. The boundary conditions for the λ's are

$$\lambda^{1i}(0,\tau) = \lambda^{2i}(0,\tau) \qquad (3.8a)$$

$$\lambda^{1i}(\pi,\tau) = \begin{cases} \lambda^{2i}(\pi,\tau) & (3.8b) \\ -\lambda^{2i}(\pi,\tau) \quad . & (3.8c) \end{cases}$$

In fact we have two choices. The first choice together with the equations of motion gives the solutions (3.3). In this sector we know that there are no tachyons. The other choice results in expansions

$$\lambda^{1i} = \sum_{r=-\infty}^{\infty} b_r{}^i\, e^{-ir(\tau-\sigma)} \qquad (3.9a)$$

$$\lambda^{2i} = \sum_{r=-\infty}^{\infty} b_r{}^i \, e^{-ir(\tau+\sigma)} , \tag{3.9b}$$

where the index r takes all half-integer values. The b's satisfy the anticommutators

$$\{b_r{}^i, b_s{}^j\} = \delta_{r+s,0} \, \delta^{ij} . \tag{3.10}$$

$$\alpha' m^2 = \sum_{n=1}^{\infty} \alpha_{-n}{}^i \alpha_n{}^i + \sum_{r=1/2}^{\infty} r \, b_{-r}{}^i b_r{}^i . \tag{3.11}$$

Computing the contributions from the zero-point fluctuations we find

$$\begin{aligned} \alpha' m_o{}^2 &= \frac{d-2}{2} \left[\sum_{n=1}^{\infty} n - \sum_{n=1}^{\infty} (n-1/2) \right] \\ &= \frac{d-2}{2} \sum_{n=1}^{\infty} n \, (1-1/2+1) \\ &\to - \frac{d-2}{16} , \end{aligned} \tag{3.12}$$

when the sum is renormalized. Again we find tachyons! Hence this sector of the model is unphysical and would lead to various inconsistencies if we insist on using it.

The states we have discovered spanned by the b- and d-oscillators together with the α's are in fact the spectrum of the Neveu-Schwarz-Ramond model. The b-sector is used to build up the bose (Neveu-Schwarz) spectrum and the d's the fermi (Ramond) spectrum. As was mentioned in the introduction the model has tachyons and we see that within this string model they are unavoidable. We could insist on using only the Ramond sector, but with those creation operators one can only have either all states being fermions or all bosons. Since we are really interested in a model with both fermions and bosons we have to discard this model.

The action (3.7) is indeed Lorentz invariant. One can construct the non-linear Lorentz transformations as in the bosonic case. However, this time we have not started from a covariant action and have no deductive way to derive the Lorentz transformations. Instead we can try to guess them. If one does that, one finds that the algebra only works when d=10. This is the critical dimension of this model.

We have discussed the spinning string model completely in the light-cone gauge. This is certainly enough, but it might be advantageous to also have a covariant formalism and in fact there is an action generalizing (2.3). It was done mimicking supergravity[12].

One considers two-dimensional supergravity (on the world-sheet σ,τ) with a "zweibein" $V_\alpha{}^a$, related to the metric $g_{\alpha\beta}$ in the usual way and a Majorana Rarita-Schwinger field ψ_α. Then the action is

$$S = -\frac{T}{2} \int d\sigma d\tau \; \eta_{\mu\nu} \, V \, \{ g^{\alpha\beta} \partial_\alpha x^\mu \partial_\beta x^\nu + i \, V_a{}^\alpha \, \bar{\lambda}^\mu \rho^a \partial_\alpha \lambda^\nu$$

$$+ 2 \, V_a{}^\alpha V_b{}^\beta \, \bar{\psi}_\alpha \rho^b \rho^a \lambda^\mu (\partial_\beta x^\nu + \frac{1}{2} \bar{\lambda}^\nu \psi_\beta) \} \,. \qquad (3.13)$$

This is an action with a lot of symmetry. Its physics is just the physics of free field theory although it contains interaction terms. It is not unlikely that this action could be used in other fields of physics if properly interpreted[25].

4. SUPERSTRINGS

The action (3.7) has a 2-dimensional global supersymmetry, which is a remnant of the local supersymmetry of (3.13). One might try to extend that symmetry and that is historically what was done[14]. The N=2 theory essentially amounts to doubling the number of fields. Still there is, of course, a tachyon-free sector but the analysis of the model shows one disgusting and at the time of the construction of the model rather sad fact; the critical dimension is d=2! This meant good-bye to the model, although its locally supersymmetric action[14] with all its beautiful symmetries might be useful for other purposes.

A better alternative is to seek other representations of the fermionic field along the string. The constraint is that these representation (of SO(d-2)) should be of the same dimension as the vector one. The natural choice is to look for a spinor representation (after all it is a fermionic field). For d=3,4,6 and 10 the lowest spinor representation is, in fact, of the same dimension as the vector one! Since SO(d-2) is a compact group the scalar product of two spinors is just the contracted sum as for vectors, and we can take over all work in section 3 up to eq.(3.8). We only make the substitution

$$\lambda^{Ai} \to \frac{1}{\sqrt{2p^+}} \, Q^{Aa} \,, \qquad (4.1)$$

where A is a two-component spinor index and a is a d-2 component spinor index. The factor $(2p^+)^{-1/2}$ is introduced for future convenience. It does not affect the SO(d-2) properties of Q^{Aa}.

The action for the superstring is then (in the light-cone gauge)[26]

$$S = -\frac{T}{2} \int_0^\pi d\sigma \int d\tau \, [\eta^{\alpha\beta} \partial_\alpha x^i \partial_\beta x^i + \frac{i}{2p^+} \bar{Q}^a \rho^\alpha \partial_\alpha Q^a] \,. \qquad (4.2)$$

I stress that we know that this action has a sector which starts with massless particles as the lowest lying states. What remains to be proven is the Lorentz invariance. Before doing that let us discuss the two sectors following from the boundary conditions (3.8). The choice we want to use is (3.8a) and (3.8b). This gives us integer modes and hence no tachyons and we can write the solutions to the equations of motion (for open strings)

$$Q_a{}^1 = \sum_{n=-\infty}^{\infty} Q_n{}^a \, e^{-in(\tau-\sigma)} \tag{4.3a}$$

$$Q_a{}^2 = \sum_{n=-\infty}^{\infty} Q_n{}^a \, e^{-in(\tau+\sigma)} , \tag{4.3b}$$

with the anticommutation rules (from (3.1))

$$\{Q_m{}^a, Q_n{}^b\} = 2p^+ \delta_{m+n,0} \, \delta^{ab} . \tag{4.4}$$

The operators $Q_{-n}{}^a$ with n positive are creation operators. They will take a bosonic state that it acts on to a fermionic one. Hence this sector will contain both bosons and fermions, in fact equally many of each kind at each mass level, building up supermultiplets at each level.

The other sector which follows by using the boundary conditions (3.8a) and (3.8c) we know has tachyons. Furthermore the fermionic oscillators will be half-integer moded and supersymmetry will be lost. Since the other sector seems sufficient we simply decree that we only use the boundary conditions (3.8a) and (3.8b) or periodic boundary conditions. This is, in fact, the way to project out the superstring states from the Neveu-Schwarz-Ramond model.

To check the symmetries of the action (4.2) we start by considering supersymmetry. In the light-cone frame supersymmetry breaks up into two generators Q_+ and Q_-. The algebra is

$$\{Q_+{}^a, Q_+{}^b\} = 2p^+ \delta^{ab} \tag{4.5a}$$

$$\{Q_-{}^{\dot{a}}, Q_-{}^{\dot{b}}\} = 2p^- \delta^{\dot{a}\dot{b}} \equiv 2H \, \delta^{\dot{a}\dot{b}} \tag{4.5b}$$

$$\{Q_+{}^a, Q_-{}^{\dot{b}}\} = \sqrt{2} \, (\gamma_i)^{a\dot{b}} \, p^i . \tag{4.5c}$$

See appendix for notations.

In an interacting field theory Q_- and H are non-linearly realized, while Q is linearly realized. In a first-quantized theory,

p^+, p^i and $Q_+^{\ +}$ are translations and hence non-linear, while p^- and Q_- are linearly realized.

Comparing (4.5a) with (4.4) we find that $Q_+^{\ a}$ is in fact $Q_o^{\ a}$. We can write this as a Noether charge using the equations of motion (4.3) as

$$Q_+^{\ a} = \frac{1}{2\pi} \int_o^\pi d\sigma (Q_a^{\ 1} + Q_a^{\ 2}) \ . \qquad (4.6)$$

This is hence a symmetry of the action. We construct $Q_-^{\ \dot{a}}$ such that it commutes to H (3.6) with the substitution (4.1). We find

$$Q_-^{\ \dot{a}} = \frac{1}{2\sqrt{2}\, p^+\pi} \int_o^\pi d\sigma \left[(\gamma^i \dot{x}^i (Q^1+Q^2))^{\dot{a}} + (\gamma^i x'^i (Q^1-Q^2))^{\dot{a}}\right]$$

$$= \frac{1}{\sqrt{2p^+}} \sum_{-\infty}^{\infty} (\gamma^i \alpha_{-n}^{\ i} Q_n)^{\dot{a}} \ . \qquad (4.7)$$

$Q_-^{\ \dot{a}}$ is also a conserved Noether charge, so the action is indeed supersymmetric.

Implementing the Q_+ generator on Q^{aA} we see that it amounts to a certain translation. However, going back to the action we see that if we can drop surface terms freely, the action possesses a bigger symmetry, an N=2 supersymmetry with separate translations in $Q_a^{\ 1}$ and $Q_a^{\ 2}$. This is the case for the closed strings with periodic boundary conditions, where the solutions to the equations of motion are

$$x^i(\sigma,\tau) = x^i + p^i\tau + \frac{i}{2} \sum_{n\neq 0} \frac{1}{n}\left(\alpha_n^{\ i}\, e^{-i2n(\tau-\sigma)} + \tilde{\alpha}_n^{\ i}\, e^{-i2n(\tau+\sigma)}\right) \qquad (4.8)$$

$$Q_a^{\ 1} = \sum_{n=-\infty}^{\infty} Q_n^{\ 1a}\, e^{-i2n(\tau-\sigma)} \qquad (4.9)$$

$$Q_a^{\ 2} = \sum_{n=-\infty}^{\infty} Q_n^{\ 2a}\, e^{-i2n(\tau+\sigma)} \ . \qquad (4.10)$$

It still remains to prove the Lorentz invariance of the action (4.2). It is an arduous task to find the representation of all the generators, especially those rotating into a minus-direction, since these ones have to be non-linear as was discussed before. The representation was found in the first paper by Green and Schwarz[27)] and here we just quote them. The representation is (in the open string case)

$$J^{ij} = \ell^{ij} - i \sum_{n=1}^{\infty} \frac{1}{n} (\alpha_{-n}{}^i \alpha_n{}^j - \alpha_{-n}{}^j \alpha_n{}^i) + K_o{}^{ij} \qquad (4.10a)$$

$$J^{+-} = \ell^{+-} \qquad (4.10b)$$

$$J^{i+} = \ell^{i+} \qquad (4.10c)$$

$$J^{i-} = \ell^{i-} - \frac{i}{p^+} \sum_{n=1}^{\infty} \frac{1}{n} (\alpha_{-n}{}^i \alpha_n{}^- - \alpha_{-n}{}^- \alpha_n{}^i) + \frac{1}{p^+} \sum_{n=-\infty}^{\infty} K_{-n}{}^{ij} a_n{}^j, \qquad (4.10d)$$

where

$$\ell^{\mu\nu} = q^\mu p^\nu - q^\nu p^\mu \qquad (4.11a)$$

$$K_m{}^{ij} = \frac{-1}{8p^+} \sum_{n=-\infty}^{\infty} Q_{m-n}{}^a (\gamma^{ij} Q_n)^a \qquad (4.11b)$$

$$\alpha_n{}^- = \frac{1}{2} \sum_{m=-\infty}^{\infty} \alpha_{n-m}{}^i \alpha_m{}^i + \frac{1}{8p^+} \sum_{m=-\infty}^{\infty} (m-n/2)\, Q_{n-m}{}^a Q_m{}^a \; . \qquad (4.11c)$$

The generators in the closed string case are essentially the ones above with each mode doubled.

To check this algebra is quite intricate. The crucial commutator is $[J^{i-}, J^{j-}]$, which develops extra terms in the quantum case which are dimension dependent and zero only if d=10. From now on we will always work in 10 dimensions.

To perform a complete analysis of the Lorentz invariance we should construct x^-. In so doing we find the supersymmetric generalization of the conditions (2.19). The condition corresponding to (2.19a) gives the mass-shell condition, while (2.19b) in the closed string case gives a further constraint (because of finite boundary conditions) saying that

$$\frac{1}{2} \int d\sigma \left(\pi p^i x^{i} - \frac{i}{4p^+} (Q^1 Q^{1\prime} + Q^2 Q^{2\prime}) \right) = N - \tilde{N} = 0 \qquad (4.12)$$

where

$$N = \sum_{n=1}^{\infty} (\alpha_{-n}{}^i \alpha_n{}^i + \frac{n}{2p^+} Q_{-n}{}^{1a} Q_n{}^{1a}) \qquad (4.13a)$$

$$\tilde{N} = \sum_{n=1}^{\infty} (\tilde{\alpha}_{-n}{}^i \tilde{\alpha}_n{}^i + \frac{n}{2p^+} Q_{-n}{}^{2a} Q_n{}^{2a}) \; . \qquad (4.13b)$$

Quantum mechanically we have to impose this condition on the physical states which hence have to satisfy

$$(N-\tilde{N}) \mid phys \rangle = 0 \; . \qquad (4.14)$$

Now we know that we have a quantum theory of free strings where the lowest lying states are massless. Next we like to know what is the spin content of these massless states. For the bosonic string one can choose a scalar vacuum state $|0\rangle$ as the ground state. Putting all fermionic oscillators in (4.10) to zero we find this state to be indeed a scalar one. However, in the superstring case there is an extra piece in the zero mode part of J^{ij} (4.10a)

$$s_o^{ij} = \frac{-1}{8p^+} Q_o^{a}(\gamma^{ij})^{ab} Q_o^{b} \; . \qquad (4.15)$$

Its effect on a vacuum state will be non-zero in general.

We should also realize that the massless level must be a supermultiplet. Such a one can be constructed from the anticommutator (4.5a) combined with the knowledge of (4.15). The generator Q_+^a is real. In the 4-dimensional case it is customary to go from an SO(2) description to a U(1) one forming complex generators (with no Lorentz index) which then build up a Clifford algebra and we can define creation and annihilation operators from which we construct the supermultiplet. For d=10 we have so far used an SO(8) covariant notation. To decompose the generators into creation and annihilation operator we must break the covariance into SU(4) x U(1). This is the formalism we will use in the next section. Here, however, I will describe an alternative formalism, which uses the full SO(8) covariance. This method can equally well be used in d=4. Consider the zero mode part of J^{ij} (4.10a) which is the relevant part on vacuum states

$$J_o^{ij} = \ell^{ij} - \frac{1}{8p^+} \; Q_o \; \gamma^{ij} \; Q_o \; \equiv \ell^{ij} + s_o^{ij} \quad . \qquad (4.16)$$

The last term in (4.13) is the spin contribution. If we try to start with a scalar vacuum $|0\rangle$ we need a constraint

$$s_o^{ij} \; |0\rangle = 0 \; . \qquad (4.17)$$

By multiplying (4.17) by another Q_o and using (4.4) and Fierz rearrangements we will find that (4.17) can only work for d=4.

For d=10 we have to try the next simplest thing. We take the vacuum to be a vector $|i\rangle$ with $\langle i|j\rangle = \delta^{ij}$. Then we must insist on a constraint

$$s_o^{ij} \; |k\rangle = \delta_{ik} \; |j\rangle - \delta_{jk} \; |i\rangle \quad . \qquad (4.18)$$

This time we find when we bang a Q_o on (4.18) that for the general state

$$\psi^{ia} \equiv Q_o{}^a \; |i\rangle \; , \qquad (4.19)$$

with the decomposition

$$\psi^{ia} = \tilde{\psi}^{ia} + \frac{1}{8}\, \gamma^i \gamma^j \psi^j \; , \qquad (4.20)$$

where $\gamma^i \tilde{\psi}^{ia} = 0$

that

$$\tilde{\psi}^{ia} = 0 \; , \qquad (4.21)$$

but that there is no constraint on

$$|\dot{a}\rangle \equiv \frac{1}{8}\, (\gamma^i Q)^{\dot{a}} \; |i\rangle \quad . \qquad (4.22)$$

with $\langle \dot{b} | a\rangle = p^+ \, \delta^{\dot{a}\dot{b}} \; |\dot{b}\rangle$.
Checking the Lorentz properties of this state we find that

$$s_o{}^{ij} \; |\dot{a}\rangle = -\, \frac{1}{2}\, (\gamma^{ij})^{\dot{a}\dot{b}} \; |\dot{b}\rangle \quad . \qquad (4.23)$$

Hence it transforms properly as a spinor. From the Fierz property (A.7)

$$Q_o{}^a Q_o{}^b = p^+ \, \delta^{ab} + \frac{1}{16}\, (\gamma^{ij})^{ab} \; Q_o \gamma^{ij} Q_o \; , \qquad (4.24)$$

we find that no other massless states can be constructed and the massless sector contains then one vector state and one spinor state which we recognize as the Yang-Mills multiplet in 10 dimensions. We can, of course, let both states transform according to some representation (such as the adjoint one) of some internal group.

It may be instructive to also check the first excited level. Here one can form 128 boson states $\alpha_{-1}{}^i \; |j\rangle$ and $Q_{-1}{}^a \; |\dot{b}\rangle$ and 128 fermionic states $\alpha_{-1}{}^i \; |\dot{a}\rangle$ and $Q_{-1}{}^a \; |i\rangle$. These form various reducible SO(8) multiplets which can be reassembled into SO(9) representation (since they are massive) using the Lorentz generators. Doing this one finds for the bosons the SO(9) representations  and  of dimensions 44 and 84 resp. The fermions form a single 128 dimensional Rarita-Schwinger SO(9) multiplet (like the $\tilde{\psi}^{ia}$ in (4.17)).

In the closed string case the massless spectrum is generated by two supersymmetry generators $Q_o{}^1$ and $Q_o{}^2$ ((4.9) and (4.10)). If both belong to the same SO(8) representation (this means that they have the same chirality in 10 dimensions (type IIb) and has been assumed in the action (4.2)), we can form complex generators

$$Q^a = \frac{1}{\sqrt{2}} (Q_o{}^1 + iQ_o{}^2)^a \ , \tag{4.25}$$

and hence construct creation and annihilation operators. For this case we can introduce a scalar vacuum, (which has to be complex, which can be seen from the supersymmetry transformations). Checking $J_o{}^{ij}$ (4.13) with the $s_o{}^{ij}$ term doubled we see that the vacuum is indeed a scalar state. Choosing Q^a as a creation operator we can form the following massless supermultiplet

$$\begin{aligned} |\,0\rangle &\sim \phi \\ Q^a\,|\,0\rangle &\sim \psi^a \\ Q^aQ^b\,|\,0\rangle &\sim \phi^{ab} \\ Q^{a_1}\ldots Q^{a_8}\,|\,0\rangle &\sim \phi^{a_1\ldots a_8} \ . \end{aligned}$$

It is easily seen that this is a reducible multiplet. To obtain an irreducible representation we impose the conditions

$$(\phi^{a_1a_2\ldots a_{2N}})^* = \frac{1}{(8-2N)!}\ \varepsilon^{a_1a_2\ldots a_8}\ \phi^{a_{2N+1}\ldots a_8} \tag{4.26}$$

$$(\psi^{a_1a_2\ldots a_{2N+1}})^* = \frac{1}{(7-2N)!}\ \varepsilon^{a_1a_2\ldots a_8}\ \psi^{a_{2N+2}\ldots a_8} \ . \tag{4.27}$$

Because of the triality properties of the representations of SO(8) we rewrite the states with vector indices. We now find the bosonic spectrum:

2 scalars ϕ

2 antisymmetric tensors $A^{ij} \sim (\gamma^{ij})^{ab}\phi^{ab}$

1 self-dual antisymmetric tensor $A^{ijk\ell}$

1 graviton g^{ij} ,

where the last two sets of states follow from $\phi^{a_1\ldots a_4}$.
The fermionic spectrum is:

2 spinors ψ^a

2 Rarita-Schwinger states $\tilde{\psi}^{ia}$.

If Q^1 and Q^2 have opposite Weyl properties (type IIa) (for this case Q^{Aa} is not a 2-dimensional spinor, but who cares?) we cannot form creation operators in an SO(8) covariant way. We then have to follow the procedure of the N=1 case. We can start with a tensor state

$$|ij\rangle \sim |i\rangle \otimes |j\rangle \sim \phi^{ij}$$

and generate the states by stuttering the N=1 procedure

$$|a\rangle \otimes |i\rangle \sim \psi^{ai}$$

$$|i\rangle \otimes |\dot{a}\rangle \sim \chi^{\dot{a}i}$$

$$|a\rangle \otimes |\dot{a}\rangle \sim \phi^{a\dot{a}} .$$

The bosonic spectrum is then

1 graviton $g^{ij} = \phi^{(ij)} - \frac{1}{8}\, \phi^{kk}\, \delta^{ij}$

1 antisymmetric tensor $A^{ij} = \phi^{[ij]}$

1 scalar $\phi = \phi^{ii}$

1 vector $A^i = (\gamma^i)^{a\dot{a}}\, \phi^{a\dot{a}}$

1 antisymmetric tensor $A^{ijk} = (\gamma^{ijk})^{a\dot{a}}\, \phi^{a\dot{a}}$.

The fermionic spectrum is

2 spinors ψ^a, $\chi^{\dot{a}}$

2 Rarita-Schwinger states $\tilde{\psi}^{ai}$, $\tilde{\chi}^{\dot{a}i}$.

This technique could, of course, also have been used in the other case above.

We can also impose constraints on the N=2 spectrum to become an N=1 spectrum. These closed strings (type I) are important since they will be able to couple to the open strings. To find this spectrum we linearly combine $Q^1+Q^2 = Q$ and use the N=1 technique. The spectrum is then clearly

$$|i\rangle \otimes |j\rangle$$

$$|i\rangle \otimes |a\rangle$$

i.e.

1 graviton	g^{ij}
1 antisymmetric tensor	A^{ij}
1 scalar	ϕ
1 spinor	ψ^a
1 Rarita-Schwinger state	$\tilde{\psi}^{ai}$.

So far we have only discussed free string states. We could anticipate that all closed strings states belong to some representation of some internal group. However, it will turn out that interaction is only possible if this representation is the trivial one.

Since the strings are extended objects in one dimension there is a possibility that they can carry an intrinsic orientation, like an "arrow" pointing in one direction along its length. When strings are oriented there are two distinct classical states for a single spatial configuration, corresponding to the two possible orientations. An open string is oriented, loosely speaking, if the end points are different, while a closed string is oriented, if one can distinguish a mode running one way around the string from a mode running the other way.

Hence the basic question is whether a string described by $x^\mu(\sigma)$, $Q^{Aa}(\sigma)$ is the same as one described by $x^\mu(\pi-\sigma)$, $Q^{Aa}(\pi-\sigma)$. Consider the closed string solution (4.8), (4.9) and (4.10). The replacement $\sigma \to \pi-\sigma$ corresponds to the interchanges $\alpha_n \leftrightarrow \tilde{\alpha}_n$, $Q_n^{\ 1} \leftrightarrow Q_n^{\ 2}$. In the type II case the two strings connected by the interchanges are different, while for type I the constraints are just such as to make the two strings the same state. Hence we conclude that type II strings are oriented while type I closed strings are non-oriented.

For open strings we will be able to allow for an internal (global) symmetry (the remnant of the gauge symmetry in a covariant formalism). The interaction allows for letting the states transform as $\phi^a{}_b$ where the index a and b run over the fundamental representation and its complex conjugate resp. Hence $\phi^a{}_b$ transforms either as the adjoint representation or the singlet one. The intuitive way to interpret this is to say that each end carry a "quark" and an "antiquark". If, hence, the "quarks" are different from the "antiquarks" the string is oriented, otherwise not. Now these strings can join their ends to form closed strings, if they are singlets. Then the SU(N) strings will form oriented closed strings, while the SO(n) and Sp(2n) strings will form non-oriented closed strings. But these strings must be of type I, since the open strings are, and hence should be non-oriented. Thus we conclude by this non-rigorous argument that only the gauge groups SO(n) or Sp(2n) are possible[28].

The notation of orientability will be important for perturbation expansions. For oriented strings the interaction must be such that the orientations match up. This will make the perturbation expansion of type I strings different from type II ones.

We have discussed the superstring so far in the light-cone gauge. This is enough to build an interacting theory as we will see in the next sections. However, it would probably be advantageous to have a covariant formalism. Such an action was finally found by Green and Schwarz[29]

$$S = \frac{T}{2} \int d\sigma \, d\tau \, (L_1 + L_2) \tag{4.28}$$

$$L_1 = -\frac{1}{2} \sqrt{-g} \, g^{\alpha\beta} \, \pi^{\mu}{}_{\alpha} \, \pi_{\mu\beta} \tag{4.29}$$

$$L_2 = -i \, \varepsilon^{\alpha\beta} \, \partial_\alpha x^\mu [\bar{\theta}^1 \gamma_\mu \partial_\beta \theta^1 - \bar{\theta}^2 \gamma_\mu \partial_\beta \theta^2]$$

$$+ \, \varepsilon^{\alpha\beta} \, \bar{\theta}^1 \gamma^\mu \partial_\alpha \theta^1 \bar{\theta}^2 \gamma_\mu \partial_\beta \theta^2 \quad , \tag{4.30}$$

where

$$\pi^{\mu}{}_{\alpha} = \partial_\alpha x^\mu - i \, \bar{\theta}^A \gamma^\mu \partial_\alpha \theta^A \quad . \tag{4.31}$$

This action is locally reparametrization invariant and has also a local as well as global supersymmetry. The local one is such that one can choose a gauge in which half of the 16-dimensional spinors θ^1 and θ^2 can be fixed. This leads very naturally to the light-cone gauge with eight-dimensional spinors. It even seems quite likely that only non-covariant gauge choices are possible since there is no way of projecting out half of the spinors in a covariant way[30]. Hence, the usefulness of the action (4.28) remains to be proven.

5. A FIELD THEORY FOR FREE SUPERSTRINGS

To describe the interaction among point-like particles we usually use a quantum field theory. Such a one can be constructed in logical steps from the quantum mechanics of free particles. However, usually we start directly with the field theory and forget the steps leading up to it. The result is that, if we have a particle described by paths $x^\mu(\tau)$, we use a field $\phi(x)$. We can think of x^μ as $x^\mu(0)$ which is a set of commuting c-numbers. Note, that we do not attempt to write fields over the whole of phase space. The operators x^μ and p^μ do not commute and we cannot simultaneously construct eigenvalues of them.

When we turn to strings we will try to follow the same lines. In the case of closed strings, they are described for a fixed τ by $x^\mu(\sigma)$, $Q_a{}^1(\sigma)$ and $Q_a{}^2(\sigma)$. However, due to the spinor nature of the Q's, this is not an (anti-)commuting set of operators. To find such a one we could either combine $Q_a{}^1$ and $Q_a{}^2$ linearly into

$$Q_a = \frac{1}{\sqrt{2}} (Q_a{}^1 + iQ_a{}^2) \tag{5.1}$$

$$\tilde{Q}_a = \frac{1}{\sqrt{2}} (Q_a{}^1 - iQ_a{}^2) \tag{5.2}$$

as was done in Eq. (4.25) to find the anticommutators

$$\{Q_a(\sigma), Q_b(\sigma')\} = 0 \tag{5.3}$$

$$\{\tilde{Q}_a(\sigma), \tilde{Q}_b(\sigma')\} = 0 \tag{5.4}$$

$$\{Q_a(\sigma), \tilde{Q}_b(\sigma')\} = 2p^+ \delta(\sigma-\sigma') . \tag{5.5}$$

A field $\phi[x^\mu(\sigma), Q_a(\sigma)]$ can now be attempted. This is what was done in Ref. 31. The problem with this method is that it is stream-lined for type IIb strings and cannot easily be used for the other string models.

The alternative way is (as was mentioned in sect. 4) to break the explicit SO(8) invariance down to SU(4) x U(1). I follow here Ref. 32. The two SO(8) spinors denoted by 8_s and 8_c decompose according to the rule

$$8_s = 4_{1/2} + \bar{4}_{-1/2} \tag{5.6}$$

$$8_c = 4_{-1/2} + \bar{4}_{1/2} , \tag{5.7}$$

where the index denotes the U(1) quantum number (which is helicity when the theory is interpreted in 4 dimensions).

The 8_s and 8_c correspond to ten-dimensional Weyl spinors of opposite handedness. Thus the chiral theories contain two 8_s Q^a-coordinates, which decompose into

$$(\theta_A, \lambda^A) \text{ and } (\tilde{\theta}_A, \tilde{\lambda}^A) . \qquad A = 1,...4$$

The notation is such that λ^A represents a 4 and θ_A a $\bar{4}$ of SU(4).

In the non-chiral IIa theory there is one 8_s coordinate and one 8_c coordinate. They decompose into

$$(\theta_A, \lambda^A) \text{ and } (\vartheta^A, \tilde{\lambda}_A) .$$

We now have the canonical anticommutation rule (in the IIb case)

$$\{\theta_A(\sigma), \lambda^B(\sigma')\} = \{\vartheta_A(\sigma), \tilde{\lambda}^B(\sigma')\} = \delta(\sigma-\sigma') \delta_A{}^B. \tag{5.8}$$

The field we then take to be a functional of $\theta_A(\sigma)$, $\vartheta_A(\sigma)$ or of the "Fourier transforms" $\lambda^A(\sigma)$, $\tilde{\lambda}^A(\sigma')$. In the open string case we can now simply use $\Theta_A(\sigma) = \frac{1}{\sqrt{2}} \left(\theta_A(+\sigma) + \vartheta_A(\sigma)\right)$ or the conjugate variable which we will call $\Lambda^A(\sigma)$.

The $x^\mu(\sigma)$ coordinate we divide up into the "time" x^+, $x^-(\sigma)$ and $x^i(\sigma)$. We can Fourier transform $x^-(\sigma)$ to let the field instead depend on p^+. Since we have broken the transverse SO(8) symmetry we sometimes use the SU(4) notation for SO(8) vectors. The decomposition rule is

$$8_v = 6_0 + 1_1 + 1_{-1} . \tag{5.9}$$

A vector A^i becomes (A^I, A^R, A^L) where $I = 1,\ldots 6$.

$$A^R = \frac{1}{\sqrt{2}} (A^7 + iA^8) \tag{5.10}$$

$$A^L = \frac{1}{\sqrt{2}} (A^7 - iA^8) . \tag{5.11}$$

It will be convenient when we discuss interactions to choose the length of the σ interval to be proportional to p^+. More specifically, the parametrization is arranged so that the p^+ density (per unit σ) is always $(2\pi)^{-1}$. This implies that the entire length of the parameter intervals is $\pi|\alpha|$, where

$$\alpha \equiv 2p^+ . \tag{5.12}$$

For the closed strings, the coordinates and momenta now have the $\tau=0$ expansions (see (4.8)-(4.10))

$$x^i(\sigma) = x^i + \frac{i}{2} \sum_{n=0} \frac{1}{n} (\alpha_n{}^i e^{2in\sigma/|\alpha|} + \tilde{\alpha}_n{}^i e^{-2in\sigma/|\alpha|}) \tag{5.13}$$

$$p^i(\sigma) = \frac{1}{\pi|\alpha|} \{p^i + \sum_{n=0} (\alpha_n{}^i e^{2in\sigma/|\alpha|} + \tilde{\alpha}_n{}^i e^{-2in\sigma/|\alpha|})\} \tag{5.14}$$

$$\theta_A(\sigma) = \frac{1}{\alpha} \Sigma\, Q_{nA}\, e^{-2in\sigma/|\alpha|} \tag{5.15}$$

$$\tilde{\theta}_A(\sigma) = \frac{1}{\alpha} \sum_{-\infty}^{\infty} \tilde{Q}_{nA}\, e^{-2in\sigma/|\alpha|} \tag{5.16}$$

$$\lambda^A(\sigma) = \frac{1}{\pi|\alpha|} \sum_{-\infty}^{\infty} Q_n{}^A\, e^{2in\sigma/|\alpha|} \tag{5.17}$$

$$\tilde{\lambda}^A(\sigma) = \frac{1}{\pi|\alpha|} \sum_{-\infty}^{\infty} \tilde{Q}_n{}^A\, e^{-2in\sigma/|\alpha|} \quad . \tag{5.18}$$

The canonical commutation rules are satisfied for

$$[\alpha_m{}^i, \alpha_n{}^j] = [\tilde{\alpha}_m{}^i, \tilde{\alpha}_n{}^j] = m\, \delta_{m+n,0}\, \delta^{ij} \tag{5.19}$$

$$\{Q_m{}^A, Q_{nB}\} = \{\tilde{Q}_m{}^A, \tilde{Q}_{nB}\} = \alpha\, \delta_{m+n,0}\, \delta^A{}_B \ . \tag{5.20}$$

All other commutators or anticommutators are zero.

For the open strings the coordinates and momenta are

$$x^i(\sigma) = x^i + i \sum_{n=1}^{\infty} \frac{1}{n} (\beta_n{}^i - \beta_{-n}{}^i) \cos \frac{n\sigma}{|\alpha|} \tag{5.21}$$

$$p^i(\sigma) = \frac{1}{\pi|\alpha|} \{p^i + \sum_{n=1}^{\infty} (\beta_n{}^i + \beta_{-n}{}^i) \cos \frac{n\sigma}{|\alpha|}\} \tag{5.22}$$

$$\Theta_A(\sigma) = \frac{1}{\sqrt{2}} (\theta_A + \tilde{\theta}_A) = \frac{1}{\alpha} \sum_{n=1}^{\infty} (R_{nA} + R_{-nA}) \cos \frac{n\sigma}{|\alpha|} \tag{5.23}$$

$$\Lambda^A(\sigma) = \frac{1}{\sqrt{2}} (\lambda^A + \tilde{\lambda}^A) = \frac{1}{\pi|\alpha|} \sum_{n=1}^{\infty} (R_n{}^A + R_{-n}{}^A) \cos \frac{n\sigma}{|\alpha|} \ . \tag{5.24}$$

where different symbols are used to avoid confusion between open and closed strings. The commutation rules are

$$[\beta_m{}^i, \beta_n{}^j] = m\, \delta_{m+n,0}\, \delta^{ij} \tag{5.25}$$

$$\{R_m{}^A, R_{nB}\} = \alpha\, \delta_{m+n,0}\, \delta^A{}_B \ . \tag{5.26}$$

All other (anti-)commutators are zero.

We can now start the field theory approach by introducing the superfields $\Psi[x(\sigma), \theta(\sigma), \tilde{\theta}(\sigma)]$ for closed strings and $\Phi^{ab}[x(\sigma), \Theta(\sigma)]$ for open strings, anticipating that the interactions allow the open string fields to have SO(n) or Sp(2n) indices a,b.

In the case of a closed string superfield, there must not be any physics in the choice of an origin for σ. That means that a shift in the σ parametrization must be immaterial, i.e.

$$\int d\sigma \left\{x'^{i} \frac{\delta}{\delta x^{I}} + \theta'_{A} \frac{\delta}{\delta\theta_{A}} + \tilde{\theta}'_{A} \frac{\delta}{\delta\tilde{\theta}_{A}}\right\} \Psi = 0 \; . \tag{5.27}$$

This is exactly equivalent to the condition (4.12) that $N=\tilde{N}$. In the case of type I strings one also has to take the non-orientation constraints into account.

$$\Psi[x(\sigma),\ \theta(\sigma),\ \tilde{\theta}(\sigma)] = \Psi[x(-\sigma),\ \tilde{\theta}(-\sigma),\ \theta(-\sigma)] \tag{5.28}$$

$$\Phi^{ab}[x(\sigma),\ \Theta(\sigma)] = -\Phi^{ba}[x(\pi|\alpha|-\sigma),\ \Theta(\pi|\alpha|-\sigma)]. \tag{5.29}$$

Furthermore there is another constraint. To understand it let us for a moment consider a superfield for the massless sector. In the type I case the generic field is $\phi(x,\ \theta_o)$. This is not an irreducible representation (compare with (4.26) and (4.27)). To form an irreducible representation define the Grassman Fourier transform

$$\hat{\phi}(\lambda_o) = \frac{1}{4\alpha^2} \int d^4\theta_o \; e^{\lambda_o{}^A\theta_{oA}} \; \phi(\theta_{oA}) \tag{5.30}$$

and impose the condition

$$\hat{\phi}(\lambda_o{}^A) = \left(\phi(\theta_{oA})\right)^* = \phi^*(\lambda_o{}^A/2\alpha) \tag{5.31}$$

where the last step makes the identification $\lambda_o \sim 2\alpha\theta_o{}^*$. This condition makes the field ϕ effectively self-conjugate.

The string generalization of these formulae is relatively straightforward. Define

$$I[\lambda,\theta] = \int_o^{\pi|\alpha|} [\lambda^A(\sigma)\ \theta_A(\sigma) + \tilde{\lambda}^A(\sigma)\ \tilde{\theta}_A(\sigma)]\ d\sigma \; , \tag{5.32}$$

where λ^A is a "c-number momentum". Next define

$$\hat{\Psi}[x,\lambda,\tilde{\lambda}] = \int D^4\theta D^4\tilde{\theta} \; e^{I[\lambda,\theta]} \; \Psi(x,\theta,\tilde{\theta}) \; . \tag{5.33}$$

If the measure is suitably normalized the condition (5.31) is generalized into

$$\hat{\Psi}[x,\lambda,\tilde{\lambda}] = \{\Psi[x,\theta,\tilde{\theta}]\}^* = \Psi^*[x,\lambda/\alpha,\tilde{\lambda}/\alpha] \; . \tag{5.34}$$

Again we have made an identification $\lambda \sim 2\alpha\theta^*$ and $\tilde{\lambda} \sim 2\alpha\tilde{\theta}^*$. This constraint hence relates Ψ^* to Ψ, which means that in a variation of fields in an action, Ψ^* should not be varied independently from Ψ.

We will treat the field theory in the light-cone gauge. In this gauge the super-Poincaré invariance will be non-linearly realized. That means that we must find a realization of this algebra on the superfields, which closes. In the interacting case all generators that take us out of the quantization plane, J^{i-}, J^{+-}, $p^- = H$ and Q_- will contain interaction terms. These operators are all called Hamiltonians by Dirac[33]. However, in the free case these extra terms are absent and it is fairly straightforward to find the realization of the algebra. To start with we do it by finding operators $j^{\mu\nu}$, q_+, q_- and p^μ, such that

$$[J^{\mu\nu}, \Psi] = j^{\mu\nu}\Psi \tag{5.35}$$

etc, with $j^{\mu\nu}$ written in terms of the phase space (5.13)-(5.18). These operators can now be taken over from the first quantized case in sect. 4, where we constructed this algebra in order to prove the Lorentz invariance. Here we rewrite them in SU(4) notations.

There is one difference with ordinary field theory, namely that some generators are implemented locally. In fact Q_+ and p^i are represented by

$$q^{1,2}_{+A} = \theta_A(\sigma),\ \bar{\theta}_A(\sigma) \tag{5.36a}$$

$$q^{1,2A}_{+} = \lambda^A(\sigma),\ \bar{\lambda}^A(\sigma) \tag{5.36b}$$

$$p^i = p^i(\sigma) \quad . \tag{5.37}$$

The Q_- generators can be written as

$$q^1{}_-{}^A = \int_0^{\pi|\alpha|} \{\sqrt{2}\,\rho^{IAB}(p^I - \tfrac{1}{\pi}x'^I)\,\theta_B + 2\pi\,\varepsilon(\alpha)\,(p^L - \tfrac{1}{\pi}x'^L)\,\bar{\lambda}^A\}\,d\sigma \tag{5.38}$$

$$q^2{}_-{}^A = \int_0^{\pi|\alpha|} \{\sqrt{2}\,\rho^{IAB}(p^I + \tfrac{1}{\pi}x'^I)\,\bar{\theta}_B + 2\pi\varepsilon(\alpha)\,(p^L + \tfrac{1}{\pi}x'^L)\,\bar{\lambda}^A\}\,d\sigma \tag{5.39}$$

$$q^1{}_{-A} = \int_0^{\pi|\alpha|} \{2(p^R - \frac{1}{\pi} x'^R)\, \theta_A + \sqrt{2}\, \pi\, \varepsilon(\alpha)\, \rho^I{}_{AB}$$

$$(p^I - \frac{1}{\pi} x'^I)\, \lambda^B\}\, d\sigma \,, \qquad (5.40)$$

$$q^2{}_{-A} = \int_0^{\pi|\alpha|} \{2(p^R + \frac{1}{\pi}\, x'^R)\, \tilde{\theta}_A + \sqrt{2}\, \pi\, \varepsilon(\alpha)\, \rho^I{}_{AB}$$

$$(p^I + \frac{1}{\pi}\, x'^I)\, \tilde{\lambda}^B\}\, d\sigma \,, \qquad (5.41)$$

where $\varepsilon(\alpha) = +1$ for $\alpha>0$ (an incoming string) and $= -1$ for $\alpha<0$ (an outgoing string). The Clebsch-Gordan coefficients $\rho^i{}_{AB}$ can be obtained from the Dirac γ-matrices in SO(8) and are described in Appendix.

The generator p^- which is the Hamiltonian is given by

$$p^- = h = \int_0^{\pi|\alpha|} \{\varepsilon(\alpha) \lfloor \pi\, p^{i^2} + \frac{1}{\pi}\, x'^{i^2} \rfloor$$

$$-2i \lfloor \theta_A \lambda^{A'} + \tilde{\theta}_A \tilde{\lambda}^{A'} \rfloor\}\, d\sigma \quad . \qquad (5.42)$$

Similarly the Lorentz generators can be constructed. We could also perform the σ-integrations in (5.38)-(5.42) and obtain

$$q^1{}_-{}^A = \frac{1}{\alpha} \sum_{n=-\infty}^{\infty} \lfloor \sqrt{2}\, \alpha_{-n}{}^I\, \rho^{IAB}\, Q_{Bn} + 2\alpha_{-n}{}^L\, Q^A{}_n \rfloor \qquad (5.43)$$

$$q^2{}_-{}^A = \frac{1}{\alpha} \sum_{n=-\infty}^{\infty} \lfloor \sqrt{2}\, \tilde{\alpha}_{-n}{}^I\, \rho^{IAB}\, \tilde{Q}_{Bn} + 2\tilde{\alpha}_{-n}{}^L\, \tilde{Q}_n{}^A \rfloor \qquad (5.44)$$

$$q^1{}_{-A} = \frac{1}{\alpha} \sum_{n=-\infty}^{\infty} \lfloor \sqrt{2}\, \alpha_{-n}{}^I\, \rho^I{}_{AB}\, Q_n{}^B + 2\alpha_{-n}{}^R\, Q_{nA} \rfloor \qquad (5.45)$$

$$q^2{}_{-A} = \frac{1}{\alpha} \sum_{n=-\infty}^{\infty} \lfloor \sqrt{2}\, \tilde{\alpha}_{-n}{}^I\, \rho^I{}_{AB}\, \tilde{Q}_n{}^B + 2\tilde{\alpha}_{-n}{}^R\, \tilde{Q}_{nA} \rfloor \qquad (5.46)$$

$$h = \frac{4}{\alpha}\, (N + \tilde{N}) \,, \qquad (5.47)$$

where

$$N = \sum_{n=-\infty}^{\infty} [\tfrac{1}{2} \alpha_{-n}{}^i \alpha_n{}^i + \frac{n}{\alpha}\, Q_{-na} Q_n{}^A] \qquad (5.48)$$

and similarly for $\tilde{N}$.

From the knowledge of the generator h (5.42) we can so form an action for the free theory

$$S = \int (\partial_+\Psi\, \partial_-\Psi)\, D^{16}z\, dx^+dx^- - \int H\, dx^+$$
$$= \int \partial_-\Psi(\partial_+\Psi - h\,\Psi)\, D^{16}z\, dx^+dx^-$$
$$= \frac{-i}{2} \int_0^\infty \alpha\, d\alpha \int \Psi_{-\alpha}(\partial_+ - h)\,\Psi_\alpha\, D^{16}z\, dx^+ \quad , \tag{5.49}$$

where in the last step we use the Fourier transform α instead of x^- and $D^{16}z$ is the functional measure over $x^i(\sigma)$, $\theta_A(\sigma)$ and $\tilde{\theta}_A(\sigma)$.

From this action we can formally derive the equal x^+ commutation rule to find for the closed strings

$$[\Psi(1), \Psi(2)] = \frac{2}{\alpha_2}\, \delta(\alpha_1 + \alpha_2) \int_0^{\pi|\alpha_2|} \frac{d\sigma_0}{\pi|\alpha_2|}$$
$$\Delta^{16}[z_1(\sigma) - z_2(\sigma+\sigma_0)] \, , \tag{5.50}$$

where Δ^{16} is the δ-functional

$$\Delta^{16}(z) \equiv \Delta^8[x^i(\sigma)]\, \Delta^4[\theta_A(\sigma)]\, \Delta^4[\tilde{\theta}_A(\sigma)] \, . \tag{5.51}$$

Given this relation a representation of the super-Poincaré algebra can be found in terms of field operators. An operator G is written as

$$G = -\frac{i}{2} \int_0^\infty \alpha d\alpha \int \Psi_{-\alpha}\, g\, \Psi_\alpha\, D^{16}z \tag{5.52}$$

in the notation of Eq. (5.35).

The expressions (5.49)-(5.52) are quite formal and must be given a precise meaning. This can be done by relating them to infinite mode expansions. A field $\Psi[x(\sigma), \theta(\sigma), \tilde{\theta}(\sigma)]$ can be expanded in component fields using complete sets of harmonic oscillator wave functions $\phi_n(x)$ and corresponding wave functions $\chi_n(\theta,\tilde{\theta})$ for the anticommuting variables.

$$\Psi[x(\sigma), \theta(\sigma), \tilde{\theta}(\sigma)] =$$
$$= \sum_{n_k, n'_\ell, m_s, m'_s} \phi_{(n_k, n'_\ell, m'_s)}(x, \theta_0, \tilde{\theta}_0)$$
$$\prod_k \phi_{n_k}(x_k) \prod_\ell \phi_{n'_\ell}(\tilde{x}_\ell) \prod_s \chi_{m_s, m'_s}(\theta_s, \tilde{\theta}_s) \, , \tag{5.53}$$

with x_k and θ_s the k'th and s'th mode of $x(\sigma)$ and $\theta(\sigma)$ resp.

The functional measure in (5.49) is then defined such that when the non-zero mode integrations are carried out one is left with

$$H = \int \alpha d\alpha\, d^8x\, d^4\theta_o d^4\tilde{\theta}_o \left[\frac{1}{2} \sum_{\{n\}} \phi_{\{n\}}(-\alpha,x,\theta_o,\tilde{\theta}_o)\right.$$

$$\frac{1}{\alpha}(\partial^{i^2} + N_\phi)\, \phi_{\{n\}}(\alpha,x,\theta_o,\tilde{\theta}_o)$$

$$\left. + \frac{1}{\alpha} \sum_{\{n\}} \psi_{\{n\}}(-\alpha,x,\theta_o,\tilde{\theta}_o)\, \frac{1}{\alpha}(\partial^{i^2} + N_\psi)\, \psi_{\{n\}}(\alpha,x,\theta_o,\tilde{\theta}_o)\right] \tag{5.54}$$

In this expression $\{n\}$ incorporates number labels for every bosonic and fermionic oscillator mode with the understanding that bosonic sets are included in the first sum and fermionic sets in the second sum. N_ϕ or N_ψ is the numerical value of $N + \tilde{N}$ (i.e. the mass squared) for this set of integers. The action is hence a sum of ordinary kinetic actions for all component fields of the string. The sum runs over infinitely many field components, which, of course, is unavoidable since the string represents an infinity of components. Some caution must be exercised in explicit computations of amplitudes to treat this infinity of modes.

At this stage let us digest the notion of the quantum string. In (5.53) and (5.54) we have given each quantum state that the string can appear in a local point-like structure. One rough way of seeing it is to consider each point along the string as a separate point-like object, which can be excited to give a specific state. However, when a string propagates all the states (or the points along the string) propagate and we must consider the extended objects. In the next section we will discuss interaction and then consider 3-string vertices. If we project each string to specific states we will obtain 3-point vertices.

A further remark is that in a superfield $\Psi[x(\sigma), \theta(\sigma), \tilde{\theta}(\sigma)]$ the Grassman coordinates are more than just book-keeping devices. We cannot Taylor-expand to get a finite number of fields $\psi[x(\sigma)]$ as could be done for ordinary superfields.

6. THE CUBIC OPEN-STRING INTERACTION

The construction of a consistent interaction for extended objects is much harder than for point-like objects. Causality will be broken unless we have a local interaction. Another problem is that if different points on the object have different times, causality can again be jeopardized. These problems are quite handedly avoided for strings in the light-cone gauge. Firstly, it is very natural to imagine a local interaction. Consider two open strings. A typical interaction will occur when their endpoints join to make one string. Similarly two closed strings can touch in one point and open up to become one closed string. (Just check under the sink in your kitchen). Secondly each point on the string carries the same (light-cone) time and the second problem above will not arise.

In this section we construct the Φ^3 interaction, that describes a process in which two open strings join their ends to give a third open string (following Ref. 32). (The complete interaction also contains the hermitian-conjugate term in which one open string breaks into two.) The parametrization of the three strings are explained in a "light-cone diagram" in fig. 1. Since strings 1 and 2 are annihilated by the vertex operator we take $\alpha_1, \alpha_2 > 0$ and $\alpha_3 = -\alpha_1 - \alpha_2 < 0$. The labeling of σ and the arrows are interpreted to mean that

$$x_r(\sigma) = x^{(r)}(\sigma_r)\,\Theta_r\,, \qquad r = 1,2,3 \tag{6.1}$$

where

$$\sigma_1 = \sigma\,, \quad \sigma_2 = \sigma - \pi\alpha_1, \quad \sigma_3 = \pi(\alpha_1 + \sigma_2) - \sigma \tag{6.2}$$

$$\Theta_1 = \begin{cases} 1 & \sigma < \pi\alpha_1 \\ 0 & \sigma > \pi\alpha_1 \end{cases} \tag{6.3a}$$

$$\Theta_2 = \begin{cases} 0 & \sigma < \pi\alpha_1 \\ 1 & \sigma > \pi\alpha_1 \end{cases} \tag{6.3b}$$

$$\Theta_3 = \Theta_1 + \Theta_2 = 1\,, \tag{6.3c}$$

and $x^{(r)}(\sigma_r)$ is given by the expansion (5.21)

$$x^{(r)}(\sigma_r) = x^{(r)} + i\sum_{n=1}^{\infty}\frac{1}{n}\left[\beta_n^{(r)} - \beta_{-n}^{(r)}\right]\cos\frac{n\sigma_r}{\alpha_r} \tag{6.4}$$

suppressing the transverse vector index. The coordinates $\theta_{rA}(\sigma)$, $\tilde{\theta}_{rA}(\sigma)$ and the momenta $p_r^{\,i}(\sigma)$, $\lambda_r^{\,A}(\sigma)$, $\tilde{\lambda}_r^{\,A}(\sigma)$ are defined in identical fashion.

The detailed assumption in the interaction diagram Fig. 1 is that at τ=0 strings 1 and 2 join to form string 3. We demand continuity in the coordinates, i.e. that

$$\Sigma\,\varepsilon_r\,x_r^{\,i}(\sigma) = 0, \qquad \Sigma\,\varepsilon_r\,\theta_{rA}(\sigma) = 0\,, \tag{6.5}$$

and local momentum conservation

$$\Sigma\,p_r^{\,i}(\sigma) = 0, \qquad \Sigma\,\Lambda_r^{\,A}(\sigma) = 0\,. \tag{6.6}$$

The essential point now is to add three-point couplings to all the Hamiltonians such that the super-Poincaré algebra closes at least to first order in the coupling constant. The procedure is performed in three steps. The first step is to work out mode expansions that are equivalent to the three-string couplings that would occur, if the entire interaction were given by the δ-functions (6.5) and (6.6). The second step consists of adding factors to the vertex that still commute with the conditions (6.5) and (6.6) such that the Hamiltonians H, Q_-^{A}, Q_{-A} are constructed leading to a correct supersymmetry algebra. In principle also the non-linear Lorentz transformations should be constructed. The third step finally is to reinterpret these oscillator expressions in functional language. This is a somewhat round-about way but reflects our limited confidence in working with functional expressions. Such methods will, of course, be worked out eventually and the construction of the vertex should then be simplified and more transparent.

Let us so go back to the starting-point, which is an expression of the functional form

$$E = g \int \prod_{r=1}^{3} d\alpha_r \, D^{12} z_r \, \delta(\Sigma\alpha_r) \, \Delta^{12}\left[\Sigma \, \varepsilon_r \, z_r(\sigma)\right]$$

$$\mathrm{tr}\left[\Phi(1) \, \Phi(2) \, \Phi(3)\right] \quad . \tag{6.7}$$

$(z = x^i, \theta_A)$

The vertex operator expressed in terms of oscillators is found by substituting the field expansion of Eq. (5.53) (in the open string case) into the expression (6.7) and writing the Δ-functionals as an infinite product of δ-functions for each mode. The coordinates for all non-zero modes can then be integrated explicitly (all integrals are Gaussian) to give a formal result

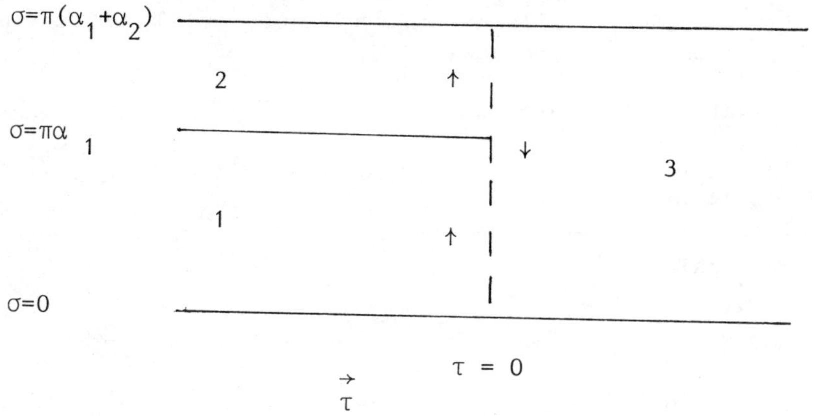

Fig. 1 Light-cone diagram depicting a vertex in which two open strings 1 and 2 join ends at time τ=o to give a single open string 3.

$$E = \int \prod_{r=1}^{3} (d\alpha_r\ d^8x_r\ d\Theta_r)\ \delta(\Sigma\alpha_r)\ \delta^{12}(z_1-z_2)\ \delta^{12}(z_2-z_3)$$

$$\sum_{\{n^{(1)},n^{(2)},n^{(3)}\}} C(\{n^{(1)},n^{(2)},n^{(3)}\}) \prod_{r=1}^{3} \phi_{\{n^{(r)}\}}(r), \quad (6.8)$$

where $\{n^{(r)}\}$ is short for the infinite set of mode numbers $\{n_k{}^{(r)}, n'_\ell{}^{(r)}, m_s{}^{(r)}, m'_s{}^{(r)}\}$ in Eq. (5.53). This is a horrendous expression and in order to avoid this morass of indices we define the number-basis vertex vector by

$$|\ V > = \sum_{\{n^{(1)},n^{(2)},n^{(3)}\}} C(\{n^{(1)},n^{(2)},n^{(3)}\})\ |\ \{n^{(1)},n^{(2)},n)\}> . \quad (6.9)$$

A coupling of three specific fields is then given by

$$C_{N_1,N_2,N_3} = < N_1,\ N_2,\ N_3\ |\ V > \quad , \quad (6.10)$$

where $|\ N_1,N_2,N_3>$ is a Fock-space vector with excitation numbers given by three sets of integers N_1, N_2, N_3. From $|V>$ one can in principle obtain the starting expression E (6.7).

To obtain $|V>$ by direct computation is quite tedious. A faster way is to notice that if we insert, say $\Sigma\ \varepsilon_r\ x_r{}^i(\sigma)$ in the integrand of expression (6.7), we get zero because of (6.5), (which is implemented by the Δ-functional). Following through the steps from (6.7) to (6.9) we deduce that $|V>$ has to satisfy

$$\Sigma\ \varepsilon_r\ x_r{}^i(\sigma)\ |V> = \Sigma\ \varepsilon_r\ \Theta_{rA}(\sigma)\ |V> = 0 \quad . \quad (6.11)$$

These conditions determine $|V>$ up to certain overall factors that do not involve the oscillators. Such factors will be determined eventually by the continued analysis.

The result of this analysis is that

$$|V> = \exp(\Delta_o+\Delta_s)\ |0>\ \delta(\Sigma\alpha_r)\ \delta^4(\Sigma\alpha_r\Theta_{rAo})\ \delta^8(\Sigma\ p_r{}^i) \ , \quad (6.12)$$

where the δ-functions represent the zero-mode Fourier components of the δ-functionals.

$$\Delta_o = \frac{1}{2} \sum_{r,s=1}^{3} \sum_{m,n=1}^{\infty} \beta_{-m}{}^{i(r)}\ \overline{N}_{mn}{}^{rs}\ \beta_{-n}{}^{i(s)}$$

$$+ \sum_{r=1}^{3} \sum_{m=1}^{\infty} \overline{N}_m{}^r\ \beta_{-m}{}^{i(r)} P^i - \frac{\tau_o}{2\alpha}\ P^2 \quad (6.13)$$

$$\Delta_s = \sum_{r,s=1}^{3} \sum_{m,n=1}^{\infty} U_{mn}^{rs} R_{-m}^{A(r)} R_{-nA}^{(s)}$$

$$+ \sum_{r=1}^{3} \sum_{m=1}^{\infty} V_m^r R_{-m}^{A(r)} \Theta_A \quad , \tag{6.14}$$

with

$$P^i = \alpha_1 p_2^i - \alpha_2 p_1^i \tag{6.15}$$

$$\Theta_A = \frac{1}{\alpha_3}(\Theta_{1Ao} - \Theta_{2Ao}) \tag{6.16}$$

$$\overline{N}_{mn}^{rs} = -\frac{mn\alpha}{n\alpha_r + m\alpha_s} \overline{N}_m^r \overline{N}_n^s \tag{6.17}$$

$$\overline{N}_m^r = \frac{1}{\alpha_r} \frac{(-1)^{m+1}}{m!} \frac{\Gamma\left(m(1 + \frac{\alpha_{r+1}}{\alpha_r})\right)}{\Gamma(1 - m \frac{\alpha_{r+1}}{\alpha_r})} \tag{6.18}$$

$$\tau_o = \sum_{r=1}^{3} \alpha_r \ln \alpha_r \tag{6.19}$$

$$\alpha = \alpha_1 \alpha_2 \alpha_3 \tag{6.20}$$

$$U_{mn}^{rs} = \frac{m}{\alpha_r} \overline{N}_{mn}^{rs} \tag{6.21}$$

$$V_m^r = -\alpha \sqrt{2} \frac{m}{\alpha_r} \overline{N}_m^r \quad . \tag{6.22}$$

The next step is to write Q_{3-A}, Q_{3-}^{A} and H_3 in the mode expansion. If we start by introducing some factor in the integrand of expression (6.7) then the resulting vector in the Fock space is a factor multiplying $|V\rangle$. The vectors corresponding to Q_{3-A}, Q_{3-}^{A} and H_3 are denoted $|Q_-^A\rangle$, $|Q_{-A}\rangle$ and $|H\rangle$. In this notation the supersymmetry algebra, at first order in the coupling constant g, takes the form

$$\sum_r q_{-r}^{A} | Q_-^{B} \rangle + (A \leftrightarrow B) = 0 \tag{6.23a}$$

$$\sum_r q_{-rA} | Q_{-B} \rangle + (A \leftrightarrow B) = 0 \tag{6.23b}$$

$$\sum_r q_{-r}^{A} | Q_{-B} \rangle + \sum_r q_{-rB} | Q_-^{A} \rangle = 2\delta^A_B | H \rangle \quad . \tag{6.23c}$$

These equations will in fact determine the factors that multiply $|V\rangle$ for each of the operators up to an overall common function of α_1, α_2 and α_3. If we insist on finding the full answer from the algebra the operators J^{i-} must be constructed. This step was circumvented by Green and Schwarz who guessed the overall factor and checked that a certain 4-point amplitude be Lorentz invariant.

Before solving the Eqs. (6.23), we first note that the operators multiplying $|V\rangle$ must not destroy the coordinate matching conditions (6.5). The unique expressions linear in oscillators with this property can be found to be

$$Z^i = |2\alpha|^{-1/2}\{P^i - \alpha \sum_{r,m} \frac{m}{\alpha_r} N_m{}^r \beta_{-m}{}^{i(r)}\} \tag{6.24}$$

$$Y_A = |\alpha/2|^{1/2}\{\Theta_A + \frac{1}{\sqrt{2}} \sum_{r,m} \frac{m}{\alpha_r} N_m{}^r R_{-mA}{}^{(r)}\} . \tag{6.25}$$

In a Yang-Mills type interaction, the cubic contributions to the supercharges do not involve spatial derivative couplings, whereas the Hamiltonian is first-order in derivatives. Therefore we expect $| Q_-{}^A \rangle$ and $| Q_{-A} \rangle$ not to involve Z^i, but $|H\rangle$ to be linear in Z^i (which is a transverse derivative for the zero mode). The simplest SU(4) covariant conjectures for the Q's are then

$$| Q_{-A} \rangle = Y_A \; | V \rangle \tag{6.26}$$

$$| Q_-{}^A \rangle = k \; \varepsilon^{ABCD} \; Y_B Y_C Y_D \; | V \rangle \tag{6.27}$$

Putting these forms into (6.23) leads to the unique solution

$$k = 2/3 \tag{6.28}$$

$$| H \rangle = \{\frac{1}{\sqrt{2}} Z^L - Z^I \rho^{IAB} Y_A Y_B + \frac{\sqrt{2}}{3} Z^R \varepsilon^{ABCD} Y_A Y_B Y_C Y_D\} \; | V \rangle \tag{6.29}$$

Of course, one has to check the whole algebra including the non-linear generators to be completely sure that (6.26)-(6.29) is a consistent result. This has so far not been done, but the uniqueness of the result and our experience with ordinary field theories in the light-cone gauge[34)] allow us to be confident that the algebra will work out.

The final step is so to transcribe the results (6.26)-(6.29) back to functional form, i.e. we have to identify the functional meaning of Z^i and Y_A. To do so consider the operator

$$p_1{}^i(\sigma) - \frac{1}{\pi} x'_1{}^i(\sigma) = \frac{1}{\pi\alpha_1} \sum \beta_m{}^{i(1)} e^{im\sigma/\alpha_1} . \tag{6.30}$$

When acting on $|V\rangle$ the effect can be calculated by commuting the lowering-operator pieces through e^{Δ_o} until they annihilate on $|0\rangle$. A number of terms arise in this process, some of them being singular as $\sigma \to \pi\alpha_1$, the interaction point. Letting

$$\varepsilon = \pi\alpha_1 - \sigma \tag{6.31}$$

be small and positive, and using

$$\varepsilon^{1/2} \sum_{m=1}^{\infty} e^{im\sigma/\alpha_1}\, m\, \overline{N}_m^{\,1} \underset{\varepsilon\to 0}{\to} \alpha_1 \eta^* (-2\alpha)^{-1/2} \quad , \tag{6.32}$$

where

$$\eta = e^{i\pi/4} \tag{6.33}$$

it follows that

$$\varepsilon^{1/2}\left(p_1{}^i(\sigma) - \tfrac{1}{\pi}\, x'{}_1{}^i(\sigma)\right) |\, V\rangle \underset{\varepsilon\to 0}{\to} \tfrac{1}{\pi}\, \eta^*\, Z^i \,|\, V\rangle \quad . \tag{6.34}$$

Similarly

$$\varepsilon^{1/2}\left(p_1{}^i(\sigma) + \tfrac{1}{\pi}\, x'{}_1{}^i(\sigma)\right) |\, V\rangle \underset{\varepsilon\to 0}{\to} \tfrac{1}{\pi}\, \eta\, Z^i \,|\, V\rangle \quad . \tag{6.35}$$

Combining these, we may define Z^i as

$$Z^i |\, V\rangle \underset{\varepsilon\to 0}{\sim} \sqrt{2\varepsilon}\, \pi\, p_1{}^i(\pi\alpha_1 - \varepsilon)\,|\, V\rangle \quad . \tag{6.36}$$

For the anticommuting modes one can define a similar formula

$$Y_A |\, V\rangle \underset{\varepsilon\to 0}{\sim} \sqrt{2\varepsilon}\, \Theta_{1A}(\pi\alpha_1 - \varepsilon)\,|\, V\rangle \quad . \tag{6.37}$$

The choice of string = 1 in these definitions is arbitrary. $|V\rangle$, Z^i and Y_A have cyclic symmetry in the coordinates of the three strings, and any string could be chosen.

The functional forms are thus

$$\begin{aligned} Q_{3-A} = g \int \Big(\prod_{r=1}^{3} D^{12} z_r\Big)\, \Delta^{12}\big[\Sigma\, \varepsilon_r z_r(\sigma)\big] \\ \int \Pi\, d\alpha_r\, \delta(\Sigma\alpha_r) \Big[\lim_{\sigma\to\pi\alpha_1} \sqrt{\pi\alpha_1 - \sigma}\, \Theta_{1A}(\sigma)\Big] \\ \mathrm{tr}\big[\Phi(1)\, \Phi(2)\, \Phi(3)\big] \quad , \end{aligned} \tag{6.38}$$

and similar expressions for $Q_{3-}{}^A$ and H using (6.36) and (6.37) together with (6.27)-(6.29).

It is natural that the interaction point enters the functional expression, and the divergence $\varepsilon^{-1/2}$ is important to single out the contributions from the interaction point.

To get a further check on the Hamiltonians, we may consider the couplings for three massless fields. They are obtained by projecting the formulae onto ground states. One can show that these expressions correspond to the cubic couplings of the super Yang-Mills theory in the light-cone gauge. This is the wanted results since that is the unique supersymmetric theory with a vector and a spinor field.

7. OTHER TYPES OF STRING INTERACTIONS

The cubic open string interaction constructed in the last section is very natural from a geometric point of view. The same will, in fact, be true for all the types of interactions. Consider first the cubic closed string interaction. Geometrically we understand this as two tubes (a closed string traces out a tube when it propagates) joining into one tube, which is easy to visualize, not only for plumbers.

The explicit construction of this vertex can be done almost immediately by a "stuttering process" if we use the SU(4) covariant formalism. (The type IIb model was treated before in an SO(8) invariant way[31].) The expression for the vertex is essentially a direct product of open-string vertices with one factor made from α's and Q's and the second factor from $\tilde{\alpha}$'s and $\tilde{Q}$'s. Let $| Q_-^A \rangle$, $| Q_{-A} \rangle$ and $|H\rangle$ represent the oscillators $\{\beta_n\}$ and $\{R_n\}$ replaced by $\{\alpha_n\}$ and $\{Q_n\}$ and let $| \tilde{Q}_-^A \rangle$, $| \tilde{Q}_{-A} \rangle$ and $|\tilde{H}\rangle$ be corresponding expressions constructed from $\{\tilde{\alpha}_n\}$ and $\{\tilde{Q}_n\}$. Then one can prove that the N=2 supersymmetry algebra is satisfied by

$$| Q_{1-}^A \rangle = | Q_-^A \rangle \otimes |\tilde{H} \rangle \tag{7.1a}$$

$$| Q_{2-}^A \rangle = | H \rangle \otimes | \tilde{Q}_-^A \rangle \tag{7.1b}$$

$$| Q_{1-A} \rangle = | Q_{-A} \rangle \otimes | \tilde{H} \rangle \tag{7.1c}$$

$$| Q_{2-A} \rangle = | H \rangle \otimes | \tilde{Q}_{-A} \rangle \tag{7.1d}$$

$$| H \rangle_{cl} = | H \rangle \otimes | \tilde{H} \rangle \tag{7.1e}$$

This can be rewritten in functional form using the formulae (6.26)-6.29), (6.36) and (6.37), and we see that the interaction vertex is quadratic in derivatives, which is typical of gravitational interaction, and by checking the couplings among three massless states, one finds the couplings of N=2 supergravity. The coupling here described is then taken to be proportional to Newton's constant κ.

In section 6 we found that two end-points of open strings can be glued together. Then there must exist an interaction where the end-points of one propagating string hook up making a closed string. By performing the same analysis as in sect. 6 one can prove that this interaction exactly coincides with the three open-string vertex (6.26)-(6.29), which is really necessary for the theory to have the local geometric interpretation we have alluded to so far. The analysis is quite long and intricate in the present formalism and it was quite gratifying when Green and Schwarz found the result.

One can also prove that there are two types of interaction which have the same topology as the cubic closed string interaction. Since the interaction is local, it does not care if the strings are open or closed. Firstly we can consider the process in which an open string and a closed string touch and exchange string segments to form a single open string, as depicted in Fig. 2. (The two possibilities in 2a and 2b must both be included, since these are unoriented type 1 strings.) The hermitian-conjugate process in which open → open + closed is also included, of course. However, there is no "crossed" interaction term with the process open + open → closed since this is not a process that occurs by an interaction in one point. The process is, of course, possible but does not correspond to a fundamental vertex. The absence of such "crossed terms" in the interaction is a fundamental difference between string theories and point-particle theories.

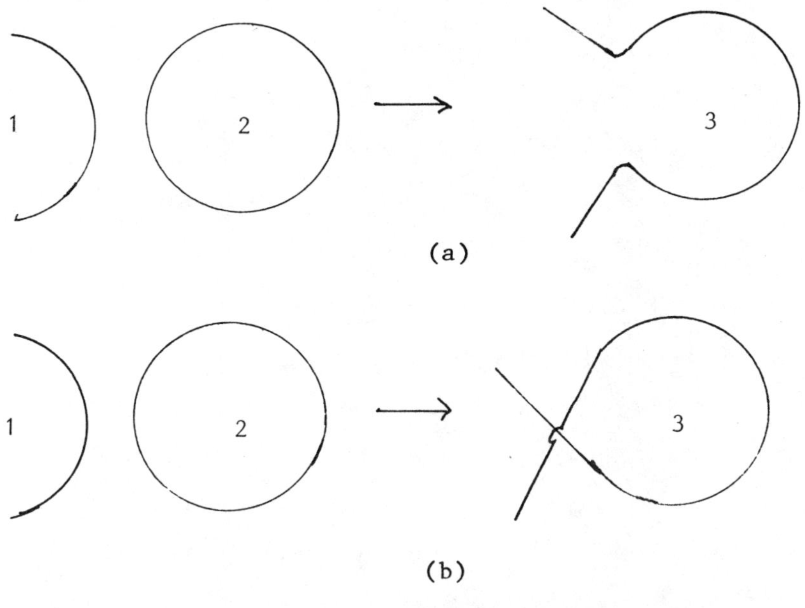

Fig. 2 Two ways in which an open string 1 can join with a type I closed string 2 to give an open string 3.

The interaction in Fig. 2, hence, must be of the form (7.1 and be of gravitational strength.

The other type of interaction which is topologically the same as the cubic closed-string vertex is the process where two open strings touch to interchange bits of the strings as in Fig. 3, which depicts the two possibilities that must be included. This Φ^4 interaction only contains terms with two initial and two final strings. Again the "crossed" process one → three and three → one are not fundamental. The interaction in Fig. 3 should again be of the form implemented by (7.1) and of gravitational strength.

We have here described five types of interactions based on two topologically different processes one with Yang-Mills strength g and one with Einstein strength κ. In the type II models only the latter one can occur, while the type I models must contain all five interactions. In fact, a really detailed check of the quantum super-Poincaré algebra (for the effective action) will show the necessity of all five interaction and will then also find a relationship $\kappa \sim g^2$ for the (10-dimensional) coupling constants.

When ordinary field theories are constructed in the light-cone gauge in the fashion used here, one finds that for Yang-Mills ,

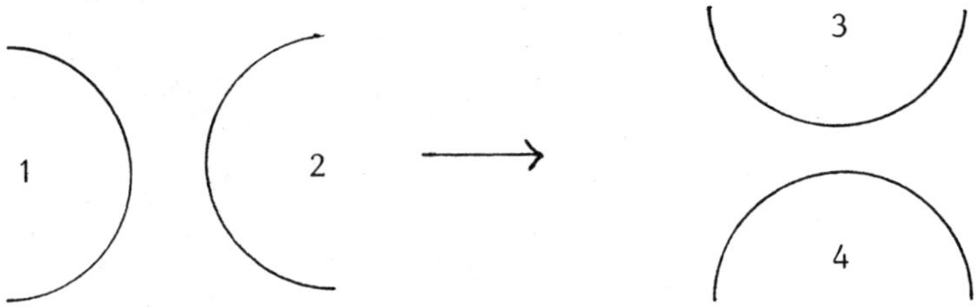

(a)

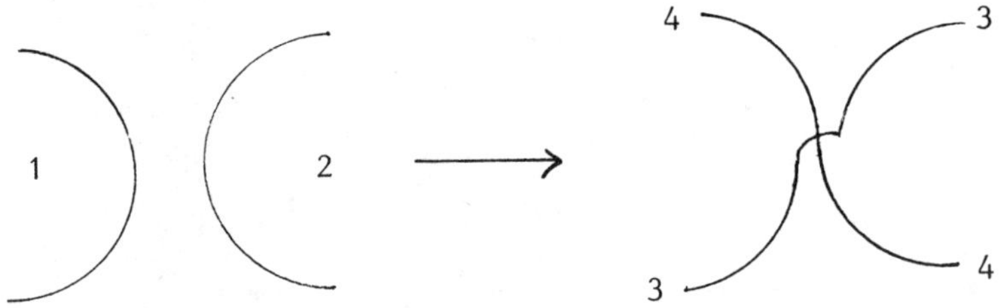

(b)

Fig. 3 Two ways in which a pair of open strings 1 and 2 can give rise to a new pair of open strings 3 and 4 by an exchange

theories it is sufficient to introduce cubic couplings in the supercharges to close the algebra, while in gravitational theories couplings to all orders are necessary. The Yang-Mills Hamiltonian, of course, has cubic and quartic couplings.

The situation turns out to be quite different in string theories. Let us imagine anticommuting Q_{3-A} as in (6.38) and the corresponding expression for Q_{3-}^{A}. Then we will formally obtain a 4-string vertex. In a field theory this gives definitely a 4-point vertex that must be included. However, in a string theory we can argue (on rather loose ground) that the 4-point vertex should be vanishingly small compared to a 4-point amplitude built up by two 3-point vertices and a propagator. In a light-cone diagram the 4-point vertex derived from the anti-commutator above is depicted in Fig. 4 while the 4-point amplitude built up by 3-point vertices is depicted in Fig. 5. In the latter case the time difference $\Delta\tau$ =

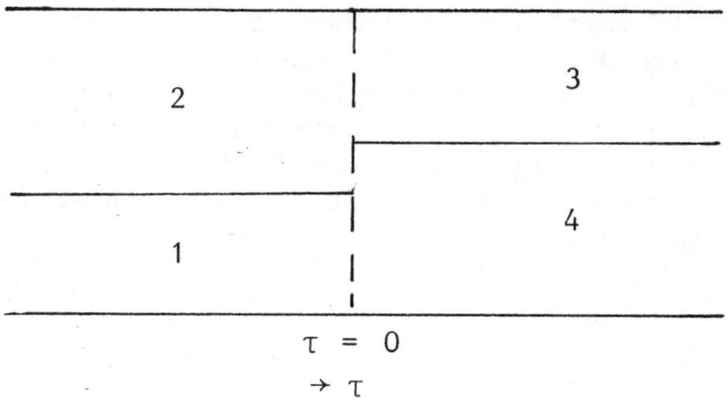

Fig. 4 Light-cone diagram depicting a 4-string vertex that would be obtained by commuting two cubic supercharges.

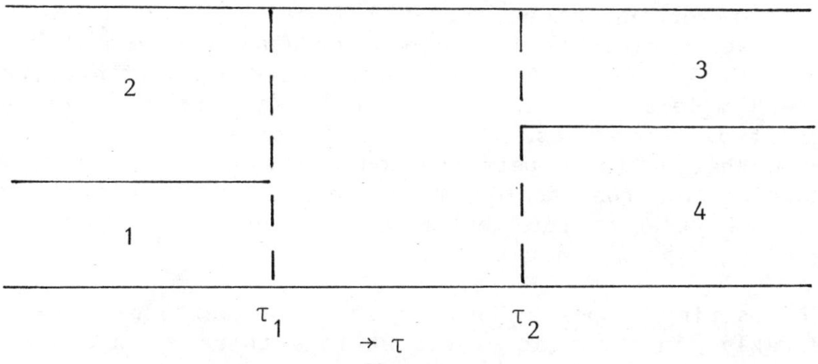

Fig. 5 Light-cone diagram depicting the process where two particles 1 and 2 scatters into particles 3 and 4.

$\tau_1 - \tau_2$ is integrated over. The 4-point vertex in Fig. 4 corresponds to just one point in the $\Delta\tau$ variable and should hence be negligible compared to the contribution from infinitely many points that the integration corresponds to. This reasoning depends quite critically on the definition of the functional measures appearing in Q_- and H. However, again resorting to mode expansions, there is an argument (which used to be called the cancelled propagator argument) that shows that the 4-point vertex should be zero. Hence, we believe we have the complete action.

The argument given here is quite qualitative and should be made much more rigorous. However, there are very good reasons to believe in the result and as a check Green and Schwarz computed the 4-point tree amplitude for external N=1 massless states just using the 3-point vertex. (They consider a channel where the "gravitational" 4-point coupling in Fig. 3 does not contribute.) By only using 3-point vertices they obtained an amplitude which in the infinite tension limit did contain also the proper 4-point interaction. The higher order contact terms of the infinite tension limit field theories are in the string theory replaced with the exchange of an infinite spectrum of massive string excitations. This is somewhat reminiscent of the way massive vector exchanges replace the four-Fermi coupling in the standard electroweak theory. This could very well be an indication that the divergence structure is improved also in the string theory compared to supergravity.

8. CONCLUSIONS

In the preceding sections we have seen how to construct interacting field theories both for strings of type I and type II. We believe we have the complete actions, although more work has to be done to prove that fact completely. It is now vital to find the methods to compute amplitudes and to check what physics the models really describe. We have some information about the quantum properties already. Green and Schwarz[21] computed the S-matrix one-loop graphs by resorting to an operator formalism, which was a method based on matrix elements in the mode expansion commonly used in old dual models. The type II models were found to be finite while for the type I models the tension had to be renormalized. As mentioned in the introduction these facts are not so surprising and there is good hope that a finite perturbation expansion does exist. With the introduction of the field theory all these questions will be tackled and this is the reason why we need an effective way of constructing higher order terms.

The string topology simplifies the perturbation expansion considerably. In fact for closed strings there is only one Feynman diagram at each order! By letting the tension tend to infinity, one gets the complete sum of all the diagrams of the underlying supergravity theory. There is a good chance that it will be easier to prove the quantum status of supergravity by first constructing the loops of the string theories and then take the infinite tension limit.

We should, of course, be somewhat cautious. The string amplitudes will be functional integrals, which we so far have only been able to define via the mode expansions, which is quite round-about. In most cases in physics so far we have not needed to solve explicitly functional integrals, but mainly used their properties as integrals. Here we will need to solve them explicitly.

Having constructed loop graphs we also have to look for anomalies. There is a standard argument[35)], which says that in a theory with both massless and massive states, it is enough to consider the massless sector to determine whether the theory has anomalies or not. The underlying ten-dimensional massless field theories are for type I: N=1 Yang-Mills coupled to N=1 supergravity and for type II: N=2 supergravity. Both N=1 Yang-Mills and N=1 supergravity contain chiral fermions and have been shown to have anomalies[36)]. The type IIa N=2 supergravity has fermions of both chiralities and is trivially anomaly-free, while the type IIb model does contain chiral fermions. It has, however, been proven to be anomaly-free due to impressive cancellations among the contributions coming from the chiral fermions and the 4-rank antisymmetric tensor field[37)]. It has even been proven to be the unique model in higher dimensions with chiral fermions that is anomaly-free! From these arguments it seems probable that the type I string models are out, but a recent calculation by Green and Schwarz[38)] indicates that if the gauge group is SO(32) also the type I model is anomaly free!

Little has so far been done to understand the physical basis for string theories. These are theories involving gravity, but we see in the present formulation no trace of a geometric theory of space-time. This is a general feature of gravity in the light-cone gauge. If the string theories are the truly fundamental theories they should be based on some fundamental physical principle. Our main guide-line so far has been finite quantum corrections, but one should be able to translate this into some more fundamental principle.

Furthermore, the string theories predict a 10-dimensional space-time. This space cannot be dimensionally reduced to 4-dimensions by shrinking the extra 6 coordinates to zero. Here a generalized Kaluza-Klein procedure must be implemented.

Finally we are facing the same difficulty as with ordinary extended supergravity, namely how do we connect it to the present day physics? Where are the quarks, leptons, gluons etc in these models? This is an urgent question, but perhaps not so urgent as the question what are really the quantum properties? If we can prove to have a consistent quantum theory including gravity that would be of such tantamount importance that it would give a great impetus for the future research into the physics of superstrings.

APPENDIX

Some Notations and Conventions

The algebra of SO(8) has three inequivalent real eight-dimensional representations, one vector and two spinors. We use 8-valued indices i,j, ... corresponding to the vector, a,b, ... corresponding to one spinor and $\dot{a}$, $\dot{b}$, ..., corresponding to the other spinor. Dirac matrices $\gamma^{i}{}_{a\dot{a}}$ may be regarded as Clebsch-Gordan coefficients for combining the three eights into a singlet. A second set of matrices $\tilde{\gamma}^{i}{}_{\dot{a}a}$ is also introduced. We choose

$$\tilde{\gamma} = \gamma^{T} \tag{A.1}$$

$$\{\gamma^{i}, \tilde{\gamma}^{j}\} = 2\,\delta^{ij} \,. \tag{A.2}$$

The 16 x 16 matrices

$$\begin{bmatrix} 0 & \gamma^{i}{}_{a\dot{a}} \\ \tilde{\gamma}^{i}{}_{\dot{b}b} & 0 \end{bmatrix}$$

form a clifford algebra. We also define

$$\gamma^{ij}_{ab} = \frac{1}{2}\left[\gamma^{i}_{a\dot{a}}\,\tilde{\gamma}^{j}_{\dot{a}b} - \gamma^{j}_{a\dot{a}}\,\tilde{\gamma}^{i}_{\dot{a}b}\right] \,. \tag{A.3}$$

These matrices are seen to be antisymmetric in a and b using (A.1)

We can also define

$$\gamma^{ij}_{\dot{a}\dot{b}} = \frac{1}{2}\left[\tilde{\gamma}^{i}_{\dot{a}a}\,\gamma^{j}_{a\dot{b}} - \tilde{\gamma}^{j}_{\dot{a}a}\,\gamma^{i}_{a\dot{b}}\right] \tag{A.4}$$

which in a similar fashion is antisymmetric in $\dot{a}$ and $\dot{b}$.

To span the whole 8 x 8 dimensional matrix spaces we also define

$$\gamma^{ijk\ell}_{ab} \equiv (\gamma^{[i}\,\tilde{\gamma}^{j}\,\gamma^{k}\,\tilde{\gamma}^{\ell]})_{ab} \,. \tag{A.5}$$

$$\gamma^{ijk\ell}_{\dot{a}\dot{b}} \equiv (\tilde{\gamma}^{[i}\,\gamma^{j}\,\tilde{\gamma}^{k}\,\gamma^{\ell]})_{\dot{a}\dot{b}} \,. \tag{A.6}$$

These matrices are symmetric.

The general Fierz formula is

$$M_{ab} = \frac{1}{8} \delta_{ab} \, \mathrm{tr}\, M - \frac{1}{16} \gamma^{ij}_{ab} \, \mathrm{tr}(\gamma^{ij} M)$$

$$+ \frac{1}{384} \gamma^{ijk\ell}_{ab} \, \mathrm{tr}(\gamma^{ijk\ell} M) \quad . \tag{A.7}$$

In the case of SU(4), the six-vector can be obtained as the antisymmetric tensor product of two 4's or two $\bar{4}$'s. The corresponding Clebsch-Gordan coefficients (or Dirac matrices) are denoted $\rho^I{}_{AB}$ and ρ^{IAB}. They are normalized as usual so that

$$\rho^{IAB} \rho^J{}_{BC} + \rho^{JAB} \rho^I{}_{BC} = 2 \, \delta^A{}_C \, \delta^{IJ} \ . \tag{A.8}$$

We also define

$$\rho^{IJ}{}_A{}^B = \frac{1}{2} (\rho^I{}_{AC} \, \rho^{JCB} - \rho^J{}_{AC} \, \rho^{ICB}) \ . \tag{A.9}$$

REFERENCES

1) G. Veneziano, Nuovo Cim. 57A (1968), 190.
2) M.A. Virasoro, Phys. Rev. D1 (1970), 2933.
3) R.C. Brower, Phys. Rev. D6 (1972), 1655.
4) P. Goddard and C.B. Thorn, Phys. Lett. 40B (1972), 235.
5) H.B. Nielsen, 15th Int. Conf. in High Energy Physics, Kiev (1970).
   L. Susskind, Nuovo Comento 69A (1970), 457.
   Y. Nambu, Proc. Int. Conf. on Symm. and Quark Models, Wayne State Univ. (1969) (Gordon and Breach, 1970).
6) P.M. Ramond, Phys. Rev. D3 (1971), 2415.
7) A. Neveu and J.H. Schwarz, Nucl. Phys. B31 (1971), 86; Phys. Rev. D4 (1971), 1109.
8) L. Brink, D.I. Olive, C. Rebbi and J. Scherk, Phys. Lett. 45B (1973, 379.
9) P. Goddard, J. Goldstone, C. Rebbi and C.B. Thorn, Nucl. Phys. B56 (1973), 109.
10) Y. Nambu, Lectures at Copenhagen Symposium, 1970 (unpublished).
   O. Hara, Progr. Theor. Phys. 46 (1971) 1549
   T. Goto, Progr. Theor. Phys. 46 (1971) 1560
11) A.M. Polyakov, Phys. Lett. 103B (1981), 207, 211.
12) L. Brink, P. Di Vecchia and P.S. Howe, Phys. Lett. 65B (1976), 471.
   S. Deser and B. Zumino, Phys. Lett. 65B (1976), 369.
13) J.-L. Gervais and B. Sakita, Nucl.Phys. B34 (1971), 477.

14) M. Ademollo, L. Brink, A. D'Adda, R. D'Auria, E. Napolitano, S. Sciuto, E. Del Giudice, P. Di Vecchia, S. Ferrara, F. Gliozzi, R. Musto, R. Pettorino, J.H. Schwarz, Nucl. Phys. B111 (1976), 77.
L. Brink and J.H. Schwarz, Nucl. Phys. B121 (1977), 285.
15) C. Lovelace, Phys. Lett. 34B (1971), 500.
16) A. Neveu and J. Scherk, Nucl. Phys. B36 (1972), 155.
17) J. Scherk and J.H. Schwarz, Nucl. Phys. B81 (1974), 118.
18) F. Gliozzi, J. Scherk and D.I. Olive, Phys. Lett. 65B (1976), 282; Nucl. Phys. B122 (1977), 253.
19) See J.H. Schwarz, Phys. Rep. 69 (1982), 223.
M.B. Green, Surveys in High Energy Physics 3 (1983), 127.
20) B. Zumino, unpublished.
21) M.B. Green and J.H. Schwarz, Nucl. Phys. B198 (1982), 441.
22) See the first reference in 12).
23) L. Brink and H.B. Nielsen, Phys. Lett 45B (1973), 332.
24) This method was suggested by F. Gliozzi, unpublished.
25) This is a point stressed by A.M. Polyakov.
26) M.B. Green and J.H. Schwarz, Phys. Lett. 109B (1982), 444.
27) M.B. Green and J.H. Schwarz, Nucl. Phys. B181 (1981), 502.
28) For a more stringent discussion, see ref. 19.
29) M.B. Green and J.H. Schwarz, Phys. Lett. (1984), 367.
30) I. Bengtsson and M. Cederwall, Institute of Theoretical Physics, Göteborg preprint (1984)
31) M.B. Green, J.H. Schwarz and L. Brink, Nucl. Phys. B219 (1983), 437.
32) M.B. Green and J.H. Schwarz, CALT-68-1090 (1984) to be published in Nucl. Phys. B.
33) P.A.M. Dirac, Rev. Mod. Phys. 26 (1949), 392.
34) M.B. Green and J.H. Schwarz, Phys. Lett. 122B (1983), 143.
A.K.H. Bengtsson, I. Bengtsson and L. Brink, Nucl. Phys. B227 (1983), 31; 41.
35) G.'t Hooft, in Recent Developments in Gauge Theories, ed. G.'t Hooft et al (Plenum, N.Y.) (1980).
36) P. Frampton and T.W. Kephart, Phys. Rev. Lett. 50 (1983), 1343; 1347.
37) L. Alvarez-Gaumé and E. Witten, Nucl. Phys. B234 (1984), 269.
38) M.B. Green and J.S. Schwarz, private communication.

SUPERGRAVITY - THE BASICS AND BEYOND

Bernard de Wit

Institute for Theoretical Physics
Princetonplein 5, P.O. Box 80.006
3508 TA Utrecht, The Netherlands

and

Daniel Z. Freedman*

Department of Mathematics and
Center for Theoretical Physics
Laboratory for Nuclear Science
Massachusetts Institute of Technology
Cambridge, MA 02139, USA

(Lecture notes written with the help of Dr. Xizeng Wu, M.I.T.)

We assume that the reader has turned to this page (as we assumed of the summer school students who came to the lectures) of his or her own free will and with some foreknowledge of what supergravity is basically about. We shall spare him both the usual motivation, of questionable inspirational value, and the conventional excuses for the lack of experimental support. Instead we will get down to business.

In the first six sections we present the basics of supergravity in concrete and deadly detail. We assume only some previous knowledge of Lagrangian field theory, mostly classical, and some previous exposure to general relativity and tensor calculus. The algebra of Dirac matrices will be studied to over-

*Supported in part by NSF Grant No 84-07109-PHY and Department of Energy Contract DE-AC-02-76 ERO 3068.

clarify the mathematical underpinnings of the charge-conjugation matrix and Majorana spinors. The free Majorana spin 1/2 field is then discussed as a quantum system. Next we discuss the free massless Rarita-Schwinger field and emphasize its gauge invariance and the fact that it describes a particle with two helicity states, $\lambda = \pm 3/2$. One of the best features of the Bonn summer school was the ice cream, especially the Bonngout Bechers available at the Café Bonngout in the center of town. We recommend them to our readers. The next topic is a discussion of tensor calculus, which emphasizes the ideas of local Lorentz invariance, torsion and the first-order formalism. Then we apply these techniques to general relativity and N = 1 supergravity, emphasizing that the equation of motion for the connection determines the torsion due to matter fields. Finally, we give a very detailed proof that N = 1 supergravity is invariant under local supersymmetry transformations.

In the next four sections we provide an introduction to a few more advanced topics necessary to understand some recent research in supergravity. First, we discuss supergravity in anti-de Sitter space, which (unfortunately) occurs as a natural background in gauged extended supergravity and Kaluza-Klein theories. Then, in the next two sections, we discuss some of the ideas of the off-shell multiplet structure which is important for the construction of phenomenological models based on matter-coupled supergravity. The last subject is a very recent Kaluza-Klein solution to 11-dimensional supergravity which clarifies the relation to gauged N = 8 supergravity in four dimensions. One final word before the the reader plunges in; don't expect any more humor.

1. GAMMA MATRICES AND MAJORANA SPINORS

In this section we will present a complete treatment of the Dirac matrices, the charge conjugation matrix and Majorana spinors[1]. The story will be told for the case of 4-dimensional Minkowski space, but things will be set up so that a generalization to N-dimensional space is easy.

Let us start with the Clifford algebra relations

$$\{\gamma^a,\gamma^b\} = 2\eta^{ab} \;,$$

$$\text{where} \qquad a,b = 0,1,2,3 \;, \qquad (1.1)$$

$$\eta_{ab} = \mathrm{diag}(1,-1,-1,-1) \;,$$

and γ^a are gamma matrices. We assume that there is one 4-dimensional representation, but work in a way which is covariant under a change of representation $\gamma^a \to \tilde{\gamma}^a = S\gamma^a S^{-1}$. Usually, we will not specify our representation, but when it is needed we use the following representation (Dirac representation)

$$\gamma^o = \begin{pmatrix} 1 & 0 \\ 0 & -1 \end{pmatrix} ,$$

$$\gamma^i = \begin{pmatrix} 0 & \sigma^i \\ -\sigma^i & 0 \end{pmatrix} , \qquad (i=1,2,3)$$

$$\gamma_5 = i\gamma^0\gamma^1\gamma^2\gamma^3 = \begin{pmatrix} 0 & 1 \\ 1 & 0 \end{pmatrix} , \tag{1.2}$$

$$C = i\gamma^2\gamma^0 = \begin{pmatrix} 0 & -i\sigma_2 \\ -i\sigma_2 & 0 \end{pmatrix} ,$$

$$\varepsilon^{0123} = 1 .$$

Consider now the set of 16 matrices

$$\begin{array}{lll} \mathbb{1} & & 1 \\ \gamma^a & & 4 \\ \gamma^{ab} & = \frac{1}{2}[\gamma^a,\gamma^b] & 6 \\ \gamma^{abc} & = \frac{1}{2}\{\gamma^a,\gamma^{bc}\} & 4 \\ \gamma^{abcd} & = \frac{1}{2}[\gamma^a,\gamma^{bcd}] & 1 \end{array} \tag{1.3}$$

These matrices have the following properties.

1. They are antisymmetric with weight one,

e.g. $\gamma^{abc} = -\gamma^{bac}$,

$\gamma^{abc} = \gamma^a\gamma^b\gamma^c$, if $a \neq b \neq c$, etc.

2. For these 16 Γ_A (A = multi-index),

$$\Gamma_A^2 = \pm \mathbb{1} ,$$

$$\Gamma_A\Gamma_B = \pm \Gamma_C , \quad (\Gamma_C = \quad \text{iff } A=B)$$

for fixed A; if B ranges over all 16 values, then so does C.

3. Trace properties:

$$\mathrm{Tr}(\mathbb{1}) = 4 ,$$

$$\mathrm{Tr}(\Gamma_A) = o \quad \text{if} \quad \Gamma_A \neq \mathbb{1} .$$

In order to prove the trace properties, one can use

$$[\tfrac{1}{2}\gamma^{ab},\gamma^c] = \eta^{bc}\gamma^a - \eta^{ac}\gamma^b . \tag{1.4}$$

Exercize: Use the Clifford algebra relation (1.1) to prove (1.4).

Note that commutator (1.4) is "isomorphic" to the algebra of the Poincaré group:

$$[M^{ab},P^c] = \eta^{bc}P^a - \eta^{ac}P^b .$$

Therefore, the Lorentz generator M^{ab} is represented in spinor space by $\frac{1}{2}\gamma^{ab}$. Finite Lorentz transformations $\Lambda = e^{\frac{1}{2}\lambda_{ab}M^{ab}}$ are represented by $L = e^{\frac{1}{4}\lambda_{ab}\gamma^{ab}}$,

$$L\gamma^a L^{-1} = \Lambda^a{}_b\gamma^b . \tag{1.5}$$

Now we can prove the trace properties. Note that

$$\mathrm{Tr}\ \gamma^a = \mathrm{Tr}\ L\gamma^a L^{-1} = \Lambda^a{}_b\ \mathrm{Tr}\ \gamma^b . \tag{1.6}$$

So $\mathrm{Tr}\ \gamma^a$ is an invariant vector; but this is impossible unless $\mathrm{Tr}(\gamma^a) = o$. Similarly, because there are no invariant antisymmetric rank 2 and 3 tensors, one has

$$\mathrm{Tr}(\gamma^{ab}) = \mathrm{Tr}(\gamma^{abc}) = o . \tag{1.7}$$

Since ε^{abcd} is a Lorentz invariant tensor, one could have

$$\mathrm{Tr}(\gamma^{abcd}) = \varepsilon^{abcd}\ p . \tag{1.8}$$

But note that

$$\gamma^{abcd} = \tfrac{1}{2}[\gamma^a,\gamma^{bcd}] ,$$

so $p = o$ because the trace of a commutator automatically vanishes.

There are some points worth mentioning. First, since the Lorentz properties are a consequence of the defining relation (1.1) of the Clifford algebra, we have proved the vanishing of certain traces by using only the Clifford algebra and no other properties of the representations. Second, all steps presented here can be easily generalized to any even dimension. For odd dimensions, one has

$$\mathrm{Tr}\left(\gamma^{a_1 a_2 \dots a_M}\right) = o \ , \text{ for } M < N \ ,$$

as a consequence of the N-dimensional Clifford algebra. Since $\gamma^{a_1 \dots a_N}$ cannot be expressed as a commutator, one has

$$\mathrm{Tr}\left(\gamma^{a_1 \dots a_N}\right) = \varepsilon^{a_1 \dots a_N} p \ ,$$

and p need not vanish. Indeed $p \neq o$ in the $2^{[N/2]}$-dimensional representation of the algebra.

4. The sixteen gamma matrices Γ_A are linearly independent as can be seen as follows. Suppose

$$\sum_A x_A \Gamma_A = o \ .$$

Then we multiply from the right by some Γ_B and take the trace

$$o = \sum_A x_A \mathrm{Tr}(\Gamma_A \Gamma_B) = \pm 4 x_B \ ,$$

where the last step follows since $\mathrm{Tr}(\Gamma_A \Gamma_B) \neq o$ only if A = B. Hence $x_B = o$ for any B.

5. The four-dimensional representation is an irreducible representation of the 4-dimensional Clifford algebra. In fact, suppose

$$\gamma^a = S \begin{pmatrix} \tilde{\gamma}^a & 0 \\ 0 & \approx\!\gamma^a \end{pmatrix} S^{-1}$$

where $\tilde{\gamma}^a$ and $\approx\!\gamma^a$ satisfies the Clifford algebra. But then $\tilde{\Gamma}_A$ (formed as in (1.3) from the submatrices $\tilde{\gamma}^a$) would be a linearly independent set of 16 matrices of dimension less than 4 x 4. This is impossible. It should be stressed that the line of arguments ($\mathrm{Tr}(\Gamma_A) = o$, $\Gamma_A \neq \mathbb{1}$ ⇒ linear independence of Γ_A ⇒ irreducibility of Γ_A) uses only the algebra (1.1) and no other properties of representation, except in the last step where the dimensionality of the representation is used. This result can also be generalized to the $2^{N/2}$ dimensional representation of N-dimensional Clifford algebra for even N. Some modifications are required for odd N.

6. The group $(\text{G-Cliff})_4$ generated by the algebra: Consider the equations

$$\Gamma_A^2 = \pm \mathbb{1} \ ,$$

$$\Gamma_A \Gamma_B = \pm \Gamma_C .$$

These "almost" define a finite group, except that multiplication by -1 is not really allowed by group axioms. Therefore consider a set of 32 elements $(\Gamma_A, -\Gamma_A)$. This is a group under matrix multiplication with multiplication table determined by (1.1) (the four-dimensional Clifford algebra).

We now list the following facts about this group. First, for every representation of the Clifford algebra, the matrices Γ_A and $-\Gamma_A$ are a representation of $(\text{G-Cliff})_4$ (but not conversely). Second, from Schur's lemma, the only matrix which commutes with an irreducible representation of $(\text{G-Cliff})_4$ and thus also the Clifford algebra (1.1) is the unit matrix. Third, consider the inequivalent irreducible representation $\mathfrak{D}^{(I)}(g)$ of $(\text{G-Cliff})_4$ with dimensionality d_I. From a fundamental result of finite group theory, one has

$$\sum_I (d_I)^2 = \text{order of group} = 32.$$

Since the trivial representation $g \to 1$ is an irreducible representation with $d = 1$, the group $(\text{G-Cliff})_4$ has one and only one inequivalent four-dimensional irreducible representation. This means that any two four-dimensional representations γ^a and $\tilde{\gamma}^a$ of (1.1) are equivalent. There is a non-singular matrix S, such that

$$\tilde{\gamma}^a = S\gamma^a S^{-1} ,$$

where S is unique up to a constant multiplier. This is because if

$$\tilde{\gamma}^a = S_1 \gamma^a S_1^{-1} = S_2 \gamma^a S_2^{-1} ,$$

then $S_2^{-1} S_1 \gamma^a = \gamma^a S_2^{-1} S_1 ,$

and $S_2^{-1} S_1 = c\,\mathbb{1}$ by Schur's Lemma.

It should be noted that all these results can be generalized to any even dimension. For odd dimensions there exist two inequivalent irreducible representations of the Clifford algebra.

Exercise for afficionados of finite group theory: find all the inequivalent irreps. of $(\text{G-Cliff})_4$.

7. The C matrix.

Consider the four matrices $-\gamma^{aT}$, which satisfy (1.1) and thus also generate $(\text{G-Cliff})_4$. Hence there exists a matrix C. Such that

$$-\gamma^{aT} = C\gamma^a C^{-1} , \qquad (1.9)$$

where the C matrix is nonsingular and unique up to a constant factor. The C matrix has the following properties:

(a) $C = p\, C^T$, for either $p = \pm 1$.

In fact, $-\gamma^a = C^{-1T} \gamma^{aT} C^T$

$$= - C^{-1T} C\gamma^a C^{-1} C^T .$$

From Schur's lemma,

$$C^{-1} C^T = p \mathbb{1} ,$$

then $C^T = pC$,

$$C = pC^T ,$$

and $p^2 = 1$.

(b) *Exercise:* show that

$$\left(C\gamma^{a_1 a_2 \ldots a_r}\right)^T = (-1)^r\, pC\gamma^{a_r \ldots a_2 a_1}$$
$$= (-1)^r\, (-1)^{\frac{1}{2}r(r-1)}\, pC\gamma^{a_1 a_2 \ldots a_r}$$

using $-\gamma^{aT} = C\gamma^a C^{-1}$ and $C^{-1} C^T = p \mathbb{1}$.

This means that each $C\Gamma_A$ is either symmetric or antisymmetric. More explicitly

C	symmetry of p	1
$C\gamma^a$	symmetry of $-p$	4
$C\gamma^{ab}$	symmetry of $-p$	6
$C\gamma^{abc}$	symmetry of p	4
$C\gamma^{abcd}$	symmetry of p	1

(c) $C = -C^T$.

This is because among complete set of 4 x 4 matrices, 10 are symmetric and 6 are antisymmetric, so $p = -1$.
These results are representation-independent. The technique presented here works also for all even N, but the symmetry of C depends on N, e.g. for $N = 6$, $C = + C^T$. For odd N, C may or may not exist, for details see ref. 1 . It should be noted that there still is a freedom to multiply C by a constant. The definition of C is representation-change covariant only if

$$\tilde{C} = S^{-1T} C S^{-1} \quad , \quad \text{for} \quad \tilde{\gamma}^a = S\gamma^a S^{-1} \ .$$

It also should be noted that everything done so far also works for Euclidean space ($\eta^{ab} = (++++)$).

8. There is a preferred class of representations of the Clifford algebra such that

$$\gamma^o = \gamma^{o\dagger}$$

$$\gamma^i = - \gamma^{i\dagger} \ .$$

Proof: The γ^a generate the finite group (G-Cliff)$_4$. For a finite group any representation is equivalent to a unitary representation. Hence one can assume

$$\gamma^{o\dagger} = (\gamma^o)^{-1} = \gamma^o \quad ,$$

$$\gamma^{i\dagger} = (\gamma^i)^{-1} = -\gamma^i \ .$$

For the preferred class the following relations hold:

$$\gamma^{a\dagger}\gamma^o = \gamma^o\gamma^a \quad ,$$

$$\gamma^{ab\dagger}\gamma^o = -\gamma^o\gamma^{ab} \ .$$

If $\quad \psi \to \bar{\psi} = \psi^\dagger\gamma^o$

then $\gamma^{ab}\psi \to \psi^\dagger(\gamma^{ab})^\dagger \gamma^o = - \bar{\psi}\gamma^{ab}$.

Hence ψ and $\bar{\psi}$ transform with opposite signs under Lorentz transformations, and $\bar{\psi}\psi$ is a scalar.

In a general representation of (1.1), there is always a Dirac adjoint matrix M with the properties that, with

$\bar{\psi} = \psi^\dagger M$, $\bar{\psi}\psi$ transforms as a scalar.

But only in the preferred class does $M = \gamma^o$. We assume the preferred class henceforth. To remain in the preferred class, one should restrict S to unitarity transformations

$$\gamma^a \to \tilde{\gamma}^a = S\gamma^a S^\dagger \ .$$

Exercise: If $\tilde{\gamma}^a = S\gamma^a S^{-1}$ is a non-unitary transformation from the preferred class to non-preferred, show that $M = (S^{-1})^\dagger S^{-1}\tilde{\gamma}^o$

9. In a preferred representation, consider the relation

$$-\gamma^{aT} = C\gamma^a C^{-1} \ .$$

The adjoint of this is

$$-\gamma^{aT} = C^{-1\dagger}\gamma^{a}C^{\dagger} ,$$

since ± signs cancel on both sides. Then

$$C^{\dagger}C\gamma^{a} = \gamma^{a}C^{\dagger}C ,$$

and $C^{\dagger}C = \ell \mathbb{1} \quad , \quad \ell > o$.

One can rescale C so that

$$C^{\dagger} = C^{-1} = - C^{*} .$$

Of course, we still have freedom to multiply C by $e^{i\theta}$.

10. Given a spinor ψ, define the charge conjugate spinor by

$$\psi_{c} = C^{-1}\bar{\psi}^{T} = C^{-1}\gamma^{oT}\psi^{*} .$$

This definition is representation covariant under unitary transformations $\psi \rightarrow S\psi$, provided C transforms as discussed above.

Exercise: Consider $\psi' = \sigma^{ab}\psi$ and show that

$$\psi'_{c} = C^{-1}\gamma^{oT}\sigma^{ab*}\psi^{*} = \sigma^{ab}\psi_{c} .$$

So ψ and ψ_c transform the same way under Lorentz transformations. Also consider the free massive Dirac equation $(i\not{\partial}-m)\psi = o$. Show that $(i\not{\partial}-m)\psi_{c} = o$.

It is worth mentioning that in d = 4 Euclidean space one can define $\psi_{c} = C^{-1}\psi^{*}$ and then the above relations still hold.

11. The facts in (10) suggests that one can impose a constraint of the form

$$\psi = \alpha\psi_{c} = \alpha C^{-1}\bar{\psi}^{T} \tag{1.10}$$

on a complex spinor. The constraint is compatible with Lorentz transformations and the equations of motion. Such a constraint is analogous to $\varphi(x) = \alpha\varphi^{*}(x)$ for a scalar field. We know that this implies half as many degrees of freedom, i.e., the particles and antiparticles associated with the field are identical. Note that the scalar field constraint is consistent only if $\varphi(x) = \alpha\alpha^{*}\varphi(x)$, i.e. α can be any phase factor. Let us explore the consistency of the spinor constraint analogously. From (1.10) one can compute

$$\bar{\psi}^T = \alpha^*\gamma^{oT}\psi^*_c = \alpha^*\gamma^{oT}C^{-1*}\gamma^{oT*}\psi^{**}$$

$$= \alpha^*\gamma^{oT}C^T\gamma^o\psi = -\alpha^*\gamma^{oT}C\gamma^o\psi \; . \qquad (1.11)$$

Substitute this in (1.10) to get the consistency condition

$$\psi = -\alpha\alpha^*C^{-1}\gamma^{oT}C\gamma^o\psi$$

$$= \alpha\alpha^*\gamma^o\gamma^o\psi = \alpha\alpha^*\psi \; . \qquad (1.12)$$

So the constraint is consitent for any phase factor α. This phase factor can be absorbed in the definition of C, so one has

$$\psi = C^{-1}\bar{\psi}^T \; , \qquad (1.13)$$

which is the condition that defines a Majorana spinor.

In some dimensions, both even and odd, the symmetry of C required by $-\gamma^{aT} = C\gamma^a C^{-1}$ does not allow any solution of the consistency condition and Majorana spinors cannot exist. They exist in Minkowski space only if d = 2,3,4 (mod 8).

Exercise: In d = 4 Euclidean space, show that the condition $\psi = \alpha\psi_c = \alpha C^{-1}\psi^*$ is inconsistent. Majorana spinors do not exist, and the Euclidean continuation of a Majorana field theory requires special treatment[2].

The question of the existence of Majorana spinors is exactly analogous to that of self-dual n forms in a space of d = 2n. For Euclidean signature, the consistency condition works in d = 4,8,12,..., but fails in d = 2,6,10. The opposite is true for Minkowski signature. For example, for d = 2 Euclidean space,

$$V^a = \varepsilon^{ab}V^b = \varepsilon^{ab}\varepsilon^{bc}V^c = -V^a \; ,$$

which is obviously inconsistent.

12. Let us introduce a further subclass of the preferred class called friendly representations (all our friends use friendly representations) in which each of the four γ^a has a definite symmetry, $\gamma^a = \pm\,\gamma^{aT}$.

Then $-\gamma^{aT} = C\gamma^a C^{-1}$

and $-\gamma^a = C\gamma^{aT}C^{-1}$.

But in general

$$-\gamma^a = C^{-1}\gamma^{aT}C$$

so $\quad C^2 = \lambda \mathbb{1}$,

and $\quad C = \lambda C^{-1} = \lambda C^{\dagger} = -\lambda C^* = \lambda\lambda^* C$,

hence $|\lambda| = 1$.

Choose the phase of C so that $\lambda = -1$ or $C = -C^{-1}$. (C is now uniquely determined up to a sign.) Given one friendly representation γ^a, then

$$\tilde{\gamma}^a = S\gamma^a S^T$$

is also a friendly representation, if S is orthogonal: $S^T = S^{-1}$.

13. In friendly representations the Majorana condition becomes

$$\psi = -\alpha C\bar{\psi}^T ,$$

but the overall phase of the four components has no physical significance, so we might as well take $\alpha = -1$ and write

$$\psi = C\bar{\psi}^T . \tag{1.14}$$

14. Majorana representations.

In these representations all four γ^a are imaginary. Hence Hermiticity implies

$$\gamma^o = -\gamma^{oT} ,$$

$$\gamma^i = \gamma^{iT} ,$$

and so

$$-\gamma^i = C\gamma^i C^{-1} ,$$

$$\gamma^o = C\gamma^o C^{-1} .$$

Therefore in this representation

$$C = \pm i\gamma^o .$$

Then with $\alpha = \pm i$, the Majorana condition is

$$\psi = \pm iC\bar{\psi}^T = -\gamma^o\gamma^{oT}\psi^* = \psi^* .$$

So ψ_α is a purely real spinor in a Majorana representation. This fact can be useful for some specialized purposes, but we will not need it. The Majorana representations are preserved by real orthogonal transformations: $\gamma^a = S\tilde{\gamma}^a S^T$. For an explicit Majorana representation the reader is referred to ref. 3.

15. Summary of the previous discussion

Representation type	Properties of C	Majorana condition
General	$C = -C^T$	not studied
Preferred: γ^o Hermitian γ^i anti-Herm.	$C = -C^T$ $C^{-1} = C^\dagger$	$\psi = C^{-1}\bar{\psi}^T$ $\bar{\psi} = -\psi^T C$
Friendly: preferred with each γ^a of definite symmetry	$C = -C^T$ $C^\dagger = C^{-1}$ $C = -C^{-1}$	$\psi = C\bar{\psi}^T$ $\bar{\psi} = \psi^T C$
Majorana repr.: preferred + friendly + imaginary γ^a	the above plus $C = \pm i\gamma^o$	$\psi = \psi^*$

General properties:

antisymmetric: C, $C\gamma^{abc}$, $C\gamma^{abcd}$

symmetric: $C\gamma^a$, $C\gamma^{ab}$.

Introduce $\gamma_5 = i\gamma^o\gamma^1\gamma^2\gamma^3$,

then $\gamma^{abcd} = -i\gamma_5\,\varepsilon^{abcd}$,

$\gamma^{abc} = -i\varepsilon^{abcd}\,\gamma_5\gamma_d$,

also

$$\sigma^{ab} = \tfrac{1}{2}\gamma^{ab} = \tfrac{1}{4}[\gamma^a,\gamma^b] .$$

Then C , $C\gamma_5$, $C\gamma_5\gamma^a$,

are antisymmetric matrices, and $C\gamma^a$, $C\sigma^{ab}$ are symmetric. From now on we will use γ_5 and σ^{ab} rather than γ^{abcd} and γ^{ab}. Further we will use a friendly representation.

16. Suppose ψ_α and χ_α are both Majorana spinors. Consider the bilinear $\bar{\chi}M\psi$ where $M = \mathbb{1}, \gamma^a, \sigma^{ab}, \gamma^5\gamma^a, \gamma^5$. Then

$$\begin{aligned}\bar{\chi}M\psi &= \chi^T CM\psi = (CM)_{\alpha\beta}\chi_\alpha\psi_\beta \\ &= \pm\ (CM)_{\beta\alpha}\psi_\beta\chi_\alpha \\ &= \pm\ \bar{\psi}M\chi\ .\end{aligned}$$

Here energy bilinear has definite symmetry; the sign depend on "statistics", the upper sign for commuting spinors, the lower sign for anticommuting spinors.

Summary:

Bilinears	Symmetry	
	Commuting ψ	Anticommuting ψ
$\bar{\chi}\psi$	−	+
$\bar{\chi}\gamma^a\psi$	+	−
$\bar{\chi}\phi^{ab}\psi$	+	−
$\bar{\chi}\gamma^5\gamma^a\psi$	−	+
$\bar{\chi}\gamma^5\psi$	−	+

2. MAJORANA SPIN-½ FIELDS

For a *Majorana* spinor field $\psi(x)$ of mass m, the action is

$$S = \tfrac{1}{2}\int d^4x\ \bar{\psi}(i\partial\!\!\!/-m)\psi\ . \tag{2.1}$$

Note that $i\bar{\psi}\partial\!\!\!/\psi$ and $\bar{\psi}\psi$ are the only Lorentz and parity invariant bilinears available. If $[\psi_\alpha,\psi_\beta] = o$, then

$$\bar{\psi}\psi = o\ ,$$

$$\bar{\psi}\partial\!\!\!/\psi = \tfrac{1}{2}\left(\bar{\psi}\gamma^\mu\partial_\mu\psi+\partial_\mu\bar{\psi}\gamma^\mu\psi\right) = \tfrac{1}{2}\ \partial_\mu\left(\bar{\psi}\gamma^\mu\psi\right)\ ,$$

and the action would vanish for commuting spinors. So it is

necessary to use anticommuting spinors even at the classical level. A factor of $\frac{1}{2}$ occurs in (2.1) because the field contains half as many degrees of freedom as a standard Dirac field. To derive the equation of motion, let us make a variation of S

$$\delta S = \frac{1}{2}\int d^4x\ [\delta\bar{\psi}(i\partial\!\!\!/-m)\psi+\bar{\psi}(i\partial\!\!\!/-m)\delta\psi]$$

$$= \int d^4x\ \delta\bar{\psi}(i\partial\!\!\!/-m)\psi + \text{surface term} . \tag{2.2}$$

This gives the equation of motion, i.e. the Dirac equation:

$$(i\partial\!\!\!/-m)\psi = o . \tag{2.3}$$

There is also a useful shortcut to obtain δS correctly (up to a surface term). One can simply vary $\bar{\psi}$ only and multiply the result by 2. This trick works whenever the derivative operator in $\mathcal{L}$ can be partially integrated. Fortunately this is the case in supersymmetric gauge theories, supersymmetric σ-models, and in supergravity in the second-order formalism with a torsion-free connection. But it does not work for the first order formalism as we will discuss later.

The Majorana field can be quantized canonically but there is a procedural difficulty. Suppose ψ_α ($\alpha=1,2,\ldots,4$) are treated as 4 independent canonical coordinates, then

$$\pi_\alpha = \frac{\partial\mathcal{L}}{\partial\dot{\psi}_\alpha} = \tfrac{1}{2}\psi^*_\alpha .$$

However ψ_α and π_α are not all independent, for they are subject to the Majorana condition. So the naive canonical anticommutation relation $\{\pi_\alpha(x),\psi_\beta(y)\}_{x_0=y_0} = -i\delta_{\alpha\beta}\ \delta(\vec{x}-\vec{y})$ is inconsistent with

$\{\psi_\alpha(x),\psi_\beta(y)\}_{x_0=y_0} = 0$. The sophisticated solution to this problem is

to use Dirac brackets instead of Poisson brackets[4]. Another simpler solution is as follows. In any representation where

$$C\gamma^o = \begin{pmatrix} 0 & M \\ M^T & 0 \end{pmatrix} \quad \text{and} \quad \psi \equiv \begin{pmatrix} \lambda \\ \chi \end{pmatrix} ,$$

one can write

$$\frac{i}{2}\bar{\psi}\dot{\psi} = \frac{i}{2}\psi^T C\gamma^o\dot{\psi} = \frac{i}{2}(\lambda M\dot{\chi}+\chi M^T\dot{\chi})$$

$$= i\lambda M\dot{\chi} + \text{a total time derivative.}$$

Then one can treat χ_1,χ_2 as canonical coordinates, and λ_1,λ_2 as canonical momentum. There is a similar and more physical approach[5] where one uses the chiral projectors $L,R = \frac{1}{2}(1\pm\gamma_5)$. Then the Majorana condition implies

$$R\psi = \overline{C(L\psi)}^T ,$$

and $\frac{i}{2}\,\bar{\psi}\partial\!\!\!/\psi = i\bar{\psi}R\partial\!\!\!/L\psi$ + total derivative.

The left-handed components of ψ are then treated as canonical coordinates, while the right-handed components of ψ are canonical momenta. This approach works because in d = 4 Minkowski space quantum field theories of Majorana and Weyl spinors are entirely equivalent. We do not have time to discuss this equivalency further.

Now we come to the momnetum-space quantization Choose spinors

$$u(\vec{p},\lambda) = \sqrt{E+m}\begin{pmatrix}\chi(\lambda) \\ \frac{\vec{\sigma}\cdot\vec{p}}{E+m}\,\chi(\lambda)\end{pmatrix}$$

$$\vec{\sigma}.\vec{p}\,\chi(\lambda) = \pm\,|\vec{p}|\,\chi(\lambda) \quad , \quad \lambda = \pm\tfrac{1}{2} , \tag{2.4}$$

which satisfies

$$(p\!\!\!/-m)u = o ,$$

$$\bar{u}u = 2m \quad , \quad \bar{u}\gamma^{\mu}u = 2p^{\mu} ,$$

$$\sum_{\lambda=\pm\frac{1}{2}} u\bar{u} = p\!\!\!/+m .$$

Define

$$v(\vec{p},\lambda) = C\,\bar{u}^T(p,\lambda) ,$$

which sat:sfies

$$(p\!\!\!/+m)v = o .$$

Then the general expansion of a free field takes the form:

$$\psi(x) = \sum_{\lambda=\pm\frac{1}{2}}\int\frac{d^3p}{(2\pi)^{3/2}\sqrt{2E}}\left[b(\vec{p},\lambda)u(\vec{p},\lambda)e^{-ipx}+b^*(\vec{p},\lambda)v(\vec{p},\lambda)e^{ipx}\right] . \tag{2.5}$$

One can easily check that the Majorana condition is satisfied. In the second-quantized theory one has the anitcommutation rules

$$\{b(\vec{p},\lambda,) \ , \ b(\vec{p}',\lambda')\} = o \ ,$$

$$\{b(\vec{p},\lambda) \ , \ b^*(\vec{p}',\lambda')\} = \delta_{\lambda\lambda'}\delta^3(\vec{p}-\vec{p}') \ ,$$

which emerge from a proper canonical quantization by the methods discussed above. Hence Majorana fields contains 2 quantum degrees of freedom, i.e. for every momentum $\vec{p}$, there are particles of helicity $\pm\frac{1}{2}$. Particles are identical with anti-particles.

Exercise: From the expression (2.5) above, compute the propagator

$$S(x,y) = -i <o|T\psi(x)\bar{\psi}(y)|o>$$

$$= -\int \frac{d^4p}{(2\pi)^4} e^{-ip.(x-y)} \frac{\not{p}+m}{p^2-m^2+i\varepsilon} \ ,$$

and also show

$$P^\mu = \int d^3p \ p^\mu \sum_\lambda b^*(\vec{p},\lambda)b(\vec{p},\lambda) \ .$$

3. FREE SPIN-3/2 FIELDS

In this section we discuss the features of free spin-3/2 fields. First, let us discuss free field theory from the view point of degrees of freedom.

The classical degrees of freedom of fields are determined by initial data for field equations; the quantum degrees of freedom of fields are the number of massless particles of a given momentum. Therefore, we have the following situation for the count of the degrees of freedom.

Field	Classical degr. of fr.	Quantum degr. of fr.
real scalar $\phi(\vec{x},t)$	2 $(\phi(\vec{x},o),\dot{\phi}(\vec{x},o))$	1
Majorana spinor $\psi_\alpha(\vec{x},t)$	4 $(\psi_\alpha(\vec{x},o))$	2
real vector $v_\mu(\vec{x},t)$	naive 8 $(v_\mu(\vec{x},o),\dot{v}_\mu(\vec{x},o))$	2
symmetric tensor $h_{\mu\nu}$	naive 16 $(h_{\mu\nu}(\vec{x},o),\dot{h}_{\mu\nu}(\vec{x},o))$	2
vector-spinor $\psi_{\mu\alpha}$	naive 16 $(\psi_{\mu\alpha}(\vec{x},o))$	2

Here we have taken into account the fact that field equations for bosonic fields are second order in $\vec{x}$ and t and those for fermionic fields first order in $\vec{x}$ and t.

It is clear from the above summary that for spin 0, $\frac{1}{2}$ fields the quantum degrees of freedom = $\frac{1}{2}\cdot$(classical degrees of freedom). But for spin $\geqslant 1$ fields, there is a mismatch between the two helicity states desired for a massless particle, and the amount of initial data required for a generic field equation. This mismatch would persist even for $m \neq o$, since

$$2s+1 < \tfrac{1}{2}\cdot(\text{classical degrees of freedom}).$$

The lessons from QED, Yang-Mills theory, general relativity and supergravity teach us that the only way to cure the mismatch is to take very special field equations with gauge invariance. In fact, gauge invariance accomplishes the following goals:

a) Relativistic covariance is maintained;
b) The Lagrangian does not contain certain "longitudinal" field components, e.g. $\partial_\mu v^\mu$;
c) Certain field equations are constraints on initial data such that

$$\tfrac{1}{2}\cdot(\text{number of independent initial data}) = 2 = \text{number of quantum degrees of freedom;}$$

d) The field describes a pure massless spin-s particle, with no lower spin admixtures (otherwise there would be some negative metric ghosts);
e) For s = 1, 3/2, 2 fields gauge invariant interactions can be introduced!

For the free complex spin-3/2 field, gauge invariance implies that the action should be invariant under the transformation

$$\psi_\mu(x) \to \psi_\mu(x) + \partial_\mu \varepsilon(x) \tag{3.1}$$

where ε_α is an arbitrary spinor. This transformation rule is modelled on $A_\mu(x) \to A_\mu(x) + \partial_\mu \Lambda(x)$ for the electromagnetic field. The only action which is Lorentz and guage invariant and of first order in derivatives is

$$S = -i \int d^d x \; \bar{\psi}_\lambda \gamma^{\lambda\nu\rho} \partial_\nu \psi_\rho \; , \tag{3.2}$$

and for d = 4,

$$S = - \int d^4 x \; \varepsilon^{\lambda\mu\nu\rho} \; \bar{\psi}_\lambda \gamma_5 \gamma_\mu \partial_\nu \psi_\rho \; . \tag{3.3}$$

Under the transformation (3.1) and its conjugate $\mathcal{L}$ transforms with a total derivative, unlike electromagnetism but like gravity. This is prescient foreknowledge of the fact that supersymmetry in a space-time transformation.

One could also have started with a more general gauge transformation rule

$$\psi'_\mu(x) \to \psi'_\mu(x) + \partial_\mu \varepsilon(x) + a\gamma_\mu \not\partial \varepsilon(x) \ . \tag{3.4}$$

Clearly, the unique invariant action is obtained from the previous by a change of variables $\psi'_\mu(x) = \psi_\mu(x) + a\gamma_\mu \gamma.\psi(x)$. Thus the more general one-parameter family of actions is physically equivalent to the simpler case a = o. Historically Rarita and Schwinger (1941) obtained an $m \neq o$ spin-3/2 field equation in a form which corresponds to $a \neq o$. (Caution: the case $a = -\frac{1}{4}$ requires special treatment since $\gamma.\psi' \equiv o$.) Note that $\partial_\nu\psi_\rho - \partial_\rho\psi_\nu$ plays the role of a gauge invariant field strength. The normalization of S is correct for a complex field as we will show below. For a Majorana field a factor of $\frac{1}{2}$ is required. Consider a more general action

$$S = \int d^4x \left[-i\bar{\psi}_\lambda \gamma^{\lambda\nu\rho} \partial_\nu \psi_\rho + m\bar{\psi}_\lambda \gamma^{\lambda\rho} \psi_\rho\right] \tag{3.5}$$

and the field equation

$$i\gamma^{\lambda\nu\rho} \partial_\nu \psi_\rho + m\gamma^{\lambda\rho} \psi_\rho = o \ . \tag{3.6}$$

We want to argue that m is the mass of the field. Use the γ-matrix identities:

$$\begin{aligned} \gamma^{\lambda\nu\rho} &= \eta^{\lambda\rho}\gamma^\nu - \eta^{\lambda\nu}\gamma^\rho - \gamma^\lambda(\eta^{\nu\rho} - \gamma^\nu\gamma^\rho) \ , \\ &= \eta^{\lambda\rho}\gamma^\nu - \eta^{\lambda\nu}\gamma^\rho + \gamma^\lambda\gamma^{\nu\rho} \ , \\ \gamma_\lambda\gamma^{\lambda\nu\rho} &= 2\gamma^{\nu\rho} \ , \\ \gamma_\lambda\gamma^{\lambda\rho} &= 3\gamma^\rho \ . \end{aligned} \tag{3.7}$$

Now contract γ_λ and ∂_λ with the field equation and obtain

$$2i\gamma^{\lambda\rho}\partial_\lambda\psi_\rho + 3m\gamma.\psi = o \ ,$$

$$m\gamma^{\lambda\rho}\partial_\lambda\psi_\rho = o \ ,$$

so $\quad m\gamma.\psi = o$.

Thus the field equation is equivalent to

$$\begin{aligned} & i\gamma^\lambda(\partial_\lambda\psi_\rho - \partial_\rho\psi_\lambda) - m\psi_\rho = o \ , \\ & m\gamma.\psi = o \ . \end{aligned} \tag{3.8}$$

The equivalence means that (3.6) implies (3.8) and (3.8) implies (3.6). The simpler equation (3.8) cannot be derived directly from any invariant Lagrangian, even for m = o, without the intermediate matrix algebra.
Now apply ∂_μ to the first equation of (3.8) and antisymmetrize it in $\mu \leftrightarrow \rho$, obtaining

$$(i\not\partial - m)(\partial_\mu\psi_\nu - \partial_\nu\psi_\mu) = o \ . \tag{3.9}$$

This suggests that the "gauge invariant components" of ψ_μ have mass m. The fact that ψ_μ entirely describes a free spin-3/2 particle of mass m is confirmed by a detailed study of (3.6) or (3.8)[6].

Exercise: Obtain the following alternate forms of the massless spin-3/2 field equation by various operations on (3.6) or (3.8):

$$\gamma_\alpha(\partial_\beta\psi_\gamma - \partial_\alpha\psi_\beta) + \gamma_\beta(\partial_\gamma\psi_\alpha - \partial_\alpha\psi_\gamma) + \gamma_\gamma(\partial_\alpha\psi_\beta \ \partial_\beta\psi_\alpha) = o \ ,$$

by contracting (3.6) with $\varepsilon_{\lambda\eta\nu\rho}$;

$$\gamma_\lambda(\partial_\alpha\psi_\beta - \partial_\beta\psi_\alpha) + i\varepsilon_{\alpha\beta}{}^{\rho\sigma}\gamma_5\gamma_\sigma(\partial_\rho\psi_\lambda - \partial_\lambda\psi_\rho) = o \ ,$$

by multiplying (3.8) by $\sigma_{\alpha\beta}$;

$$\partial_\alpha\psi_\beta - \partial_\beta\psi_\alpha - i\gamma_5\varepsilon^{\mu\nu}{}_{\alpha\beta}\partial_\mu\psi_\nu = o \ ,$$

by contracting (3.6) with $\gamma_\lambda\sigma_{\alpha\beta}$.

Exercise: In d = 3, show that the massless spin-3/2 field equation $\gamma^{\mu\nu\rho}\partial_\nu\psi_\rho = o$ implies $\partial_\nu\psi_\rho - \partial_\rho\psi_\nu = o$, which means that the system contains no degrees of freedom. This is the supersymmetric counterpart of the statement that $R_{\mu\nu} = o$ implies $R_{\mu\nu\lambda\rho} = o$, implying no degrees of freedom for gravity in d = 3.

Since a massless spin-3/2-field is a gauge field, we must fix a gauge. The physical content can easily be seen in the non-covariant gauge $\gamma^i\psi_i = \vec{\gamma}.\vec{\psi} = o$. If $\vec{\gamma}.\vec{\psi} \neq o$ initially, we can regauge to

$$\psi'_\mu = \psi_\mu + \partial_\mu\varepsilon$$

with $\vec{\gamma}.\vec{\psi}' = \vec{\gamma}.\vec{\psi} + \vec{\gamma}.\vec{\nabla}\varepsilon = o$.

This fixes uniquely $\varepsilon(x)$ as

$$\varepsilon(\vec{x},t) = -\vec{\gamma}.\vec{\nabla}\int \frac{d^3y}{4\pi|\vec{x}-\vec{y}|}\,\gamma.\psi(\vec{y},t) \ ,$$

i.e. $\vec{\gamma}.\vec{\nabla}$ is an invertible operator on the space of square integrable functions on $\mathbb{R}^3$. The $\vec{\gamma}.\vec{\psi} = o$ gauge is the analogue of the Coulomb gauge $\vec{\nabla}.\vec{A} = o$ in electromagnetism. The field equation in the form (3.8) is then

$$(\gamma^o\partial_o - \vec{\gamma}.\vec{\nabla})\psi_\rho - \partial_\rho(\gamma^o\psi_o + \vec{\gamma}.\vec{\psi}) = o .$$

For $\rho = o$ component, this is

$$\vec{\gamma}.\vec{\nabla}\psi_o = o .$$

Since $\vec{\gamma}.\vec{\nabla}$ is invertible, one has

$$\psi_o(\vec{x},t) = o . \tag{3.10}$$

For the remaining space components of the field equation, one has

$$(\gamma_o\partial_o - \vec{\gamma}.\vec{\nabla})\vec{\psi} = o .$$

Contract with $\vec{\gamma}$ giving

$$\vec{\nabla}.\vec{\psi} = o . \tag{3.11}$$

Thus from the gauge condition and equation of motion we derive the 3 independent constraints on $\psi_\rho(\vec{x},t)$ on the initial surface.

$$\begin{aligned} \vec{\gamma}.\vec{\psi}(\vec{x},t) &= o , \\ \psi_o(\vec{x},t) &= o , \\ \vec{\nabla}.\vec{\psi}(\vec{x},t) &= o . \end{aligned} \tag{3.12}$$

This means that only 4 out of the 16 original components of ψ_ρ constitute independent initial data. That is

$\frac{1}{2}\cdot$(# of classical degr. fr.) = desired quantum degrees of freedom.

Our system is now reduced to the Dirac equation:

$$i\partial\!\!\!/\psi_\mu = o$$

plus the constraints above.

Let us use the Fourier analysis to write the general solution to this system. $\psi_\mu(\vec{x},t)$ can be expanded in plane wave "vector-spinors" as

$$\psi_\mu(\vec{x},t) = e^{-ip.x}\, v_\mu(\vec{p})u(\vec{p},\pm) ,$$

where

$$u(\vec{p},\pm) = \sqrt{E}\begin{pmatrix} \chi_\pm(\vec{p}) \\ \pm\ \chi_\pm(\vec{p}) \end{pmatrix} ,$$

and p is a positive null 4-momentum. So far $v_\mu(\vec{p})$ is arbitrary, but it can be expanded in the complete set:

$$v_\mu(\vec{p}) = ap_\mu + b\bar{p}_\mu + c\varepsilon_\mu(\vec{p},+) + d\varepsilon_\mu(\vec{p},-)$$

where

$$p_\mu = (|\vec{p}|,\vec{p})$$

$$\bar{p}_\mu = (-|\vec{p}|,\vec{p})$$

$\varepsilon_\mu(\vec{p},\pm)$ = transverse polarization vectors for the direction $\vec{p}$.

Now impose constraints

$$\psi_o = o \quad \Rightarrow \quad (a-b)u(\vec{p},\pm) = o \quad \Rightarrow \quad a = b$$

$$\vec{\nabla}.\vec{\psi} = o \quad \Rightarrow \quad \vec{p}^{\,2}(a+b)u(\vec{p},\pm) = o \ \Rightarrow \quad a = b = o .$$

So we have reduced ψ_μ to

$$\psi_\mu(\vec{x},t) \sim e^{-ip.x}\Big\{a_{++}\varepsilon_\mu(\vec{p},+)u(\vec{p},+) + a_{--}\varepsilon_\mu(\vec{p},-)u(\vec{p},-)$$
$$+ a_{+-}\varepsilon_\mu(\vec{p},+)u(\vec{p},-) + a_{-+}\varepsilon_\mu(\vec{p},-)u(\vec{p},+)\Big\} .$$

Now enforce the constraint $\vec{\gamma}.\vec{\psi} = o$ by the use of spin-$\frac{1}{2}$ angular momentum properties. One finds

$$\vec{\sigma}.\vec{\varepsilon}(p,\pm)\chi(\vec{p},\pm) = o \quad \Rightarrow \quad a_{++}, a_{--} \text{ unconstrained,}$$

and $\vec{\sigma}.\vec{\varepsilon}(\vec{p},\pm)\chi(\vec{p},\mp) \sim \chi(\vec{p},\pm) \quad \Rightarrow \quad a_{+-} = a_{-+} = o$.

Thus we are left with 2 independent wave functions of helicity ± 3/2.

The plane wave expansion for ψ_μ is then

$$\psi_\mu(x) = \sum_{\lambda=\pm}\int \frac{d^3p}{(2\pi)^{3/2}\sqrt{2E}}\Big[b(\vec{p},\lambda)\varepsilon_\mu(\vec{p},\lambda)u(\vec{p},\lambda)e^{-ipx}$$
$$+ d^*(\vec{p},\lambda)\varepsilon^*_\mu(\vec{p},\lambda)C\bar{u}^T(\vec{p},\lambda)e^{ipx}\Big] , \qquad (3.13)$$

where $b(\vec{p},\lambda)$, $b^*(\vec{p},\lambda)$, $d(\vec{p},\lambda)$ and $d^*(\vec{p},\lambda)$ are creation and annihilation operators for particles and antiparticles. For Majorana spinors, $\psi_\mu = C\bar{\psi}^T_\mu$, the expansion is very similar, but

$d(\vec{p},\lambda) = b(\vec{p},\lambda)$, i.e. particles = antiparticles.

Let us now give a heuristic discussion of the normalization of action. The part of S containing time derivatives is

$$S = -i \int d^4x \;\bar{\psi}_i\gamma^{ioj}\dot{\psi}_j$$

$$= i \int d^4x \;\bar{\psi}_i\gamma^o\gamma^{ij}\dot{\psi}_j$$

$$= i \int d^4x \;\bar{\psi}_i\gamma^o(\gamma^i\gamma^j-\eta^{ij})\dot{\psi}_j$$

$$= i \int d^4x \;\bar{\psi}_i\gamma^o\dot{\psi}_i \;,$$

since $\gamma^i\dot{\psi}_j = o$ in our gauge and $\eta^{ij} = -\delta^{ij}$. Therefore the normalization of the ψ_i is the same as for a standard fermion field.

Exercise: Compute the canonical stress tensor for the action (3.3) and show that

$$P^\mu = \sum_{\lambda=\pm 3/2} \int d^3p \; p^\mu\Big(b^*(\vec{p},\lambda)b(\vec{p},\lambda)+d^*(\vec{p},\lambda)d(\vec{p},\lambda)\Big) \;.$$

The last topic of this section is a brief treatment of the covariant path integral quantization of the massless spin-3/2 field. The only justification will be by analogy with Yang-Mills theory. This is admittedly somewhat dangerous. For a more rigorous treatment readers are referred to refs. [7,8] in which the BRS invariance of supergravity is discussed.

Introduce the action with conserved sources $J^\mu_\alpha(x)$ $(\partial_\mu J^\mu=o)$:

$$S[\psi,J] = \int d^4x \left[-i\bar{\psi}_\lambda\,\gamma^{\lambda\nu\rho}\partial_\nu\psi_\rho+\bar{\psi}_\lambda J^\lambda+\bar{J}^\lambda\psi_\lambda\right] .$$

The generating functional is then

$$Z[J] = \int \prod_x db(x)d\bar{b}(x)d\psi_\rho(x)d\bar{\psi}_\lambda(x)$$

$$\exp\, i\left\{S[\psi,J] - \zeta \int d^4x \;\bar{b}(x)\partial\!\!\!/ b(x)\right\}$$

$$\prod_x \delta(F(\psi)-b(x))\delta(\bar{F}(\bar{\psi})-\bar{b}(x)) \quad \det{}^{-1}\frac{\delta F}{\delta\varepsilon}\det{}^{-1}\frac{\delta\bar{F}}{\delta\varepsilon} \;, \quad (3.14)$$

where $F_\alpha(\psi)$ is a nonsingular gauge condition, $b_\alpha(x)$ is a dummy fermion weighting field, and ζ is gauge parameter. Do the b, $\bar{b}$ integration and obtain

$$Z[J,\bar{J}] = \int \prod_x d\psi_\rho(x) d\bar{\psi}_\lambda(x) \exp i S'[\psi,J] \det{}^{-1} \frac{\delta F}{\delta \varepsilon} \det{}^{-1} \frac{\delta \bar{F}}{\delta \varepsilon} \tag{3.15}$$

where

$$S'[\psi,J] = S[\psi,J] - \zeta \int d^4x\, \bar{F}(\bar{\psi}(x)) i\partial\!\!\!/ F(\psi(x)) \ .$$

If one chooses a covariant gauge condition $F = i\gamma.\psi$, then

$$S' = -i \int d^4x\, \bar{\psi}_\lambda(x) \left[\gamma^{\lambda\nu\rho}\partial_\nu + \zeta\gamma^\lambda \partial\!\!\!/ \gamma^\rho\right] \psi_\rho(x) \ . \tag{3.16}$$

This defines an effective wave operator

$$W^{\lambda\rho} = -i\gamma^{\lambda\nu\rho}\partial_\nu + i\zeta\gamma^\lambda \partial\!\!\!/ \gamma^\rho \ . \tag{3.17}$$

Exercise: Use some γ-algebra to show that the ζ-independent part can be written as (in momentum space)

$$-\gamma^{\lambda\nu\rho} p_\nu = - p\!\!\!/ Q^{\lambda\rho} = Q^{\lambda\rho} p\!\!\!/$$

where

$$Q^{\lambda\rho} = \left(\eta^{\lambda\tau} - \frac{p^\lambda p^\tau}{p^2}\right) \left(\eta_{\tau\sigma} - \gamma_\tau\gamma_\sigma\right) \left(\eta^{\sigma\rho} - \frac{p^\sigma p^\rho}{p^2}\right) .$$

Clearly the gauge-invariant part of $W^{\lambda\rho}$ is not invertible, and it should not be for any gauge field. However $W^{\lambda\rho}$ is invertible and we can define the propagator $D_{\mu\nu}$ as the inverse

$$W^{\lambda\mu} D_{\mu\nu} = \delta^\lambda{}_\nu \ .$$

Exercise: Show that

$$D_{\mu\nu} = \frac{d_{\mu\nu}(p)}{p^2 + i\varepsilon}$$

with $-d_{\mu\nu} = \frac{1}{2}\, \gamma_\nu p\!\!\!/ \gamma_\mu + (2+\zeta^{-1}) \dfrac{p_\mu p\!\!\!/ p_\gamma}{p^2}$.

It is curious that the index order is reversed in the first term. For $\zeta = -\frac{1}{2}$, the propagator simplifies to one term. This is the gauge used in the early calculations of 1-loop finiteness in supergravity[9].

A proof of unitarity of a gauge field system requires as a necessary condition that the pole term of the propagator has a positive residue for conserved sources. In momentum space, we have

$$\frac{\gamma(p)}{p^2+i\varepsilon} = \frac{\bar{J}^\mu d_{\mu\nu} J^\nu}{p^2+i\varepsilon} .$$

One can show that the residue $\gamma(p)$ is positive and of rank 2 using a helicity decomposition of the conserved source plus some γ-matrix algebra.

In general, gauge theories are consistent only if the gauge fields couple to sources which are conserved. When interactions are present the sources are not external but are currents made up of other fields of the system. They are conserved because they are the Noether currents of the global symmetry which is gauged by the gauge field. Our discussion again confirms this general feature of gauge theories. In interacting supergravity theories, path integral quantization has qualitatively new features: the most striking of which are the 4-ghost interaction terms (in addition to the usual 2-ghost term. For details, see refs. 7 and 8.

4. TENSOR CALCULUS AND TORSION

In this section we assume that students are familiar with "ordinary" tensor calculus in Riemannian geometry. Prior exposure to the vierbein formalism will be useful. What we will stress are the subjects of local Lorentz invariance, torsion and first-order formalism which are important in the treatment of spinors in relativity and for the presentation and proof of invariance in supergravity.

In general relativity, the universe is viewed as a manifold M, usually four-dimensional, and the entire formalism is set up to be covariant under reparametrizations or diffeomorphisms. Specifically under an infinitesimal change of coordinates

$$x^\mu \to x'^\mu = x^\mu - \xi^\mu(x) , \qquad (4.1)$$

scalars, vectors and tensors transform as

$$\begin{aligned} \phi &\to \phi + \xi^\mu \partial_\mu \phi , \\ V^\mu &\to V^\mu + \xi^\rho \partial_\rho V^\mu - \partial_\rho \xi^\mu V^\rho , \\ V_\mu &\to V_\mu + \xi^\rho \partial_\rho V_\mu + \partial_\mu \xi^\rho V_\rho , \\ T_\mu{}^\nu &\to T_\mu{}^\nu + \xi^\rho \partial_\rho T_\mu{}^\nu + \partial_\mu \xi^\rho T_\rho{}^\nu - \partial_\rho \xi^\nu T_\mu{}^\rho , \end{aligned} \qquad (4.2)$$

respectively. In every case the transformation rule defines a so-called Lie derivative of the corresponding tensor, e.g.

$$\delta_G T_\mu{}^\nu(x) = \mathcal{L}_\xi T_\mu{}^\nu = \xi^\rho \partial_\rho T_\mu{}^\nu + \partial_\mu \xi^\rho T_\rho{}^\nu - \partial_\rho \xi^\nu T_\mu{}^\rho . \tag{4.3}$$

Lie derivatives are defined only in terms of the manifold structure of M, no connection is required. The Lie derivative of a tensor is still a tensor with the same rank.

Among the tensors defined on M, the covariant totally antisymmetric tensors ϕ, ω_μ, $\omega_{\mu\nu} = -\omega_{\nu\mu}$, etc. have a particularly useful structure. One can define differential forms

$$\begin{aligned} \omega^{(1)} &= \omega_\mu dx^\mu , \\ \omega^{(2)} &= \tfrac{1}{2}\omega_{\mu\nu} dx^\mu \wedge dx^\nu , \end{aligned} \tag{4.9}$$

and their wedge products and exterior derivatives

$$\begin{aligned} d\omega^{(1)} &= \partial_\mu \omega_\nu dx^\mu \wedge dx^\nu \\ &= \tfrac{1}{2} (\partial_\mu \omega_\nu - \partial_\nu \omega_\mu) dx^\mu \wedge dx^\nu , \\ d\omega^{(2)} &= \tfrac{1}{2} \partial_\rho \omega_{\mu\nu} dx^\rho \wedge dx^\mu \wedge dx^\nu \\ &= \tfrac{1}{6} (\partial_\rho \omega_{\mu\nu} + \partial_\mu \omega_{\nu\rho} + \partial_\nu \omega_{\rho\mu}) dx^\rho \wedge dx^\mu \wedge dx^\nu . \end{aligned} \tag{4.5}$$

We will not develop the exterior algebra and calculus of forms here, but one application of the formulae above will be needed. It should be stressed that in the definition of exterior derivative no connection is used. The d operation maps a p-form into a p+1-form, or a rank p antisymmetric tensor into a rank (p+1) antisymmetric tensor.

One of the quantities of primary importantce in general relativity is the metric tensor $g_{\mu\nu}$, in terms of which the line element and the inner product of vectors are defined as

$$ds^2 = g_{\mu\nu} dx^\mu dx^\nu \tag{4.6}$$

$$V.W = g_{\mu\nu} V^\mu W^\nu . \tag{4.7}$$

The metric is covariant, symmetric ($g_{\mu\nu} = g_{\nu\mu}$) and non-singular. Its inverse is $g^{\mu\nu}$ and

$$g^{\mu\nu} g_{\nu\rho} = \delta^\mu{}_\rho = g_{\rho\nu} g^{\nu\mu} . \tag{4.8}$$

The signature of $g_{\mu\nu}$ is (+ − − −). This means that by matrix algebra, g can be diagonalized by an orthogonal transformation

$$g_{\mu\nu} = O_{\mu a} D_{ab} O_{\nu b} ,$$

where $O^{-1}_{b\nu} = O_{\nu b}$, and

$$D_{ab} = \begin{bmatrix} \lambda^{(0)} & & & 0 \\ & -\lambda^{(1)} & & \\ & & -\lambda^{(2)} & \\ 0 & & & -\lambda^{(3)} \end{bmatrix} .$$

The nonsingularity implies $\lambda^{(a)} \neq o$, and signature $(+ - - -)$ means that $\lambda^{(a)} > o$, for all $a = 0,1,2,3$.

One also can define the vierbein:

$$e^a_{\ \mu} = \sqrt{\lambda^{(a)}}\, O_{\mu a} . \tag{4.9}$$

Then $g_{\mu\nu}$ can be expressed as

$$g_{\mu\nu} = e^a_{\ \mu} \eta_{ab} e^b_{\ \nu} . \tag{4.10}$$

where η_{ab} is the Minkowski metric.
Further $e^a_{\ \mu}$ has an inverse $e^\mu_{\ a}$ which satisfies

$$e_a^{\ \mu} e^b_{\ \mu} = \delta_a^{\ b} \quad \text{and} \quad e_a^{\ \mu} e^a_{\ \nu} = \delta^\mu_{\ \nu} . \tag{4.11}$$

One can easily show that

$$\begin{aligned} e_a^{\ \mu} &= g^{\mu\nu} \eta_{ab} e^b_{\ \nu} , \\ e_a \cdot e_b &= e_a^{\ \mu} g_{\mu\nu} e_b^{\ \nu} = \eta_{ab} . \end{aligned} \tag{4.12}$$

Geometrically, the $e_a^{\ \mu}$ form an orthonormal set of vectors in the tangent space of M at x. Any vector or tensor can be referred to an orthonormal frame, e.g.

$$\begin{aligned} V^\mu &= e_a^{\ \mu} V^a , & V^a &= e^a_{\ \mu} V^\mu , \\ W_\mu &= e^a_{\ \mu} W_a , & W_a &= e_a^{\ \mu} W_\mu , \\ V^\mu W_\mu &= V^a W_a . \end{aligned} \tag{4.13}$$

One can use $e^a_{\ \mu}$ and $e_a^{\ \mu}$ to go back and forth between the coordinate bases with indices $\mu, \nu, \rho \ldots$ and local Lorentz bases with indices $a, b, c \ldots$. One can also use $g_{\mu\nu}$ and $g^{\mu\nu}$ to raise and lower coordinates and use η_{ab} and η^{ab} to raise and lower Lorentz indices.

Given $g_{\mu\nu}(x)$ the vierbein $e^a_{\ \mu}$ is not uniquely determined. Any local Lorentz transformation

$$e'^a{}_\mu(x) = \Lambda^a{}_b(x) e^b{}_\mu(x) \tag{4.14}$$

leads to an equally good vierbein. Thus we extend the principle of general covariance, by requiring that the equations of a gravitational theory be covariant not only under changes of coordinate bases in the tangent space of M, b·t also under local Lorentz transformations of orthonormal bases of M. The local Lorentz invariance is an optional complication if we de·l only with tensors and vectors, but it is necessary to treat spi... s because spinors are defined by their transformation properties under Lorentz transformations.

Let us now summarize the local Lorentz transformation rules of the various fields we will deal with. A local Lorentz transformation is represented by

$$\Lambda(x) = \exp\left[\tfrac{1}{2}\lambda_{ab}(x) M^{ab}\right] , \tag{4.15}$$

and in any representation of the proper Lorentz group

$$\left[M^{ab}, M^{cd}\right] = \eta^{bc} M^{ad} - \eta^{ac} M^{bd} - \eta^{bd} M^{ac} + \eta^{ad} M^{bc} . \tag{4.16}$$

An infinitesimal transformation induces the variation δ_L in any field. For vectors,

$$\begin{aligned} (M^{ab})^c{}_d &= \eta^{bc}\delta^a{}_b - \eta^{ac}\delta^b{}_d , \\ \delta_L V^a(x) &= \lambda^a{}_b(x) V^b(x) , \\ \delta_L V_a(x) &= \lambda_a{}^b(x) V_b(x) , \end{aligned} \tag{4.17}$$

where

$$\begin{aligned} \lambda^a{}_b(x) &= \eta^{ac}\lambda_{cb}(x) , \\ \lambda_a{}^b(x) &= \eta^{bc}\lambda_{ac}(x) . \end{aligned}$$

For spinors,

$$\begin{aligned} M^{ab} &= \sigma^{ab} = \tfrac{1}{4}[\gamma^a, \gamma^b] , \\ \delta_L \psi(x) &= \tfrac{1}{2}\lambda_{ab}(x)\sigma^{ab}\psi(x) , \\ \delta_L \bar{\psi}(x) &= -\tfrac{1}{2}\bar{\psi}(x)\lambda_{ab}(x)\sigma^{ab} . \end{aligned} \tag{4.18}$$

Every field referred to local Lorentz frames transforms as a coordinate scalar, e.g.

$$\delta_G V^a(x) = \xi^\mu(x)\partial_\mu V^a(x)$$
$$\delta_G \psi(x) = \xi^\mu(x)\partial_\mu \psi(x) \quad . \tag{4.19}$$

The local Lorentz invariance is implemented like Yang-Mills gauge invariance with gauge group SO(3,1). In fact, one can define the spin connection $\omega_{\mu ab}$ as a set of 24 independent quantities which transform as

$$\delta_L \omega_{\mu ab} = \partial_\mu \lambda_{ab} + \lambda_a{}^c \omega_{\mu cb} + \lambda_b{}^c \omega_{\mu ac}$$
$$\delta_G \omega_{\mu ab} = \xi^\rho \partial_\rho \omega_{\mu ab} + \partial_\mu \xi^\rho \omega_{\rho ab} = \mathcal{L}_\xi \omega_{\mu ab} \ . \tag{4.20}$$

The connection defines the notion of parallel transport, so that one has Lorentz covariant derivatives

$$D_\mu \psi = (\partial_\mu + \tfrac{1}{2}\omega_{\mu ab}\sigma^{ab})\psi \ ,$$
$$(D_\mu V)^a = \partial_\mu V^a + \omega_\mu{}^a{}_b V^b \ , \tag{4.21}$$
$$(D_\mu V)_a = \partial_\mu V_a + \omega_{\mu a}{}^b V_b \ .$$

The SO(3,1) field strength is the curvature tensors:

$$R_{\mu\nu ab} = \partial_\mu \omega_{\nu ab} - \partial_\nu \omega_{\mu ab} + \omega_{\mu a}{}^c \omega_{\nu cb} - \omega_{\nu a}{}^c \omega_{\mu cb} \ . \tag{4.22}$$

From this definition one can derive the Ricci identity:

$$[D_\mu, D_\nu] = \tfrac{1}{2} R_{\mu\nu ab} M^{ab} \ ,$$
$$\text{e.g.}\ [D_\mu, D_\nu] V^a = R_{\mu\nu}{}^a{}_b V^b \ ,$$
$$[D_\mu, D_\nu] V_a = R_{\mu\nu}{}^a{}_b V^b \ , \tag{4.23}$$
$$[D_\mu, D_\nu]\psi = \tfrac{1}{2} R_{\mu\nu ab}\sigma^{ab}\psi ,$$

and the Bianchi identity

$$D_{[\mu} R_{\nu\rho]ab} = 0 \ . \tag{4.24}$$

All these objects above transform either as covariant vectors or as antisymmetric tensors under δ_G. (Note: $D_\mu D_\nu V_a$ is not a tensor but $D_{[\mu} D_{\nu]} V_a$ is.)

For vectors and tensors we can readily convert between coordinate and local Lorentz indices. For example, $D_\mu V^a$ is a local Lorentz vector and a world covariant vector. Hence

$$D_b V^a \equiv e_b^{\ \mu} D_\mu V^a$$

is a local Lorentz tensor ($T_b^{\ a}$) and a world scalar. One also defines the second rank covariant tensor

$$\begin{aligned}
\mathcal{D}_\mu V_\nu &\equiv e_{a\nu} D_\mu V^a \\
&= e_{a\nu} D_\mu (e^{a\rho} V_\rho) \\
&= e_{a\nu} (\partial_\mu e^{a\rho} + \omega_\mu{}^a{}_b e^{b\rho}) V_\rho + e_{a\nu} e^{a\rho} \partial_\mu V_\rho \\
&= \partial_\mu V_\nu + e_{a\nu} (D_\mu e^{a\rho}) V_\rho \\
&= \partial_\mu V_\nu - (e^{a\rho} D_\mu e_{a\nu}) V_\rho \ ,
\end{aligned} \tag{4.25}$$

since $\partial_\mu M^{-1} = -M^{-1} \partial_\mu M M^{-1}$. Since $\mathcal{D}_\mu V_\nu$ is a local Lorentz scalar and a world tensor, it can be identified with the world covariant derivative

$$\mathcal{D}_\mu V_\nu = \partial_\mu V_\nu - \Gamma_{\mu\nu}^{\ \ \rho} V_\rho \ , \tag{4.26}$$

with

$$\Gamma_{\mu\nu}^{\ \ \rho} \equiv e^{a\rho} D_\mu e_{a\nu} \ .$$

In general

$$\Gamma_{\mu\nu}^{\ \ \rho} \neq \Gamma_{\nu\mu}^{\ \ \rho} \ .$$

Similarly one can define

$$\mathcal{D}_\mu V^\nu \equiv e_a^{\ \nu} D_\mu V^a = \partial_\mu V^\nu + \Gamma_{\mu\rho}^{\ \ \nu} V^\rho \ . \tag{4.27}$$

Exercise: Show directly from the definitions that

(a) $\Gamma_{\mu\nu}^{\ \ \rho}$ transforms as the usual affine connection under δ_G

(b) $\mathcal{D}_\mu V_\nu$ transforms as $T_{\mu\nu}$

(c) $\mathcal{D}_\mu V^\nu$ transforms as $T_\mu^{\ \nu}$.

One can generalize this discussion to any tensor $T_{\mu\nu}^{\ \ \rho}$. One can define its local Lorentz derivatives $D_\lambda T_{ab}^{\ \ c}$ as above, and then its world covariant derivative

$$\mathcal{D}_\lambda T_{\mu\nu}^{\ \ \rho} \equiv e^a_{\ \mu} e^b_{\ \nu} e_c^{\ \rho} D_\lambda T_{ab}^{\ \ c} \ . \tag{4.28}$$

One finds that $\Gamma_{\mu\nu}^{\ \ \rho}$ appears in $\mathcal{D}_\lambda T_{\mu\nu}^{\ \ \rho}$ in the expected way, and that $\mathcal{D}_\lambda T_{\mu\nu}^{\ \ \rho}$ is a tensor.

It should be noted that $\Gamma_{\mu\nu}^{\rho}$ and $\omega_{\mu ab}$ really are equivalent. They express the same rule for parallel transport in the coordinate basis and local Lorentz basis, respectively.

For "mixed" quantities it is useful to preserve the two distinct derivatives, e.g. for a vector spinor $\psi_\mu(x)$,

$$D_\mu\psi_\nu(x) = \left(\partial_\mu + \tfrac{1}{2}\omega_{\mu ab}\sigma^{ab}\right)\psi_\nu \tag{4.29}$$

is a local Lorentz spinor, not a tensor, but

$$\mathfrak{D}_\mu\psi_\nu(x) = D_\mu\psi_\nu(x) - \Gamma_{\mu\nu}^{\ \ \nu}\psi_\rho \tag{4.30}$$

is a tensor.

The definition of $\Gamma_{\mu\nu}^{\ \ \rho}$ implies that

$$D_\mu e_{a\nu} = \partial_\mu e_{a\nu} + \omega_{\mu a}^{\ \ \ b} e_{b\nu} - \Gamma_{\mu\nu}^{\ \ \rho} e_{a\rho} = o\ . \tag{4.31}$$

This relation is also called the vierbein postulate. Defining

$$\Gamma_{\mu\nu,\rho} \equiv \Gamma_{\mu\nu}^{\ \ \sigma} g_{\sigma\rho} = e^a_{\ \rho} D_\mu e_{a\nu}\ ,$$

$$\Gamma_{\mu\rho,\nu} = \Gamma_{\mu\rho}^{\ \ \sigma} g_{\sigma\rho} = e^a_{\ \nu} D_\mu e_{a\rho}\ ,$$

we find that

$$\Gamma_{\mu\nu,\rho} + \Gamma_{\mu\rho,\nu} = \partial_\mu g_{\nu\rho}\ ,$$

or equivalently,

$$\mathfrak{D}_\mu g_{\nu\rho} \equiv \partial_\mu g_{\nu\rho} - \Gamma_{\mu\nu}^{\ \ \sigma} g_{\sigma\rho} - \Gamma_{\mu\rho}^{\ \ \sigma} g_{\nu\sigma} = o\ .$$

The last equation is called the metric postulate. This equation means that lengths and scalar products of vectors are preserved under parallel transport. Length preservation means that the tangent space group is O(3,1). The metric postulate also tells us that covariant differentiation commutes with index raising

$$\mathfrak{D}_\mu V^\nu = g^{\nu\rho}\mathfrak{D}_\mu V_\rho\ . \tag{4.32}$$

There exists an important identity: the generalized Ricci identity. By direct computation one finds

$$[\mathfrak{D}_\mu\ \mathfrak{D}_\nu]V^\rho = R_{\mu\nu\ \sigma}^{\ \ \ \rho} V^\sigma - S_{\mu\nu}^{\ \ \sigma} D_\sigma V^\sigma \tag{4.33}$$

where

$$R_{\mu\nu\ \sigma}^{\ \ \ \rho} = \partial_\mu\Gamma_{\nu\sigma}^{\ \ \rho} - \partial_\nu\Gamma_{\mu\sigma}^{\ \ \rho} + \Gamma_{\mu\tau}^{\ \ \rho}\Gamma_{\nu\sigma}^{\ \ \tau} - \Gamma_{\nu\tau}^{\ \ \rho}\Gamma_{\mu\sigma}^{\ \ \tau}\ ,$$

$$S_{\mu\nu}{}^{\sigma} = \Gamma_{\mu\nu}{}^{\sigma} - \Gamma_{\nu\mu}{}^{\sigma} . \qquad (4.34)$$

In fact, $R_{\mu\nu}{}^{\rho}{}_{\sigma}$ is simply the previous curvature tensor, expressed in a coordinate basis,

$$R_{\mu\nu}{}^{\rho}{}_{\sigma} = e^{a\rho} e^{b}{}_{\sigma} R_{\mu\nu ab} . \qquad (4.35)$$

Exercise: Prove this by evaluation of $[\mathfrak{D}_\mu, \mathfrak{D}_\nu] e_a{}^{\rho} = 0$.
$S^{\sigma}_{\mu\nu}$ is also a tensor, called the torsion tensor of the connection ω or Γ.

Exercise: Show that $\Gamma_{[\mu\nu]}{}^{\sigma}$ transforms as a tensor under δ_G.

The generalized Ricci identity can also be extended to all tensorial and local Lorentz quantities.

Exercise: Show that

$$\begin{aligned} [D_\mu, D_\nu] V_a &= R_{\mu\nu a}{}^{b} V_b - S_{\mu\nu}{}^{\sigma} D_\sigma V_b \\ [\mathfrak{D}_\mu, \mathfrak{D}_\nu] V_\rho &= R_{\mu\nu\rho}{}^{\sigma} V_\sigma - S_{\mu\nu}{}^{\sigma} \mathfrak{D}_\sigma V_\rho \\ [D_\mu, D_\nu] \psi &= \tfrac{1}{2} R_{\mu\nu ab} \sigma^{ab} \psi - S_{\mu\nu}{}^{\sigma} D_\sigma \psi . \end{aligned} \qquad (4.36)$$

The torsion tensor is an important geometric object. It may be written as

$$S_{\mu\nu}{}^{\sigma} = \Gamma_{\mu\nu}{}^{\sigma} - \Gamma_{\nu\mu}{}^{\sigma} = e^{a\sigma} (D_\mu e_{a\nu} - D_\nu e_{a\mu}) . \qquad (4.37)$$

Define

$$S_{\mu\nu\rho} = S_{\mu\nu}{}^{\sigma} g_{\sigma\rho}$$

and $\omega_{\mu\nu\rho} \equiv \omega_{\mu ab} e^{a}{}_{\nu} e^{b}{}_{\rho}$.

Then the above can be written as

$$S_{\mu\nu\rho} = \Omega_{\mu\nu\rho} + \omega_{\mu\rho\nu} - \omega_{\nu\rho\mu} , \qquad (4.38)$$

where

$$\Omega_{\mu\nu\rho} = e^{a}{}_{\rho} (\partial_\mu e_{a\nu} - \partial_\nu e_{a\mu}) .$$

One can write similar expressions for $-S_{\nu\rho\mu}$ and $S_{\rho\mu\nu}$ and add them up and solve to obtain

$$\begin{aligned} \omega_{\mu\nu\rho} &= \tfrac{1}{2} [\Omega_{\mu\nu\rho} - \Omega_{\nu\rho\mu} + \Omega_{\rho\mu\nu}] + \tfrac{1}{2} [S_{\nu\rho\mu} - S_{\mu\nu\rho} - S_{\rho\mu\nu}] \\ &\equiv \omega_{\mu ab}(e) e^{a}{}_{\nu} e^{b}{}_{\rho} + K_{\mu\nu\rho} , \end{aligned} \qquad (4.39)$$

where $\omega_{\mu ab}(e)$ is called the tensor-free or Riemannian spin connection and $K_{\mu\nu\rho}$ is the contortion tensor. Then one has

$$\omega_{\mu ab} = \omega_{\mu ab}(e) + K_{\mu ab} \,. \tag{4.40}$$

Exercise: From the definition of $\Gamma_{\mu\nu}{}^{\rho}$ show that

$$\Gamma_{\mu\nu}{}^{\rho} = \Gamma_{\mu\nu}{}^{\rho}(g) - K_{\mu\nu}{}^{\rho} \tag{4.41}$$

where $\Gamma_{\mu\nu}{}^{\rho}(g)$ is the Christoffel symbol. Also derive

$$K_{\mu\nu\rho} - K_{\nu\mu\rho} = -S_{\mu\nu\rho} \,. \tag{4.42}$$

In mathematical textbooks[10] the torsion and curvature of a connection are related to the frame 1-forms:

$e^a = e_\mu{}^a dx^\mu$, and the connection 1-forms:

$\omega^{ab} = \omega_\mu{}^{ab} dx^\mu$, by the Cartan structure equations:

$$\begin{aligned} de^a + \omega^a{}_b \wedge e^b &= S^a \equiv \tfrac{1}{2} S_{\mu\nu}{}^a dx^\mu \wedge dx^\nu \\ d\omega^a{}_b + \omega^a{}_c \wedge \omega^c{}_b &= R^a{}_b \equiv \tfrac{1}{2} R_{\mu\nu}{}^a{}_b dx^\mu \wedge dx^\nu \end{aligned} \tag{4.43}$$

where S^a and $R^a{}_b$ are the torsion 2-form and the curvature 2-form, respectively.

Exercise: Show that the first equation above is equivalent to

$S_{\mu\nu}{}^a = (D_\mu e_\nu{}^a - D_\nu e_\mu{}^a)$ and the second equation equivalent to $R_{\mu\nu}{}^a{}_b =$ $\partial_\mu \omega_\nu{}^a{}_b - \partial_\nu \omega_{\mu ab} + \omega_{\mu a}{}^c \omega_{\nu cb} - \omega_{\nu a}{}^c \omega_{\mu cb}$.

Thus our definitions are equivalent to those of mathematicians (for the tangent space group SO(3,1). In fact,

$$\omega_{\mu ab} = \omega_{\mu ab}(e) + K_{\mu ab}$$

is the general "solution" of $S^a = de^a + \omega^a{}_b \wedge e^b$.

Important exercise: Derive the Bianchi identities by applying the exterior derivative d to the equations (4.43) above for S^a and $R^a{}_b$. Obtain

$$\begin{aligned} R^a{}_b \wedge e^b &= dS^a + \omega^a{}_b \wedge S^b \\ dR^a{}_b + \omega^a{}_c \wedge R^c{}_b - R^a{}_c \wedge \omega^c{}_b &= o \,. \end{aligned} \tag{4.44}$$

In terms of components the results are:

$$R_{[\mu\nu\rho]}{}^{a} = - D_{[\mu}S_{\nu\rho]}{}^{a} ,$$
$$D_{[\mu}R_{\nu\rho]}{}^{a}{}_{b} = o , \qquad (4.45)$$

where $[\mu\nu\rho]$ = sum of three cyclic permutations. One should note that the usual first Bianchi identity for $R_{\mu\nu\rho}{}^{\sigma}$ is modified here by the torsion. Also the automatic covariantization of the Lie and exterior derivatives

$$\mathcal{L}_{\xi}V_{\mu} = \xi^{\rho}(\mathcal{D}_{\rho}V_{\mu}) + (\mathcal{D}_{\mu}\xi^{\rho})V_{\rho}$$
$$\partial_{\mu}V_{\nu} - \partial_{\nu}V_{\mu} = \mathcal{D}_{\mu}V_{\nu} - \mathcal{D}_{\nu}V_{\mu}$$

is true if torsion vanishes, but must be modified if $S_{\mu\nu}{}^{\rho} \neq o$. The symmetry property of the Ricci term $R_{\mu\nu} = R_{\mu\rho\nu}{}^{\rho} = R_{\nu\mu}$ and the pair exchange symmetry $R_{\mu\nu\lambda\rho} = R_{\lambda\rho\mu\nu}$ who fail when there is torsion.

Finally, we mention that the presentation given here is somewhat different from that of mathematicians. Motivated by physics, we introduced the metric at an early stage and thus prejudiced ourselves in favor of an SO(3,1) valued connection which preserves length under parallel transport. Mathematicians[10] typically organize their basic material as follows. First, a manifold M is abstractly defined, also definitions of vectors and tensors on M are given. Then one defines the Lie derivatives and exterior algebra and calculus, using only the manifold structure. A connection is introduced as an extra structure on M, which defines the notion of parallel transport. The connection is, in general, GL(4) valued and $\Gamma_{\mu\nu}{}^{\rho}$ and $\omega_{\mu}{}^{a}{}_{b}$ are the same object viewed from a coordinate frame and a general frame in T(M), respectively. The torsion and curvature are abstract properties of the connection, and the Cartan structure equations hold at this level. Finally, the metric is introduced as an additional structure, which defines the length of tangent vectors. One can then show that there is a unique torsion-free connection which preserves lengths under parallel transport. This is equivalent to $\Gamma_{\mu\nu}{}^{\rho}(g)$ and $\omega_{\mu ab}(e)$ above.

5. THE FIRST-ORDER FORMALISM FOR N = 0 AND N = 1 SUPERGRAVITY

In this section we apply the previous mathematical formalism to describe pure gravity and N = 1 supergravity. The action for pure gravity is

$$S_2[e,\omega] = \frac{-1}{4\kappa^2}\int d^4x\, eR$$
$$= \frac{-1}{4\kappa^2}\int d^4x\, e\, e^{a\mu}e^{b\nu}R_{\mu\nu ab}(\omega) \qquad (5.1)$$

where

$$\kappa^2 = 4\pi G ,$$

$$e = \det(e^c{}_\rho)$$

$$R_{\mu\nu ab} = \partial_\mu\omega_{\nu ab} - \partial_\nu\omega_{\mu ab} + \omega_{\mu a}{}^c\omega_{\nu cb} - \omega_{\nu a}{}^c\omega_{\mu cb} .$$

We assume only the transformation rules of the previous lecture for δ_G and δ_L which imply that S is a Lorentz and world scalar. Here $\omega_{\mu ab}$ and $e_{a\mu}$ are independent field variables, and the dynamics will determine the relation between them. We will work in four dimensions and use some tricks which simplify things there.

Let us define $\varepsilon^{\mu\nu\lambda\rho}$ as the usual totally antisymmetric tensor density ($\varepsilon^{0123} = 1$). It is a local Lorentz scalar. We also define $\hat{\varepsilon}^{abcd}$ (with $\hat{\varepsilon}^{0123} = 1$) as a totally antisymmetric frame tensor. It is a world scalar. Indices are lowered as one would expect, namely

$$\hat{\varepsilon}_a{}^{bcd} = \eta_{af}\,\hat{\varepsilon}^{fbcd} , \text{ etc.,}$$

and it is easy to show that

$$\hat{\varepsilon}^{abef}\,\hat{\varepsilon}_{cdef} = -2\left(\delta^a{}_c\delta^b{}_d - \delta^a{}_d\delta^b{}_c\right) . \tag{5.2}$$

The Greek and Latin ε's are related by the definition of the determinant

$$\det(e_f{}^\tau)\,\varepsilon^{\mu\nu\lambda\rho} = e_a{}^\mu e_b{}^\nu e_c{}^\lambda e_d{}^\rho\,\hat{\varepsilon}^{abcd} ,$$

$$\varepsilon^{\mu\nu\lambda\rho} = e\, e_a{}^\mu e_b{}^\nu e_c{}^\lambda e_d{}^\rho\,\hat{\varepsilon}^{abcd} , \tag{5.3}$$

since $\quad e = 1/\det(e_f{}^\tau) \quad .$

Exercise: Derive the identiy

$$\varepsilon^{\mu\nu\lambda\rho}\,\hat{\varepsilon}_{cdef}\, e_\lambda{}^e e_\rho{}^f = e\hat{\varepsilon}^{abef}\,\varepsilon_{cdef}\, e_a{}^\mu e_b{}^\nu = -2e\left(e^\mu{}_c e^\nu{}_d - e^\mu{}_d e^\nu{}_c\right) .$$

This leads to

$$\tfrac{1}{4}\varepsilon^{\mu\nu\lambda\rho}\,\hat{\varepsilon}_{abcd}\, e^c{}_\lambda e^d{}_\rho R_{\mu\nu}{}^{ab} = -e\, e^{a\mu}e^{b\nu}R_{\mu\nu ab} \tag{5.4}$$

which allows us to write the action $S_2[e,\omega]$ either in "eR form" or "$\varepsilon\hat{\varepsilon}$ form".

Now let us compute the variation of $S_2[e,\omega]$. Using the eR form of the action for $\delta e_{a\mu}$ terms and the $\varepsilon\hat{\varepsilon}$ form of the action for $\delta\omega_{\mu ab}$ terms. Then we have

$$\delta S_2[e,\omega] = \frac{1}{2\kappa^2}\int d^4x\, e\left\{R^{a\mu}(\omega)-\tfrac{1}{2}e^{a\mu}R(\omega)\right\}\delta e_{a\mu}$$

$$+\frac{1}{16\kappa^2}\int d^4x\, \varepsilon^{\mu\nu\lambda\rho}\,\hat{\varepsilon}_{abcd}\, e^c{}_\lambda\, e^d{}_\rho\left(D_\mu\delta\omega_\nu{}^{ab}-D_\nu\delta\omega_\mu{}^{ab}\right) \tag{5.5}$$

where we have used

$$\begin{aligned}
\delta e^{a\mu} &= -e^{a\rho}\,\delta e_{c\rho}\, e^{c\mu} \quad ,\\
\delta e &= e\, e^{a\mu}\,\delta e_{a\mu} \quad ,\\
R^{a\mu}(\omega) &= e^{a\lambda}\, R_{\lambda\gamma cd}(\omega) e^{c\mu} e^{d\gamma} \quad ,\\
R(\omega) &= e_{a\mu}\, R^{a\mu}(\omega) \quad .
\end{aligned} \tag{5.6}$$

Thus the equation of motion for $e_{a\mu}$ is a nonsymmetric form of the Einstein equation

$$R^{a\mu}(\omega) - \tfrac{1}{2}e^{a\mu}R(\omega) = 0 \ . \tag{5.7}$$

In the second term of (5.5) the D_μ can be integrated by parts, since $\hat{\varepsilon}_{abcd}$ is an invariant proper Lorentz tensor and $\varepsilon^{\mu\nu\lambda\rho}$ is a Lorentz scalar. Thus

$$\delta S_2[e,\omega]_{\text{2-nd term}} = \frac{1}{4\kappa^2}\int d^4x\, \varepsilon^{\mu\nu\lambda\rho}\,\hat{\varepsilon}_{abcd}\, e^c{}_\lambda\left(D_\nu e^d{}_\rho\right)\delta\omega_\mu{}^{ab} \quad ,$$

and

$$\frac{\delta S_2}{\delta\omega_\mu{}^{ab}} = \frac{1}{4\kappa^2}\,\varepsilon^{\mu\nu\lambda\rho}\,\hat{\varepsilon}_{abcd}\, e^c{}_\lambda\, D_\nu e^d{}_\rho = 0 \ . \tag{5.8}$$

is the equation of motion for the connection. Multiply this by $\varepsilon_{\mu\alpha\beta\gamma}$ and $\hat{\varepsilon}^{ab}{}_{ef}$, use ε identities and change indices to get

$$e^{[a}_{[\rho}\left(D_\nu e^{b]}_{\nu]}-D_\nu e^{b]}_{\mu]}\right) = 0 \ . \tag{5.9}$$

In this way we get 24 equations, which is what we expect for the Euler-Lagrange equations of $\delta\omega_\mu^{ab}$. They are all independent linear equations, hence we get 24 conditions

$$D_\mu e^a_\mu - D_\nu e^a_\mu = 0 \ . \tag{5.10}$$

Exercise: Derive (5.10) algebraically from (5.9) by contracting (5.9) with e^ρ_b and then contracting the resulting equation with e^ν_a.

From the work of the previous section we see that the unique solution of (5.10) is

$$\omega_{\mu ab} = \omega_{\mu ab}(e) \ . \tag{5.11}$$

Therefore, the first-order formalism of pure gravity leads to a torsion-free and length preserving connection.

Exercise: Insert $\omega(e)$ in the Einstein equation (5.7), multiply by e^{ν}_{a} and derive the conventional Einstein equation

$$R^{\mu\nu}(g) - \tfrac{1}{2}g^{\mu\nu}R(g) = 0 \ , \tag{5.12}$$

where $R^{\mu\nu}(g)$ is the usual symmetric Ricci tensor.

Now we add matter to the system in the form of a Majorana spin-3/2 field $\psi_{\rho}(x)$ coupled to gravity. The action is

$$S_{3/2}[e,\omega,\psi] = -\tfrac{1}{2}\int d^4x \ \varepsilon^{\mu\nu\lambda\rho} \ \bar{\psi}_{\lambda}\gamma_5\gamma_{\mu}D_{\nu}\psi_{\rho} \ . \tag{5.13}$$

This is a "minimal coupling" of the free spin-3/2 action of Section 3 to gravity. Here

$$\gamma_{\mu} = \gamma^{a}e_{a\mu}$$

$$\text{and} \quad D_{\nu}\psi_{\rho} = \partial_{\nu}\psi_{\rho} + \tfrac{1}{2}\omega_{\nu ab}\sigma^{ab}\psi_{\rho} \ . \tag{5.14}$$

Note that $D_{\nu}\psi_{\rho}$ gives a tensorial action because of $[\nu\rho]$ antisymmetrization. We might have started with

$$\mathcal{D}_{\mu}\psi_{\rho} = D_{\mu}\psi_{\rho} - \Gamma_{\mu\rho}^{\ \ \sigma}\psi_{\sigma}$$

which gives an action that differs by the torsion contribution $S_{\nu\rho}^{\sigma} = \Gamma^{\sigma}_{[\nu\rho]}$. This choice is also covariant, but it does not lead to a correct theory of supergravity.

Let us concentrate on the equation of motion for the connection which gets a contribution

$$\frac{\delta S_{3/2}(e,\omega,\psi)}{\delta\omega_{\mu}^{\ ab}} = \tfrac{1}{4}\varepsilon^{\mu\nu\lambda\rho} \ \bar{\psi}_{\lambda}\gamma_5\gamma_{\nu}\sigma_{ab}\psi_{\rho} \ . \tag{5.15}$$

Now use some γ-matrix algebra,

$$\begin{aligned}\gamma_5\gamma_c\sigma_{ab} &= \gamma_5 \ \tfrac{1}{2}[\gamma_c,\sigma_{ab}] + \tfrac{1}{2} \ \gamma_5 \ \{\gamma_c,\sigma_{ab}\} \\ &= \tfrac{1}{2} \ \gamma_5(\eta_{ac}\gamma_b - \eta_{bc}\gamma_a) + \tfrac{1}{2} \ \gamma_5\gamma_{abc} \ .\end{aligned} \tag{5.16}$$

$$\text{But} \quad \tfrac{1}{2} \ \gamma_5\gamma_{abc} = -\frac{i}{2} \ \hat{\varepsilon}_{abcd}\gamma^{d}$$

and $\bar{\psi}_{[\lambda}\gamma_5\gamma_a\psi_{\rho]} = o$, for a Majorana spinor.

Hence

$$\frac{\delta S_{3/2}}{\delta\omega_\mu^{ab}} = \frac{i}{8}\,\varepsilon^{\mu\nu\lambda\rho}\,\hat{\varepsilon}_{abcd}\,e^c_\lambda\bar{\psi}_\gamma\gamma^d\psi_\rho \;. \tag{5.17}$$

Combine (5.17) with the previous variation (5.8) of the gravity action to get

$$\frac{\delta(S_2+S_{3/2})}{\delta\omega_\mu^{ab}} = \frac{1}{8\kappa^2}\,\varepsilon^{\mu\nu\lambda\rho}\,\hat{\varepsilon}_{abcd}\,e^c_\lambda\left(D_\nu e^d_\rho - D_\rho e^d_\nu + i\kappa^2\,\bar{\psi}_\nu\gamma^d\psi_\rho\right) = o.$$

By the same arguments as previous, these 24 equations imply

$$D_\mu e^a_\nu - D_\nu e^a_\mu = -i\kappa^2\,\bar{\psi}_\mu\gamma^a\psi_\nu \;. \tag{5.19}$$

By the definition of torsion, we find the torsion tensor

$$S_{\mu\nu}{}^a = -i\kappa^2\,\bar{\psi}_\mu\gamma^a\psi_\nu \;, \tag{5.20}$$

and the connection

$$\omega_{\mu ab}(e,\psi) = \omega_{\mu ab}(e) + K_{\mu ab} \;, \tag{5.21}$$

where $K_{\mu\nu}{}^a$ is the contortion tensor computed from $S_{\mu\nu}{}^a$ as in the previous section.

Therefore, we see that in the first-order formalism, the presence of matter, usually spinor matter, automatically gives a connection with torsion.

The other equations of motion of supergravity system can be derived from $S_2 + S_{3/2}$. The vierbein variation is

$$\frac{\delta(S_2+S_{3/2})}{\delta e_{a\mu}} = \frac{1}{2\kappa^2}\,e[R^{a\mu}(\omega)-\tfrac{1}{2}e^{a\mu}R(\omega)] -$$

$$- \tfrac{1}{2}\varepsilon^{\mu\nu\lambda\rho}\,\bar{\psi}_\lambda\gamma_5\gamma^a D_\nu\psi_\rho = o \;. \tag{5.22}$$

Insert $\omega = \omega(e) + K$ into (5.22) and multiply it by $e_a{}^\nu$ to get the Einstein equation with source $T^{\nu\mu}$, which is quadratic and quartic in ψ_ρ.

For the equations of motion of ψ_ρ we just compute

$$\delta(S_2+S_{3/2}) = -\tfrac{1}{2}\int d^4x\ \varepsilon^{\lambda\mu\nu\rho}\,[\delta\bar\psi_\lambda\gamma_5\gamma_\mu D_\nu\psi_\rho+\bar\psi_\lambda\gamma_5\gamma_\mu D_\nu\delta\psi_\rho]\ . \tag{5.23}$$

Keeping in mind the anti-symmetrization enforced by $\varepsilon^{\lambda\mu\nu\rho}$, the second term may be manipulated as follows:

$$\begin{aligned}
\bar\psi_\lambda\gamma_5\gamma_\mu D_\nu\delta\psi_\rho &= -\delta\bar\psi_\lambda\overleftarrow{D}_\nu\gamma_5\gamma_\mu\psi_\rho \\
&= \delta\bar\psi_\lambda\gamma_5 D_\nu\gamma_\mu\psi_\rho\ \text{ + total derivatives} \\
&= \delta\bar\psi_\lambda\gamma_5(\gamma_\mu D_\nu\psi_\rho\ +\ (D_\nu e_{a\mu})\gamma^a\psi_\rho) \\
&= \delta\bar\psi_\lambda\gamma_5(\gamma_\mu D_\nu\psi_\rho-\tfrac{1}{2}S_{\mu\nu}{}^{\sigma}\,\gamma_\sigma\psi_\rho)\ .
\end{aligned} \tag{5.24}$$

Inserting this we find

$$\delta(S_2+S_{3/2}) = \int d^4x\ \delta\bar\psi_\lambda\ \varepsilon^{\lambda\mu\nu\rho}\ \gamma_5(\gamma_\mu D_\nu\psi_\rho-\tfrac{1}{2}S_{\mu\nu}{}^{\sigma}\,\gamma_\sigma\psi_\rho)\ . \tag{5.25}$$

This shows explicitly that caution must be used when integrating local Lorentz derivatives D_ν by parts. The same is true for the total covariant derivative $\mathcal{D}_\nu$ in the first order formalism. In general there are extra torsion terms. When $S_{\mu\nu}{}^{\sigma} \sim \bar\psi_\mu\gamma^\sigma\psi_\nu$ is inserted in (5.25) the explicit torsion term vanishes by a Fierz rearrangement (Fierz transformations will be discussed in the next section) and the equation of motion becomes

$$\varepsilon^{\lambda\mu\nu\rho}\ \gamma_5\gamma_\mu D_\nu(\omega(e)+K)\psi_\rho = o\ , \tag{5.26}$$

which includes linear and trilinear terms in ψ.

In the final form of the field equation, namely (5.22) and (5.26) only the fields $e_{a\mu}$ and ψ_μ appear. There is also an action which leads directly to these equations (with some rearrangement of terms). This is the so-called "second-order" form of the supergravity action (and the form first obtained in 1976). One simply substitutes the explicit connection $\omega_{\mu ab}(e,\psi)$ of (5.21) into the first order action to get

$$\begin{aligned}
S[e,\psi] &= S_2[e,\omega(e,\psi)] + S_{3/2}[e,\omega(e,\psi),\psi] \\
&= -\frac{1}{4\kappa^2}\int d^4x\, e\, e^{a\mu}e^{b\nu}R_{\mu\nu ab}(\omega(e)) \\
&\quad -\tfrac{1}{2}\int d^4x\ \varepsilon^{\mu\nu\lambda\rho}\ \bar\psi_\lambda\gamma_5\gamma_\mu D_\nu\psi_\rho \\
&\quad -\frac{1}{16}\kappa^2\int d^4x\ e\Big[(\bar\psi^\lambda\gamma^\mu\psi^\rho)(\bar\psi_\lambda\gamma_\mu\psi_\rho+2\bar\psi_\mu\gamma_\lambda\psi_\rho)-4(\bar\psi\cdot\gamma\psi_\sigma)^2\Big].
\end{aligned} \tag{5.27}$$

The first term is the standard action of general relativity, and the second is a minimal coupling of ψ_μ with

$$D_\mu\psi_\rho = \left(\partial_\mu + \tfrac{1}{2}\omega_{\mu ab}(e)\sigma^{ab}\right)\psi_\rho .$$

Including the Christoffel term is optional here because $\Gamma_{\mu\rho}{}^{\tau}(e)$ is symmetric. The last expression contains the fourth order contact terms which are the residual physical effects of the torsion. These terms lead to Feynman graph vertices which are absolutely necessary to include. Without them the S-matrix elements in flat-space would not be Lorentz invariant. Indeed the second order form of the action is better suited than the first-order form for computing Feynman amplitudes.

6. GAUGE INVARIANCE OF THE N = 1 SUPERGRAVITY ACTION

In supergravity the spin-3/2 field ψ_ρ is treated as the gauge field for supersymmetry transformations. In this lecture we will give a proof of the gauge invariance of the N = 1 supergravity action.

First, let us see why a gauge interaction of ψ_ρ is needed. Suppose one takes the free kinetic Lagrangian for spin-3/2 field and adds some fairly standard interaction in flat space, e.g. a Yukawa interaction with a scalar field,

$$\mathcal{L} = -\varepsilon^{\lambda\mu\nu\rho}\,\bar{\psi}_\lambda\gamma_5\gamma_\mu\partial_\nu\psi_\rho + \tfrac{1}{2}(\partial_\mu\varphi)^2 + g\bar{\psi}^\mu\psi_\mu\varphi .$$

Then it is well known that the resulting theory is pathological: there is acausal propagation[11] and the canonical commutation relations are inconsistent with the constraint structure[12]. The heuristic reason for the problems is that the highest derivative terms of wave operator are "non-invertible", because of the gauge invariance of the free field, but the interaction is not gauge invariant. This suggests that if we can introduce interactions which are consistent with the free field gauge invariance, then after gauge fixing we will have a system of field equations, with an invertible wave operator and $1/p^2$ propagators and thus causal propagation.

In supergravity, since the supersymmetry algebra $\{Q,\bar{Q}\} \sim P$ brings in translations, it is natural that the full gauge system must involve at least ψ_ρ and $e_{a\mu}$. This is consistent with the fact that the algebra has an irreducible representation involving massless particles of spin-3/2 and spin-2.

We will present a proof of gauge invariance, which takes advantage of some of the modern improvements that have been developed since 1976 and also takes advantage of the machinery developed in connection with our treatment of torsion and the equations of motion.

Let us write the action again as

$$S_{SG}[e,\psi,\omega] = \int d^4x\ \varepsilon^{\mu\nu\lambda\rho} \left[\frac{1}{16\kappa^2} \varepsilon_{abcd}\ e^c_\lambda e^d_\rho R_{\mu\nu}{}^{ab}(\omega) - \right.$$

$$\left. - \tfrac{1}{2}\ \bar{\psi}_\lambda \gamma_5 \gamma_\mu D_\nu \psi_\rho \right] . \tag{6.1}$$

In this action we envisage substituting

$$\omega_{\mu ab} = \omega_{\mu ab}(e) + K_{\mu ab}$$

with $S_{\mu\nu}{}^a = -i\kappa^2\ \bar{\psi}_\mu \gamma^a \psi_\nu$,

so that we will really have only two independent fields, $e_{a\mu}$ and ψ_μ, and we really are dealing with the second-order action (5.27). However, do *not* substitute explicitly and work out the torsion terms as was done in the previous section. We will show that the action is invariant under the following local supersymmetry transformation rules of the independent fields:

$$\delta e_{a\mu} = -i\kappa\bar{\varepsilon}\gamma_a \psi_\mu$$
$$\delta\psi_\rho = \frac{1}{\kappa} D_\mu(\omega)\varepsilon = \frac{1}{\kappa}\left(\partial_\mu + \omega_{\mu ab}\sigma^{ab}\right)\varepsilon , \tag{6.2}$$

where $\varepsilon(x)$ is an arbitrary anticommuting spinorial gauge parameter. Again we envisage substituting $\omega = \omega(e) + K$ in $\delta\psi$.

To start the proof of invariance, write the formal variation as

$$\delta S[e,\psi,\omega(e,\psi)] = \int d^4x \left\{\delta e_{a\mu}(x) \frac{\delta S}{\delta e_{a\mu}(x)} + \delta\bar{\psi}_\lambda \frac{\delta S}{\delta\bar{\psi}_\lambda}\right.$$

$$\left. + \frac{\delta S}{\delta\omega_{\gamma cd}} \left(\frac{\delta\omega_{\gamma cd}}{\delta e_{a\mu}} \delta e_{a\mu} + \delta\bar{\psi}_\lambda \frac{\delta\omega_{\gamma cd}}{\delta\bar{\psi}_\lambda}\right)\right\} . \tag{6.3}$$

When we substitute $\omega = \omega(e) + K$, the last term vanishes, since

$\left.\frac{\delta S}{\delta\omega}\right|_{\omega=\omega(e)+K} = o$ by definition. Thus we need consider the first two terms, which are the explicit variations of the first order action with respect to $e_{a\mu}$ and $\bar{\psi}_\lambda$.

This device of using the connection equation of motion to eliminate the third term is the so-called "1.5-order" formalism. It is really second-order, since only the physical fields $e_{a\mu}$ and ψ_ρ are taken as independent variables. But it uses the fact that the action has a simple first-order form, and it retains the grouping of terms ($\omega=\omega(e)+K$) suggested by that form[13]. It should be stressed that it is correct to use the equation of motion only for an auxiliary field such as ω in a test of invariance.

Now use the variations $\frac{\delta S}{\delta e}$ and $\frac{\delta S}{\delta \bar{\psi}}$ already calculated in the last lecture:

$$\frac{\delta S}{\delta e_{a\mu}} = \frac{1}{2\kappa^2}\, e\left[R^{a\mu}(\omega)-\tfrac{1}{2}e^{a\mu}R(\omega)\right] - \tfrac{1}{2}\varepsilon^{\mu\nu\lambda\rho}\ \psi_\lambda\gamma_5\gamma^a D_\nu(\omega)\psi_\rho\ ,$$
$$\frac{\delta S}{\delta\bar{\psi}_\lambda} = -\varepsilon^{\mu\nu\lambda\rho}\ \gamma_5\left[\gamma_\mu D_\nu(\omega)\psi_\rho - \tfrac{1}{4}S_{\mu\nu}{}^{\sigma}\gamma_\sigma\psi_\rho\right]. \tag{6.4}$$

Substitute these results and the transformation rules in the formal variations and get

$$\delta S = \int d^4x \left\{(-i\kappa\bar{\varepsilon}\gamma_a\psi_\mu)\right\}\left\{\frac{e}{2\kappa^2}\underbrace{\left[R^{a\mu}-\tfrac{1}{2}e^{a\mu}R\right]}_{①} -\right.$$
$$- \tfrac{1}{2}\varepsilon^{\mu\nu\lambda\rho}\ \underbrace{\bar{\psi}_\lambda\gamma_5\gamma^a D_\nu\psi_\rho}_{②} \tag{6.5}$$
$$\left.- \frac{1}{\kappa}\varepsilon^{\lambda\mu\nu\rho}\ \bar{\varepsilon}\overleftarrow{D}_\lambda\gamma_5\left[\underbrace{\gamma_\mu D_\nu\psi_\rho}_{③} - \underbrace{\tfrac{1}{4}S_{\mu\nu}{}^{\sigma}\gamma_\sigma\psi_\rho}_{④}\right]\right\}_{\omega=\omega(e)+K}$$

where we have labelled four terms ① - ④ which will be discussed separately. Our task is to show that (6.5) vanishes which will take several pages of exhausting manipulations.

Use the partial integration technique discussed above (5.11) to obtain

$$③ = \frac{1}{\kappa}\varepsilon^{\lambda\mu\nu\rho}\ \bar{\varepsilon}\gamma_5 D_\lambda(\gamma_\mu D_\nu\psi_\rho) + \text{total derivative}$$
$$= \underbrace{\frac{1}{2\kappa}\varepsilon^{\lambda\mu\nu\rho}\ \bar{\varepsilon}\gamma_5\gamma_\mu[D_\lambda,D_\nu]\psi_\rho}_{③a} + \underbrace{\frac{1}{2\kappa}\varepsilon^{\lambda\mu\nu\rho}\ S_{\lambda\mu}{}^{a}\bar{\varepsilon}\gamma_5\gamma_a D_\nu\psi_\rho}_{③b}\ . \tag{6.6}$$

If we now use the Ricci identity of Section 4, and the same γ-matrix algebra for $\gamma_\mu\sigma^{ab}$ used in Section 5, we can write the first term as

$$③a = \frac{1}{4\kappa}\varepsilon^{\lambda\mu\nu\rho}\ \bar{\varepsilon}\gamma_5\gamma_\mu\sigma^{ab}\psi_\rho R_{\mu\nu ab}\psi_\rho$$
$$= \frac{1}{4\kappa}\varepsilon^{\lambda\mu\nu\rho}\left[\underbrace{\tfrac{1}{2}\gamma_5\left(e_\mu{}^a\gamma^b - e_\mu{}^b\gamma^a\right)}_{③a-A} - \underbrace{\tfrac{i}{2}\hat{\varepsilon}_\mu{}^{abc}\gamma_c}_{③a-V}\right]_{\mu\nu ab}\psi_\rho\ . \tag{6.7}$$

Furthermore (with $\hat{\varepsilon}_{\mu abc} = e^d{}_\mu\,\varepsilon_{dabc}$)

$$③a\text{-}V = \frac{-i}{8\kappa}\varepsilon^{\mu\nu\lambda\rho}\,\hat{\varepsilon}_{\mu abc}\,\bar{\varepsilon}\,\gamma^{c}\psi_{\rho}R_{\lambda\nu}{}^{ab}$$

$$= \frac{i}{8\kappa}e\left(e_a{}^{\nu}e_b{}^{\lambda}e_c{}^{\rho} + e_b{}^{\nu}e_c{}^{\lambda}e_d{}^{\rho} + e_c{}^{\nu}e_a{}^{\lambda}e_b{}^{\rho}\right)\bar{\varepsilon}\gamma^{c}\psi_{\rho}R_{\lambda\nu}{}^{ab}$$

$$= \frac{ie}{4\kappa}\left(-e_c{}^{\rho}R + R_c{}^{\rho} + R_c{}^{\rho}\right)\bar{\varepsilon}\,\gamma^{c}\psi_{\rho}$$

$$= \frac{i}{2\kappa}e\left(R^{a\mu} - \tfrac{1}{2}e^{a\mu}\right)\bar{\varepsilon}\,\gamma^{a}\psi_{\mu} = -\;①\;. \tag{6.8}$$

To get from the first to the second line we use an ε-contraction identity similar to one discussed in section 5. Notice that the first significant cancellation, namely between (3a-V) and (1), has now occurred.

$$③a\text{-}A = \frac{1}{4\kappa}\varepsilon^{\lambda\mu\nu\rho}\,\bar{\varepsilon}\,\gamma_5\gamma_b\psi_{\rho}R_{\lambda\nu\mu}{}^{b}$$

$$= \frac{1}{4\kappa}\varepsilon^{\lambda\mu\nu\rho}\left(\bar{\varepsilon}\gamma_5\gamma_b\psi_{\rho}\right)D_{\lambda}S_{\mu\nu}{}^{b}$$

$$= -\frac{i}{2}\kappa\;\varepsilon^{\lambda\mu\nu\rho}\left(\bar{\varepsilon}\gamma_5\gamma_b\psi_{\rho}\right)\bar{\psi}_{\mu}\gamma^{b}D_{\lambda}\psi_{\nu}\;. \tag{6.9}$$

Here we have used the modified first Bianchi identity presented in Section 4 and

$$S_{\mu\nu}{}^{b} = -i\kappa^2\;\bar{\psi}_{\mu}\gamma^{b}\psi_{\nu}$$

$$D_{\lambda}S_{\mu\nu}{}^{b} = -2i\kappa^2\;\bar{\psi}_{\mu}\gamma^{b}D_{\lambda}\psi_{\nu}\;. \tag{6.10}$$

There are now three terms left with the general structure $\varepsilon\, D\psi\,\psi\psi$, namely (2) of (6.5), (3b) of (6.6) after insertion of the torsion, and (3a-A) of (6.9). We will show that the sum of these three terms vanishes because of the Fierz rearrangement properties of the Dirac matrices. There is also one term, namely (4) of (6.5) of the structure $\varepsilon\,\overleftarrow{D}\,\psi\psi\psi$, and this may be shown to actually vanish by the Fierz technique.

The Fierz rearrangement technique works as follows. The 16 matrices Γ_A of Section 1 are a complete set of 4x4 matrices. Therefore the quantity $\delta_{\alpha\beta}\delta_{\gamma\delta}$ can be viewed as a "matrix in γβ" and expanded as

$$\delta_{\alpha\beta}\delta_{\gamma\delta} = \sum_A b_A(\alpha,\delta)\Gamma_{A\gamma\beta}\;.$$

Of course the coefficients $b_A(\alpha,\delta)$ can be viewed as "matrices in

$\alpha\delta$" and expanded in the Γ_A again. The full expansion takes the form

$$\delta_{\alpha\beta}\delta_{\gamma\delta} = \tfrac{1}{4}\Big[a\delta_{\alpha\delta}\delta_{\gamma\beta}+b\gamma^p{}_{\alpha\delta}\gamma_{p\gamma\beta}+c\sigma^{pq}{}_{\alpha\delta}\sigma_{pq\gamma\beta}$$
$$+ d\left(\gamma_5\gamma^p\right)_{\alpha\delta}\left(\gamma_5\gamma_p\right)_{\gamma\beta}+e\gamma_{5\alpha\delta}\gamma_{5\gamma\beta}\Big] \qquad (6.11)$$

where $\frac{1}{4}$ is put in for convenience and a–e must be numerical coefficients. By trace techniques, i.e. by multiplying both sides of (6.11) by matrices $\delta_{\beta\gamma}$, $\gamma^q{}_{\beta\gamma}$ etc., one finds the coefficients a=1, b=1, c=−2, d=−1, e=1. When (6.11) is used to reorder spinor expressions an extra (−) sign appears due to anticommutativity of the spinor fields.

Let we apply (6.11) to term (3a−A) of (6.9) and reorder so that the bilinears $\bar\psi_\mu\Gamma_A\psi_\rho$ occur where Γ_A is a generic matrix in (6.11). Due to the anti-symmetrization in μ and ρ and to Majorana property, only $\bar\psi_\mu\gamma^p\psi_\rho$ and $\bar\psi_\mu\sigma^{pq}\psi_\rho$ and constitute. However the expressions $\gamma_b\gamma_p\gamma^b = -2\gamma_p$ and $\gamma_b\sigma_{pq}\gamma^b = o$ (valid in four dimensions only!) then occur between $\bar\varepsilon\gamma_5$ and $D_\lambda\psi_\nu$ and kill the tensor term. Hence one can write

$$\text{(3a−A)} = \frac{-i}{4}\kappa\,\varepsilon^{\lambda\mu\nu\rho}\left(\bar\varepsilon\gamma_5\gamma_p D_\lambda\psi_\nu\right)\left(\bar\psi_\mu\gamma^p\psi_\rho\right)$$
$$= \frac{i}{4}\kappa\,\varepsilon^{\lambda\mu\nu\rho}\left(\bar\varepsilon\gamma_5\gamma_p D_\lambda\psi_\mu\right)\left(\bar\psi_\nu\gamma^p\psi_\rho\right)\,. \qquad (6.12)$$

Similar remarks apply to (2) of (6.5), and it rearranges as

$$\text{(2)} = \frac{i}{4}\kappa\,\varepsilon^{\lambda\mu\nu\rho}\left(\bar\varepsilon\gamma_5\gamma_p D_\nu\psi_\rho\right)\left(\bar\psi_\lambda\gamma^p\psi_\mu\right)$$
$$= \frac{i}{4}\kappa\,\varepsilon^{\lambda\mu\nu\rho}\left(\varepsilon\gamma_5\gamma_p D_\nu\psi_\rho\right)\left(\psi_\nu\gamma^p\psi_\rho\right)\,. \qquad (6.13)$$

Now we insert the explicit torsion tensor of (6.10) in term 3b of (6.6) and observe, after some dummy index changes, that it is equal and opposite to the sum of (6.12) and (6.13). Hence

$$\text{(3a−A)} + \text{(3b)} + \text{(2)} = o\,. \qquad (6.14)$$

The final step is to make that term (4) may be reordered as

$$\text{(4)} \sim \varepsilon^{\lambda\mu\nu\rho}\,\gamma_b\psi_\rho\bar\psi_\mu\gamma^b\psi_\nu$$

$$= -\tfrac{1}{4}\varepsilon^{\lambda\mu\nu\rho}\ \gamma_b\gamma_p\gamma^b\psi_\nu\bar{\psi}_\mu\gamma^p\psi_\rho$$

$$= \tfrac{1}{2}\varepsilon^{\lambda\mu\nu\rho}\ \gamma_p\psi_\nu\bar{\psi}_\mu\gamma^p\psi_\rho$$

$$= -\tfrac{1}{2}\varepsilon^{\lambda\mu\nu\rho}\ \gamma_p\psi_\rho\bar{\psi}_\mu\gamma^p\psi_\nu\ . \qquad (6.15)$$

Since the last term is equal to $-\frac{1}{2}$ times what we started with, it must vanish. Hence ④ = o, and the proof of gauge invariance is complete!

7. SUPERSYMMETRY IN ANTI-DE SITTER SPACE

The simple pure N=1 supergravity theory discussed in Secs. 5 and 6 is probably unrealistic; it seems that it requires more than an elementary graviton and gravitinofield to describe our world. The simple theory can be generalized in various ways, by coupling to N≡1 matter multiplet, by adding appropriate fields to obtain N-extended supergravity for $2 \leqslant N \leqslant 8$, by gauging the SO(N) internal symmetry in the previous theories, and by formulating supergravity in dimensions $d > 4$ with the Kaluza-Klein program in view.

In this section we briefly discuss N=1 supergravity and the supersymmetric chiral multiplet in anti-de Sitter space. This is quite relevant to gauged N-extended supergravity and to Kaluza-Klein theories, where anti-de Sitter space occurs as a natural background. The isometry group of anti-de Sitter space in SO(3,2), which is locally isomorphic to Sp(4). Its graded extensions are the super algebras OSp(4). Thus one view of the subject is that we are explaining field theories which are invariant under these super algebras (rather than their Wigner-Inönü contractions which give flat-space supersymmetry). Another view is that we are exploring quantum field theory in a space-time background with a cosmological constant, perhaps with the hope of learning why this constant vanishes.

We will approach this subject by considering the addition of a gravitino mass term as in (3.5) to the N=1 supergravity action (5.1) and (5.13) in "1.5-order form". We would not expect the gauge invariance to be maintained for the same reason it is not maintained in (3.5). However, we regain full gauge invariance if we add a cosmological term to the action and modify the gravitino transformation rule. We present the theory and then show how it leads to anti-de Sitter space.

In four space-time dimensions, we have the Lagrangian density[14]

$$\mathcal{L} = -\frac{1}{4\kappa^2}\, eR(\omega) - \frac{i}{2}\,\bar{\psi}_\mu\gamma^{\mu\nu\rho}D_\nu\psi_\rho - \tfrac{1}{2}\, ae\;\bar{\psi}_\mu\gamma^{\mu\nu}\psi_\nu + \frac{3}{2}\,\kappa^{-2}\, a^2 e \ , \qquad (7.1)$$

and transformation rules

$$\begin{aligned} \delta e^a{}_\mu &= -i\kappa\bar{\varepsilon}\gamma^a\psi_\mu \ , \\ \delta\psi_\mu &= \frac{1}{\kappa}\left[D_\mu(\omega)+\tfrac{1}{2}ia\gamma_\mu\right]\varepsilon \ , \end{aligned} \qquad (7.2)$$

where a is a new coupling constant of dimension [1/length]. The action obtained by integrating (7.1) is invariant under (7.2). To see this one notes that the gravitino mass term transforms into a term proportional to $\bar{\varepsilon}\gamma^{\mu\nu}D_\mu\psi_\nu$ which is cancelled by the variation of the gravitino kinetic action under the new term in $\delta\psi_\mu$. Further there is a term proportional to $\bar{\varepsilon}\gamma^\mu\psi_\mu$ which is cancelled by the variation of the vierbein determinant in the cosmological term. For $d > 4$ or $N > 1$ there are additional $\bar{\varepsilon}\psi\bar{\psi}\psi$ terms in the variation which are cancelled only when one adds additional fields to (7.1) to construct a full supergravity theory. In this section we will be concerned with the implications of (7.1-2) and we will ignore the extra terms.

The existence of a supersymmetric gravitino mass term is surprising, because mass terms for gauge fields usually signal symmetry breaking. Therefore let us first investigate whether the lagrangian (7.1) has supersymmetric solutions. Because of the cosmological term flat space is no longer a solution of the gravitational field equations. Purely bosonic solutions of (7.11) are characterized by

$$\begin{aligned} R_{\mu a} - \tfrac{1}{2}e_{\mu a}R &= -3a^2 e_{\mu a} \ , \\ \psi_\mu &= o \ . \end{aligned} \qquad (7.3)$$

Obviously $e_\mu{}^a$ is invariant under supersymmetry in view of the fact that $\psi_\mu = o$. The remaining condition for having a supersymmetric solution is therefore

$$\delta\psi_\mu = \frac{1}{\kappa}\,(D_\mu+\tfrac{1}{2}ia\gamma_\mu)\,\mathring{\varepsilon} = o \ . \qquad (7.4)$$

Spinor parameters $\mathring{\varepsilon}$ that satisfy this restriction are called *Killing spinors*. To have Killing spinors requires restrictions on the gravitational field, as follows by applying $D_\mu + \tfrac{1}{2}ia\gamma_\mu$ twice, i.e.

$$(D_\mu+\tfrac{1}{2}ia\gamma_\mu)\,(D_\nu+\tfrac{1}{2}ia\gamma_\nu)\,\mathring{\varepsilon} = o \ . \qquad (7.5)$$

Antisymmetrizing over μ and ν, and using the Ricci identity (4.33) and the absence of torsion in this background, one finds an algebraic condition

$$\left(R_{\mu\nu}{}^{ab}-a^2 e_{[\mu}{}^{a}\, e_{\nu\,]}{}^{b}\right)\gamma_{ab}\,\overset{o}{\varepsilon}= o \; . \tag{7.6}$$

Hence, supersymmetry requires the existence of Killing spinors, which in turn implies a condition on the Riemann tensor

$$R_{\mu\nu}{}^{ab} = a^2\, e_{[\mu}{}^{a}\, e_{\nu\,]}{}^{b} \; . \tag{7.7}$$

This is consistent with the field equations (7.3), since (7.7) implies

$$\begin{aligned} R_{\mu a} &= 3a^2 e_{\mu a} \; , \\ R &= 12a^2 \; . \end{aligned} \tag{7.8}$$

The curvature (7.7) describes a so-called anti-de Sitter space, and we have thus shown that an anti-de Sitter background is a supersymmetric solution of the theory described by (7.1). Note that a de Sitter space is obtained if we replace a by ia. However, this replacement is not consistent with the Majorana property of ψ_μ and ε [15].

However, the interpretation of the gravitino mass-like term is more subtle in anti-de Sitter space. In spite of the mass term the gravitino field only describes the number of degrees of freedom appropriate for a massless gravitino field[16]. The masslike term can also be absorbed in the definition of the covariant derivative, because

$$\begin{aligned} &-\tfrac{1}{2}i\bar{\psi}_\mu\gamma^{\mu\nu\rho}(D_\nu+\tfrac{1}{2}ia\gamma_\nu)\psi_\rho \\ &= -\tfrac{1}{2}i\bar{\psi}_\mu\gamma^{\mu\nu\rho}D_\nu\psi_\rho-\tfrac{1}{2}a\bar{\psi}_\mu\gamma^{\mu\nu}\psi_\nu \; . \end{aligned} \tag{7.9}$$

This modified covariant derivative, which coincides with the expression that appears in $\delta\psi_\mu$ and thus in the Killing condition (7.4), has a well-defined meaning. An anti-de Sitter space is invariant under an SO(3,2) isometry group. The transformations of this group consist of the SO(3,1) Lorentz transformation and 4 translations. In a flat space the translations would commute, but since the space is not flat here the commutator of two infinitesimal translations yields an infinitesimal Lorentz transformation. For spinors the SO(3,2) anti-de Sitter algebra is generated by the generators of the Lorentz group, $\Sigma_{ab} = \frac{1}{2}\gamma_{ab}$, and by 4 additional generators $\Sigma_{a5} = -\frac{1}{2}i\gamma_a = -\Sigma_{5a}$. To exhibit the nature of the anti-de Sitter algebra we can summarize the SO(3,2) commutation relation in the following way

$$\left[\Sigma_{AB},\Sigma_{CD}\right] = -\eta_{AC}\Sigma_{BD} + \eta_{AD}\Sigma_{BC} + \eta_{BC}\Sigma_{AD} - \eta_{BD}\Sigma_{AC} , \qquad (7.10)$$

where the indices A,B,... run from 0 to 4 and η_{AB} is given by

$$\eta_{AB} = \mathrm{diag}(1,-1,-1,-1,1) . \qquad (7.11)$$

Indices a,b always run from 0 to 3 and we identify $\Sigma_{AB} = \Sigma_{ab}$ for A,B $\leqslant$ 3, and $\Sigma_{AB} = \Sigma_{a5}$ if A = a and B = 5.

The modified covariant derivative $(\partial_\mu+\frac{1}{2}\omega_\mu{}^{ab}\sigma_{ab}+\frac{1}{2}ia\gamma_a e^a{}_\mu)$ has therefore the form of an SO(3,2) covariant derivative with $\omega_\mu{}^{ab}$ and $e_\mu{}^a$ playing the role of gauge fields. Rewriting the condition for Killing spinors as

$$\partial_\mu \varepsilon = \left(-\tfrac{1}{2}\omega_\mu{}^{ab}\sigma_{ab}-\tfrac{1}{2}ia\gamma_\mu\right)\varepsilon , \qquad (7.12)$$

it follows that the infinitesimal translation of a Killing spinor in anti-de Sitter space simply corresponds to an infinitesimal SO(3,2) rotation of the spinors. The precise form of this rotation is determined by the anti-de Sitter connection and vierbein fields $\omega_\mu{}^{ab}$ and $e_\mu{}^a$.

Consider now the commutator of two successive supersymmetry transformations on the vierbein

$$[\delta_1,\delta_2]e_\mu{}^a = -i\kappa\left\{\bar\varepsilon_2\gamma^a\delta_1\psi_\mu - \bar\varepsilon_1\gamma^a\delta_2\psi_\mu\right\}$$

$$= -i\left\{\bar\varepsilon_2\gamma^a(D_\mu+\tfrac{1}{2}ia\gamma_\mu)\varepsilon_1-\bar\varepsilon_1\gamma^a(D_\mu+\tfrac{1}{2}ia\gamma_\mu)\varepsilon_2\right\}$$

$$= -i\left\{D_\mu(\bar\varepsilon_2\gamma^a\varepsilon_1)+ia(\bar\varepsilon_2\gamma^{ab}\varepsilon_1)e_{\mu b}\right\} , \qquad (7.13)$$

where we made use of $\bar\varepsilon_1\gamma^a\varepsilon_2 = -\bar\varepsilon_2\gamma^a\varepsilon_1$ and $\bar\varepsilon_1\gamma^a\gamma_\mu\varepsilon_2 = \bar\varepsilon_2\gamma_\mu\gamma^a\varepsilon_1$ which holds for any pair of anticommutating Majorana spinors. Since supersymmetry is one of the invariances of the original action, the commutator (7.13) should yield an infinitesimal transformation that again leaves the action invariant. The second term in (7.13) obviously satisfies that requirement, because it represents an infinitesimal Lorentz transformation on the vierbein with parameter $\varepsilon^{ab} = \bar\varepsilon_2\gamma^{ab}\varepsilon_1$. After a few manipulations it turns out that also the first term in (7.13) is an infinitesimal invariance transformation. To see this we write

$$D_\mu\xi^a = D_\mu(e_\nu{}^a\xi^\nu)$$

$$= \partial_\mu\xi^\nu e_\nu{}^a+\xi^\nu D_\nu e_\mu{}^a+\xi^\nu S_{\mu\nu}{}^a . \qquad (7.14)$$

After using (5.20) one obtains

$$D_\mu \xi^a = \partial_\mu \xi^\nu e_\mu{}^a - \xi^\nu \partial_\nu e_\mu{}^a - (\xi^\nu \omega_\mu{}^{ab}) e_{\mu b} - i\kappa^2 \xi^\nu \bar{\psi}_\nu \gamma^a \psi_\mu \ , \tag{7.15}$$

which decomposes into a general coordinate transformation with parameter $\xi^\mu = i\bar{\varepsilon}_1\gamma^\mu\varepsilon_2$, a local Lorentz transformation with parameter $\lambda^{ab} = \xi^\nu\omega_\nu{}^{ab}$, and a supersymmetry transformation with parameter $\varepsilon = \xi^\nu\psi_\nu$. Because (7.14) is manifestly covariant this particular combination of (field-dependent) symmetry transformation is sometimes called a "covariant translation"[17]. Let us now return to the anti-de Sitter background, where (7.13) should vanish, i.e.

$$D_\mu(\bar{\varepsilon}_2\gamma^a\varepsilon_1) + ia(\bar{\varepsilon}_2\gamma^{ab}\varepsilon_1)e_{\mu b} = o \ . \tag{7.16}$$

This equation is indeed satisfied if one assumes that the supersymmetry parameters are Killing spinors. Furthermore, it shows that there is a particular combination of a general coordinate transformation and a local Lorentz transformation that leaves the vierbein invariant, i.e.

$$\xi^\mu = i\bar{\varepsilon}_1\gamma^\mu\varepsilon_2 \quad , \quad \lambda^{ab} = ia\bar{\varepsilon}_2\gamma_{ab}\varepsilon_1 \ . \tag{7.17}$$

Since the local Lorentz transformations do not act on the metric tensor, the general coordinate transformation in (7.17) leaves $g_{\mu\nu}$ invaraint

$$\delta g_{\mu\nu} = \mathcal{D}_\mu\xi_\nu + \mathcal{D}_\nu\xi_\mu = o \ . \tag{7.18}$$

This is the defining condition for a *Killing vector*. By general theorems[18] a d-dimensional maximally symmetric space has $\frac{1}{2}d(d+1)$ isometries, and there are thus at most $\frac{1}{2}d(d+1)$ linearly independent Killing vectors. As shown by (7.17) quantities that carry Lorentz indices are not inert under the general coordinate transformation generated by a Killing vector, but they can be invariant under a combined transformation of a general coordinate transformation and a local Lorentz transformation. So the vielbein is invariant under

$$\delta e_\mu{}^a = \partial_\mu\xi^\nu \, e_\nu{}^a + \xi^\nu \, D_\nu e_\mu{}^a + \lambda^{ab} e_{\mu b} \ , \tag{7.19}$$

as demonstrated by (7.16). Under the isometry group, a spinor will thus transform as

$$\delta\psi = \xi^\nu \, D_\nu\psi + \tfrac{1}{2}\lambda^{ab}\gamma_{ab}\psi \ . \tag{7.20}$$

The Killing spinors therefore transform linearly under the isometry group according to

$$\delta\psi = (\tfrac{1}{4}\varepsilon^{ab}\Gamma_{ab} + \tfrac{1}{2}ia\xi^\mu\gamma_\mu)\psi \ . \tag{7.21}$$

To elucidate the above results it might be worthwhile to consider the analogous situation in flat Minkowski space (i.e. a=o). Here the vielbein can be chosen as

$$(e_\mu{}^a)^{flat} = \delta_\mu{}^a \ , \tag{7.22}$$

and the isometry group is the Poincaré group consisting of constant Lorentz transformations and translations. Hence the corresponding Killing vectors can be parametrized by

$$\xi^\mu = \Lambda^{\mu\nu} x_\nu + a^\mu \tag{7.23}$$

where $\Lambda^{\mu\nu}$ and a^μ are the constant transformation parameters of the Poincaré group ($\Lambda^{\mu\nu}=-\Lambda^{\nu\mu}$). The flat metric $g_{\mu\nu}{}^{flat} = \delta_{\mu\nu}$ is obviously invariant under reparametrizations defined by (7.23), and indeed $D_\mu\xi_\nu = \partial_\mu\xi_\nu = \Lambda_{\mu\nu}$ is antisymmetric so that (7.18) is satisfied. The vielbein (7.22) is only invariant after including a compensating Lorentz transformation acting on the local Lorentz indices, with parameters $\lambda^{ab} = \Lambda^{ab}$. Indeed it is easy to verify that the result of the combined transformation vanishes, i.e.

$$\partial_\mu\xi^\nu e_\nu{}^a + \xi^\nu\partial_\nu e_\mu{}^a + \lambda^{ab} e_{\mu b} = \partial_\mu(\Lambda^{\nu\rho}x_\rho + a^\nu)e_\nu{}^a + \Lambda^{ab} e_{\mu b} = o \ , \tag{7.24}$$

for ξ^μ given by (7.23). Spinors are subject to the compensating "local" Lorentz transformation as well and transform as

$$\delta\psi = - (\Lambda^{\mu\nu}x_\nu + a^\mu)\partial_\mu\psi + \tfrac{1}{2}\Lambda^{ab}\Gamma_{ab}\psi \ , \tag{7.25}$$

which is the familiar flat space result. In this context a Killing spinor is a constant spinor, so that only the last term in (7.25) remains, which is precisely the result (7.21) with a = o.

Now that we have defined the supersymmetry parameter in an anti-de Sitter space it is possible to also consider supermultiplets. As an example consider the N=1 chiral multiplet in anti-de Sitter space. Hence, we have a scalar field A, a pseudoscalar field B and a Majorana spinor ψ, which transform under supersymmetry as

$$\delta A(x) = \frac{1}{\sqrt{2}}\, \bar{\varepsilon}\psi$$

$$\delta B(x) = \frac{1}{\sqrt{2}}\, \bar{\varepsilon} i\gamma_5\psi$$

$$\delta\psi(x) = - \frac{1}{\sqrt{2}} \left[i\partial\!\!\!/(A+i\gamma_5 B) + pA + iq\gamma_5 B\ \varepsilon(x)\right] . \tag{7.26}$$

The supersymmetry parameter $\varepsilon(x)$ is a Killing spinor, i.e.

$$\left(\partial_\mu + \tfrac{1}{2}\omega_\mu{}^{ab}\sigma_{ab} + \frac{i}{2}a\gamma_\nu\right)\varepsilon(x) = o \ . \tag{7.27}$$

Note that we are not in supergravity here, so that $e_\mu{}^a$ corresponds to a given anti-de Sitter background. The constant parameters p and q in (7.26) will be determined by subsequent considerations.

In order that (7.26) is a supermultiplet the supersymmetry transformations must close under commutation. For A and B this is easy to verify

$$[\delta_1,\delta_2]A = -\ \tfrac{1}{2}\ \left\{\bar{\varepsilon}_2[i\partial\!\!\!/(A+i\gamma_5B)+pA+iq\gamma_5B]\varepsilon_1 - (1\leftrightarrow 2)\right\} \ . \tag{7.28}$$

Because ε_1 and ε_2 are Majorana spinors the only term that remains is

$$[\delta_1,\delta_2]A = \xi^\mu\partial_\mu A \ , \tag{7.29}$$

where $\xi^\mu = i\bar{\varepsilon}_1\gamma^\mu\varepsilon_2$. This result agrees with the commutation on the vierbein (7.13), because the second term is a Lorentz transformation which does not act on the scalar field A.

The evaluation of the commutator on B is completely analogous, so we proceed to examine the commutator on ψ. Therefore consider first

$$\delta_1\delta_2\psi = -\ \frac{i}{2}\left\{i\gamma^\mu\left[\partial_\mu(\bar{\varepsilon}_1\psi)-\gamma_5\partial_\mu(\bar{\varepsilon}_1\gamma_5\psi)\right] + p(\bar{\varepsilon}_1\psi)-q\gamma_5(\bar{\varepsilon}_1\gamma_5\psi)\right\}\varepsilon_2 \ . \tag{7.30}$$

Because ε is not constant in anti-de Sitter space, we use the Killing condition (7.27) to write

$$\begin{aligned} \partial_\mu(\bar{\varepsilon}\psi) &= \tfrac{1}{2}ia\ \bar{\varepsilon}\gamma_\mu\psi + \bar{\varepsilon}\,\overset{\circ}{D}_\mu\psi \ , \\ \partial_\mu(\bar{\varepsilon}\gamma_5\psi) &= \tfrac{1}{2}ia\ \bar{\varepsilon}\gamma_\mu\gamma_5\psi + \bar{\varepsilon}\gamma_5\,\overset{\circ}{D}_\mu\psi \ , \end{aligned} \tag{7.31}$$

where the derivative $\overset{\circ}{D}_\mu$ is the covariant derivative in anti-de Sitter space. Thus (7.30) becomes

$$\begin{aligned} \delta_1\delta_2\psi = {} & \tfrac{1}{4}a\left[\gamma^\mu\varepsilon_2\ \bar{\varepsilon}_1\gamma_\mu\psi-\gamma^\mu\gamma_5\varepsilon_2\ \bar{\varepsilon}_1\gamma_\mu\gamma_5\psi\right] \\ & - \frac{i}{2}\left[\gamma^\mu\varepsilon_2\ \bar{\varepsilon}_1\ \overset{\circ}{D}_\mu\psi-\gamma^\mu\gamma_5\varepsilon_2\ \bar{\varepsilon}_1\ \gamma_5\overset{\circ}{D}_\mu\psi\right] \\ & - \frac{i}{2}\left[p\varepsilon_2\ \bar{\varepsilon}_1\psi-q\gamma_5\varepsilon_2\ \ \bar{\varepsilon}_1\gamma_5\psi\right] \ . \end{aligned} \tag{7.32}$$

One must now Fierz reorder this expression according to the procedure of the previous section. In the first term one finds that only $\bar{\varepsilon}_1\varepsilon_2$ and $\bar{\varepsilon}_1\gamma_5\varepsilon_2$ occur, which are symmetric in ε_1 and ε_2 and therefore cancel in the commutator. The second term involves both $\bar{\varepsilon}_1\gamma^\rho\varepsilon_2$ and $\bar{\varepsilon}_1\gamma^\rho\gamma_5\varepsilon_2$, but only the first is antisymmetric and leads to

$$[\delta_1,\delta_2]\psi = \tfrac{1}{2}i\bar{\varepsilon}_1\gamma^\rho\varepsilon_2\ \gamma^\mu\gamma^\rho\ \overset{\circ}{D}_\mu\psi\ . \tag{7.33}$$

The last term in (7.32) contains all possible bilinears in ε_1 and ε_2. Keeping only the antisymmetric vector and tensor bilinear, one finds

$$[\delta_1,\delta_2]\psi = \tfrac{1}{4}\Big[(p+q)(\bar{\varepsilon}_1\gamma^\rho\varepsilon_2)\gamma_\rho\psi-\tfrac{1}{2}(p-q)(\bar{\varepsilon}_1\gamma^{ab}\varepsilon_2)\gamma_{ab}\psi\Big]\ . \tag{7.34}$$

Combining (7.33) and (7.34) the net result is

$$[\delta_1,\delta_2]\psi = \xi^\mu\ \overset{\circ}{D}_\mu\psi - \frac{1}{8}(p-q)\bar{\varepsilon}_1\gamma^{ab}\varepsilon_2\gamma_{ab}\psi - \tfrac{1}{2}\xi^\rho\gamma_\rho[\overset{\circ}{\not{D}} + \frac{i}{2}(p+q)\psi]\ . \tag{7.35}$$

The first term in (7.35) is precisely the covariant translation that was also found in (7.29). The second term is a Lorentz transformation of the spinor, whose form coincides with the Lorentz transformation found in (7.13) provided we choose

$$p-q = 2a\ . \tag{7.36}$$

The third term has no obvious interpretation as a symmetry transformation, but it takes the form of a Dirac equation in anti-de Sitter space with mass

$$2m = p+q\ . \tag{7.37}$$

Hence we have found that (7.26) defines a supermultiplet provided that the coefficients p and q are given by

$$\begin{aligned} p &= a+m\ , \\ q &= -a+m\ , \end{aligned} \tag{7.38}$$

and the fields satisfy the field equations. By taking the limit $a \to o$ one recovers the corresponding result in a flat Minkowski space, where one must also impose the field equations.

At first instance one finds that only the fermion field equation

$$\overset{\circ}{\not{D}}\psi + im\psi = o \tag{7.39}$$

must be imposed, but by supersymmetry this equation varies into the bosonic equations which must then be satisfied as well. By explicitly varying (7.39) with respect to $\delta\psi$ as given in (7.26) one finds the bosonic equations

$$\square A + (m^2 - ma - 2a^2)A = o \; ,$$
$$\square B + (m^2 + ma - 2a^2)B = o \; . \qquad (7.40)$$

In order to derive (7.40) one uses again the Killing condition for the supersymmetry parameter. Hence the fields A, B and ψ, which form a supermultiplet, satisfy field equations in anti-de Sitter space with unequal mass terms[19]. This surprising result confirms our earlier statement that the interpretation of mass terms differs from the usual one in flat Minkowski space. The reason that the mass terms can be different is directly related to the fact that the mass operator $P_\mu^{\,2}$ is not a Casimir operator in anti-de Sitter space.

The massless limit of the multiplet is given by m=o, where the field equations can be expressed as

$$\mathring{\not{D}}\psi = o \; , \; (\square - 1/6 \; R)A = o \; , \; (\square - 1/6 \; R)B = o \; , \qquad (7.41)$$

where R is the scalar curvature which is equal to $R = 12a^2$ (cf (7.8)). The equations (7.41) are the proper field equations for massless free fields in a general gravitational background; it can be shown that they are in fact conformally invariant, as we will exploit in Sections 8 and 9.

8. OFF-SHELL ASPECTS OF SUPERGRAVITY AND CONFORMAL (SUPER)GRAVITY

The formulation of supergravity that we have been using so far suffers from serious deficiences. To see this let us consider the change in the number of degrees of freedom that takes place when the fields move off the mass shell. In general the graviton and gravitino field equations impose restrictions on various components of these fields, and imply that the momenta associated with the nonzero components are on the mass shell. Since the fields in pure supergravity are massless, the mass shell is the light cone. If one now considers the fields for momentum values that are not on the light cone then the number of field components must change if we insist on a Lorentz-covariant formulation. For instance, a graviton has 2 physical degrees of freedom, but if we move off shell in a manifestly Lorentz-invariant way then at least 5 degrees of freedom are required because the graviton fields must then have the number of degrees of freedom of a massive spin-2 particle. Actually, after subtraction of the 4 gauge degrees of freedom associated with general coordinate transformations, the graviton field contains 6 degrees of freedom corresponding to a symmetric two-rank tensor. Therefore we have one more degree of freedom than is necessary according to the above "off-shell counting" argument[20]; the extra degree of freedom is related to the scale factor of the metric tensor, and is put to zero by the gravitational field equation (5.12).

Table 1. Counting on- and off-shell degrees of freedom described by graviton and gravitino fields

	m=0	m≠0	fields
graviton	2	5	6
gravitino	2	4+4	12

The change in the number of degrees of freedom is even more drastic for spinor fields; a real spinor subject to the Dirac equation describes only half as many degrees of freedom as the number of components of the spinor field. However, when moving off shell all field components will contribute, so that the field describes the number of degrees of freedom corresponding to 2 fermions. For a d=4 gravitino field this implies that at least 2×4 = 8 field components are required to have a Lorentz-covariant off-shell formulation, because the number of degrees of freedom for a massive spin-3/2 particle is equal to 4. Again we note that the actual number contained in the gravitino field is larger, namely 16-4 = 12, where we have subtracted the 4 gauge degrees of freedom associated with supersymmetry. The exta 4 degrees of freedom that are not required on the basis of off-shell counting reside in $\gamma^{\mu}\psi_{\mu}$; these auxiliary degrees of freedom are indeed put to zero by the gravitino field equation. The situation that we describe is rather general, and is summarized for the supergravity fields in Table 1.

The crucial observation is that on the one hand one has too many degrees of freedom, since $\det(g_{\mu\nu})$ and $\gamma^{\mu}\psi_{\mu}$ are not necessary on the basis of off-shell counting, whereas on the other hand there are not yet enough field components off-shell to constitute a full supermultiplet, because a massive spin-2 and two massive spin-3/2 particles do not form a supermultiplet. In order to have a supermultiplet the states corresponding to a massive spin-1 particle must be included. We have already encountered the same phenomenon in Section 7 where the definition of a chiral supermultiplet (in anti-de Sitter space) forced us to impose the field equations. The reason for this can be traced back to the fact that we had based ourselves on 2 bosonic and 4 fermionic degrees of freedom, which only form a supermultiplet on shell where half the number of fermionic degrees of freedom is suppressed.

Let us attempt to construct a field representation for supergravity that is also supersymmetric off-shell. The strategy is to first remove the graviton and gravitino degrees of freedom that are not necessary for having an off-shell description. Subsequently we will then attempt to introduce new fields in order to complete the supermultiplet. To restrict the graviton and gravitino fields to 5

Table 2. The massive supermultiplet associated with the fields of conformal N=1 supergravity

fields	number of degrees of freedom	off-shell spin content
$e_\mu{}^a$	5	massive spin-2
ψ_μ	4+4	2× massive spin-3/2
A_μ	3	massive spin-1

and 8 degrees of freedom, respectively, one could simply impose a condition such as det g=1 and $\gamma^\mu\psi_\mu = 0$. However, it is more advantageous to exploit conformal gauge transformations for this purpose. The relevant transformation rules are

$$\delta e_\mu{}^a(x) = -\Lambda(x)\, e_\mu{}^a(x)\ , \qquad \text{(dilatation)}$$

$$\delta\phi_\mu(x) = -i\gamma_\mu\eta(x)\quad . \qquad \text{(S supersymmetry)} \tag{8.1}$$

The dilatations are local scale transformations which remove the degrees of freedom associated with the scale factor of the metric tensor. The spinor $\eta(x)$ characterizes a new supersymmetry transformation which can be used to remove the $\gamma^\mu\psi_\mu$ component of the gravitino field. With these extra gauge transformations we have now succeeded in reducing the number of degrees of freedom to 5 for the graviton, namely 16 components of the vierbein field from which we subtract 6 gauge degrees of freedom associated with local Lorentz transformations, 4 with general coordinate transformations and 1 with the local scale transformations; for the gravitino we count 8 degrees of freedom, namely 16 components contained in ψ_μ from which we subtract 4 + 4 gauge degrees of freedom associated with ordinary (Q) and special (S) supersymmetry. By introducing a single gauge field A_μ it is now easy to see that the numbers of bosonic and fermionic degrees of freedom match; the full massive supermultiplet corresponding to the resulting field representation is shown in Table 2.

The gauge theory based on $e_\mu{}^a$, ψ_μ and A_μ, consisting of 8+8 degrees of freedom, is called conformal N=1 supergravity[21]. According to the above counting arguments it represents the minimal field representation that is needed for an off-shell description of gravity that is Lorentz covariant and supersymmetric. The gauge field A_μ turns out to be associated with local chiral transformations of the gravitino field, i.e.

$$\delta A_\mu = \partial_\mu \Lambda_A(x) ,$$

$$\delta \sigma_\mu = \tfrac{1}{2} i \gamma_5 \Lambda_A(x) \psi_\mu(x) . \tag{8.2}$$

Conformal supergravity theories have been constructed for $N \leqslant 4$ in d=4, and for N=1 in d=10 dimensions. For a further discussion we refer to the original literature and to lectures given elsewhere[22,23].

In spite of the fact that we are introducing a theory with conformal (super)symmetries, the goal is still to describe and construct supergravity theories that are not necessarily invariant under all gauge invariances of the (super)conformal algebra. This can be done by means of a (super)conformal multiplet calculus which employs so-called compensating fields. We prefer to exhibit this first in the context of a rather simple example, which goes back to Stueckelberg[24]. Consider a free Proca field described by the lagrangian

$$\mathcal{L} = -\tfrac{1}{4}(\partial_\mu V_\nu - \partial_\nu V_\mu)^2 + \tfrac{1}{2} m^2 V_\mu{}^2 . \tag{8.3}$$

Because of the mass term this lagrangian is not gauge invariant under the standard gauge transformations $\delta V_\mu = \partial_\mu \Lambda$. Nevertheless, it is possible to introduce gauge invariance by simply redefining V_μ according to

$$V_\mu = A_\mu - m^{-1} \partial_\mu \phi . \tag{8.4}$$

This linear combination is invariant under the combined transformations

$$\delta A_\mu(x) = \partial_\mu \Lambda(x) \quad , \quad \delta\phi(x) = m\Lambda(x) , \tag{8.5}$$

and consequently also the lagrangian (8.3) is invariant when expressed in terms of the new fields A_μ and ϕ.

Since we have introduced an extra scalar field ϕ and local gauge transformations characterized by the parameter $\Lambda(x)$ at the same time, we have not altered the total number of degrees of freedom. But what we have achieved is that the 4 degrees of freedom contained in V_μ can now be treated on the basis of two separate fields, A_μ and ϕ. The field A_μ contains 3 degrees of freedom, the minimal number that is required for a Lorentz covaraint off-shell description. The field V_μ transforms as a reducible representation of the Poincaré group, and the gauge transformations (8.5) allow us to decompose V_μ into irreducible representations without the need for nonlocal projection operators as one encounters when decomposing straightforwardly into transversal and longitudinal components:

$$V_\mu = V_\mu{}^T + V_\mu{}^L ,$$

$$V_\mu{}^T = \left(\delta_\mu{}^\nu - \frac{\partial_\mu \partial^\nu}{\Box}\right) V_\nu \quad , \quad V_\mu{}^L = \frac{\partial_\mu \partial^\nu}{\Box} V_\nu . \tag{8.6}$$

The field ϕ is called a *compensating* field; such a field enables one to redefine any other field transforming under the gauge transformation into a gauge invariant quantity; in such a redefinition the compensating field enters as the parameter of a field-dependent gauge transformation, whose variation compensates for the effect of a gauge transformation on the original field. The expression (8.4) is an example of such a redefinition.

The Lagrangian (8.3) when expressed in terms of the fields A_μ and ϕ takes the form

$$\mathcal{L} = -\tfrac{1}{4}(\partial_\mu A_\nu - \partial_\nu A_\mu)^2 + \tfrac{1}{2}(D_\mu \phi)^2 ,$$
$$D_\mu \phi = \partial_\mu \phi - m A_\mu , \tag{8.7}$$

which is just the Lagrangian of a scalar field coupled to an abelian gauge field. However, the theory remains gauge equivalent to the original theory: the field ϕ can simply be reabsorbed into A_μ through a field-dependent gauge transformation (cf. (8.4)) after which ϕ decouples from the theory because of gauge invaraince. Alternatively one may impose a gauge condition such as $\phi = 0$, which again leads back to (8.3).

Compensators can also be used in gravity. Here one may consider local scale transformations (dilatations) under which the vierbein field transforms as

$$e_\mu{}^a(x) \to e^{-\Lambda(x)} e_\mu{}^a(x) . \tag{8.8}$$

The Einstein action is not invariant under (8.8), but the following action is[25]

$$S_{\mathrm{Weyl}} = \int d^4x \, e \left[R_{\mu\nu}{}^2 - \tfrac{1}{3} R^2 \right] , \tag{8.9}$$

where $R_{\mu\nu}$ is the (symmetric) Ricci tensor. To write down the Einstein action in a way that is invariant under dilatations we introduce a compensating scalar field ϕ, which transforms under dilatations as

$$\phi(x) \to e^{\Lambda(x)} \phi(x) . \tag{8.10}$$

In order to have a scale invariant Lagrangian we recall the conformally invariant wave equation (cf. (7.41):

$$(\Box - \tfrac{1}{6}R)\phi = 0 \ . \tag{8.11}$$

Here the d'Alembertian is defined by

$$\begin{aligned}\Box\phi &= \eta^{ab}D_a D_b \phi \\ &= g^{\mu\nu}\mathcal{D}_\mu \partial_\nu \phi \\ &= e^{-1}\partial_\mu(e g^{\mu\nu}\partial_\nu\phi) \ ,\end{aligned} \tag{8.12}$$

as follows from direct application of the formulae in Section 4. From (8.11) it is straightforward to construct a scale invariant action for the scalar field

$$S = -\int d^4x \ e\phi(\Box - \tfrac{1}{6}R)\phi \ . \tag{8.13}$$

To prove the invariance of (8.13) under the combined transformations (8.8) and (8.10) is somewhat tedius but essentially straightforward.

Although (8.13) has the form of a scalar field coupled to gravity with a so-called improvement term it is really Einstein gravity in disguise. To see this one uses the scale transformations (8.10) to put the scalar field equal to a dimensionful constant κ^{-1}. Hence we impose a gauge condition

$$\phi(x) = \kappa^{-1} \ . \tag{8.14}$$

In that case it is easy to see that (8.13) reduces to the Einstein action modulo a factor

$$S = \frac{1}{6\kappa^2}\int d^4x \ eR \ . \tag{8.15}$$

At this point it is useful to repeat the counting arguments given at the beginning of this section. The vierbein field $e_\mu{}^a$ contains 16 components, from which we subtract 4 gauge degrees of freedom associated with general coordinate transformations, 6 associated with Lorentz transformations, and 1 associated with scale transformation. Hence we are left with 5 degrees of freedom, which is just the minimal number for describing spin-2. As it turns out, 5 degrees of freedom suffice to write down the action (8.9), but for the Einstein action an extra auxiliary degree of freedom is required. This degree of freedom is provided by the compensating field $\phi(x)$.

Although we are thus working within the context of conformal gravity the compensating field enables one to couple matter to gravity without any restrictions. We give two typical examples:

(1) *Cosmological term.* The lagrangian

$$L_{cosm} = g\kappa^{-4} e \ , \tag{8.16}$$

where g is a dimensionless constant, is obviously not scale invariant. However, (8.6) is simply gauge equivalent to

$$L_{cosm} = ge\phi^4 \ . \tag{8.17}$$

This is verified directly by imposing (8.14).

(2) *Scalar fields.* The Weyl weight w of a field characterizes its transformation under dilatations according to

$$S(x) \rightarrow e^{w\Lambda(x)} S(x) \ . \tag{8.18}$$

If we introduce extra scalar fields besides the compensator then we can adjust its Weyl weight by multiplying these fields with a suitable power of the compensating field. In particular we may make it zero by replacing S by

$$S \rightarrow \phi^{-w} S \ . \tag{8.19}$$

The kinetic term for a scalar field S can then be written in scale invariant form as

$$L_{kin} = \tfrac{1}{2} e\phi^2 g^{\mu\nu} \partial_\mu S \partial_\nu S \ , \tag{8.20}$$

whereas the scale invariant version of the potential is

$$L_{pot} = -e\phi^4 V(S) \ . \tag{8.21}$$

The above approach has been used extensively to couple matter to N=1 and N=2 supergravity. The first step in this so-called superconformal tensor calculus[26] is the construction of N=1 and N=2 conformal supergravity. For N=1 the field content has been specified earlier in this section[21]. The field content of N=2 conformal supergravity is considerably more complicated, and we will not consider it here[27]. Instead we will proceed and discuss the coupling of scalar fields to N=1 and N=2 supergravity based on some general features of conformal supergravity and certain supermultiplets.

9. SPINLESS FIELDS COUPLED TO N=1 AND N=2 SUPERGRAVITY

In the previous section we have explained how to couple spinless fields to gravity in the framework of conformal gravity (cf. (8.20)). In order to couple n fields ϕ^A (A=1,...,n) to gravity one starts from n+1 fields. The extra field then acts as a compensator for the local scale transformations. However, according to (8.20) the n+1 fields were not treated on equal footing because they were chosen such that only the compensator field transforms under scale transformations. The other fields were defined with zero Weyl weight factor. For reasons that will become clear later, we prefer a framework in which all fields are treated uniformly, so that they all transform under dilations according to (this formalism has been introduced in the first ref. of 28)

$$\phi^I(x) \rightarrow e^{\Lambda(x)}\phi^I(x) \qquad (I=0,1,\ldots,n) \tag{9.1}$$

This can always be achieved by taking products and/or powers of the various fields. Note that the compensator is now included in the fields ϕ^I.

A conformally invariant Lagrangian is now

$$\mathcal{L} = e\,\eta_{IJ}\,\phi^I(\Box - 1/6R)\phi^J \ , \tag{9.2}$$

where η_{IJ} is a constant symmetric tensor, and the conformally invariant d'Alembertian $(\Box - 1/6R)$ was given in the previous section. The coupling (9.2) will be called the "minimal case". More general couplings can be constructed by taking linear combinations of Lagrangians

$$\mathcal{L} = e\,g(\phi)(\Box - 1/6R)f(\phi) \ , \tag{9.3}$$

where g and f are arbitrary homogeneous functions of the fields ϕ^I. The homogeneity is required because they must scale under dilatations just as the original fields (cf. (9.1)); only in that case one obtains a conformally invariant action. Hence

$$f(\lambda\phi) = \lambda f(\phi) \tag{9.4}$$

from which one proves

$$f(\phi) = f_I(\phi)\phi^I \ , \tag{9.5}$$

$$f_{IJ}(\phi)\phi^J = o \quad , \tag{9.6}$$

and similar relations for $g(\phi)$. Here f_I and f_{IJ} denote the first- and second-order derivatives of $f(\phi)$ with respect to the fields ϕ^I.

Using (9.5) one may rewrite (9.3) modulo a total divergence as

$$\mathcal{L} = e\,N_{IJ}(\phi)\left\{-1/6\phi^I\phi^J R - \partial_\mu\phi^I\partial^\mu\phi^J\right\} \ , \tag{9.7}$$

where $N_{IJ} = \frac{1}{2}(f_I g_J + f_J g_I)$ is a symmetric homogeneous function. Taking linear combinations of Lagrangians (9.3), characterized by arbitrary pairs of functions f and g leaves the structure of (9.7) unaltered. Exploiting the invariance under local dilatations, we adjust the coefficient of the Riemann scalar to a constant

$$N_{IJ}(\phi)\phi^I\phi^J = \kappa^{-2} \ . \tag{9.8}$$

The Lagrangian (9.7) now describes Einstein gravity, coupled to n+1 scalar fields ϕ^I subject to the nonlinear constraint (9.8).

To elucidate this result let us return to the minimal case (9.2), where N_{IJ} is just a constant tensor η_{IJ}. By redefining the fields ϕ^I one can obviously diagonalize η_{IJ} with eigenvalues ±1 (zero eigenvalues can be excluded, because they imply that certain fields are simply not present in (9.2)). In order to have the correct sign for the kinetic term of the scalar fields one must choose

$$\eta_{IJ} = \mathrm{diag}(+1,-1,-1,\ldots,-1) \ . \tag{9.9}$$

Therefore, the minimal case corresponds to an SO(n,1) nonlinear sigma model. The fact that the simplest coupling yields a *non-compact* sigma model is directly related to the conformal aspects of this approach. Note that the size of the manifold of this sigma model is given in terms of Newton's constant, as specified by (9.8). It is, however, possible to write any coupling of scalar fields to gravity in this framework. For example, scalar fields without self-interaction correspond to a tensor N_{IJ} decomposed as

$$N_{IJ} = \begin{pmatrix} 1 - \sum\limits_{C=1}^{n} (\phi^C/\phi^o)^2 & \phi^B/\phi^o \\ \phi^A/\phi^o & -\delta^{AB} \end{pmatrix} , \tag{9.10}$$

where A,B = 1,...,n. Clearly, as a physical system this case is much simpler than (9.2), but in the formalism of this section it takes a more complicated form.

We now turn to chiral multiplets coupled to N=1 supergravity. A chiral multiplet consists of a complex scalar X, a Majorana spinor Ψ and a complex auxiliary scalar B. To construct a kinetic term it is convenient to define another chiral multiplet, called the "kinetic multiplet". This multiplet, denoted by $T\phi$ has components proportional to the complex conjugate of the original chiral multiplet ϕ. In superfield notation it is simply

$$T\phi = D^2\bar{\phi} \ , \tag{9.11}$$

where $D\phi = o$, so its components are proportional to (in rigid superspace)

$$T\phi \propto (\bar{B}, \not\partial\bar{\Psi}, \Box\bar{X}) \ . \tag{9.12}$$

In a conformal supergravity background chiral multiplets exist with arbitrary Weyl weight, but in order to define a kinetic multiplet one must have w=1. The kinetic multiplet itself has then w=3. Its components now contain many nontrivial modifications that depend on the superconformal background fields. For our purpose, it suffices to know that $\Box\bar{X}$ generalizes to the conformally invariant d'Alembertian with derivatives that are also covariant with respect to the chiral U(1) gauge group of conformal supergravity.

A supersymmetric action follows from the product of a chiral multiplet with the kinetic multiplet

$$\mathcal{L} \propto \phi T\phi \rightarrow X(\Box + \ldots)\bar{X} + \ldots \ . \tag{9.13}$$

Since the expression is not yet real one adds the hermitean conjugate. A more general Lagrangian for chiral multiplets coupled to conformal supergravity can be obtained by introducing arbitrary homogeneous functions of first degree, just as before, and writing

$$\mathcal{L} \propto g(\phi)Tf(\phi) + \text{h.c.} \Rightarrow g(x)(\Box + \ldots)\bar{f}(\bar{X}) + \ldots + \text{h.c.} \ . \tag{9.14}$$

Repeating the same steps as for ordinary gravity, the coupling of scalar fields belonging to n+1 chiral multiplets to N=1 conformal supergravity can be written as

$$\mathcal{L} = e\, N_{I\bar{J}}(X,\bar{X}) \left\{- 1/6\, X^I\bar{X}^J\, R - (\partial_\mu - iA_\mu)X^I(\partial^\mu + iA^\mu)\bar{X}^J\right\} . \tag{9.15}$$

In terms of the functions f(X) and g(X), the matrix $N_{I\bar{J}}$ reads

$$N_{I\bar{J}}(X,X) = g_I(X)\bar{f}_j(\bar{X}) + f_I(X)\bar{g}_J(\bar{X}) \ , \tag{9.16}$$

so $N_{I\bar{J}}$ is now hermitean and homogeneous in X^I and $\bar{X}^I$ *separately*. Note that $N_{I\bar{J}}$ is just the second-order derivative of a real homogeneous function,

$$N_{I\bar{J}} = \frac{\partial}{\partial X^I}\frac{\partial}{\partial \bar{X}^J} Y(X,\bar{X}) \ , \tag{9.17}$$

$$\text{with } Y(X,\bar{X}) = f(X)\bar{g}(\bar{X}) + \text{c.c.} \ . \tag{9.18}$$

The Lagrangian (9.15) and correspondingly (9.16) and (9.18) may be further generalized by taking a sum over Lagrangians each characterized by a pair of arbitrary homogeneous functions. The symmetry property

$$N_{IK\bar{J}} \equiv \frac{\partial}{\partial X^K} N_{I\bar{J}} = N_{KI\bar{J}} \ , \tag{9.19}$$

follows directly from (9.17) and

$$N_{IK\bar{J}}\, X^K = 0 \qquad (9.20)$$

follows from the homogeneity of $Y(X,\bar{X})$ in X and $\bar{X}$ separately.

The main difference with the case of ordinary gravity is that we are now dealing with complex fields, which transform under the local U(1) gauge transformation of conformal supergravity. The conformal invariance of (9.15) allows us to impose the gauge condition

$$N_{I\bar{J}}(X,\bar{X})\, X^I \bar{X}^J = \kappa^{-2} \,. \qquad (9.21)$$

Because the chiral gauge field A_μ has no kinetic term, this field acts as an auxiliary field which may be eliminated by means of its field equations. The Lagrangian then takes the form

$$\mathcal{L} = -\frac{1}{6\kappa^2}\, eR - e\left\{N_{IJ}\,\partial_\mu X^I \partial^\mu \bar{X}^J - \tfrac{1}{4}\left(N_{I\bar{J}} X^I \overleftrightarrow{\partial}_\mu \bar{X}^J\right)^2\right\} . \qquad (9.22)$$

Before discussing the implications of this result let us first consider the case of N=2 supergravity. Scalar fields are contained in vector, tensor or scalar supermultiplets, and here we only discuss the vector multiplets. These can be described by N=2 reduced chiral superfields[29]. The physical fields are a complex scalar X, an SU(2) doublet of Majorana spinors Ω_i and a gauge field W_μ. In addition there is an SU(2) triplet of auxiliary fields Y_{ij}. The independent components of the corresponding chiral superfield are X, Ω_i, Y_{ij}, and the field strength F(W) associated with W_μ. The other components are not independent; the reason is that this is a *reduced* chiral superfield[30]. In rigid superspace the components are

$$\phi \propto \left(X, \Omega_i, F(W), Y_{ij}, \not{\partial}\Omega^i, \Box\bar{X}\right) . \qquad (9.23)$$

An arbitrary function of ϕ gives rise to a general chiral superfield, whose highest component leads to a supersymmetric action.

The above results still hold in a superconformal background, but there are restrictions on the Weyl weights. A reduced chiral superfield must have w=1, and a supersymmetric action can only be written down if the chiral superfield has w=2. Consequently all invariant actions are based on a homogeneous function of second degree, i.e.

$$F(\lambda\phi) = \lambda^2 F(\phi) \,, \qquad (9.24)$$

from which the following identities follow

$$F(\phi) = \tfrac{1}{2}F_I(\phi)\phi^I \,,$$
$$F_I(\phi) = F_{IJ}(\phi)\phi^J \,, \qquad (9.25)$$
$$F_{IJK}(\phi)\phi^K = 0 \quad .$$

The highest component of a function F of the reduced chiral superfield contains the term

$$F(\phi) \rightarrow F_I(X)\Box\bar{X}^I \;. \qquad (9.26)$$

In a superconformal background $\Box\bar{X}$ contains modifications that depend on the superconformal fields, but the relevant terms coincide with those found for N=1 conformal supergravity, namely a term $R\bar{X}$ and derivatives that are covariant with respect to chiral U(1) transformations. Adding these modifications to (9.26) the derivation of the Lagrangian is straightforward. Modulo a total divergence, one finds

$$\mathcal{L} = -\frac{1}{6}\, eR\, F_I(X)\bar{X}^I - e(\partial_\mu - iA_\mu)F_I(X)(\partial^\mu + iA^\mu)\bar{X}^I \;. \qquad (9.27)$$

Using the identities (9.25), and adding the hermitean conjugate this result takes the same form as (9.15), but now with a function $N_{I\bar{J}}$ given by

$$N_{I\bar{J}} = F_{IJ}(X) + \bar{F}_{IJ}(\bar{X}) \;. \qquad (9.28)$$

Hence in N=2 supergravity $N_{I\bar{J}}$ is real and symmetric. Again $N_{I\bar{J}}$ can be written in the form (9.17), with $Y(X,\bar{X})$ given by

$$Y(X,\bar{X}) = F_{IJ}(X)X^I\bar{X}^J + \text{c.c.} \qquad (9.29)$$

The identities (8.17) and (8.18) still hold, but take a stronger form

$$\frac{\partial}{\partial X^K} N_{I\bar{J}}(X,\bar{X}) = F_{IJK}(X) \;. \qquad (9.30)$$

Modulo a slight complication that leads to a change in the coefficient of the Riemann scalar[31] the same Lagrangian (9.22) is then obtained after imposing the dilatational gauge condition (9.21) and eliminating the U(1) gauge field A_μ.

In the minimal case the tensor $N_{I\bar{J}}$ is constant. After field redefinitions it can be brought in diagonal form, and in order to have the correct sign for the scalar kinetic term one chooses

$$N_{I\bar{J}} = \eta_{IJ} = \text{diag}\,(+1,-1,\ldots,-1) \;. \qquad (9.31)$$

The Lagrangian (9.15) now describes $n+1$ complex scalar fields coupled to gravity and subject to an U(n,1) invariant condition

(cf. 9.21)

$$\eta_{IJ}\, X^I \bar{X}^J = \kappa^{-2} \;. \tag{9.32}$$

Furthermore, the overall phase of X^I is irrelevant in view of the chiral U(1) invariance. Therefore, we are dealing with only 2n independent degrees of freedom, and the model in question is just a noncompact version of the $\mathbb{C}P^n$ model. The size of the manifold $\mathbb{C}P^n$ is given by Newton's constant.

In the general case the scalar fields in these Lagrangians describe a Kähler manifold[32]. To show this one first chooses inhomogeneous coordinates

$$Z^I = \frac{X^I}{X^o} \;, \tag{9.33}$$

so that $Z^o=1$ and the condition (8.19) reads

$$|X^o|^{-2} = N_{I\bar{J}}\, Z^I \bar{Z}^J \;. \tag{9.34}$$

Because $N_{I\bar{J}}$ is a homogeneous function of zeroth degree in both X^I and $\bar{X}^I$ it depends only on ratios of fields such as (9.33). In terms of the coordinate Z^I and $\bar{Z}^I$ the function $N_{I\bar{J}}$ satisfies the following conditions

$$\begin{aligned} \frac{\partial}{\partial Z^A}\left(N_{I\bar{J}}\, Z^I \bar{Z}^J\right) &= N_{A\bar{J}} \bar{Z}^J \;, \\ \frac{\partial}{\partial \bar{Z}^B} N_{I\bar{J}} \bar{Z}^J &= N_{I\bar{B}} \;, \end{aligned} \tag{9.35}$$

which follow straightforwardly from (9.20) and (9.33). Since $Z^o=1$ we only differentiate with respect to Z^A and $\bar{Z}^A$ (A=1,...,n).

It is now straightforward to write the scalar kinetic term in (9.22) in terms of the fields Z^I, where it is helpfull to recall that the overall phase of X^I does not enter because of local U(1) invariance. The result is

$$\mathcal{L}_{\text{spin-o}} = e\, \frac{\partial_\mu Z^I \partial^\mu \bar{Z}^J}{(N_{P\bar{Q}} Z^P \bar{Z}^Q)^2} \left(N_{I\bar{J}} N_{K\bar{L}} - N_{I\bar{L}} N_{K\bar{J}}\right) Z^K \bar{Z}^L \;. \tag{9.36}$$

Note that $\partial_\mu Z^o = o$, so that the indices I,J in (9.36) run effectively over A,B=1,...,n. It is now easy to verify from (9.35) that

$$\frac{\partial}{\partial Z^A} \frac{\partial}{\partial \bar{Z}^B} \ln\left(N_{I\bar{J}}(Z,\bar{Z}) Z^I \bar{Z}^J\right) = \frac{1}{(N_{P\bar{Q}} Z^P \bar{Z}^Q)^2} \left(N_{A\bar{B}} N_{K\bar{L}} - N_{A\bar{L}} N_{K\bar{B}}\right) Z^K \bar{Z}^L \;, \tag{9.37}$$

which shows that the metric $g_{A\bar{B}}$ of the nonlinear sigma model defined by (9.36) can be expressed as the second-order derivative of a scalar potential. This is the defining property of a Kähler manifold, and

$$K(Z,\bar{Z}) = \ell n\left(N_{I\bar{J}}(Z,\bar{Z})Z^I\bar{Z}^J\right) \tag{9.38}$$

is thus identified as the Kähler potential corresponding to (9.36). It is possible that the Kähler manifold has a certain invariance. In that case there are complex reparametrizations

$$Z^A \to Z^A + \xi^A(Z) \tag{9.39}$$

that leave the kinetic term (9.36) invariant. The variations $\xi^A(Z)$ are then so-called Killing vectors (cf. (7.18)). It is possible but by no means necessary that these invariance transformations of the scalar fields can be extended to an invariance of the full supergravity field equations. In N=2 supergravity these transformations act then on the spin-1 fields of the vector multiplets by means of generalized duality transformations[28].

10. GAUGED N=8 SUPERGRAVITY IN d=11 DIMENSIONS

In this last section we turn to the relation between supergravity theories in d=4 and d=11 dimensions. By spontaneous compactification of the extra dimensions[33], a possible framework for unifying the fundamental interactions, one obtains an effective "low-energy" theory consisting of all the massless states. For example, it is well-known that N=8 supergravity can be obtained by a truncation of d=11 supergravity[34] in which the fields do not depend on the extra dimensions[35]. This corresponds to retaining only the massless modes in a solution where the extra dimensions are compactified to a torus T^7. However, d=11 supergravity has a large variety of solutions. They are all of the Freund-Rubin type[36] where the antisymmetric field strength F_{MNPQ} (M,N,P,Q = 1,...,11) acquires a nonzero value if all indices are in the 4-dimensional subspace corresponding to ordinary space-time

$$F_{\mu\nu\rho\sigma}(x,y) = if\eta_{\mu\nu\rho\sigma} . \tag{10.1}$$

Here x^μ denotes the coordinates of 4-dimensional space-time and y^m are the extra coordinates; $\eta_{\mu\nu\rho\sigma}$ is the invariant Levi-Civita tensor of the 4-dimensional subspace. If we assume that this subspace is maximally symmetric then the only nonzero components of F_{MNPQ} besides (10.1) are those where all indices are in the 7-dimensional subspace. Because of that the Bianchi identity for F_{MNPQ},

$$\partial_{[M}F_{NPQR]} = 0 , \tag{10.2}$$

implies that f is y-independent. Furthermore it is customary to decompose the d=11 metric tensor as

$$g_{MN}(x,y) = \begin{pmatrix} g_{\mu\nu}(x) & 0 \\ 0 & g_{mn}(y) \end{pmatrix} . \qquad (10.3)$$

There are only two fully supersymmetric compactifications of this type[37]. In the first solution (10.1) vanishes and the 7- and 4-dimensional subspaces are flat. This corresponds to the compactification of the extra dimensions to the torus T^7. In the other supersymmetric solution[38], the 4-dimensional subspace is an anti-de Sitter space and the extra 7 dimensions are compactified to the sphere S^7. The corresponding Riemann tensors are

$$\begin{aligned} R_{mnpq} &= -m_7^{\,2}(g_{mp}g_{nq}-g_{mq}g_{np}) , \\ R_{\mu\nu\rho\sigma} &= m_4^{\,2}(g_{\mu\rho}g_{\nu\sigma}-g_{\mu\sigma}g_{\nu\rho}) , \end{aligned} \qquad (10.4)$$

where $m_4^{\,2} = 4m_7^{\,2}$ and $f^2 = 18m_7^{\,2}$, with $|m_7|$ the inverse radius of S^7. Hence the anti-de Sitter ground state is fully supersymmetric and invariant under the SO(8) isometries of S^7. Furthermore, the full spectrum of small fluctuations about this ground state has been determined, and consists of one massless supermultiplet[37,38] and an infinite tower[39] of massive N=8 anti-de Sitter supermultiplets[40] (the decomposition into supermultiplets confirms that the ground state is supersymmetric). There is one d=4 supergravity theory with the same features, namely gauged N=8 supergravity[41]. This theory has a supersymmetric anti-de Sitter ground state which leaves the local SO(8) group invariant (there are other versions of gauged N=8 supergravity where the gauge group is not SO(8), but these do not have a supersymmetric solution[42]). Therefore it seems obvious that the S^7 compactification of d=11 supergravity must correspond to gauged N=8 supergravity coupled to an infinite tower of massive supermultiplets.

If the above reasoning is correct one should be able to write down a d=4 field theory of supergravity coupled to matter which is equivalent to the full d=11 theory. Subsequently the matter multiplets can be put to zero, after which one is left with pure gauged N=8 supergravity. It is thus possible to truncate d=11 supergravity directly to N=8 supergravity by suitably restricting the y-dependence of the d=11 fields. The x-dependence is then characterized in terms of a finite number of x-dependent functions which should correspond to the d=4 fields. While such a restriction was straightforward on T^7, where the fields were taken constant in y, it is extremely complicated on S^7. A complete solution for the latter case is in fact not known. However, further knowledge of this truncation is highly desirable. It would rigorously establish the complete relationship between the S^7 compactification and N=8 supergravity, which may shed more light on the remarkable symmetry structure of the latter. Moreover, since solutions of pure supergravity are in general also solutions of a theory of supergravity coupled to matter, one would be able to formulate all d=4 solutions of pure N=8 supergravity[43] as solutions of d=11 supergravity. Note that the reverse is not always true! Solutions of d=11 supergravity do not necessarily coincide with solutions of pure N=8 supergravity, just as solutions of matter coupled to supergravity are not always solutions of pure supergravity. In fact, most of the known d=11 solutions are not related to pure N=8 supergravity.

The embedding of the N=8 supergravity fields into d=11 supergravity is partially known. The previous correspondence has been verified for a number of field configurations with $SO(7)^{\pm}$, G_2 and SU(4) invariance[44,45]. To explain how these results have been obtained, let us return to the analysis of the small fluctuations about some classical solution. The d=11 fields are then conveniently expanded in terms of a suitable set of eigenfunctions $Y^{(n)}(y)$ of the relevant mass operator associated with the background

$$\phi(x,y) = \sum_n \phi^{(n)}(x)Y^{(n)}(Y) \ . \qquad (10.5)$$

At first instance one might expect that restricting the decomposition (10.5) to the modes on S^7 corresponding to the massless supermultiplet defines the truncation of d=11 supergravity to gauged N=8 supergravity. However, the y-dependence specified in (10.5) is not free of ambiguity because the functions $Y^{(n)}(y)$ are subject to y-dependent gauge transformations which are a subset of the gauge transformations of the full d=11 theory. Furthermore the proper identification of the d=4 fields at a finite distance away from the background involves nonlinear modifications, as has been pointed out in ref. 46. Instead of specifying the y-dependence of $\phi(x,y)$ one may as well be specifying the y-dependence of $f(\phi(x,y))$ where f is some unknown function such that $f(\phi(x,y)) \propto \phi(x,y)$ in the linear approximation. Therefore from the result of the linearized analysis one cannot infer much about the full y-dependence of the fields.

There is no doubt that the above complications play a role, because already for the T^7 reduction nonlinear field redefinitions and gauge choices are required for a proper identification of the d=4 fields. Moreover, one can show directly that the supersymmetry transformation rules for the modes in (10.5) are not consistent upon truncation to the massless sector, in the sense that modes belonging to massive supermultiplets transform into modes belonging to the massless supermultiplet. Therefore the massive supermultiplets cannot be put equal to zero, because they reappear through the supersymmetry transformations. In deriving this result one uses supersymmetry parameters that leave the S^7 background invariant; the y-dependence of these parameters is restricted by the Killing condition

$$\overset{\circ}{D}_m \overset{\circ}{\varepsilon}(x,y) = -\tfrac{1}{2} m_7 i \overset{\circ}{\Gamma}_m \overset{\circ}{\varepsilon}(x,y) \quad . \tag{10.6}$$

To make the transformation rules consistent one must exploit non-linear field redefinitions of the type mentioned above. Also the supersymmetry parameters $\overset{\circ}{\varepsilon}(x,y)$ are then modified accordingly. However, in the S^7 background all modifications vanish. In ref. 47 it has been demonstrated that redefinitions of this type are sufficient to make the linearized transformations consistent upon truncation to the massless multiplet.

Our strategy is to identify the fields of pure N=8 supergravity in d=11 supergravity by requiring that the field transformations are consistent upon truncation to the massless supermultiplet. To do this it is important that we first bring the transformation rules in a convenient form. Therefore we redefine the fields and the transformation parameters in such a way that the d=11 transformation rules resemble as closely as possible the transformation rules of pure N=8 supergravity. Somewhat surprisingly, this can be done without specifying either the background or the y-dependence of the fields[44]. Consequently these redefinitions are a direct generalization of the procedure followed for T^7 in ref. 35.

We do not present the explicit transformation rules here, but refer the reader to ref. 44. The important observation is that only a restricted set of (nonlinear) field redefinitions is allowed if one wishes to preserve the qualitative features of these transformation rules. It turns out that for the fermions those redefinitions must take the form of field-dependent chiral SU(8) rotations. Therefore field-dependent chiral SU(8) rotations are the only possibility for making the supersymmetry transformations consistent upon truncation to the massless supermultiplet. In the S^7 solution the SU(8) rotations are zero, and when the fields move away from this solution one expects the fields to rotate under SU(8). Since the rotation depends on the fields, which in turn depend on y, the y-dependence of the modes thus changes continuously and differs from the y-dependence that was previously found for the small fluctuations about the supersymmetric solution.

From these considerations one can also deduce the full embedding of the scalar and pseudoscalar fields of N=8 supergravity into the 7-dimensional metric. It is given by[45]

$$\left[\frac{\det\, g_{mn}(x,y)}{\det\, \overset{\circ}{g}_{mn}(y)}\right]^{-\frac{1}{2}} g^{mn}(x,y) = \frac{1}{16}\Big(K^{mIJ}(y)K^{nKL}(y)+K^{nIJ}(y)K^{mKL}(y)\Big)$$
$$\times \left(u_{ij}{}^{IJ}(x)+v_{ijIJ}(x)\right)\left(u^{ij}{}_{KL}(x)+v^{ijKL}(x)\right) \qquad (10.7)$$

where $g_{mn}(x,y)$ is the 7-dimensional metric, $\overset{\circ}{g}_{mn}(y)$ the metric of the round S^7, and the $K^{mIJ}(y)$ are the 28 Killing vectors of S^7, which can be defined in terms of Killing spinors, i.e.

$$K^{mIJ}(y) = i\bar{\eta}^I(y)\Gamma^a\eta^J(y)\,\overset{\circ}{e}_a{}^m(y)\,. \qquad (10.8)$$

The indices I,J,... = 1,...,8 label the 8 independent Killing spinors that satisfy (10.6), the Γ^a are 7-dimensional gamma matrices, and $\overset{\circ}{e}_a{}^m(y)$ is the S^7 siebenbein. The quantities $u_{ij}{}^{IJ}$ and v_{ijIJ} characterize the scalar and pseudoscalar fields of N=8 supergravity[41]. For infinitesimal values of these fields $u_{ij}{}^{IJ}$ and $v_{ij}{}^{IJ}$ acquire the form (i,j=1,...,8 are local SU(8) indices)

$$u_{ij}{}^{IJ} = \delta_{ij}{}^{IJ} + \mathcal{O}(\phi^2)\,,$$
$$v_{ijIJ} = \tfrac{1}{4}\sqrt{2}\,\phi_{ijIJ} + \mathcal{O}(\phi^3)\,. \qquad (10.9)$$

This result holds in the so-called symmetric gauge where there is no distinction between i,j,... and I,J,... indices. The field ϕ_{IJKL} contains 35 scalar and 35 pseudoscalar fields and satisfies the constraint

$$\phi^{IJKL} = (\phi_{IJKL})^* = \frac{1}{24}\,\varepsilon^{IJKLMNPQ}\phi_{MNPQ}\,. \qquad (10.10)$$

Substituting (10.9) into (10.7) gives

$$g^{mn}(x,y) = \overset{\circ}{g}{}^{mn}(y)$$
$$+\frac{\sqrt{2}}{32}\,(\phi(x)+\phi^*(x))_{IJKL}\Big(K^{mIJ}(y)K^{nKL}(y) - \frac{1}{9}\,\overset{\circ}{g}{}^{mn}(y)K^{pIJ}(y)K_p{}^{KL}(y)\Big)$$
$$+\,\mathcal{O}(\phi^2)\,. \qquad (10.11)$$

The first-order term coincides precisely with the y-dependence found for the metric fluctuations about the S^7 background[37,38]. These fluctuations were identified with the scalar fields of gauged N=8 supergravity, and as we see clearly from (10.11) this identification is correct for *infinitesimal* fluctuations, but *not* for finite deviations. The form of (10.7) is also in accord with a result from ref. 44, where it was argued that a purely pseudoscalar background entails metric deviations in general.

As mentioned above (10.7) has been verified for a variety of scalar and pseudoscalar field configurations. Much is known about configurations with G_2 invariance, where the scalar (pseudoscalar) fields are proportional to a selfdual (antiselfdual) tensor, which satisfies

$$C_{\pm}^{IJKP} C_{\pm LMNP} = 6\,\delta^{IJK}_{LMN} + 9\,\delta^{[I}_{[L}\, C_{\pm}{}^{JK]}{}_{MN]} \ . \qquad (10.12)$$

Both tensors are invariant under inequivalent subgroups of SO(8), which we denote by $SO(7)^{\pm}$. Their common subgroup is G_2.

According to (10.11) a vacuum expectation value of the scalar proportional to C_+ involves the tensor

$$\xi_{ab}(y) = \frac{1}{16}\, C^{+}{}_{IJKL}\ \bar{\eta}^{I}(y)\Gamma_a \eta^{J}(y)\ \bar{\eta}^{K}(y)\Gamma_b \eta^{L}(y) \ , \qquad (10.13)$$

which is related to an $SO(7)^+$ invariant vector

$$\xi_a(y) = \frac{i}{16}\, C^{+}{}_{IJKL}\ \bar{\eta}^{I}(y)\Gamma_{ab}\eta^{J}(y)\ \bar{\eta}^{K}(y)\Gamma^{b}\eta^{L}(y) \ . \qquad (10.14)$$

This vector satisfies

$$\begin{aligned}
&\xi_a\xi_a = (21+\xi)(3-\xi) \ , \quad \xi = \xi_{aa} \ , \\
&\xi_a\xi_b - \delta_{ab}\xi_c^2 = 6(\xi-3)(\xi_{ab}+3\delta_{ab}) \ , \\
&\mathring{D}_a\xi = 2\xi_a \ , \quad \mathring{D}_c\xi_{ab} = \frac{1}{3}\,\delta_{ab}\xi_c - \frac{1}{3}\,\xi_{(a}\delta_{b)c} \ , \\
&\mathring{D}_a\xi_b = (3-\xi)\delta_{ab} - \frac{1}{3-\xi}\,\xi_a\xi_b \ ,
\end{aligned} \qquad (10.15)$$

where $\mathring{D}_a$ is the derivative on the round S^7 background. The length of the vector (10.14) vanishes for $\xi = -21$ and $\xi = 3$, which correspond to the poles and the equator of S^7, respectively. In standard polar coordinates on S^7 one proves

$$\xi = -\,21 + 24\,\sin^2\phi \ , \qquad (10.16)$$

where $o \leqslant \phi \leqslant \pi$ is the lattitude on S^7.

Metrics g^{mn} that correspond to a G_2 invariant configuration of scalar and pseudoscalar N=8 supergravity fields take the form

$$g^{mn} = f^{-1}(\xi)\left((\mathring{g}^{mn} - \hat{\xi}^m\hat{\xi}^n) + H^2(\xi)\,\hat{\xi}^m\hat{\xi}^n\right) , \qquad (10.17)$$

where $\hat{\xi}_a$ is the vector ξ_a normalized to unit length; tangent space indices on ξ_a are converted to world indices by means of the S^7 siebenbein. Furthermore, the functions $f(\xi)$ and $H(\xi)$ are given by

$$f(\xi) = \gamma^{-1/9}(\alpha+\beta\xi)^{-1/3} \quad ,$$
$$H(\xi) = (\alpha+\beta\xi)^{-1/2} \quad , \tag{10.18}$$

where α, β, and γ are real parameters that satisfy the conditions

$$\alpha,\gamma > o \ , \ -\frac{\alpha}{3} < \beta < \frac{\alpha}{21} \ . \tag{10.19}$$

Vacuum expectation values of the pseudoscalars proportional to C_- involve the so-called Cartan-Schouten torsion tensor[48]

$$S_{abc}(y) = \frac{i}{16} C^-{}_{IJKL} \ \bar{\eta}^I(y)\Gamma_{[ab}\eta^J(y) \ \bar{\eta}^K(y)\Gamma_{c]}\eta^L(y) \ , \tag{10.20}$$

which satisfies

$$\overset{\circ}{D}_a S_{bcd} = \frac{1}{6} \varepsilon_{abcdefg} \ S^{efg} \quad ,$$
$$S^{[abc}S^{d]ef} = -\tfrac{1}{4} \varepsilon^{abcd[e}{}_{gh} \ S^{f]gh} \quad ,$$
$$S^{a[bc}S^{de]f} = -\frac{1}{6} \varepsilon^{bcde(a}{}_{gh} \ S^{f)gh} \quad ,$$
$$S^{abc}S_{cde} = 2\delta^{ab}_{cd} - \frac{1}{8} \varepsilon^{ab}{}_{cdefg} \ S^{efg} \ . \tag{10.21}$$

Note that the pseudoscalar vacuum expectation values proportional to C^- do not change the form of (10.17), although, through the field equations, the presence of the pseudoscalars affects the actual values for the parameters α, β and γ.

In order to embed d=4 solutions of pure N=8 supergravity into d=11 supergravity one must also specify the 4-dimensional metric. In general $g_{\mu\nu}(x,y)$ is not independent of y; in order to be consistent with N=8 supergravity with constant values for scalar and pseudoscalar fields the 4-dimensional metric must take the form

$$g_{MN}(x,y) = \begin{pmatrix} \Delta^{-1}(y) g_{\mu\nu}(x) & 0 \\ 0 & g_{mn}(y) \end{pmatrix} , \tag{10.22}$$

where

$$\Delta(y) = \frac{\det g_{mn}(y)}{\det \overset{\circ}{g}_{mn}(y)} \tag{10.23}$$

and $g_{\mu\nu}(x)$ is the anti-de Sitter metric.

So far, no general formula is known for the embedding of the scalar and pseudoscalar fields into the field strength F_{mnpq}. For the G_2 invariant field configurations this tensor can be decomposed into the three independent structures that can be constructed in terms of (10.14) and (10.20). In this way G_2 invariant solutions of d=11 supergravity have been found. There are just the corresponding solutions of the N=8 potential, namely one SO(8) invariant supersymmetric solution[38] (the round S^7), a (degenerate) $SO(7)^-$ invariant solution without supersymmetry[49] (the parallelized S^7), an $SO(7)^+$ invariant solution without supersymmetry[50], and a (degenerate) G_2 invariant solution with N=1 residual supersymmetry[45]. It is outside the scope of these lectures to discuss further details of these solutions and of the embedding of pure N=8 supergravity into d=11 supergravity. We refer to the original literature for more explicit results.

REFERENCES

1. Some useful references are:
   J.J. Sakurai, Advanced Quantum Mechanics (Addison-Wesley, Reading, 1968).
   J. Scherk, in Recent Developments in Gravitation, Cargèse, 1978 (ed. by M. Lévy and S. Deser, Plenum Press, 1979) p. 479.
   P. van Nieuwenhuizen, Phys. Rep. 68 (1981) 192; in Relativity, Groups and Topology II (ed. by B.S. DeWitt and R. Stora, North-Holland, 1984) p. 825.
2. S.W. Hawking, in Recent Developments in Gravitation, Cargèse 1978, (ed. by M. Lévy and S. Deser, Plenum Press, 1979) p. 145.
   See also, J. Fröhlich and K. Osterwalder, Helv. Phys. Acta 47 (1974) 781; H. Nicolai, Nucl. Phys. B140 (1978) 294.
3. C. Itzykson and J.B. Zuber, Quantum Field Theory (McGraw-Hill, 1980).
4. G. Senjanović, Phys. Rev. D16 (1977) 307.
   P. van Nieuwenhuizen, Phys. Rep., cited in ref. 1, p. 264.
5. A. Das and D.Z. Freedman, Nucl. Phys. B114 (1976) 271.
6. D. Lurié, Particles and Fields (Interscience Publishers, 1968).
7. E.S. Fradkin and M.A. Vasiliev, Phys. Lett. 72B (1977) 70.
   E.S. Fradkin and T.E. Fradkina, Phys. Lett. 72B (1978) 343.
   R.E. Kallosh, Zh. Eksp. Teor. Fiz Pis'ma 26 (1977) 575, Nucl. Phys. B141 (1978) 141.
   G. Sterman, P.K. Townsend and P. van Nieuwenhuizen, Phys. Rev. D17 (1978) 1501.
   B. de Wit and M.T. Grisaru, Phys. Lett. 74B (1978) 57.
   B. de Wit and J.W. van Holten, Phys. Lett. 79B (1978) 389.
8. P. van Nieuwenhuizen, Phys. Rep., cited in ref. 1.
9. M.T. Grisaru, P. van Nieuwenhuizen and J. Vermaseren, Phys. Rev. Lett. 37 (1976) 75.
10. S. Helgason, Differential Geometry and Symmetric Spaces (Ac.

Press, 1962).
11. G. Velo and D. Zwanziger, Phys. Rev. 186 (1969) 1337.
12. K. Johnson and E.C. Sudarshan, Ann. Phys. (NY) 13 (1961) 126.
13. P.K. Townsend and P. van Nieuwenhuizen, Phys. Lett. 67B (1977) 439.
See also A.H. Chamseddine and P.C. West, Nucl. Phys. 129B (1977) 39.
E.S. Fradkin and M.A. Vasiliev, Lebedev Institute preprint 1976.
14. P.K. Townsend, Phys. Rev. D15 (1977) 2808.
15. It has recently been argued that a supersymmetric de Sitter groundstate is possible for complex (N=2) supergravity, J. Lukierski and A. Nowicki, Wroclaw preprint 609, 1984.
16. S. Deser and B. Zumino, Phys. Rev. Lett. 38 (1977) 1433.
17. B. de Wit and D.Z. Freedman, Phys. Rev. D12 (1975) 2286.
18. See, for instance, S. Weinberg, Gravitation and Cosmology (Wiley, 1972).
19. See, for instance, E.A. Ivanov and A.S. Sorin, J. Phys. A13 (1980) 1159; P. Breitenlohner and D.Z. Freedman, Phys. Lett. 115B (1982) 197; H. Nicolai, in Supersymmetry and Supergravity '84 (ed. by B. de Wit, P. Fayet and P. van Nieuwenhuizen, World Scientific, 1984), p. 368.
20. B. de Wit and S. Ferrara, Phys. Lett. 81B (1979) 317.
21. M. Kaku, P.K. Townsend and P. van Nieuwenhuizen, Phys. Rev. D17 (1978) 3179; P.K. Townsend and P. van Nieuwenhuizen, Phys. Rev. D19 (1979) 3166.
22. E. Bergshoeff, M. de Roo and B. de Wit, Nucl. Phys. B182 (1981) 173, B217 (1983) 489.
23. B. de Wit, in Supergravity '81, eds. S. Ferrara and J.G. Taylor (Cambridge Univ. Press, 1982).
24. E.C.G. Stueckelberg, Helv. Phys. Acta 11 (1938) 225.
25. H. Weyl, Sitzungsber. K. Preuss. Akad. Wiss. (1918) 465 (reprinted in "The Principle of Relativity", Dover, 1952), Math. Z. 2 (1918) 384.
26. This has been reviewed by B. de Wit, in Supersymmetry and Supergravity '82 (ed. by S. Ferrara, J.G. Taylor and P. van Nieuwenhuizen, World Scientific, 1983); A. van Proeyen, in Supersymmetry and Supergravity 1983 (ed. by B. Milweski, World Scientific, 1983), where an extensive list of references is given.
27. B. de Wit, J.W. van Holten and A. Van Proeyen, Nucl. Phys. B167 (1980) 186 (E: B172 (1980) 543).
28. B. de Wit and A. Van Proeyen, Nucl. Phys. B245 (1984) 89; see also E. Cremmer, C. Kounnas, A. Van Proeyen, J.P. Derendinger, S. Ferrara, B. de Wit, and L. Giradello, Cern preprint TH 3965/84; B. de Wit, P.G. Lauwers and A. Van Proeyen, Bonn preprint, 1984.
29. R. Grimm, M.F. Sohnius and J. Wess, Nucl. Phys. B133 (1978) 275.

30. R.J. Firth and J.D. Jenkins, Nucl. Phys. B85 (1975) 525; M. de Roo, J.W. van Holten, B. de Wit and A. Van Proeyen, Nucl. Phys. B173 (1980) 175.
31. B. de Wit, J.W. van Holten and A. Van Proeyen, Nucl. Phys. B184 (1981) 77; B. de Wit, R. Philippe and A. Van Proeyen, Nucl. Phys. B219 (1983) 143.
32. The Kähler property in N=1 supergravity was first shown in E. Witten and J. Bagger, Phys. Lett. 115B (1982) 202. For N=2 this result was established in B. de Wit, P.G. Lauwers, R. Philippe, Su S.-Q., and A. Van Proeyen, Phys. Lett. 134B (1984) 37. Another class of scalar fields follows from the coupling of hypermultiplets. In this case one obtains quaternionic manifolds, as was shown in P. Breitenlohner and M.F. Sohnius, Nucl. Phys. B187 (1981) 409; E. Witten and J. Bagger, Nucl. Phys. B222 (1983) 1.
33. Th. Kaluza, Sitzungsber. Preuss. Akad. Wiss. Berlin Kl (1921) 966; O. Klein, Z. Phys. 37 (1926) 895; B.S. DeWitt, Dynamical theory of groups and fields (Gordon and Breach, New York, 1965); R. Kerner, Ann. Inst. H. Poincaré 9 (1968) 143; A. Trautman, Rep. Math. Phys. 1 (1970) 29; Y.M. Cho and P.G.O. Freund, Phys. Rev. D12 (1975) 1711; J. Scherk and J.H. Schwarz, Phys. Lett. B57 (1975) 463; E. Cremmer and J. Scherk, Nucl. Phys. B103 (1976) 393; B108 (1976) 409; A. Salam and J. Strathdee, Ann. of Phys. (NY) 141 (1982) 316.
34. E. Cremmer, B. Julia and J. Scherk, Phys. Lett. 76B (1978) 409.
35. E. Cremmer and B. Julia, Phys. Lett. 80B (1978) 48, Nucl. Phys. B159 (1979) 409.
36. P.G.O. Freund and M.A. Rubin, Phys. Lett. 97B (1980) 233.
37. B. Biran, F. Englert, B. de Wit and H. Nicolai, Phys. Lett. 124B (1983) 45.
38. M.J. Duff and C.N. Pope, in Supersymmetry and Supergravity '82 (ed. by S. Ferrara, J.G. Taylor and P. van Nieuwenhuizen, World Scientific, 1983).
39. B. Biran, A. Casher, F. Englert, M. Rooman and P. Spindel, Phys. Lett. 134B (1984) 179; A. Casher, F. Englert, M. Rooman and H. Nicolai, Preprint TH3794-CERN; E. Sezgin, Trieste preprint IC-83-220.
40. H. Nicolai and D.Z. Freedman, Nucl. Phys. B237 (1984) 342.
41. B. de Wit and H. Nicolai, Phys. Lett. 108B (1981) 285, Nucl. Phys. B208 (1982) 323.
42. C.M. Hull, MIT preprints.
43. N.P. Warner, Phys. Lett. 128B (1983) 169.
44. B. de Wit and H. Nicolai, Nucl. Phys. B243 (1984) 91.
45. B. de Wit, H. Nicolai and N.P. Warner, CERN preprint TH.4052/84.
46. B. de Wit and H. Nicolai, Nucl. Phys. B231 (1984) 506.
47. M.A. Awada, B.E.W. Nilsson and C.N. Pope, Phys. Rev. D29 (1984) 334.
48. E. Cartan and J.A. Schouten, Proc. K. Akad. Wet. Amsterdam 29 (1926) 933.

49. F. Englert, Phys. Lett. 119B (1982) 339.
50. B. de Wit and H. Nicolai, Phys. Lett. 148B (1984) 60.

$d = 4$ SUPERGRAVITY FROM $d = 11$: CONJECTURES REVISITED

M.J. Duff[1]
Institute for Theoretical Physics
University of California
Santa Barbara, California, 93106

ABSTRACT

We examine the relationship between the massless sector of the theory obtained by Kaluza-Klein compactification of $d = 11$ supergravity on the round S^7 and the $N = 8$ gauged $SO(8)$ supergravity in $d = 4$. Some time ago it was conjectured that (I) the two theories are the same and (II) non-symmetric extrema of the $d = 4$ effective potential correspond to other solutions in $d = 11$ with the same S^7 topology; non-zero VEV's for scalars and/or pseudoscalars corresponding to distortion of and/or "torsion" on S^7. As yet neither conjecture is rigorously proved. Here we review the evidence in their favor, which is now overwhelming. As an interesting spin-off, one discovers the criterion for the consistency of the Kaluza-Klein ansatz.

1. d = 11 Supergravity

1.1 The Lagrangian, its symmetries, and the equations of motion

The Lagrangian of $N = 1$ supergravity in $d = 11$ is given with signature $(- + + + + + + + + + +)$ by [1]

---

[1]On leave from the Theory Division, CERN, Geneva and from the Blackett Laboratory,- Imperial College, London.

$$\mathcal{L} = \frac{1}{4} e\, e_B{}^N e_A{}^M R_{MN}{}^{AB}(\omega)$$
$$- \frac{i}{2} e \bar{\Psi}_M \Gamma^{MNP} D_N[\frac{1}{2}(\omega + \tilde{\omega})] \Psi_P$$
$$- \frac{1}{48} e F_{MNPQ} F^{MNPQ} + \frac{2}{(12)^4} e \epsilon^{M_1 \dots M_{11}} F_{M_1 \dots M_4} F_{M_5 \dots M_8} A_{M_9 \dots M_{11}}$$
$$+ \frac{3}{4(12)^4} e [\bar{\Psi}_M \Gamma^{MNWXYZ} \Psi_N + 12 \bar{\Psi}^W \Gamma^{XY} \Psi^Z](F_{WXYZ} + \tilde{F}_{WXYZ}) \qquad (1.1.1)$$

where $e = \det e_M{}^A$, $\bar{\Psi} = \Psi^+ \Gamma_0$, $D_M(\omega) = \partial_M - \frac{1}{4}\omega_M{}^{AB}\Gamma_{AB}$ and

$$\omega_{MAB} = \frac{1}{2}(-\Omega_{MAB} + \Omega_{ABM} - \Omega_{BMA}) + K_{MAB}, \qquad (1.1.2)$$

$$K_{MAB} = \frac{i}{4}[-\bar{\Psi}_N \Gamma_{MAB}{}^{NP} \Psi_P + 2(\bar{\Psi}_M \Gamma_B \Psi_A - \bar{\Psi}_M \Gamma_A \Psi_B + \bar{\Psi}_B \Gamma_M \Psi_A)], \qquad (1.1.3)$$

$$\Omega_{MN}{}^A = 2\partial_{[N} e_{M]}^A \qquad (1.1.4)$$

$$\tilde{\omega}_{MAB} = \omega_{MAB} + \frac{i}{4} \bar{\Psi}_N \Gamma_{MAB}{}^{NP} \Psi_P \qquad (1.1.5)$$

$$F_{MNPQ} = 4\partial_{[M} A_{NPQ]} \qquad (1.1.6)$$

$$\tilde{F}_{MNPQ} = F_{MNPQ} - 3\bar{\Psi}_{[M} \Gamma_{NP} \Psi_{Q]} \qquad (1.1.7)$$

We are using the convention that $M, N, P...$ refer to $d = 11$ world indices and $A, B, C...$ refer to $d = 11$ tangent space indices. $\epsilon^{MNP\dots}$ is a tensor, rather than a tensor density, and

$$\epsilon_{12\dots 11} = e \qquad (1.1.8)$$

The Γ-matrices satisfy

$$\{\Gamma_A, , \Gamma_B\} = -2\eta_{AB} \tag{1.1.9}$$

where η_{AB} is the metric in the tangent space and $\Gamma_{A_1 \ldots A_p} \equiv \Gamma_{[A_1} \ldots \Gamma_{A_p]}$. The spinors appearing in (1.1.1) are anticommuting and satisfy the Majorana condition

$$\bar{\Psi} = \Psi^T C \tag{1.1.10}$$

where the charge conjugation matrix C is antisymmetric and defined by

$$C^{-1}\Gamma_A C = -\Gamma_A{}^T \tag{1.1.11}$$

Upon variation with respect to $e_M{}^A$, Ψ_M, and A_{MNP} we obtain the field equations [2]

$$\begin{aligned} &R_{MN}(\tilde{\omega}) - \frac{1}{2} g_{MN} R(\tilde{\omega}) \\ &= \frac{1}{3}[\tilde{F}_{MNPQ}\tilde{F}_N{}^{PQR} - \frac{1}{8} g_{MN}\tilde{F}_{PQRS}\tilde{F}^{PQRS}] \end{aligned} \tag{1.1.12}$$

$$\Gamma^{MNP}\tilde{D}_N(\tilde{\omega})\Psi_P = 0 \tag{1.1.13}$$

$$\nabla_M(\tilde{\omega})\tilde{F}^{MPQR} = -\frac{1}{576}\epsilon^{PQRM_1 \ldots M_4 M_5 \ldots M_8}\tilde{F}_{M_1 \ldots M_4}\tilde{F}_{M_5 \ldots M_8} \tag{1.1.14}$$

where

$$\tilde{D}_M(\tilde{\omega})\Psi_N = D_M(\tilde{\omega})\Psi_N + T_M{}^{PQRS}\tilde{F}_{PQRS}\Psi_N \tag{1.1.15}$$

and

$$T^{SMNPQ} = \frac{1}{144}(\Gamma^{SMNPQ} - 8\Gamma^{[MNP}g^{Q]S}) \tag{1.1.16}$$

We see that the appearance of $\frac{1}{2}(\omega+\tilde{\omega})$ and $\frac{1}{2}(F+\tilde{F})$ in the Lagrangian ensures that only the supercovariant $\tilde{\omega}$ and $\tilde{F}$ enter the field equations.

The action and equations of motion are invariant under the following symmetries

a) $d = 11$ general covariance, with parameterξ^M

$$\delta e_M{}^A = e_N{}^A \partial_M \xi^N + \xi^N \partial_N e_M{}^A \tag{1.1.17}$$
$$\delta \Psi_M = \Psi_N \partial_M \xi^N + \xi^N \partial_N \Psi_M \tag{1.1.18}$$
$$\delta A_{MNP} = 3A_{Q[NP}\partial_{M]}\xi^Q + \xi^Q \partial_Q A_{MNP} \tag{1.1.19}$$

b) Local $SO(1,10)$ Lorentz transformations with parameter $\alpha_{AB} = -\alpha_{BA}$

$$\delta e_M{}^A = -e_M{}^B \alpha_B{}^A \tag{1.1.20}$$
$$\delta \Psi_M = -\frac{1}{4}\alpha_{AB}\Gamma^{AB}\Psi_M \tag{1.1.21}$$
$$\delta A_{MNP} = 0 \tag{1.1.22}$$

c) $N = 1$ supersymmetry transformations with anticommuting parameter ϵ

$$\delta e_M{}^A = -i\bar{\epsilon}\Gamma^A \Psi_M \tag{1.1.23}$$
$$\delta \Psi_M = \tilde{D}_M(\tilde{\omega})\epsilon \tag{1.1.24}$$
$$\delta A_{MNP} = \frac{3}{2}\bar{\epsilon}\Gamma_{[MN}\Psi_{P]} \tag{1.1.25}$$

d) abelian gauge transformations with parameter $\Lambda_{MN} = -\Lambda_{NM}$

$$\delta e_M{}^A = 0 \tag{1.1.26}$$

$$\delta \Psi_M = 0 \tag{1.1.27}$$

$$\delta A_{MNP} = \partial_{[M}\Lambda_{NP]} \tag{1.1.28}$$

e) odd number of space or time reflections together with

$$A_{MNP} \rightarrow -A_{MNP} \tag{1.1.29}$$

This summarizes the Lagrangian and transformation rules to be used in subsequent sections. For possible Chern-Simons modifications, see [3]. The ramifications of adding such a term and its effect on the $d = 4$ theory are very interesting but will not be pursued here.

1.2 The Freund-Rubin ansatz: $M_4 \times M_7$ ground state.

We now see how $d = 11$ supergravity admits spontaneous compactification to $d = 4$. We start by looking for solutions of (1.1.12) to (1.1.14) which might be candidates for a ground state, at least at the tree level.

We are eventually interested in obtaining a four dimensional theory which admits maximal spacetime symmetry. With signature $(-+++)$ this means that the vacuum should be invariant under $SO(1,4)$, Poincaré, or $SO(2,3)$ according as the cosmological constant is positive, zero, or negative, corresponding to de Sitter, Minkowski or anti-de Sitter space (AdS), respectively.

The first requirement of maximal symmetry is that the VEV of any fermion field should vanish and accordingly we set

$$< \psi_M >= 0\,, \tag{1.2.1}$$

and focus our attention on the equations for g_{MN} and A_{MNP} which from (1.1.12) and (1.1.14) are

$$\begin{aligned} R_{MN} - \frac{1}{2} g_{MN} R &= \frac{1}{3}\left[F_{MNQP} F_N{}^{PQR} - \frac{1}{8} g_{MN} F_{PQRS} F^{PQRS}\right], && (1.2.2)\\ \nabla_M F^{MNPQ} &= -\frac{1}{576}\epsilon^{M_1 \ldots M_8 NPQ} F_{M_1 \ldots M_4} F_{M_5 \ldots M_8} && (1.2.3)\end{aligned}$$

We look for solutions of the direct product form $M_4 \times M_7$, compatible with maximal spacetime symmetry. Thus we set

$$\begin{aligned} < g_{\mu\nu} > &= \overset{\circ}{g}_{\mu\nu}(x) \quad , \quad < F_{\mu\nu\rho\sigma} >= \overset{\circ}{F}_{\mu\nu\rho\sigma}(x), \\ < g_{mn} > &= \overset{\circ}{g}_{mn}(y) \quad , \quad < F_{mnpq} >= \overset{\circ}{F}_{mnpq}(y) \end{aligned} \tag{1.2.4}$$

but

$$< g_{\mu n} >= 0\,,\ < F_{\mu\nu\rho q} >=< F_{\mu\nu p q} >=< F_{\mu n p q} >= 0, \tag{1.2.5}$$

where x^μ are the spacetime coordinates and y^m the extra coordinates. The y independence of $\overset{\circ}{g}_{\mu\nu}$ and the x independence of $\overset{\circ}{g}_{mn}$ in (1.2.4) are necessary if $\overset{\circ}{g}_{MN}$ is to be a product metric, whereas the y independence of $F_{\mu\nu\rho\sigma}$ and the x independence of F_{mnpq} are consequences of the Bianchi identity

$$\partial_{[M} F_{NPQR]} = 0 \tag{1.2.6}$$

and (1.2.5).

It is important to note that maximal spacetime symmetry alone would not rule out a "warped-product" ansatz for which $< g_{\mu\nu}(x,y) >= f(y) g_{\mu\nu}(x)$.

Note however that since the Bianchi identity does not involve the metric, the y-independence of $F_{\mu\nu\rho\sigma}$ is still guaranteed. In what follows we shall confine our attention to "warp-factor one" *i.e.* $f(y) = 1$. The case $f(y) \neq 1$ will be postponed until Section 3.

The ansatz of Freund and Rubin [4] is to set

$$\mathring{F}_{\mu\nu\rho\sigma} = 3m\mathring{\epsilon}_{\mu\nu\rho\sigma}, \qquad (1.2.7)$$

$$\mathring{F}_{mnpq} = 0, \qquad (1.2.8)$$

where m is a real constant and the factor of 3 is chosen for future convenience. The superscript zero on the Levi-Cevita tensor $\epsilon_{\mu\nu\rho\sigma}$ means that $\epsilon_{0123} = \sqrt{-\mathring{g}}$. Substituting (1.2.7) and (1.2.8) into the field equaitons we find that (1.2.3) is trivially satisfied while (1.2.2) yields the product of a four-dimensional Einstein spacetime

$$\mathring{R}_{\mu\nu} = -12m^2\mathring{g}_{\mu\nu} \qquad (1.2.9)$$

with Minkowski signature $(- + ++)$ and a seven-dimensional Einstein space

$$\mathring{R}_{mn} = 6m^2\mathring{g}_{mn} \qquad (1.2.10)$$

with Euclidean signature $(+ + + + + + +)$.

For future reference we also record the form taken by the supercovariant derivative $\tilde{D}_M$ appearing in (1.1.15), when evaluated in the Freund-Rubin background geometry. First we decompose the $d = 11$ $\hat{\Gamma}$ matrices

$$\hat{\Gamma}_A = (\gamma_\alpha \otimes \mathbf{1}\,,\ \gamma_5 \otimes \Gamma_a) \qquad (1.2.11)$$

where

$$\{\gamma_\alpha, \gamma_\beta\} = -2\eta_{\alpha\beta} \qquad (1.2.12)$$

$$\{\Gamma_a, \Gamma_b\} = -2\delta_{ab} \qquad (1.2.13)$$

and where $\alpha, \beta...$ are spacetime indices for the tangent space group $SO(1,3)$ and $a, b...$ are extra dimensional indices for the tangent space group $SO(7)$.

Substituting (1.2.4), (1.2.5), (1.2.7), (1.2.8) and (1.2.11) into (1.1.15) we find that

$$\overset{\circ}{\tilde{D}}_\mu = D_\mu + m\gamma_\mu\gamma_5 \tag{1.2.14}$$

$$\overset{\circ}{\tilde{D}}_m = D_m - \frac{m}{2}\Gamma_m, \tag{1.2.15}$$

where $\overset{\circ}{\gamma}_\mu = \overset{\circ}{e}_\mu{}^\alpha\gamma_\alpha$ and $\overset{\circ}{\Gamma}_m = e_m{}^a\overset{\circ}{\Gamma}_a$.

Several comments are now in order. The constancy of m in the ansatz (1.2.7) is necessary in order to satisfy the field equations. However the ansatz (1.2.8) is not necessary and indeed we shall consider solutions with non-zero F_{mnpq} in Section 2. The maximally symmetric solution to (1.2.9) is, in fact, anti-de Sitter space since the cosmological constant $\Lambda = -12m^2$ is negative. There are infinitely many seven-dimensional Einstein spaces M_7 satisfying (1.2.10) and, for the moment, M_7 will be left arbitrary. The important point is that Einstein spaces of positive curvature and Euclidean signature are automatically compact [5] and hence spontaneous compactification to $d = 4$ has indeed been achieved.

2. Relation between d = 11 and d = 4 supergravity

2.1 The de Wit-Nicolai theory

In 1981 de Wit and Nicolai [6] constructed a four-dimensional supergravity theory by gauging the $SO(8)$ subgroup of E_7 in the global $E_7\times$ local $SU(8)$ supergravity of Cremmer and Julia [7]. In common with the Cremmer-Julia theory, this theory described the self-interaction of a single massless $N = 8$ supermultiplet of spin (2, 3/2, 1, 1/2, 0^+, 0^-) but with local $SO(8)\times$ local $SU(8)$ invariance. In addition to the gravitational constant, there was a new parameter; the $SO(8)$ gauge coupling constant e, and it was necessary to modify the Cremmer-Julia Lagrangian and transformation rules by other e dependent terms in order to preserve the $N = 8$ supersymmetry. In particular there were Yukawa-like interactions between the fermion and spin 0 fields, and a non-trivial effective potential for the scalars. As in the Cremmer-Julia theory, the $SU(8)$ gauge potentials were composite fields with no kinetic term of their own and were therefore expressible in terms of the spin 0 fields and their derivatives. The transformation of the various fields under $SO(8) \times SU(8)$ is summarized in Table I. We note, in particular that the fermions are *chiral* under $SU(8)$. The number of spin 0 *fields* is reduced to 133 by assigning the u's and v's to a "sechsundfunfsigbein" which is an element of E_7. The number of physical spin-0 states is $70 = 133 - 63$ since 63 fields may be gauged away by an $SU(8)$ rotation.

TABLE I

spin	2	3/2	1	1/2	0	0
field	$e_\mu{}^a$	$\psi_\mu{}^i$	$B_\mu{}^{IJ}$	χ^{ijk}	U^{ijIJ}	$V_{ij}{}^{IJ}$
SO(8)	1	1	28	1	28	28
SU(8)	1	8	1	56	28	$\overline{28}$

Since E_7 is no longer a symmetry, however, we cannot parametrize these 70 scalars by the coset $E_7/SU(8)$ as Cremmer-Julia did. They are no longer Goldstone bosons.

The existence of the effective potential naturally prompted the question of its extrema. As discussed by de Wit and Nicolai [6], the obvious symmetric extremum for which both the 35 scalars and 35pseudoscalars have zero VEV's yields an $SO(8)$ vacuum state with $N = 8$ supersymmetry. ($SU(8)$ is not, or course, a symmetry of the vacuum.) The physical states belong to the familiar massless $N = 8$ supermultiplet with $(1, 8_s, 28, 56_s, 35_v, 35_c)$ representations of $SO(8)$. The effective potential then yields a non-zero cosmological constant for this vacuum given by $4\pi G\Lambda = -3e^2$, and the Yukawa couplings yield a "mass term" for the gravitino. The relevant algebra is therefore the $OSp(4|8)$ de Sitter supersymmetry rather than the $N = 8$ Poincaré sypersymmetry of the Cremmer-Julia vacuum. Note, however, that the de Wit-Nicolai theory goes smoothly over to the Cremmer-Julia theory in the limit $e \to 0$. We shall return to this point shortly.

For full details of the de Wit-Nicolai theory, see [6]. By truncation, one may also obtain gauged $SO(N)$ supergravities for $N < 8$, in particular $N = 5$ and 6. The gauged $N \leq 4$ theories were already known to exist [8,9,10]. The existence of gauged extended supergravity for $N > 4$ was already strongly suggested, before the explicit construction of de Wit and Nicolai, by the discovery that such theories would necessarily have vanishing one-loop β function [11]. This was strongly reminiscent of the vanishing one-loop β function in $N > 2$ super Yang-Mills theories, and indeed both were subsequently explained by the device of spin-moment sum rules [12,13].

2.2 Conjectures on the seven-sphere

Shortly after the de Wit-Nicolai theory was constructed, Duff and Pope [13,14,16,17] observed that $d = 11$ supergravity admitted vacuum solutions of the form $AdS \times S^7$, and that since S^7 has isometry group $SO(8)$ and admits 8 Killing spinors, this gives rise via the Kaluza-Klein mechanism to an effective four-dimensional theory with local $SO(8)$ invariance and $N = 8$ supersymmetry. It followed automatically that this $d = 4$ theory described the interaction of one massless $N = 8$ supermultiplet of spins (2, 3/2, 1, 1/2, 0^+, 0^-) in $SO(8)$ reps (1, 8_s, 28, 56_s, 35_v, 35_c) with an infinite tower of massive $N = 8$ supermultiplets. It also followed that in this $N = 8$ vacuum Λ was proportional to $-Ge^2$ with a calculable, but at the time not yet calculated, coefficient.

All this led to the conjecture [13,14,16,17] that the dynamics of the massless $N = 8$ supermulitplet of this Kaluza-Klein theory was nothing but the $N = 8$ theory of de Wit and Nicolai. Let us call this conjecture I. The arrival of the Englert solution [15] led to the observation [16] that the round S^7 and the round S^7 with "torsion" (*i.e.* non-vanishing A_{mnp}) were merely different phases of the same four-dimensional theory, and that "torsion" corresponded to a non-zero VEV for the 35_c pseudoscalars. This led to conjecture II: Just as the symmetric $N = 8$ extremum of the de Wit-Nicolai effective potential in $d = 4$ corresponds to the round S^7 without torsion, so non-symmetric extrema correspond to other solutions of the $d = 11$ theory with different geometry but the same S^7 topology, and that non-zero VEVs for scalars and pseudoscalars in $d = 4$ correspond to a deformation of and/or torsion on S^7 in $d = 11$ [16,17] (Of course, the deformation or torsion in question would have to correspond to the 35_v scalars and/or the 35_c pseudoscalars. This means, in particular, that the squashed S^7 of Awada, Duff and Pope [19] would not admit any such de Wit-Nicolai interpretation, since here it is the 300 of $SO(8)$ which acquires the non-zero VEV. This latter fact was not appreciated until later [21], however). Thus one had a $d = 11$ geometrical Kaluza-Klein interpretation of the $d = 4$ Brout-Englert-Higgs-Kibble effect. This relationship was further explored in [18,20,22].

While some held the truth of these conjectures to be self-evident, others raised objections. Their verification would require a complete non-linear analysis of the $d = 4$ equations of motion obtained from $d = 11$ and, to date, only a partial results are known mainly because of the complicated non-polynomial dependence on the scalars and pseudoscalars. Let us therefore list some of these objections and see how far they have been overcome to date. First of all, implicit in conjecture I is the assumption that it is in fact possible to truncate the full Kaluza-Klein theory to its massless sector in a consistent fashion. As it turns out, such fears are in general well founded. It is a remarkable property of S^7 that, with the exception of T^7, it provides the only examples of M_7 compactifications of $d = 11$ supergravity to admit such consistent trucations. This is the subject of Section 3.

A second objection to conjecture I concerned the cosmological constant. A necessary condition for agreement is that the spacetime cosmological constant Λ in the round S^7 vacuum must be related to the $S0(8)$ gauge coupling constant e by the same formula $4\pi G\Lambda = -3e^2$ as obtained in the $N = 8$ phase of the de Wit-Nicolai theory. Weinberg [23] has shown how the coupling constant may be calculated starting from pure Einstein gravity in higher dimensions. Applied to the round S^7 of radius m^{-1}, the formula gives $e^2 = 64\pi Gm^2$ which, combined with $\Lambda = -12m^2$ from (1.2.9) yields $16\pi G\Lambda = -3e^2$ which disagrees by a factor of 4 with the de Wit-Nicolai value: an apparent disaster for conjecture I. However, it turns out that the presence of the A_{MNP} field in $d = 11$ supergravity leads to a modification of Weinberg's pure gravity calculation [24]. As described in Sections 2.3 and 3.2 when this is taken into account it changes the relation between e^2 and m^2 to $e^2 = 16\pi Gm^2$ which yields precisely the de Wit-Nicolai relation.

A third objection to conjecture I, closely related to the other two, involved the massless ansatz and the supersymmetry transformation rules. One important consistency check, pointed out in [16], was whether the $d = 11$ supersymmetry tranformation rules consistently yielded the correct $d = 4$ transformation rules when the ansatz for the massless supermultiplet was substituted in. This massless ansatz was known in full at the linearized level [17,20] and is given in Section 2.3, so one could check the transformation rules to the same order of approximation. In fact, as we shall see in Section 2.3, everything works out well [25,26] despite earlier claims that there was a mismatch in the y dependence.

A surprising result, not anticipated at the time conjecture I was made, is the appearance of the 294_v massless 0^+ in the round S^7 spectrum not belonging to the massless $N = 8$ supermultiplet [27]. The inclusion of these fields in the massless ansatz would clearly violate supersymmetry and lead to inconsistencies. When discussing conjecture I it is always important therefore to distinguish the ansatz for the mass *supermultiplet* from the ansatz for the massless *sector*.

At the time conjecture II was made, only one non-symmetric extremum of the de Wit-Nicolai theory was known [28]. This was Warner's $SO(3) \times SO(3)$ extremum with $N = 0$. Since then Warner has discovered many more and has classified all those which have at least $SU(3)$ symmetry [29]. See Section 2.4.

Up until recently, these results posed the gravest threat to conjecture II. For although all comparisons of the $SO(7)_c$ extremum in $d = 4$ with the Englert solution in $d = 11$ proved positive [30,31], the other non-symmetric extrema defied all attempts to yield to a $d = 11$ origin. It was even proved that no Freund-Rubin type solution with or without "torsion" could have G_2 symmetry unless it was the round S^7 [32]. The resolution of this problem is the subject Section 2.4 of and turns out to have far-reaching consequences for Kaluza-Klein.

2.3. Linearized massless ansatz and transformation rules

The ansatz for the massless supermultiplet at the linearized level can be obtained by retaining only the lowest modes in the Fourier expansion with the harmonics specialized to S^7. Making the field redefinitions

$$h'_{MN} = h_{MN} + \frac{1}{2} g_{MN} h^q_q \tag{2.3.1}$$

$$\Psi'_M = \Psi_M - \frac{1}{2}\hat{\Gamma}_M \hat{\Gamma}^q \Psi_q \tag{2.3.2}$$

the ansatz is [17,20];

$$h'_{\mu\nu}(x,y) = h_{\mu\nu}(x) \tag{2.3.3}$$

$$h'_{\mu n}(x,y) = \frac{1}{2} B^{IJ}_\mu(x)\eta^{IJ}_n \tag{2.3.4}$$

$$h'_{mn}(x,y) = S^{IJKL}(x)\eta^{IJKL}_{mn} \tag{2.3.5}$$

$$f_{\mu\nu\rho\sigma}(x,y) = \frac{3m}{2}\epsilon_{\mu\nu\rho\sigma}(h'^{\tau}_{\tau} - \frac{2}{9} h'^{t}_{t}) \tag{2.3.6}$$

$$f_{\mu\nu pq}(x,y) = \frac{1}{24m}\epsilon_{\mu\nu\rho\sigma}\nabla^\sigma\nabla_q h'^{t}_{t} \tag{2.3.7}$$

$$f_{\mu\nu pq}(x,y) = -\frac{1}{2m}\epsilon_{\mu\nu\rho\sigma}\nabla^\rho\nabla_{[p} h'_{q]}{}^{\sigma} \tag{2.3.8}$$

$$f_{\mu npq}(x,y) = -\frac{1}{2}\partial_\mu P^{IJKL}(x)\eta^{IJKL}_{mnpq} \tag{2.3.9}$$

$$f_{mnpq}(x,y) = 2mP^{IJKL}(x)\eta^{IJKL}_{mnpq} \tag{2.3.10}$$
$$\Psi'_{\mu}(x,y) = \psi^{I}_{\mu}(x)\eta^{I} \tag{2.3.11}$$
$$\Psi'_{m}(x,y) = \chi^{IJK}(x)\eta^{IJK}_{m} \tag{2.3.12}$$

The y-dependent tensors η are defined in terms of "Killing spinors." A Killing spinor η on M^7 is defined to be a solution of the equation

$$\overset{o}{\tilde{D}}_m \eta = 0$$

where $\overset{o}{\tilde{D}}_m$ is the $d=7$ supercovariant derivative of (1.2.15). On the round S^7 there exists the maximum number (8) of Killing spinors, denoted $\eta^I (I = 1...8)$ where I denotes the 8_s representation of $SO(8)$. From these we may construct

$$\eta^{IJ}_m = \bar{\eta}^I \Gamma_m \eta^J \quad , \quad \eta^{IJKL}_{mn} = \eta^{[IJ}_m \eta^{KL]}_n$$
$$\eta^{IJKL}_{mnp} = \eta^{IJ}_{[mn}\eta^{KL}_{p]} \quad , \quad \eta^{IJKL}_{mnpq} = \eta^{IJ}_{[mn}\eta^{KL}_{pq]} \tag{2.3.13}$$

where we also define

$$\eta^{IJ}_{mn} = \bar{\eta}^I \Gamma_{mn} \eta^J \quad , \quad \eta^{IJK}_m = \eta^{[I}\Gamma^{JK]}_m \tag{2.3.14}$$

The vectors η^{IJ}_m are in fact the 28 Killing vectors on the round S^7.

The expansion of the supersymmetry parameter, which we shall modify later, will for now be taken to be

$$\epsilon(x,y) = \epsilon^I(x)\eta^I \tag{2.3.15}$$

The idea now is to substitute these ansätze into the $d=11$ supersymmetry transformation rules (1.1.23), (1.1.24) and (1.1.25) in order to verify to first order in fields, four dimensional supersymmetry rules are obtainable *i.e.* that the y dependence of the left and right hand sides of each of these equations matches. Here we follow the paper of Awada, Nilsson and Pope [25]. Starting with (1.1.23) we first convert it into transformation law for the linearized metric

$$\delta h_{MN} = -2i\bar{\epsilon}\hat{\Gamma}^{(0)}_{(M}\Psi_{N)} \tag{2.3.16}$$

where $\hat{\Gamma}^{(0)}_M$ is obtained from $\hat{\Gamma}_M$ using the background vielbein. The calculations here are straightforward, and indeed yield equations in which the y dependence matches.

There are three cases to consider, corresponding to $MN = \mu\nu$, μn, and mn, and so dropping the y dependence these give, after some algebra,

$$\begin{aligned}
\delta h_{\mu\nu} &= -2i\bar{\epsilon}^I \gamma_{(\mu} \psi^I_{\nu)} \\
\delta B^{IJ}_\mu &= -2i\bar{\epsilon}^{[I} \gamma_5 \psi^{J]}_\mu - 2i\bar{\epsilon}^K \gamma_\mu \chi^{KIJ} \\
\delta S^{IJKL} &= -2i\bar{\epsilon}^{[I} \gamma_5 \chi^{JKL]_+}
\end{aligned} \tag{2.3.17}$$

where in the final equation $[IJKL]_+$ denotes the self-dual projection of $[IJKL]$. This projection follows from the duality property of η^{IJKL}_{mn} defined in (2.3.13).

We now convert (1.1.25) into a transformation law for F_{MNPQ} and so after linearization we obtain

$$\delta f_{MNPQ} = 6D_{[M}(\bar{\epsilon}\hat{\Gamma}^{(0)}_{NP}\Psi_{Q]}). \tag{2.3.18}$$

On substituting the ansätze into (2.3.18), one again finds after some straightforward but tedious algebra that the y dependence matches. There are five cases to consider, corresponding to the various possible combinations of spacetime and internal indices. All except the cases of 3 or 4 internal indices reproduce the previously obtained transformation laws (2.3.17). The remaining cases produce the last bosonic transformation law

$$\delta P^{IJKL} = -2\bar{\epsilon}^{[I}\chi_-^{JKL]} \tag{2.3.19}$$

We now turn to the fermion transformation law (1.1.24). This is the equation which at first sight appears to give rise to a mismatch of the y dependence, and hence an inconsistent result, and so here we shall describe the calculation in greater detail. Linearizing (1.1.24), we obtain

$$
\begin{aligned}
\delta\Psi_M = \overset{o}{\tilde{D}}_M\epsilon - \frac{1}{4}(2e_A{}^N \overset{o}{D}_{[M}V_{N]B} \\
- e_A{}^P e_B{}^Q e_M{}^D \overset{o}{D}_P V_{QD})\hat{\Gamma}^{AB}\epsilon \\
- \frac{i}{144}(\hat{\Gamma}_M^{(0)NPQR} + 8\delta_M^N\hat{\Gamma}^{(0)PQR})f_{NPQR}\epsilon \\
- \frac{i}{144}(\hat{\Gamma}_M^{(1)\,NPQR} + 8\delta_M^N\hat{\Gamma}^{(1)PQR})F^{(0)}_{NPQR}\epsilon
\end{aligned}
\tag{2.3.20}
$$

where $V_M{}^A$ is the fluctuation around the background vielbein $e_M{}^A$ corresponding to the metric fluctuation h_{MN}. It is convenient to impose Lorentz gauge condition $V_m{}^\alpha = 0$. The matrices $\hat{\Gamma}_M^{(1)NPQR}$ and $\hat{\Gamma}^{(1)PQR}$ are the terms linear in $V_M{}^A$ which result from converting tangent space to world indices on the Dirac matrices $\hat{\Gamma}_A{}^{BCDE}$ and $\hat{\Gamma}^{CDE}$.

We now substitute the ansätze into (2.3.20), and consider first the case $M = m$. On general grounds we expect the variation $\delta\Psi_m$ to involve terms containing B_μ, $\nabla_{[\mu}B_{\nu]}$, $\partial_\mu S$, $\partial_\mu P$, S, and P. Detailed calculation shows that the term in B_μ vanishes, while after a Fierz transformations the y dependence of the term in $\nabla_{[\mu}B_{\nu]}$ is seen to match with the left hand side of (2.3.20). For the scalar and pseudoscalar terms, we require the Fierz transformation

$$
\begin{aligned}
& - 144(\delta^{I[J}\eta_m^{KLM]} + \frac{1}{9}\Gamma_m\delta^{I[J}\eta^{KLM]}) \\
& = \Gamma_m{}^{npq}\eta^I\eta_{npq}^{JKLM} + 6\Gamma^{np}\eta^I\eta_{mnp}^{JKLM} + 18\Gamma^n\eta^I\eta_{mn}^{JKLM} \\
& - 2\Gamma_m\eta^I\eta_{pq}^{JKLM}g^{pq} + \Gamma_m{}^p\eta^I\eta_{np}^{[JK}\eta_q^{LM]}g^{nq} \\
& - 2\eta^I\eta_{mn}^{JK}\eta_p^{LM}g^{np}
\end{aligned}
\tag{2.3.21}
$$

where we have dropped the $^{(0)}$ superscript on the $d = 7$ Γ matrices. It is crucial to note that the first two terms on the right hand side of (2.3.21) are anti-self-dual in $JKLM$, since they are associated with the 35_c of three-forms of the pseudoscalar ansatz, while the remaining terms are self-dual in $JKLM$, since they are associated with the 35_v of Killing tensors of the scalar ansatz. The terms in $\delta\Psi_m$ involving $\partial_\mu S$ and $\partial_\mu P$ are precisely the self-dual and anti-self-dual parts of (2.3.21), respectively, and hence the y dependence of these terms matches with $\delta\Psi_m$; they occur in the combination $\partial_\mu(S + i\gamma_5 P)$.

Repeating the procedure for the S and P terms in $\delta\Psi_m$, one runs into an apparent inconsistency. The relevant terms in the transformation rule are

$$\delta\chi^{IJK}\big|_{S,P}(\eta_m^{IJK}+\frac{1}{9}\Gamma_m\eta^{IJK})$$
$$=\frac{m}{72}S^{JKLM}\epsilon^I(-5\Gamma_m{}^p\eta^I\eta_{np}^{[JK}\eta_q^{LM]}g^{nq}-2\Gamma_m\eta^I\eta_{pq}^{JKLM}g^{pq}-18\Gamma^n\eta^I\eta_{mn}^{JKLM})$$
$$+\frac{im}{72}P^{JKLM}\gamma_5\epsilon^I(-8\Gamma_m^{npq}\eta^I\eta_{npq}^{JKLM}-12\Gamma^{pq}\eta^I\eta_{mpq}^{JKLM}) \qquad (2.3.22)$$

Using (2.3.21), one can now show that the y dependence on each side of (2.3.22) does not match. To remedy this, we note that the ansatz (2.3.15) for $\epsilon(x,y)$ implies that $\tilde{D}_m\epsilon=0$, but that the ansatz can be modified by the addition of S- and P-dependent terms in such a manner that $\tilde{D}_m\epsilon$ now produces precisely the required extra terms in (2.3.22) in order that the y dependence coincides with that of the Fierz equation (2.3.21). The modified ansatz is

$$\epsilon(x,y)=\big[1+\frac{1}{72}S^{JKLM}(-2\eta_{mn}^{JKLM}g^{mn}+3\eta_{mn}^{[JK}\eta_p^{LM]}g^{np}\Gamma^m)$$
$$+\frac{i}{72}\gamma_5P^{JKLM}(-3\eta_{mnp}^{JKLM}\Gamma^{mnp})\big]\epsilon^I(x)\eta^I \qquad (2.3.23)$$

Note that this does not upset any of the previously derived linearized transformation rules, since these ϵ always appear in terms already linear in fields. Summarizing, the transformation rule for $\delta\Psi_m$ gives, in terms of $F_{\mu\nu}^{IJ}=2\nabla_{[\mu}B_{\nu]}^{IJ}$,

$$\delta\chi^{IJK}=\frac{3}{8}F_{\mu\nu}^{[IJ}\gamma^{\mu\nu}\epsilon^{K]}-2\gamma_5\partial\mu(S^{IJKL}+i\gamma_5P^{IJKL})\gamma^\mu\epsilon^L$$
$$-4m(S^{IJKL}+i\gamma_5P^{IJKL})\epsilon^L. \qquad (2.3.24)$$

Finally, substituting the ansätze for the massless fields, and the modified ansatz (2.3.23) for $\epsilon(x,y)$, into the transformation rule for $\delta\Psi_\mu$, one finds that the y dependence matches on each side of the equation, and

$$\delta\psi_\mu^I=\bar{D}_\mu\epsilon^I-2mB_\mu^{IJ}\epsilon^J-[\frac{(1-\gamma_5)}{2}F_{\mu\nu}^{+IJ}-\frac{(1+\gamma_5)}{2}F_{\mu\nu}^{-IJ}]\gamma^\nu\epsilon^J, \qquad (2.3.25)$$

where $F^{\pm}_{\mu\nu} = \frac{1}{2}(F_{\mu\nu} \pm {}^*F_{\mu\nu})$ and ${}^*F_{\mu\nu} = \frac{i}{2}\epsilon_{\mu\nu\rho\sigma}F^{\rho\sigma}$. The derivative $\tilde{D}_\mu$ is the anti-de Sitter covariant derivative $D_\mu + m\gamma_\mu\gamma_5$ including terms up to first order in fluctuations. The transformation rules must, of course, be $SO(8)$ gauge covariant, and defining an $SO(8)$ gauge covariant derivative $\mathcal{D}_\mu$ by

$$\mathcal{D}_\mu\epsilon^I = D_\mu\epsilon^I - eB^{IJ}_\mu\epsilon^J \tag{2.3.26}$$

then (2.3.25) may be written as

$$\delta\psi^I_\mu = \tilde{\mathcal{D}}_\mu\epsilon^I - [\frac{(1-\gamma_5)}{2}F^{+IJ}_{\mu\nu} - \frac{(1+\gamma_5)}{2}F^{-IJ}_{\mu\nu}]\epsilon^J \tag{2.3.27}$$

provided the $SO(8)$ gauge coupling constant e is chosen to be

$$e = 2m. \tag{2.3.28}$$

Here we have defined $\tilde{\mathcal{D}}_\mu$ in the same manner as $\tilde{D}_\mu$, *i.e.* $\tilde{\mathcal{D}}_\mu = \mathcal{D}_\mu + m\gamma_\mu\gamma_5$. (As we shall see in Section 3.2, the normalization chosen for the massless ansatz of this section is such that $4\pi G = 1$ where G is the $d = 4$ Newton's constant. This implies $e^2 = 16\pi G m^2$ *i.e.* $4\pi G\Lambda = -3e^2$ as required for consistency with conjecture I.)

It is important to check that the redefinition of $\epsilon(x, y)$ in (2.3.23) does not upset any of our previous results. In particular, the criterion for unbroken supersymmetry $< \delta\Psi_M >= 0$ remains unaltered. This is because $< S >=< P >= 0$ in the round S^7 background.

Thus we see that the truncation of the $d = 11$ theory to include just the massless supermultiplet is indeed consistent with the transfromation rules at this order of approximation. Moreover, to the same order of approximation, the $d = 4$ transformation rules coincide with those of the de Wit-Nicolai theory as may be seen after some trivial rescalings of the fields. In particular, note that the $SO(8)$ covariantization of the derivative in (2.3.27) and the appearance throughout of the combination of fields $(S + i\gamma_5 P)$. When one considers the very different $d = 11$ origins of S and P, this latter fact is truly remarkable and is suggestive of some deeper connection between g_{MN} and A_{MNP} in $d = 11$ which may explain the $SU(8)$ symmetry in $d = 4$.

One may wonder how the *linearized* ansatz can give information about couplings constants and covariant derivatives. The answer is that, although we worked to only linear order in *fields*, the transformation rules involve products of fields and supersymmetry parameters. Hence they yield information about cubic terms in the Lagrangian. Thus consistency of the above transformation rules is already probing the non-linear structure. It is to the non-linear theory that we now turn.

2.4 New S^7 solutions and Warner's extrema

Warner [29] has now completed the list of all the extrema of the de Wit-Nicolai potential in $d = 4$ which induce a breaking of $SO(8)$ to some smaller group containing $SU(3)$. It is not yet known how many other extrema there are, but the original example of $SO(3) \times SO(3)$ is known as mentioned in Section 2.1. The known extrema are shown in Table 2.

Let us now reexamine conjecture II in the light of these extrema. All comparisons which have been made to date regarding gauge symmetries, supersymmetries, cosmological constants and partial mass spectra are consistent with the identification of : E_0 with the Duff-Pope S^7 solution [13,14,16,17]; E_1 with the Englert S^7 solution [15]; and E_4 with the Pope-Warner S^7 solution [33].

TABLE 2

Known extrema of the scalar potential

Extremum	Gauge Symmetry	Supersymmetry	Non-zero VEVs
E_0	SO(8)	N=8	-
E_1	$SO(7)_c$	N=0	P
E_2	$SO(7)_s$	N=0	S
E_3	G_2	N=1	S,P
E_4	SU(4)	N=0	P
E_5	SU(3)xU(1)	N=2	S,P
E_6	SO(3)xSO(3)	N=0	S,P

In all cases, non-zero VEVs for the pseudoscalars P in $d = 4$ correspond to non-zero F_{mnpq} in $d = 11$. The fact that $< S >= 0$ does not necessarily imply the *round* S^7 geometry since the seven-sphere is "stretched" in the Pope-Warner solution. This reflects the fact that the pseudoscalar will enter at the non-linear level in the ansatz for $g_{mn}(x, y)$.

Much more mysterious from the point of view of conjecture II are those extrema with $< S > \neq 0$, namely E_2, E_3, E_5 and E_6. Indeed, it can be proved that with the usual product ansatz of Section 1.2, there is no Freund-Rubin type solution, with or without "torsion", capable of reproducing these extrema [32]. This led to claims that conjecture II was wrong. It was realized in [34] that the resolution to the problem was the "warp-factor" $f(y)$ [35,32,36] discussed in Section 1.2, which we have ignored up until now. In other words, one should generalize the ground state ansatz for the metric $\hat{g}_{MN}$ to be

$$< \hat{g}_{MN}(x, y) >= \begin{pmatrix} f(y) g_{\mu\nu}(x) & 0 \\ 0 & g_{mn}(y) \end{pmatrix} \tag{2.4.1}$$

One can now find an S^7 solution with $SO(7)_s$ symmetry and $N = 0$ which can be identified with the E_2 extremum [34]. In [37], it was shown that this de Wit-Nicolai solution can be written in the simple form

$$\begin{aligned} f(y) &= a\left(1 - \frac{4}{5} \sin^2 \alpha\right)^{2/3} \\ g_{mn} dy^m dy^n &= b f^{-1/2} \sin^2 \alpha d\Omega_6^2 + c f d\alpha^2 \end{aligned} \tag{2.4.2}$$

where $d\Omega_6^2$ is the metric on the round S^6, and α is the seventh coordinate with $0 \leq \alpha \leq \pi$ and a, b, and c are constants. The E_3 extremum can be explained in a similar way [38], as can presumably the E_5 and E_6. It is also interesting to note that all the known solutions with $f(y) \neq 1$ correspond to *inhomogeneous* deformations of S^7 *i.e.* they are not coset spaces. This is because any extremum with $< S > \neq 0$ corresponds to a deformation h_{mn} of S^7 induced by the 35_v Killing tensors [17]. As explained in [21,22], the trace $h^m{}_m$ is non-constant for the 35_v (in contrast to the 1 and 300) and hence the deformed S^7 is inhomogeneous.

As for stability, E_0, E_3, and E_5 are automatically stable; E_1 and E_2 are known to be unstable [26]; the rest have not yet been analyzed.

3.1 Generic Kaluza-Klein Theories

In attempting to establish the truth of conjecture I and II, we are forced to re-examine some of the basic assumptions of Kaluza-Klein theories in general. Implicit in much of the Kaluza-Klein literature is the assumption that it is in fact possible to truncate the full theory to its massless sector in a consistent fashion. This is the subject of this section and is based on the paper by Duff, Nilsson, Pope and Warner [39], and a forthcoming paper by Duff, Nilsson, and Pope [40].

Let us begin by recalling that in modern approaches to Kaluza-Klein theories, the extra (k) dimensions are treated as physical and are not to be regarded merely as a mathematical device. In this framework, therefore, it is essential that at every stage in the derivation of the effective four-dimensional field theory one maintains consistency with the higher-dimensional field equations. To derive this effective theory one selects the ground state of (spacetime) $\times$ (compact manifold M_k) and performs a generalized Fourier expansion of all the fields in terms of harmonics on M_k. Provided one retains all the modes in this expansion (*i.e.* all the massive states) then no such problems of inconsistency can arise. Moreover, it is well-known that if M_k has isometry group G then the theory includes massless Yang-Mills bosons with gauge group G. (Assuming, as we shall for the time being, that any other matter fields non-zero in the ground-state are singlets under G.)

In practice, however, one is often interested in extracting an effective "low energy" theory by discarding all but a finite number of states including the massless graviton, the massless gauge fields and other matter fields which are usually (but not necessarily) massless. For historical reasons this procedure is often called the "Kaluza-Klein ansatz". Despite the name, this should not be an ad hoc procedure but should correspond to retaining only the appropriate Fourier modes, for example, the zero eigenvalue modes for the mass operators in the case of the massless particles. It is generally believed that the correct ansatz for the metric tensor $\hat{g}_{MN}(x, y)$ is

$$\begin{aligned}
\hat{g}_{\mu\nu}(x,y) &= g_{\mu\nu}(x) + A^{\alpha}_{\mu}(x)A^{\beta}_{\nu}K^{m\alpha}(y)K^{n\beta}(y)\mathring{g}_{mn}(y) \\
\hat{g}_{\mu n}(x,y) &= A^{\alpha}_{\mu}(x)K^{m\alpha}(y)\mathring{g}_{mn}(y) \\
\hat{g}_{mn}(x,y) &= \mathring{g}_{mn}(y)
\end{aligned} \tag{3.1.1}$$

where the coordinates $x^{\mu}(\mu = 1...4)$ refer to spacetime and $y^{m}(m = 1...k)$ to the extra dimensions and $\mathring{g}_{mn}(y)$ is the metric on M_k. The quantities $K^{m\alpha}(y)$ are the Killing vecores corresponding to the isometries of this metric and α runs over the dimension of the isometry group G. The claim that this is the correct ansatz is based on the observation that substituting (3.1.1) into the higher dimesional action and integrating over y, one obtains the four dimensional Einstein-Yang-Mills action with metric $g_{\mu\nu}(x)$ and gauge potential $A_\mu^\alpha(x)$.

But as we have already emphasized the correct Kaluza-Klein ansatz must be consistent with the higher dimensional *field equations* and, as we shall now demonstrate, (3.1.1) does not in general satisfy this criterion. As has already been stressed elsewhere [16,17] one obvious source of inconsistency is the neglect of scalar fields in (3.1.1). For example, setting $\hat{g}_{55} = 1$ in the $d = 5$ pure gravity theory is inconsistent with the $\hat{R}_{55} = 0$ components of the Einstein equation which would force $F_{\mu\nu}F^{\mu\nu}$ to vanish. In this case, the remedy is simple: one includes the single massless scalar field ϕ via $\hat{g}_{55} = \phi(x)$, then after a suitable Weyl rescaling the $\hat{R}_{55}$ equation becomes

$$\Box(log\phi) \sim \phi F_{\mu\nu}F^{\mu\nu}.$$

In this example, the process of restoring consistency stops here; there is no need to include any higher massive modes in the Fourier expansion.

The purpose of this section is to point out a new, and much more serious, source of inconsistency arising from (3.1.1) . To illustrate it, let us consider the field equations of pure gravity with a positive cosmological constant Λ in $d = 4 + k$ dimensions

$$\hat{R}_{MN} = \Lambda \hat{g}_{MN} \tag{3.1.2}$$

This theory admits a classical ground state solution of de Sitter spacetime $\times$ compact manifold M_k. (The fact that spacetime is de Sitter rather than anti-de Sitter or Minkowski space need not concern us here since we are concerned only with the mathematical consistency of the ansatz.) Substituting (3.1.1) into (3.1.2) we find the $d = 4$ Einstein equation

$$R_{\mu\nu} - \frac{1}{2}g_{\mu\nu}R + \Lambda g_{\mu\nu} = \frac{1}{2}(F_{\mu\rho}{}^{\alpha}F_{\nu}{}^{\rho\beta} - \frac{1}{4}g_{\mu\nu}F_{\rho\sigma}{}^{\alpha}F^{\rho\sigma\beta})K_n{}^{\alpha}K^{n\beta} \tag{3.1.3}$$

where $F_{\mu\nu}{}^{\alpha}$ is the Yang-Mills field strength. The inconsistency is now apparent. The left hand side of (3.1.3) is independent of y while the right-hand side depends on y via the Killing vector combination $K_m{}^\alpha K^{m\beta}$. For example, when $M_k = S^k$ with its $SO(k+1)$ invariant metric,

$$K_m^\alpha(y) K^{m\beta}(y) = \delta^{\alpha\beta} + Y^{\alpha\beta}(y) \tag{3.1.4}$$

where $Y^{\alpha\beta}(y)$ is that harmonic of the scalar Laplacian with next to lowest non-vanishing eigenvalue $2\Lambda(k+1)/(k-1)$ belonging to the $k(k+3)/2$ dimensional representation of $SO(k+1)$.

How can consistency be restored? It is not difficult to see that including scalars in the ansatz (3.1.1) is not sufficient to cure the problem with the gauge field's stress tensor. One might try including the next-to-lightest massive graviton which involves the harmonic $Y^{\alpha\beta}(y)$ in an attempt to provide the correct y dependence on the left hand side of (3.1.3). However, it is well known that only with zero [41] or an infinite number of massive gravitons is the theory consistent. In other words, one would end up having to put back all the massive states.

If one insists on making a truncation to a finite number of states including gauge bosons, one may alternatively select a subgroup G' of G with Killing vectors $K_m^{\alpha'}$, where α' runs over the dimension of G', for which

$$K_m^{\alpha'} K^{m\beta'} = \delta^{\alpha'\beta'} \tag{3.1.5}$$

One obvious way to achieve this arises when M_k is itself the non-abelian group manifold G' which, with its bi-invariant metric, has isometry group $G = G' \times G'$. Then the left-invariant or right-invariant vector fields are Killing vectors which separately obey (3.1.5), but taken together the cross terms generate a y-dependent piece as in (3.1.4). Contrary to many claims in the literature, therefore, the Kaluza-Klein ansatz for group manifolds yields the gauge bosons only of G' and not $G' \times G'$. The full invariance of $G' \times G'$ is restored only by including the massive modes. For abelian group manifolds G', the isometry group is only G' and all Killing vectors automatically satisfy (3.1.5). So for compactification on the k-torus T^k, the ansatz is always consistent.

Another way to achieve (3.1.5) arises when M_k is a principle bundle with G' as the fibre. For example $M_k = S^{4n+3}$ (n = positive integer) is an $SU(2)$ bundle over HP^n. In this case there are two Einstein metrics: the round sphere with $G = SO(4n+4)$ and the squashed sphere with $G = Sp(n+1) \times SU(2)$. In either case, the $SU(2)$ Killing vectors on the fibre satisfy (3.1.5). Once again, the residual gauge group is much smaller than the isometry group.

Note that (3.1.5) implies the existence of everywhere non-vanishing vector fiels and so (3.1.5) can never be satisfied on spaces not admitting such fields *i.e.* on spaces with non-zero Euler number χ (odd dimensional spaces always have $\chi = 0$). For example S^{2n} has $\chi = 2$, and CP^n has $\chi = n + 1$. In these cases, the Kaluza-Klein ansatz would involve no gauge bosons at all. Since (3.1.5) is a sum of k terms, it follows that for a general space, the number of Killing vectors satisfying (3.1.5) is less than or equal to k, the dimension of the space, with equality for group manifolds.

Of course, having cured the inconsistency of the $\mu\nu$ components of the Einstein equation, one must also ensure the consistency of the remaining components. For $\hat{R}_{mn}$, this would involve at least the inclusion of a scalar singlet in the ansatz for $\hat{g}_{mn}$.

3.2 $d = 11$ supergravity and the seven -sphere

The situation we have described so far changes radically when we turn our attention to $d = 11$ supergravity. The reason for this difference is the presence of the three-index gauge field $\hat{A}_{MNP}$ in addition to the metric $\hat{g}_{MN}$. Recall that the bosonic field equations are

$$\hat{R}_{MN} - \frac{1}{2}\hat{g}_{MN}\hat{R} = \frac{1}{3}(\hat{F}_{MNPQ}\hat{F}_N{}^{PQR} - \frac{1}{8}\hat{g}_{MN}\hat{F}^2) \tag{3.2.1}$$

$$\nabla_M \hat{F}^{MPQR} = -\frac{1}{576}\epsilon^{M_1\ldots M_8 PQR}\hat{F}_{M_1\ldots M_4}\hat{F}_{M_5\ldots M_8} \tag{3.2.2}$$

where $\hat{F}_{MNPQ} = 4\partial_{[M}\hat{A}_{NPQ]}$. It is this $\hat{A}_{MNP}$ field, rather than an explicit cosmological constant, which is responsible for the spontaneous compactificaiton to $d = 4$. In particular, let us consider the Freund-Rubin type compactifications of Section I with ground state $AdS \times M_7$ obtained from

$$< \hat{g}_{\mu\nu} > = \overset{\circ}{g}_{\mu\nu}(x) \qquad < \hat{g}_{mn} >= \overset{\circ}{g}_{mn}(y)$$
$$< \hat{F}_{\mu\nu\rho\sigma} > = 3m\epsilon_{\mu\nu\rho\sigma} \tag{3.2.3}$$

with all other components vanishing, where $\overset{\circ}{g}_{\mu\nu}$ is the metric on AdS and $\overset{\circ}{g}_{mn}(y)$ the Einstein metric on M_7 satisfying $\overset{\circ}{R}_{mn} = 6m^2\overset{\circ}{g}_{mn}$.

The critical observation is that the standard Kaluza-Klein ansatz for the gauge bosons must now be augmented by the additional ansatz [17,25]

$$\hat{F}_{\mu\nu pq} = -\frac{1}{4}m^{-1}\epsilon_{\mu\nu\rho\sigma}F^{\rho\sigma\alpha}\nabla_{[p}K^{\alpha}_{q]}. \tag{3.2.4}$$

(In the case of the round S^7 this can be seen from (2.3.4) and (2.3.8)). Since we are concerned primarily with possible inconsistencies in the gauge field sector, we shall postpone the inclusion of the scalar fields. Substituting (3.1.1) and (3.2.4) into (3.2.1) we find the $d = 4$ Einstein equation [24]

$$\begin{aligned} &R_{\mu\nu} - \frac{1}{2}g_{\mu\nu}R - 12m^2 g_{\mu\nu} = \\ &\frac{1}{2}(F_{\mu\nu}{}^{\alpha}F_{\nu}{}^{\rho\beta} - \frac{1}{4}g_{\mu\nu}F_{\rho\sigma}{}^{\alpha}F^{\rho\sigma\beta})(K_m{}^{\alpha}K^{m\beta} \\ &+ \frac{1}{2}m^{-2}\nabla_m K^{\alpha}_n \nabla^m K^{n\beta}) \end{aligned} \tag{3.2.5}$$

Comparing (3.2.5) with (3.1.3) we see that instead of the consistency condition (3.1.5) which permits at most 7 Killing vectors we now have

$$K^{\alpha'}_m K^{m\beta'} + \frac{1}{2}m^{-2}\nabla_m K^{\alpha'}_n \nabla^m K^{n\beta'} = \delta^{\alpha'\beta'} \tag{3.2.6}$$

which permits at most 28 since it is the sum of 28 terms. Under what circumstances is (3.2.6) satisfied? Let us begin by examining the maximally symmetric round S^7 solution. Since S^7 admits the maximum of 8 unbroken supersymmetries i.e. 8 Killing spinors $\eta^I (I = 1...8)$ satisfying $D_m\eta^I = \frac{m}{2}\Gamma_m\eta^I$, we may form all 28 Killings vectors of $SO(8)$ via [17]

$$K^{IJ}_m = \bar{\eta}^I\Gamma_m\eta^J \tag{3.2.7}$$

and for convenience, we choose the η's to be orthonomal $\bar{\eta}^I\eta^J = \delta^{IJ}$. Using the Fierz identity for $d = 7$ commuting Majorana spinors [17]

$$\psi\bar{\phi}\chi = \frac{1}{8}(\chi\bar{\phi}\psi - \Gamma_m\chi\bar{\phi}\Gamma^m\psi - \frac{1}{2}\Gamma_{mn}\chi\bar{\phi}\Gamma^{mn}\psi + \frac{1}{6}\Gamma_{mnp}\chi\bar{\phi}\Gamma^{mnp}\psi) \tag{3.2.8}$$

one may verify that condition (3.2.6) is indeed satisfied for all 28 Killing vectors of $SO(8)$ and that the ansatz (3.1.1) and (3.2.4) are therefore consistent. Of course one must also include the scalars and pseudoscalars as well in order to verify complete consistency, not only of the $\mu\nu$ components of (3.2.1) but also the other components of (3.2.1) and all components of (3.2.2). One must also check the fermion ansatz. On the round S^7 the total ansatz for the massless $N = 8$ supermultiplet of spins $(2, 3/2, 1, 1/2, 0^+, 0^-)$ is not yet known to all orders owing to the complicated non-polynomial dependence on the scalars and pseudoscalars. However, to the order to which it is known, everything is consistent. This seems to be guaranteed by the fact that the ansatz for all spins can be expressed in terms of the Killing spinors η^I [7].

Indeed, whenever M_7 admits $N \leq 8$ Killing spinors one may always find Killing vectors satisfying (3.2.6) namely those constructed as in (3.2.7). For such compactifications the total gauge group G is always of the form $G = SO(N) \times K$ for some group K [42]. The construction (3.2.7) singles out $G' = SO(N)$ as the subgroup surviving in the consistent ansatz. These are the gauge bosons of the N-extended gravity supermultiplet.

In fact the round S^7 seems to be the only M_7 which exploits the extra freedom permitted by (3.2.6) as compared with (3.1.5). To see this, we first rewrite (3.2.6) as

$$\frac{1}{4m^2}(\Box + 16m^2) K_m^{\alpha'} K^{m\beta'} = \delta^{\alpha'\beta'}. \tag{3.2.9}$$

Thus whenever M_7 admits Killing vectors $K_m^{\alpha'}$ satisfying (3.1.5) then $\frac{1}{2}K_m^{\alpha'}$ will also satisfy (3.2.9). In order to see whether there are any more Killing vectors satisfying (3.2.9) but not (3.1.5), let us observe that in the case of a general coset space the appropriately normalized Killing vectors satisfy

$$K_m^{\alpha'} K^{m\beta'} = \frac{1}{4}\delta^{\alpha'\beta'} + \phi^{\alpha'\beta}(y) \tag{3.2.10}$$

where $\phi^{\alpha'\beta'}(y)$ is tracefree. The extra freedom of (3.2.9) over (3.1.5) is that $\phi^{\alpha'\beta'}(y)$ can now be non-zero, provided that is satisfies

$$-\Box\phi^{\alpha'\beta'} = 16m^2\phi^{\alpha'\beta'} \tag{3.2.11}$$

Moreover, a case by case analysis of all homogeneous M_7's reveals that, with the exception of the round S^7, none of them permits eigenfunctions of the Laplacian in the right representation and with the right eigenvalues to satisfy (3.2.11). Interestingly enough, condition (3.2.11) implies the existence of massless scalars in the $O^{+(1)}$ tower of [43]. In the case of the round S^7, these are just the 35_v massless scalars of the $N = 8$ supermultiplet. None of the other M_7's has massless scalars of this kind. The surviving gauge groups G' compatible with the consistent ansatz are listed in Table 3 for some typical M_7 space.

TABLE 3

Gauge groups surviving in the consistent massless ansatz

M_7	G	G'
round S^7	SO(8)	SO(8)
squashed S^7	SO(5)xSU(2)	SU(2)
M(m,n)	SU(3)xSU(2)xU(1)	U(1)
N(k,1)	SU(3)xU(1)	U(1)
N(1,1)	SU(3)xSU(2)	SU(2)
Q(p,q,r)	$[SU(2)]^3$xU(1)	U(1)
T^7	$[U(1)]^7$	$[U(1)]^7$

We have seen that all gauge bosons appearing in the consistent ansatz for pure gravity in $d = 11$ will also appear in the consistent ansatz for supergravity but with a normalization differing by a factor of 2. We remark incidentally that this is responsible for the modification of Weinberg's [23] formula relating the gauge coupling constant to the radius of the extra dimensions. One finds [24]

$$e^2(supergravity) = \frac{1}{4}e^2(pure\ gravity). \tag{3.2.12}$$

Note that by integrating (3.1.3) and (3.2.5) over y, one may verify that Wienberg's formula and (3.2.12) remain valid for all the gauge bosons of G and not merely those of G'. Note also that the ansatz (3.1.1) together with the normalization of the Killing vector chosen in (3.2.6) establishes the convention $4\pi G = 1$ in the Einstein equation (3.2.5). This result was used in Section 2.3 in deriving the relation $e^2 = 16\pi G m^2$ on the round S^7, a result which could also be obtained from (3.2.12) without reference to the transformation rules. To obtain the coupling constant directly, one can calculate the Yang-Mills structure constants from the algebra of the Killing vectors having once established their normalization.

3.3 Significance of consistency

We conclude this section with some comments on the significance of Kaluza-Klein consistency.

(i) The consistency condition (3.2.6) was derived under the assumption of Freund-Rubin compactifications given by (3.2.3), but we might also consider more general AdS solutions with $< \hat{g}_{\mu\nu} >= f(y) g_{\mu\nu}(x)$ and/or $< \hat{F}_{mnpq} > \neq 0$. We have little to say about the consistency of the Kaluza-Klein ansatz for general solutions of this kind but we can make some plausible conjectures in the case of those which are topologically S^7. These conjectures are related to conjectures I and II of Section 2. It seems highly plausible that just as the round S^7 ansatz for the massless fields of the de Wit-Nicolai theory was already completely consistent without any reduction of the vacuum symmetry G, so the ansatze for the other S^7 vacua corresponding to Table 3 will also be completely consistent provided we retain the same de Wit-Nicolai fields, some of which will now be massive. Indeed, it seems that S^7 and $d = 11$ supergravity go hand-in-hand.

(ii) For a generic Kaluza-Klein theory, it would seem that our consistency condition is equivalent to the requirement that the entire Kaluza-Klein ansatz (and not merely the ground state metric) is left invariant under a transitively acting subgroup of the isometry group. This is a version of what is sometimes called "K-invariance" [44] of the ansatz. For example, this would yield only the gauge bosons of G' in the case of a group manifold G' and only the gauge bosons of $SU(2)$ in the case of S^{4n+3}. The reason for this equivalence is that the consistency condition amounts to requiring that the Lagrangian obtained by substituting the ansatz into the higher dimensional Lagrangian be independent of y. Although this is guaranteed by "K-invariance", it is clearly too strong a condition as the example of S^7 and $d = 11$ supergravity illustrates. [The smallest subgroup of $SO(8)$ which acts transitively on S^7 is $SO(5)$ and so "K-invariance" would yield only the gauge bosons of $SU(2)$ This gives rise to new consistent truncations of $d = 11$ supergravity with $N \leq 3$, not contained in the $N = 8$ truncation [45].]

(iii) In summary we have seen that the standard Kaluza-Klein ansatz for massless gauge bosons is in general inconsistent with the higher dimensional field equations. Consistency can be restored either by putting back the infinite tower of massive states or else by reducing the gauge group G to some subgroup G'. Exeptions to this rule are provided by certain S^7 compactifications of $d = 11$ supergravity. An interesting question, to which we do not know the answer, is whether there are other Kaluza-Klein theories with consistent truncations. We emphasize that we are not disputing the Yang-Mills content of the complete Kaluza-Klein theory but only the ability to truncate consistently while retaining the full gauge symmetry.

So far we have been concerned with mathematical consistency. Does inconsistency matter from a physical point of view? Here it is important to distinguish between Kaluza-Klein theories which predict a Minkowski spacetime ground state and those, of the kind discussed in this paper, which predict AdS with a Planck-sized cosmological constant. In the former case the answer is probably no if we are interested in energies small compared with the compactification scale, since the inconsistency will only be relevant for operators of dimension > 4. (We are grateful to E. Witten and S. Weinberg for pointing this out.) In the case of a cosmological constant $\Lambda \sim e^2/G$ (typical of $d = 11$ supergravity and all those Kaluza-Klein theories for which Λ is not fine-tuned to zero), then this is no longer the case. For example dimension six operators like $GR\ F_{\mu\nu}F^{\mu\nu}$ may be converted to dimension four operators like $e^2\ F_{\mu\nu}F^{\mu\nu}$ on using the field equations $R = e^2/G + ...$ (This observation was also relevant when computing the $\beta(e)$ function in gauged supergravity and its relation to the renormalization of $G\Lambda$ [11].) In this case, of course, it is no longer clear that the massless sector is in any case a good approximation to the full theory.

In our opinion, until such time as the cosmological constant problem in Kaluza-Klein theories has been solved (as opposed to being merely fine-tuned away), the physical significance of the Kaluza-Klein ansatz remains obscure.

In summary, it seems that conjectures I and II are almost certainly true but we are as far away from understanding their *physical* significance as we were when they were made. For example, the squashed S^7 of [19] is a perfectly good solution of the $d = 11$ theory but has nothing to do with the de Wit-Nicolai theory. This S^7 solution has $< S > \neq 0$ and $< P > = 0$ and it is interesting to compare it with the $< S > \neq 0$, $< P > = 0$ S^7 solutions in Table 2. Warner [38] has pointed out that these S^7 geometries can all be embedded in flat $\mathbf{R}^8$ in contrast to the squashed S^7 [19,21].

Recent Results

The problem of consistent truncations in Kaluza-Klein theories in general, and $d = 11$ supergravity in particular, is treated in a recent paper by Duff and Pope [45]. Here it is stressed that if the truncation to the $N = 8$ massless supermultiplet is consistent in the sense described in the paper, then conjecture II is an automatic consequence of conjecture I. It is also made clear that the problem of inconsistency is not one of non-linearity: an inconsistent ansatz goes wrong already at the linear level! Further progress in demonstrating the truth of these conjectures *to all orders* has recently been made by Nilsson [46] and by de Wit, Nicolai and Warner [47]. The need for a final proof is amply demonstrated by a recent paper of Boucher [48] which claims to prove that there is no consistent $N = 8$ truncation on S^7! In our opinion, Boucher has failed to realize that although the K-invariance of Section 3.3 is a sufficient condition for consistency, it is not necessary [45].

The final proof should also clear up on or two other naggging points. Does the $N = 8$ theory obtained from $d = 11$ on S^7 exhibit the $SU(8)$ structure? (The Cremmer-Julia theory obtained from T^7 does not. Further field redefinitions are necessary [7]). Let us bear in mind that in this formulation the de Wit-Nicolai theory has anomalies [49] even though the $d = 11$ theory is anomaly-free. What role, if any, does the $d = 11$ Chern-Simons term [3] play in all of this?

ACKNOWLEDGEMENTS

I am grateful to my collaborators M. Awada, B. Nilsson, C. Pope and N. Warner. Conversations with C. Pope and N. Manton are especially acknowledged. This work was supported in part by the National Science Foundation under Grant No. PHY77-27084, supplemented by funds from the National Aeronautics and Space Administration. I am grateful for the hospitality of the Institute for Theoretical Physics, Santa Barbara.

REFERENCES

1. E. Cremmer, B. Julia and J. Scherk, "Supergravity Theory in 11 Dimensions," Phys. Lett. **76**B, 409 (1978).

2. E. Cremmer and S. Ferrara, "Formulation of $d = 11$ Supegravity in Superspace," Phys. Lett. **91**B, 61 (1980).

3. L. Alvarez-Gaumé and E. Witten "Gravitational Anomalies," Nucl. Phys. B**234**, 269 (1984); L. Alvarez-Gaumé, S. Della Pietra, G. Moore, HUTP-84/A028.

4. P.G.O. Freund and M.A. Rubin, "Dynamics of Dimensional Reduction," Phys. Lett. B**97**, 233 (1980).

5. This is Myers' theorem.

6. B. de Wit and H. Nicolai, "$N = 8$ Supergravity," Nucl. Phys. B**208**, 323 (1982).

7. E. Cremmer and B. Julia, "The $SO(8)$ Supergravity," Nucl. Phys. B**159**, 141 (1979).

8. D.Z. Freedman and A. Das, "Gauge Internal Symmetry in Extended Supergravity," Phys. Lett. **74**B, 333 (1977).

9. E.S. Fradkin and M. Vasiliev, unpublished.

10. A. Das, M. Fischler and M. Rocek, "Super-Higgs Effect in a new Class of Scalar Models and a Model of Super QED," Phys. Rev. D**16**, 3427 (1977).

11. S.M. Christensen, M.J. Duff, G.W. Gibbons and M. Rocek, "Vanishing One-Loop β Function in Gauged $N > 4$ Supergravity," Phys. Rev. Lett. **45**, 161 (1980).

12. T. Curtwright, "Charge Renormalization and High Spin Fields," Phys. Lett **102**B, 17 (1981).

13. M.J. Duff, "Ultraviolet Divergences in Extended Supergravity" in : Supergravity, 81, eds. S. Ferrara and J.G. Taylor, 257 (C. U. P., 1982).

14. M.J. Duff and D.J. Toms, "Kaluza-Klein Kounterterms," in "Unification of the Fundamental Interactions II," eds. J. Ellis and S. Ferrara (Plenum 1982).

15. F. Englert, "Spontaneous Compactification of 11-Dimensional Supergravity," Phys. Lett. **119**B, 339 (1982).

16. M.J. Duff, "Supergravity, the Seven-Sphere and Spontaneous Symmetry Breaking," Proceedings of the Marcel Grossman Meeting on General Relativity, Shanghai August 1982, ed. Hu Ning (Science Press and North Holland, 1983) and Nucl. Phys. B**219**, 389 (1983).

17. M.J. Duff and C.N. Pope, "Kaluza-Klein Supergravity and the Seven Sphere," in Supersymmetry and Supergravity '82 eds. S. Ferrara, J.G. Taylor and P. van Nieuwenhuizen (World Scientific Publishing, 1983).

18. R. D'Auria, P. Fré and P. van Nieuwenhuizen, "Symmetry Breaking in $d = 11$ Supergravity on the Parallelized Seven-Sphere," Phys. Lett. **122**B, 225 (1983).

19. M.A. Awada, M.J. Duff, and C.N. Pope, "$N = 8$ Supergravity Breaks Down to $N = 1$," Phys. Rev. Lett. **50**, 294 (1983).

20. B. Biran, F. Englert, B. de Wit, and H. Nicolai, "Gauged $N = 8$ Supergravity and its Breaking from Spontaneous Compactification," Phys. Lett. **124**B, 45 (1983).

21. M.J. Duff, B.E.W. Nilsson and C.N. Pope, "Spontaneous Supersymmetry Breaking by the Squashed Seven-Sphere," Phys. Rev. Lett. **50**, 2043 (1983); Erratum **51**, 846 (1983).

22. M.J. Duff, B.E.W. Nilsson and C.N. Pope, "Superunification from Eleven Dimensions," Nucl. Phys. B**233**, 433 (1984).

23. S. Weinberg, "Charges from Extra Dimensions," Phys. Lett. **124**B, 265 (1983).

24. M.J. Duff, C.N. Pope and N.P. Warner, "Cosmological and Coupling Constants in Kaluza-Klein Supergravity," Phys. Lett. **130**B, 254 (1983).

25. M.A. Awada, B.E.W. Nilsson and C.N. Pope, "The Supersymmetry Transformation Rules for Kaluza-Klein Supergravity on the Seven-Sphere," Phys. Rev. (rapid communication) D**29**, 335 (1984).

26. B. de Wit and H. Nicolai, "On the Relation Between $d = 4$ and $d = 11$ Supergravity," Nucl. Phys. B**243**, 91 (1984).

27. B. Biran, A. Casher, F. Englert, M. Rooman and P. Spindel, "The fluctuating Seven Sphere in Eleven Dimensional Supergravity," Phys. Lett. **134**B, 179 (1984).

28. N.P. Warner, "Some New Extrema of the Scalar Potential of Gauged $N = 8$ Supergravity," Phys. Lett. **128**B, 169 (1983).

29. N.P. Warner, "Some Properties of the Scalar Potential in Gauged Supergravity Theories," Nucl. Phy. B**231**, 250 (1984).

30. B. de Wit and H. Nicolai, "The Parallelizing S^7 Torsion in Gauged $N = 8$ Supergravity," Nucl. Phys. B**231**, 506 (1984).

31. B. Biran and Ph. Spindel, "Instability of the Parallelized Seven-Sphere An Eleven-Dimensional Approach," Bruxelles-Mons preprint ULB-TH 84/00 (March 1984).

32. M. Gunaydin and N.P. Warner, Nucl. Phys. B (to appear).

33. C.N. Pope and N.P. Warner, "An $SU(4)$ Invariant Compactification of $d = 11$ Supergravity on a Stretched Seven-Sphere," Phys. Lett. B, to appear.

34. B. de Wit, and H. Nicolai, "A New $SO(7)$ Invariant Solution of $d = 1$ Supergravity," TH-3956-CERN (1984).

35. V.A. Rubakov and M.E. Shaposhnikov, "Extra Space-Time Dimensions Toward a Solution to the Cosmological Constant Problem," Phys. Lett. **125**B 139 (1983).

36. P. van Nieuwenhuizen, "An Introduction to Simple Supergravity and the Kaluza-Klein Program" in Relativity, Groups and Topology II, eds. B.S. de Witt and R. Stora, (North Holland, 1984).

37. P. van Nieuwenhuizen and N.P. Warner, ITP-SB-84-75 (unpublished).

38. N.P. Warner (unpublished).

39. M.J. Duff, B.E.W. Nilsson, C.N. Pope and N.P. Warner, "On the Consistency of the Kaluza-Klein Ansatz," Phys. Lett. B, to appear.

40. M.J. Duff, B.E.W. Nilsson, and C.N. Pope, "Kaluza-Klein Supergravity" Phys. Reports (to appear).

41. D. Boulware and S. Deser, Ann. Phys. (N.Y.) **81**, 193 (1975).

42. L. Castellani, R. D'Auria and P. Fré, "$SU(3) \otimes SU(2) \otimes U(1)$ From $D = 11$ Supergravity," Nucl Phys. B**239**, 610 (1984).

43. M.J. Duff, B.E.W. Nilsson and C.N. Pope, "The Criterion for Vacuum Stability in Kaluza-Klein Supergravity," Phys. Lett. **139**B, 154 (1984).

44. P. Forgacs and N.S. Manton, "Space-Time Symmetries in Gauge Theories," Commun. Math Phys. **72**, 15 (1980).

45. M.J. Duff and C.N. Pope, "Consistent Truncations in Kaluza-Klein Theories," ITP preprint NSF-ITP-84-166.

46. B.E.W. Nilsson, preprint, Gothenburg 84-52.

47. B. de Wit, H. Nicolai and N. Warner (to appear).

48. W. Boucher, D.A.M.T.P. preprint.

49. S. Ferrara, P. di Vecchia, and L. Girardello, CERN preprint TH 4026/84.

SUPERSYMMETRY AND SUPERGRAVITY FOR MODEL BUILDERS: GENERAL COUPLINGS AND SYMMETRY BREAKING

Sergio Ferrara

CERN, Geneva, Switzerland

1. INTRODUCTION

In the present lectures, supersymmetric gauge theories of fundamental particle interactions will be described at an introductory level. The past few years have shown us how crucial is the role played by supergravity[1] effects in describing a physically meaningful theory encompassing the fundamental electroweak and strong 'low-energy' gauge interactions in a supersymmetric setting.

These theories[2] nowadays offer one of the most attractive alternatives to the so-called Standard Model, and will have important consequences for particle physics at the present available energies and at those soon to be achieved. In particular, investigation of spontaneous supersymmetry breaking in Lagrangian field theory, and its potential applications to models for elementary particle interactions, have revealed that the so-called supersymmetry-breaking scale M_{SB}, defined through the goldstino decay constant[3] can possibly be as large as

$$M_{SB} = O[(m_W m_P)^{\frac{1}{2}}] , \tag{1.1}$$

i.e. intermediate between the Planck scale $m_P = O(10^{19}$ GeV) and the weak scale $m_W = O(100$ GeV). Under these circumstances supergravity corrections to globally supersymmetric models can no longer be neglected since the gravitino mass growth implies, through the super-Higgs mechanism, the relation[4,5]

$$m_{3/2} = \sqrt{\frac{8\pi}{3}}\,\frac{M_{SB}^2}{m_P} = O(m_W) \ , \tag{1.2}$$

and the intramultiplet splitting of particle masses, due to supergravity effects, is typically[6,7]

$$\Delta m^2 = O\left(m_{3/2}^2\right) \ . \tag{1.3}$$

Equation (1.3) reflects the fact that the Goldberger-Treiman relation for supersymmetry,

$$\Delta m_i^2 = g_i M_{SB}^2 \ , \tag{1.4}$$

can be fulfilled with a large value of M_{SB} provided the goldstino coupling to ordinary matter fields is very small:

$$g_i = O(m_W/m_P) \ . \tag{1.5}$$

This is what naturally happens when the spontaneous supersymmetry breaking is caused by a 'hidden' sector which is gravitationally coupled to ordinary matter[5].

We refer the reader to recent reviews[2] for the phenomenological motivations for considering an effective supersymmetry-breaking scale $\Delta m^2 = O(m_W^2)$. One of the advocated motivations is related to the so-called hierarchy problem of grand unified theories (GUTs) of electroweak and strong interactions[7].

From a more fundamental point of view, the interest for supersymmetry in particle physics is at least threefold:

i) It offers a theoretical framework in which ultraviolet divergences are milder owing to mutual cancellations of boson and fermion loops. One aspect of these cancellations is the fact that supersymmetry avoids quadratic divergences for elementary scalar masses, thus avoiding unnatural fine-tuning for the Higgs scalar v.e.v.'s . These fine-tunings are essential, in the Standard Model, in order for the theory to be meaningful beyond the 1 TeV scale.

ii) It is the only symmetry principle which enlarges, in a non-trivial way, the space-time Poincaré symmetry through the mathematical concept of graded Lie algebra.

iii) Supersymmetry provides a non-trivial link between particle physics and gravitation. This link is operative both in N-extended supergravity theories in which all elementary fields are in the same N-extended supermultiplet as well as in 'low-energy' supergravity in which the gravitino mass controls the particle mass splitting and the electroweak gauge symmetry breaking.

The present lectures are intended for physicists willing to practice in the field of 'low-energy' supersymmetry, and contain a useful dictionary for constructing models for elementary particle interactions.

The lectures are organized as follows: In Sections 2, 3, and 4 we derive the final form of locally supersymmetric N = 1 supergravity-matter coupled systems from general invariance principles and tensor calculus for supersymmetric field theories.

In Section 5 the super-Higgs effect is described, and in Section 6 the low-energy limit of locally supersymmetric Lagrangians, with m_P (Planck mass) $\to \infty$ and $m_{3/2}$ (gravitino mass) kept fixed, is obtained. In Section 7, new examples of the super-Higgs sector of the theory are given: one example uses the idea of 'flat' potentials; the other, a locally supersymmetric generalization of the Fayet-Iliopoulos mechanism.

Section 8 reports simple tree-level examples of the generalization of the Standard Model for the electroweak theory to local supersymmetry. In Section 9, N = 2 extended supergravity theories are considered, and the relation between the N = 1 and N = 2 scalar potential is derived. The connection with possible truncation of higher-N extended supergravity is briefly discussed. In Section 10 the Higgs and super-Higgs effect for N = 2 pure Yang-Mills theories 'minimally coupled' to N = 2 supergravity is studied. All gauge internal symmetry-breaking patterns for unitary groups which also break local supersymmetry are found.

Finally, in Section 11 the N = 2 super-Higgs effect without cosmological constant is discussed, and N = 2 flat potentials are introduced.

2. TENSOR CALCULUS

The action and transformation laws of locally supersymmetric systems are constructed by the use of the N = 1 tensor calculus of supergravity[8]. A major simplification occurs by employing the so-called conformal tensor calculus which uses the Weyl multiplet for the gravitational sector, i.e. the multiplet which contains only the gauge fields of conformal supergravity[9].

The N = 1 Weyl multiplet contains the following set of fields,

$$(e_{m\mu},\ \psi_{\mu a},\ A_{\mu},\ b_{\mu})\ , \tag{2.1}$$

which are respectively the gauge fields for general coordinate transformations (vierbein), for Q supersymmetry (gravitino), for chiral U(1) (R-symmetry), and dilatations.

The additional gauge fields $\omega_{\mu mn}$, $\Phi_{\mu a}$, and $f_{m\mu}$ for local Lorentz rotations, special supersymmetry, and special conformal transformations M_{mn}, S_a, and K_m, respectively, can be solved in terms of the gauge fields in Eq. (2.1) by means of the conventional constraints*)

$$R^m_{\mu\nu}(P) = 0 \ , \qquad (2.2a)$$

$$\gamma^\mu R_{\mu\nu}(Q) = 0 \ , \qquad (2.2b)$$

$$\hat{R}_{\mu\nu mn}(M)\, e^{\nu n} + \frac{i}{3}\,\hat{R}_{\mu m}(A) = 0 \ , \qquad (2.2c)$$

$$\left(\tilde{T}_{mn} = \frac{1}{2}\,\varepsilon_{mnpq}T_{pq}\right) , \qquad (2.2d)$$

where $\hat{R}_{\mu\nu mn}$ denotes the fully convariant Lorentz connection (Riemann tensor). The curvatures are covariant, but since the constraints in Eqs. (2.2) are not Q-invariant, the variations $\delta_Q\omega_{\mu nn}$, $\delta_Q\Phi_{\mu a}$ and $\delta_Q f_{m\mu}$ are not simply given by the group rule. In the conformal tensor calculus we use two basic multiplets other than the one of Eq. (2.1). The scalar (or chiral) multiplet $S = (z, \chi_L, h)$, with Weyl weight w (w is the Weyl weight of the first component z of S) transforming as

$$\delta z = \bar{\varepsilon}_L\chi_L - \frac{1}{3}\, iw\alpha z \ , \qquad (2.3a)$$

$$\delta\chi_L = \frac{1}{2}\,\hat{\not{D}}z\varepsilon_R + \frac{1}{2}\,h\varepsilon_L + wz\eta_L + i\alpha\left(\frac{1}{2} - \frac{1}{3}\,w\right)\chi_L \ , \qquad (2.3b)$$

$$\delta h = \bar{\varepsilon}_R\hat{\not{D}}\chi_L + 2\bar{\eta}_L\chi_L(1-w) + i\alpha\left(1 - \frac{1}{3}\,w\right)h \ , \qquad (2.3c)$$

where ε is a space-time dependent anticommuting Weyl (or Majorana) spinor corresponding to Q transformations, and α is the space-time dependent parameter of the U(1) chiral group of the superconformal algebra, normalized such that for the U(1) gauge field A_μ

$$\delta_\alpha A_\mu = \partial_\mu\alpha \ . \qquad (2.4)$$

---

*) We use the conventions of Ref. 2. We also use the following conventions: $a = 1,2$, two-component spinor indices; $m = 1, \ldots, 4$, flat Lorentz indices; $\mu = 1, \ldots, 4$, world indices; $\alpha = 1 \ldots \dim G$, gauge group indices; $i,j = 1 \ldots \dim R$, G-group representation indices, R being a finite unitary representation of G. We also set, in most of our formulae, the gravitational constant $k = 1$; k is related to the Planck mass m_P as follows: $k = \sqrt{8\pi}/m_P$.

In Eq. (2.3) η is a space-time dependent anticommuting Weyl (or Majorana) spinor corresponding to special S_a supersymmetry transformations, and $\hat{D}_\mu$ denotes the superconformal covariant derivative.

The second multiplet is a real vector multiplet V: $V = (C, S_L, H, N_m, \lambda_R, D)$. Its transformation is

$$\delta G = \frac{1}{2} i\bar{\varepsilon}_L S_L - \frac{1}{2} i\bar{\varepsilon}_R S_R , \tag{2.5a}$$

$$\delta S_L = \frac{1}{2} iH\varepsilon_L - \frac{1}{2} \not{B}\varepsilon_R - \frac{1}{2} i\hat{\not{D}}C\varepsilon_R - iwC\eta_L + \frac{1}{2} i\alpha S_L , \tag{2.5b}$$

$$\delta H = -i\bar{\varepsilon}_R \hat{\not{D}} S_L - i\varepsilon_R \lambda_R + i(-2 + w)\bar{\eta}_L S_L + i\alpha H , \tag{2.5c}$$

$$\delta B_m = \left[- \frac{1}{2} \bar{\varepsilon}_L \hat{D}_m S_L - \frac{1}{2} \bar{\varepsilon}_L \gamma_m \lambda_R + \frac{1}{2}(1 + w)\bar{\eta}_L \gamma_m S_R\right] + \text{h.c.} , \tag{2.5.d}$$

$$\delta\lambda_R = \frac{1}{2}(\sigma\cdot\hat{F} - iD)\varepsilon_R + \frac{1}{2} w(iH\eta_R - \not{B}\eta_L + i\hat{\not{D}}C\eta_L) +$$

$$+ \frac{1}{2} i\alpha\lambda_R - w\not{\Lambda}_K S_L , \tag{2.5e}$$

$$\delta D = \left[\frac{1}{2} i\bar{\varepsilon}_L \hat{\not{D}}\lambda_R + iw\bar{\eta}_L\left(\lambda_L + \frac{1}{2} \hat{\not{D}}S_R\right) - w\Lambda_{Km}\hat{D}_m C\right] + \text{h.c.} , \tag{2.5f}$$

with

$$\hat{F}_{mn} = 2\hat{D}_{[m}B_{n]} + i\varepsilon_{mnpq}\hat{D}_p\hat{D}_q C , \quad [mn] = \frac{1}{2}(mn - nm) , \tag{2.5g}$$

where the symbol [ ] denotes antisymmetrization, and Λ_{Km} is the space-time dependent parameter for special conformal transformations.

The scalar multiplet S, for Weyl weight $w = 3$, has its last component h with Weyl weight $w = 4$, and it can be extended to a superconformal density

$$e^{-1} L_{S(w=3)} = h + \bar{\psi}_R\cdot\gamma\chi_L + \bar{\psi}_{\mu R}\sigma_{\mu\nu}\psi_{\nu R} z . \tag{2.6}$$

Analogously we can embed a vector multiplet V with $w = 2$ into a scalar multiplet S_V with $w = 3$, whose components are

$$(-H^*, \ -i\lambda_L - i\hat{\not{D}}S_R, \ D + \hat{\Box} C + I\hat{D}_m B^m) . \tag{2.7}$$

Then we can apply Eq. (2.6) to the multiplet S_V to obtain a density formula for the vector multiplet with $w = 2$:

$$e^{-1}L_{V(w=2)} = D + \hat{\Box} C + \left[\frac{1}{2}\, i\bar{\psi}\cdot\gamma(\lambda_R + \hat{\not{D}}S_L) - \frac{1}{2}\,\psi_{\mu L}\sigma_{\mu\nu}\psi_{\nu L}H + \text{h.c.}\right] . \tag{2.8}$$

We can also get a vector embedding of a scalar multiplet S with $w = 0$ into a vector multiplet V_S with $w = 0$, whose components are

$$V_S = \left[i(z - z^*),\ \ 2\chi_L,\ \ -2ih,\ \ -\hat{D}_m(z + z^*),\ 0,\ 0\right] . \tag{2.9}$$

The multiplication rules of scalar and vector multiplets are given respectively by

$$S_1S_2 = (z_1z_2,\ \ z_1\chi_{2L} + z_2\chi_{1L},\ \ z_1h_2 + z_2h_1 - 2\chi_{1L}\chi_{2L}) \ ; \tag{2.10a}$$

$$\begin{aligned} V_1V_2 = \Big[& C_1C_2,\ \ C_1S_{2L} + C_2S_{1L},\ \ C_1H_2 + C_2H_1 - \bar{S}_{1L}S_{2L}, \\ & C_1B_{m2} + \frac{1}{2}\, i\bar{S}_{1L}\gamma_m S_{2R} + (1 \leftrightarrow 2), \\ & C_1\lambda_{2R} + \frac{1}{2}(-\hat{\not{D}}C_1S_{2L} + H_1S_{2R} - i\not{B}_1S_{2L}) + (1 \leftrightarrow 2) , \\ & C_1D_2 - \bar{S}_{1L}\lambda_{2L} - \bar{S}_{1R}\lambda_{2R} + \frac{1}{2}\, H_1H_2^* - \frac{1}{2}\, B_1\cdot B_2 - \frac{1}{2}\, \hat{D}_mC_1\hat{D}^mC_2 \\ & - \frac{1}{2}\, \bar{S}_{1L}\hat{\not{D}}S_{2R} - \frac{1}{2}\, \bar{S}_{1R}\hat{\not{D}}S_{2L} + (1 \leftrightarrow 2)\Big] \ ; \end{aligned} \tag{2.10b}$$

$$\begin{aligned} \frac{1}{2}(S_1\bar{S}_2 + \bar{S}_1S_2) = \Big[& \frac{1}{2}\, z_1z_2^*,\ \ -iz_2^*\chi_{1L},\ \ -z_2^*h_1\ , \\ & \frac{1}{2}\, i\left(z_2^*\hat{D}_mz_1 - z_1\hat{D}_mz_2^*\right) - i\bar{\chi}_{2R}\gamma_m\chi_{1L}\ , \\ & -ih_1\chi_{2R} + i\hat{\not{D}}z_2^*\chi_{1L}\ , \\ & h_1h_2^* - \hat{D}_mz_1\hat{D}^mz_2^* - \chi_{1L}\overleftrightarrow{\hat{\not{D}}}\chi_{1R}\Big] + (1 \leftrightarrow 2) . \end{aligned} \tag{2.11}$$

The resulting multiplets have Weyl weight $w_1 + w_2$.

The rules we have given so far are sufficient for constructing all superconformal invariant actions. However, if we want to restrict our tensor calculus to Poincaré supersymmetry only, we have to introduce Poincaré rules which will enable us to construct Poincaré supersymmetric local densities for arbitrary Weyl weight of the fields involved. This is most easily achieved by introducing a new compensating chiral multiplet S_0 with Weyl weight $w = 1$, and then using it to fix a special superconformal gauge. This choice of compensating multiplet corresponds to the so-called old minimal formulation of $N = 1$ supergravity originally proposed in Refs. 7. The connection of this formulation with the new minimal formulation of Sohnius and West[10], which uses a different compensating multiplet to fix the superconformal gauge, has been given in general elswehere[11].

The chiral compensating multiplet has components

$$S_0 = \left(a,\ \xi,\ \frac{1}{3}\,u\right)\ , \tag{2.12}$$

where $u = S - iP$ is the complex auxiliary field of Poincaré supergravity defined by Cremmer et al.[5].

To get Poincaré supergravity from the superconformal tensor calculus we fix a superconformal gauge by fixing the gauges of dilatation, U(1) chiral rotation, special S-supersymmetry transformation and conformal boosts, by the conditions

$$a = 1,\quad \xi = 0\ ,\quad b_\mu = 0\ . \tag{2.13}$$

These conditions are preserved under a newly defined Poincaré Q local supersymmetry with the parameter

$$\begin{aligned}\delta^P(\varepsilon) = \delta_Q(\varepsilon) &+ \delta_S\left(\eta_L = -\frac{1}{6}\,u\varepsilon_L - \frac{1}{6}\,i\not{A}\varepsilon_R\right)\\ &+ \delta_K\left(\Lambda_{K\mu} = \frac{1}{4}\,\bar{\varepsilon}\phi_\mu - \frac{1}{4}\,\bar{\eta}\psi_\mu\right)\ .\end{aligned} \tag{2.14}$$

Equation (2.14) provides the rule for computing a Poincaré local supersymmetry transformation in terms of the local superconformal rules.

To compute the density formula for a chiral multiplet S of Weyl weight $w = 0$ we first construct the new chiral multiplet SS_0^3 with Weyl weight $w = 3$. Since in the Poincaré gauge defined by Eq. (2.13) using Eq. (2.10), we have

$$S_0^n = \left(1,\ 0,\ \frac{1}{3}\,nu\right), \tag{2.15}$$

we obtain

$$SS_0^n = \left(z,\ \chi_L,\ h + \frac{1}{3}\,nuz\right). \tag{2.16}$$

Then the Poincaré density formula for S is given by

$$\begin{aligned}[SS_0^3]_{F,SC} &= [S]_{F,\text{Poincaré}} \\ &= L_{S(w=0)} = eh + euz + e\bar{\psi}_R\cdot\gamma\chi_L + e\bar{\psi}_{\mu R}\sigma_{\mu\nu}\psi_{\nu R}z\,.\end{aligned} \tag{2.17}$$

For the vector multiplet V, we take a Weyl weight w = 0 and then introduce the compensating vector multiplet $S_0\bar{S}_0$ with w = 2 having components

$$S_0\bar{S}_0 = \left[1,\ 0,\ -\frac{2}{3}\,u,\ -\frac{2}{3}\,A_m,\ 0,\ \frac{2}{9}\left(uu^* - A_mA^m\right)\right]. \tag{2.18}$$

We then get a Poincaré density formula for V by using a superconformal density formula for $VS_0\bar{S}_0$, with w = 2, and using the multiplication rule given by Eq. (2.11) in the Poincaré gauge [Eq. (2.13)]:

$$\begin{aligned}[VS_0\bar{S}_0]_{D,SC} &= [V]_{D,\text{Poincaré}} = e^{-1}\,L_{V(w=0)} \\ &= D - \frac{1}{2}\,i\bar{\psi}\cdot\gamma\gamma_5\lambda - \frac{1}{3}(u^*H + uH^*) + \frac{2}{3}\,B_mA^m \\ &\quad - \frac{1}{3}\,i\bar{S}\gamma_5\gamma R^P + \frac{1}{4}\,i\varepsilon^{mnrs}\bar{\psi}_m\gamma_n\psi_r\left(B_s - \frac{1}{2}\,\bar{\psi}_sS\right) - \frac{2}{3}\,C\,e^{-1}L_{SC}\,,\end{aligned} \tag{2.19}$$

where

$$\begin{aligned}R^P_\mu &= e^{-1}\,\varepsilon_{\mu\nu}{}^{\rho\sigma}\gamma_5\gamma^\nu\Big[\partial_\rho + \frac{1}{2}\,\omega_{\rho mn}(e_1\psi)\sigma^{mn} \\ &\quad + \frac{1}{2}\,i\gamma_5A_\rho + \frac{1}{6}\,\gamma_\rho(\omega - i\not{A}\gamma_5)\Big]\psi_0\,,\end{aligned} \tag{2.20a}$$

$$\begin{aligned}L_{SC} &= -\frac{1}{2}\,eR[e,\omega(e,\psi)] \\ &\quad - \frac{1}{2}\,\varepsilon^{\mu\nu\rho\sigma}\bar{\psi}_\mu\gamma_5\gamma_\nu D_\rho[\omega(e,\psi)]\psi_\sigma - \frac{1}{3}\,euu^* + \frac{1}{3}\,eA_mA^m\,.\end{aligned} \tag{2.20b}$$

We remark that the Poincaré density formula as well as the Poincaré local supersymmetry transformations are w-independent. If S has weight w, then $S' = SS_0^{-w}$ has w = 0. Equation (2.15) defines the components of S' with $h' = h - \frac{1}{3}wuz$. This defines the w-independent transformation rules of Ref. 6. Analogously for a vector multiplet V with Weyl weight w, we define the new multiplet $V' = V(S_0\bar{S}_0)^{-w/2}$ with w = 0.

The components of V' are easily computed to be

$$V' = (C,\ S_L,\ \mathcal{H}',\ B'_m,\ \lambda'_R,\ D')\ , \tag{2.21}$$

with

$$\mathcal{H}' = \mathcal{H} + \frac{1}{3}wCu,\quad B'_m = B_m + \frac{1}{3}wCA_m\ , \tag{2.22a}$$

$$\lambda'_R = \lambda_R + \frac{1}{6}w(uS_R - i\not{A}S_L)\ , \tag{2.22b}$$

$$D' = D + \frac{1}{6}w\left(\mathcal{H}u^* + \mathcal{H}^*u - 2B_mA^m + \frac{1}{2}i\bar{S}\gamma_5\gamma\cdot R^P\right)$$

$$+ \frac{1}{18}w^2C(uu^* - A_mA^m)\ . \tag{2.22c}$$

The density formulae for S' and V' as given by Eq. (2.17) (with h = h') and Eq. (2.19) (with $\mathcal{H} = \mathcal{H}'$, $B_m = B'_m$, $\lambda_R = \lambda'_R$, $D = D'$) are local supersymmetric densities with respect to the following Poincaré local supersymmetry transformations:

$$\delta^P z = \bar{\varepsilon}_L\chi_L\ , \tag{2.23a}$$

$$\delta^P\chi_L = \frac{1}{2}\hat{\not{\mathcal{D}}}^P z\varepsilon_R + \frac{1}{2}h'\varepsilon_L\ , \tag{2.23b}$$

$$\delta^P h' = \bar{\varepsilon}_R\hat{\not{\mathcal{D}}}^P\chi_L - \frac{1}{3}\bar{\varepsilon}_L\chi_L u - \frac{1}{6}i\bar{\varepsilon}_R\not{A}\chi_L\ , \tag{2.23c}$$

$$\delta^P C = \frac{1}{2}i\bar{\varepsilon}_L S_L + \text{h.c.}\ , \tag{2.23d}$$

$$\delta^P S_L = \frac{1}{2}i\mathcal{H}'\varepsilon_L - \frac{1}{2}\not{B}'\varepsilon_R - \frac{1}{2}i\hat{\not{\mathcal{D}}}^P C\varepsilon_R\ , \tag{2.23e}$$

$$\delta^P\mathcal{H}' = -i\bar{\varepsilon}_R\hat{\not{\mathcal{D}}}^P S_L - i\bar{\varepsilon}_R\lambda'_R + \frac{1}{3}i\bar{S}_L\varepsilon_L u + \frac{1}{6}\bar{S}_L\not{A}\varepsilon_R\ , \tag{2.23f}$$

$$\delta^P B'_m = \Big[-\frac{1}{2}\,\bar{\varepsilon}_L \Big(\hat{D}^P_m - \frac{1}{2}\, iA_m \Big) S_L - \frac{1}{2}\,\bar{\varepsilon}_L \gamma_m \lambda'_R$$

$$+ \frac{1}{12}\,\bar{S}_R \gamma_m (u\varepsilon_L + i A\!\!\!/ \varepsilon_R) \Big] + \text{h.c.}, \qquad (2.23g)$$

$$\delta^P \lambda'_R = \frac{1}{2}(\sigma \cdot \hat{F}^P - iD')\varepsilon_R \qquad (2.23h)$$

$$\delta^P D' = \Big[\frac{1}{2}\, i\bar{\varepsilon}_L \Big(\hat{D\!\!\!/}^P - \frac{1}{2}\, i A\!\!\!/ \Big) \lambda'_R \Big] + \text{h.c.}\ , \qquad (2.23i)$$

with

$$\hat{F}^P_{mn} = 2D_{[m}B'_{n]} + \bar{\psi}_{[m}\hat{D}^P_{n]}\, S + \bar{\psi}_{[m}\gamma_{n]}\lambda' + \bar{S}e^\mu_{[m}D_{n]}\psi_\mu\ , \qquad (2.24)$$

We now extend our rules of tensor calculus by introducing a Yang-Mills group[6]. To maintain both supersymmetry an d gauge invariance, the space-time dependent parameters of Yang-Mills transformations must be assigned to an entire chiral multiplet.

The Lie algebra-valued transformation parameters are

$$\Lambda = \tilde{g}^\alpha \Lambda^\alpha T^{\alpha j}_{i}\ , \qquad (2.25)$$

where $\tilde{g}^\alpha$ are gauge coupling constants, T^α are representation matrices for the generators of the group G, and Λ^α are scalar multiplets with Weyl weight w = 0. For a simple group, $\tilde{g}^\alpha = \tilde{g}$. We will often use the simpler notation $\tilde{g}$ for the set of gauge coupling constants $\tilde{g}^\alpha$ in the case of semi-simple or not semi-simple gauge groups G.

The components of the parameter multiplet Λ^α are denoted by

$$\Lambda^\alpha = (y^\alpha,\ \rho^\alpha,\ v^\alpha)\ . \qquad (2.26)$$

For the Yang-Mills connection (gauge potential) we introduce a Lie algebra-valued vector multiplet

$$V = \tilde{g} V^\alpha T^{\alpha j}_{i}$$

which, under a finite Yang-Mills variation, transforms as

$$e^{2V} \to e^{-i\bar{\Lambda}}\, e^{2V}\, e^{i\Lambda}\ . \qquad (2.28)$$

In the infinitesimal, Eq. (2.28) reads

$$\delta V = \frac{i}{2}\, i(\Lambda - \bar{\Lambda}) + \frac{1}{2}\, i[V,\Lambda + \bar{\Lambda}] + \frac{1}{6}\, i[V,[V,\Lambda - \bar{\Lambda}]] + O(V^3)\ . \tag{2.29}$$

If we are interested in discussing a supersymmetric Yang-Mills theory, we can take gauge choices for all Λ transformations except the real scalar Re y, which actually corresponds to the standard Yang-Mills parameter in ordinary space-time.

The Wess-Zumino gauge is defined by the following gauge choice:

$$C^\alpha = S^\alpha = H^\alpha = 0 \tag{2.30}$$

for the components of V^α.

The constraints in Eq. (2.30) are not invariant under supersymmetry and gauge transformations, since under these combined transformations we get

$$\delta C^\alpha = \frac{1}{2}\, i(y - y^*)\ , \tag{2.31a}$$

$$\delta S_L^\alpha = \rho_L^\alpha - \frac{1}{2}\, \not{B}\varepsilon_R\ , \tag{2.31b}$$

$$\delta H^\alpha = -iv^\alpha - \frac{1}{2}\, i\bar{\varepsilon}_R\gamma^\mu \not{B}^\alpha \psi_{\mu R} - i\bar{\varepsilon}_R\lambda_R^\alpha\ . \tag{2.31c}$$

Therefore the Wess-Zumino gauge is only invariant for $y = y^*$, and under a redefined supersymmetry transformation

$$\delta(\varepsilon) = \delta_\Omega(\varepsilon) + \delta\left(\rho_L^\alpha = \frac{1}{2}\, \not{B}^\alpha \varepsilon_R\right) + \delta\left(v^\alpha = -\frac{1}{2}\, \bar{\varepsilon}_R\gamma^\mu \not{B}^\alpha \psi_{\mu R} - \bar{\varepsilon}_R\lambda_R^\alpha\right)\ . \tag{2.32}$$

By applying this rule to the components of V^α in the Wess-Zumino gauge

$$V_{WZ}^\alpha = (0,\ 0,\ 0,\ \lambda_R^\alpha,\ B_\mu^\alpha,\ D^\alpha)\ , \tag{2.33}$$

we get

$$\delta B_\mu = -\frac{1}{2}\, \bar{\varepsilon}_L\gamma_\mu\lambda_R - \frac{1}{2}\, \bar{\varepsilon}_R\gamma_\mu\lambda_L - D_\mu y\ , \tag{2.34a}$$

$$\delta\lambda_R = \frac{1}{2}(\sigma\cdot\hat{F} - ID)\varepsilon_R + i[\lambda_R, y] \ , \tag{2.34b}$$

$$\delta D = \frac{1}{2} i\bar{\varepsilon}_L\hat{\not{D}}\lambda_R - \frac{1}{2} i\bar{\varepsilon}_R\hat{\not{D}}\lambda_L + i[D, y] \ , \tag{2.34c}$$

where $\hat{F}_{\mu\nu}$ and $\hat{D}_\mu$ are now covariant derivatives with respect to local supersymmetry and Yang-Mills transformations.

For chiral multiplets S^i transforming according to some (generally complex) representation of G, we have the finite transformation

$$S \to e^{-i\Lambda} S \tag{2.35}$$

or in the infinitesimal,

$$\delta S = -i\Lambda S \ . \tag{2.36}$$

Equation (2.36), when written in components, reads

$$\delta S' = -i(yz,\ y\chi_L + \rho_L z,\ yh + vz - \bar{\rho}_L\chi_L) \ . \tag{2.37}$$

Therefore in the transformation law given by Eq. (2.3) all derivatives are now replaced by Yang-Mills covariant derivatives, and in δh there is an extra term due to the last term in Eq. (2.32):

$$\delta' h = i\bar{\varepsilon}_R\lambda_R z = i\tilde{g}\bar{\varepsilon}_R\lambda_R^\alpha T_i^{\alpha j} z_j \ , \tag{2.38}$$

We end this section by giving the vector multiplication of two scalar multiplets which can actually be converted in a chiral multiplication of a chiral multiplet S_1 with the "kinetic multiplet" of S_2, $T(S_2)$, which is the curved generalization of the flat space superfield $\bar{D}\bar{D}\bar{S}$.

The components of the multiplet T(S) are

$$T(S_i) = \left(h^{*i} + \frac{1}{3} u^* z^{*i},\ \tilde{\psi}_L^i,\ \tilde{H}^i\right) \ , \tag{2.39}$$

where

$$\tilde{\psi}_L^i = \hat{\not{D}}^P \chi_R^i + \frac{1}{6} i\not{A}\chi_R^i + \frac{1}{6}\gamma\cdot R_R^P z^{*i} - i\tilde{g} z^{*j} T_j^{\alpha i}\lambda_L^\alpha \ , \tag{2.40}$$

$$\tilde{H}^i = \hat{\Box}^c z^{*i} - \frac{2}{3} u\left(h^{*i} + \frac{1}{3} u^* z^{*i}\right) + \tilde{g} z^{*j} T_j^{\alpha i} D^\alpha - 2i\tilde{g}\bar{\chi}_R^j T_j^{\alpha i} \lambda_R^\alpha , \quad (2.41)$$

and $\hat{\Box}^c$ denotes the conformal d'Alembertian with Yang-Mills covariant derivatives.

Using the chiral density formula for the new chiral multiplet

$$C_j^i S_{1i} T(S_{2j}) + C_j^{i*} S_{2j} T(S_{1i}) \quad (2.42)$$

we get a new local density corresponding to the vector multiplication of chiral multiplets instead of their chiral multiplication as given by Eq. (2.10). The final result for the local density associated with the chiral multiplet Eq. (2.42) is

$$\begin{aligned}
&e^{-1} L\ C_j^i S_{1i} T(S_{ij}) + C_j^{i*} S_{2j} T(S_{1i}) = \\
&C_j^i\Big[\frac{1}{6}(R + e^{-1}\,\varepsilon^{\mu\nu\rho\sigma}\bar{\psi}_\mu\gamma_5\gamma_\nu D_\rho\psi_\sigma) z_{1i} z_2^{*j} - \hat{D}_\mu z_{1i}\hat{D}^\mu z_2^{*j} \\
&- \bar{\chi}_{1Li}\not{D}\chi_{2R}^j - \bar{\chi}_{2R}^j\not{D}\chi_{1Li} + \left(h_{1i} + \frac{1}{3} u z_{1i}\right)\left(h_2^{*j} + \frac{1}{3} u^* z_2^{*j}\right) \\
&- \frac{1}{9} A_m^2 z_{1i} z_2^{*j} + \frac{1}{3} iA^m\left(-\bar{\chi}_{1Li}\gamma_m\chi_{2R}^j + z_2^{*j}\hat{D}_m z_{1i} - z_{1i}\hat{D}_m z_2^{*j}\right) \\
&+ \frac{1}{8} e^{-1}\,\varepsilon^{\mu\nu\rho\sigma}\bar{\psi}_\mu\gamma_\nu\psi_\rho\left(z_{1i}\hat{D}_\sigma z_2^{*j} - z_2^{*j}\hat{D}_\sigma z_{1i}\right) \\
&+ \frac{1}{3}\bar{\chi}_{1Li}\sigma^{\nu\mu} D_\mu\psi_\nu z_2^{*j} + \frac{1}{3}\bar{\chi}_{2R}^j\sigma^{\nu\mu} D_\mu\psi_\nu z_{1i} \\
&+ \bar{\chi}_{1Li}\sigma^{\nu\mu}\psi_\nu\hat{D}_\mu z_2^{*j} + \bar{\chi}_{2R}^j\sigma^{\nu\mu}\psi_\nu\hat{D}_\mu z_{1i} - D_\mu\bar{\chi}_{1Li}\sigma^{\nu\mu}\psi_\nu z_2^{*j} \\
&- D_\mu\bar{\chi}_{2R}^j\sigma^{\nu\mu}\psi_\nu z_{1i} + 2i\tilde{g} T_i^{\alpha k}\bar{\lambda}_L^\alpha\chi_{1Lk} z_2^{*j} - 2i\tilde{g}\bar{\chi}_{2R}^k T_k^{\alpha j}\lambda_R^\alpha z_{1i} \\
&+ \tilde{g} D^\alpha T_i^{\alpha k} z_{1k} z_2^{*j} - \frac{1}{2} i\tilde{g}\bar{\psi}\cdot\gamma\gamma_5\lambda^\alpha T_i^{\alpha k} z_{1k} z_2^{*j}\Big] + \text{h.c.}\ ,
\end{aligned} \quad (2.43)$$

where $C_j^i z_{1i} z_2^{*j}$ is invariant under G, i.e.

$$C_j^i T_i^{\alpha k} z_{1k} z_2^{*j} - C_j^i T_k^{\alpha j} z_{1j} z_2^{*k} = 0 . \quad (2.44)$$

The other (antisymmetric) combination of the chiral multiplets,

$$C_j^i S_{1i} T(S_{2j}) - C_j^{i*} S_{2j} T(S_{1i}) , \quad (2.45)$$

gives rise to a density formula which is a total derivative.

Equation (2.43) produces the kinetic terms of the chiral fields and their non-polynomial generalizations.

3. MANIFESTLY INVARIANT ACTIONS AND TRANSFORMATION LAWS

In this section we give the matter-coupled supergravity Lagrangian in its superspace form and its component form together with the transformation laws of the component fields under local Poincaré supersymmetry transformation[6].

Let us start with the most general supersymmetric superspace action for a set of n chiral multiplets. This si given in flat superspace by

$$\int d^4x\, d^4\theta\, \Phi(S,\bar{S}) + \int d^4x\, \mathrm{Re} \int d^2\theta\, g(S) \; , \tag{3.1}$$

where $\Phi(S,\bar{S})$ is an arbitrary real function of the chiral multiplets S_i and their complex conjugate $\bar{S}^i$, and g(S) is a chiral function constructed out of the chiral multiplets S^i. The first component of g(S), $g(z_i)$, is often called the superpotential, The first term in Eq. (3.1) gives rise to a D-type density, whilst the second term gives rise to an F-type density. They are, respectively, the non-polynomial generalizations of the scalar kinetic term $S_i\bar{S}^i$ of global supersymmetry and the self-interaction term (including the mass term) of the chiral multiplets S^i. If the chiral multiplets S_i transform according to some representation of a compact Lie group G, the action is assumed to be G-invariant. This requires the following group properties on the scalar functions $\Phi(z,z^*)$ and g(z):

$$z^{*j} T^{\alpha i}_{j} \Phi_i = \Phi^j T^{\alpha i}_{j} z_i \; , \tag{3.2a}$$

$$g^j T^{\alpha i}_{j} z_i = 0 \; , \tag{3.2b}$$

$$\left(\Phi^i = \frac{\partial \Phi}{\partial z_i} \; , \quad \Phi_i = \frac{\partial \Phi}{\partial z^{*i}} \; , \quad g^i = \frac{\partial g}{\partial z_i} \right) \tag{3.2c}$$

The extension of the action [Eq. (3.1)] to local supersymmetry and to a local Yang-Mills group G needs the following modification to Eq. (3.1) [12]:

$$\int d^4x\, d^4\theta\, E\left\{\Phi(S,\bar{S}\, \bar{e}^{2V}) + \mathrm{Re}\left[\frac{1}{R}\, g(S)\right]\right\}, \tag{3.3}$$

where V is the gauge vector multiplet of G, E is the superspace determinant, and R is the chiral scalar curvature superfield[12]. The component form of Eq. (3.3) can be computed with the rules of the tensor calculus developed in the previous section.

The remaining part of the Yang-Mills supergravity coupling corresponds to the supergravity extension of the flat Yang-Mills action[12]:

$$\int d^4x\, \mathrm{Re} \int d^2\theta\, W^{\alpha}_{a}\varepsilon^{ab}W^{\alpha}_{b}\, , \tag{3.4}$$

where W^{α}_{a} is the field-strength chiral multiplet.

We can extend Eq. (3.4), including non-polynomial interactions with the chiral multiplets S_i, as follows[6]:

$$\int d^4x\, \mathrm{Re} \int d^2\theta\left[f_{\alpha\beta}(S)W^{\alpha}_{a}\varepsilon^{ab}W^{\beta}_{b}\right], \tag{3.5}$$

where $f_{\alpha\beta}(S)$ is a chiral superfield transforming as the symmetric product of the adjoint representation of G.

The curved space generalization of Eq. (3.5) is

$$\int d^4x\, d^4\theta\, E\, \mathrm{Re}\left[\frac{1}{R}\, f_{\alpha\beta}(S)W^{\alpha}_{a}\varepsilon^{ab}W^{\beta}_{b}\right] \tag{3.6}$$

and we can compute Eq. (3.6) using the local density formula for chiral multiplets obtained in Section 2. In principle, under the requirement that the field-Lagrangian contains only first derivatives in fermion fields and second derivatives in the scalar fields, we could also have new invariants whith higher power of the field strength multiplet W^{α}_{a}, such as $f_{\alpha_1 \ldots \alpha_n}(S)W^{\alpha_1} \cdots W^{\alpha_n}$. We will not consider these terms here. All other possible modifications of Eqs. (3.3) and (3.6) would necessarily contain higher derivatives of boson and fermion fields. These terms cannot appear in the tree-level Lagrangian if we require normal propagation of the physical fields, but they may well appear in the effective action as a result of radiative corrections, in exact analogy to the high curvature terms which appear in the quantum loop expansion of the Einstein theory.

In global supersymmetry, if the gauge group G contains an Abelian factor U(1), there is an extra possible invariant -- the so-called Fayet-Iliopoulos term[13]

$$\int d^4x \ d^4\theta \ V . \tag{3.7}$$

The naive extension of Eq. (3.7) to curved space,

$$\int d^4x \, d^4\theta \, EV , \tag{3.8}$$

is not gauge invariant, since under a U(1) transformation $E \to E$ and $V \to V + (\Lambda - \bar{\Lambda})$. Then we have to modify both Eq. (3.8) and the U(1) gauge transformation in order to preserve a U(1) gauge invariance[14].

The Fayet-Iliopoulos term can be introduced in curved superspace as the result of the gauging of an R-symmetry[15], which is defined as follows

$$V_R \to V_R + i/g_R(\Lambda - \bar{\Lambda}) , \tag{3.9a}$$

$$S_0 \to e^{i\Lambda} S_0 , \quad \bar{S}_0 \to e^{-i\bar{\Lambda}} \bar{S}_0 , \tag{3.9b}$$

$$S_i \to e^{-in_i\Lambda} S_i , \quad \bar{S}^i \to e^{in_i\Lambda} \bar{S}_i , \tag{3.9c}$$

and V_R is the gauge multiplet for this U(1) gauge group. The extension of Eqs. (3.3) and (3.5) to local R-symmetric interactions requires the following condition on the functions $\Phi(S,\bar{S})$, $g(S)$, and $f_{\alpha\beta}(S)$[11]:

$$\Phi(S_i,\bar{S}^i) = \Phi \ e^{-in_iq} S_i , \ e^{in_iq} \bar{S}^i , \tag{3.10a}$$

$$g(S_i) = e^{3iq} \, e^{-in_iq} S_i , \tag{3.10b}$$

$$f_{\alpha\beta}(S_i) = f_{\alpha\beta} \, e^{-in_iq} S_i , \tag{3.10c}$$

where q is a real constant parameter.

Equations (3.10) imply that Φ and $f_{\alpha\beta}$ are R-symmetric, but the superpotential g(S) must transform with a given phase under R-symmetry. This is due to the non-invariance of the chiral measure E/R in Eq. (3.3) or, equivalently, to the transformation property of the compensating multiplet S_0 in the density formula given by Eq. (2.17).

Under the restriction (3.10) the most general Lagrangian density invariant under the Yang-Mills group $G \times U_R(1)$ is[11]

$$\int d^4x\, d^4\theta\, E\Big[e^{-g_R V_R}\Phi\Big(S_i,\, \bar{S}^i\, e^{n_i g_R V_R}\Big)_e^{2V}\Big]$$

$$+ \mathrm{Re}\left[\frac{1}{2}\, g(S_i) + f_{\alpha\beta}(S_i)\, W^\alpha W^\beta + f_R(S_i) W_R^2\right] \tag{3.11}$$

or equivalently, using the density formulae of Section 2,

$$L = -\frac{1}{2}\Big[S_0\, e^{-g_R V_R}\, \bar{S}_0\, \Phi\Big(S_i,\, \bar{S}^i\, e^{n_i g_R V_R}\, e^{2V}\Big)\Big]_D$$

$$+ \Big[g(S_i) S_0^3\Big]_F - \Big[f_{\alpha\beta}(S_i) W^\alpha W^\beta\Big]_F - \Big[f_R(S_i) W_R^2\Big]_F\ , \tag{3.12}$$

where W_R is the field-strength multiplet for V_R, $g_R = k^2\xi$, where ξ is a dimensional parameter (dim $\xi = 2$), and g_R is the R-gauge coupling constant. We will now derive the component expression of Eqs. (3.3) and (3.6), and comment at the end on their generalization to the case of a gauged R-symmetry [Eq. (3.12)].

In order to compute the following density formulae,

$$\Big[S_0 \bar{S}_0 \Phi(S, \bar{S}\, e^{2V})\Big]_{\hat{D}}\ , \tag{3.13}$$

$$\Big[g(S) S_0^3\Big]_F\ , \tag{3.14}$$

$$\Big[f_{\alpha\beta} W^\alpha W^\beta\Big]_F = \Big[f W^2\Big]_F, \tag{3.15}$$

we have to work out the components of the vector multiplet $\Phi(S, \bar{S}\, e^{2V})$ and of the chiral multiplets g(S) and fW^2, respectively.

This can be done using the multiplication rules (2.10), (2.11), and (2.42), and the Wess-Zumino gauge choice for the vector multiplet V^{α}. In this gauge $V^{\alpha}V^{\beta}V^{\gamma} = 0$, and we get

$$\Phi(S,\bar{S}e^{2}) = \Phi(S,\bar{S}) + 2\tilde{g}\Phi_{i}\bar{S}^{j}T^{\alpha i}_{j}V^{\alpha} + 2\tilde{g}^{2}(\Phi_{ij}\bar{S}^{k}T^{\alpha i}_{k}\bar{S}^{\ell}T^{\beta j}_{\ell} + \Phi_{i}\bar{S}^{k}T^{\alpha j}_{k}T^{\beta i}_{j})V^{\alpha}V^{\beta} . \tag{3.16}$$

The components $C(\Phi)$, $S_L(\Phi)$, $H(\Phi)$, $B_m(\Phi)$, $\lambda_R(\Phi)$, and $D(\Phi)$ of the local vector multiplet given in Eq. (3.16) are given by[6]

$$C(\Phi) = \Phi(z,z^{*}) , \tag{3.17a}$$

$$S_L(\Phi) = -2i\Phi^{i}\chi_{Li} , \tag{3.17b}$$

$$H(\Phi) = -2\Phi^{i}h_{i} + 2\Phi^{ij}\bar{\chi}_{Li}\chi_{Lj} , \tag{3.17c}$$

$$B_m(\Phi) = i\Phi^{i}\hat{D}_{m}z_{i} - i\Phi_{i}\hat{D}_{m}z^{*i} - 2i\Phi^{j}_{i}\bar{\chi}^{i}_{R}\gamma_{m}\chi_{Lj} , \tag{3.17d}$$

$$\begin{aligned}\lambda_R(\Phi) = &-2i\Phi^{j}_{i}h_{j}\chi^{i}_{R} + 2i\Phi^{ij}_{k}\chi^{k}_{R}\bar{\chi}_{Li}\chi_{Lj} \\ &+2i\Phi^{j}_{i}\hat{\not{D}}z^{*i}\chi_{Lj} + 2\tilde{g}\lambda^{\alpha}_{R}\tilde{D}^{\alpha} ,\end{aligned} \tag{3.17e}$$

$$\begin{aligned}D(\Phi) = &\, 2\Phi^{i}_{j}h_{i}h^{*j} - 2\Phi^{ij}_{k}\bar{\chi}_{Li}\chi_{Lj}h^{*k} - 2\Phi^{k}_{ij}\bar{\chi}^{i}_{R}\chi^{j}_{R}h_{k} \\ &+ 2\Phi^{ij}_{k\ell}\bar{\chi}_{Li}\chi_{Lj}\bar{\chi}^{k}_{R}\chi^{\ell}_{R} - 2\Phi^{j}_{i}\hat{D}_{\mu}z^{*i}\hat{D}^{\mu}z_{j} \\ &- 2\Phi^{j}_{i}\bar{\chi}_{Lj}\overleftrightarrow{\hat{\not{D}}}\chi^{i}_{R} - 2\,\Phi^{i}_{jk}\hat{D}_{\mu}z^{*k} - \Phi^{ik}_{j}\hat{D}_{\mu}z_{k}\,\bar{\chi}_{Li}\gamma^{\mu}\chi^{j}_{R} \\ &+ 2\tilde{g}D^{\alpha}\tilde{D}^{\alpha} + 4i\tilde{g}\bar{\lambda}^{\alpha}_{L}\tilde{D}^{\alpha i}\chi_{Li} - 4i\tilde{g}\bar{\lambda}^{\alpha}_{R}D^{\alpha}_{i}\chi^{i}_{R} ;\end{aligned} \tag{3.17f}$$

where

$$\tilde{D}^{\alpha} = \Phi_{i}T^{\alpha i}_{j}z^{*j} = \Phi^{i}T^{\alpha j}_{i}z_{j} , \qquad \tilde{D}^{\alpha i} = \frac{\partial\tilde{D}^{\alpha}}{\partial z}\ \tilde{D}^{\alpha}_{i} = \frac{\partial\tilde{D}^{\alpha}}{\partial z^{*i}}$$

(the derivatives $\hat{D}_{\mu}$ are supersymmetric and Yang-Mills covariant).

Analogously, the components of the two chiral multiplets g(S) amd fW^{2} are given respectively by[6]

$$g(S) = \left[g(z), \quad \chi_L(g) = g^i\chi_{Li}, \quad h(g) = g^i h_i - g^{ij}\bar{\chi}_{Li}\chi_{Lj}\right], \tag{3.18}$$

$$fW^2 = \left[z(fW^2), \ \chi_L(fW^2), \ h(fW^2)\right] \tag{3.19}$$

with

$$z(fW^2) = -\frac{1}{2} f_{\alpha\beta}\bar{\lambda}^{\alpha}_L\lambda^{\beta}_L, \tag{3.20a}$$

$$\chi_L(fW^2) = \frac{1}{2} f_{\alpha\beta}(\sigma\cdot\hat{F}^{-\alpha} - iD^{\alpha})\lambda^{\beta}_L - \frac{1}{2} f^i_{\alpha\beta}\chi_{Li}\bar{\lambda}^{\alpha}_L\lambda^{\beta}_L, \tag{3.20b}$$

$$\begin{aligned} h(fW^2) &= f_{\alpha\beta}\left(-\bar{\lambda}^{\alpha}_L\hat{\not{D}}\lambda^{\beta}_R - \frac{1}{2}\hat{F}^{-\alpha}_{\mu\nu}F^{-\beta}_{\mu\nu} + \frac{1}{2}D^{\alpha}D^{\beta}\right) \\ &+ f^i_{\alpha\beta}\bar{\chi}_{Li}(-\sigma\cdot\hat{F}^{-\alpha} + iD^{\alpha})\lambda^{\beta}_L \\ &- \frac{1}{2} f^i_{\alpha\beta}h_i\bar{\lambda}^{\alpha}_L\lambda^{\beta}_L + \frac{1}{2} f^{ij}_{\alpha\beta}\bar{\chi}_{Li}\chi_{Lj}\bar{\lambda}^{\alpha}_L\lambda^{\beta}_L . \end{aligned} \tag{3.20c}$$

Using the density formulae (2.17) and (2.19), we finally get

$$\begin{aligned} e^{-1}L(\Phi) = &-\frac{1}{6}\Phi\, e^{-1} L_{SC} + \Phi^i_j\left(-\frac{1}{2}D_\mu z_i D^\mu z^{*j} - \bar{\chi}_{Li}\not{D}\chi^j_R + \frac{1}{2}h_i h^{*j}\right) \\ &- \Phi^{ij}_k\bar{\chi}_{Li}\chi_{Lj}h^{*k} + \Phi^{ij}_k\bar{\chi}_{Li}\not{D}z_j\chi^k_R \\ &+ \frac{1}{2}\Phi^{ij}_k\bar{\chi}_{Li}\chi_{Lj}\bar{\chi}^k_R\chi^{\ell}_R + \frac{1}{3}u^*(\Phi^i h_i - \Phi^{ij}\chi_{Li}\chi_{Lj}) \\ &+ \frac{1}{3}iA^\mu\left[\frac{1}{2}\Phi^{i}_j\bar{\chi}^j_R\gamma_\mu\chi_{Li} + \Phi^i(D_\mu z_i - \bar{\psi}_{\mu L}\chi_{Li})\right] \\ &+ \Phi^i_j\bar{\psi}_{\mu L}\not{D}z^{*j}\gamma_\mu\chi_{Li} - \frac{4}{3}\Phi^i\bar{\chi}_{Li}\sigma^{\mu\nu}D_\mu\psi_{\nu L} \\ &- \frac{1}{8}e^{-1}\varepsilon^{\mu\nu\rho\sigma}\bar{\psi}_\mu\gamma_\nu\psi_\rho\left(\Phi^i D_\sigma z_i + \frac{1}{2}\Phi^j_i\bar{\chi}^i_R\gamma_\sigma\chi_{Lj}\right) \\ &- \frac{1}{2}\Phi^i_j\bar{\psi}_{\mu R}\chi^j_R\bar{\psi}^\mu_L\chi_{Li} + \frac{1}{6}\Phi^i\bar{\chi}_{Li} \\ &\times\left(\bar{\psi}_{\mu L}\bar{\psi}\cdot\gamma\psi^\mu + \frac{1}{2}\sigma^{\mu\nu}\psi^\rho_L\bar{\psi}_\nu\gamma_\rho\psi_\mu + \sigma^{\mu\nu}\psi_{\nu L}\bar{\psi}_\mu\gamma\cdot\psi\right) \\ &+ \frac{1}{2}\tilde{g}\Phi^i T^{\alpha j}_i z_j(D^\alpha + i\bar{\psi}_L\cdot\gamma\lambda^\alpha_R) - 2i\tilde{g}\Phi^j_i T^{\alpha k}_j z_k\bar{\lambda}^\alpha_R\chi^i_R + \text{h.c.}, \end{aligned} \tag{3.21}$$

$$\begin{aligned} e^{-1}L(g) = &-\frac{1}{2}g^{ij}\bar{\chi}_{Li}\chi_{Lj} + \frac{1}{2}g^i h_i + \frac{1}{2}gu + \frac{1}{2}\bar{\psi}_R\cdot\gamma\chi_{Li} \\ &+ \frac{1}{2}g\bar{\psi}_{\mu R}\sigma^{\mu\nu}\psi_{\nu R} + \text{h.c.}, \end{aligned} \tag{3.22}$$

$$e^{-1}\,L(fW^2) = \frac{1}{2}\,f_{\alpha\beta}\Big[-\frac{1}{4}\,F^{\alpha}_{\mu\nu}F^{\beta}_{\mu\nu} - \frac{1}{2}\,\bar{\lambda}^{\alpha}\tilde{\not{D}}\lambda^{\beta} + \frac{1}{2}\,D^{\alpha}D^{\beta}$$

$$+ \frac{1}{4}\,F^{\alpha}_{\mu\nu}\bar{\psi}_{\rho}\sigma^{\mu\nu}\gamma^{\rho}\lambda^{\beta} + \frac{1}{4}\,F^{\alpha}_{\mu\nu}\tilde{F}^{\beta}_{\mu\nu} - \frac{1}{2}\,D_{\mu}(\bar{\lambda}^{\alpha}_{L}\gamma^{\mu}\lambda^{\beta}_{R})\Big]$$

$$+ \frac{1}{2}\,f^{i}_{\alpha\beta}\Big[\bar{\chi}_{Li}(-\sigma\cdot\hat{F}^{-\alpha} + iD^{\alpha})\lambda^{\beta}_{L} - \frac{1}{2}\,h_{i}\bar{\lambda}^{\alpha}_{L}\lambda^{\beta}_{L}$$

$$- \frac{1}{2}\,\bar{\psi}_{R}\cdot\gamma\chi_{Li}\bar{\lambda}^{\alpha}_{L}\lambda^{\beta}_{L}\Big] + \frac{1}{4}\,f^{ij}_{\alpha\beta}\bar{\chi}_{Li}\chi_{Lj}\bar{\lambda}^{\alpha}_{L}\lambda^{\beta}_{L} + \text{h.c.} \tag{3.23}$$

Here $\tilde{D}_{\mu}\lambda^{\alpha}$ is a supersymmetric and Yang-Mills covariant derivative but without the $\psi_{\mu}D^{\alpha}$ term. The other derivatives are Yang-Mills and general coordinate covariants but they have no ψ_{μ} torsion.

The over-all Lagrangian corresponding to the superspace expressions of Eqs. (3.3) and (3.6) is therefore given by[6]

$$L = L(\Phi) + L(g) + L(fW^2)\ . \tag{3.24}$$

The three terms in Eq. (3.24) are separately invariant under local supersymmetry transformations.

These transformations have the following form, for the several fields involved in Eq. (3.24):

$$\delta B^{\alpha}_{\mu} = -\frac{1}{2}\,\bar{\varepsilon}_{L}\gamma_{\mu}\lambda^{\alpha}_{R} - \frac{1}{2}\,\bar{\varepsilon}_{R}\gamma_{\mu}\lambda^{\alpha}_{L}\ , \tag{3.25a}$$

$$\delta\lambda^{\alpha}_{R} = \frac{1}{2}\,\sigma^{\mu\nu}\hat{F}^{\alpha}_{\mu\nu}\varepsilon_{R} - \frac{1}{2}\,iD^{\alpha}\varepsilon_{R}\ , \tag{3.25b}$$

$$\delta D^{\alpha} = \frac{1}{2}\,i\varepsilon_{L}\Big(\hat{\not{D}}^{P} - \frac{1}{2}\,i\not{A}\Big)\lambda_{R} - \frac{1}{2}\,i\bar{\varepsilon}_{R}\Big(\hat{\not{D}}^{P} + \frac{1}{2}\,i\not{A}\Big)\lambda^{\alpha}_{L}\ , \tag{3.25c}$$

$$\delta z_{i} = \bar{\varepsilon}_{L}\chi_{Li}\ , \tag{3.25d}$$

$$\delta\chi_{Li} = \frac{1}{2}\,\hat{\not{\partial}}z_{i}\varepsilon_{R} + \frac{1}{2}\,h_{i}\varepsilon_{L}\ , \tag{3.25e}$$

$$\delta h_{i} = \bar{\varepsilon}_{R}\Big(\hat{\not{D}}^{P} - \frac{1}{2}\,i\not{A}\Big)\chi_{Li} - \frac{1}{3}\,\bar{\chi}_{Li}(u\varepsilon_{L} + i\not{A}\varepsilon_{R} + i\tilde{g}\bar{\varepsilon}_{R}\lambda^{\alpha}_{R}T^{\alpha j}_{i}z_{j})\ , \tag{3.25f}$$

$$\delta e^{m}_{\mu} = \frac{1}{2}\,\bar{\varepsilon}_{L}\gamma^{m}\psi_{\mu R} + \frac{1}{2}\,\bar{\varepsilon}_{R}\gamma^{m}\psi_{\mu L}\ , \tag{3.25g}$$

$$\delta\psi_{\mu L} = \left[\partial_\mu + \frac{1}{2}\,\omega_{\mu mn}(e,\psi)\sigma^{mn} + \frac{1}{2}\,iA_\mu\right]\varepsilon_L$$

$$+ \frac{1}{6}\,\gamma_\mu(u^*\varepsilon_R - i\not{A}\varepsilon_L)\ , \tag{3.25h}$$

$$\delta A_\mu = \frac{3}{4}\,i\bar{\varepsilon}_L\left(R^P_\mu - \frac{1}{3}\,\gamma_\mu\gamma\cdot R^P\right)_L + \text{h.c.}\ , \tag{3.25i}$$

$$\delta u = \frac{1}{2}\,\bar{\varepsilon}_R\gamma\cdot R^P_L\ . \tag{3.25j}$$

The over-all Lagrangian given by Eq. (3.24) contains auxiliary field components of the multiplets S^i, V^α, and the supergravity multiplet

$$(e_{\mu m},\ \psi_{\mu a},\ u,\ A_m)\ . \tag{3.26}$$

In addition, an appropriate Weyl rescaling of the vierbein field, as well as redefinitions of the fermion fields, have to be performed in the first term of Eq. (3.24), in its standard form[5]. These redefinitions, as well as the elimination of the auxiliary fields h_i, D^α, u, and A_m, will be done in the next section.

These manipulations will allow us to recast the final Lagrangian and transformation rules in a simple form in terms of the particle fields

$$(B^\alpha_m,\ \lambda^\alpha_R),\quad (z_i,\ \chi_{Li}),\quad (e_{\mu m},\ \psi_{\mu a})\ . \tag{3.27}$$

We end this section by showing (using simple superspace arguments) that the action given by Eq. (3.24) depends only on a particular combination of the function $\Phi(S,\bar{S})$ and $g(S)$[5].

To prove this, we make a super-Weyl rescaling on the supervierbein so that its superdeterminant transforms as

$$E \to E'\ \exp\,(\Sigma + \bar{\Sigma})\ , \tag{3.28}$$

where Σ is a chiral parameter superfield. Under the same rescaling the second term in Eq. (3.3) transforms as

$$\int d^4x\ d^4\theta\ \text{Re}\left[\frac{E'}{R'}\,(\exp\,3\Sigma)g(S)\right]\ , \tag{3.29}$$

whilst the Yang-Mills part [Eq. (3.6)] keeps the same form:

$$\int d^4x\, d^4\theta \, \mathrm{Re}\, \frac{\bar{E}'}{\bar{R}'} f_{\alpha\beta}(S) W^{\alpha} W^{\beta} \; . \tag{3.30}$$

If we now choose Σ such that

$$\exp 3\Sigma g(S) = 1, \quad \text{i.e. } \exp \Sigma = g(S)^{-1/3} \; , \tag{3.31}$$

we finally obtain that Eq. (3.3) becomes

$$\int d^4x\, d^4\theta\, E' \exp \frac{1}{3}\, G(S, \bar{S}\, e^{2V}) + \int d^4x\, d^4\theta \, \mathrm{Re} \left(\frac{E'}{R'}\right) , \tag{3.32}$$

where the function G is given by

$$G(z,z^*) = 3 \log \Phi(z,z^*) - \log |g(z)|^2 \; , \tag{3.33}$$

so that

$$\exp \frac{1}{3}\, G(z,z^*) = \frac{\Phi(z,z^*)}{|g(z)|^{2/3}} \; . \tag{3.34}$$

The use of the function $G(S,\bar{S})$ makes the introduction of the Fayet-Iliopoulos term in formula (3.11) trivial[11,16]. In fact, in this case we simply get

$$\exp \left[\frac{1}{3}\, G(S,\bar{S} \exp n_i g_R V_R \exp 2V)\right] =$$

$$\frac{(\Phi(S_i,\bar{S} \exp n_i g_R V_R \exp 2V)}{\left[g^*(\bar{S} \exp n_i g_R V_R \exp 2V) g(S)\right]^{1/3}} \; . \tag{3.35}$$

This means that in the G superfield we simply have to replace $\bar{S}\, e^{2V}$ with $\bar{S}\, e^{n_i g_R V_R}\, e^{2V}$, i.e. we have to covariantize with respect to the full gauge group $G \times U_R(1)$.

We note that this simple rule does not hold in the case of a vanishing superpotential, $g(S_i) = 0$, since in this case the rescaling given by formula (3.31) can no longer be performed[11]. At the end of the next section we will give simple substitution rules for the Lagrangian in terms of physical fields which will enable us to go from the general form of $G(S,\bar{S})$ to the limiting case corresponding to $g(S) = 0$.

4. FINAL FORM OF THE LAGRANGIAN, TRANSFORMATION LAWS, AND GAUGED R-SYMMETRY

The over-all action of the coupled-matter Yang-Mills supergravity multiplets is given in Eq. (3.24).

In order to obtain the final form of the Lagrangian, we have to eliminate, by means of their field equations, the auxiliary fields h_i, D^α, u, and A_m of the multiplets involved.

It is important to notice that, after their elimination, only the total Lagrangian (3.24) -- but not the three separate pieces given by Eqs. (3.21) to (3.23) -- will be invariant under local supersymmetry transformations of the physical fields.

The part of the Lagrangian (3.24) which contains the auxiliary fields is

$$\begin{aligned} e^{-1} L_{aux} &= \frac{1}{18}\,\Phi(uu^* - A_m A^m) + \frac{1}{2}\,\Phi^i_j h_i h^{*j} \\ &- \Phi^{ij}_k h^{*k} \bar{\chi}_{Li}\chi_{Lj} + \frac{1}{2}\, g^i h_i - \frac{1}{4}\, f^i_{\alpha\beta} h_i \bar{\lambda}^\alpha_L \lambda^\beta_L \\ &+ \frac{1}{3}\left(\Phi^i h_i - \Phi^{ij}\bar{\chi}_{Li}\chi_{Lj} + \frac{3}{2}\, g^*\right) \\ &+ \frac{1}{3}\, iA^\mu\left[\frac{1}{2}\,\Phi^i_j \bar{\chi}^j_R \gamma_\mu \chi_{Li} + \Phi^i (D_\mu z_i - \psi_{\mu L}\chi_{Li}) + \frac{3}{4}\,\bar{\lambda}^\alpha_L \gamma_\mu \lambda^\beta_R f_{\alpha\beta}\right] \\ &+ \frac{1}{2}\,\tilde{g}\Phi^i T^{\alpha j}_i z_j D^\alpha + \frac{1}{4}\, f_{\alpha\beta} D^\alpha D^\beta + \frac{i}{2}\, f^i_{\alpha\beta}\bar{\chi}_{Li}\lambda^\alpha_L D^\alpha + \text{h.c.} \end{aligned} \tag{4.1}$$

If we define the new expressions

$$J = 3 \log - \frac{\Phi}{3} \tag{4.2}$$

and

$$\tilde{u} = u + J^i h_i \ , \tag{4.3}$$

then Eq. (4.1) reads

$$\begin{aligned} e^{-1} L_{aux} &= \frac{1}{18}\,\Phi(\tilde{u}\tilde{u}^* - A_m A^m) + \frac{1}{6}\,\Phi J^i_j h_i h^{*j} \\ &+ \frac{1}{2}\, g\left[\tilde{u} - h_i\left(J^i - \frac{g^i}{g}\right)\right] - \frac{1}{g}\, u^*\left(J^{ij} + \frac{1}{3}\, J^i J^j\right)\Phi\bar{\chi}_{Li}\chi_{Lj} \end{aligned}$$

(continued)

$$- \frac{1}{3} \Phi h^{*k} \bar{\chi}_{Li} \chi_{Lj} \left(J_k^{ij} + \frac{2}{3} J_k^i J^j \right) - \frac{1}{4} f^i_{\alpha\beta} h_i \bar{\lambda}^\alpha_L \lambda^\beta_L$$

$$+ \frac{1}{3} iA^\mu \left[\frac{1}{2} \Phi^{i}_{j} \bar{\chi}^j_R \gamma_\mu \chi_{Li} + \Phi^i (D_\mu z_i - \bar{\psi}_{\mu L} \chi_{Li}) + \frac{3}{4} \bar{\lambda}^\alpha \gamma_\mu \lambda^\beta_R f_{\alpha\beta} \right]$$

$$+ \frac{1}{2} \tilde{g} \Phi^i T^{\alpha j}_i z_j D^\alpha + \frac{1}{4} f_{\alpha\beta} D^\alpha D^\beta + \frac{i}{2} f^i_{\alpha\beta} \bar{\chi}_{Li} \lambda^\alpha_L D^\beta + \text{h.c.} \qquad (4.4)$$

The field equations for the auxiliary fields are

$$\tilde{u} = \frac{2}{\Phi} \left[- \frac{1}{2} g^* + \frac{1}{g} \left(J^{ij} + \frac{1}{3} J^i J^j \right) \Phi \bar{\chi}_{Li} \chi_{Lj} \right] , \qquad (4.5a)$$

$$\frac{1}{3} \Phi h_i J^i_k = - \frac{1}{2} g^* \left(\frac{g^*_k}{g^*} - J_k \right) + \frac{1}{3} \Phi \left(J^{ij}_k + \frac{2}{3} J^i_k J^j \right) \bar{\chi}_{Li} \chi_{Lj}$$
$$+ \frac{1}{4} f^*_{\alpha\beta k} \bar{\lambda}^\alpha_R \lambda^\beta_R , \qquad (4.5b)$$

$$\frac{2}{3} \Phi A_\mu = \frac{1}{2} i \Phi^i_j \bar{\chi}^j_R \gamma_\mu \chi_{Li} + i \Phi^i (D_\mu z_i - \bar{\psi}_{\mu L} \chi_{Li})$$
$$+ \frac{3}{4} i f_{\alpha\beta} \bar{\lambda}^\alpha_L \gamma_\mu \lambda^\beta_R + \text{h.c.} , \qquad (4.5c)$$

$$-\text{Re } f_{\alpha\beta} D^\beta = \tilde{g} \Phi^i T^{\alpha j}_i z_j + \frac{1}{2} i f^i_{\alpha\beta} \bar{\chi}_{Li} \lambda^\beta_L - \frac{1}{2} i f^*_{\alpha\beta i} \bar{\chi}^i_R \lambda^\beta_R . \qquad (4.5d)$$

The insertion of Eqs. (4.5) into Eq. (4.4) finally gives

$$e^{-1} L = e^{-1} L_{\text{aux}} + e^{-1} \hat{L} , \qquad (4.6)$$

where

$$e^{-1} L_{\text{aux}} = - \frac{g}{\Phi} \left| \frac{1}{2} g^* - \frac{1}{g} \left(J^{ij} + \frac{1}{3} J^i J^j \right) \Phi \bar{\chi}_{Li} \chi_{Lj} \right|^2 - \frac{3}{\Phi} (J^{-1})^k_\ell$$
$$\times \left[\frac{1}{2} g^* \left(\frac{g^*_k}{g^*} - J_k \right) - \frac{1}{3} \Phi \left(J^{ij}_k + \frac{2}{3} J^i_k J^j \right) \bar{\chi}_{Li} \chi_{Lj} - \frac{1}{4} f^*_{\alpha\beta} \bar{\lambda}^\alpha_R \lambda^\beta_R \right]$$
$$\times \left[\frac{1}{2} g \left(\frac{g^\ell}{g} - J^\ell \right) - \frac{1}{3} \Phi \left(J^\ell_{mn} + \frac{2}{3} J^\ell_m J_n \right) \bar{\chi}^m_R \chi^n_R - \frac{1}{4} f^\ell_{\gamma\delta} \bar{\lambda}^\gamma_L \lambda^\delta_L \right]$$
$$- \frac{1}{4\Phi} \left[\Phi^i_j \bar{\chi}^j_R \gamma_\mu \chi_{Li} + \Phi^i (D_\mu z_i - \bar{\psi}_{\mu L} \chi_{Li}) - \Phi_i (D_\mu z^{*i} - \bar{\psi}_{\mu R} \chi^i_R) \right.$$
$$\left. + \frac{3}{2} \text{Re } f_{\alpha\beta} \bar{\lambda}^\alpha_L \gamma_\mu \lambda^\beta_R \right]^2 - \frac{1}{2} (\text{Re } f)^{-1}_{\alpha\beta} \left(\tilde{g} \Phi^i T^{\alpha j}_i z_j + \frac{i}{2} f^i_{\alpha\gamma} \bar{\chi}_{Li} \lambda^\gamma_L \right.$$
$$\left. - \frac{1}{2} i f^*_{\alpha\gamma i} \bar{\chi}^i_R \lambda^\gamma_R \right) \times \left(\tilde{g} \Phi^k T^{\beta\ell}_k z_\ell + \frac{1}{2} i f^k_{\beta\delta} \bar{\chi}_{Lk} \lambda^\delta_L - \frac{1}{2} i f^*_{\beta\delta k} \bar{\chi}^k_R \lambda^\delta_R \right)$$

(4.7a)

$$e^{-1}\,\hat{L} = \frac{1}{12}\,\Phi\Big\{R[\omega(e)] + \bar{\psi}_\mu\gamma_5\gamma_\nu D_\rho\psi_\sigma\varepsilon^{\mu\nu\rho\sigma} + \frac{1}{16}\,(\bar{\psi}_\mu\gamma_\lambda\psi_\rho$$

$$+ 2\bar{\psi}_\lambda\gamma_\mu\psi_\rho)\bar{\psi}^\mu\gamma^\lambda\psi^\rho - \frac{1}{4}\,(\bar{\psi}_\mu\gamma\cdot\psi)^2 + e^{-1}\,\partial_\mu(e\bar{\psi}\cdot\gamma\psi^\mu)\Big\}$$

$$+ \Phi^i_j\left(-\frac{1}{2}\,D_\mu z_i D^\mu z^{*j} - \bar{\chi}_{Li}\not{D}\chi^j_R\right) + \Phi^{ij}_k\bar{\chi}_{Li}\not{D}z_j\chi^k_R$$

$$+ \Phi^i_j\bar{\psi}_{\mu L}\not{D}z^{*j}\gamma^\mu\chi_{Li} - \frac{4}{3}\,\Phi^i\bar{\chi}_{Li}\sigma^{\mu\nu}D_\mu\psi_{\nu L}$$

$$- \frac{1}{8}\,e^{-1}\,\varepsilon^{\mu\nu\rho\sigma}\bar{\psi}_\mu\gamma_\nu\psi_\rho\left(\Phi^i D_\sigma z_i + \frac{1}{2}\,\Phi^j_i\bar{\chi}^i_R\gamma_\sigma\chi_{Lj}\right)$$

$$- \frac{1}{2}\,\Phi^i_j\bar{\psi}_{\mu L}\chi_{Li}\bar{\psi}^\mu_R\chi^j_R + \frac{1}{2}\,\Phi^{ij}_{k\ell}\bar{\chi}_{Li}\chi_{Lj}\bar{\chi}^k_R\chi^\ell_R$$

$$+ \frac{1}{6}\,\Phi^i\chi_{Li}\left(\psi_{\mu L}\bar{\psi}\cdot\gamma\psi^\mu + \frac{1}{2}\,\sigma^{\mu\nu}\psi^\ell_L\bar{\psi}_\nu\gamma_\rho\psi_\mu + \sigma^{\mu\nu}\psi_{\nu L}\bar{\psi}_\mu\gamma\cdot\psi\right)$$

$$+ \frac{1}{2}\,i\tilde{g}\Phi^i T^{\alpha j}_i z_j\bar{\psi}_L\gamma\lambda^\alpha_k - 2i\tilde{g}\Phi^i_k T^{\alpha j}_i z_j\bar{\lambda}^\alpha_R\chi^k_R$$

$$- \frac{1}{2}\,g^{ij}\bar{\chi}_{Li}\chi_{Lj} + \frac{1}{2}\,g^i\bar{\psi}_R\cdot\gamma\chi_{Li} + \frac{1}{2}\,g\bar{\psi}_{\mu R}\sigma^{\mu\nu}\psi_{\nu R}$$

$$+ \frac{1}{2}\,f_{\alpha\beta}\Big[-\frac{1}{4}\,F^\alpha_{\mu\nu}F^\beta_{\mu\nu} - \frac{1}{2}\,\bar{\lambda}^\alpha\not{D}\lambda^\beta + \frac{1}{2}\,\bar{\lambda}^\alpha\gamma^\mu\sigma^{\rho\sigma}\psi_\mu F^\beta_{\rho\sigma}$$

$$+ \frac{1}{4}\,\bar{\lambda}^\alpha\gamma^\mu\sigma^{\rho\sigma}\psi_\mu\bar{\psi}_\rho\gamma_\sigma\lambda^\alpha - \frac{1}{16}\,\bar{\lambda}^\alpha\gamma^\mu\sigma^{mn}\lambda^\beta(2\bar{\psi}_\mu\gamma_m\psi_n$$

$$+ \bar{\psi}_m\gamma_\mu\psi_n) + \frac{1}{4}\,F^\alpha_{\mu\nu}\tilde{F}^\beta_{\mu\nu} - \frac{1}{2}\,D_\mu(\bar{\lambda}^\alpha_L\gamma^\mu\lambda^\beta_R)\Big]$$

$$+ \frac{1}{2}\,f^i_{\alpha\beta}\Big[-\bar{\chi}_{Li}\sigma_{\mu\nu}\lambda^\beta_L\left(F^{-\alpha}_{\mu\nu} - \frac{1}{2}\,\bar{\psi}_{\rho L}\sigma_{\mu\nu}\gamma^\rho\lambda^\alpha_R + \frac{1}{2}\,\bar{\psi}_R\cdot\gamma\sigma_{\mu\nu}\lambda^\alpha_L\right)$$

$$- \frac{1}{2}\,\bar{\psi}_R\cdot\gamma\chi_{Li}\bar{\lambda}^\alpha_L\lambda^\beta_L\Big] + \frac{1}{4}\,f^{ij}_{\alpha\beta}\bar{\chi}_{Li}\chi_{Lj}\bar{\lambda}^\alpha_L\lambda^\beta_L + \text{h.c.} \qquad (4.7b)$$

In order to recast the Einstein term and the Rarita-Schwinger action in canonical form, as well as to keep the gaugino and chiral fermion terms in a quasi-canonical form, the following Weyl rescaling on the vierbein and the fermion fields must be performed:

$$e_{m\mu} \to e_{m\mu}e^\sigma\,, \qquad e \to e^{4\sigma}\,e\,, \qquad (4.8a)$$

$$\lambda \to e^{-3\sigma/2}\,\lambda\ , \tag{4.8b}$$

$$\chi \to e^{-\sigma/2}\,\chi\ , \tag{4.8c}$$

$$\psi_\mu \to e^{\sigma/2}\,\psi_\mu\ , \tag{4.8d}$$

with

$$e^{2\sigma} = -\frac{3}{\Phi}\ , \qquad \sigma = -\frac{1}{6}\,J\ . \tag{4.9}$$

Under the Weyl rescaling of the vierbein we get the standard changes for the Einstein-curvature term and of the Lorentz connection:

$$\frac{1}{6}\,e\Phi R \to -\frac{1}{2}\,eR - \frac{3}{4}\,e(\partial_\mu \log \Phi)^2 + \text{4-div.}\ , \tag{4.10a}$$

$$\omega_{\mu mn} \to \omega_{\mu mn} - 2e^\nu{}_{[m}e_{n]\mu}\partial_\nu\sigma\ . \tag{4.10b}$$

After the substitutions (4.8), the total Lagrangian becomes the sum of a pure bosonic part and a fermionic part:

$$e^{-1}\,L = e^{-1}\,L_B + e^{-1}\,L_F\ , \tag{4.11}$$

where

$$\begin{aligned} e^{-1}\,L_B = & -\frac{1}{2}\,R + J^i_j D_\mu z_i D_\mu z^{*j} - \frac{1}{4}\,\mathrm{Re}\,f_{\alpha\beta}F^\alpha_{\mu\nu}F^\beta_{\mu\nu} + \frac{1}{4}\,i\,\mathrm{Im}\,f_{\alpha\beta}F^\alpha_{\mu\nu}\tilde{F}^\beta_{\mu\nu} \\ & + \frac{1}{4}\,|g|^2\,e^{-J}\left[3 + (J^{-1})^k_\ell\left(\frac{g^*_k}{g^*} - J_k\right)\left(\frac{g^\ell}{g} - J^\ell\right)\right] \\ & - \frac{1}{2}\,\tilde{g}^2\,\mathrm{Re}\,f^{-1}_{\alpha\beta}(J^i T^{\alpha j}_i z_j)\,(J^k T^{\beta\ell}_k z_\ell) \end{aligned} \tag{4.12}$$

and

$$\begin{aligned} e^{-1}\,L_F = & \frac{1}{2}\,e^{3\sigma}\left(J^{ij} + \frac{1}{3}\,J^i J^j\right)g\bar{\chi}_{Li}\chi_{Lj} \\ & + \frac{1}{2}\,g\left(\frac{g^\ell}{g} - J^\ell\right)e^{3\sigma}\,\bar{J}^{-1k}_\ell\left(J^{ij}_k + \frac{2}{3}\,J^i_k J^j\right)\bar{\chi}_{Li}\chi_{Lj} \\ & + \frac{1}{2}\,g\left(\frac{g^\ell}{g} - J^\ell\right)\frac{3}{4\Phi}\,e^\sigma\,J^{-1k}_\ell f^*_{\alpha\beta k}\bar{\lambda}^\alpha_R\lambda^\beta_R - \frac{1}{18}\,\Phi\,e^{2\sigma} \\ & \times\left(J^{ij} + \frac{1}{3}\,J^i J^j\right)\bar{\chi}_{Li}\chi_{Lj}\left(J_{k\ell} + \frac{1}{3}\,J_k J_\ell\right)\bar{\chi}^k_R\chi^\ell_R - \frac{1}{6}\,\Phi\,e^{2\sigma}\,J^{-1k}_\ell \end{aligned}$$

$$\times \left(J^{ij}_{k} + \frac{2}{3} J^{i}_{k}J^{j}\right)\bar{\chi}_{Li}\chi_{Lj}\left(J^{\ell}_{mn} + \frac{2}{3} J^{\ell}_{m}J_{n}\right)\bar{\chi}^{m}_{R}\chi^{n}_{R}$$

$$- \frac{1}{4} f^{\ell}_{\alpha\beta}\bar{\lambda}^{\alpha}_{L}\lambda^{\beta}_{L}J^{-1k}_{\ell}\left(J^{ij}_{k} + \frac{2}{3} J^{i}_{k}J^{j}\right)\bar{\chi}_{Li}\chi_{Lj}$$

$$- \frac{3}{32} \frac{1}{\Phi} e^{-2\sigma} J^{-1k}_{\ell} f^{\ell}_{\alpha\beta}\bar{\lambda}^{\alpha}_{L}\lambda^{\beta}_{L}f^{*}_{\gamma\delta k}\bar{\lambda}^{\gamma}_{R}\lambda^{\delta}_{R}$$

$$- \frac{1}{2\Phi} e^{2\sigma} \Phi^{i}_{j}\bar{\chi}^{j}_{R}\gamma^{\mu}\chi_{Li}\Phi^{k}D_{\mu}z_{k} - \frac{1}{2\Phi} e^{2\sigma} \Phi^{i}D_{\mu}z_{i}$$

$$\times(\Phi_{j}\bar{\psi}_{\mu R}\chi^{j}_{R} - \Phi^{j}\bar{\psi}_{\mu L}\chi_{Lj}) - \frac{3}{4\Phi} \mathrm{Re}\ f_{\alpha\beta}\Phi^{i}D_{\mu}z_{i}\bar{\lambda}^{\alpha}\gamma^{\mu}\lambda^{\beta}_{R}$$

$$- \frac{1}{8\Phi}(\Phi^{i}_{j}\bar{\chi}^{j}_{R}\gamma_{m}\chi_{Li})^{2} + \frac{1}{2\Phi} \Phi^{i}_{j}\bar{\chi}^{j}_{R}\gamma^{\mu}\chi_{Li}\Phi^{k}\bar{\psi}_{\mu L}\chi_{Lk}\ e^{2\sigma}$$

$$- \frac{3}{8\Phi} \Phi^{i}_{j}\bar{\chi}^{j}_{R}\gamma^{m}\chi_{Li}\ \mathrm{Re}\ f_{\alpha\beta}\bar{\lambda}^{\alpha}_{L}\gamma_{m}\lambda^{\beta}_{R}$$

$$+ \frac{3}{4\Phi} \Phi^{i}\bar{\psi}_{\mu L}\chi_{Li}\ \mathrm{Re}\ f_{\alpha\beta}\bar{\lambda}^{\alpha}_{L}\gamma^{\mu}\lambda^{\beta}_{R}$$

$$+ \frac{1}{4\Phi} e^{2\sigma} \Phi^{j}\bar{\psi}_{\mu L}\chi_{Lj}(\Phi_{i}\bar{\psi}_{\mu R}\chi^{i}_{R} - \Phi^{i}\bar{\psi}_{\mu L}\chi_{Li})$$

$$- \frac{9}{32\Phi} \mathrm{Re}\ f_{\alpha\beta}\ \mathrm{Re}\ f_{\gamma\delta}\bar{\lambda}^{\alpha}_{L}\gamma^{m}\lambda^{\beta}_{R}\bar{\lambda}^{\gamma}_{L}\gamma_{m}\lambda^{\delta}_{R}\ e^{-2\sigma}$$

$$- \frac{1}{2}\ i\ \mathrm{Re}\ f^{-1}_{\alpha\beta}\tilde{g}\Phi^{i}T^{\alpha j}_{i}z_{j}f^{k}_{\beta\gamma}\bar{\chi}_{Lk}\lambda^{\gamma}_{L}\ e^{2\sigma}$$

$$+ \frac{1}{8}\ \mathrm{Re}\ f^{-1}_{\alpha\beta}f^{i}_{\alpha\gamma}\chi_{Li}\lambda^{\gamma}_{L}(f^{j}_{\beta\delta}\bar{\chi}_{Lj}\lambda^{\delta}_{L} - f^{*}_{\beta\delta j}\bar{\chi}^{j}_{R}\lambda^{\delta}_{R})$$

$$- \frac{1}{4}\ \bar{\psi}_{\mu}\gamma_{5}\gamma_{\nu}D_{\rho}\psi_{\sigma}\sigma^{\mu\nu\rho\delta}\ e^{-1} - \frac{1}{64}\ (\bar{\psi}_{\mu}\gamma_{\lambda}\psi_{\rho} + 2\bar{\psi}_{\lambda}\gamma_{\mu}\psi_{\rho})\bar{\psi}^{\mu}\gamma^{\lambda}\psi^{\rho}$$

$$+ \frac{1}{16}\ (\bar{\psi}_{\mu}\cdot\gamma\cdot\psi)^{2} + 3\ \frac{\Phi^{i}_{j}}{\Phi}\ \bar{\chi}_{Li}\not{\partial}\chi^{j}_{R} - 3\ \frac{\Phi^{ij}_{k}}{\Phi}\ \bar{\chi}_{Li}\not{\partial}z_{j}\chi^{k}_{R}$$

$$- 3\ \frac{\Phi^{i}_{j}}{\Phi}\ \bar{\psi}_{\mu L}\not{\partial}z^{*j}\gamma_{\mu}\chi_{Li} + \frac{4\Phi^{i}}{\Phi}\ \chi_{Li}\sigma^{\mu\nu}D_{\mu}\psi_{\nu}$$

$$- \frac{3\Phi^{i}}{\Phi}\ \bar{\chi}_{Li}\gamma^{\nu}\gamma^{\mu}\psi_{L\nu}D_{\mu}\sigma + \frac{3}{8}\ e^{-1}\ \varepsilon^{\mu\nu\rho\sigma}\psi_{\mu}\gamma_{\nu}\psi_{\rho}$$

$$\times\left(\frac{\Phi^{i}}{\Phi}\ D_{\sigma}z_{i} + \frac{1}{2}\ \frac{\Phi^{j}_{i}}{\Phi}\ \bar{\chi}^{i}_{R}\gamma_{\sigma}\chi_{Lj}\right) + \frac{3}{2}\ \frac{\Phi^{i}_{j}}{\Phi}\ \psi_{\mu L}\bar{\chi}_{Li}\bar{\psi}_{\mu R}\chi^{j}_{R}$$

$$- \frac{3}{2} \frac{\Phi^{ij}_{k}}{\Phi} \bar{\chi}_{Li}\chi_{Lj}\bar{\chi}^{k}_{R}\chi^{\ell}_{R} - \frac{1}{2} \Phi^{i}\bar{\chi}_{Li}$$

$$\times \left(\chi_{\mu L}\bar{\psi}\cdot\gamma\psi^{\mu} + \frac{1}{2} \sigma_{\mu\nu}\psi_{\rho L}\bar{\psi}^{\nu}\gamma^{\rho}\psi^{\mu} + \sigma^{\mu\nu}\psi_{\nu L}\bar{\psi}_{\mu}\gamma\cdot\psi \right)$$

$$- \frac{3}{2} i\tilde{g} \frac{\Phi^{i}}{\Phi} T^{\alpha j}_{i} z_{j}\bar{\psi}_{L}\cdot\gamma\lambda^{\alpha}_{R} + 6i\tilde{g} \frac{\Phi^{j}_{i}}{\Phi} T^{\alpha k}_{j} z_{k}\bar{\lambda}^{\alpha}_{R}\chi^{i}_{R}$$

$$- \frac{1}{2} g^{ij} e^{3\sigma} \bar{\chi}_{Li}\chi_{Lj} + \frac{1}{2} g^{i} e^{3\sigma} \psi_{R}\cdot\gamma\chi_{Li}$$

$$+ \frac{1}{2} g\, e^{3\sigma} \bar{\psi}_{\mu R}\sigma^{\mu\nu}\psi_{\nu R} + \frac{1}{2} \mathrm{Re}\, f_{\alpha\beta}\Big[- \frac{1}{2} \bar{\lambda}^{\alpha}\not{D}\lambda^{\beta} + \frac{1}{2} \bar{\lambda}^{\alpha}\gamma^{\mu}\sigma^{\rho\sigma}\psi_{\mu}$$

$$\times \left(F^{\beta}_{\rho\sigma} + \frac{1}{2} \bar{\psi}_{\rho}\gamma_{\sigma}\lambda^{\beta} \right) - \frac{1}{32} \bar{\lambda}^{\alpha}\gamma_{5}\gamma_{\nu}\lambda^{\beta} e^{-1} \varepsilon^{\mu\nu\rho\sigma}\bar{\psi}_{\mu}\gamma_{\rho}\psi_{\sigma}\Big]$$

$$- \frac{1}{4} i\, \mathrm{Im}\, f_{\alpha\beta} D^{\mu}(\bar{\lambda}^{\alpha}\gamma^{\mu}\lambda^{\beta}_{R}) + \frac{1}{2} f^{i}_{\alpha\beta}\Big[-\bar{\chi}_{Li}\sigma^{\mu\nu}\lambda^{\beta}_{L}$$

$$\times \left(F^{-\alpha}_{\mu\nu} - \frac{1}{2} \bar{\psi}_{\rho L}\sigma_{\mu\nu}\gamma^{\rho}\lambda^{\alpha}_{R} + \frac{1}{2} \bar{\psi}_{R}\cdot\gamma\sigma_{\mu\nu}\lambda^{\alpha}_{L} \right)$$

$$- \frac{1}{2} \bar{\psi}_{R}\cdot\gamma\chi_{Li}\bar{\lambda}^{\alpha}_{L}\lambda^{\beta}_{L}\Big] + \frac{1}{4} f^{ij}_{\alpha\beta}\bar{\chi}_{Li}\chi_{Lj}\bar{\lambda}^{\alpha}_{L}\lambda^{\beta}_{L} + \text{h.c.} \qquad (4.13)$$

The kinetic part of Eq. (4.13) can be further diagonalized through the redefinition

$$\psi_{\mu L} \to \psi_{\mu L} - \gamma_{\mu} \frac{\Phi_{i}}{\Phi} \chi^{i}_{R} \,. \qquad (4.14)$$

If, in addition, we perform the following chiral redefinitions on the fermion fields,

$$\psi_{\mu L} \to \left(\frac{g}{g^{*}}\right)^{\frac{1}{4}} \psi_{\mu L} \,, \quad \chi_{Li} \to \left(\frac{g}{g^{*}}\right)^{\frac{1}{4}} \chi_{Li} \,, \quad \lambda^{\alpha}_{L} \to \left(\frac{g}{g^{*}}\right)^{\frac{1}{4}} \lambda^{\alpha}_{L} \qquad (4.15)$$

and we define the function

$$G = J - \log \frac{1}{4} |g|^{2} \qquad (4.16)$$

so that

$$G_{i} = J^{i} - \frac{g^{i}}{g}, \quad G^{i}_{j} = J^{i}_{j}, \tag{4.17}$$

the final Lagrangian becomes

$$L = L_{B} + L_{FK} + L_{FM} + L_{(4)F}, \tag{4.18}$$

where

$$e^{-1} L_{B} = -\frac{1}{2} R + G^{i}_{j} D_{\mu} z_{i} D^{\mu} z^{*j} - \frac{1}{4} \mathrm{Re}\, f_{\alpha\beta} F^{\alpha}_{\mu\nu} F^{\beta}_{\mu\nu}$$
$$+ \frac{1}{4} i\, \mathrm{Im}\, f_{\alpha\beta} F^{\alpha}_{\mu\nu} \tilde{F}^{\beta}_{\mu\nu} + e^{-G}(3 + G_{k} G^{-1k}_{\ell} G^{\ell})$$
$$+ \frac{1}{2} \tilde{g}^{2}\, \mathrm{Re}\, f^{-1}_{\alpha\beta} (G^{i} T^{\alpha j}_{i} z_{j}) (G^{k} T^{\beta\ell}_{k} z_{\ell}) ; \tag{4.19}$$

L_{FK} is the curved generalization of the fermionic terms; L_{FM} is the quadratic part in the fermion fields without derivative terms; $L_{(4)F}$ is the part of the Lagrangian which contains four-fermion interaction terms [not coming from the spin-3/2 torsion contribution already contained in the Lorentz connection $\omega_{\mu mn}(e,\psi)$] and also other interactions of dimension greater than four containing covariant derivatives of boson fields:

$$e^{-1} L_{FK} = -\frac{1}{4} \mathrm{Re}\, f_{\alpha\beta} \bar{\lambda}^{\alpha} \not{D} \lambda^{\beta} - \frac{1}{4} e^{-1} \bar{\psi}_{\mu} \gamma_5 \gamma_{\nu} D_{\rho} \psi_{\sigma} \varepsilon^{\mu\nu\rho\sigma}$$
$$+ G^{i}_{j} \bar{\chi}_{Li} \not{D} \chi^{j}_{R} - \frac{i}{8} \mathrm{Im}\, f_{\alpha\beta}\, e^{-1} D_{\mu} (e \bar{\lambda}^{\alpha} \gamma_5 \gamma_{\mu} \lambda^{\beta}) + \text{h.c.} , \tag{4.20}$$

$$e^{-1} L_{F,M} = e^{-G/2} \bar{\psi}_{\mu R} \sigma^{\mu\nu} \psi_{\nu R} - \bar{\psi}_{R} \cdot \gamma \Big(e^{-G/2} G^{i} \chi_{Li}$$
$$- \frac{i}{2} \tilde{g} G^{i} T^{\alpha j}_{i} z_{j} \lambda^{\alpha}_{L} \Big) + e^{-G/2} (G^{ij} - G^{i} G^{j}$$
$$- G^{\ell} G^{-1k}_{\ell} G^{ij}_{k}) \bar{\chi}_{Li} \chi_{Lj} - 2i\tilde{g} G^{k}_{i} z^{*j} T^{\alpha i}_{j} \bar{\lambda}^{\alpha}_{L} \chi_{Lk}$$
$$+ \frac{1}{2} f^{k}_{\alpha\beta} \Big(\frac{1}{2} e^{-G/2} G_{\ell} G^{-1\ell}_{k} \bar{\lambda}^{\alpha}_{L} + i\tilde{g}\, \mathrm{Re}\, f^{-1}_{\alpha\gamma} G^{i} T^{\gamma j}_{i} z_{j} \bar{\chi}_{Lk} \Big) \lambda^{\beta}_{L} + \text{h.c.} , \tag{4.21}$$

$$
\begin{aligned}
e^{-1} L_{(4)F} &= \frac{1}{2} \operatorname{Re} f_{\alpha\beta}\left(\frac{i}{2} \bar{\lambda}^{\alpha}\gamma^{\mu}\sigma^{\rho\sigma}\psi_{\mu}F^{\beta}_{\rho\sigma} - \frac{1}{2} \bar{\lambda}^{\alpha}_{L}\gamma_{\mu}\lambda^{\beta}_{R}G^{i}D_{\mu}z_{i}\right) \\
&- \frac{1}{2} f^{i}_{\alpha\beta}\bar{\chi}_{Li}\sigma\cdot F^{\alpha}\lambda^{\beta}_{L} + \frac{1}{8} e^{-1} \varepsilon^{\mu\nu\rho\sigma}\bar{\psi}_{\mu}\gamma_{\nu}\psi_{\rho}G^{i}D_{\sigma}z_{i} \\
&- G^{j}_{i}\bar{\psi}_{\mu L}\not{D}z^{*i}\gamma^{\mu}\chi_{Lj} - \bar{\chi}_{Li}\not{D}z_{j}\chi^{k}_{R}\left(G^{ij}_{k} + \frac{1}{2} G^{i}_{k}G^{j}\right) \\
&+ \frac{1}{32} G^{-1k}_{\ell} f^{\ell}_{\alpha\beta}f^{*}_{\gamma\delta k}\bar{\lambda}^{\alpha}_{L}\lambda^{\beta}_{L}\bar{\lambda}^{\gamma}_{R}\lambda^{\delta}_{R} + \frac{3}{32}(\operatorname{Re} f_{\alpha\beta}\bar{\lambda}^{\alpha}_{L}\gamma_{m}\gamma^{\beta}_{R})^{2} \\
&+ \frac{1}{8} \operatorname{Re} f_{\alpha\beta}\bar{\lambda}^{\alpha}\gamma^{\mu}\sigma^{\rho\sigma}\psi_{\mu}\bar{\psi}_{\rho}\gamma_{\sigma}\lambda^{\beta} + \frac{1}{2} f^{i}_{\alpha\beta}\left(\bar{\chi}_{Li}\sigma^{\mu\nu}\lambda^{\alpha}\cdot\bar{\psi}_{\nu L}\gamma_{\mu}\lambda^{\beta}_{R}\right. \\
&\left.+ \frac{1}{4} \bar{\psi}_{R}\cdot\gamma\chi_{Li}\bar{\lambda}^{\alpha}_{L}\lambda^{\beta}_{L}\right) + \frac{1}{8} G^{j}_{i}\bar{\chi}^{i}_{k}\gamma_{d}\chi_{Lj}(\varepsilon^{abcd}\bar{\psi}_{a}\gamma_{b}\psi_{c} \\
&- \bar{\psi}_{a}\gamma_{5}\gamma^{d}\psi_{a}) + \frac{1}{16} \bar{\chi}_{Li}\gamma^{\mu}\chi^{j}_{R}\bar{\lambda}^{\delta}_{R}\gamma_{\mu}\lambda^{\gamma}_{L}(-2G^{i}_{j} \operatorname{Re} f_{\gamma\delta} \\
&+ \operatorname{Re} f^{-1}_{\alpha\beta}f^{i}_{\alpha\gamma}f^{*}_{\beta\delta j}) - \frac{1}{16} \bar{\chi}_{Li}\sigma_{\mu\nu}\chi_{Lj}\bar{\lambda}^{\gamma}_{L}\sigma^{\mu\nu}\lambda^{\delta}_{L} \\
&\times \operatorname{Re} f^{-1}_{\alpha\beta}f^{i}_{\alpha\gamma}f^{j}_{\beta\delta} + \frac{1}{16} \bar{\chi}_{Li}\chi_{Lj}\bar{\lambda}^{\gamma}_{L}\lambda^{\delta}_{L} \\
&\times (-4G^{ij}_{k}G^{-1k}_{\ell}f^{\ell}_{\gamma\delta} + 4f^{ij}_{\gamma\delta} - \operatorname{Re} f^{-1}_{\alpha\beta}f^{i}_{\alpha\gamma}f^{j}_{\beta\delta}) \\
&+ \left(- \frac{1}{2} G^{ij}_{k\ell} + \frac{1}{2} G^{ij}_{m}G^{-1m}_{n}G^{n}_{k\ell} - \frac{1}{4} G^{i}_{k}G^{j}_{\ell}\right) \\
&\times \bar{\chi}_{Li}\chi_{Lj}\bar{\chi}^{k}_{R}\chi^{\ell}_{R} + \text{h.c.}
\end{aligned}
\tag{4.22}
$$

The final form of the Lagrangian, as given by Eq. (4.18), is invariant under the following local supersymmetry transformations:

$$
\delta e^{m}_{\mu} = \frac{1}{2} \bar{\varepsilon}_{L}\gamma^{m}\psi_{\mu R} + \text{h.c.} , \tag{4.23a}
$$

$$
\begin{aligned}
\delta\psi_{\mu L} &= \left[\partial_{\mu} + \frac{1}{2} \omega_{\mu mn}(e,\psi)\sigma^{mn}\right]\varepsilon_{L} + \frac{1}{2} \sigma_{\mu\nu}\varepsilon_{L}G^{j}_{i}\bar{\chi}^{i}_{R}\gamma^{\nu}\chi_{Lj} \\
&+ \frac{1}{2} \gamma_{\mu}\varepsilon_{R} e^{-G/2} + \frac{1}{4} \psi_{\mu L}(G^{i}\bar{\varepsilon}_{L}\chi_{Li} - G_{i}\bar{\varepsilon}_{R}\chi^{i}_{R}) \\
&- \frac{1}{4} \varepsilon_{L}(G^{i}D_{\mu}z_{i} - G_{i}D_{\mu}z^{*i}) + \frac{1}{4} (\sigma_{\nu\mu} + g_{\mu\nu})\varepsilon_{L}\bar{\lambda}^{*}_{L}\gamma^{\nu}\lambda^{\beta}_{R} \operatorname{Re} f_{\alpha\beta} ,
\end{aligned}
\tag{4.23b}
$$

$$\delta B_\mu^\alpha = -\frac{1}{2}\bar{\varepsilon}_L\gamma_\mu\lambda_R^\alpha + \text{h.c.}\ , \tag{4.23c}$$

$$\delta\lambda_R^\alpha = \frac{1}{2}\sigma^{\mu\nu}\hat{F}_{\mu\nu}^\alpha\varepsilon_R + \frac{1}{2}i\varepsilon_R \operatorname{Re} f_{\alpha\beta}^{-1}\Big(-\tilde{g}G^i T_i^{\beta j}z_j + \frac{1}{2}if_{\beta\gamma}^i\bar{\chi}_{Li}\lambda_L^\gamma$$

$$- \frac{1}{2}if_{\beta\gamma i}^*\bar{\chi}_R^i\lambda_R^\gamma\Big) - \frac{1}{4}\lambda_R^\alpha(G^i\bar{\varepsilon}_L\chi_{Li} - G_i\bar{\varepsilon}_R\chi_R)\ , \tag{4.23d}$$

$$\delta z_i = \bar{\varepsilon}_L\chi_{Li}\ , \tag{4.23e}$$

$$\delta\chi_{Li} = \frac{1}{2}\hat{\not{\partial}}z_i\varepsilon_R - \frac{1}{2}\varepsilon_L\, e^{-G/2}\, G_i^{-1j}G_j - \frac{1}{8}\varepsilon_L\bar{\lambda}_R^\alpha\lambda_R^\beta G^{-1k}f_{\alpha\beta k}^*$$

$$+ \frac{1}{2}\varepsilon_L G_i^{-1k}G_k^{j\ell}\bar{\chi}_{Lj}\chi_{L\ell} + \frac{1}{4}\chi_{Li}(G_j\bar{\varepsilon}_R\chi_R^j - G^j\bar{\varepsilon}_L\chi_{Lj})\ . \tag{4.23f}$$

The transformations (4.23) are directly obtained from the original transformations (3.25), with the substitution of Eqs. (4.5) in the fermionic field variations and with the redefinitions (4.8), (4.14), and (4.15) for the vierbein field and for the fermionic fields. In addition, the local supersymmetry parameter in Eqs. (4.23) has also been redefined, compared with Eqs. (3.25), as follows:

$$\varepsilon_L \to e^{-\frac{1}{2}J}\left(\frac{g}{g^*}\right)^{\frac{1}{4}}\varepsilon_L\ . \tag{4.24}$$

We note that, by using conformal tensor calculus, we could avoid the redefinitions given by Eqs. (4.8), (4.9), (4.14), (4.15), and (4.24), by suitably choosing a special conformal gauge. This has been shown elsewhere[17].

Finally, we discuss the case in which the gauge group is $G \times U_R(1)$, corresponding to the superspace action formulae given in Eqs. (3.11) and (3.12). As a consequence of Eq. (3.35), the presence of a gauged R-symmetry is most easily discussed in the final form of the Lagrangian and transformation laws as given by Eqs. (4.18) and (4.23).

In fact, the Lagrangian has precisely the same expression as in Eq. (4.18) with suitable gauge covariantization with respect to the $U_R(1)$ gauge field. The only difference now is the fact that the D-term of the R-multiplet, defined as

$$D_R = g_R G^i n_i z_i \ , \tag{4.25}$$

can have a constant term, owing to the non-invariance of the superpotential under R-transformations:

$$g^i(z) n_i g_R z_i = 3 g_R g(z) \ , \tag{4.26}$$

where $n_i g_R = q_i$ are the U(1) charges of the scalar fields z_i.

It is obvious that these considerations are no longer valid for a vanishing superpotential function $g(z) = 0$. In this case the final form of the Lagrangian and transformation laws comes directly from Eqs. (4.18) and (4.25) with the following substitution rules[11,18]

i) $e^{-G}\ , \quad e^{-G/2} \to 0\ ,$

$G^i \nabla_m z_i \Rightarrow J^i \nabla_m z_i + \frac{3}{2}\, i g_R V^R_m\ ,$

ii) $\left(G_i \nabla_m z^{*i} \Rightarrow J_i \nabla_m z^{*i} - \frac{3}{2}\, i g_R V^R_m\right)\ ,$

iii) $\tilde{g}_\alpha G^i T^{\alpha j}_i z_j \Rightarrow \tilde{g}_\alpha J^i T^{\alpha j}_i z_j - \frac{3}{2}\, g_R \delta^\alpha_R\ ,$

iv) for other derivatives of G not in the form given by (ii) and (iii), replace $G \to J$.

We note, in particular, that the substitution rule (ii) produces new minimal gauge couplings of the R-gauge field to the gravitino, gauginos, and chiral fermions (but not to the scalars) which are usually eliminated through the chiral redefinitions (4.15) when $g(z) \neq 0$.

It is important to note that some models discussed in the literature fall within this class[19]. However, these models can only have spontaneously broken supersymmetry with a non-vanishing cosmological constant.

In the more interesting physical situation of spontaneously broken R-symmetry and supersymmetry with a vanishing cosmological constant, the superpotential cannot vanish and the final form of the Lagrangian as given by Eq. (4.18) applies.

5. THE SUPER-HIGGS EFFECT IN N = 1 SUPERGRAVITY

Spontaneous breaking of local supersymmetry implies the super-Higgs effect, i.e. the gravitino mass generation and the disappearance of the massless goldstino which is absorbed by the 'massive' spin-$^3/_2$ gravitino.

This phenomenon may or may not occur with a vanishing cosmological constant, i.e. in a Minkowski or de Sitter space-time.

The goldstino field is uniquely defined by the spin-$\frac{1}{2}$ fermion which couples to the spin-$^3/_2$ gravitino in $L_{F,M}$ given by Eq. (4.21):

$$-\bar{\psi}_R\cdot\gamma\eta_L\ , \qquad \eta_L = e^{-G/2}\, G^i\chi_{Li} - \frac{i}{2}\, D^\alpha\lambda_L^\alpha\ , \tag{5.1}$$

where

$$D^\alpha = \tilde{g}G^i T^{\alpha j} z_j\ . \tag{5.2}$$

A necessary and sufficient condition for spontaneous supersymmetry breaking therefore implies that one of the quantities[6],

$$e^{-G/2}\, G^i, \qquad D^\alpha\ , \tag{5.3}$$

is different from zero at the minimum of the potential given in Eq. (4.19):

$$V(z,z^*) = -e^{-G}(3 + G^i G_i^{-1j} G_j) + \frac{1}{2}\,\mathrm{Re}\, f^{-1}_{\alpha\beta} D^\alpha D^\beta\ . \tag{5.4}$$

In the more interesting situation of a Minkowski space-time, we also demand the vanishing of the cosmological constant

$$V(z_0,z_0^*) = 0 \quad \text{at} \quad \left.\frac{\partial V}{\partial z}\right|_{z=z_0} = 0\ . \tag{5.5}$$

Under condition (5.5), the gravitino mass has a precise meaning, and this is given by the universal formula[5]:

$$m_{3/2} = e^{-G\,2} . \tag{5.6}$$

We now discuss the particularly interesting case of 'minimal' coupling of the Yang-Mills system to supergravity. This is defined by the two conditions

$$G^{i}_{j} = -\frac{1}{2}\delta^{i}_{j} , \qquad f_{\alpha\beta} = \delta_{\alpha\beta} , \tag{5.7}$$

in which case all the kinetic terms are canonical and the Lagrangian depends only on the superpotential g(z). In this case we have

$$G(z,z^{*}) = -\frac{1}{2}|z|^{2} - \log\frac{1}{4}|g(z)|^{2} . \tag{5.8}$$

The scalar potential (5.4) becomes

$$V(z,z^{*}) = e^{-G}(2G^{i}G_{i} - 3) + \frac{1}{2}(D^{\alpha})^{2} , \tag{5.9}$$

with

$$D^{\alpha} = -\frac{1}{2}\tilde{g}_{\alpha}z^{*i}T^{\alpha j}_{i}z_{j} + \xi\delta^{\alpha}_{R} .$$

The term $L_{F,M}$ given by Eq. (4.21), which determines the fermion mass matrix is

$$e^{-1}L_{F,M} = e^{-G/2}\left(\bar{\psi}_{\mu R}\sigma^{\mu\nu}\psi_{\nu R} - \bar{\psi}_{R}\cdot\gamma\tilde{\eta}_{L} - \frac{2}{3}\bar{\tilde{\eta}}_{L}\tilde{\eta}_{L}\right)$$
$$+ \bar{\chi}_{Li}M^{ij}\chi_{Lj} + 2\bar{\chi}_{Li}M^{i\alpha}\lambda^{\alpha}_{L} + \bar{\lambda}^{\alpha}_{L}M^{\alpha\beta}\lambda^{\beta}_{L} + \text{h.c.} \tag{5.10}$$

where

$$\tilde{\eta}_{L} = G^{i}\chi_{Li} - \frac{i}{2}e^{G/2}D^{\alpha}\lambda^{\alpha}_{L} . \tag{5.11}$$

The spin-½ fermion mass matrix has the form

$$M^{ij} = e^{-G/2}\left(G^{ij} - \frac{1}{3}G^{i}G^{j}\right) , \tag{5.12a}$$

$$M^{i\alpha} = -\frac{1}{3}iG^{i}D^{\alpha} - ID^{\alpha i} , \tag{5.12b}$$

$$M^{\alpha\beta} = -\frac{1}{6}\, e^{G/2} D^{\alpha} D^{\beta} \;. \qquad (5.12c)$$

We can now compute the quadratic mass relation which generalizes the result of spontaneously broken, globally supersymmetric Yang-Mills theories. The scalar contribution to the trace of the square mass is

$$\mathrm{Tr}\, M_0^2 = -(N+4) D^{\alpha} D^{\alpha} + 2G^{i} G_{i} D^{\alpha} D^{\alpha} + \eta\, e^{-G} (4G_{ij} G^{ij} + N)$$
$$+ 4D^{\alpha i} D^{\alpha}_{i} + 4D^{\alpha} D^{\alpha i}_{i} \;. \qquad (5.13)$$

The spin-1 contribution is

$$\mathrm{Tr}\, M_1^2 = 12 D^{\alpha}_{i} D^{i\alpha} \;. \qquad (5.14)$$

The spin-½ contribution is

$$\mathrm{Tr}\, M_{\frac{1}{2}}^2 = 8\, e^{-G} \left(G^{ij} - \frac{1}{3} G^{i} G^{j} \right) \left(G_{ij} - \frac{1}{3} G_{i} G_{j} \right)$$
$$+ 16 \left(D^{\alpha i} + \frac{1}{3} G^{i} D^{\alpha} \right) \left(D^{\alpha}_{i} + \frac{1}{3} G_{i} D^{\alpha} \right) + \frac{2}{g}\, e^{G} (D^{\alpha})^2 (D^{\beta})^2 \qquad (5.15)$$

and the gravitino contribution is

$$\mathrm{Tr}\, M_{3/2}^2 = 4\, e^{-G} \;. \qquad (5.16)$$

Therefore we finally get the mass formula

$$\text{super-trace } M^2 = \sum_{J=0}^{3/2} (-)^{2J} (2J+1) m_J^2 = (N-1)(2m_{3/2}^2 - K^2 D^{\alpha} D^{\alpha})$$
$$- 2\tilde{g}_{\alpha} D^{\alpha}\, \mathrm{Tr}\, T^{\alpha} \;. \qquad (5.17)$$

Equation (5.17) gives back the result of global supersymmetry[20] if we set $m_{3/2} = 0$, $k = 0$. It also reproduces the result of Ref. 11 for $N = 1$ and $\tilde{g} = 0$. We also note that the last term in Eq. (5.17) is only possible for Abelian U(1) factors of G with $\mathrm{Tr}\, T^{\alpha} \neq 0$. Equation (5.17) summarizes the mass relations of spontaneously broken supersymmetry when the simultaneous occurrence of the Higgs and super-Higgs effects takes place.

6. LOW-ENERGY LIMIT OF SPONTANEOUSLY BROKEN, LOCALLY SUPERSYMMETRIC LAGRANGIANS

In the practical current applications of spontaneously broken supergravity models to particle interactions, the Lagrangians described in the previous sections will contain, beyond the Planck scale m_P, some lower energy scales.

In particular they will contain the weak interaction scale m_W and other possible fundamental scales, such as the GUT scale $m_X = O(10^{15}\text{-}10^{16}\ \text{GeV})$.

Although our previous formalism is valid for any value of the gravitino mass, which is related to the supersymmetry-breaking scale M_{SB} as follows[4,5],

$$m_{3/2} = \sqrt{\frac{8\pi}{3}}\,\frac{M_{SB}^2}{m_P}\ , \tag{6.1}$$

we will consider the particular situation of an intermediate supersymmetry breaking scale

$$M_{SB} = \sqrt{\mu m_P}\ , \tag{6.2}$$

where μ is a scale parameter $O(m_W)$. Under this circumstance

$$m_{3/2} = \sqrt{\frac{8\pi}{3}} = O(m_W) \tag{6.3}$$

and supergravity corrections to low-energy physics play a crucial role. This is already implied by the modification of the mass spectrum through the quadratic mass relation given by Eq. (5.17).

The purpose of the present section is to derive all possible supergravity-induced breaking terms which survive in the low-energy limit defined by letting $m_P \to \infty$ with $m_{3/2}$ kept fixed.

The final result will be that in the limit $m_{3/2}/m_P \ll 1$, spontaneously broken supergravity will manifest itself as an explicitly softly broken global supersymmetric theory with well-defined renormalizable interactions[21].

In particular, the soft-breaking terms, at most of dimension three in the particle fields, are such that no quadratic divergence arises through radiative corrections if gravitational interactions are neglected[22]. This circumstance has the resulting effect of not destroying the characteristic feature of supersymmetric theories, i.e. of keeping radiative corrections of light particles, including Higgs, generally small, compared to some other large scale of the theory such as the GUT scale m_X of the Planck scale m_P.

The important point in obtaining the low-energy limit of a spontaneously broken supergravity model is to realize that the cancellation of the cosmological constant [see Eq. (5.5)] implies the presence, in the theory, of some scalar fields x_i acquiring v.e.v.'s of order m_P which break supersymmetry[5]. It is natural, although not compulsory, that these fields are singlets under the Yang-Mills group G introduced in the previous section, especially if we do not want the GUT scale $m_X < m_P$ to be related to the Planck scale. The full set of scalar fields z_i belonging to the chiral multiplets S^i of the theory therefore split into

$$z_i = (x_i, y_i) \; , \tag{6.4}$$

where $x_i = O(m_P)$ and $\langle y_i\rangle/\langle x_j\rangle << 1$.

The remaining fields y_i are called low-energy fields and generally transform non-trivially under G.

The requirement that the low-energy effective Lagrangian indeed describes light fields with renormalizable interactions can be fulfilled under the condition that a decoupling occurs between x-type and y-type multiplets as well as gauge vector multiplets. Furthermore, the gravitino mass $m_{3/2}$ must be finite for $m_P \to 0$ [21].

We define the dimensionful G function to be (dim G = 2):

$$G(z,z^*) = J(z,z^*) - \frac{1}{k^2} \log |k^3 g(z)|^2 \; . \tag{6.5}$$

The y-dependent part of the Kähler potential is

$$J\left(\frac{1}{k}\zeta, \frac{1}{k}\zeta^*; y, y^*\right) = -\frac{1}{2} y_i \Lambda^i_j(\zeta,\zeta^*) y^{*j} + B^i(\zeta,\zeta^*) y_i + B^*_i(\zeta,\zeta^*) y^{*i} , \quad (6.6)$$

where $\zeta_i = kx_i$. Analogously, the y-dependent part of the superpotential $g(z) = g(x,y)$ must be a finite polynomial, at most of degree three in the y-variables with ζ-dependent coefficients.

Equation (3.5) defines the Kähler potential $J(z,z^*)$ up to a Kähler transformation

$$J(z,z^*) \to J(z,z^*) + f(z) + f^*(z^*) , \quad (6.7a)$$

$$g(z) \to g(z) \exp k^2 f(z) . \quad (6.7b)$$

The behaviour of the y-dependent part of the superpotential $g(z)$ for $m_P \to \infty$, consistent with Eq. (6.6), must be

$$g(\zeta,0) = O(m_P^2) , \quad k \to 0 . \quad (6.8)$$

In addition, the dimensionless function $f_{\alpha\rho}(z)$ must behave as

$$f_{\alpha\beta}(z) = \delta_{\alpha\beta}(1 + i\theta_\alpha) , \quad k \to 0 \quad (6.9)$$

for $m_P \to \infty$ and x_i, y_i fixed.

Let us now consider the general form of the Lagrangian (4.18) in the limit $m_P \to \infty$ $(k \to 0)$ in the y_i fields, after the x_i fields have acquired v.e.v.'s of $O(m_P)$ which break supersymmetry.

Equation (6.8) implies a finite gravitino mass,

$$m_{3/2} = \lim_{k\to 0} k^2 |\langle g\rangle| \exp - \frac{k^2}{2} \langle J\rangle . \quad (6.10)$$

The low-energy fields also include (coming from the x-type fields) the goldstino, which provides the missing degrees of freedom of the massive spin-$\frac{3}{2}$ gravitino and two G-singlet scalar fields, the scalar partners of the would-be goldstino, with masses $O(m_{3/2})$[6]. However, these fields are not included in the low-energy Lagrangian of the y fields, because in the limit $m_P \to \infty$ they decouple from the y-type multiplets.

The gravitino mass defined by Eq. (6.10) provides the scale of all explicit supersymmetry-breaking terms appearing in the limiting Lagrangian for the low-energy fields y_i. This has the general form[23]

$$L = L(\mathrm{SUSY};f) + m_{3/2}(c_\alpha \bar{\lambda}_L^\alpha \lambda_L^\alpha + \mathrm{h.c.}) - m_{3/2}\left[h(y) + h^*(y^*)\right] - \frac{1}{2} m_{3/2}^2 y^{*j} \Lambda^i_j y_i \; . \tag{6.11}$$

The first term is a globally supersymmetric Lagrangian with superpotential

$$f(y) = \lim_{k \to 0} g(\langle x \rangle, y) \exp\left(-\frac{1}{2} k^2 \langle J \rangle\right). \tag{6.12}$$

The function h(y) is another analytic polynomial, invariant under G, of (at most) degree three in the y fields; the $c_{\alpha,s}$ in the gaugino mass terms are complex dimensionless coefficients of order 1; and Λ^i_j is a numerical Hermitian matrix which depends on the Kähler potential J defined by Eq. (6.6).

In addition, the D-term of the supersymmetric part of Eq. (6.11) can contain dimensionful constants (Fayet-Iliopoulos terms), corresponding to U(1) factors of G:

$$D^2_{U(1)} = (\tilde{g} y_i Q^i_j y^{*j} + \xi)^2 \; , \tag{6.13}$$

this even if the original Lagrangian given by Eq. (4.18) does not contain any gauged R-symmetry, as defined in Section 3. The reason is that if the scalar fields x_i have v.e.v.'s $\langle x_i \rangle = O(m_P)$, then we can get a purely gravitationally induced effective D-term coming from terms linear in the U(1) gauge multiplet which couples to the x-fields*),

$$k^2 D^{\alpha} W_{\alpha} \Psi(S,\bar{S})|_D \;, \tag{6.14}$$

producing an effective interaction,

$$D_{U(1)} \psi^j_i h^i h^*_j \rightarrow c D_{U(1)} m^2_{3/2} \;, \tag{6.15}$$

for $\psi^j_i \sim \delta^j_i$ and $|h_x| = O(m_{3/2})$.

This term should not be confused with a possible intrinsic ξ term in the original Lagrangian (4.18) which could be introduced by the gauging of R-symmetry as explained in Section 3. The term in Eq. (6.13) rather produces a soft-breaking term of dimension two in the scalar fields, of the type

$$L_{soft} \simeq c m^2_{3/2} \, y_i (Q)^i_j y^{*j} \tag{6.16}$$

where Q is the U(1) charge matrix of the y fields and which looks like a Fayet-Iliopoulos term since it can be rewritten in the form (6.13).

It is interesting to note that the general form (6.11) can also be obtained by the coupling of a locally supersymmetric Lagrangian for the matter fields y_i, to a Volkov-Akulov field, i.e. the non-linearly realized goldstino field. This has recently been shown by Samuel and Wess[24].

The reason why this approach gives the same result as the linear realization is due to the decoupling which occurs, for $m_P \rightarrow \infty$, between the goldstino multiplet and the light-matter fields y_i. In particular, the scalar degrees of freedom of the goldstino multiplet, which acquire v.e.v.'s of $O(m_P)$, do not play any role in the final softly broken Lagrangian we have considered. Then the entire

*) For a related argument, in the framework of non-linear realization, see Samuel and Wess[24].

goldstino multiplet can be simply replaced by a Volkov-Akulov field θ transforming, for constant ε, as[24]:

$$\delta_\varepsilon \theta(x) = M_{SB}^2 \varepsilon + \frac{1}{M_{SB}^2} \xi^\mu(x) \partial_\mu \theta(x) , \tag{6.17}$$

with

$$\xi^\mu(x) = \frac{1}{2} \bar{\varepsilon}_L \gamma^\mu \theta_R(x) - \frac{1}{2} \bar{\theta}_L(x) \gamma^\mu \varepsilon_R . \tag{6.18}$$

The Lagrangian (6.11) shows that the low-energy effects due to spontaneously broken supergravity, with $m_{3/2} = O(m_W)$, result in explicit soft-breaking terms in the global limit which do not alter the ultraviolet behaviour of the purely globally supersymmetric part of the Lagrangian. In particular, they do not introduce quadratic divergences since these terms are soft in a supersymmetric sense, according to the analysis of Ref. 22.

To understand the origin of the various soft-breaking terms in Eq. (6.11), it is useful to consider special forms for the functions $G(z,z^*)$ and $f_{\alpha\beta}(z)$ which appear in the general form of the matter-Yang-Mills supergravity action given by Eq. (4.18).

In the case of minimal coupling, defined by Eq. (5.7), we get[21,25]

$$h(y) = (A-3)f(y) + \sum_I y_i \frac{\partial f}{\partial y_i} , \tag{6.19a}$$

$$c_\alpha = 0 , \tag{6.19b}$$

$$\Lambda_i^j = \delta_i^j , \tag{6.19c}$$

where A is a parameter depending on the particular way the x fields enter into the superpotential $g(z) = g(x,y)$.

For example, if

$$g(x,y) = g(x) + g(y) , \tag{6.20}$$

with

$$g(x) = ax + b , \tag{6.21}$$

then

$$A = 3 - \sqrt{3} \, . \tag{6.22}$$

In the minimal case we may obtain an arbitrary function h(y) for the soft-breaking term in Eq. (6.11), which is linear in $m_{3/2}$, by relaxing the property (Eq. 6.19)*).

If we introduce general mixing terms of the y fields to the x fields, then h(y) receives an additional contribution given by

$$\Delta h(y) = \lim_{k\to 0} \frac{\sqrt{3}}{k} \left.\frac{\langle \partial g(x,y)\rangle}{\partial x}\right|_{x=x_0} \exp\left(-\frac{k^2}{2}\langle J\rangle\right) , \tag{6.23}$$

where $\partial g/\partial x$ is the derivative of g in the direction of the would-be goldstino.

Note that in order to get Eq. (6.22), what is needed is to introduce a superpotential of the form**)

$$g(x,y) = g(x) + g(y) + xg'(y) + O\left(\frac{1}{m_P}\right) + \dots \, . \tag{6.24}$$

The reason for introducing non-vanishing gaugino mass terms in Eq. (6.11) is because we can have an x-dependence in the function $f_{\alpha\beta}(z)$ which appears in the Yang-Mills part of the original Lagrangian[6]. The coefficients c_α in Eq. (6.11) are related to the first derivative of the function $f_{\alpha\beta}$ in the x-direction:

$$\begin{aligned} m_{3/2} c_\alpha \delta_{\alpha\beta} &= \lim_{k\to 0} \frac{1}{k} f^{i}_{\alpha\beta} G_j (-J^{-1})^j_i \exp\left(-\frac{k^2\langle G\rangle}{2}\right) \\ &= \lim_{k\to 0} \frac{1}{k} f^{x}_{\alpha\beta} G_x (-J^{-1})^x_x \exp\left(-\frac{k^2\langle G\rangle}{2}\right). \end{aligned} \tag{6.25}$$

---

*) An example of the soft-breaking term given by an analytic function h(y) is obtained by integrating out the heavy degrees of freedom of GUTs, which in general modify the soft-breaking terms of light fields[26].

**) An example which is a particular case of Eq. (6.23) has been considered by Cremmer et al.[27].

Finally, to get mass terms for the y fields of the form $y_i \Lambda^i_j y^{*j}$ with $\Lambda^i_j \neq \delta^i_j$, we need a non-minimal Kähler potential[23]:

$$J(z,z^*) = -\tfrac{1}{2}|x_i|^2 - \tfrac{1}{2}|y_i|^2 - \tfrac{1}{2} k^2 |x_i|^2 y^{*\ell} A^m_\ell y_m \,, \qquad (6.26)$$

where A^m_ℓ is a numerical Hermitian matrix.

Then one gets in this case the matrix relation

$$\Lambda = \mathbb{1} + 3A \,. \qquad (6.27)$$

It turns out that Eqs. (4.24) and (4.26) give a correction to the general y-independent mass sum rule for minimal coupling regarding the y multiplets[6]:

$$\sum_{J=0}^{1} (-)^{2J}(2J + 1)m_J^2 = 2(\dim T)m^2_{3/2} - 2\tilde{g}^\alpha D^\alpha \operatorname{Tr} T^\alpha . \qquad (6.28)$$

This is related to the modification of the general quadratic mass sum rule in the case of non-minimal coupling[16].

For instance, a non-minimal Kähler potential gives an additional contribution to Eq. (6.27) of the form[16]:

$$e^{-G} G_a (G^{-1})^a_b G^c (G^{-1})^d_c R^b_d \,, \qquad R^b_d = \partial^b \partial_d \log \operatorname{Det} G^m_n \,, \qquad (6.29)$$

when R^ℓ_i is the contracted Riemann tensor (Ricci tensor) constructed out of the Kähler metric G^j_i.

For the particular example given by Eq. (6.25), the term in Eq. (6.28) gives a contribution to the quadratic mass sum rule proportional to $3m^2_{3/2}$ Tr A.

Further examples of a super-Higgs sector of the theory with non-trivial Kähler curvature and non-trivial mass relations will be given in the next section.

The Lagrangian given by Eq. (6.11) is a tree-level result. If we start with a minimal coupling of matter to supergravity, then $c_\alpha = 0$ and $\Lambda^j_i = \delta^j_i$. It is clear that for general values of Λ^j_i and

c_α the low-energy Lagrangian (6.11) is stable under radiative corrections. However, if we start with the minimal coupling [Eq. (5.7)] then radiative corrections, including gravitational ones, will induce corrections which at most will be reabsorbed in definitions of the coefficients c_α and the matrix Λ_i^j as well as in a possible Fayet-Iliopoulos term for Abelian factors of G. These corrections, depending on the particular model under consideration, can turn out to be small. In this situation the low-energy spectrum can be described by Eq. (6.18), in terms of a few parameters, and we can build simple extensions of the standard model for weak interactions as well as GUT models with well-defined predictions for particle phenomenology in the 100 GeV to 1 TeV energy range.

Simple examples of the generalization of the Standard Model, based on low-energy supergravity, will be given in Section 8.

7. FURTHER EXAMPLES OF A SUPER-HIGGS SECTOR: FLAT POTENTIALS AND GAUGED R-SYMMETRY

In this section we will consider the super-Higgs effect in situations that are different from the flat Kähler potential situation

$$\mathcal{G}_i^j = -\frac{1}{2}\,\delta_i^j \tag{7.1}$$

as considered in Sections 5 and 6.

We again consider the situation of a single chiral multiplet (z,χ_L). The z-potential is given, in general, by the first term of Eq. (5.4) for i = 1:

$$V(z,z^*) = -e^{-\mathcal{G}}(\mathcal{G}_z\mathcal{G}_{z^*}\mathcal{G}^{-1}_{zz^*} + 3)\ . \tag{7.2}$$

The gravitino mass is given by Eq. (5.6), provided $V(z,z^*)$ vanishes at the point $V_z = 0$. The Kähler curvature, for a signle scalar field, is given by[16]

$$R_{zz^*} = \partial_z\partial_{z^*}\ \log \mathcal{G}_{zz^*} \tag{7.3}$$

and it vanishes for the 'minimal kinetic term' (7.1). This corresponds to the super-Higgs sector described by a Polony field[28]. For a non-trivial, non-constant metric G_{zz^*}, the mass-sum rule of the super-Higgs sector becomes[16]:

$$m_A^2 + m_B^2 - 4m_{3/2}^2 = -2\ e^{-G}\ \frac{G_z G_{z^*}}{(G_{zz^*})^2}\ R_{zz^*} \tag{7.4}$$

There are interesting cases where the scalar potential given by Eq. (7.2) is identically equal to zero, although the supersymmetry may be broken. Rewriting the scalar potential of Eq. (7.2) as[5]

$$V = -g\ e^{-(4/3)G}\ G^{-1}_{zz^*}\partial_z\partial_{z^*}\ e^{(1/3)G}\ , \tag{7.5}$$

it follows immediately that the potential is identically equal to zero[29] if*)

$$G = \frac{3}{2}\log\left[\phi(z) + \phi^*(z^*)\right]^2\ . \tag{7.6}$$

---

*) To better emphasize the geometric relevance of global flatness it is worth noting that Eq. (7.6) is equivalent to the unique solution

$$G = \frac{3}{2}\log\ (z + z^*)$$

up to field redefinition $z \to f(z)$. The scalar Lagrangian takes the form

$$3\sqrt{-g}\ g^{\mu\nu}\ \partial_\mu z \partial_\nu z^* / (z + z^*)^2$$

This Lagrangian is the same as the scalar sector of the N = 4 supergravity theory[30] and it describes a non-linear σ-model with an SU(1,1) non-compact symmetry

$$z \to \alpha z + i\beta / i\gamma z + \delta\ ,$$

with α, β, γ, δ real, and $\alpha\beta + \beta\gamma = 1$.

We also note that through the particular redefinition

$$z \to z + \sqrt{3}/z - \sqrt{3}$$

the kinetic term takes the form of the conformal scalar coupling[15,31] with $J = +3\log\ (1 - \frac{1}{3}\ zz^*)$.

Contrary to the minimal kinetic term case, the Kähler manifold is not flat ($R_{zz^*} = 0$), since using Eq. (7.3) it is found that

$$R_{zz^*} = -\frac{2}{3} G_{zz^*} . \tag{7.7}$$

Therefore, this defines an Einstein-Kähler manifold. However, relation (7.7) is not sufficient to ensure the vanishing of the potential.

Note that in the positive energy domain defined by

$$\mathcal{D} : G_{zz^*} < 0 \tag{7.8}$$

supersymmetry is spontaneously broken and the gravitino mass is different from zero:

$$m^2_{3/2} = e^{-G} = \frac{1}{|\phi(z) + \phi^*(z^*)|^3} , \qquad \text{for all } z \in \mathcal{D} \tag{7.9}$$

From the explicit expression for G_{zz^*}, if follows that

$$\mathcal{D} : |\phi_z| \neq 0, \qquad (\phi + \phi^*)^2 < \infty . \tag{7.10}$$

So the 'gravitino mass' is zero only at the boundary points $(\phi + \phi^*) = \infty$. The fact that the scalar potential is degenerate ($V \equiv 0$) reflects an arbitrariness for the values of the gravitino mass. This ambiguity can easily be removed if we demand the flatness condition ($V \equiv 0$) only locally, around a point $z_0 \in \mathcal{D}$.

$$V_{z \to z_0} \simeq C|z-z_0|^{2n} , \ n > 2 . \tag{7.11}$$

In fact we can build a large class of positive definite potentials demanding the relation

$$\partial_z \partial_{z^*} e^{(1/3)G} = \Phi_{zz^*}(z,z^*) , \tag{7.12}$$

where the real function $\Phi(z,z^*)$ satisfies the conditions

$$\Phi_{zz^*}(z,z^*) > 0, \qquad \text{for all } z \in \mathcal{D} , \tag{7.13}$$

with $\mathcal{D}$ the positive kinetic energy domain ($G_{zz^+} < 0$), and

$$\Phi_{zz^*}(z_0, z_0^*) = 0 \; .$$

The general solution of Eq. (7.12) is then

$$G = \frac{3}{2} \log \, (\phi + \phi^* + \Phi)^2 \; , \tag{7.15}$$

with

$$G_{zz^*} = 3 \, \frac{\Phi_{zz^*}(\phi + \phi^* + \Phi) - |\phi_z + \Phi_z|^2}{(\phi + \phi^* + \Phi)^2} < 0 \; . \tag{7.16}$$

Equation (7.16) defines the positive kinetic energy domain $\mathcal{D}$.

The corresponding scalar potential is positive definite in $\mathcal{D}$, provided that $\Phi_{zz^*} > 0$ and $\phi + \phi^+ + \Phi > 0$, as can easily be seen from its analytic expression

$$V_0 = 3 \, \frac{\Phi_{zz^*}(\phi + \phi^* + \Phi)}{|\phi + \phi^* + \Phi|^3 \left[|\phi_z + \Phi_z|^2 - \Phi_{zz^*}(\phi + \phi^* + \Phi) \right]} \; . \tag{7.17}$$

At the minimum ($z = z_0$) the potential vanishes identically and the gravitino mass is well defined. If the potential is locally flat at z_0, the fourth derivative of Φ at $z = z_0$ must also vanish:

$$\Phi_{zzz^*z^*}(z_0, z_0^*) = 0 \; . \tag{7.18}$$

In that case, the relation (7.7) between the curvature R_{zz^*}, and the Kähler metric G_{zz^*} is still valid locally ($z \sim z_0$), and the curvature contribution to the (mass)^2 sum rule[16] is minus four times the gravitino mass squared:

$$\Delta m^2 = -2 \, e^{-G} \, \frac{G_z G_{z^*}}{(G_{zz^*})^2} \, R_{zz^*} \Bigg|_{z=z_0} = -4 m_{3/2}^2 \; . \tag{7.19}$$

We may understand how the (mass)^2 sum rule can be satisfied with a non-vanishing gravitino mass (signal of supersymmetry breaking) and vanishing scalar masses, as is obvious from the local flatness of the potential at the minimum.

Let us now examine the interesting physical case where the usual matter fields are also coupled to supergravity. In general, the positivity properties of the potential are destroyed when the matter fields are coupled in an arbitrary way. The solution to this problem is to assume that G is the sum of two uncoupled functions $G(z,z^*)$, as in Eq. (7.15), and $\bar{G}(y^i,y_i^*)$, so that the potential becomes

$$V = -e^{-G_T}\left[\frac{G_z G_{z^*}}{G_{zz^*}} + 3\right] - e^{-G_T}\,\bar{G}_i(\bar{G}^{-1})^i_j\bar{G}^j + \frac{1}{2}D^\alpha D^\alpha , \qquad (7.20)$$

with

$$G_T = G(z,z^*) + \bar{G}(y_i,y^{*i}) , \qquad (7.21)$$

and, therefore, remains positive definite provided that $-\bar{G}^i_j$ is a Hermitian positive matrix, the latter being a necessary condition for ensuring a meaningful kinetic term.

The vanishing minimum arises for $z = z_0$ as before, and

$$G_i(\bar{G}^{-1})^i_j\bar{G}^j = 0 , \qquad (7.22)$$

$$D^\alpha = 0 .$$

The choice of minimal kinetic terms for the matter sector (y^i) leads to the following positive definite potential:

$$V(z,y_i) = e^{-\bar{G}}V_0(z,z^*) + e^{-G_T}\left(\sum_i \left|h_i + y_i\right|^2\right) + \frac{1}{2}(D^\alpha)^2 , \qquad (7.23)$$

where $V_0(z,z^*)$ is given by Eq. (7.17). The total G_T function reads

$$G_T = \frac{3}{2}\log(\phi + \phi^* + \Phi)^2 - y_i y^{i*} - h(y_i) - h^*(y^{*i}) . \qquad (7.24)$$

The absolute minimum of the potential (7.23) is zero and occurs at $z = z_0$ and $y_i = -h_i^*$ (if there are solutions). Note that the condition $y_i = -h_i^*$ automatically implies the vanishing of D terms because of the invariance of h under the internal group[6]. The vanishing of the cosmological constant is thus automatically satisfied without any unnatural fine tuning. The gravitino mass, in that case, is given by

$$m^2_{3/2} = e^{-G_T(z,z^*,y_i,y^{i*})}\Big|_{\text{at the minimum}} \quad . \tag{7.25}$$

In the flat limit, the hidden sector decouples and the scalar particle z remains massless for locally flat potentials. The potential for the matter fields becomes

$$V(y_i,y^{*i}) = \sum_i \left| \frac{\partial g(y_i)}{\partial y_i} + m_{3/2} y^{*i} \right|^2 + \frac{1}{2} D^\alpha D^\alpha \ , \tag{7.26}$$

where we have rescaled $h(y^i) = (1/M^2 m_{3/2}) g(y^i)$ in order to have a residual effect. The D term in that case is given by

$$D^\alpha = g^\alpha y^{*i} T^{\alpha j}_i y_j \ . \tag{7.27}$$

The form of the potential presented in Eq. (7.26) no longer depends on the specific form of the hidden sector, and gives rise to soft global supersymmetry-breaking terms of the same form as in the A = 3 case or the case of a factorized form of the superpotential[27], and therefore the supertrace formula for the matter fields remains the same as in the minimal case[6,27]

$$\sum_B m^2_B - 2\sum_F m^2_F = 2N_y m^2_{3/2} - 2r_i m^2_{\lambda_i} \ , \tag{7.28}$$

where N_y is the number of chiral multiplets in the y sector, r_i is the dimension of the internal gauge group, and m_{λ_i} are the gaugino masses coming from a non-minimal choice of f_{AB}[6,27]. In the gravity-hidden sector, the bosonic degrees of freedom are massless, although their fermionic partners obtain masses equal to the gravitino mass ($m_{3/2}$). To appreciate this fact better, let us generalize our results to the case of N chiral superfields (z_a,χ_a) in the hidden sector. In that case, the corresponding scalar potential of Eq. (7.2) becomes

$$V = -e^{-G}\left[G_a (G^{-1})^a_b G^b + 3\right] \tag{7.29}$$

and may be rewritten as

$$V = -\frac{g}{N^2}\, e^{-[(N+3)/N]G}\,((G^{-1})^a_b\,\partial_a\,\partial^b\, e^{(N/3)G}\ . \tag{7.30}$$

The flatness of the potential in this case implies a particular G, so that

$$(G^{-1})^a_b\,\partial_a\,\partial^b\; e^{(N/3)G} = 0\ . \tag{7.31}$$

For simplicity, we examine only the case of chiral superfields which are singlets under the internal group. An obvious particular solution of Eq. (7.31) is

$$G = \sum_{a=1}^{N} \frac{3}{2N} \log\left[\phi_a(z_a) + \phi^*_a(z^*_a)\right]\ , \tag{7.32}$$

where $\phi_a(z^a)$ is a function of the z^a field only. The potential is identically zero, as in the one-superfield case. Here, also, we can realize the locally flat potential requirement ($z_a \sim z^0_a$) by modifying G to

$$G = \sum_{a=1}^{N} \frac{3}{2N} \log\left[\phi_a(z_a) + \phi^*_a(z^*_a) + \Phi_a(z_a, z^*_a)\right]\ , \tag{7.33}$$

with

$$\begin{aligned} &\partial_a \partial^a \Phi_a > 0 && \text{for all } z_a \in \mathcal{D}\ , \\ &\partial_a \partial^a \partial_a \partial^a \Phi_a = 0 && \text{at } z^0_a\ . \end{aligned} \tag{7.34}$$

All scalar fields are then massless, contrary to their fermionic partners and the gravitino whose masses satisfy the following $(\text{mass})^2$ formula[16]:

$$-2\sum_{F=1}^{N-1} m_F^2 - 4m^2_{3/2} = -2\, e^{-G}\, G_a (G^{-1})^a_b G^c (G^{-1})^d_c R^b_d + 2(N-1)m^2_{3/2}\ , \tag{7.35}$$

where the curvature tensor R^b_a is given by[16]

$$R^b_a = \partial^b \partial_a \log \text{Det}\, G^m_n\ . \tag{7.36}$$

The particular structure of the function G in Eq. (7.33) implies the following non-trivial R_a^b:

$$R_a^b = -\frac{2N}{3} G_a^b . \tag{7.37}$$

Here also the Kähler space is an Einstein manifold.

Using Eq. (7.37) we find

$$e^{-G} G_a (G^{-1})^a_b G^c (G^{-1})^d_c R^b_d = 2Nm^2_{3/2} \tag{7.38}$$

and, using Eqs. (7.5) and (7.7), we finally obtain the following mass formula:

$$2 \sum_{F=1}^{N-1} m_F^2 = 2(N-1)m^2_{3/2} , \tag{7.39}$$

which means that every fermionic degree of freedom acquires a mass equal to the gravitino mass. Their bosonic partners are massless. Note the reverse role of bosonic and fermionic degrees of freedom. In fact, contrary to the minimal kinetic term case case[6], the extra mass contribution originating in the supersymmetry breaking is distributed among the fermionic degrees of freedom, whilst their bosonic partners remain massless.

Just as in the case of one chiral superfield in the hidden sector, the matter fields must be coupled in such a way that the properties of the hidden sector are not destroyed. We must choose the total G function as the sum of the hidden sector $G(z^a, z^*_a)$ and the matter one $G(y^i, y^*_i)$. The resulting potential for the y sector is the same as before.

In the second part of this section we consider a super-Higgs sector implemented by local R-symmetry[15]. The introduction of local R-symmetry in supergravity has been extensively explained in Section 6 and it allows us to generalize the Fayet-Iliopoulos mechanism[13] for spontaneous supersymmetry breaking to local supersymmetry.

We consider a class of 'minimal' models in which the super-Higgs sector is due to spin $(0^{\pm}, \frac{1}{2})$ chiral multiplet carrying non-trivial R-charge and a spin $(1, \frac{1}{2})$ vector multiplet gauging R-symmetry[32]. This is the minimal multiplet content for a super-Higgs sector with local R-symmetry if we demand the vanishing of the cosmological constant with positive definite potential. We exhibit a simple model in which the supersymmetry breaking receives equal contributions from the D-term due to the vector multiplet and the f-term due to the chiral multiplet.

The final mass spectrum consists of a real scalar, a chiral spinor, and a massive vector, all with the same mass $m = 2m_{3/2}$. In an R-symmetric theory with a single chiral multiplet coupled to N = 1 supergravity, the most general form of the scalar potential is[11]

$$V(z,z^*) = -e^{-G}(G_z G_{z^*} G^{-1}_{zz^*} + 3) + \frac{1}{2} g^2 D^2 , \tag{7.40}$$

where $G = G(z,z^*)$ and $D = -G_z z$; g is the R-gauge coupling constant. Using the identity

$$G_z = z^* G'(\rho) , \quad G_{z^*} = zG'(\rho) , \quad G_{zz^*} = G'(\rho) + \rho G''(\rho) =$$
$$= -D'(\rho) = -\frac{1}{\rho} D'(\omega) , \tag{7.41}$$

$$\rho = zz^* , \quad \omega = \log \rho ,$$

we can rewrite (7.40) as follows:

$$V(\omega) = e^{-G}\left(\frac{D^2(\omega)}{D'(\omega)}\right) - 3 + \frac{1}{2} g^2 D^2(\omega) , \tag{7.42}$$

with $D' = f(D)$, $dD/f(D) = d\omega$.

The G function is then given by the following integral:

$$G(\omega) = - \int d\omega D(\omega) = - \int dD \frac{D}{F(D)} . \tag{7.43}$$

If we now define

$$e^{-G} = \phi(D) \ , \tag{7.44}$$

then

$$-G' \ e^{-G} = \phi'(D)D' = \phi(D)D \ ; \quad \frac{1}{D'} = \frac{\phi'(D)}{\phi(D)D} \ , \tag{7.45}$$

so we finally get

$$\omega = \int \frac{dD}{D} \frac{\phi'(D)}{\phi(D)}$$

and the potential can be rewritten in the D-variable as follows[32]:

$$\begin{aligned} V &= \phi(D)\left(\frac{D^2\phi'(D)}{D\phi(D)} - 3\right) + \frac{1}{2} g^2D^2 \\ &= -3\phi(D) + D\phi'(D) + \frac{1}{2} g^2D^2 \ . \end{aligned} \tag{7.47}$$

Equation (7.47) defines the most general R-symmetric potential in terms of the function $\phi(D)$, which is related to G and D through Eqs. (7.44) and (7.45).

Positive definite potentials correspond to different choices of $\phi(D)$, so that Eq. (7.47) is semipositive definite.

The simplest case is to take $\phi(D)$ linear in D[32]:

$$\phi(D) = \alpha D - \beta/3 \ . \tag{7.48}$$

Then Eq. (7.47) becomes

$$V = \frac{1}{2} g^2\left(D^2 - \frac{4\alpha}{g^2} D + \frac{2\beta}{g^2}\right) \ . \tag{7.49}$$

Positivity means

$$V = \frac{1}{2} g^2(D-\xi)^2 \ , \tag{7.50}$$

which demands

$$\frac{2\alpha}{g^2} = \xi \ , \quad \frac{2\beta}{g^2} = \xi^2 \ ; \quad \phi(D) = \frac{1}{2} g^2\xi D - \frac{1}{6} g^2\xi^2 \ . \tag{7.51}$$

At the minimum

$$D = \xi \ , \quad \phi(\xi) = \frac{1}{3} g^2\xi^2 \ , \quad m_{3/2}^2 = \frac{1}{3} g^2\xi^2 \ . \tag{7.52}$$

Using Eq. (7.46) we can rewrite the potential in terms of ρ:

$$D = \frac{\xi}{3} \frac{1}{1 - (\rho \ \rho_0)^{\xi/3}} \ . \tag{7.53}$$

So we finally get

$$V(\rho) = \frac{2}{3} m_{3/2}^2 \frac{\left[1 - (\rho/\rho_{min})^{\xi/3}\right]^2}{\left[1 - \frac{2}{3} (\rho/\rho_{min})^{\xi/3}\right]^2} \ , \quad \rho_{min} = \frac{2m_{3/2}^3}{g^2} \ , \tag{7.54}$$

where the following conditions have been used to compute ρ_{min} and ρ_0:

$$D(\rho_{min}) = \xi \ , \quad D'(\rho_{min}) = 1 \ . \tag{7.55}$$

From Eqs. (7.55) we get

$$\rho_{min} = \frac{2}{3} \xi^2 \ , \quad (\rho_{min}/\rho_0)^{\xi/3} = \frac{2}{3} \ . \tag{7.56}$$

Note that the potential (7.54) becomes singular for $\rho/\rho_{min} = (3/2)^{3/\xi}$, so the variable ρ is constrained in the region $0 < \rho < \rho_{min}(3/2)^{3/\xi}$. From Eq. (7.47) we also get, at the minimum,

$$-e^{-G} G_z G_{z^*} G_{z^*} G^{-1}_{zz^*} = \frac{1}{2} g^2 D^2 = \frac{1}{2} g^2\xi^2 = \frac{2}{3} m_{3/2}^2 \ ,$$

which means that the gravitino mass gets an equal contribution from the 'D'-breaking and the 'f'-breaking terms. As we will see below, this means that the would-be goldstino is an equal mixture of the χ and λ spinors of the chiral and vector multiplets. It is now straight-forward to compute the three-level particle masses of the theory.

The scalar square-mass matrix is

$$M_B^2 = g^2\rho_{min}\begin{pmatrix} 1 & 1 \\ 1 & 1 \end{pmatrix} , \quad \text{i.e. } M_1^2 = 0, \quad M_2^2 = 2g^2\rho_{min} = 4m_{3/2}^2 \ . \tag{7.57}$$

The massless mode is the would-be Goldstone boson of spontaneously broken R-symmetry; the physical massive mode has mass $M_2 = 2m_{3/2}$.

The vector boson mass is

$$m_V^2 = 2g^2\rho_{min} = 4m_{3/2}^2 . \tag{7.58}$$

For the fermion mass matrix we notice that the $\chi\chi$ term of the N = 1 supergravity Lagrangian[6],

$$+\bar{\chi}_L\chi_L\left(G_{zz} - G_zG_z - G_zG_{zzz^*}G^{-1}_{zz^*}\right) , \tag{7.59}$$

is absent owing to the vanishing of the expression in brackets in our model.

The remaining term of the spin-½ mass matrix is

$$\lambda_L\chi_L(2igz^*) + \text{h.c.} , \tag{7.60}$$

The goldstino mode is defined by the term coupled to the spin-½ gauge field in the supergravity Lagrangian[6],

$$\bar{\psi}_R\cdot\gamma\eta_L + \text{h.c.}, \tag{7.61}$$

$$\eta_L = m_{3/2}G_z\chi_L - \frac{i}{2}\,gzG_z\lambda_L = i\sqrt{\frac{3}{2}}\,m_{3/2}\left(\frac{i\sqrt{2}\chi_L + \lambda_L}{\sqrt{2}}\right) . \tag{7.62}$$

Therefore, Eq. (7.60) can be rewritten as

$$\bar{\lambda}_L(i\sqrt{2}\chi_L)\sqrt{2}\,gz^* + \text{h.c.} , \qquad \sqrt{2}\,gz^*_{min} = 2m_{3/2} . \tag{7.63}$$

Equation (7.63) shows that the orthogonal combination to the would-be Goldstone fermion

$$\psi_L = \frac{i\sqrt{2}\cdot\chi_L - \lambda_L}{\sqrt{2}} \tag{7.64}$$

has mass $m_\psi = 2m_{3/2}$. Therefore, the physical spectrum consists of a massless spin-2 graviton, a massive spin-3/2 gravitino, and a scalar, a spinor, and a vector, all with the same mass:

$$m_S = m_V = m_\psi = 2m_{3/2} . \tag{7.65}$$

The square-mass sum rule gives in this case

$$\mathrm{St}M^2 = \sum_{J=0}^{3/2} (-)^{2J} m_J^2 = 4m_{3/2}^2 . \qquad (7.66)$$

To understand this result, we have to apply the general mass formula in N = 1 supergravity coupled to matter in the presence of non-vanishing Kähler curvature and the non-vanishing D-term[16]:

$$\mathrm{St}M^2 = 2(g^2 D + g^2 z\Gamma_z D) = 2R_{zz^*} \frac{G_z G_{z^*}}{(G_{zz^*})^2} e^{-G} , \qquad (7.67)$$

where $z\Gamma_z$ is the Kähler connection

$$z\Gamma_z = z\partial_z \log G_{zz^*} . \qquad (7.68)$$

In our case,

$$z\Gamma_z\Big|_{min} = z\partial_z \log D'(\rho)\Big|_{min} = \left(-1 + \frac{5}{3}\xi\right) , \qquad (7.69)$$

$$R_{zz^*} = -2G_{zz^*} , \qquad (7.70)$$

which at the minimum gives

$$R_{zz^*}\Big|_{min} = 2 . \qquad (7.70')$$

Therefore we get

$$\mathrm{St}M^2 = 2\left[g^2\xi + g^2\xi\left(-1 + \frac{5}{3}\xi\right)\right] + 4m_{3/2}^2 \frac{G_z G_{z^*}}{G_{zz^*}}$$

$$= \frac{10}{3} g^2\xi^2 - 4m_{3/2}^2 \frac{\xi^2}{\rho} = 10m_{3/2}^2 - 4 \times \frac{3}{2} m_{3/2}^2 = 4m_{3/2}^2 , \qquad (7.71)$$

which agrees with the left-hand side given by Eq. (7.66). Incidentally, we note that Eq. (7.71) shows that the Kähler manifold is an Einstein space.

We now consider the potential given by Eq. (7.54) in the low-energy limit $m_P \to \infty$, with $m_{3/2}$ fixed[32]. By means of Eq. (7.52) we can express the dimensionless variable ξ as

$$\xi = \sqrt{24\pi}\,\frac{m_{3/2}}{g m_P} \tag{7.72}$$

so that two different limits are possible.

If $g \to 0$ with $m_P \to \infty$, then it is easy to see that

$$V_{m_P\to\infty, g\to 0} = 2m_{3/2}^2\,\phi^2 + O\left(\frac{1}{m_P}\right) , \tag{7.73}$$

where ϕ is the physical scalar degree of freedom. The vector, spinor, and scalar degrees of freedom decouple in this limit, and we get a free theory of massive particles of spin 0, ½, and 1, respectively. This situation is entirely analogous to the normal super-Higgs effect without local chiral symmetry when the scalar field of the hidden sector just decouples in the limit $m_P \to \infty$, $m_{3/2}$ fixed[21].

However, in the presence of a gauged R-symmetry a second non-trivial limit exists for the supergravity Lagrangian with $M_P \to \infty$, $m_{3/2}$ fixed, and g fixed[32]. In this case the gauge-R interaction is non-gravitational in the sense that g does not vanish with $M_P \to \infty$.

From a physical point of view, this second limit is less interesting: it corresponds to non-renormalizable interactions at low energy, since non-renormalizable terms in the effective Lagrangian appear, which are scaled by inverse powers of $m_{3/2}$.

8. GENERALIZATION OF THE STANDARD MODEL TO LOCAL SUPERSYMMETRY: MINIMAL MODELS

'Minimal' models for the low-energy gauge theory $SU(3)_C \times SU(2)_L \times U(1)$ based on low-energy supergravity (SUGRA) with tree-level breaking of $SU(2)_L \times U(1)$ have been constructed[21,25,27,33–35].

The ordinary quarks and leptons are embedded in chiral spin (½, 0) multiplets with the following $SU(3) \times SU(2)_L \times U(1)$ quantum numbers:

$$Q = (3, 2, 1/3) , \quad u^c = (\bar{3}, 1, -4/3) , \quad d^c = (\bar{3}, 1, 2/3) , \tag{8.1a}$$

$$L = (1, 2, -1) , \quad e^c = (1, 1, 2) . \tag{8.1b}$$

The Higgs sector contains three chiral multiplets, two SU(2) doublets

$$H = (1,\ 2,\ -1)\ ,\qquad H^c = (1,\ 2,\ 1)\ , \tag{8.2}$$

and an over-all singlet Y,

$$Y = (1,\ 1,\ 0)\ . \tag{8.3}$$

After $SU(2)_L \times U(1)$ breaking down to $U_{em}(1)$, the scalar Higgs sector contains two charged and five neutral scalars.

Equation (8.1) defines an ordinary fermion family. The present fermion spectrum, with the inclusion of the t-quark, is obtained by repeating the chiral multiplets three times as defined by Eq. (8.1). Extension of these models to SU(5) with the same set of particle states has been considered[36-38]. This requires the following structure of SU(5) representations for the basic spin (½, 0) chiral multiplets:

$3(10 + \bar{5})$	quarks and leptons,
$5,\ \bar{5},\ 1,\ 24$	Higgs sector,
24	gauge multiplet.

Models have been constructed with a tree-level breaking SU(5) → → $SU_L(3) \times SU(2) \times U(1) \times SU(3)_C \times U(1)_{em}$ with the hierarchy $m_W[O(m_{3/2})] < m_X < m_P$. Unfortunately, unless there are further modifications, this class of models has an unstable hierarchy. In fact the very existence of the Y-SU(5) singlet spoils, at the one-loop level, the tree-level hierarchy $m_W << m_X$. To understand this point, it is sufficient to observe that because of the $YH\bar{H}$ coupling and of the heavy triplet mass splitting $\delta m^2_{3/2} = O(m_X m_{3/2})$, Figure 1 destabilizes the tree-level hierarchy; it induces Higgs doublet radiative square masses $\delta m_2^2 \simeq O(m_X, m_{3/2})$ [39]. This is the sliding singlet problem and it was first observed, in the context of global

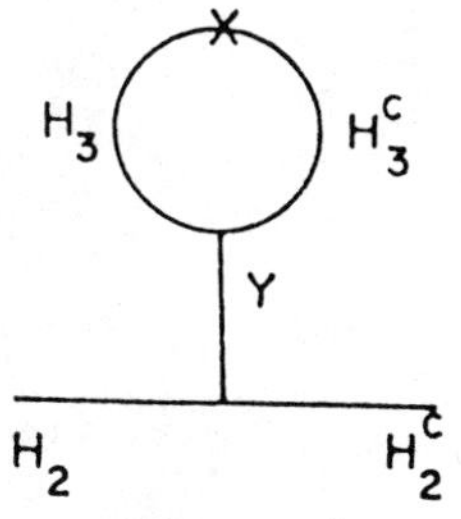

Figure 1.

SUSY GUTs, by Polchinski and Susskind[3]. The sliding singlet problem in local SUSY GUTs can be partially cured by introducing additional 50, $\overline{50}$, and 75 representations for the chiral multiplets[38]. The missing partner scenario[40] can then avoid $H_3H_3^c$ direct mass terms as well as triplet direct mass terms contained in the 50 and $\overline{50}$ representations. Upon integration on the superheavy fields, the low-energy SUGRA theory reduces to the form given by the first two lines of Eq. (6.11) and involves

i) the chiral multiplets given by Eqs. (8.1) and (8.2);

ii) the spin (1, ½) gauge multiplets of SU(3) × SU(2) × U(1).

The complete particle spectrum of the minimal low-energy SUGRA theory with tree-level breaking is given by Table 1.

Table 1

Ordinary particles		Superpartners
12×3 (quarks)$_{\text{Left}}$	→	24×3 (squarks)
3×3 (leptons)$_{\text{Left}}$	→	6×3 (sleptons)
$H^{\pm}$, 3 neutral (Higgs)	→	H_1^0 (shiggs)
2 neutral (Y-Higgs)	→	λ_Y (Y-Majorana)
3 ($W^{\pm}$, Z^0)	→	$\lambda_1^{\pm}, \lambda_2^{\pm}, \lambda_1^0, \lambda_2^0$ {W-inos, Z-inos}
1 (γ)	→	λ_γ (photino)
8 (g)	→	(8) λ_g (gluinos)
Gravitational sector		
1 (graviton)	→	(1) gravitino
	→	2 scalars (goldstino partners)

An example of tree-level mass spectrum for the superpartners of Table 1 is given above[38]: it corresponds to the choice A = 3 of Eq. (6.19a) and the following Higgs superpotential.

$$h(Y)_{\text{Higgs}} = \alpha Y H H^c + \frac{1}{3} \beta Y^3 \ . \tag{8.5}$$

<u>Fermions (gauginos and Higgses)</u>

$$m^2_{W(1,2)} = m^2_W + \frac{1}{2} m^2_{3/2} \left[1 \pm \left(1 + \frac{4m^2_W}{m^2_{3/2}} \right)^{\frac{1}{2}} \right] , \tag{8.6a}$$

$$m^2_{Z(1,2)} = m^2_Z + \frac{1}{2} m^2_{3/2} \left[1 \pm \left(1 + \frac{4m^2_Z}{m^2_{3/2}} \right)^{\frac{1}{2}} \right] , \tag{8.6b}$$

$$m^2_{0(3)} = m^2_{3/2} \ , \quad m^2_{3/2}(2-x)^2 \ , \tag{8.6c}$$

$$m_g, \ m_\gamma \ \text{arbitrary.} \tag{8.6d}$$

<u>Scalar Higgs sector</u>

$$m^2_\pm = m^2_W + 4m^2_{3/2} \ , \quad m^2_0 = m^2_Z + 4m^2_{3/2} \ , \tag{8.7a}$$

$$m^2_{0(2,3)} = \frac{1}{2}(1-x)m^2_{3/2}\left[5 - x \pm (9 - 10x + x^2)^{\frac{1}{2}}\right] , \tag{8.7b}$$

$$m^2_{0(4,5)} = \frac{1}{2} m^2_{3/2}\left[9 - 2x + x^2 \pm (3+x)(9 - 10x + x^2)^{\frac{1}{2}}\right] . \tag{8.7c}$$

<u>Squarks and sleptons</u>

$$M^\pm_{sq(s\ell)} = \left| m_{3/2} \pm m_{q(\ell)} \right| \ . \tag{8.8}$$

The entire spectrum of the superpartners of quarks, leptons, and gauge vector bosons is predicted in terms of two parameters: the gravitino mass $m_{3/2}$ and a dimensionless (Higgs coupling) constant $x = \beta/\alpha$.

The particle spectrum satisfies the tree-level mass sum rules[38]:

$$\sum_{J=0}^{1} (-)^{2J}(2J+1)m_J^2(\text{W sector}) = 4m_{3/2}^2 \ , \tag{8.9a}$$

$$\sum_{J=0}^{1} (-)^{2J}(2J+1)m_J^2(\text{Z sector}) = 2m_{3/2}^2 \ , \tag{8.9b}$$

$$\sum_{0}^{1/2} (-)^{2J}(2J+1)m_J^2(\text{quark, lepton}) = 2m_{3/2}^2 \ , \tag{8.9c}$$

$$\sum_{J=0}^{1/2} (-)^{2J}(2J+1)m_J^2(\text{4 neutral}) = 4m_{3/2}^2 \ . \tag{8.9d}$$

There is a remarkable class of tree-level models considered in Refs. 27 and 34, in which the tree-level mass spectrum is further simplified. This is obtained when the bare gaugino masses in Eq. (6.11) are set equal to the gravitino masses, and the $W^{\pm}$-ino and Z^0-ino masses are found to be

$$m_{W^{\pm}(1,2)} = |m_{3/2} \pm m_W| \ , \quad m_{Z(1,2)} = |m_{3/2} \pm m_Z| \ ,$$

whilst for the associated Higgs sector the same mass relations as given by Eq. (8.7a) hold. Also, the squark and slepton masses are given by Eq. (8.8). In this situation the first two mass formulae of Eqs. (8.9) are modified accordingly:

$$\sum_{J=0}^{1} (-)^{2J}(2J+1)m_J^2(\text{W sector}) = 0 \ , \tag{8.10a}$$

$$\sum_{J=0}^{1} (-)^{2J}(2J+1)m_J^2(\text{Z sector}) = 0 \ . \tag{8.10b}$$

The (s) particle mass spectrum (superpartners of known particles) can be plotted in this case as a function of $m_{3/2}$ and is given by Fig. 2. (For definiteness, a value of the top-quark mass $m_t \simeq 20$ GeV has been assumed).

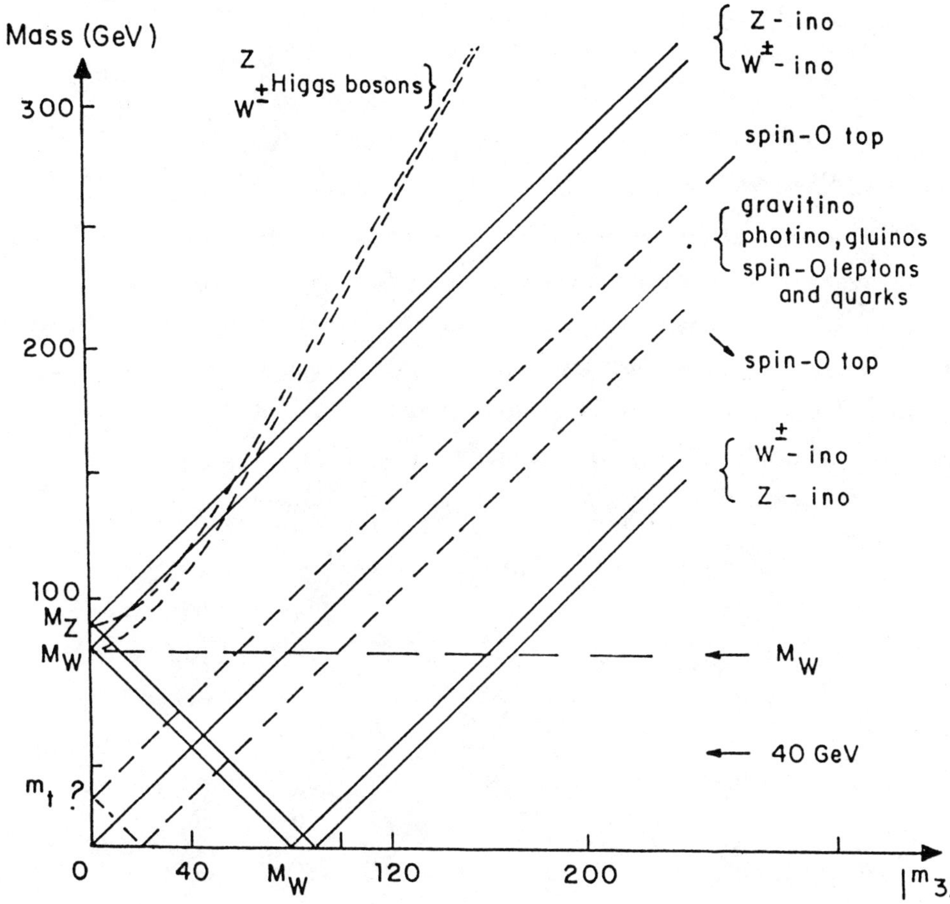

Figure 2.

Recently, a class of SUGRA GUT models based on radiative $SU(2)_L \times \times U(1)$ breaking have been constructed without the sliding singlet problem[41].

These models are somewhat similar to the globally supersymmetric models with induced radiative breaking, with the only (but important) difference that the radiative breaking is now induced by the soft-breaking terms due to the supergravity rather than to the Fayet-O'Raifeartaigh hidden sector.

These models fall into two classes[41]:

i) very large top-quark mass: $m_t \geq 60$ GeV and $m_{3/2} \simeq O(100$ GeV);

ii) small top-quark mass: $m_t \geq 20$-40 GeV, and a light gravitino: $m_{3/2} \geq O(20$ GeV).

This second class of model is particularly interesting because it can have a rather light mass spectrum for the superpartners of ordinary particles (near the experimental bound). Moreover, they predict a light neutral scalar with mass of a few GeV. A typical mass spectrum (in GeV) for the superpartners (Table 2) is reported here for $m_{3/2} = 20$ GeV and different values of the top-quark mass. The (s)particle mass spectrum in GeV is from Kounnas et al. in Ref. 41.

Table 2

Top		25	35	50
All families	$(\text{sleptons})_L$	27	36	25
	$(\text{sleptons})_R$	20	23	19
1st and 2nd family	$(\text{squark})_L$	77	108	67
	$(\text{squark})_R$	74 74	104 103	65 65
3rd family	$(\text{sbottom})_L$	76	106	66
	$(\text{sbottom})_R$	74	104	65
	$(\text{stop})_L$	96	132	106
	$(\text{stop})_R$	54	78	37
Charged Higgses		93	94	83
Neutral Higgses		104 3 46	105 5.3 48	95 5 19
Gluinos		84	118	72
Photino		4.6	7.3	3
W-inos		87 82	94 79	90 75
Z-inos		108 85	116 80	106 83
Neutral Majorana		23	24	9

9. N = 2 SUPERGRAVITY COUPLED TO VECTOR MULTIPLETS: GENERAL COUPLINGS AND THE SCALAR POTENTIAL

Lagrangians based on spontaneously broken N = 1 supergravity provide phenomenologically acceptable models for particle physics with interesting new predictions in the 100 to 1000 GeV energy range where new particles, the superpartners of the known elementary particles, should soon be discovered[2,41]. However, from a more fundamental point of view, N = 1 supergravity, being non-renormalizable, cannot be the final answer for the unification of fundamental forces, and new symmetries, perhaps extended supersymmetry, should possibly manifest themselves at higher energies.

In the present section, we focus on the simplest extension of N = 1 local supersymmetry, namely N = 2 local supersymmetry[42]. For computational reasons, it is important to have the fundamental Lagrangians written in the simplest possible way. In particular, if a possible scenario is envisaged, in which N = 2 supergravity is broken down to N = 1 or N = 0 supersymmetry, it is relevant to have a formulation of N = 2 Lagrangians in terms of N = 1 variables[42].

Our formulation is particularly useful in connection with the σ-model structure of $N > 4$ extended supergravities. In the N = 2 language, the scalar sector of these theories is given by an interaction of a Kählerian σ-model related to N = 2 vector multiplets[44] and a quaternionic σ-model related to N = 2 hypermultiplets[45,46], these multiplets having a unique embedding in the $N > 4$ supergravity multiplets. We claim that in the case in which the quaternionic manifold of the hypermultiplet sector is also Kählerian these theories have a simple description in terms of N = 1 invariant functions. This is the case for N = 4, 5, and 6 extended supergravity. Remarkably enough, the general form of the scalar potential given in this section, as shown below, does indeed apply to these theories.

Our task is to formulate the most general N = 2 coupling of matter and Yang-Mills multiplets to supergravity in terms of the N = 1 invariant functions[5,6]:

$$G(z,z^*) = +J(z,z^*) - \ln|h(z)|^2 \tag{9.1}$$

and

$$\tilde{f}_{ab}(z) = \tilde{f}_{ba}(z) , \tag{9.2}$$

which determine the most general coupling of a Yang-Mills matter system to N = 1 supergravity. The z_i variables in Eqs. (9.1) and (9.2) denote the complex scalar fields of chiral multiplets, which in N = 1 supergravity can belong to arbitrary (generally complex) representations of the gauge group G.

Let us recall the properties of the functions $G(z,z^*)$ and $\tilde{f}_{ab}(z)$. The function $G(z,z^*)$ is a G-invariant function*),

$$G^i T^{aj}_i z_j = z^{*i} T^{aj}_i G_j , \tag{9.3}$$

containing the Kähler potential $J(z,z^*)$ and the superpotential h(z). The scalar-kinetic terms are[6]:

$$G^j_i \partial_\mu z_j \partial_\mu z^{*i} , \quad G^j_i = +J^i_i , \tag{9.4}$$

whilst the chiral part of the scalar potential is given by

$$-e^{-G}\left[G^i (G^{-1})^i_j G_j + 3\right] . \tag{9.5}$$

The function $\tilde{f}_{ab}(z)$ is related to the Yang-Mills part of the Lagrangian[6]. For instance, the part of the Lagrangian quadratic in the Yang-Mills field strengths is:

$$-\frac{1}{4}\,\mathrm{Re}\,\tilde{f}_{ab} F^a_{\mu\nu} F^b_{\mu\nu} + \frac{i}{4}\,\mathrm{Im}\,\tilde{f}_{ab} F^a_{\mu\nu} \tilde{F}^b_{\mu\nu} , \tag{9.6}$$

whilst the gauge contribution to the scalar potential is

$$\frac{1}{2}\,(\mathrm{Re}\,\tilde{f})^{-1\,ab}\, D_a D_b$$

---

*) In the present section we use the following conventions for gauge indices: i, j, k, ... refer to arbitrary representations of G, whilst a, b, c, ... always refer to the adjoint representation. Moreover, $G^i = \partial G\, \partial z_i$; $G_i = \partial G/\partial z^{*i}$ and $G^i_j = \partial^2 G/\partial z_i \partial z^{ij}$.

(9.7a)

$$D_a = gG^i T^j_{ai} z_j = -gJ^i T^j_{ai} z_j , \tag{9.7b}$$

g denoting the Yang-Mills gauge coupling constant. Spontaneous symmetry-breaking in local supersymmetry occurs whenever the vacuum state corresponds to[5,6]

$$\left\{ \left\langle e^{-G/2}\, G_i \right\rangle ,\ \left\langle D_a \right\rangle \right\} \neq 0 . \tag{9.8}$$

We now consider the case of N = 2 supergravity coupled to matter. We use three different multiplets of physical fields:

i) the supergravity multiplet[42]

$$(e_{m\mu},\ \Psi_{\mu\pm},\ B_\mu) \tag{9.9}$$

containing the vierbein field, two Majorana gravitinos, and a U(1) gauge field (graviphoton);

ii) the Yang-Mills multiplets[47,48]

$$(A^a_\mu,\ \Omega^a_\pm,\ z^a) , \qquad a = 1, \ldots, n = \dim G \tag{9.10}$$

containing the Yang-Mills gauge bosons, two gauginos, and a complex scalar field;

iii) the hypermultiplet[47,48] (9.11)

$$(\xi^i,\ A^i_\pm) , \qquad i = 1, \ldots, 2r$$

transforming as a 2r-dimensional representation of G and containing 2r Majorana spinors and 4r scalars satisfying a reality property which should be respected by the representation of the gauge group G.

The N = 2 system has an over-all SO(2) × G gauge invariance. We identify the SO(2) with the gauge invariance of the B_μ field*).

*) In this section we consider only the case where the SO(2) group is gauged by the vector field B_μ sitting in the graviton multiplet. More general situations have been considered in Ref. [49].

We denote by g' its gauge coupling constant. This gauge invariance acts on the ± indices in formulae (9.9) to (9.11), and the only scalars which are not SO(2) inert are thus the scalar fields of the hypermultiplet (9.11).

Couplings of Abelian vector multiplets[50] or hypermultiplets[50] to N = 2 supergravity have been constructed first using the Noether method. The development of the N = 2 tensor calculus[44,45,51–53] allowed more general actions to be obtained. In particular, for the coupling of vector multiplets to supergravity, it was shown by de Wit et al.[44] that there exist more general actions, whose scalar sector describe[44] a σ-model with a Kähler structure. The Kähler potential is in this case

$$J = \ln y \; ; \quad y = z^{*I} N_{IJ} z^{J} , \tag{9.12}$$

where the index I runs over the values 0 to n = dim G ($z^0 = 1$). The action is constructed from a function $f(z^a)$ which is a G-invariant analytic function in the complex variables z^a. In Eq. (9.12) N_{IJ} and y are then given by

$$N_{00} = \frac{1}{2}\left(f - f_a z^a + \frac{1}{2} f_{ab} z^a z^b\right) + \text{c.c.} , \tag{9.13a}$$

$$N_{0a} = \frac{1}{4}\left(f_a - f_{ab} z^b\right) + \text{c.c.} , \tag{9.13b}$$

$$N_{ab} = \frac{1}{4} f_{ab} + \text{c.c.} , \tag{9.13c}$$

$$y = \frac{1}{2}\left[f + f^* + \frac{1}{2}\left(f_a - f^*_a\right)\left(z^{*a} - z^a\right)\right] ; \tag{9.13d}$$

$$\left(f_a = \partial f/\partial z^a , \quad f_{ab} = \partial^2 f/\partial z^a \partial z^b\right) . \tag{9.13e}$$

If we neglect the vector B_μ [as we omit a multiplet of spin (3/2, 1) in the comparison between N = 2 and N = 1], we obtain kinetic terms as in formula (9.6) with

$$\tilde{f}_{ab} = -\frac{1}{4} f_{ab} + \frac{y_a y_b}{z^{*I} y_I} , \tag{9.14a}$$

$$y_a = \frac{\partial y}{\partial z^a} = N_{aI} z^{*I} , \quad y_0 = N_{0I} z^{*I} . \tag{9.14b}$$

From Eqs. (9.14) we notice that the N = 2 Yang-Mills metric $\tilde{f}_{ab}$ has a non-analytic part $y_a y_b/(z^{*I} y_I)$. This term comes form the elimination of an (antisymmetric tensor) auxiliary field[49,53] of the spin (3/2, 1) multiplet so it cannot be directly derived from the N = 1 superspace Lagrangian of Ref. 6.

The 'minimal coupling' is the case where

$$f = 1 - z^a z^a , \tag{9.15}$$

leading to

$$N_{00} = 1 , \quad N_{0a} = 0 , \quad N_{ab} = -\delta_{ab} . \tag{9.16}$$

The linear invariance of f is now extended to SO(n,). The complex form of G, G , is a maximal subgroup of SO(n,). The corresponding function y is

$$y = 1 - z^a z^{*a} ,$$

whose linear invariance is now U(n). The scalar kinetic terms derived from the Kähler potential,

$$J = \ln (1 - z^a z^{*a}) , \tag{9.18}$$

have the structure of a CP^n σ-model with scalar fields living on a Grassmannian manifold U(n,1)/U(n) × U(1).

For the couplings of hypermultiplets to N = 2 supergravity, Bagger and Witten[46] used the Noether method to show that the scalar sector always describes a σ-model with a quaternionic structure. Some of these models were already obtained by the tensor calculus[44,45,52]). Combined general couplings of hypermultiplets and vector multiplets have been studied by de Wit et al.[44].

We temporarily neglect hypermultiplets and confine ourselves to the general local coupling of vector multiplets described by the G-invariant analytic function f(z):

$$f_b T_a{}^b{}_c z^c = 0 \ , \tag{9.19}$$

where $iT_a{}^b{}_c$ are the (real) structure constants of G. The over-all scalar potential of the coupled N = 2 Yang-Mills supergravity system has been established by de Wit et al.[44,49,53], and it reads:

$$V = -8g'^2 (N^{-1})^{00} - 16g'^2 y^{-1} - g^2 y^{-2} N_{ab} \left(z^{*c} T_c{}^a{}_d z^d \right) \left(z^{*e} T_e{}^b{}_m z^m \right) \ , \tag{9.20}$$

where g is the coupling constant of the gauge group G, whilst g′ is the coupling constant of the SO(2) local symmetry of gauged N = 2 supergravity[54]. It is apparent that the gauge part of V has the D^2 structure which is familiar from N = 1 supergravity.

In the case of 'minimal coupling' [Eqs. (9.16) to (9.18)], one finds for the potential

$$V_{min} = -8g'^2 \left(1 + \frac{2}{1 - z^a z^{*a}} \right) + \frac{g^2}{(1 - z^a z^{*a})^2} \left| z^{*c} T_c{}^b{}_d z^d \right|^2 . \tag{9.21}$$

All stationary points, supersymmetry, and symmetry-breaking patterns for this potential have been studied in Ref. 55.

Our aim is to recast formula (9.20) in the form given by formulae (9.5) and (9.7), namely to find the gauge-invariant function $\mathcal{G}$ given by formula (9.1). The answer turns out to be extremely simple: we claim that the function $\mathcal{G}$ which reproduces the Yang-Mills sector of N = 2 supergravity coupled to gauge multiplets is simply given by

$$\mathcal{G} = -\ln \left(\frac{8g'^2}{y} \right) = \ln (8g'^2) + \ln y \ . \tag{9.22}$$

This means that the superpotential is fixed to be a constant

$$h(z) = 2\sqrt{2}\, g' \ , \tag{9.23}$$

proportional to the SO(2) gauge coupling constant g'. As we will show below, the same expression for G will give the potential for N = 4 and N = 5 supergravities.

To derive this result, we need to calculate the potential of the N = 1 theory specified by the two functions $\tilde{f}_{ab}$ [Eq. (9.14)] for the coupling of the N = 1 Yang-Mills multiplet and G [Eq. (9.22)] for the chiral multiplet in the adjoint representation. The chiral part of this potential is given by Eq. (9.5). From the logarithmic form (9.22) for G we derive*)

$$G_{\bar{a}b} = +\frac{y_{\bar{a}b}}{y} - \frac{y_{\bar{a}}y_b}{y^2}$$

$$(G^{-1})^{a\bar{b}} = +y(y^{-1})^{a\bar{b}} + \frac{(y^{-1})^{a\bar{c}}}{1-w}\, y_{\bar{c}}y_d(y^{-1})^{d\bar{b}} , \qquad (9.24)$$

where

$$w = \frac{1}{y}\, y_a(y^{-1})^{a\bar{b}}y_{\bar{b}} . \qquad (9.25)$$

The chiral potential is then

$$V = 8g'^2\,\frac{1}{y}\left(\frac{w}{w-1} - 3\right) . \qquad (9.26)$$

This form of the N = 1 potential is valid without any assumption for y. Notice, however, that in our N = 2 case

$$\frac{\partial^2 y}{\partial z^a \partial z^{*b}} = N_{ab} . \qquad (9.27)$$

The expression for w [Eq. (9.25)] in N = 2 is

$$y(1-w) = z^{*I}N_{IJ}z^J - z^{*I}N_{Ia}(y^{-1})^{ab}N_{bJ}z^J . \qquad (9.28)$$

The only non-vanishing contribution comes from I = J = 0. The identity

$$\left(N_{00} - N_{0a}y^{-1\;ab}N_{b0}\right)(N^{-1})^{00} = 1 , \qquad (9.29)$$

---

*) $\bar{a}$ denotes derivatives with respect to z^{*a}.

which can be proved using

$$N_{00}(N^{-1})^{00} + N_{0a}(N^{-1})^{a0} = 1 \,, \tag{9.30a}$$

$$N_{b0}(N^{-1})^{00} + N_{bc}(N^{-1})^{c0} = 0 \,, \tag{9.30b}$$

$$y^{-1\ ab}N_{bc} = \delta^a_c \,, \tag{9.30c}$$

then implies

$$y(1-w) = \frac{1}{(N^{-1})^{00}} \,. \tag{9.31}$$

The form (9.26) is now the first two terms of the N = 2 potential (9.20).

We now turn to the gauge part, to identify the last term of Eq. (9.20) with Eq. (9.7). First, we note that the equation for the gauge invariance of f [Eq. (9.19)] can be derived with respect to z^d, giving

$$f_{db}T_a{}^b{}_c z^c + f_b T^b{}_{ad} = 0 \,. \tag{9.32}$$

The expression for G [Eq. (9.22)] yields

$$D_a = -g\frac{1}{y}\,y_b T_a{}^b{}_c z^c \,. \tag{9.33}$$

We then use Eq. (9.14) for y_b, the definition of N [Eqs. (9.13)], and Eq. (9.32) to obtain

$$D_a = -\frac{g}{4y}\quad z^{*d} f_{db}T_a{}^b{}_c z^c - f^*_{cb}T_a{}^b{}_d z^{*d} z^c \quad . \tag{9.34}$$

Up to now, the fact that the scalar fields span the adjoint representation has not been used: the matrix $(T_a)^b{}_c$ in Eq. (9.34) could be replaced by any representation. For the case of the adjoint representation, however, these matrices are antisymmetric in their lower indices. Hence, from Eq. (9.32), the quantity

$$f_{db}T_a{}^b{}_c z^c \tag{9.35}$$

is antisymmetric in the a, d indices. We then obtain

$$D_a = \frac{g}{y} N_{ab}\left(z^{*d} T_d{}^b{}_c z^c\right) . \tag{9.36}$$

Using analogous manipulations we can prove that

$$y_a z^{*b} T_b{}^a{}_c z^c = 0 , \tag{9.37}$$

which allows us to replace N_{ab} in Eq. (9.36) by $-2 \operatorname{Re} \tilde{f}_{ab}$ [see Eq. (9.14a)]. Equation (9.7a) then yields

$$V_g = 2 \frac{g^2}{y^2} \operatorname{Re} \tilde{f}_{ab}\left(z^{*d} T_d{}^a{}_c z^c\right)\left(z^{*e} T_e{}^b{}_m z^m\right) , \tag{9.38}$$

which can indeed be identified with the last term in Eq. (9.20) using once more Eqs. (9.14) and (9.37).

To summarize, we have shown that the complete bosonic Lagrangian for N = 2 gauge multiplets coupled to N = 2 supergravity is completely specified in N = 1 formulation by the analytic function f(z). Of course, we omitted the field B_μ in the N = 1 formulation since this field belongs to the spin (3/2, 1) multiplet which we neglected. As mentioned before, the elimination of an auxiliary field of this same multiplet is also responsible for the non-analytic term in the metric $\tilde{f}_{ab}$ as given in Eq. (9.14a). This term does not allow the complete Yang-Mills part of the Lagrangian to be recast in an N = 1 superfield form as given in Ref. 6.

When the SO(2) symmetry of supergravity is gauged, a chiral potential is induced, deriving from a fixed constant N = 1 superpotential proportional to the SO(2) gauge coupling constant g'. Notice that there is no SO(2) gauge potential, simply because the scalar fields of the Yang-Mills multiplet are neutral under SO(2). On the contrary, an SO(2) gauge potential would appear for the scalars of a hypermultiplet. The latter situation is explicitly realized in N = 5 supergravity.

Let us now consider some implications of our results for extended supergravities with $N > 4$. All these theories, in terms of $N = 2$ supersymmetry representations, contain vector multiplets and thus scalar fields whose potential should fit in with our description. For $N > 5$, these theories will also contain hypermultiplets. But we will show that, at least for $N = 5$, the scalars in these hypermultiplets also follow our prescription owing to the largest invariance of the manifold which contains all the scalar fields of the theory.

If we decompose $N > 4$ extended supergravity multiplets into $N = 2$ supermultiplets, we can regard the scalar sector of these theories, which describes a σ-model over the manifold G/H, as a particular interaction of a Kählerian σ-model[56] (related to the scalar fields of the $N = 2$ vector multiplets) with a quaternionic σ-model (related to the scalar fields of the $N = 2$ hypermultiplets). The same procedure also applies to $N = 4$ supergravity coupled to $N = 4$ Yang-Mills multiplets.

We call coset disintegration the decomposition of a σ-model associated with G/H in terms of σ-models associated with G_1/H_1 and G_2/H_2, with $G_1, G_2 \subset G$, and $H_1, H_2 \subset H$.

Consider first the gravity multiplet of $N = 4$ supergravity. It contains a complex scalar field, obviously inert under both SO(4) and the gauge group. This multiplet, when considered in terms of $N = 2$ supersymmetry, splits into the $N = 2$ gravity multiplet, an SU(2) doublet of gravitino multiplets, and one vector multiplet which contains the scalar field. Thus the scalar potential of the $N = 4$ theory has to follow from our results. Moreover, it is known that this scalar field is a coordinate of the coset space SU(1,1)/U(1). The corresponding Kähler potential (up to a field redefinition) is then exactly the same as in the 'minimal' coupling case, $f(z) = 1 - z^2$ [Eq. (9.16)]. and the potential is then obtained from

$$G = -\ln\left[8g'^2(1 - zz^*)^{-1}\right] , \tag{9.39}$$

which gives

$$V = -8g'^2 \left(\frac{2zz^*}{1 - zz^*} + 3 \right) \tag{9.40}$$

with vanishing gauge contributions for an SO(4) singlet complex field z. This result agrees with the result of Cremmer et al.[57] for the gauging of SO(4) supergravity*).

The N = 5 supergravity with local SO(5) has been studied by de Wit and Nicolai[58]. The scalar fields, transforming like a complex $\underset{\sim}{5}$ of SO(5), have kinetic terms given by a non-linear σ-model based on the manifold SU(5,1)/SU(5) × U(1). Their Kähler potential (up to Kähler transformation and field redefinition) is

$$-J(z,z^*) = \ln(y) = -\ln(1 - z_i z_i^*) , \qquad i = 1, \ldots, 5 . \tag{9.41}$$

When the N = 5 gravity multiplet is decomposed into N = 2 multiplets, the scalar fields enter an SO(3) triplet of vector multiplets containing three complex scalars which are SO(2) singlets, the two remaining complex scalars being part of a hypermultiplet. First we can show that if we add to the Kähler potential $J(z,z^*)$ a constant superpotential such that the G function reads

$$G(z,z^*) = -\ln(8g'^2) + \ln(1 - z_i z_i^*) \tag{9.42}$$

as usual, according to our rules we recover the N = 5 potential. It is very simple, using the SO(5) transformation properties of the scalar fields to establish that the gauge part of the potential is

$$2g'^2 \frac{1}{y^2}\left[\left(z_i z_i^*\right)^2 - z_i z_i\right)\left(z_j z_j^*\right)\right] , \tag{9.43}$$

while the chiral part of the potential defined by Eqs. (9.5) and (9.42) is

$$-8g'^2 - 16g'^2 y^{-1} \tag{9.44}$$

---

*) Other gauging of N = 4 supergravity can be obtained in N = 2 by gauging the SO(2) using a physical vector multiplet[49].

as is usual in the 'minimal' case. The sum of formulae (9.43) and (9.44) is precisely the potential found in Ref. 58. An obvious way to make clear the relation with N = 2 is to split the function G according to

$$J = \ln \left[\exp(J_1) + \exp(J_2) - 1\right] , \qquad (9.45)$$

where

$$J_1 = \ln \left(1 - z_a z_a^*\right) , \qquad a = 1, 2, 3 \qquad (9.46)$$

gives the Kähler potential for the scalar fields of the three vector multiplets which are coordinates of an SU(3,1)/SU(3) × U(1) manifold, whilst

$$J_2 = \ln (1 - z_4 z_4^* - z_5 z_5^*) \qquad (9.47)$$

is the Kähler potential for the scalars of the hypermultiplet, in the manifold SU(2,1)/SU(2) × U(1), which is both Kählerian and quaternionic according to Bagger and Witten[46].

It should be possible to extend our considerations to N = 6 and N = 8 supergravities. The N = 6 theory contains 15 complex scalar fields which are coordinates of the Kählerian manifold $SO^*(12)/U(6)$. When decomposed in terms of N = 2 multiplets, the N = 6 theory contains seven vector multiplets with scalars living in the manifold $O^*(8)$ × SU(1,1)/U(4) × U(1), and four hypermultiplets, the corresponding manifold being SU(4,2)/SU(4) × SU(2) × U(1), which is Kählerian and quaternionic. Consequently, our prescription should also apply to the hypermultiplet sector of N = 6. The analogue for the N = 8 theory is the following: the 70 scalars parametrizing the coset $E_{7,7}/SU(8)$ are split into two pieces. First, 15 complex scalars belonging to 15 N = 2 vector multiplets span the Kähler manifold $SO^*(12)/U(6)$. Then 40 real scalars, which are coordinates of the quaternionic manifold $E_{6,2}/SU(6) \times SU(2)$, are components of 20 N = 2 hypermultiplets. This last coset space is indeed the only non-Kählerian manifold present in the decomposition of all N > 5 extended supergravities. This is related to the fact that $E_{7,7}/SU(8)$ is not a Kähler manifold, and is a consequence of PCT self-conjugation of the N = 8 theory and of the 20 hypermultiplets contained in this theory.

10. HIGGS AND SUPER-HIGGS EFFECT IN N = 2 SUPERGRAVITY YANG-MILLS SYSTEMS

In the present section we consider the first example of the super-Higgs effect in N = 2 supergravity coupled to pure Yang-Mills matter multiplets.

For definiteness we consider the 'minimal' case, which corresponds to the particular choice of the G-invariant analytic function f(z) given by Eq. (9.15). The scalar potential is given by Eq. (9.21).

We can summarize the analysis of the present section as follows[55].

If $g' = 0$ there are no extrema which break supersymmetry. We find a set of degenerate vacua which can induce the Brout-Englert-Higgs mechanism and break G in Minkowski space, as was the case of global supersymmetry.

If $g' \neq 0$, the situation is completely different. There is a unique stationary point (a local maximum) which breaks neither supersymmetry nor the gauge group G. It corresponds to massless multiplets in anti de Sitter space. All other stationary points break both N = 2 supersymmetry and the gauge group G. They correspond to the combined Higgs and super-Higgs effects of N = 2 supergravity. We recall that since the gauge multiplets are not charged with respect to the SO(2) gauge symmetry of N = 2 supergravity, this latter symmetry remains unbroken; consequently both supersymmetries are simultaneously broken. In the absence of matter multiplets with non-vanishing vacuum expectation values, the cosmological constant is always negative; so the spontaneous breakdown takes place in anti de Sitter space[59]. A situation with vanishing cosmological constant will be considered in the next section.

For $G = SU(M)$, we give all gauge symmetry-breaking patterns. They are characterized by all sets of (non-negative) integers m_1, m_2, ..., m_M so that

$$\sum_{n=1}^{M} n m_n = M \ . \tag{10.1}$$

The residual gauge symmetry which corresponds to such a set is

$$H = SU(m_1) \times SU(m_2) \times \ldots \times SU(m_M) \times U(1)^{p-1} \ ,$$

where p is the number if non-zero m_n's. The N = 2 supersymmetry is broken, except in the case $m_1 = M$, where H = G. The maximal unbroken symmetry is then $H_{max} = SU(M-2) \times U(1)$.

To check the stability of these non-trivial stationary points, we computed the scalar masses for $H = H_{max}$ from the part of the potential quadratic in the fields. We then used the criterion of stability in anti de Sitter space obtained by Breitenlohner and Freedman[60]. It turns out that these stationary points are stable if the ratio g'^2/g^2 is large enough.

For M > 7, the breaking pattern $SU(M) \to SU(3) \times SU(2) \times U(1)$ is possible*). Interestingly enough, such an unconventional embedding of the usual low-energy gauge group, obtained from the N = 2 supergravity potential, still allows us [for instance in the case of SU(7) and SU(8)] to obtain spin-½ states with the correct assignments of colour and electroweak quantum numbers when matter hypermultiplets are added. These fields, which could be identified with quarks and leptons, are unavoidably accompanied by mirror particles and by exotic states, making the fermions not completely realistic at the present stage.

We now proceed to derive and discuss our results.

---

*) The N = 10 case is an exception as will later become clear from the study of the general pattern of SU(M).

The N = 2 vector multiplet consists of a vector field A_μ, two Majorana spinors $\Omega_\pm$, and a complex scalar field z, all transforming according to the adjoint representation of G. The usual (global) D-term reads

$$D = g(z,z^*) \tag{10.2}$$

(we adopt a matrix notation $z = a + ib$, with a and b Hermitian). The scalar potential, given by Eq. (9.20), is

$$V(z,z^*) = x_0^4 \,\mathrm{Tr}\,(D^2) - \frac{16g'^2 x_0^2}{k^4} - \frac{8g'^2}{k^4}\,, \tag{10.3}$$

where

$$x_0 = \frac{1}{\sqrt{1 - k^2\,\mathrm{Tr}(zz^*)}}\,. \tag{10.4}$$

This non-polynomial function of z may be rewritten as a fourth-order polynomial by defining

$$x = x_0 z\,. \tag{10.5}$$

Then

$$V(x,x^*) = g^2\,\mathrm{Tr}\,\left[(x,x^*)^2\right] - 16\,\frac{g'^2}{k^2}\,\mathrm{Tr}\,(xx^*) - 24\,\frac{g'^2}{k^4}\,. \tag{10.6}$$

The first term of this potential corresponds to the gauge potential of N = 1 supersymmetry; it is positive definite and vanishes in directions for which $[x,x^*] = 0$, i.e. D is zero[61]. Thus, in the absence of g', supersymmetry remains unbroken since the vacuum expectation value (v.e.v.) of D is zero. The scale of gauge symmetry-breaking is, however, arbitrary, and the cosmological constant is zero.

When the g' coupling constant is introduced, $V(x,x^*)$ becomes unbounded from below. The v.e.v.'s are given by stationary points of the potential which satisfy*)

---

*) We use the following conventions: $x = A + iB$ (with A and B Hermitian) is developed on a set of Hermitian generators T_i, $x = (1/\sqrt{2})x_i T^i$, normalized with $\mathrm{Tr}(T_i T_j) = 2\delta_{ij}$ and $[T_j, T_k] = 2if_{jkl}T_l$. We then have: $\mathrm{Tr}(xx^*) = x_i x_i = A_i A_i + B_i B_i$ and $[x,x\ ] = if_{jkl}x_j x_k T_l = 2if_{jkl}A_j B_k T_l$. From now on, we set k = 1.

$$0 = \frac{dV}{Dx_i} = -4g^2 f_{ijk} f_{\ell mk} x^*_j x_\ell x^*_m - 16g'^2 x^*_i \ . \tag{10.7}$$

These stationary points correspond to locally stable solutions in anti de Sitter space if the following stability condition is fulfilled[60]:

$$\frac{m^2}{\langle V\rangle} < 3/4 \ , \tag{10.8}$$

where m^2 is the most negative eigenvalue of the scalar mass matrix given by the quadratic part of the potential, and $\langle V\rangle$ is the value of the potential for this solution. Equation (10.7) leads immediately to

$$\mathrm{Tr}\ (\langle A\rangle\langle B\rangle) = 0 \ , \tag{10.9}$$

$$\mathrm{Tr}\ (\langle A\rangle^2) = \mathrm{Tr}\ (\langle B\rangle^2) \ , \tag{10.10}$$

$$\langle x_0\rangle^4\ \mathrm{Tr}(\langle D\rangle^2) = 8g'^2\ \mathrm{Tr}\ (\langle x\rangle\langle x\rangle^*) = 16g'^2\ \mathrm{Tr}\ (\langle A\rangle^2) \ , \tag{10.11}$$

$$\langle V\rangle = -24g'^2\left[1 + \frac{2}{3}\ \mathrm{Tr}\langle A\rangle^2)\right] \tag{10.12}$$

for the v.e.v.'s. Two kinds of solutions can occur. First, all v.e.v.'s are zero. Supersymmetry and gauge symmetry remain unbroken and $\langle V\rangle = -24g'^2$. The quadratic term in the potential gives a universal squared 'mass', $m_u^2 = -16g'^2$, to all scalar fields, which verifies that $m_u^2 = (2/3)\langle V\rangle$. This result is precisely what is expected from an actually massless particle in an anti de Sitter space. Moreover, unbroken supersymmetry and gauge symmetry indicate that all scalars have to remain massless. This stable solution is the only stationary point of V in the case where there is no D^2 term either because g is zero or because the gauge group is Abelian ($f_{ijk} = 0$).

If, however, some v.e.v.'s are non-zero, supersymmetry and gauge symmetry will be broken. From Eq. (10.7) it is apparent that non-zero v.e.v.'s of scalar fields will be of order g'/g. The supersymmetry-breaking scale $\sqrt{D}$ will then be of order $\sqrt{g'^2/g}$, and the cosmological constant is always negative.

The Goldstone fermions, which after the super-Higgs effect are eaten up by the two gravitinos, are given by

$$\Omega_{G\pm} = \mathrm{Tr}\,(\langle D\rangle \Omega_{\pm})$$

($\Omega_{\pm}$ are the two multiplets of Majorana fermions belonging to the $N = 2$ Yang-Mills vector multiplet).

Equation (10.7) allows us to work out the general stationary points of $V(x,x^*)$, for a given gauge group G. In the case where $G = SU(M)$ we obtain the following solution: $\langle x\rangle = \langle A\rangle + i\langle B\rangle$ can be written in block-diagonal form, with m_n blocks of dimension n, n being an arbitrary integer, and with the obvious constraint [Eq. (10.1)] $M = \sum_{n=1}^{M} nm_n$. Each possible v.e.v. is then characterized by the M integers

$$m_1,\ m_2,\ \ldots,\ m_M\ . \tag{10.13}$$

Each of these blocks of dimension n, in the basis where $\langle A\rangle$ is diagonal, reads

$$\langle A\rangle = \frac{g'}{g}\,\mathrm{diag}\,[n-1,\ n-3,\ \ldots,\ -(n-1)]$$

$$\langle B\rangle = \begin{pmatrix} 0 & B_1 & 0 & \cdots & 0 & 0 \\ B_1^* & 0 & B_2 & \cdots & 0 & 0 \\ 0 & B_2^* & 0 & \cdots & 0 & 0 \\ \vdots & \vdots & \vdots & & \vdots & \vdots \\ 0 & 0 & 0 & & 0 & B_{n-1} \\ 0 & 0 & 0 & \cdots & B_{n-1}^* & 0 \end{pmatrix} \tag{10.14}$$

with

$$|B_j|^2 = \frac{g'^2}{g^2}\,j(n-j)\ ; \tag{10.15}$$

$\langle A \rangle$ and $\langle B \rangle$ are obviously traceless in each block and verify that

$$\mathrm{Tr}\,(\langle A\rangle^2) = \mathrm{Tr}\,(\langle B\rangle^2) = \frac{g'^2}{3g^2}\,(n+1)n(n-1)\ . \tag{10.16}$$

The invariance of a given solution characterized by $m_1, m_2, \ldots, m_M$ is

$$SU(m_1) \times SU(m_2) \times \ldots \times SU(m_L) \times U(1)^{p-1}\ ,$$

where p is the number of non-zero m_n and $L = [M/2]$ is the largest value for which m_L can be larger than 1. Let us give two simple examples.

<u>G = SU(2):</u> There are two solutions:

I: $m_1 = 2,\ m_2 = 0$: $\langle x \rangle = 0$

SU(2) unbroken.

II: $m_1 = 0,\ m_2 = 1$:

$$\langle A \rangle = \frac{g'}{g}\begin{pmatrix} 1 & 0 \\ 0 & -1 \end{pmatrix},\quad \langle B \rangle = \frac{g'}{g}\begin{pmatrix} 0 & e^{i\alpha} \\ e^{-i\alpha} & 0 \end{pmatrix},$$

SU(2) completely broken.

<u>G = SU(3):</u> There are three solutions:

I: $m_1 = 3,\ m_2 = 0,\ m_3 = 0$: $\langle x \rangle = 0$

SU(3) unbroken.

II: $m_1 = 1,\ m_2 = 1,\ m_3 = 0$:

$$\langle x \rangle = \frac{g'}{g}\begin{pmatrix} 1 & ie^{i\alpha} & 0 \\ ie^{-i\alpha} & -1 & 0 \\ 0 & 0 & 0 \end{pmatrix},\quad SU(3) \to U(1)$$

III: $m_1 = 0,\ m_2 = 0,\ m_3 = 1$

$$\langle x \rangle = 2\,\frac{g'}{g}\begin{pmatrix} 1 & ie^{i\alpha}/\sqrt{2} & 0 \\ ie^{-i\alpha}/\sqrt{2} & 0 & ie^{i\beta}/\sqrt{2} \\ 0 & ie^{-i\beta}/\sqrt{2} & -1 \end{pmatrix},\quad \text{SU(3) completely broken}$$

Notice that the solution $0 = m_1 = m_2 = \ldots = m_{M-1}$, $m_M = 1$, always breaks SU(M) completely. Supersymmetry breaking is thus compatible with a large variety of gauge symmetry-breaking patterns. However, the embedding of the unbroken $SU(m_1) \times SU(m_2) \times \ldots SU(m_L)$ subgroup into SU(M) is restricted to be such that

$$\begin{aligned} M = {} & (\underset{\sim}{m_1}, \underset{\sim}{1}, \ldots, \underset{\sim}{1}) \\ & + 2(\underset{\sim}{1}, \underset{\sim}{m_2}, \underset{\sim}{1}, \ldots, \underset{\sim}{1}) \\ & + \ldots \\ & + L(\underset{\sim}{1}, \underset{\sim}{1}, \ldots, \underset{\sim}{1}, \underset{\sim}{m_L}) \ . \end{aligned} \tag{10.17}$$

It is tempting at this point to look for SU(M) breaking into $G_0 = SU(3) \times SU(2) \times U(1)$. It is clear, however, that all embeddings of G_0, as defined by Eqs. (10.13) and (10.14), will lead to exotic quantum numbers when matter hypermultiplets are considered. These multiplets contain a doublet of Weyl fermions and complex scalars transforming under G in conjugate representations. Let us consider the simplest candidate model which breaks into G_0, based on SU(7), with $m_1 = 3$ and $m_2 = 2$. The unbroken $SU(3) \times SU(2) \times U(1)$ subgroup is defined through the embedding

$$\begin{aligned} \underset{\sim}{7} &= (\underset{\sim}{3}, \underset{\sim}{1}, \tfrac{2}{3}) + 2(\underset{\sim}{1}, \underset{\sim}{2}, -\tfrac{1}{2}) \ , \\ \underset{\sim}{21} &= 2(\underset{\sim}{3}, \underset{\sim}{2}, \tfrac{1}{6}) + 3(\underset{\sim}{1}, \underset{\sim}{1}, -1) \\ &\quad + (\underset{\sim}{\bar{3}}, \underset{\sim}{1}, \tfrac{4}{3}) + (\underset{\sim}{1}, \underset{\sim}{3}, -1) \ , \\ \underset{\sim}{35} &= 3(\underset{\sim}{3}, \underset{\sim}{1}, -\tfrac{1}{3}) + 2(\underset{\sim}{\bar{3}}, \underset{\sim}{2}, \tfrac{5}{6}) + (\underset{\sim}{3}, \underset{\sim}{3}, -\tfrac{1}{3}) \\ &\quad + 2(\underset{\sim}{1}, \underset{\sim}{2}, -\tfrac{3}{2}) + (\underset{\sim}{1}, \underset{\sim}{1}, 2) \ . \end{aligned} \tag{10.18}$$

The antisymmetric tensors of SU(7) contain all the fields necessary to classify quarks and leptons (with mixed chirality) and additional SU(2) and U(1) exotic states. Since $\underset{\sim}{7}$ contains only one SU(3) triplet, there will be no colour exotics. Moreover, the electric charge of all leptonic states [i.e. SU(3) singlets] is an integer, whilst coloured states have charges $\pm\tfrac{4}{3}$, $\pm\tfrac{1}{3}$, $\mp\tfrac{2}{3}$. It is not a trivial result that symmetry-breaking precisely selects a U(1) group which is a candidate for a weak hypercharge. This fact is indeed not a peculiarity of SU(7). The same occurs using SU(8) with, however, a few differences. The embedding of SU(3) × SU(2) × U(1) is the following:

$$\begin{aligned}
\underset{\sim}{8} &= 2(\underset{\sim}{3}, \underset{\sim}{1}, \tfrac{1}{6}) + (\underset{\sim}{1}, \underset{\sim}{2}, -\tfrac{1}{2}) \, , \\
\underset{\sim}{28} &= 3(\underset{\sim}{\bar{3}}, \underset{\sim}{1}, \tfrac{1}{3}) + (\underset{\sim}{1}, \underset{\sim}{1}, -1) \\
&\quad + (\underset{\sim}{6}, \underset{\sim}{1}, \tfrac{1}{3}) + 2(\underset{\sim}{3}, \underset{\sim}{2}, -\tfrac{1}{3}) \, , \\
\underset{\sim}{56} &= 3(\underset{\sim}{\bar{3}}, \underset{\sim}{2}, -\tfrac{1}{6}) + 2(\underset{\sim}{8}, \underset{\sim}{1}, \tfrac{1}{2}) + 4(\underset{\sim}{1}, \underset{\sim}{1}, \tfrac{1}{2}) \\
&\quad + 2(\underset{\sim}{3}, \underset{\sim}{1}, -\tfrac{5}{6}) + (\underset{\sim}{6}, \underset{\sim}{2}, -\tfrac{1}{6}) \, , \\
\underset{\sim}{70} &= 3(\underset{\sim}{3}, \underset{\sim}{1}, \tfrac{2}{3}) + 3(\underset{\sim}{\bar{3}}, \underset{\sim}{1}, -\tfrac{2}{3}) + (\underset{\sim}{6}, \underset{\sim}{1}, -\tfrac{2}{3}) \\
&\quad + (\underset{\sim}{\bar{6}}, \underset{\sim}{1}, \tfrac{2}{3}) + 4(\underset{\sim}{1}, \underset{\sim}{2}, 0) + 2(\underset{\sim}{8}, \underset{\sim}{2}, 0) \, .
\end{aligned} \tag{10.19}$$

We obtain states with quantum numbers of quarks and leptons, and colour and electric charge exotics. Leptonic and coloured states can have electric charges $\pm\tfrac{1}{2}$, whilst there are some SU(3) triplets with charges $\pm\tfrac{1}{6}$ and $\pm\tfrac{5}{6}$. The weak hypercharge obtained from SU(8)-breaking seems less attractive than in the SU(7) case. The extension of this analysis to larger SU(M) groups is straightforward.

This discussion of course does not give any indication of the possibility of separating left-handed and right-handed fermions inside an N = 2 matter multiplet, whose coupling to N = 2 supergravity can certainly modify the analysis of both the symmetry-breaking pattern and the cosmological constant problem.

We would now like to discuss the problem of the stability of the symmetry-breaking stationary points of the potential. To check the stability we need the 'mass' spectrum*) of the scalar fields (or at least the most negative squared mass) obtained from the quadratic

---

*) Recall that in anti de Sitter space, these 'masses' do not correspond to the physical masses of states.

terms in the Taylor expansion of $V(z,z^*)$, and from the structure of the scalar kinetic terms which read:

$$L_{kin} = -(M_{kin})_i^j \left(\partial_\mu z^i\right)\left(\partial^\mu z_j^*\right), \tag{10.20}$$

with

$$(M_{kin})_i^j = \left\langle 2 \frac{\partial}{\partial z^i} \frac{\partial}{\partial z_j^*} \ln x_0 \right\rangle . \tag{10.21}$$

Equivalently,

$$(M_{kin})_i^j = \langle x_0 \rangle^2 \left(\delta_i^j + \langle x \rangle^j \langle x \rangle_i^*\right) . \tag{10.22}$$

Thus, apart from an over-all $\langle x_0 \rangle^2$ factor, the kinetic terms for directions with non-zero v.e.v.'s are not canonically normalized and the mass matrix has to be correspondingly rescaled.

However, the mass spectrum shows several universal features which do not depend on the choice of the gauge group G. Scalar fields which correspond to generators T^i of the little group of $\langle z \rangle$, $[T^i, \langle z \rangle] = 0$, do not receive any mass contribution from the D^2 part of the potential. They have only the 'universal squared mass' $-16g'^2$ and always fulfil the stability requirement since

$$\frac{m^2}{V} = \frac{2}{3\left[1 + \frac{2}{3}\,\mathrm{Tr}(A^2)\right]} < \frac{2}{3} . \tag{10.23}$$

Notice that the non-zero v.e.v.'s shift the ratio from $2/3$, so that the scalars are no longer massless states. Other general results can be obtained. The 'Higgs particle', i.e. the scalar state collinear with the v.e.v. $\langle z \rangle = \langle a \rangle + i\langle b \rangle$, has a squared mass $+32g'^2\langle x_0 \rangle^2$, whilst the orthogonal state spanned by $\langle b \rangle - i\langle a \rangle$ [see Eqs. (10.9) and (10.10)] is as usual a Goldstone state. The states collinear with $\langle b \rangle + i\langle a \rangle$ and $\langle a \rangle - i\langle b \rangle$ have squared 'masses' $-32g'^2$, which satisfies the stability criterion if

$$\mathrm{Tr}\,(\langle A \rangle^2) > 7/6 . \tag{10.24}$$

This inequality leads to a lower bound of order 1 on the ratio g'^2/g^2 since

$$\mathrm{Tr}\ (\langle A\rangle^2) = 0(1)\left[g'^2/g^2\right] .$$

These last general results completely solve the G = SU(2) [or SO(3)] case for which Eq. (10.7) requires

$$\langle x\rangle_{SU(2)} = \frac{g'}{g}\begin{pmatrix} 1 & i \\ i & -1 \end{pmatrix}$$

i.e. $\mathrm{Tr}\ (\langle A^2\rangle) = \mathrm{Tr}\ (\langle B\rangle^2\ ) = 2g'^2/g^2$. This v.e.v. breaks SU(2) completely; we thus obtain three Goldstone states, a Higgs particle with positive squared mass $32g'^2(1 + 4g'^2/g^2)$, and two scalar states with squared mass $-32g'^2$. Stability is ensured as long as

$$g'^2/g^2 > 7/12\ . \tag{10.26}$$

The largest unbroken gauge invariance with broken supersymmetry in the case G = SU(M) is SU(M − 2) × U(1). It is obtained by breaking the SU(2) factor in the maximal subgroup SU(M − 2) × SU(2) × U(1) of SU(M), by using the $\langle x\rangle_{SU(2)}$ solution [Eq. (10.25)]. The scalar mass spectrum is given in Table 3 [states are classified according to SU(M − 2) × SU(2)]. Since the most negative squared mass is the same as in the G = SU(2) case, stability will be ensured when the condition (10.26) holds.

We have shown that all solutions have a negative cosmological constant. The only way to make the cosmological constant possibly vanishing is to add hypermultiplets whose scalars take a non-vanishing vacuum expectation value and (or) to depart from the minimal coupling of supergravity to the Yang-Mills action. The latter situation will be considered in the next section. In this case the breaking pattern of the gauge group could be rather different from the one discussed in the present section.

Table 3

$SU(M-2) \times SU(2)$ states	Squared masses
$(\underset{\sim}{1}, \underset{\sim}{3})$	3 Goldstone states 3 states with squared masses $+32g'^2\ (1+4g'^2/g^2)$ $-32g'^2$ $-32g'^2$
(Adjoint, $\underset{\sim}{1}$) + $(\underset{\sim}{1}, \underset{\sim}{1})$	$2(M-2)^2$ real states with squared mass $-16g'^2$
$(M-2, \underset{\sim}{2})$ + $(\overline{M-2}, \underset{\sim}{2})$	$4(M-2)$ Goldstone states $4(M-2)$ states with squared masses $-24g'^2$

11. SUPER-HIGGS EFFECT IN N = 2 SUPERGRAVITY WITHOUT COSMOLOGICAL CONSTANT: N = 2 FLAT POTENTIALS

In the present section, examples of N = 2 extended gravity theories with vanishing cosmological constant and spontaneous supersymmetry breaking will be considered.

We will confine ourselves to only those situations in which the super-Higgs effect is induced by[43,49] coupling supergravity to vector multiplets. The O(2) group which rotates the two gravitinos can in general by gauged by a linear combination of the vector partner of the graviton B_μ and the vector fields A_μ:

$$g'B_\mu + g_i A^i_\mu \ . \tag{11.1}$$

If the vectors A^i_μ gauge a group G, then g_i can be non-zero for only U(1) Abelian factors of G. The scalar potential for n vector multiplets is given, in N = 1 language, by[44]

$$e^{\mathcal{G}}\mathcal{G}^{i}\mathcal{G}_{j}(\mathcal{G}^{-1})^{j}_{i} - 3 \ , \tag{11.2}$$

with the function $\mathcal{G}$ given by

$$\mathcal{G} = \ln y - \ln h(z)|^{2} \ , \tag{11.3}$$

with y as given by Eq. (9.13d), and superpotential h(z),

$$h(z) = 2(g^{T}z) \ , \tag{11.4}$$

where $g^{T}z = g' + g_{i}z^{i}$.

Equation (11.4) generalizes Eq. (9.22) for $g_{i} \neq 0$. In terms of the matrix N_{IJ} defined by Eq. (9.13) the over-all potential (11.2) reads

$$V = -\frac{1}{2}\, g^{T}N^{-1}g - \frac{|g^{T}z|^{2}}{z^{T}Nz^{*}} \ , \tag{11.5}$$

where g is the constant n + 1 dimensional vector (g', g_{i}) and z stands for $(1, z_{i})$. The invariance of V under the global transformation[62,63]

$$x \simeq Ux, \quad N \to (U^{T})^{-1}Nu^{-1}, \quad g \to (U^{T})^{-1}g \tag{11.6}$$

$$(z_{i} = x_{i}/x_{0}), \quad U \in GL(n+1, R) \ ,$$

allows us to write the potential in the $g_{i} = 0$ gauge by performing a general linear transformation. Under such transformation the analytic function f(z) which defines the vector multiplet coupling to N = 2 supergravity transforms as follows:

$$f(z) \to g(z) = \left(U^{0}_{0} + U^{j}_{0}z_{j}\right)^{2} f\left(\frac{U^{0}_{i} + U^{j}_{i}z_{j}}{U^{0}_{0} + U^{j}_{0}z_{j}}\right) \tag{11.7}$$

In the $g_{i} = 0$ gauge the scalar potential reads

$$\frac{V}{g'^{2}} = -\frac{1}{2}\left(N^{-1}\right)^{0}_{0} - \frac{1}{z^{T}Nz^{*}} \tag{11.8}$$

as given by Eq. (9.22).

We consider first the coupling of a single vector multiplet to the N = 2 supergravity action. In this case it can be shown that the trace of the scalar mass-matrix satisfied the relation

$$V_{zz^*} = 2G_{zz^*}V$$

at the extremum $V_z = 0$.

Therefore the absence of a cosmological constant implies

$$m_A^2 + m_B^2 = 0, \quad \text{i.e.} \quad m_A = m_B = 0 \; . \tag{11.10}$$

In Ref. 44 it was shown that the point

$$V_z = V = 0 \tag{11.11}$$

can never be a local minimum of V for any choice of f(z).

Therefore the only locally stable solution of Eq. (11.11) with $e^G \neq 0$ is a flat potential, i.e.[62]

$$V \equiv 0 \; , \qquad e^G \neq 0 \; . \tag{11.12}$$

The most general solution for a flat potential of a single complex scalar field z was given, for N = 1 supergravity, in Section 7 by Eq. (7.6)

$$G = \log \left[\phi(z) + \phi^*(z^*)\right]^3 \; . \tag{11.13}$$

If we demand G as given by Eq. (11.13) to be of the form (11.7), we find two solutions corresponding to the following choices of the N = 2 invariant function f(z):

$$f_1(z) = -2i(\alpha z + \beta)^3 + iP_2(z) \; , \tag{11.14a}$$

$$f_2(z) = -i(\alpha z + \beta)^{3/2} + iP_2(z) \; , \qquad \alpha,\beta \text{ real} \; , \tag{11.14b}$$

where $P_2(z)$ is a second-degree polynomial with real coefficients. The solutions (11.14) correspond to the special values for $\phi(z)$ as given by Eq. (11.13)

$$\phi_1(z) = -i(\alpha + \beta z) + i\gamma \ , \qquad \phi_2(z) = (\alpha z + \beta)^{\frac{1}{2}} + i\gamma \ , \tag{11.15}$$

where γ is an arbitrary real constant. Equations (11.14) give the first example of an N = 2 supergravity theory with vanishing cosmological constant and spontaneous supersymmetry breaking.

The gravitino mass in these models is undetermined at the tree level since the vacuum expectation value of z is arbitrary. Coupling to other multiplets or radiative corrections may remove this degeneracy, as has been proposed in the litetature[29,63].

Vanishing potentials can be generalized in different ways to the case of many vector multiplets.

One obvious way, which is only possible if the n-vector multiplets gauge Abelian group, is simply given by choosing the $f(z^i)$ to be

$$f(z) = \sum_i f_{(i)}(z^i) \ , \tag{11.16}$$

in which $f_{(i)}(z^i)$ depends only on z^i and is given by Eqs. (11.14). A more interesting way is obtained by the generalization of $f_1(z)$ to many fields, namely

$$f(z^i) = iC_{ijk} z^i z^j z^k \ , \tag{11.17}$$

in which C_{ijk} is an arbitrary (totally symmetric) constant.

If $g_i \neq 0$ the function $f(z^i)$ is given by

$$f(z^i, g^i) = \frac{1}{g^T z} iC_{ijk} z^i z^j z^k \tag{11.18}$$

and still gives a vanishing potential for arbitrary g_i. It is interesting to observe that for $g_i = 0$, Eq. (11.17) corresponds to a Kähler potential log y, with

$$y = -iC_{ijk}\left(z_i^i - z^{*i}\right)(z^j - z^{*j})(z^k - z^{*k}) \ . \tag{11.19}$$

This is the most general form of y which can be obtained from the coupling of n-vector multiplets to N = 2 supergravity in D = 5 dimensions[64]. In that case the most general scalar potential[65] corresponds to a subclass of D = 4 potentials defined through Eq. (11.3), with h(z) as given by Eq. (11.4) but y given by Eq. (11.19).

Note that for $g_i \neq 0$ the D = 5 vanishing potentials are different from the gauge-transformed vanishing potentials defined by the f-function as given by Eq. (11.18). The simplest way to understand this fact is by noticing that the D = 5 potentials correspond to the Kähler metric given by Eq. (11.19), whilst the D = 4 vanishing potentials correspond to the Kähler metric given by Eq. (9.13) in terms of f(z) as defined by Eq. (11.18). The two Kähler metrics only coincide when $g_i = 0$. We would like to conclude this final section by a comment on the positivity domain for the kinetic-energy terms defined by the Kähler metric (11.17). If the n-vector multiplets gauge the unitary groups SU(n) (in this case $n = n^2 - 1$) then C_{ijk} can be identified with the d_{ijk} symbols of SU(M). In this case the positivity-energy domain for the scalar fields is empty[62].

There is a non-empty domain in the case of G' = U(M) with

$$f(z) = i\left(z^3 + zz^{i2} + d_{ijk}z^i z^j z^k\right) \tag{11.20}$$

and $\alpha > 0$.

Of course there is a non-empty domain for $G = U(1)^n$ and for all C_{ijk} coefficients obtained from the D = 4 dimensional reductions of D = 5 supergravity Maxwell coupled systems[64,65].

REFERENCES

1. For a review see, for example, P. van Nieuwenhuizen, Phys. Rep. 68:191 (1981).
2. For recent reviews see, for example, R. Barbieri and S. Ferrara, Surveys in High-Energy Physics 4:33 (1983).
   B. Zumino, Berkeley preprint UCB-PTH-83/2 (1983), LBL-15819 (1983), to appear in the Proc. Solvay Conf., Austin, Texas, 1982.
   H.-P. Nilles, Proc. Conf. on Problems of Unification and Supergravity, La Jolla, Calif., 1983 (AIP Conference Proceedings No. 116, 1983), p. 109.
   J. Polchinski, Harvard preprint HUTP-83/A036 (1983).
   S. Ferrara, Proc. 4th Silarg Symp. on Gravitation, Caracas, 1982, ed. C. Aragone (World Scientific, Singapore, 1982), p. 11.
   D.V. Nanopoulos, CERN preprint TH.3699 (1983), to appear in Proc. Europhysics Study Conference on Electroweak Effects at High Energies, Erice, 1983.
   J. Ellis, preprint CERN TH.3718 (1983), to appear in Proc. Int. Symp. on Lepton and Photon Interactions at High Energies, Cornell, 1983.
   R. Barbieri, Unconventional weak interactions, Univ. Pisa preprint (1983), to appear in Proc. Int. Symp. on Lepton and Photon Interactions at High Energies, Cornell, 1983.
3. R. Barbieri, S. Ferrara and D.V. Nanopoulos, Z. Phys. C13:267 (1982) and Phys. Lett. 116B:6 (1982).
   J. Ellis, L. Ibanez and G. Ross, Phys. Lett. 113B:283 (1982).
   S. Dimopoulos and S. Raby, Nucl. Phys. B219:479 (1983).
   J. Polchinski and L. Susskind, Phys. Rev. D26:3661 (1982).
4. S. Deser and B. Zumino, Phys. Rev. Lett. 38:1433 (1977).
5. E. Cremmer, B. Julia, J. Scherk, S. Ferrara, L. Girardello and P. van Nieuwenhuizen, Phys. Lett. 79B:231 (1978) and Nucl. Phys. B147:105 (1979).
6. E. Cremmer, S. Ferrara, L. Girardello and A. van Proeyen, Phys. Lett. 116B:231 (1982) and Nucl. Phys. B212:413 (1983).
7. E. Gildener, Phys. Rev. D14:1667 (1976).
   E. Gildener and S. Weinberg, Phys. Rev. D15:3333 (1976).
   L. Maiani, Proc. Summer School on Weak Interactions, Gif-sur-Yvette, 1979 (IN2P3, Paris, 1980), p. 3.
   M. Veltman, Acta Phys. Pol. B12, 437 (1981).
   E. Witten, Nucl. Phys. B188:513 (1981).
   S. Dimopoulos and S. Raby, Nucl. Phys. B199:353 (1981).
8. S. Ferrara and P. van Nieuwenhuizen, Phys. Lett. 74B:333 (1978), and Phys. Lett. 76B:404 (1978).
   K.S. Stelle and P.C. West, Phys. Lett. 74B:330 and 77B:376 (1978).
9. M. Kaku, P.K. Townsend and P. van Nieuwenhuizen, Phys. Rev. D17: 3179 (1978).
   B. de Wit, in Supergravity 1982, eds. S. Ferrara, J.G. Taylor and P. van Nieuwenhuizen (World Scientific Publishing, Singapore, 1982), p. 85.
   T. Kugo and S. Uehara, Nucl. Phys. B222:125 (1983).

10. M.F. Sohnius and P.C. West, Phys. Lett. 105B:353 (1981).
11. S. Ferrara, L. Girardello, T. Kugo and A. van Proeyen, Nucl. Phys. B223:191 (1983).
12. See, for example, J. Wess and J. Bagger, Supersymmetry and supergravity (Princeton Univ. Press, 1983).
13. P. Fayet and J. Iliopoulos, Phys. Lett. 51B:461 (1974).
14. K.S. Stelle and P. C. West, Nucl. Phys. B145:175 (1978).
15. R. Barbieri, S. Ferrara, D.V. Nanopoulos and K.S. Stelle, Phys. Lett. 113B:219 (1982).
16. M.T. Grisaru, M. Rocek and A. Karlhede, Phys. Lett. 120B:189 (1982).
17. T. Kugo and S. Uehara, Kyoto preprint KUNS 646 (1982), to be published in Nuclear Physics B.
18. J. Bagger, Nucl. Phys. B211:302 (1983).
19. D.Z. Freedman, Phys. Rev. D15:1173 (1977).
B. de Wit and P. van Nieuwenhuizen, Nucl. Phys. B139:531 (1979).
20. S. Ferrara, L. Girardello and F. Palumbo, Phys. Rev. D20:403 (1979).
21. R. Barbieri, S. Ferrara and C.A. Savoy, Phys. Lett. 119B:343 (1982).
22. L. Girardello and M.T. Grisaru, Nucl. Phys. B194:65 (1982).
23. S.K. Soni and H.A. Weldon, Phys. Lett. 126B:215 (1983).
R. Barbieri and S. Ferrara, see Ref. 2.
24. S. Samuel and J. Wess, Columbia Univ. preprints CU-TP-258 (1982) and CU-TP-260 (1983).
25. H.-P. Nilles, M. Srednicki and D. Wyler, Phys. Lett. 120B:346 (1983).
26. L. Hall, J. Likken and S. Weinberg, Phys. Rev. D27:2359 (1983).
27. E. Cremmer, P. Fayet and L. Girardello, Phys. Lett. 122B:41 (1983).
28. J. Polony, Budapest report KFKI-(1977) (unpublished).
29. E. Cremmer, S. Ferrara, C. Kounnas and D.V. Nanopoulos, Phys. Lett. B113:61 (1983).
S. Ferrara and A. van Proeyen, Phys. Lett. 138B:77 (1984).
30. E. Cremmer, J. Scherk and S. Ferrara, Phys. Lett. 74B:61 (1978).
31. Ngee-Poug Chang, S. Ouvry and Xizeng Wu, Phys. Rev. Lett. 51:327 (1983).
32. E. Cremmer, S. Ferrara, L. Girardello, C. Kounnas and A. Masiero, Phys. Lett. 137B:62 (1984).
33. J.M. Frere, D.R.T. Jones and S. Raby, Nucl. Phys. B222:11 (1983).
34. P. Fayet, Phys. Lett. 125B:178 (1983).
35. R. Barbieri and L. Maiani, Rome Univ. preprint No. 343 (1983).
36. R. Arnowitt, A.H. Chamseddine and P. Nath, Phys. Rev. Lett. 49:970 (1982).
L. Ibanez, Phys. Lett. 118B:73 (1972).
37. J. Ellis, J.S. Hagelin, D.V. Nanopoulos and K. Tamvakis, Phys. Lett. 124B:484 (1983).
B.A. Ovrut and S. Raby, Phys. Lett. 125B:270 (1983).
L. Hall and M. Suzuki, Berkeley preprint LBL-16150, UCB-PTH-83/8 (1983).

38. S. Ferrara, D.V. Nanopoulos and C.A. Savoy, Phys. Lett. 123B:214 (1983).
39. H.-P. Nilles, M. Srednicki and D. Wyler, Phys. Lett. 124B:337 (1983).
A.B. Lahanas, Phys. Lett. 124B:341 (1983).
40. A. Masiero, D.V. Nanopoulos, K. Tamvakis and T. Yamagida, Phys. Lett. 115B:380 (1982).
B. Grinstein, Nucl. Phys. B206:387 (1982).
S. Dimopoulos and F. Wilczek, Santa Barbara preprint (1982).
C. Kounnas, D.V. Nanopoulos, M. Quiros and M. Srednicki, Phys. Lett. 127B:82 (1983).
41. J. Ellis, D.V. Nanopoulos and K. Tamvakis, Phys. Lett. 121B:123 (1983).
H.-P. Nilles, Nucl. Phys. B214:366 (1983).
L. Alvarez-Gaumé, J. Polchinski and M. Wise, Nucl. Phys. B221: 495 (1983).
J. Ellis, J.S. Hagelin, D.V. Nanopoulos and K. Tamvakis, Phys. Lett. 125B:275 (1983).
L.E. Ibanez and C. Lopez, Phys. Lett. 126B:54 (1983).
L.E. Ibanez, Phys. Lett. 126B:196 (1983).
M. Claudson, L.J. Hall and I. Hinchliffe, Nucl. Phys. B228:501 (1983).
C. Kounnas, A.B. Lahanas, D.V. Nanopoulos and M. Quiros, Phys. Lett. 132B:95 (1983).
L. Ibanez and C. Lopez, Nucl. Phys. B233:511 (1984).
J.P. Derendinger and C.A. Savoy, Nucl. Phys. B237:307 (1984).
C. Kounnas, A.B. Lahanas, D.V. Nanopoulos and M. Quiros, Nucl. Phys. B236:438 (1984).
42. S. Ferrara and P. van Nieuwenhuizen, Phys. Rev. Lett. 37:1669 (1976).
43. J.P. Derendinger, S. Ferrara, A. Masiero and A. van Proeyen, Phys. Lett. 140B:307 (1984).
44. B. de Wit, P.G. Lauwers, R. Philippe, S.Q. Su and A. van Proeyen Phys. Lett. 143B:37 (1984).
45. P. Breitenlohner and M. Sohnius, Nucl. Phys. B187:409 (1981).
46. J. Bagger and E. Witten, Nucl. Phys. B222:1 (1983).
47. P.H. Dondi and M. Sohnius, Nucl. Phys. B81:317 (1974).
R.J. Firth and O.J. Jenkins, Nucl. Phys. B85:525 (1975).
R. Grimm, M. Sohnius and J. Wess, Nucl. Phys. B133:275 (1978).
48. P. Fayet, Nucl. Phys. B113:135 (1976) and B149:137 (1979).
49. B. de Wit and A. van Proeyen, NIKHEF preprint-H/84-4 (1984).
50. J.F. Luciani, Nucl. Phys. B132:325 (1978).
51. C.K. Zachos, Phys. Lett. 76B:329 (1978).
52. M. de Roo, J.W. van Holten, B. de Wit and A. van Proeyen, Nucl. Phys. B173:175 (1980).
P. Breitenlohner and M. Sohnius, Nucl. Phys. B178:151 (1980).
B. de Wit, J.W. van Holten and A. van Proeyen, Phys. Lett. 95B: 51 (1980) and Nucl. Phys. B184:77 (1981) [E: B122:516 (1983)
B. de Wit, R. Philippe and A. van Proeyen, Nucl. Phys. B219:149 (1983).

53. B. de Wit, P.G. Lauwers, R. Philippe and A. van Proeyen, Phys. Lett. 135B:295 (1984).
54. D.Z. Freedman and A. Das, Nucl. Phys. B120:221 (1977).
E.S. Fradkin and M.A. Vasiliev, Lebedev Institute preprint, No. 197 (1976).
55. J.P. Derendinger, S. Ferrara, A. Masiero and van Proeyen, Phys. Lett. 136B:354 (1984).
56. B. Zumino, Phys. Lett. 87B:203 (1979).
J. Bagger and E. Witten, Phys. Lett. 118B:103 (1982).
57. E. Cremmer, S. Ferrara and J. Scherk, unpublished.
A. Das, M. Fischler and M. Rocek, Phys. Rev. D16:3427 (1977).
58. B. de Wit and H. Nicolai, Nucl. Phys. B188:98 (1981).
59. B. Zumino, Nucl. Phys. B127:189 (1977),
60. P. Breitenlohner and D.Z. Freedman, Phys. Lett. 115B:197 (1982), and Ann. Phys. 144:249 (1982).
61. F. Buccella, J.P. Derendinger, S. Ferrara and C.A. Savoy, in Unification of the fundamental particle interactions, II (eds. J. Ellis and S. Ferrara) (Plenum Press, New York and London, 1983), p. 349 and Phys. Lett. 115B:375 (1982).
P.H. Frampton and T. Kephart, Phys. Rev. Lett. 48:1237 (1982) and Nucl. Phys. B211:239 (1983).
62. E. Cremmer, J.P. Derendinger, B. de Wit, S. Ferrara, L. Girardello, C. Kounnas, A. van Proeyen, preprint CERN TH.3965 (1984);
see also J.P. Derendinger and S. Ferrara, preprint CERN TH.3903 (1984), Proc. Spring School on Supergravity and Supersymmetry, Trieste, 1984 (World Scientific, Singapore, 1984), p. 159.
63. J. Ellis, A.B. Lahanas, D.V. Nanopoulos and K. Tamvakis, Phys. Lett. 134B:429 (1984).
J. Ellis, C. Kounnas and D.V. Nanopoulos, CERN preprint TH.3773 (1983).
64. J. Gunaydin, G. Sierra and P.K. Townsend, Phys. Lett. 133B:72 (1983).
65. J. Gunaydin, G. Sierra and P.K. Townsend, Nucl. Phys. B242:244 (1984) and Phys. Lett. 144B:41 (1984).

SUPERGRAPHS

M. T. Grisaru

Physics Department
Brandeis University
Waltham, MA 02254

I. INTRODUCTION

From the early days of supersymmetry it has been clear that the quantum properties of supersymmetric theories are best investigated and exhibited by using superfield perturbation theory and supergraphs. In this way the results of calculations are manifestly supersymmetric, cancellations due to supersymmetry are automatic and the calculations themselves are generally simpler since one evaluates simultaneously bosonic and fermionic contributions.

The original superfield Feynman rules[1] were not in optimal form. An improved version,[2] along with better calculational techniques was presented in 1979, just preceeding the renewed interest in supersymmetry, and since then we have made rapid advances in developing the formalism and tools, for both global and local supersymmetry. In these lectures I shall attempt to review the methods and summarize the results. I will divide my subject into two parts: ordinary supergraphs, and background field supergraphs. Many details that I omit can be found in the original references and in "Superspace".[3] Newer results can be found in some recent papers.[4]

I use the notation and conventions of "Superspace" with two component spinors $\psi\alpha$, $\chi_{\dot{\alpha}}$, and four-vectors $A_a \equiv A_{\alpha\dot{\alpha}}$. Superspace coordinates for N = 1 supersymmetry are $x^{\alpha\dot{\alpha}}$, θ^{α}, $\bar{\theta}^{\dot{\alpha}}$. Superfields are denoted by $\Psi(x,\theta,\bar{\theta})$. The derivatives

$$D_A = \{\partial_{\alpha\dot{\alpha}}, D_\alpha, \bar{D}_{\dot{\alpha}}\}$$

$$D_\alpha = \partial_\alpha + \frac{i}{2}\bar{\theta}^{\dot{\alpha}}\partial_{\alpha\dot{\alpha}} \qquad \bar{D}_{\dot{\alpha}} = \bar{\partial}_{\dot{\alpha}} + \frac{i}{2}\theta^\alpha\partial_{\alpha\dot{\alpha}} \tag{1.1}$$

$$\{D_\alpha, \bar{D}_{\dot{\alpha}}\} = i\partial_{\alpha\dot{\alpha}}$$

are covariant with respect to global supersymmetry. We deal primarily with "chiral" and "antichiral" scalar superfields $\phi, \bar{\phi}$, satisfying the differential constraint

$$\bar{D}_{\dot{\alpha}}\phi = 0 \quad , \quad D_\alpha\bar{\phi} = 0 \tag{1.2}$$

and real scalar superfields satisfying $V = V^\dagger$. We define $D^2 = \frac{1}{2}D^\alpha D_\alpha = iD_+D_-$, where D_+, D_- are the two independent components of the covariant derivative. Higher powers of D_α, or any other spinor, vanish since it is a Grassman (anticommuting) variable.

Integrals over superspace are defined using the concept of Berezin integration of Grassman variables $\int d\theta\, 1 = 0$, $\int d\theta\, \theta = 1$, which implies $\int d\theta f(\theta) = f'(0)$. We shall use only the integrals

$$\int d^4x\, d^2\theta \quad , \quad \int d^4x\, d^2\bar{\theta} \quad , \quad \int d^4x\, d^4\theta = \int d^4x\, d^2\theta d^2\bar{\theta} \tag{1.3}$$

The first one defines a supersymmetry invariant only if the integrand is chiral. The last one vanishes if the integrand is chiral. It is easy to verify, using the properties of the Berezin integral, that

$$\int d^4x\, d^2\theta \ldots = \int d^4x\, D^2 \ldots$$
$$\int d^4x\, d^4\theta = \int d^4x\, D^2\bar{D}^2 \tag{1.4}$$

The supersymmetry invariance of integrals over superspace is simply a consequence of the fact that supersymmetry transformations of superfields are just coordinate transformations.

II. GLOBAL SUPERGRAPHS

We use functional methods to obtain the Feynman rules. The procedure is exactly the same as for ordinary Feynman rules. A typical action describing supersymmetric Yang-Mills and a chiral multiplet is (ignoring group theory details)

$$\int d^4x\, d^2\theta\, W^\alpha W_\alpha + \int d^4x\, d^4\theta\, \bar{\phi}\, e^V \phi$$
$$+ \int d^4x\, d^2\theta\, [\tfrac{m}{2}\phi^2 + \tfrac{\lambda}{6}\phi^3] + \text{h.c.} \tag{2.1}$$

where

$$W^\alpha = i\bar{D}^2 e^{-V} D^\alpha e^V = i\bar{D}^2 D^\alpha V + \ldots \tag{2.2}$$

is the field strength superfield. The gauge transformations are described by chiral and antichiral superfield parameters. A chiral superfield transforms by

$$\phi \to e^{i\Lambda}\phi \qquad\qquad \bar{D}_{\dot\alpha}\Lambda = 0 \tag{2.3}$$

while the gauge transformations of the Yang-Mills superfield are

$$e^V \to e^{i\bar{\Lambda}} e^V e^{-i\Lambda} \tag{2.4}$$

It is easy to verify that the action (2.1) is invariant under these transformations provided the mass and chiral self-interaction terms are chosen appropriately.

As gauge-fixing functions we use

$$F = \bar{D}^2 V \quad , \quad \bar{F} = D^2 V \tag{2.5}$$

These are chiral and antichiral quantities, and will lead to Faddeev-Popov ghosts that are also chiral and antichiral. To fix the gauge we start with a functional integral over V and ϕ, including factors $\delta(\bar{D}^2V - a)\delta(D^2V - \bar{a})$ and a Faddeev-Popov determinant. We then gauge average with a factor $\exp \frac{1}{\alpha}\int d^4x\, d^4\theta\, \bar{a}a$ which gives a gauge fixing action $\frac{1}{\alpha}\int D^2V\bar{D}^2V$, and replace the Faddeev-Popov determinant by a ghost action, according to the standard prescription

$$S_{FP} = \int [c'\delta_{c,\bar{c}}(D^2V) + \bar{c}'\delta_{c,\bar{c}}(\bar{D}^2V)] \tag{2.6}$$

where $\delta_{c,\bar{c}}$ indicates a gauge variation with parameters replaced by ghost fields.

The quadratic part of the gauge action has the form $\frac{1}{2}\int VD^\beta\bar{D}^2D_\beta V$. If we choose the gauge parameter $\alpha = 1$ and use the identity

$$D^2\bar{D}^2 + \bar{D}^2 D^2 - D^\beta \bar{D}^2 D_\beta = \Box \tag{2.7}$$

we are led to the gauge and ghost action

$$\int d^4x\, d^4\theta \Big[-\tfrac{1}{2} V \Box V + V D^\alpha V \bar{D}^2 D_\alpha V + \ldots + \bar{c}' c - c' \bar{c} + \ldots\Big] \tag{2.8}$$

where higher order V terms come from the expansion of the exponentials in (2.2), and higher order ghost and V interaction terms arise because of the nonlinear nature of the gauge transformation (2.4).

We derive the Feynman rule by introducing real and chiral sources for the fields and defining the generating functional in standard fashion. By Legendre transformation we obtain the effective action $\Gamma(V,\phi,\bar{\phi})$ that is given by a sum of one particle irreducible graphs with external field factors V, ϕ, $\bar{\phi}$. We define

$$\begin{aligned} Z(J,j,\bar{j}) &= \int \mathcal{D}V\, \mathcal{D}\phi \ldots \exp\Big[S(V,\phi,\bar{\phi}) + \int d^4x\, d^4\theta\, JV \\ &\qquad + \int d^4x\, d^2\theta\, j\phi + \text{h.c.}\Big] \\ &= \exp S_{int}\Big[\frac{\delta}{\delta J}, \frac{\delta}{\delta j}, \frac{\delta}{\delta \bar{j}} \ldots\Big) \cdot \\ &\quad \cdot \int \mathcal{D}V \ldots \exp\Big[\int\Big[-\tfrac{1}{2} V \Box V + JV\Big] + \int \bar{\phi}\phi + \int\Big[\tfrac{m}{2}\phi^2 + j\phi\Big] + \text{h.c.}\Big] \end{aligned} \tag{2.9}$$

and doing the Gaussian integrals we are led to introduce propagators (see Ref. 2, 3)

$$\begin{aligned} \langle V(x,\theta) V(x',\theta') \rangle &= \frac{1}{\Box}\, \delta^4(x-x')\delta^4(\theta-\theta') \\ \langle \bar{\phi}(x,\theta)\phi(x',\theta') \rangle &= -\frac{1}{\Box - m^2}\delta^4(x-x')\delta^4(\theta-\theta') \\ \langle \phi(x,\theta)\phi(x',\theta') \rangle &= -\frac{m\bar{D}^2}{\Box(\Box - m^2)} \end{aligned} \tag{2.10}$$

The vertices are obtained from S_{int}, using

$$\begin{aligned} \frac{\delta J(x,\theta)}{\delta J(x',\theta')} &= \delta^4(x-x')\delta^4(\theta-\theta') \\ \frac{\delta j(x,\theta)}{\delta j(x',\theta')} &= \bar{D}^2 \delta^4(x-x')\delta^4(\theta-\theta') \end{aligned} \tag{2.11}$$

The rules for functional differentiation of an unconstrained superfield are the same as for ordinary functions. For a chiral superfield, e.g. j, the variation must be taken with due regard to the chirality constraint $\bar{D}_{\dot\alpha} j = 0$ and this accounts for the $\bar{D}^2$ factor in the second equation of (2.11). Therefore, in general we have an operator $\bar{D}^2$ acting on each chiral propagator leaving a vertex and similarly D^2 acting on an antichiral propagator.

At the vertices we also have $\int d^4x\, d^4\theta$, $\int d^4x\, d^2\theta$ or $\int d^4x\, d^2\bar\theta$ integrations, the latter if the integrands in the original action where purely chiral, e.g ϕ^n, or antichiral.

However, since we get for each ϕ line a $\bar{D}^2$, we can use one such operator to convert to a full superspace integral since

$$\int d^4x\, d^2\theta\, \bar{D}^2 \ldots = \int d^4x\, d^4\theta \ldots \qquad (2.12)$$

Therefore our vertices always carry a $d^4\theta$ integration but we have the rule that at a chiral vertex we omit a factor $\bar{D}^2$ from one of the lines.

From the V self-interaction terms in (2.8) we also get operators D^α, $\bar{D}^2 D_\alpha$, etc. acting on the propagators. One of the main activities in supergraph calculation is the manipulation of these operators, so-called D-algebra. We explain this now.

From a Feynman graph, we obtain a contribution that consists of a product of propagators $G(x_i - x_{i+1}, \theta_i - \theta_{i+1})$, spinorial derivatives D_α, $\bar{D}_{\dot\alpha}$ acting on them and integrals over vertices, as well as a number of external line factors $\phi(x,\theta)$, $\phi(x',\theta')$, $V(x'',\theta'')$, etc. The D_α's satisfy

$$D_\alpha G(x - x', \theta - \theta') = - D'_\alpha G(x - x', \theta - \theta') \qquad (2.13)$$

Furthermore, from the explicit expressions (1.1), it can be seen that they can be integrated by parts when inside integrals. Thus, for any internal line, we can move the D's from one end to the other, and then integrate them by parts so that the propagator is free of any derivatives. This procedure can be successively repeated on the lines of a given loop with vertices 1, 2,...n, so that one generates a sum of terms containing an expression

$$G_{1,2}\, G_{2,3} \ldots G_{n-1,n} (DD\bar{D}D\bar{D}\bar{D} \ldots G_{n,1}) \qquad (2.14)$$

We use two facts: since $(D)^3 = 0$, we can use the anticommutation relations in (1.1) to reduce the number of spinor derivatives in (2.14) to at most 2 D's and 2 $\bar{D}$'s, times a number of space-time derivatives. Also, since $G_{ij} = \frac{1}{\Box}\delta^4(x_i - x_j)\delta^4(\theta_i - \theta_j)$, which is always a δ-function in the θ variables since $\Box$ is independent of θ, in (2.10) we have a factor

$$\delta(\theta_1 - \theta_2)\ldots\ \delta(\theta_{n-1} - \theta_n)\ DD\ldots\delta(\theta_n - \theta_1)$$
$$\sim \delta(\theta_1 - \theta_n)\ DD\ldots\delta(\theta_n - \theta_1) \tag{2.15}$$

and this is zero ($\delta(\theta) = \theta^2\bar{\theta}^2$ so that, e.g., $\delta(\theta)\delta(\theta) = 0$) unless we have exactly 2 D's and 2 $\bar{D}$'s, in which case

$$\delta(\theta_1 - \theta_n)D^2\bar{D}^2\delta(\theta_n - \theta_1) = \delta(\theta_1 - \theta_n) \quad . \tag{2.16}$$

(each D removes a θ from the second δ-function). This fact embodies many of the cancellations due to supersymmetry.

Repeating this procedure loop by loop, we generate factors $\partial_{\alpha\dot{\alpha}}$ (from $\{D_\alpha, \bar{D}_{\dot{\alpha}}\} = i\partial_{\alpha\dot{\alpha}}$), and derivatives acting on the external field factors, but we can use the $\delta(\theta_i - \theta_j)$ to do all but one of the θ interactions, ending up with a contribution to the effective action of the form

$$\int d^4\theta\ d^4x_1\ldots.d^4x_n \quad F(x_1\ldots.x_n) \quad \phi(x_1,\theta)D^a\phi(x_2\theta)\ldots. \tag{2.17}$$

containting in general a nonlocal function F, but local in θ, with a $d^4\theta$ integral.

As an example consider the action $\int \bar{\phi}e^V\phi$ and a one loop contribution from a massless chiral field to the V self-energy. According to our rules we have the diagram

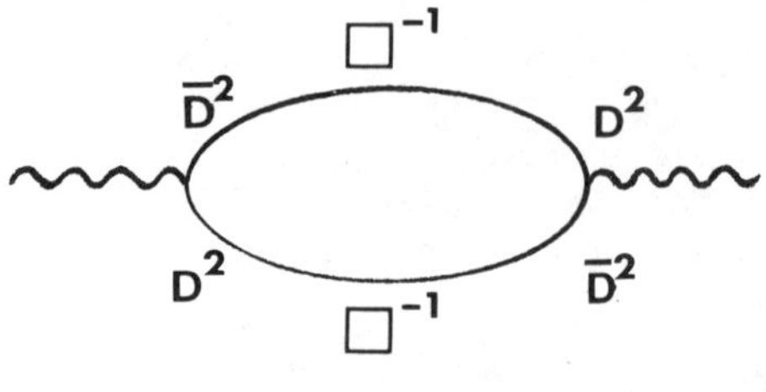

Figure 1: A self-energy supergraph

Integrating by parts the D's as described above and going to momentum space we find

$$\frac{1}{2}\int \frac{d^4k}{(2\pi)^4} d^4\theta\, V(-k,\theta) \int \frac{d^4P}{(2\pi)^4} \frac{-p^2 - p^{\alpha\dot{\alpha}}\bar{D}_{\dot{\alpha}}D_{\alpha} + \bar{D}^2D^2}{p^2(k+p)^2} V(k,\theta) \quad (2.18)$$

where the D's act on $V(k,\theta)$.

After D-algebra, one is left with ordinary momentum space integrals to perform. The D-algebra itself becomes routine after a while. There exist many tricks that can be used to simplify the procedure and some of them can be found in the papers that describe higher loop calculations in the Wess-Zumino model and SSYM.[5] In the remainder of this section I will discuss a number of general results.

1) The method leads to an effective action that is expressed as an integral over superspace of superfields and their covariant derivatives.

2) If one does dimensional regularization <u>after</u> the D-algebra has been completed one is essentially using the dimensional reduction scheme of Siegel.[6] While the scheme may run into ambiguities at higher loops, it or any other regularization scheme that maintains translational invariance will be supersymmetric if done after D-algebra, simply as a consequence of 1).

3) The effective action is local in θ.

4) The effective action involves a $\int d^4\theta$ integral so that <u>local</u> terms with $\int d^2\theta$ do not get generated. Therefore, mass $\int d^2\theta\phi^2$ or chiral interaction $\int d^2\theta\ \phi^n$ terms, which are present in the original action, do not receive finite or infinite radiative corrections. This is the simplest nonrenormalization theorem of supersymmetry. Note that the result is true even in nonrenormalizable theories, e.g. supergravity. New divergences are generated, but they always involve $d^4\theta$ integrals.

5) Power counting is very simple: $d^2\theta$ has dimension 1, D_α has dimension $\frac{1}{2}$, while for the superfields dim $\phi = 1$, dim $V = 0$. From the form of the effective action in (2.17) it follows that at each order of perturbation theory, to get new divergences involving necessarily local function F with dimension ≥ 0, we can have at most a factor $\bar{\phi}\phi$ (and a logarithmic divergence) but an arbitrary number of V lines. Thus the chiral kinetic term requires (wave function) renormalization in general. Since V is a gauge field, the divergences are controlled by the Slavnov-Taylor identities so that in fact only the 2- and 3-point functions require renormalization. This result becomes clearer in the background field method to which

I now turn. The actual proof of renormalizability of the theory is complicated by the presence of infrared divergences. Rigorous results on this subject have been proven recently by Piguet and Sibold.[7]

Supergraph techniques may be extended to higher N supersymmetric theories provided an unconstrained superfield formalism exists. This is necessary in order to have a definition of functional integration and differentiation (N = 1 chiral superfields are constrained, but the chirality constraint can be explicitly solved so this presents no problem). For N = 2,3 recent work[8] is very promising.

III. BACKGROUND FIELD SUPERGRAPHS

Conventional quantization of gauge theories, both in ordinary space and in superspace, generally leads to an effective action that is not gauge invariant (but it has BRS invariance and, of course, physical quantities are gauge invariant). In the background field method one constructs an effective action that is gauge invariant. This is achieved by introducing two kinds of fields, quantum and background. The background field only appears in external lines of Feynman graphs and requires no gauge fixing. For the quantum field we still have to fix the gauge in order to define propagators, but this can be done in a manner that respects the gauge invariance of the effective action.

The procedure has been reviewed by Abbott.[9] Instead of the usual gauge Lagrangian $L(A_a)$, one considers the Lagrangian $L(A_a+B_a) - L(B_a) - L'(B_a)A_a$ where B_a is the background field. The quantization is carried out in conventional fashion for the field A_a except that the gauge-fixing function is chosen background covariant:

$$\partial^a A_a \rightarrow D^a(B)A_a \tag{3.1}$$

$$D^a(B) = \partial^a + B^a \times \tag{3.2}$$

It is easy to see that in this fashion one maintains invariance under $\delta B_a = \partial_a\lambda + \lambda\times B_a$ provided the quantum field transforms as $\delta A_a = \lambda\times A_a$. The ghost action is also covariantized with respect to the background field. One can show that Feynman graphs with A internal and B external lines give a suitable effective action $\Gamma(B_a)$ for constructing Green's functions.

In this example the background-quantum splitting is linear: in the Lagrangian we substitute $A \rightarrow A + B$. This is not always the case. For our purpose the correct splitting in superspace does match that of ordinary Yang-Mills provided the latter is interpreted as a covariantization with respect to B of the A covariant derivative:

$$D_a(A) = \partial_a + A_a\times \rightarrow \partial_a + (B_a + A_a)\times = D_a(B) + A_a\times \tag{3.3}$$

since $[D_a, D_b] = F_{ab}$, the field strength, the quantum-background splitting of the Lagrangian is recovered by substituting in $-\frac{1}{4}(F_{ab})^2$ the expression for the covariant derivative in (3.3).

The same recipe works for SSYM. The covariant derivatives (in "chiral" representation[3]) are

$$\nabla_\alpha = e^{-V}D_\alpha e^V \equiv D_\alpha - i\Gamma_\alpha = D_\alpha + D_\alpha V + \ldots$$

$$\nabla_{\dot\alpha} = \bar{D}_{\dot\alpha} \tag{3.4}$$

$$\nabla_{\alpha\dot\alpha} = -i\,\{\nabla_\alpha, \nabla_{\dot\alpha}\} \equiv \partial_{\alpha\dot\alpha} - i\,\Gamma_{\alpha\dot\alpha}$$

and the field strength W_α can be expressed as

$$W_\alpha = \frac{i}{2}[\nabla^{\dot\alpha}, \{\nabla_{\dot\alpha}, \nabla_\alpha\}] \tag{3.5}$$

In (3.4) we have introduced the connection superfields defined by $\nabla_A = D_A - i\Gamma_A$. The splitting (nonlinear this time)

$$V \rightarrow V + V_B \tag{3.6}$$

is realized by

$$\nabla_\alpha \rightarrow e^{-V}\nabla_\alpha(V_B)e^V$$

$$\nabla_{\dot\alpha} \rightarrow \nabla_{\dot\alpha}(V_B)$$

$$\nabla_{\alpha\dot\alpha} = -i\{\nabla_\alpha, \nabla_{\dot\alpha}\}$$

In other words, we have replaced the usual derivatives in (3.4) by background covariant derivatives. Substituting into the action $\int d^4x\, d^2\theta\, W^2$ the expression (3.5), but with the derivatives of (3.7) leads to an expression containing the quantum fields explicitly and the background fields implicitly through the covariant derivatives.

Gauge fixing is done with a term $\nabla^2 V \bar{\nabla}^2 V$ that is background covariant. We note that since the gauge fixing function $\bar{\nabla}^2 V$ satisfies a <u>covariant</u> chirality condition $\bar{\nabla}_{\dot{\alpha}}(\bar{\nabla}^2 V) = 0$, the Faddeev-Popov ghosts c, c' will be covariantly chiral fields: $\bar{\nabla}_{\dot{\alpha}} c = 0$. The gauge averaging $\exp \frac{1}{\alpha} \int \bar{a}a$ that takes us from $\delta(\bar{\nabla}^2 V - a)$ to the gauge fixing Lagrangian introduces an additional, "Nielsen-Kallosh" ghost[2,3]: since $a = \bar{\nabla}^2 V$ is also covariantly chiral the integral $\int \mathcal{D}a\, \mathcal{D}\bar{a}[\exp \int \bar{a}a]$ produces a determinant with a nontrivial dependence on the background field, and this can be cancelled by an additional factor $\int \mathcal{D}b\, \mathcal{D}\bar{b}[\exp \int \bar{b}b]$ where the covariantly chiral superfield b has abnormal statistics.

Covariantly chiral superfields have an implicit dependence on the gauge field. This dependence can be made explicit: In the representation of (3.4) if covariantly chiral and antichiral fields are defined by

$$\nabla_{\dot{\alpha}} \phi_c = 0 \quad , \quad \nabla_{\alpha} \tilde{\phi}_c = 0 \tag{3.8}$$

we can write

$$\begin{aligned} \phi_c &= \phi_o & \bar{D}_{\dot{\alpha}} \phi_o &= 0 \\ \tilde{\phi}_c &= e^{-V} \phi_o & D_{\alpha} \bar{\phi}_o &= 0 \end{aligned} \tag{3.9}$$

in terms of ordinary chiral and antichiral fields. The usual chiral action in the presence of a gauge field can be rewritten in terms of covariantly chiral fields:

$$\int \bar{\phi}_o e^V \phi_o = \int e^{-V} \bar{\phi}_o \phi_o = \int \tilde{\phi} \phi \tag{3.10}$$

After quantum-background splitting it goes into

$$\int \tilde{\phi} e^V \phi \tag{3.11}$$

where now V is the quantum field and the ϕ's are background covariantly chiral.

After gauge fixing the Yang-Mills action becomes

$$-\frac{1}{2} \int d^4x\, d^4\theta\, \{V[\Box - iW^{\alpha}\nabla_{\alpha} - i\bar{W}^{\dot{\alpha}}\bar{\nabla}_{\dot{\alpha}}]V - V\{\nabla^{\alpha} V, \bar{\nabla}^2 \nabla_{\alpha} V\} + \ldots\} \tag{3.12}$$

involving background covariant derivatives and field strengths, and we have additional ghost and chiral superfield terms.

To compute supergraphs we proceed as follows. We separate out $\nabla_\alpha = D_\alpha - i\Gamma_\alpha$, and also in $\Box = \frac{1}{2}\nabla^a\nabla_a$, $\nabla_a = \partial_a - i\Gamma_a$ and rewrite the action as

$$- \frac{1}{2} \int d^4x\, d^4\theta\, \{V[\Box_o - i\Gamma^a\partial_a - \frac{i}{2}(\partial^a\Gamma_a) - \frac{1}{2}\Gamma^a\Gamma_a$$

$$- iW^\alpha(D_\alpha - i\Gamma_\alpha) - i\bar{W}^{\dot{\alpha}}(\bar{D}_\alpha - i\bar{\Gamma}_{\dot{\alpha}})]V \tag{3.13}$$

$$- VD^\alpha V\bar{D}^2 D_\alpha V - iV[[\Gamma^\alpha, V], \bar{D}^2 D_\alpha V] + \ldots\}$$

We have now quantum background vertices, and also quantum-quantum vertices, but the rules are as before. We draw supergraphs with only external background lines, do D-algebra as usual, and evaluate momentum integrals. As compared to ordinary methods we have achieved the following:

1) As guaranteed by the background field method we have manifest gauge invariance: the complete result, to any order of perturbation theory is expressible in terms of gauge invariant quantities.

2) Improved power counting: by contrast with ordinary Yang-Mills theory, SSYM has the property that the superfield V does not appear bare in the covariant derivatives. These depend on the connections Γ_A, and from the definitions in (3.4) it follows that the Γ_A involve derivatives of V. Since in the present method the background fields appear in the action of (3.12) only through the connections supergraphs will give results that depend explicitly only on the $\Gamma_{\alpha\dot{\alpha}}$ (and their derivatives) and these have dimensions $\geqslant \frac{1}{2}$. Therefore, by the same power counting arguments of the previous section, we conclude that any divergent graphs can contain at most four Γ_α. (Actually the complete effective action, by background gauge invariance, can only have a divergence proportional to the original action $\int W^2$).

3) Simplifications of the calculations: this is already evident in ordinary Yang-Mills.[9] Not only are the individual graphs more convergent, but they are also fewer in number.

If chiral superfields are also present it would seem that the background V field would appear explicitly, since in order to carry out calculations one has to express the background covariantly chiral fields in (3.11) in terms of ordinary chiral fields. However, with the exception of certain one-loop contributions this is not the case, and the Feynman rules for covariantly chiral fields can also be formulated in terms of covariant derivatives[3]: at a vertex we have now factors $\bar{\nabla}^2$, ∇^2 for chiral or antichiral lines leaving the vertex, while in the propagators the d'Alembertians are replaced by covariant d'Alembertians $\Box_\pm$ with, for example

$$\Box_+ = \Box - iW^\alpha\nabla_\alpha - \frac{i}{2}(\nabla^\alpha W_\alpha) \quad , \quad \Box = \frac{1}{2}\nabla^a\nabla_a \tag{3.14}$$

The propagators can then be expanded in terms of $\Box_0$, spinor derivatives D_α, and connections Γ_α, Γ_a.

We have used background field methods to calculate the two-loop β-function in SSYM and have indeed found them simpler than ordinary methods.[10] Another example of the power of the method can be seen in the calculation of the one-loop S-matrix in N = 4 SSYM.[11]

Recently, further progress has been made by the introduction of covariant supergraphs.[4] These supergraphs are useful for computing higher loops in Yang-Mills theory and in supergravity. The new rules for supergraph calculations start with the background field method and actions such as that of (3.12). However, instead of separating the covariant spinor derivative into an ordinary spinor derivative and a connection $\nabla_\alpha = D_\alpha - i\Gamma_\alpha$ we write the supergraphs in terms of covariant propagators

$$\Box^{-1}_{cov} = [\Box - iW^\alpha\nabla_\alpha - i\bar{W}^{\dot{\alpha}}\nabla_{\dot{\alpha}}]^{-1} \tag{3.12}$$

and covariant vertices such as $V\ \nabla^\alpha V\ \bar{\nabla}^2\nabla_\alpha V$ and do the D-algebra directly with the covariant spinor derivatives. We now expand

$$\Box^{-1}_{cov} = \Box^{-1} + \Box^{-1}(iW^\alpha\nabla_\alpha + i\bar{W}^{\dot{\alpha}}\bar{\nabla}_{\dot{a}})\Box^{-1} + \ldots \tag{3.13}$$

which gives us additional vertices, and, in pushing ∇_α's around a loop we must take into account commutators that arise, e.g. $[\nabla_\alpha, \Box] \neq 0$. The commutators generate additional W_α factors as well as covariant space-time derivatives $\nabla_{\alpha\dot{\alpha}}$. Eventually the algebra leads to expressions such as (2.15) or (2.16) with covariant ∇, $\bar{\nabla}$, for which the same rules apply. We must have exactly two ∇_α and two $\bar{\nabla}_{\dot{\alpha}}$ to get a nonzero result, and if we start with more than this we reduce the number by using $\{\nabla_\alpha, \bar{\nabla}_{\dot{\alpha}}\} = i\nabla_{\alpha\dot{\alpha}}$, generating this covariant space-time derivatives. We again do the θ-integration loop-by-loop and eventually we are led to x-space Feynman diagrams with factors $\nabla_a = \partial_a - i\Gamma_a$ in the numerator and propagators $\Box^{-1}=[\frac{1}{2}\nabla^a\nabla_a]^{-1}$, as well as explicit W_α, $\nabla_\beta W_\alpha$ external field factors, but no bare ∇_α that would give rise to pure Γ_α vertices. Thus, supergraphs computed in this fashion do not depend explicitly on the dimension $\frac{1}{2}$ spinor connection Γ_α but only on the dimension 1 vector connection Γ_a and therefore, again by simple power counting are more convergent. The calculations themselves are simpler especially since the number of graphs one must compute is smaller than in ordinary background field calculations.[4]

In the unconstrained field formalism of Gates and Siegel[11] supergraphs can also be used for calculations in supergravity. The technical details, rather complicated, are described for example in "Superspace".[3] For actual calculations the background field method is preferable and covariant D-algebra can also be used to great advantage.[4] Although the calculations are complicated they are considerably simpler than calculations using component Feynman graphs. We hope supergraph techniques will allow us to investigate some of the quantum properties of supergravity.

REFERENCES

1. A. Salam and J. Strathdee, Phys. Rev. D11 (1975) 1521,
   J. Honerkamp, M. Schlindwein, F. Krause and M. Scheunert, Nucl. Phys. B95 (1975) 397,
   S. Ferrara and O. Piguet, Nucl. Phys. B93 (1975) 261.
2. M. T. Grisaru, W. Siegel and M. Rocek, Nucl. Phys B159 (1979) 429.
3. S. J. Gates, M. T. Grisaru, M. Rocek and W. Siegel, "Superspace", Benjamin-Cummings, Reading, MA (1983).
4. M. T. Grisaru and D. Zanon, Phys. Lett. 142B (1984) 359,
   M. T. Grisaru and D. Zanon, "Covariant Supergraphs: I and II", Brandeis preprints, BRX-TH-163, 169 (1984).
5. L. F. Abbott and M. T. Grisaru, Nucl. Phys. B169 (1980) 415,
   W. E. Caswell and D. Zanon, Nucl. Phys. B182 (1981) 125,
   M. T. Grisaru, M. Rocek and W. Siegel, Nucl. Phys. B183 (1981) 141.
6. W. Siegel, Phys. Lett. 84B, (1979) 193, 94B (1979) 37.
7. O. Piguet and K. Sibold, "The off-shell infrared problem in N = 1 supersymmetric Yang-Mills theories", Université de Genève UGVA-DPT 04-423 (1984).

8. A. Galperin, E. Ivanov, S. Kalitzin, V. Ogievetsky and E. Sokatchev, "Unconstrained N = 2 matter, Yang-Mills and supergravity theories in harmonic superspace", Trieste preprint IC/84/43 (1984).
9. L. F. Abbott, Nucl. Phys. B185 (1981) 189.
10. L. F. Abbott, M. T. Grisaru and D. Zanon, Nucl. Phys. B244 (1984) 454.
11. M. T. Grisaru and W. Siegel, Phys. Lett. 110B (1982) 49.
12. W. Siegel and S. J. Gates, Nucl. Phys. B147 (1979) 77.

HOW CAN WE DECIDE IF NATURE IS SUPERSYMMETRIC?

Gordon L. Kane

Randall Laboratory of Physics
The University of Michigan
Ann Arbor, Michigan 48109

ABSTRACT

These lectures emphasize the possible connections of Supersymmetry to experiment. Particular concern is given to considering how (a) we would recognize evidence for supersymmetry if we saw it, and (b) if nature is not supersymmetric on the weak scale how we can establish that. Anomalous events with missing P_T at present and future colliders are discussed in some detail, including a derivation of the expected rates starting from the Lagrangian, giving Feynman rules, calculating cross sections, and considering experimental cuts. A few remarks are made about light Higgs bosons in supersymmetric theories. The emphasis throughout the lectures is pedagogical.

Introduction and Motivation

At a school where a great deal of the subject matter is related to supersymmetry, it is hardly necessary to discuss the motivation for pursuing the subject in much detail. In this matter as well as a lot of others, I will also rely on the recent review [1] with Haber, where an extensive discussion of much of the subject of these lectures occurrs [2].

The general motivation to study supersymmetry arises from a number of factors. It is a powerful gauge field theory, and may allow, through supergravity, a connection between gravity and he other forces in nature. More in line with our purposes is the possibility that supersymmetry can complement the standard model. Although the standard model gives a full description of all low energy interactions in nature up to a scale of about 100 GEV, as far as we are aware today, it is incomplete in many ways. Most clearly, we do not understand the physics of the Higgs sector, and how to make sense of the need for scalar bosons. If a light Higgs boson exists, it may even be a strong indication that nature is supersymmetric, since in the framework of supersymmetry scalar particles are incorporated into the fundamental theory without ad hoc interactions, and quantum corrections to their masses may not be large.

If nature is supersymmetric on the weak scale, we will observe both new particles, the superpartners of the normal particles, and also new interactions, since supersymmetry provides a new source of parity, charge conjugation and, perhaps, of CP violation, and predicts deviations from the standard model for $(g-2)\mu$, the ρ parameter, and other observables. There will not be time to cover all of these subjects here [in particular, I will not discuss the indirect results such as those just mentioned], and we will concentrate on new particle effects since there are a number of good ways to observe superpartners, both at present and at future machines.

Spectrum

Before we deal explicitly with the spectrum, it is useful to narrow down the possibilities by some general observations. There could be a number N of generators of supersymmetric transformations, some carrying additional quantum numbers besides raising or lowering the spin. However, if $N \geq 2$, there would be multiplets with a left-handed electron changed into a right-handed electron by lowering the helicity. The right-handed electron cannot be the usual one since e_L and e_R have different SU(2) properties and cannot be in the same Susy multiplet if Susy transformations commute with SU(2) ones. Consequently, if $N \geq 2$, the supersymmetry must be broken in stages, with the only remaining symmetry at the weak scale where SU(2) invariance holds being an $N = 1$ supersymmetry. For our purposes, it is sufficient to consider $N = 1$. If there is a higher supersymmetry, additional particles may exist besides those we consider, but all the ones we discuss will be present. So we consider the situation with one supersymmetry generator that does not carry any additional quantum numbers.

Then all superpartners have quantum numbers identical to those of their partners (except for spin), and the known particles cannot be partners of each other.

It is interesting to consider a possible exception to this situation. Could the neutral Higgs $H°$ be the neutrino superpartner $\tilde{\nu}$ which gets a vacuum expectation value $\langle\tilde{\nu}\rangle$? The answer is no, for several reasons:

(1) One needs two doublets in the theory for two reasons, (a) to cancel anomalies, and (b) to give mass to both up and down quarks and leptons.

(2) If one writes the most general hypercharge conserving Lagrangian under these assumptions, it gives an effective Hamiltonian which allows proton decay at typical weak interactive rates.

(3) When $\langle\tilde{\nu}\rangle \neq 0$ lepton number conservative is violated. That is not as excluded by data as one might imagine [3], but constraints do exist and probably $\langle\tilde{\nu}\rangle$ cannot be as large as 250 GEV.

Thus the spectrum of particles in a supersymmetric world requires a doubling, and gives a rich set of possibilites for experimental detection. The usual particles are

u_L, u_R, d_L, ...

e^-_L, e^-_R, μ^-_L, ...

ν_e, ν_μ, ...

g

γ

$W^\pm$, Z°

$$\begin{pmatrix} H_1^+ \\ H_1^0 \end{pmatrix}, \quad \begin{pmatrix} H_2^\circ \\ H_2^- \end{pmatrix}$$

where each helicity state is treated separately. Then the supersymmetric weak interaction eigenstates are the same states displaced by half a unit of spin, denoted by $\tilde{x}$.

spin 0: $\tilde{u}_L$, $\tilde{u}_R$, $\tilde{d}_L$, ----

$\tilde{e}^-_L$, $\tilde{e}^-_R$, $\tilde{\mu}^-_L$, ----

$\tilde{\nu}_e$, $\tilde{\nu}_\mu$, ----

spin 1/2: $\tilde{\gamma}$, $\tilde{g}$, $\tilde{w}^\pm$, $\tilde{z}$, $\tilde{H}_1^+$, $\tilde{H}_1^0$, $\tilde{H}_2^0$, $\tilde{H}_2^-$

Choosing a set of names which also denote properties, we will use an -ino suffix for the partners of gauge bosons--photino, gluino, wino, zino, higgsino. Instead of $\tilde{\gamma}$ and $\tilde{Z}$ we could have used the partners of W^3 and B, $\tilde{w}^3$ and b, wino and bino. The name is useful because it also tells how the states interact and what their quantum numbers are. For the partners of fermions, we use the names scalar-quarks, scalar-leptons, and scalar-neutrinos. For the partners of gauge bosons and Higgs, we use the collective names charginos when they are electrically charged, and neutralinos when they are electrically neutral.

The states listed above are weak-interaction eigenstates. We will see below that the usual electroweak symmetry breaking mixes them, so the physical mass eigenstates will be linear combinations of the weak eigenstates.

If supersymmetry were unbroken there would be a scalar-proton $\tilde{p} = ud\tilde{u} + u\tilde{d}u$, a femionic hydrogen atom made of $\tilde{e} + p$ and $e + \tilde{p}$, weak interactions mediated by $\tilde{w}$ and $\tilde{z}$, etc. Since $\tilde{e}$ and e, p and p, etc, would be degenerate, all these states and interactions would be easy to observe. None of them are observed. This means that either supersymmetry is not relevant to nature, or that supersymmetry is a broken symmetry and the masses of the superpartners are large compared to those of the ordinary particles.

We will, of course, assume that the masses of the partners are heavy. Although ways to break supersymmetry have been studied, there is so far no compelling procedure or pattern of masses, and we will simply treat the masses as arbitrary in the following. That is all right, as we will see how to search for superpartners over the entire range of masses up to the kinematic limits accessible at machines. Eventually, a search will be carried out over the full range relevant to supersymmetry if it is a symmetry to help understand the weak scale.

Coupling strengths

Although the masses are due to the symmetry breaking and are unknown, the couplings are mainly gauge couplings such as g_s, g_2, or e and are the same measured ones as for the normal particles. This is extremely important as it allows us to calculate cross sections, decay branching ratios, and interactions fully except for kinematical factors depending on the unknown masses. A full set of Feynman rules is given in ref 1, and we will exhibit some below.

Basically, every vertex of the standard theory converts into several in the supersymmetric theory, all with the same strength, obtained by replacing particles by their superpartners in pairs. For example,

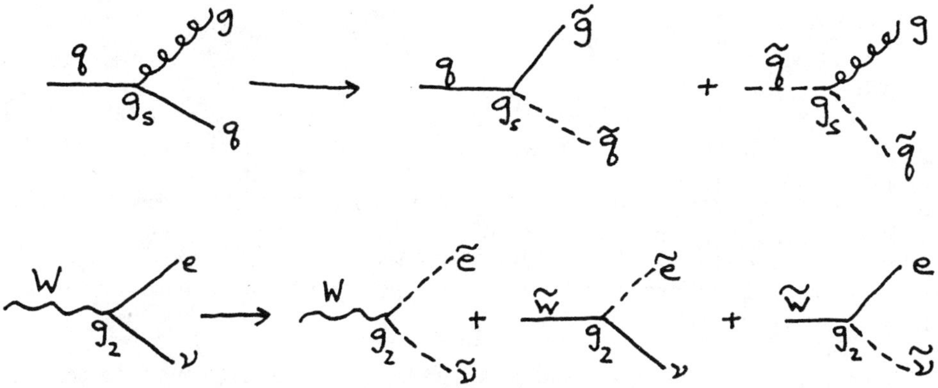

The space-time structure is always the minimal one. All vertices obtained in this way will be present, and up to an occasional factor of $\sqrt{2}$ they will be the right size, so one can think qualitatively about the perturbative predictions of supersymmetry by just drawing the appropriate diagrams. Cross sections, decays, etc., can be estimated without detailed knowledge of the Lagrangian. A few additional vertices occur in the full theory, mainly four-point functions or Higgs-scalar-quark couplings.

Mixing

In several sectors of the theory, it turns out that the weak eigenstates having all quantum numbers the same except for SU(2) will mix with one another, by SU(2) breaking. For example, in the scalar sector the left-handed scalar-electron $\tilde{e}_L$ is in an SU(2) doublet while $\tilde{e}_R$ is an SU(2) singlet. The subscripts L,R remind us of how $\tilde{e}$ interacts. So there is a term $H\tilde{e}_L\tilde{e}_R$ in the Lagrangian, which gives an off diagonal mass term when H^0 gets a vacuum expectation valve. In addition, supersymmetry breaking will give $\tilde{e}_L$ and $\tilde{e}_R$ masses, perhaps different ones (say L,R). Thus, the $\tilde{e}$ mass matrix will be of the form

$$\begin{pmatrix} L & g\langle H^0\rangle \\ g\langle H^0\rangle & R \end{pmatrix}$$

Diagonalizing this will give mass eigenstates $\tilde{e}_1,\tilde{e}_2$ whose couplings are mixtures of the electroweak doublet and singlet couplings.

Charginos

Similarly, the interaction Lagrangian for the charginos contains terms

$$L \sim \frac{ig_2}{\sqrt{2}} \left(H_1^0 \tilde{W}^+ \tilde{H}_1^- + H_2^0 \tilde{W}^- \tilde{H}_1^+ \right) + M\vec{\tilde{W}}\cdot\vec{\tilde{W}} - \mu\tilde{H}_1^+\tilde{H}_2^-.$$

The first two terms are normal interactions and give an off diagonal term in the wino-higgsino mass matrix when H_1°, H_2° get vacuum expectation valves. The latter two terms break supersymmetry softly (i.e., do not introduce quadratic divergences) and give masses to superpartners. We will see these same parameters M,μ below in the neutralino mass matrix, and they will play an important role in tests of the relevance of supersymmetry to understanding the weak scale.

One can form the chargino mass matrix, with diagonal terms M,μ and off-diagonal ones $g_2\langle H_1^\circ\rangle$ or $g_2\langle H_2^\circ\rangle$. Depending on the relative valves of M,μ, $\langle H_1^\circ\rangle$, $\langle H_2^\circ\rangle$ the chargino mass eigenstates could be any mixture of wino states (which couple like objects with weak isospin of one) and higgsino states (which couple as weak isodoublets).

Neutral Current Couplings

It is instructive to examine the neutral current couplings of possible chargino mass eigenstates. In the language we are using, a physical electron is a four-component spinor

$$e^- = \begin{pmatrix} e_L^- \\ e_R^{+*} \end{pmatrix}$$

constructed of two-component spinors with definite electroweak couplings. Similarly, here the physical particles will be states $X^\pm$ of the form

$$X^+ = \begin{pmatrix} a\tilde{w}^+ + b\tilde{h}^+ \\ c\tilde{h}^{-*} + d\tilde{w}^{-*} \end{pmatrix}$$

where a,b,c,d are determined by the mass matrix. The neutral current is

$$J_\mu^Z = T_\mu^3 - \sin^2\theta_W \, J_\mu^Q$$

$$T_\mu^3 = (V_\mu - A_\mu)\,\tau^3$$

$$J_{\mu^+}^Q = QV_\mu$$

In this form, the coupling for $Z^\circ \to e^+e^-$ is obtained from

$$T^3_\mu e^- = -\frac{1}{2}(V\mu - A\mu)$$

$$J^Q_\mu e^- = -V_\mu$$

$$T^3_\mu e^+ = 0$$

$$J^Q_\mu e^+ = +V_\mu$$

so (remembering $V^*_\mu = -V_\mu$, $A^*_\mu = A_\mu$) we get

$$J^Z_\mu e = \left(-\frac{1}{2}(V_\mu - A_\mu) + \sin^2\theta_W V_\mu\right) + \left(+\sin^2\theta_W V\mu\right)$$

$$= \left(-\frac{1}{2} + 2\sin^2\theta_W\right) V_\mu + \frac{1}{2} A_\mu$$

$$= g_V V_\mu + g_a A_\mu$$

For $X^\pm$ the eigenvalves of T^3, Q are

$$T^3_\mu \tilde{w}^\pm = \pm (V_\mu - A_\mu)$$

$$T^3_\mu \tilde{h}^\pm = \pm \frac{1}{2} (V_\mu - A_\mu)$$

$$J^Q_\mu \tilde{w}^\pm = \pm V_\mu$$

$$J^Q_\mu \tilde{h}^\pm = \pm V_\mu$$

For any mass matrix one can write the neutral current couplings. It is interesting to note that they can be quite large. With $\sin^2\theta_W = 1/4$ for simplicity, one has $g^e_V = 0$, $g^e_a = -1/2$, while for some reasonable a,b,c,d one can have $g_V, g_a \gtrsim 1$. Then $g^2_V + g^2_a$ can be much larger for $\tilde{w}^\pm$ or $\tilde{h}^\pm$ than for electrons.

Another interesting property of the chargino mass matrix emerges if we consider what happens when the supersymmetry breaking is such that $M \cong 0$ or $\mu \cong 0$. Then the determinant of the chargino mass matrix is just the product of the off-diagonal entries, but they are determined by Higgs vacuum expectation valves, so we know something about them. In the simple case where the two vev's are equal, the determinant of the mass matrix is just M_W^2. Since the determinant is also the product of the eigenvalves, $M_+M_- = M_W^2$. Then one of the eigenvalves is

$$M_- \leqslant M_W,$$

so we can expect to find a chargino lighter than the W. However, if both M, μ are non-negligible compared with M_W this is no longer true, so it is a model dependent result.

Neutralinos

There are at least four neutralinos, which can be chosen to be $\tilde{B}$, $\tilde{W}^0$, $\tilde{H}_1^0$, $\tilde{H}_2^0$ or $\tilde{Z}$, $\tilde{\gamma}$, $\tilde{H}_1^0$, $\tilde{H}_0^2$. The Lagrangian will contain terms

$$M\ \vec{\tilde{W}}\cdot\vec{\tilde{W}} - \mu\ \tilde{H}_1^+\ \tilde{H}_2 + M'\tilde{B}\tilde{B}$$

where M, μ are the same as for the charginos, plus ordinary SU(2) breaking terms. Then choosing a basis

$$\psi^T = (\tilde{B}, \tilde{W}^0, \tilde{H}_1^0, \tilde{H}_2^0)$$

the mass matrix is

$$-\frac{1}{2}\psi^T \begin{pmatrix} M' & 0 & M_Z s_V s_W & -M_Z c_V s_W \\ 0 & M & -M_Z s_V c_W & M_Z c_V c_W \\ M_Z s_V s_W & -M_Z s_V c_W & 0 & -\mu \\ -M_Z c_V s_W & M_Z c_V c_W & -\mu & 0 \end{pmatrix}$$

where $s_W = \sin\theta_W$, $c_W = \cos\theta_W$, $s_V = \sin\theta_V$, $c_V = \cos\theta_V$, $\tan\theta_V = v_1/v_2$. The zeros in the upper left hand part come because SU(2) invariance does not allow a $\tilde{W}\tilde{B}$ term, and the zeros in the lower right hand part because a diagonal mass term from a doublet would be of the form $\varepsilon_{ij}\, H_i\, H_j$ which vanishes by the antisummetry of ε_{ij}. To reduce the number of parametors we note that in a grand unified theory the B and W° couplings are related, which gives $M' = \frac{5}{3} g'^2\, M/g^2$, and we will assume that some such relation holds. Also, one can vary v_1/v_2, but presumably not too far from unity, so we will assume $v1/v2 \simeq 1$ (but in any practical application one must check). Then M, μ are the remaining parameters and one can study the neutralino masses as a function of M, μ.

One particularly important consequence that emerges is that the two lightest of the neutralinos have masses much smaller than M, μ. It happens because the determinant of the mass matrix is (with $v_1 = v_2$).

$$M_1 M_2 M_3 M_4 = \mu M M_W^2 \tan^2\theta_W \left(1 - \frac{5}{3}\frac{\mu M}{M_W^2}\right).$$

Then if $\mu \simeq M \simeq M_W$, some of the eigenvalves must be small compared to M_W. The range of m_1+m_2 for μ,M up to 800 GeV is shown in Fig. 1.

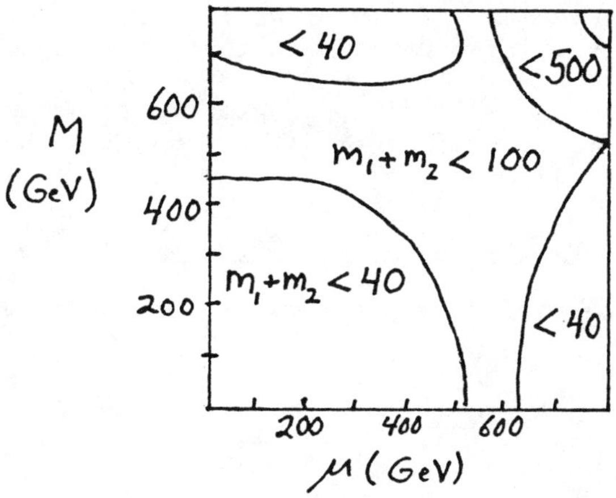

Figure 1.

We will use this later when we discuss neutralino production.

R-parity

In a normal supersymmetric theory there is a conserved quantum number called R-parity. It is a discrete symmetry that almost always remains even after all supersymmetry breaking, giving a multiplicatively conserved quantum number. All ordinary particles have R=+1, and all supersymmetric partners are assigned R=-1. Formally, one can define the R-parity of any particle of spin j, baryon number B, and lepton number L to be

$$R=(-1)^{2j+3B+L}$$

Then several important practical consequences follow. (1) Super partners are always produced in pairs if one starts from normal particles. (2) There will be a superpartner in the decay of any superpartner. (3) The lightest superpartner will be stable. These results greatly affect how we look for superpartners in practice.

The lightest superpartner?

What is the lightest superpartner [4]? Presumably by cosmological and abundance arguments, it is not charged. It could either be the scalar-neutrino ($\tilde{\nu}_e$, $\tilde{\nu}_\mu$, $\tilde{\nu}_\tau$), or the lightest of the neutralinos. Although people often speak of the lightest neutralino as the photino, in practice that need not be the case, and its coupling could easily be that of a Higgsino or intermediate. Cosmological restrictions force the lightest neutralino to be in the GeV range, while they do not restrict how light the scalar-neutrino can be, so there may be some prejudice in favor of the latter. Models, on the other hand, tend to have the neutralino lightest. For our purposes it will not matter much--we will generally call the lightest state a photino, but it should be kept in mind that it might well be otherwise.

Photino (or lightest superpartner) interactions [5]

Suppose a pair of superpartners are produced, e.g.

$$e^+e^- \rightarrow \tilde{\ell}^+\tilde{\ell}^-$$

$$\bar{p}p \rightarrow \tilde{g}\tilde{g}X$$

Then the produced superpartners decay, e.g.

$$\tilde{\ell} \rightarrow \ell\tilde{\gamma}$$

$$\tilde{g} \rightarrow q\bar{q}\tilde{\gamma},$$

giving normal particles plus photinos. Then the key to detecting supersymmetry is to understand $\tilde{\gamma}$ (or $\tilde{\nu}$,...) interactions. In general the answer depends somewhat on masses and details, but basically we can give a useful answer which qualitatively covers the situations of interest.

The incoming $\tilde{\gamma}$ will hit a hadron in the target and excite it. In the parton model approximation,

$$\sigma(s) = \sum_q \int dx\, q(x)\, \hat{\sigma}(\hat{s})$$

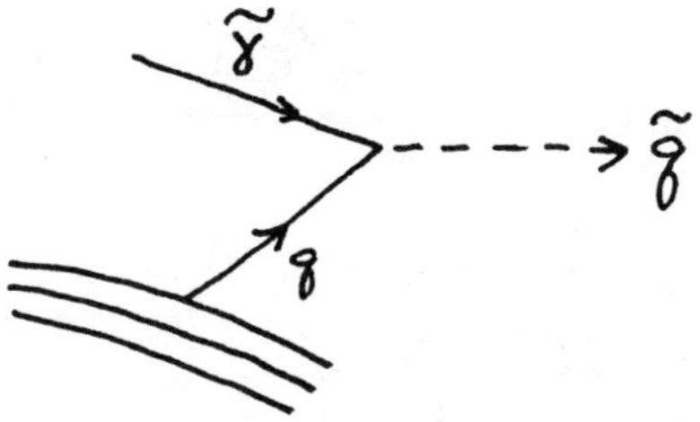

The $\tilde{\gamma}$ of momentum k and proton have a c.m. energy squared s, and the $\tilde{\gamma}$ and quark q a c.m. energy squared $\hat{s}$. The quark has a momentum q=xp, carrying a fraction x of the proton momentum p. The quark structure function q(x) is a measured function telling the probability of the quark carrying a momentum fraction x. Let us assume for simplicity that a real scalar-quark $\tilde{q}$ of momentum $\tilde{p}$ and mass $\tilde{m}$ is excited. Then

$$\hat{\sigma} = \frac{(2\pi)^4}{4(\hat{s}/2)} \overline{|M|^2}\, \delta^4 (\tilde{p} - q - k)\, d^3\tilde{p}/(2\pi)^3\, 2\tilde{E}$$

$$= \frac{\pi}{\tilde{s}} \overline{|M|^2}\, \delta (\tilde{p} - \tilde{m}^2),$$

and $\tilde{s} = \tilde{p}^2 = (k+q)^2 \approx 2\, x\, k\cdot q = xs$ neglecting the initial quark and photino masses for simplicity. The matrix element is

$$M = \sqrt{2}\, e_q e\, \overline{u}\, P_R\, V$$

where P_R is a right handed projection operator, $\overline{u}$ and v the quark and photino spinors, and e_q the quark charge. We will see below how to derive this; one could look it up in the Feynman rules of ref 1. If one simply used $\overline{u}v$ or $\overline{u}\gamma_5 v$ the qualitative answer would be the same.

Then

$$\frac{1}{4} \sum_{\text{spins}} |M|^2 = \frac{e_q^2 e^2}{2} \text{Tr}\, (\not{k}\, \not{q}\, P_R) = e_q^2 e^2\, k\cdot q = e_q^2 e^2\, \tilde{m}^2/2$$

$$\hat{\sigma} = \frac{\pi}{2} \frac{e_q^2 e^2 \tilde{m}^2}{\hat{s}}\, \delta\, (sx - \tilde{m}^2)$$

$$\sigma = \sum_q \int dx\, q(x)\, \frac{1}{x}\, \delta\,(x-\tilde{m}^2/s)\, \frac{\pi}{2}\, \frac{e^2 e_q^2 \tilde{m}^2}{s^2} = \frac{2\pi^2\alpha}{\tilde{m}^2} \sum_q e_q^2\; x\, q(x)$$

where we have assumed that all the quarks that contribute have degenerate $\tilde{m}^2$ so we can factor it out. The sum is just the structure function $F_2(x)$ from deep inelastic process, so finally the photino interaction cross section is

$$\sigma \simeq \frac{2\pi\alpha}{\tilde{m}^2} F_2\,(\tilde{m}^2/s).$$

How big is this? For $\tilde{m} > 20$ GeV, and $\tilde{m}^2/s \ll 1$, we can look up F_2 (x) and find $F_2 \lesssim 0.15$. Then $\sigma \lesssim 10^{-32}$ cm^2. Since this is below 10^{-6} of a typical total cross section, we see that normally photinos (or $\tilde{\nu}$) simply escape a detector. The usual signature of supersymmetry is a particle which escapes detectors.

In practice the above estimate is probably an overestimate. The full calculation involves a number of diagrams,

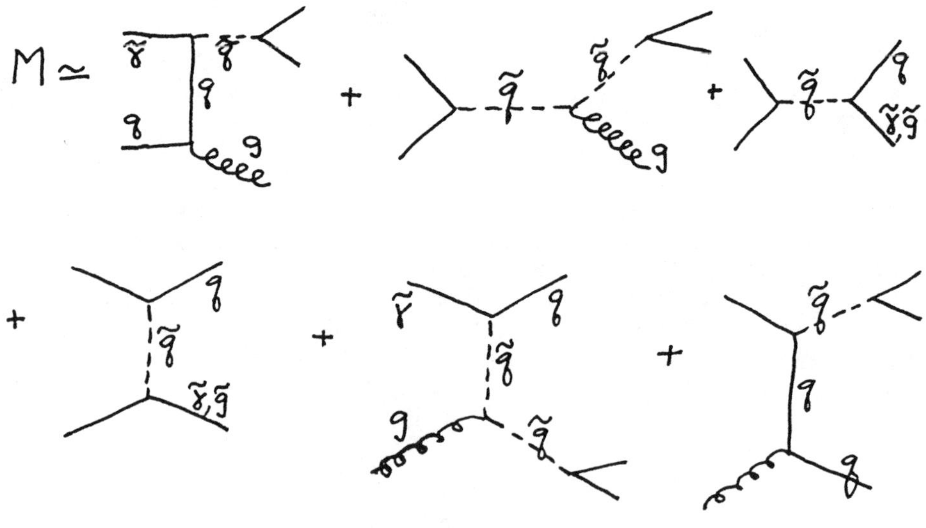

Figure 2.

and one finds

$$\sigma \simeq 2\text{x}10^{-37}\, \tilde{E}_\gamma\, \left(\frac{m_W}{\tilde{m}_q}\right)^4 \tilde{F}\ \text{cm}^2$$

where diagrams c and d give

$$\tilde{F} \simeq \sum_q \int_{\tilde{m}_g^2/s}^{1} e_q^2 \, dx \, q(x)(1 - \tilde{m}_g^2/xs)^2 \; (1 + m_g^2/8xs)$$

and the others have not been included in any analysis yet.

One sees that $\sigma_{\tilde{\gamma}} \gtrsim \sigma_\nu$ so photinos (or scalar-neutrinos) would interact in typical neutrino detectors but not in typical collider detectors. If they were detected the characteristic x and y distributions are different from those of neutrinos, and one could distinguish.

Decays of Superpartners and present mass limits

To understand how supersymmetric particles might have been detected already if their masses are in certain ranges, and how they might be detected at future machines, we want to consider each partner, note its dominant decays, and see how it might have been produced and detected. This is a major part of the subject matter of ref 1, and here we will just summarize the results. Even more recent results and limits at e^+e^-colliders are summarized by Chen [6].

Scalar-neutrinos

If $\tilde{\nu}$ are not the lightest supersymmetric partners, they will decay into others. They decays have been studied in ref. 7. Possible decays are shown in fig. 2, with additional diagrams being possible.

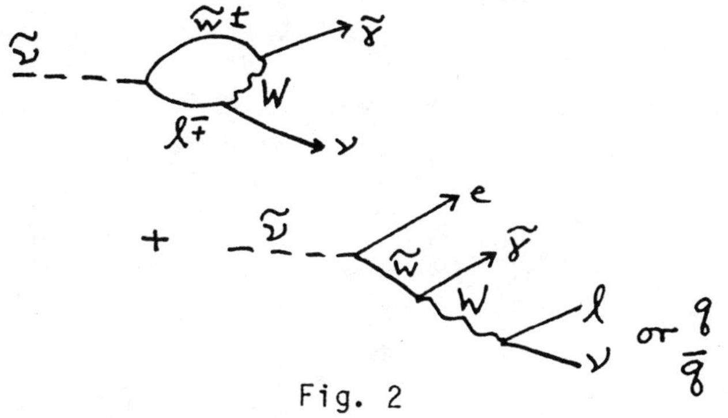

Fig. 2

From the calculation of ref. 7 we can define an effective effective $\tilde{\nu}\tilde{\gamma}\nu$ vertex,

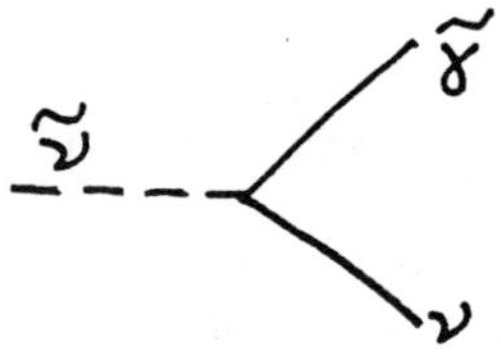

$g_{eff}\ \overline{u}P_R u$

with

$$g_{eff} \simeq \frac{g_2^2 e}{16\sqrt{2}\pi} F(\tilde{m}_\nu, \tilde{m}_\gamma).$$

F is a function defined in ref. 7; for a large range of relevant masses, F≃1/4, so $g_{eff} \simeq 5\times10^{-4}e$. This gives a decay width of order $\Gamma \simeq 10^{-10}\ \tilde{m}_\nu$ so a decay lifetime of $\tilde{\tau} < 10^{-15}$ sec, too short to ever be detected directly. (The same numbers approximately describe $\Gamma(\tilde{\gamma}\to\tilde{\nu}\nu)$ if $\tilde{m}_\gamma > \tilde{m}_\nu$, since the same effective vertex describes the decay.)

Possible flavor effects should also be kept in mind. If $m(\tilde{\nu}_\mu) > m(\tilde{\nu}_e)$, $\tilde{\nu}_\mu \to \mu e \tilde{\nu}_e$.

At present the only limits arise from the absence of any deviations from the standard model predictions for τ or μ decay. If $\tilde{\nu}$ masses were small enough, the decay

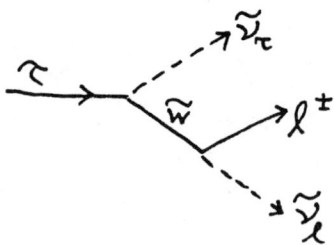

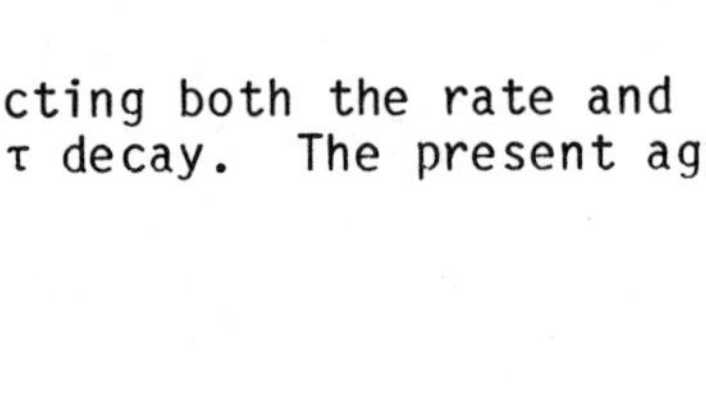

would occur, affecting both the rate and the lepton energy distribution for τ decay. The present agreement of these with the standard model predictions can set restrictions [8], which very approximately are

$$\tilde{m}_{\nu_\tau} + \tilde{m}_{\nu_\ell} \gtrsim m_\tau \qquad \ell = e,\mu.$$

Similarly, from the absence of $\mu \to e\tilde{\nu}_\mu\bar{\tilde{\nu}}_e$ we expect $\tilde{m}_{\nu_\mu} + \tilde{m}_{\nu_e} > m_\mu$.

One can produce $\tilde{\nu}$ various ways, such as (1) $e^+e^- \to \tilde{\nu}\bar{\tilde{\nu}}$, (2) $e^+e^- \to \gamma\tilde{\nu}\bar{\tilde{\nu}}$, (3) $Z^0 \to \tilde{\nu}\bar{\tilde{\nu}}$, (4) $W \to \ell\tilde{\nu}$. To observe a signature in the first of these one would need to get a one-sided event where one $\tilde{\nu}$ escaped or decayed into $\tilde{\gamma}\nu$ which escape, while the other decayed to $\ell^+\ell^-\tilde{\gamma}\nu$ or $\tilde{\gamma}\ell q\bar{q}$. The second, "photon-counting", we will discuss below. The third can increase the rate for Z° to give undetected decays, or give a signature such as the first. The last will give a single $\ell^\pm$ plus $\not{p}_T$ as a signature, and may be detectable.

Scalar-lepton

If $\tilde{\ell}^\pm$ are produced they will decay rapidly via $\tilde{\ell}^\pm \to \ell^\pm\tilde{\gamma}$. In $e^+e^- \to \tilde{\ell}^+\tilde{\ell}^-$, events would occur with missing energy from the escaping photinos, and an acolinear, acoplanar e^+e^- or $\mu^+\mu^-$ (but not μe) pair. There are no such events detected, so $\tilde{m}_{\ell^\pm} > 17$GeV for $\ell = e,\mu,\tau$; if nothing is detected the limits will go up to 23 Gev at PETRA. For $\tilde{e}^\pm$ a stronger result is obtainable.

Photon-counting

One expects $e^+e^- \to \nu\bar{\nu}$ to occur at present day colliders, via Z^0 and W contributions. But there is nothing to observe. It was emphasized some time ago [9] that one could observe $e^+e^- \to \gamma\nu\bar{\nu}$ where the γ was radiated off an initial $e^\pm$, with an approximate rate

$$\frac{d\sigma}{dx_\gamma d\cos\phi_\gamma} \simeq \frac{2\alpha}{\pi}\,\frac{\sigma(e^+e^- \to \nu\bar{\nu})}{x_\gamma \sin^2\theta_\gamma}$$

where θ_γ is the angle of the photon, and $x_\gamma = 2E_\gamma/\sqrt{s}$.

Clearly, the presence of any other light particle which escapes collider detectors and is coupled to W or Z gives a similar contribution. Thus one can have [10]

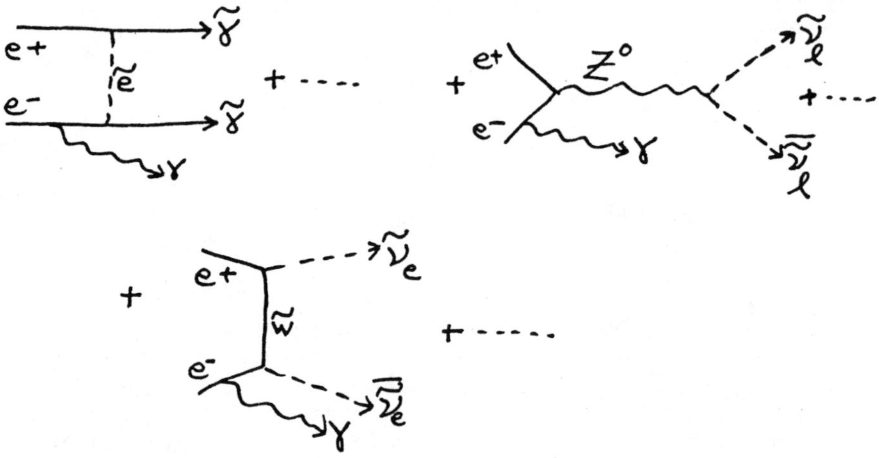

Two experiments are currently underway [11] at PEP to look for the rate of single hard γ's. Note that $\nu\bar{\nu}$ and $\tilde{\nu}\bar{\tilde{\nu}}$ will track the Z as the energy changes, as will any higgsino component in the so-called photino (see discussion above). If $\tilde{m}_W/m_W = 1/2$, $\sigma(e^+e^-\to\tilde{\nu}\bar{\tilde{\nu}})/\sigma(e^+e^-\to\nu\bar{\nu}) = 4$.

If there are no candidates a limit is obtained, such as shown in figure 3.

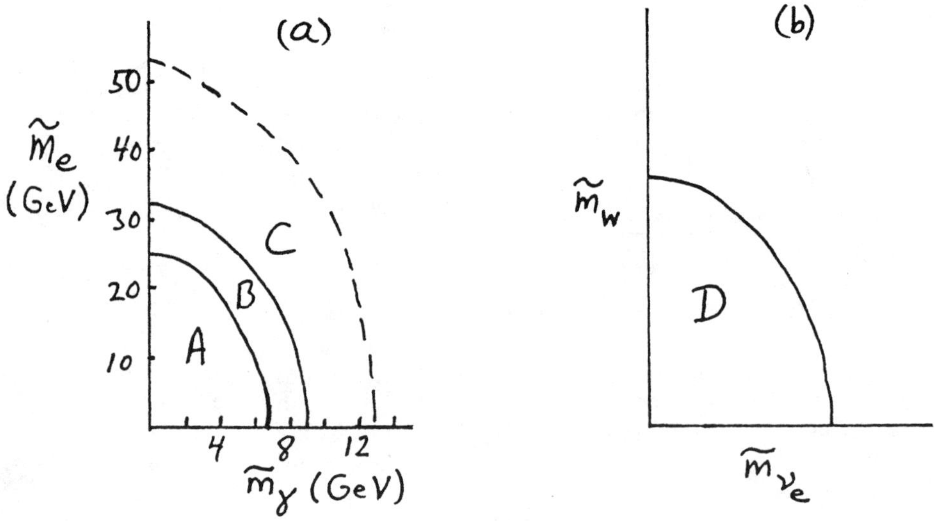

Figure 3.

In part a, section A shows approximately the region excluded by the present data from the MAC collaboration; if $\tilde{e}_R$ and $\tilde{e}_L$ have the same mass the excluded region is A+B. If the masses are within region B+C, a signal should be observed within the next year or two. The same data could be interpreted as in fig 3b, with an excluded region D. Since the analysis has not yet been done for part b, I will not put numbers on, but they would be about the same as the numbers in part a.

<u>Charginos</u>

A full treatment of charginos must take the mixing of winos and higgsinos into account, with their very different couplings. For simplicity here I will mainly discuss the wino part, but ref 1 and original literature should be consulted for a complete discussion.

Winos can have a number of important decays, depending on the other superpartner masses. Some of them are shown in fig 4.

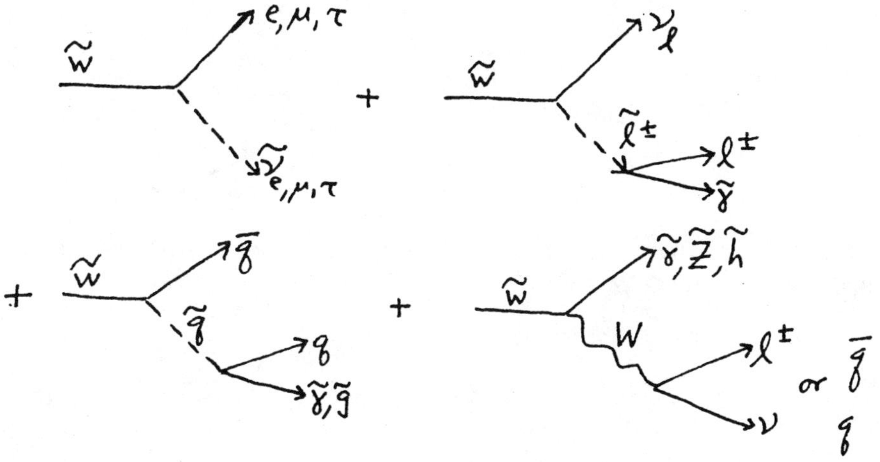

Any two body decays which are kinematically allowed, such as $\ell\tilde{\nu}_\ell$, $\nu_\ell\tilde{\ell}$, $q\tilde{q}$, would dominate.

Limits on $\tilde{w}$ come from $e^+e^-\to\gamma\tilde{\nu}_e\bar{\tilde{\nu}}_e$ which was discussed above, from the absence of $t\to\tilde{w}\tilde{b}$ (see below), and from $e^+e^-\to\tilde{w}^+\tilde{w}^-$,

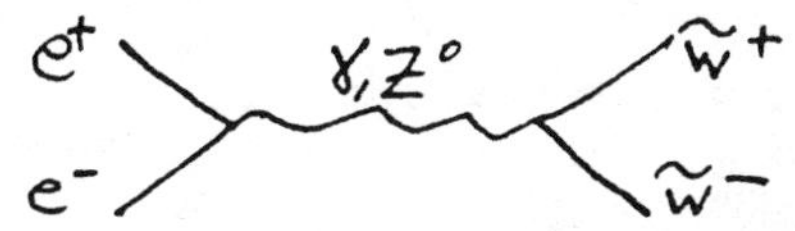

The γ coupling is just the charge, and we derived the Z° couplings of winos earlier. The limits will depend on the decay modes. If $\tilde{m}_\gamma \lesssim 4$ GeV, the MARK J detector has published limits [6], so long as $\tilde{w}$ decays with a significant leptonic branching ratio, of $\tilde{m}_{\chi^\pm} > 17$ GeV.

Eventually detection or strong limits will come from future CERN collider experiments or FNAL (see below). If there is a changino with mass below $m_Z/2$, the large neutral current couplings for most of the mass eigenstates will allow detection and extensive study on the Z°.

Neutralinos

We have discussed above the ways in which neutralinos mix to give mass eigenstates. At present there are no limits on the existence of neutralinos which constrain the theory. The lightest neutralino is probably stable and escapes collider detectors, as explained above. The other neutralinos will mainly decay by the modes

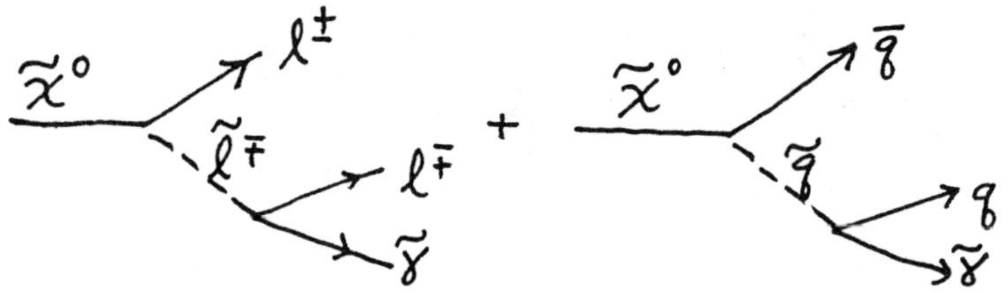

Which mode(s) dominate depends on the masses, both explicitly and through the missing angles coming from the mass matrix. As before, we write $\tilde{\gamma}$ for the lightest neutralino though the state in question may not couple like a photino.

The best way to search for neutralinos is by their associated production, $f\bar{f}\to\tilde{\chi}_2\tilde{\chi}_1$, where $\tilde{\chi}_1$ escapes and $\tilde{\chi}_2$ decays as above, giving a one-sided event. We will discuss them explicitly below.

Scalar-quarks

If scalar quarks are kinematically able to decay via $\tilde{q}\to q\tilde{g}$ without considerable phase space suppression, that mode will dominate. Since it does not give a useful signature, detection and limits are both difficult if $\tilde{m}_g<\tilde{m}_q$.

If $\tilde{m}_q < \tilde{m}_g$, then presumably $\tilde{q}\to q\tilde{\gamma}$ dominates, although in some cases other modes such as $\tilde{t}\to\tilde{b}\tilde{w}$ could be important. In many models the contribution to scalar-quark masses from supersymmetry breaking is flavor independent, and most scalar-quarks are approximately degenerate, in which case $\tilde{q} \to q\tilde{\gamma}$ will dominate.

A complicated analysis, based on the absence [12] of scalar-quarkonium states in the few GeV range, the absence of events with a coplanar jet pairs and missing energy at PETRA [13], and isospin invariance, probably implies that $\tilde{m}_q > 17$ GeV for both Q = 2/3 and Q = -1/3 quarks if $\tilde{m}_q < \tilde{m}_g$. Analysis of the $S\bar{p}pS$ UA1 data [14] probably implies [15] $\tilde{m}_q \gtrsim 40\text{-}50$ GeV if $\tilde{m}_q < \tilde{m}_g$. If $\tilde{m}_q > \tilde{m}_g$ there are no published restrictions on larger m_q; the absence of scalar-quarkonium states only implies [12] $\tilde{m}_q \gtrsim 3$ GeV for Q = 2/3.

Given that there is a heavy quark observed in a semileptonic decay, such as $t \to b\ell\nu$, one can conclude that $m_t < \tilde{m}_W + \tilde{m}_b$, $m_t < \tilde{m}_t + \tilde{m}_\gamma$, $m_t < \tilde{m}_g + \tilde{m}_t$, $m_t < \tilde{m}_t + \tilde{m}_Z$ since otherwise one of the two body decays would have dominated.

Gluinos

Since gluinos are color octets they couple with large Clebsch-Gordan coefficients to quarks and gluons, and can be produced copiously in hadronic collisions. This allows strong limits to be placed on their masses if they are not detected.

If the lightest neutralino is lighter than the gluino, the latter will always decay via

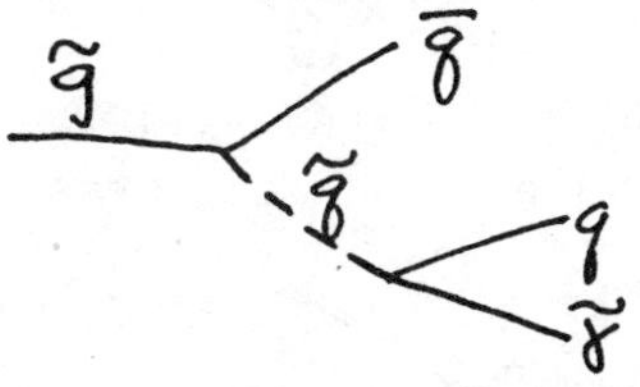

Detailed study of limits on $\tilde{m}_g$ has been done in ref. 1 and 16. Using various arguments one can conclude that if $\tilde{g}$ decays it would have been detected if $\tilde{m}_g < 2\text{-}5$ GeV(depending on its lifetime, which in turn depends on $\tilde{m}_q$). The recent CERN $S\bar{p}pS$ UA1 data on monojet events allow one to conclude that if $\tilde{g} \to q\bar{q}\tilde{\gamma}$ than $\tilde{m}_g \gtrsim 40$ GeV [15].

If $\tilde{g}$ were stable (or very long-lived) these limits probably still apply. The apparent absence of gluinoball states in $\psi \to \gamma\chi$ implies $\tilde{m}_g \gtrsim 1$ GeV even for stable $\tilde{g}$. Since $\tilde{g}$ would be shielded by gluons and form color singlet hadrons, they spend most of their life in a bound state and are off-mass shell, so that even massless $\tilde{g}$ can decay [17]. The lifetime is lengthened, but the decay still probably occurs rapidly enough that effects would have been observed. At present the only window open for gluinos seems to be one where gluinos are heavier than about a GeV, and are considerably lighter than photinos (of order a few GeV) so they form heavy, neutral, stable hadrons. When the Υ decays $\Upsilon \to \gamma\chi$ are fully analyzed for extra states a limit $\tilde{m}_g > 4$ GeV will be obtained [18] for stable $\tilde{g}$ if no signal is found, and experiments looking for heavy, neutral, stable hadrons at the FNAL Tevatron will be able to detect them up to a Kinematic limit of 5-10 GeV. Eventually collider experiments with time-of-flight capability can look for much heavier stable neutral hadrons.

$\tilde{g} \to g\tilde{\gamma}$

Partly for pedagogical purposes, to see the rich structure of supersymmetric theories, it is interesting to discuss the decay [19] $\tilde{g} \to g\tilde{\gamma}$. Since all three particles are odd eigenstates of charge conjugation, the decay appears to be forbidden; indeed, it does not appear at tree level in the theory.

However, at one loop it is induced, in an amount which depends on the mass matrices. The process

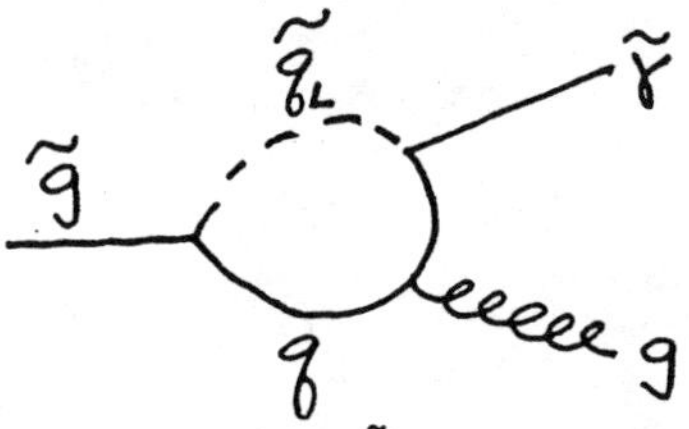

contributes, where we write $\tilde{q}_L$ for the scalar-quark--in fact we mean here the associated mass eigenstate nearest to $\tilde{q}_L$. So does

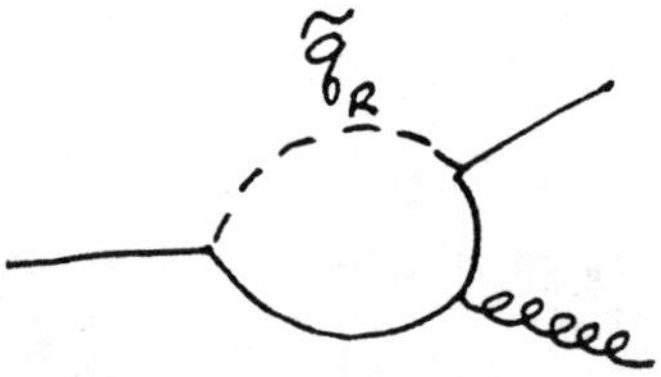

contribute. If the theory were invariant under parity then these two (gauge-invariant) contributions would cancel, and we would have $\tilde{g} \not\to g\tilde{\gamma}$.

But the electroweak contributions to the mixing of scalars guarantee that $\tilde{m}_L \neq \tilde{m}_R$, so P and C are separately violated (the Lagrangian is CP invariant so the P,C violations cancel to give no CP violation). Then necessarily $\Gamma(\tilde{g} \to g\tilde{\gamma}) \neq 0$. The size of $\Gamma(\tilde{g} \to g\tilde{\gamma})$ is model dependent since it is a function of the entries in the scalar-quark mass matrix, which in turn depends on the supersymmetry breaking. Computing the above diagram, the branching ratio comes out to be

$$BR(\tilde{g}\to g\tilde{\gamma}) \simeq \sum_{\text{flavors}} \frac{\alpha_s}{4} \frac{(\tilde{m}_R^2 - \tilde{m}_L^2)^2}{\tilde{m}_R^4 + \tilde{m}_L^4},$$

where one has to sum over all quark flavors in the loops. Duncan [20], and Suzuki [21], have shown that the absence of parity-violation in nuclei restricts $\tilde{m}_R - \tilde{m}_L$ for u,d quarks to be rather small, but there are no restrictions for heavier flavors. Typically in models the above branching ratio is of order 1%, and it can be as large as 10%.

When $BR(\tilde{g}\to g\tilde{\gamma})$ is large it can provide a useful signature for gluinos, since it gives a two body decay with large P_T. If $\tilde{g}$ are ever detected, the size of this BR will provide a useful constraint on models.

Promising Ways to Look for Superpartners

A. The Z° Width

Almost all superpartners couple to the Z° and would appear in its decay if they were sufficiently light. Possible pairs are $\tilde{\ell}^+\tilde{\ell}^-$, $\tilde{q}\tilde{q}$, $\tilde{\nu}\tilde{\nu}$, $\tilde{w}^+\tilde{w}^-$, $\tilde{h}^+\tilde{h}^-$, ... When mixing is present we have seen above how to write the neutral current couplings. The Feynman rules for all couplings are given in ref. 1.

To estimate the effect on the Z° width of such additional channels, we have to include separately the partners of the fermions and the gauginos. Each scalar partner pair gives a contribution

$$\Gamma/\Gamma_{SM} = 1 + \frac{2}{4}\left(1 - 4\tilde{m}^2/m_Z^2\right)^{3/2}$$

where Γ_{SM} is the standard model width of the Z°, the 1/4 is due to spin suppression of 1/4 for spin zero states, the 2 comes from having 2 scalar partners (e.g. $\tilde{L}$,R), and the remaining factor is a phase space correction. Using $\tilde{m}$ = 20 GeV as a reasonable lower limit, we get a possible increase $\Gamma/\Gamma_{SM} \leqslant 1.3$. For the gauginos, depending on mixing, we could have (using the couplings we derived earlier)

$$\Gamma_{\tilde{w}^+\tilde{w}^-}/\Gamma_{\ell^+\ell^-} \leq \frac{(2 - 2\sin^2\theta_w)^2}{(\frac{1}{2} - 2\sin^2\theta_w)^2 + \frac{1}{4}} = 10$$

which gives

$$\Gamma_{\tilde{w}^+\tilde{w}^-} \leq 0.3\ \Gamma_{SM}.$$

A phase space correction is s-wave here and rather small.

Thus, finally we could get effects as large as

$$\Gamma_{susy} \leq 1.6\ \Gamma_{SM}$$

where Γ_{susy} includes Γ_{SM} + additional supersymmetric contributions. To reach this 60% increase in $\Gamma_{Z°}$, all the partners would have to be light, and all mixings favorable. Probably the real correction would be rather small. In any specific model it can be explicitly calculated.

B. <u>One-Sided Events</u>

1. $W \to \tilde{w}\tilde{\gamma}$

A number of authors [25] have studied the possibility of finding the decay $W^{\pm} \to \tilde{w}^{\pm}\tilde{\gamma}$ (as usual, we denote $\tilde{w}, \tilde{\gamma}$ by the simplest names but the mass eigenstates could couple differently). The rate could be large, with

$$\Gamma(W \to \tilde{w}\tilde{\gamma})/\Gamma(W \to e\nu) \leq 1/2.$$

If $\tilde{w} \to \ell\nu\tilde{\gamma}$ the event has a simple one-sided signature as shown,

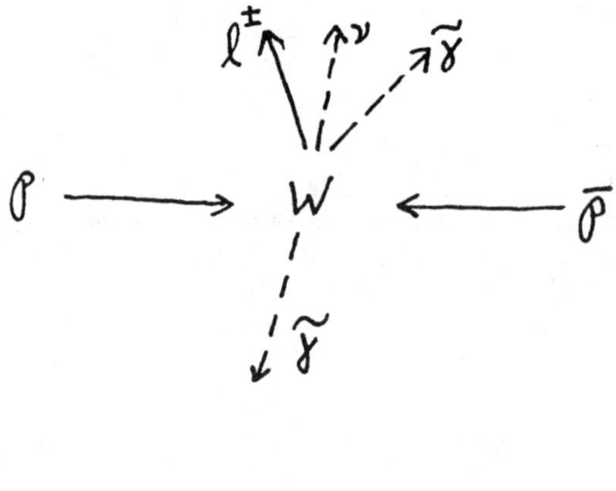

since only $\ell^{\pm}$ would be detected. However, the background from $W \to e\nu$, $W \to \tau\nu \to e\bar{\nu}\bar{\nu}\nu$, and from semileptonic decays of c, b is large, and it would require very good statistics to detect such a signal.

The prospects may be better from $\tilde{w} \to q\bar{q}\tilde{\gamma}$. The event is again one-sided since $q+\bar{q}$ will be in one hemisphere,

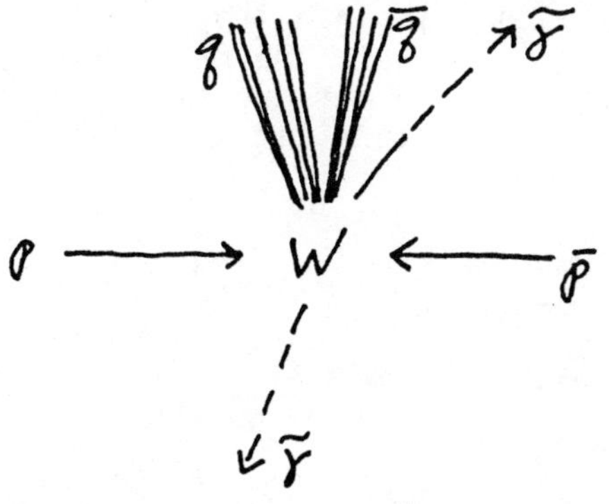

and the background may be less, especially if one can veto events with one or two hard $\ell^{\pm}$ and missing ν. There is also background from $\bar{p}p \to Z°gX$, $Z° \to \nu\bar{\nu}$, since the $q+\bar{q}$ above could appear as a single fat jet; probably this could be separated because the gluons give a jet of low effective mass, while the mass of the $q+\bar{q}$ would be of order $2/3\ \tilde{m}_W$. The most difficult background may be $pp \to W^{\pm}gX$, $W^{\pm} \to \tau^{\pm}\nu$, $\tau^{\pm} \to$ hadrons, giving two jets ($g+\tau$) and $\not{p}_T$. So far no estimates of signal/background have been done for the $q\bar{q}\tilde{\gamma}$ signature.

In many models one chargino is lighter than the W (as discussed in the section on the chargino mass matrix) and $m_W > m_{\tilde{w}} + m_{\tilde{\gamma}}$ so $\tilde{w}\tilde{\gamma}$ occurs at a useful rate. When many W's are available, at CERN in 1985 or 1987 and at FNAL in 1987-1988, it may be possible to detect $W \to \tilde{w}\tilde{\gamma}$.

2. Associated Production of $\tilde{z}\tilde{\gamma}$ at e+e- or Hadron Colliders [23,24]

If supersymmetric partners are not found, eventually this process will provide the most likely way, and ultimately it will provide the most definitive test of whether supersymmetry is relevant to understanding the weak scale.

Consider the process

$$f\bar{f} \to \tilde{z}\tilde{\gamma}$$

where $f\bar{f} = q\bar{q}$ or e^+e^-, and $\tilde{z}$ and $\tilde{\gamma}$ are two neutralinos, probably the two lightest ones at a machine limited by the available energy to produce new particles, and perhaps the two most strongly coupled ones at a higher energy machine limited by luminosity. As usual, these are mass eigenstates which will in general couple to fermions in a way not precisely like z or γ but with some higgsino mixed in. As we saw above, the sum of the masses of the two lightest neutralinos is much less than the scale of the supersymmetry breaking parameters in the mass matrix, so that if the supersymmetry breaking parameters are even well above the weak scale one can still have neutralinos light enough to produce at present machines.

The cross sections arise from the diagrams

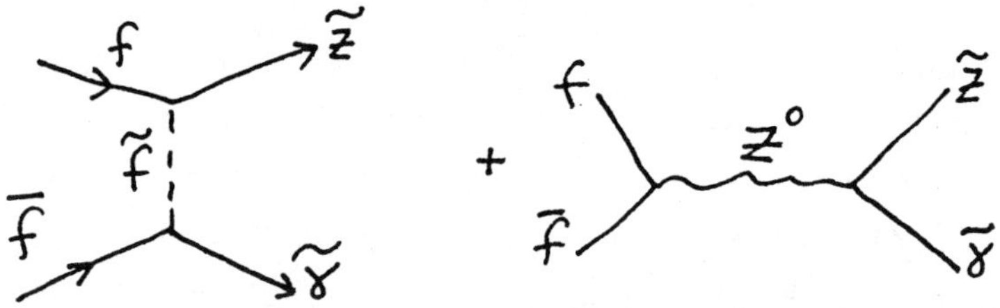

The off-diagonal coupling of Z° in the second term is not zero in a supersymmetric theory (it is like a flavor changing neutral current, and arises whenever the mass eigenstates are a mixture of Dirac and Majorara particles), and it can be large if much higgsino is mixed in. The full cross sections for e^+e^- are computed in ref. 24 and those for $q\bar{q}$ can be obtained by small changes. Some cross sections are given in ref. 16 for the case of unmixed $\tilde{\gamma}$ and $\tilde{z}$, so the rates obtained there are probably somewhat large, and neglect the second contribution above, which affects distributions. The first diagram gives a rate that grows with s for $s \lesssim \tilde{m}_f^2$, and the second obviously has the Z° pole and then falls. Typical cross sections are of order 0.1 pb at PEP or PETRA energies.

The signatures are quite good, which is partly why we emphasize these events. The lightest neutralino escapes the detector. The heavier one decays via

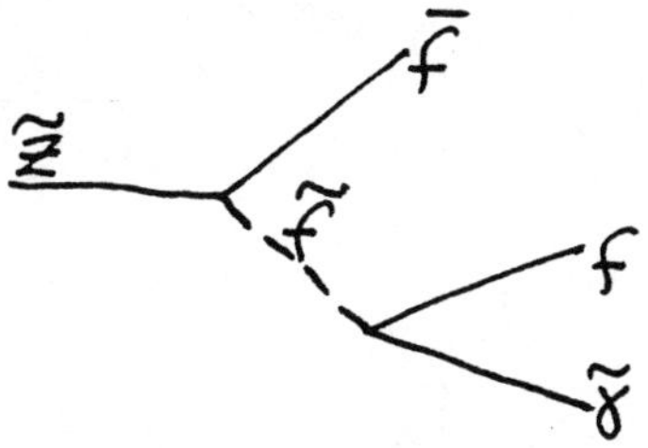

where f is any lepton or quark. Thus again one-sided events such as

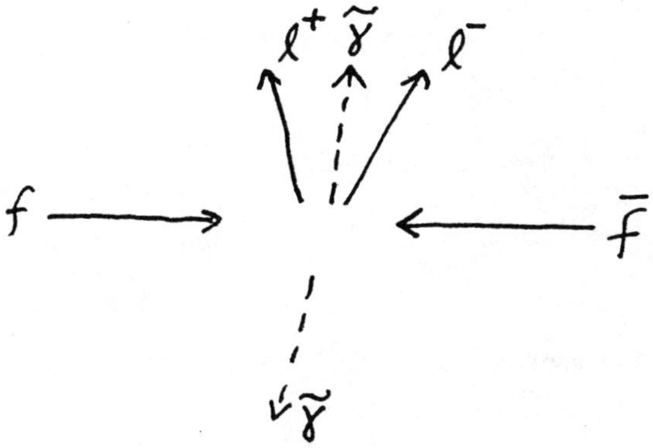

or

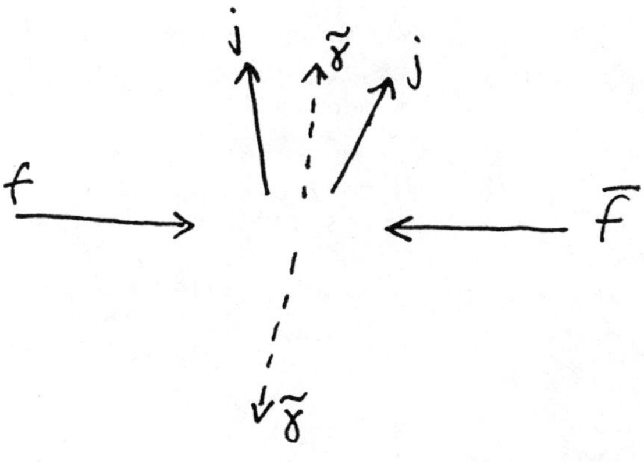

appear. There is no standard model background for such events. (Such events can originate from other new physics, so if they are ever detected it will be very interesting to work out what their origin is; methods exist for doing that.)

C. Anomalous Events With Missing P_T

Associated production of various pairs of superpartners is possible. Just as one can produce qq, gg, $W^{\pm}g$, ... pairs at a hadron collider [1,28-32], one can produce $\tilde{w}\tilde{g}$, $\tilde{q}\tilde{g}$, $\tilde{g}\tilde{g}$, $\tilde{q}\tilde{q}$, $\tilde{q}\tilde{\gamma}$, etc. Since each superpartner gives one particle that escapes the detector, these events produce the characteristic supersymmetry signature of missing energy and P_T.

To calculate the cross section at a hadron collider, we must start with the hadrons, and express the full cross section in terms of the cross section for the constituents. Pictorially, to produce $\tilde{w}\tilde{g}$ one has

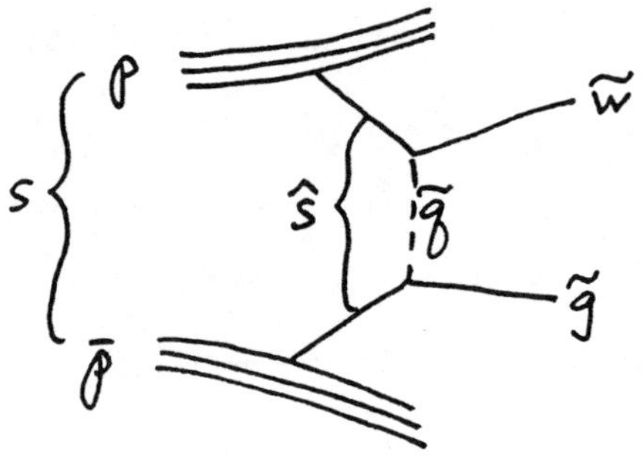

and one writes in the parton model,

$$\sigma_{\tilde{w},\tilde{g}}(s) = \int dx_1 dx_2 \; q(x_1)\bar{q}(x_2)\hat{\sigma}_{\tilde{w}\tilde{g}}(\hat{s})$$

where x_1 and x_2 are the fraction of the hadron momentum carried by q and $\bar{q}$, $q(x_1)$ and $\bar{q}(x_2)$ are the structure functions that give the probability of a quark q carrying momentum fraction x_1, etc. $\hat{\sigma}_{\tilde{w}\tilde{g}}$ is the constituent cross section for $q\bar{q}\to\tilde{w}\tilde{g}$ at a constituent center of mass energy $\sqrt{\hat{s}}$, which is related to s by $\hat{s}=x_1x_2s$. For a given process one adds all constituent processes which give that final state. Other diagrams with gluons in hadrons contribute too. Such a procedure is known to work approximately at the CERN collider for production of $W^{\pm}$,$Z°$, and jet pairs ($q\bar{q}$,qg,gg); if anything it somewhat underestimates cross sections, but appears to be accurate to a factor of 2 or so. In addition, various experimental cuts and restrictions must be imposed on theoretical cross sections before comparison with experiment.

Here we will carry out a sample calculation including mixing effects, etc. We start from the two-component Lagrangian, derive the Feynman rules in four-component form, and calculate the constituent cross section. Then we discuss the effects of structure functions and experimental cuts on rates, and finally the experimental signatures.

Consider the example

$q\bar{q}\to\tilde{w}\tilde{g}$.

There are two diagrams to make $\tilde{w}^+\tilde{g}$ (if the experimental apparatus cannot distinguish $\tilde{w}^+$ from $\tilde{w}^-$, one must double the final rate)

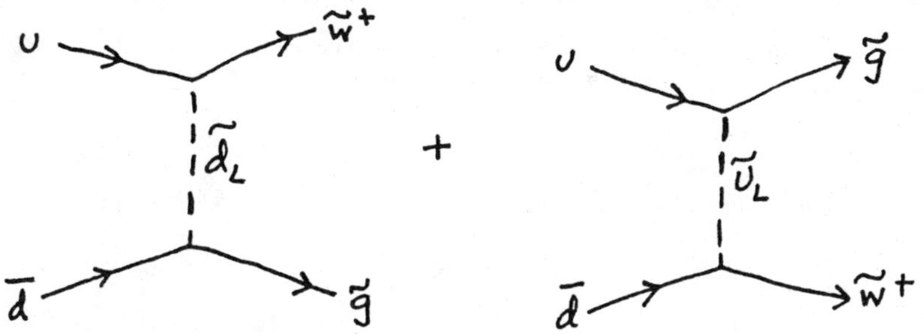

To find the Feynman rules we start with the lagrangian in two component form,

$$L_{int} = ig\sqrt{2}\, T^a_{ij}(\lambda^a \psi_j A_i^* - \bar{\lambda}^a \bar{\psi}_j A_i).$$

T^a_{ij} is the group generator in the representation under which ψ,A transform, λ^a is a two-component gaugino, ψ a matter fermion, and A a complex scalar. This is what arises in component form from the covariant derivative applied to superfields. Using $\lambda^\pm=(\lambda^1\pm i\lambda^2)/\sqrt{2}$ and putting in explicit particles we have

$$L_{int} = ig\,(\lambda^- u_L\, \tilde{d}_L^* + \lambda^+ d_L \tilde{u}_L^* + h.c.)$$

Since the wino and higgsino mix to form the mass eigenstates, they will be of the form

$$\chi_1^+ = V_{11}\lambda^+ + V_{12}\tilde{h}^+$$

with χ_2^+ defined similarly. The mixing matrix V will be of the form

$$V = \begin{pmatrix} \cos\phi_+ & \sin\phi_+ \\ \sin\phi_+ & -\cos\phi_+ \end{pmatrix}$$

Similarly, we will have

$$\chi_1^- = U_{11}\lambda^- + U_{12}\tilde{h}^-,$$

etc., with

$$U = \begin{pmatrix} \cos\phi_- & \sin\phi_- \\ -\sin\phi_- & \cos\phi_- \end{pmatrix}.$$

The signs come out as above when one proceeds naively. If one wants to make U,V to be identical in form, one can write $\chi_2 \rightarrow \gamma_5 \chi_2$ which changes the signs of the mass eigenvectors. If the two vacuum expectation values v_1 and v_2 are equal, then $\phi_+ = \phi_-$.

The states χ_1^+, χ_1^-, are like the two component states e_L^-, e_R^+. They can be combined to form a 4-component mass eigenstate

$$\tilde{\chi}_1^+ = \begin{pmatrix} \chi_1^+ \\ \overline{\chi_1^-} \end{pmatrix}$$

etc. Then $\tilde{\chi}_1^\pm$ are antiparticles of mass m_1, $\tilde{\chi}_2^\pm$ are antiparticles of mass m_2.

We proceed by first rewriting the Lagrangian in terms of 4-component spinors. Some useful identities, which can easily be verified by explicit calculation, are as follows.

We define 4-component states ψ of the form

$$\psi = \begin{pmatrix} \psi_L \\ \psi_R \end{pmatrix}$$

by

$$\psi = \begin{pmatrix} \xi \\ \overline{\eta} \end{pmatrix}$$

where ξ, η are two-component spinors satisfying

$$(\overline{\sigma}_\mu p^\mu)\xi = m\overline{\eta}$$

$$(\sigma_\mu p^\mu)\overline{\eta} = m\xi$$

with $\sigma^\mu = (1,\vec{\sigma})$ and $\overline{\sigma}^\mu = (1,-\vec{\sigma})$. Then

$$(\gamma_\mu p^\mu - m)\psi = 0,$$

and

$$\psi^C = C\bar{\psi}^T = \binom{\eta}{\bar{\xi}},$$

where C is the charge conjugation matrix. Products of ξ,η are

$$\eta\xi = \eta^\alpha\xi_\alpha = \xi\eta$$

$$\overline{\eta\xi} = \overline{\xi\eta}$$

$$\eta_2\sigma^\mu\eta_1 = -\eta_1\sigma^\mu\eta_2.$$

Then by direct calculations,

$$\bar{\psi}_1\psi_2 = \eta_1\xi_2 + \bar{\eta}_2\bar{\xi}_1$$

$$\bar{\psi}_1\gamma_5\psi_2 = -\eta_1\xi_2 + \bar{\eta}_2\bar{\xi}_1$$

$$\bar{\psi}_1\gamma^\mu\psi_2 = \bar{\xi}_1\bar{\sigma}^\mu\xi_2 - \bar{\eta}_2\bar{\sigma}^\mu\eta_1$$

$$\bar{\psi}_1\gamma^\mu\gamma_5\psi_2 = -\bar{\xi}_1\bar{\sigma}^\mu\xi_1 - \bar{\eta}_2\bar{\sigma}^\mu\eta_1$$

So one can solve for the products of two-component spinors:

$$\eta_1\xi_2 = \bar{\psi}_1 P_L\psi_2$$

$$\bar{\eta}_2\bar{\xi}_2 = \bar{\psi}_1 P_R\psi_2$$

$$\bar{\xi}_1\bar{\sigma}^\mu\xi_2 = \psi_1\gamma^\mu P_L\psi_2$$

$$\bar{\eta}_2\bar{\sigma}^\mu\eta_1 = -\bar{\psi}_1\gamma^\mu P_R\psi_2.$$

For our example we only need some of these, but to do several processes they would all be necessary.

For Majorana spinors we put $\eta=\xi$ so that

$$\bar{\psi}_1\psi_2 = \bar{\psi}_2\psi_1$$

$$\bar{\psi}_1\gamma_5\psi_2 = \bar{\psi}_2\gamma_5\psi_1$$

$$\bar{\psi}_1\gamma_\mu\psi_2 = -\bar{\psi}_2\gamma_\mu\psi_1 \qquad \text{Majorana}$$

$$\bar{\psi}_1\gamma_\mu P_L\psi_2 = -\bar{\psi}_2\gamma_\mu P_R\psi_1$$

etc.

Now we can complete the process. We go to 4-component form by replacing the various 2-component products by their 4-component forms, with projection operators, and we go to mass eigenstates by replacing $\tilde{w},\tilde{w}^c$ with $\tilde{x}_i$. The 4-component form is

$$L_{int} = -g[\bar{\tilde{w}}P_L u\tilde{d}_L^* + \bar{u}P_R\tilde{w}\tilde{d}_L + \bar{\tilde{w}}^c P_L d\tilde{u}_L^* + \bar{d}P_R\tilde{w}^c\tilde{u}_L].$$

The $\tilde{w}^c$ terms arise naturally when all 2-component products are needed. Then using

$$\tilde{w} = U_{11}\tilde{x}_1 + U_{21}\tilde{x}_2$$

$$\tilde{w}^c = V_{11}\tilde{x}_1^c + V_{21}\tilde{x}_2^c$$

we get

$$L_{int} = -g[(U_{11}^*\bar{\tilde{x}}_1 + U_{21}^*\bar{\tilde{x}}_2)P_L u\tilde{d}_L^* + \ldots]$$

from which we can directly read the Feynman rules. For our case, we need

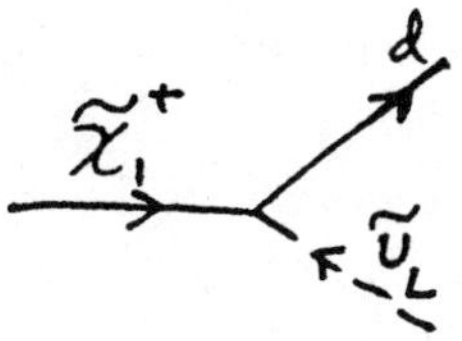

$-ig\ U_{11}P_R$

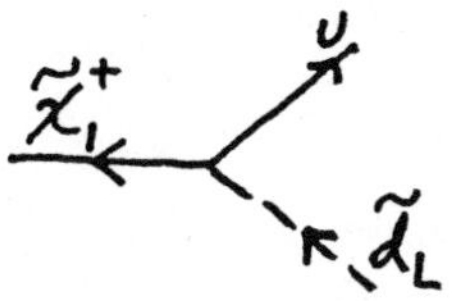

$-ig\ V_{11}P_RC$

The details of the mixing angles could be obtained by diagonalizing the mass matrix. This has been done if ref. 1, and the results are, for $v_1 = v_2$,

$$\phi_+ = \phi_- = \frac{1}{2}\tan^{-1}\frac{2M_W}{M-\mu} .$$

When $M=\mu$, $\phi_+=\phi_-=45°$ and the masses of the eigenstates are

$$\tilde{M}_{1,2} = M_W \pm \mu .$$

We do not actually use these valves, but they give one some intuition about the role of V_{11}, U_{11}.

By a similar procedure we can obtain the other rules we need:

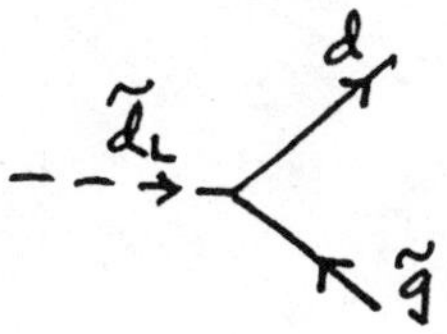

$ig_s\sqrt{2}T^aP_R$

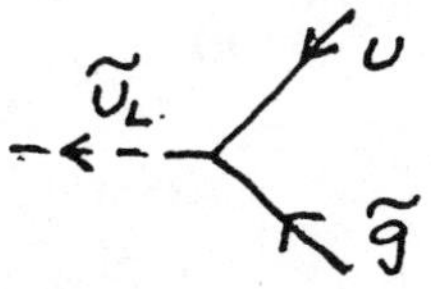

$-ig_s\sqrt{2}T^aC^{-1}P_L$

For any Dirac or Majorana spinor,

$u(k,s) = C\bar{v}^T(k,s)$

$v(k,s) = C\bar{u}^T(k,s)$.

The usual propogators are

$\sum_s u^s(p)\bar{u}^s(p) = \not{p} + m$

$\sum_s v^s(p)\bar{v}^s(p) = \not{p} - m$

Using the above spinor identities in these one can obtain the propogators which arise in the Majorana case,

$\sum_s u^s(p)v^{sT}(p) = (\not{p}+m)C^T$

etc. Combining all these relations one can use normal trace techniques to compute cross sections.

To simplify the formulas as we proceed, let us write $\tilde{m}_d=\tilde{m}_u=\tilde{q}$, $\tilde{m}_g=\tilde{g}$, $\tilde{m}_w=\tilde{w}$. It will be obvious when we mean the mass or the particle.

Then the amplitudes are (momenta are in parentheses)

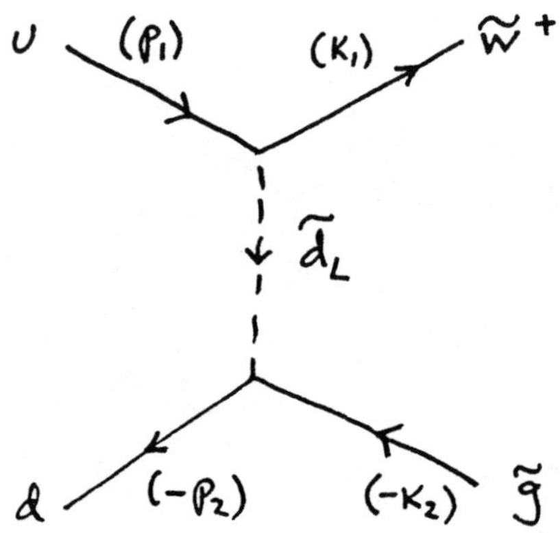

$$-iM_a = \frac{-i}{\hat{t}-\tilde{q}^2}[\bar{u}(k_1)ig\sin\phi_- P_L u(p_1)][\bar{v}(p_2)ig_s \sqrt{2}P_R v(k_2)]$$

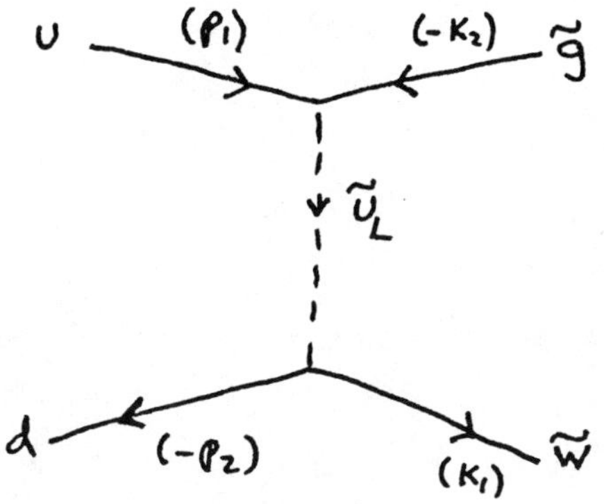

$$-iM_b = \frac{-i}{\hat{u}-\tilde{q}^2}\,[v^T(k_2)(-ig_s\sqrt{2})C^{-1}P_L u(p_1)][\bar{v}(p_2)ig\,\sin\phi_+\,P_R C\bar{u}^T(k_1)].$$

where $\hat{t}$, $\hat{u}$ are the constituent process momentum transfer variables, $\hat{t} = (p_1-k_1)^2$, $\hat{s}+\hat{t}+\hat{u} = \tilde{g}^2+\tilde{w}^2$. We have not written the color factors since they give an overall factor of $\mathrm{Tr}T^aT^a = 4$ in $|M|^2$.

Then, summing over final spins and colors and averaging over initial spins and colors,

$$\frac{d\hat{\sigma}}{d\hat{t}} = \frac{1}{16\pi\hat{s}^2}\,\frac{1}{4}\,\frac{1}{9}\sum_{\substack{\text{spins,}\\ \text{colors}}} |M|^2$$

$$= \frac{8g^2g_s^2}{16\mathrm{x}36\pi\hat{s}^2}\Big[\sin^2\phi_-\,\frac{(\tilde{w}^2-\hat{t})(\tilde{g}^2-\hat{t})}{(\tilde{q}^2-\hat{t})^2} + \sin^2\phi_+\,\frac{(\tilde{w}^2-\hat{u})(\tilde{g}^2-\hat{u})}{(\tilde{q}^2-\hat{u})^2}$$

$$+ \frac{2\hat{s}\tilde{w}\tilde{g}\,\sin\phi_+\sin\phi_-}{(\tilde{q}^2-\hat{t})(\tilde{q}^2-\hat{u})}\Big].$$

If we compute for the other mass eigenstate, from the form of the mixing matrices, $\sin\phi_\pm \rightarrow \cos\phi_\pm$.

Next, we can integrate over angles, using

$$\hat{t},\hat{u} = [-(\hat{s}-\tilde{w}^2-\tilde{g}^2) \pm \sqrt{\lambda}\cos\theta]/2$$

$$\lambda = \hat{s}^2+\tilde{w}^4+\tilde{g}^4 - 2\hat{s}\tilde{w}^2 - 2\hat{s}\tilde{g}^2 - 2\tilde{w}^2\tilde{g}^2.$$

Then the full cross section is

$$\hat{\sigma}(u\bar{d}\to\tilde{g}\tilde{w}^+) = \frac{2\pi\alpha_2\alpha_s}{9\hat{s}^2}\{(\sin^2\phi_+ + \sin^2\phi_-)[\sqrt{\lambda} + (\tilde{g}^2-\tilde{w}^2)\ell n\, r]$$

$$+ \frac{4swg}{\hat{s}^2+\tilde{w}^2-\tilde{g}^2}\sin\phi_+\sin\phi_-\ \ell n\, r\}$$

where

$$r = (\hat{s} + \tilde{w}^2 - \tilde{g}^2 + \sqrt{\lambda})/(\hat{s} + \tilde{w}^2 - \tilde{g}^2 - \sqrt{\lambda}).$$

Note that the interference term sign is important. To check this calculation one can make the changes that would convert it into another one that exists and compare. E.g., for $e^+e^-\to\tilde{\gamma}\tilde{\gamma}$, one should put (i) $g_s\to e$, (ii) $1/9\ \mathrm{Tr}T^aT^a\to 1$, (iii) $\pm g\sin\phi_\pm\to\sqrt{2}e$, (iv) $\tilde{m}_w,\tilde{m}_g\to\tilde{m}_\gamma$, (v) 1/2 for identical particles.

This result is sufficiently complicated that it is not obvious what the form is. To see some simple forms we give cross sections for several processes when all superpartner masses are set equal to $\tilde{m}$

<u>$q\bar{q}\to\tilde{w}\tilde{g}$</u> (summed over $\tilde{w}^+ + \tilde{w}^-$)

$$\hat{\sigma} \simeq \frac{4\pi}{9}\frac{\alpha_2\alpha_s}{\hat{s}}\ \ell n\frac{1+\beta}{1-\beta}\left(\sin^2\phi_+ + \sin^2\phi_-\right)$$

<u>$qg\to\tilde{q}\tilde{g}$</u> (per flavor)

$$\hat{\sigma} \simeq \frac{\pi\ \alpha_s^2}{\hat{s}}\ \ell n\frac{1+\beta}{1-\beta}$$

<u>$q\bar{q}\to\tilde{\gamma}\tilde{g}$</u>

$$\hat{\sigma} \simeq \frac{\pi\alpha\alpha_s}{\hat{s}}\{1 - \frac{\tilde{m}^2}{\hat{s}}(1 + \ell n\frac{\hat{s}}{\tilde{m}^2})\}$$

where

$$\beta = \sqrt{1 - 4\tilde{m}^2/\hat{s}}.$$

Each of them is a number of order one times the appropriate product of α's, divided by $\hat{s}$.

To estimate them numerically we can use $\alpha_2 \simeq 1/30$, $\alpha_s \approx 0.1$. To set the scale, note that if we put $\hat{s} = (150\ \text{GeV})^2$,

$$\frac{\pi\alpha_2\alpha_s}{\hat{s}} \simeq 20\times10^{-35}\ \text{cm}^2.$$

The current CERN experiment [14,33] has an integrated luminosity such that one event corresponds to a cross section of about 10^{-35} cm^2.

However, before one can compare experiment and theory, one must convolute with the structure functions, as discussed above, and impose experimental cuts. Where one is at the edge of phase space, as here, that can sharply reduce the expected number of events. We will not go into details here since the prediction of supersymmetric constituent cross sections, and the signatures, are our main concern.

The rate for $\tilde{w}\tilde{g}$ production at the CERN collider is in the right region to produce such events if they are kinematically allowed. Next we need to discuss the possible signatures. If we produce $\tilde{w}\tilde{g}$, then $\tilde{g} \to q\bar{q}\tilde{\gamma}$ assuming $\tilde{m}_g > \tilde{m}_\gamma$. If the masses are appropriate, $\tilde{w} \to e\tilde{\nu} \to \tilde{e}\nu$, giving $e + \not{p}_T$ in both cases. Then the full signature is $\tilde{w}\tilde{g} \to ejj\not{p}_T$, where the $\not{p}_T$ is due to two or three particles. The jets should be light quark ones, with low effective mass and normal multiplicity. If jets below a certain $\not{p}_T$ are not counted, sometimes these events would look like $ej\not{p}_T$. Similar events with μ and τ will occur.

If $\tilde{g}\tilde{q}$ (which is the largest cross section for non-electroweak processes) are produced, $\tilde{g} \to jj\tilde{\gamma}$ and $\tilde{q} \to j\tilde{\gamma}$ (assuming $\tilde{m}_q > \tilde{m}_g$) so $\tilde{g}\tilde{q} \to jjj\not{p}_T$ where the $\not{p}_T$ is due to two missing particles. For light $\tilde{g}$ some jets can coalesce or be below cuts so $jj\not{p}_T$ or $j\not{p}_T$ can appear. Similar remarks hold for $\tilde{g}\tilde{g}$ or $\tilde{q}\tilde{q}$, with somewhat lower cross sections.

At each machine from now on one must look for such events as the signal for production of supersymmetric partners. If events such as $j\not{p}_T$, $jj\not{p}_T$, $jjj\not{p}_T$, $\ell^\pm j\not{p}_T$, $\ell^\pm jj\not{p}_T$ are ever observed <u>at the</u>

right rate it could be the observation of supersymmetry. If they are not observed, the limits on such masses will be increasingly large, until eventually it will be clear that supersymmetry is not relevant to understanding phenomena at the weak scale.

Comments on Light Higgs Bosons in Supersymmetric Theories

In a supersymmetric theory there are two Higgs doublets. The two major modifications occur to the minimal Standard Model. (1) The physical Higgs particle spectrum has a pair of charged Higgs boson and three neutral bosons, one pseudoscalar and two scalars. (2) There are two vacuum expectation values, which can be different. The coupling to fermions goes inversely as the vacuum expectation values, and can be enhanced for either down-type quarks or up-type quarks or both, subject to the condition $v_u^2 + v_d^2 = v^2$ where v^2 is the usual vev. See the lectures of P. Nilles for an extensive discussion of the Higgs masses in a supersymmetric theory.

Because of the enhanced coupling, it may be possible to find a light Higgs experimentally at a higher branching ratio than in the SM, e.g. through $\Upsilon \to \gamma H^0$. In that case, some constraints [22] must be satisfied before a supersymmetric Higgs interpretation of a possible signal is allowed. First, if the branching ratio for $\Upsilon \to \gamma H^0$ were greatly enhanced (so $v_d << v_u$), then one would expect a similar enhancement for other down-type objects (s,τ,μ,e,d) but not for up-type objects. Therefore the dominant decays of such an H^0 should be to τ's, not to hadrons. The only way to evade that would be if the decay to gluons could be large; it goes through fermion loops [23]. But if the gluon pair branching ratio were larger, then H^0 would be produced from hadrons. One can show [22] that $BR(H^0 \to gg) < 6\%$ from existing data [24] on μ pair production for $7\ \text{GeV} \lesssim m_H \lesssim m_\Upsilon$. Thus, if a light H^0 were to be found "early" via a larger-than-SM branching ratio in Υ decays (or Υ', Υ'') it could only be consistent with the Higgs expected in a supersymmetric theory with two doublets if H^0 decayed almost entirely into τ pairs.

Acknowledgements

I would like to thank my various collaborators in this field, J. Leveille, R. Rolnick, J. Ellis, J. Hagelin, S. Raby, S. Petcov, and especially J.-M. Frere and H. Haber, for many instructive discussions and enjoyable interactions.

REFERENCES

1. Howard E. Haber and G. L. Kane, "The Search for Supersymmetry: Probing Physics Beyond the Standard Model", UM HE TH 83-17, to be published in Physics Report.

2. A complete set of references is given in ref. 1. Recent reviews along the lines of ref 1 and the present lectures by Fayet; by Ellis; by Nanopoulos, Savoy-Navarro, and Tao; by Savoy; and by Arnovitt, Nath, and Chamseddine are cited in ref 1. The 1983 SLAC Summer Institute lectures of J. Ellis are closest in their approach to the present ones (preprint CERN TH 3747).

3. G. L. Kane and J.-M. Frere, unpublished, L. J. Hall and M. Suzuki, Nucl. Phys. B231 (1984) 491; I-Hsiu Lee, preprint UCB-PTH-83/25.

4. J. S. Hagelin, G. L. Kane, and S. Raby, Los Alamos preprint LA-UR-83-3711; L. Ibanez, preprint FTUAM-83-28.

5. P. Fayet, Phys. Lett. 86B (1979) 272.

6. Min Chen, Lab. for Nuc. Sci. Technical Report 139.

7. See R. M. Barnett, H. E. Haber, and K. S. Lackner, Phys. Rev. D29 (1984) 1381 and further references there.

8. G. L. Kane and W. Rolnick, Nucl Phys. B232 (1984) 21.

9. E. Ma and J. Okada, Phys. Rev. D18 (1978) 4219; G. Barbiellini, B. Richter, and J. L. Siegrist, Phys. Lett. 106B (1984) 414.

10. P. Fayet, Phys. Lett. 117B (1982) 460; J. Ellis and J. S. Hagelin, Phys. Lett. 122B (1983) 303; J. D. Ware and Marie E. Machacek, Northeastern preprint NUB#2626.

11. MAC Collaboration, and D. Burke and R. Hollebeck et al.

12. C. Nappi, Phys. Rev. D25 (1982) 84.

13. H. Takeda, Proceedings of the Brighton European Physicist Society Meeting, August 1983.

14. UA1 Collaboration, G. Arnison et al., CERN-EP/84-42.

15. J. Ellis and J. Kowalski, CERN preprint; R. M. Barnett, H. E. Haber, and G. L. Kane, in preparation.

16. Sally Dawson, E. Eichten, and C. Quigg, Fermilab - PUB 83/82-THY.

17. E. Franco, Phys. Lett. 124B (1983) 271.

18. T. Goldman and H. E. Haber, Los Alamos preprint LA-UR-83-2323; W.-Y. Keung and A. Khare, BNL preprint; J. H. Kuhn and S. Ono, preprint PITHA 83/25.

19. H. E. Haber and G. L. Kane, Nucl. Phys. B232 (1984) 333.

20. M. J. Duncan, Nucl. Phys. B214 (1983) 21.

21. M. Suzuki, Phys. Lett. 115B (1982) 40.

22. For a general discussion of such constraints, see Howard E. Haber and G.L. Kane, "Implications of a Higgs Interpretation of the $\zeta(8.J)$," preprint UM TH 84-26.

23. F. Wilczek, Phys. Rev. Lett. 39 (1977) 1303.

24. K. Ueno et al., Phys. Rev. Lett. 42 (1979) 486.

25. S. Weinberg, Phys. Rev. Lett. 50 (1983) 387; R. Arnowitt, A. H. Chamseddine, and P. Nath, Phys. Rev. Lett. 50 (1983) 1232; G.L. Kane and J.-M. Frere, Nucl. Phys. B223 (1983) 331; R. Fayet, Phys. Lett. 133B (1983) 363; J. Ellis, J. S. Hagelin, D. V. Nanopoulos, and M. Srednicki, Phys. Lett. 127B. (1983) 233.

26. G. L. Kane and J.-M. Frere, Nucl. Phys. B223 (1983) 331.

27. J. Ellis, J.-M. Frere, J. S. Hagelin, G. L. Kane, and S. T. Petkov, Phys. Lett. 132B (1983) 436.

28. H. E. Haber and G. L. Kane, Santa Cruz preprint.

29. J. Ellis and L. Kowalski, CERN preprint.

30. G. Altarelli and B. Mele, Rome preprint.

31. E. Reya and D. P. Roy, Dortmund preprint DO-TH 84/03.

32. V. Barger et. al, Wisconsin preprint.

33. UA2 Collaboration, P. Bagnaia et al., CERN-EP/84-40.

SUPERSYMMETRIC FUNCTIONAL INTEGRATION MEASURES*

Hermann Nicolai

CERN

1211 Geneva 23

ABSTRACT

Supersymmetric theories can be directly characterized in terms of the associated bosonic functional integration measures where fermions have been integrated out. This method and its various recent developments are reviewed.

1. INTRODUCTION

Supersymmetric theories (for reviews and further references, see e.g.[1]) are usually characterized by their invariance properties with respect to transformations that involve anticommuting parameters. The latter play an essential role in the formulation of supersymmetric theories and their use sometimes facilitates complex calculations, for example in perturbation theory. On the other hand, they are not as wieldy as ordinary numbers and many arguments based on their use remain formal. It is therefore of interest that there exists an alternative characterization of (a subclass of) supersymmetric theories directly in terms of the functional integration measure by means of which expectation values in a supersymmetric theory are defined [2] . This result, which enables us to reconstruct supersymmetric theories and to study their properties without recourse to anticommuting quantities, is essentially

*Lectures delivered at the 1984 NATO Advanced Study Institute on Supersymmetry, Bonn, Federal Republic of Germany, August 1984

a consequence of the fact that, in (flat space) supersymmetric theories, the vacuum energy vanishes if the groundstate is supersymmetric [3]. In the absence of interactions, this just means that fermionic and bosonic degrees of freedom cancel (fermions being counted as negative). For interacting supersymmetric theories, the vanishing of the vacuum energy density is expressed by

$$\langle 0 | H(\lambda_1, \dots, \lambda_n) | 0 \rangle = 0 \tag{1.1}$$

for arbitrary interaction parameters $\lambda_1, \dots, \lambda_n$. By virtue of the well known relation between the vacuum energy and the (Euclidean) vacuum functional, (1.1) implies that

$$Z(\lambda_1, \dots, \lambda_n) = \int d\phi \exp[-S(\lambda_i, \phi)] = \text{const.} \tag{1.2}$$

as a function of the parameters $\lambda_1, \dots, \lambda_n$ (we will see later on that $Z(\lambda)$ may exhibit jump discontinuities on certain "critical surfaces" in parameter space). For an interacting theory, the vacuum functional (1.2) is given by a non-trivial integral and the fact that it is piecewise constant as a function of $\lambda_1, \dots, \lambda_n$ puts stringent constraints on the form of the interaction in (1.2). One may therefore expect that supersymmetry can be alternatively characterized as a property of the associated functional integration measure.

To set up the necessary conventions and notation, let us assume that the supersymmetric model under consideration is based on one (or several) supermultiplet(s) consisting of one (or several) bosonic field(s) $A(x)$ together with their fermionic partner(s) $\psi(x)$ which are usually Majorana spinors. In sections 2 and 3 of these lectures, it will furthermore be assumed that the maximum spin is 1/2; supersymmetric gauge theories are treated in section 4. The number of dimensions is left unspecified for the moment. More significantly, all models to be discussed will be used in their "on-shell" version: it appears to be one of the advantages of the present approach that it works independently of the existence or non-existence of an "off-shell" formulation. A central input is the fact that the fermions can be "integrated out" [4]. To do so, one decomposes the action into a bosonic and a fermionic part, i.e.

$$S(\lambda, \phi) = S_B(\lambda, A) + \frac{1}{2}\bar{\psi} M(\lambda, A) \psi \tag{1.3}$$

with the fermionic part

$$\frac{1}{2}\bar{\psi} M(\lambda, A)\psi \equiv$$

$$\equiv \frac{1}{2}\int dx\, dy\, \bar{\psi}(x)\, M(x, y; \lambda, A)\, \psi(y) \tag{1.4}$$

Since the expression (1.4) is quadratic in the fermions, its exponential can be integrated in closed form with the result

$$\int d\psi \, \exp\left[-\tfrac{1}{2}\bar{\psi} M(\lambda, A)\psi\right] = \left\{\det M(\lambda, A)\right\}^{1/2} = \tag{1.5}$$

$$\equiv \left\{\det M(\lambda = 0, A)\right\}^{1/2} D(\lambda, A)$$

for Majorana fermions (which obey $\bar{\psi}_\alpha = \psi_\beta C_{\beta\alpha}$). $D(\lambda, A)$ is the Matthews Salam determinant [4] . The restriction to actions, which are quadratic in the fermions is mainly for practical reasons, although it remains to be explored which of the results presented in these lectures would survive if this restriction were dropped. In the same vein as (1.5), the fermionic propagator in a scalar background may be obtained by explicit integration. It is given by the formula (I omit Dirac and possible internal indices)

$$\underbrace{\psi(x)\,\bar{\psi}(y)} \equiv M^{-1}(x, y; \lambda, A) =$$

$$= (\det M)^{-1/2} \int d\psi \, \psi(x)\bar{\psi}(y) \exp\left[-\tfrac{1}{2}\bar{\psi} M \psi\right] \tag{1.6}$$

Analogous formulas hold for all n-point fermionic expectation values ("Wick's theorem"). Eliminating the fermionic variables by means of (1.5) or (1.6), one can transform the functional integration measure (1.2) into a purely bosonic functional integration measure

$$d\mu(\lambda, A) \equiv \exp\left[-S_B(\lambda, A)\right] \det M(\lambda, A)^{1/2} dA \tag{1.7}$$

The central result to be elaborated in these lectures is summarized in the following

Theorem [2]: Supersymmetric theories are characterized by the existence of a (generally non-linear and non-local) transformation T of the bosonic fields

$$T_\lambda : A(x) \longrightarrow A'(x, \lambda, A) \tag{1.8}$$

with the following properties

(i) T_λ is invertible in the sence of formal power series at least.

(ii) $S(\lambda; A) = S_o(A'(\lambda, A))$ where S is the full bosonic part of the action and S_o its quadratic part (including mass terms)

(iii) The Jacobi determinant of the transformation T equals the Matthews-Salam determinant, i.e.

$$\det \frac{\delta A'(x,\lambda,A)}{\delta A(y)} = D(\lambda,A) \tag{1.9}$$

For supersymmetric gauge theories, the presence of a gauge fixing term which violates supersymmetry requires some modifications of these statements which will be extensively discussed in section 4.

As a prelude, it is instructive to see how the theorem works for supersymmetric quantum mechanics ([5], see also [6]) where the transformation (1.8) can be easily constructed in closed form [7,8]. This example may serve as the prototype for the more complicated models to be discussed later since it displays all the relevant features. The relevant variables in a Lagrangian formalism are $q(t) \in \mathbb{R}$ and two anticommuting variables $\psi(t)$ and $\bar{\psi}(t)$. The (Euclidean) Lagrangian reads

$$\mathcal{L} = \frac{1}{2}\dot{q}^2 + \frac{1}{2}V(q)^2 + \bar{\psi}\left(\frac{d}{dt} + V'(q)\right)\psi \tag{1.10}$$

The Matthews-Salam determinant for this model is

$$\begin{aligned} M &= \det\left[\left(\frac{d}{dt} + V'(q(t))\right)\delta(t-t')\right] = \\ &= D(V(q)) \det \frac{d}{dt}\delta(t-t') \end{aligned} \tag{1.11}$$

The variable substitution which makes the bosonic part of the action Gaussian and whose Jacobian equals D, is given by

$$(Tq)(t) = q(t) \pm \int \theta(t-t')V(q(t'))dt' \tag{1.12}$$

where $\theta(t-t')$ is the "free fermionic propagator", i.e.

$$\frac{d}{dt}\theta(t-t') = \delta(t-t') \tag{1.13}$$

Note that (1.12) is usually presented in the equivalent form [7,8]

$$\begin{aligned} \xi(t) &= \frac{d}{dt}(Tq)(t) = \\ &= \dot{q}(t) \pm V(q(t)) \end{aligned} \tag{1.14}$$

The sign ambiguity in (1.12) and (1.14) arises from the fact that the only term which is sensitive to it is the topological invariant

$$\int_{-\infty}^{+\infty} \dot{q}\, V(q(t))\, dt = W(\infty) - W(-\infty) \tag{1.15}$$

where W(q), V(q) = W'(q), is the "superpotential"; in fact, (1.15) is just the instanton number of the configuration q(t) [8]. (1.14) is a stochastic differential equation to which well-developed techniques may be applied [9] (the connection between supersymmetry and stochastic differential equations was first pointed out in [10]). However, the mapping (1.8) is non-local in general, and therefore the stochastic structure is lost unless one admits the possibility of "stochastic integro-differential equations". Nonetheless, there are some models for which (1.8) becomes local just as for (1.14), namely the N=2 Wess-Zumino model in two dimensions [7,10] and the N=1 Yang-Mills theory in four dimensions, see refs. [11-14] and section 4 of these lectures. Such theories are sometimes called "stochastic".

In these lectures, I will not be overly concerned with questions of mathematical rigour although many of the results described can probably be made more rigorous (e.g. for supersymmetric quantum mechanics, see [15]), and the existence of the mapping (1.8) may even facilitate a more rigorous investigation of supersymmetric theories (for instance, one may possibly employ (1.8) to regularize supersymmetric theories [16]). A Euclidean signature will be used throughout these lectures. Although this is a common practice, one here has to deal with Euclidean Majorana spinors. These are sometimes professed not to exist, but in fact do, if one renounces the usual hermiticity properties [17]. This causes no harm since in Euclidean field theory, the relevant requirement is Osterwalder Schrader positivity [18] not hermiticity. For the construction of the transformation (1.8), I will mainly rely on the recently developed method of ref. [19], which is much more transparent than the original one of [2]. Moreover, as a new result, a simple proof of (1.9) is given; this closes a gap that had occured in the original proof [20]. An alternative method is based on the use of "stochastic identities" [12]. The idea is to determine the combinations of bosonic fields that satisfy free field vacuum expectation values by a careful analysis of the supersymmetry transformation rules. This approach is not discussed in detail here since it has been reviewed in a recent paper [14]. Finally, although it is somewhat contrary in spirit to the approach expounded in these lectures, it should be noted that the theorem has a superspace analog: the Jacobian for the change of variables in an integral over unconstrained superfields is always equal to one [21].

2. PROOF OF MAIN THEOREM

For the sake of definiteness and notational convenience, only the N=1 Wess-Zumino model in two dimensions will be considered in this section as the extension of the proof given below to other cases without additional gauge invariances is immediate. In two space-time dimensions, the minimal supermultiplet contains a scalar A and a two component Majorana spinor ψ_α [22]. The most general Lagrangian reads

$$\mathcal{L} = \frac{1}{2}(\partial_\mu A)^2 + \frac{1}{2}(mA + \lambda p(A))^2 + \\ + \frac{1}{2}\bar{\psi}\gamma^\mu\partial_\mu\psi + \frac{1}{2}(m + \lambda p'(A))\bar{\psi}\psi \tag{2.1}$$

where auxiliary fields have been eliminated and the interaction part has been split off. Thus, p(A) is an arbitrary polynomial of degree ≧ 2. (2.1) transforms as a total divergence under the supersymmetry transformations

$$\delta\psi_\beta = \varepsilon_\alpha\delta_\alpha\psi_\beta \quad , \quad \delta A = \varepsilon_\alpha\delta_\alpha A \tag{2.2}$$

where

$$\delta_\alpha A = \mathcal{C}_{\alpha\beta}\psi_\beta = -\bar{\psi}_\alpha \\ \delta_\alpha\psi_\beta = \gamma^\mu_{\beta\alpha}\partial_\mu A - \delta_{\alpha\beta}(mA + \lambda p(A)) \tag{2.3}$$

with the two dimensional charge conjugation matrix $\mathcal{C}_{\alpha\beta}$.

The transformation (1.8) can be constructed iteratively by exploiting [19] the fact that the derivative of $\mathcal{L}$ with respect to any of its coupling parameters can be written as a supervariation [3]. The relevant result in our case reads

$$\frac{\partial\mathcal{L}}{\partial\lambda} = -\delta_\alpha\hat{\Delta}_\alpha \quad \text{with} \quad \hat{\Delta}_\alpha \equiv \frac{1}{2}\psi_\alpha p(A) \tag{2.4}$$

We also need the integrated operator insertion

$$\Delta_\alpha \equiv \int \hat{\Delta}_\alpha(x)\, d^2x \tag{2.5}$$

The validity of (2.4) may be verified directly from (2.1) and (2.3). The vanishing of the vacuum energy is a consequence of

$$\frac{\partial}{\partial \lambda} \langle 1 \rangle = \langle \delta_\alpha \Delta_\alpha \rangle \tag{2.6}$$

and the supersymmetry Ward identity which states that

$$\langle \delta_\alpha Y \rangle = 0 \tag{2.7}$$

for an arbitrary string of operators Y. If X is some (not necessarily local) product of <u>bosonic</u> operators, one gets in a similar manner

$$\begin{aligned} \frac{\partial}{\partial \lambda} \langle X \rangle &= \left\langle \frac{\partial X}{\partial \lambda} \right\rangle + \langle X \, \delta_\alpha \Delta_\alpha \rangle = \\ &= \left\langle \frac{\partial X}{\partial \lambda} \right\rangle - \langle \delta_\alpha X \cdot \Delta_\alpha \rangle \end{aligned} \tag{2.8}$$

where the Ward identity (2.7) has been used to "integrate by parts". The expectation values in (2.6), (2.7), and (2.8) signify integration over <u>both</u> fermions and bosons. Since the second term on the right side of (2.8) contains two fermion operators, we can make use of (1.6) to integrate out the fermions before integrating over bosons. This motivates the definition of the "R-operator" [19,13]

$$RX \equiv \frac{\partial X}{\partial \lambda} - \underbrace{\delta_\alpha X \cdot \Delta_\alpha} \tag{2.9}$$

which is now a purely bosonic expression and defines a nonlocal and nonlinear mapping of the functional $X[A]$ onto another functional $RX[A]$ (it is, in some sense, the generator of the transformation to be constructed). The fermion propagator occurring in (2.9) is defined by

$$[\not\partial + m + \lambda p'(A(x))]_{\alpha\gamma} \underbrace{\psi_\gamma(x) \bar{\psi}_\beta(y)} = \delta_{\alpha\beta} \delta(x-y) \tag{2.10}$$

or, equivalently,

$$\begin{aligned} \underbrace{\psi_\alpha(x) \bar{\psi}_\gamma(y)} [-\overleftarrow{\not\partial} + m + \lambda p'(A(y))]_{\gamma\beta} &= \\ = \delta_{\alpha\beta} \delta(x-y) \end{aligned} \tag{2.11}$$

Both terms in (2.9) act like derivatives, i.e.

$$R(X_1 X_2) = RX_1 \cdot X_2 + X_1 RX_2 \text{ , etc.} \tag{2.12}$$

Since the fermionic integration has already been performed in (2.9), the expectation value can now be represented as a purely bosonic expectation value. One then takes higher order derivatives of $\langle X \rangle$ with respect to λ. For the iteration of this procedure, it is crucial that the expectation value (2.9) can be computed by "undoing" the fermionic integration for the Matthews-Salam determinant but not for the fermionic propagator in (2.9) before applying the R operator again. Thus,

$$\begin{aligned} \langle X \rangle &= \langle X \rangle_{\lambda=0} + \sum_{s=1}^{\infty} \frac{\lambda^s}{s!} \frac{\partial^s}{\partial \lambda^s} \langle X \rangle \Big|_{\lambda=0} = \\ &= \int d\mu(\lambda=0, A) \sum_{s=0}^{\infty} \frac{\lambda^s}{s!} R^s X \Big|_{\lambda=0} \end{aligned} \tag{2.13}$$

The bosonic measure is defined as in (1.7) with $S_B(\lambda, A)$ from (2.1). From (2.12) and (2.13), it is obvious that

$$\begin{aligned} \langle \prod_n A(x_n) \rangle &= \int d\mu(\lambda, A) \prod_n A(x_n) = \\ &= \int d\mu(\lambda=0, A') \prod_n A(x_n, \lambda, A') \end{aligned} \tag{2.14}$$

where

$$\begin{aligned} A(x, \lambda, A') &= (T_\lambda^{-1} A')(x) \equiv \\ &\equiv \sum_{s=0}^{\infty} \frac{\lambda^s}{s!} (R^s A')(x) \Big|_{\lambda=0} \end{aligned} \tag{2.15}$$

Consequently, the prescription (2.15) enables one to rewrite the vacuum expectation values of the interacting theory as vacuum expectation values of transformed fields in a free theory. As (2.14) is true for arbitrary monomials of the field operators, it follows that

$$d\mu(\lambda, A) = d\mu(\lambda = 0, A'(\lambda, A)) \tag{2.16}$$

To prove parts (ii) and (iii) of the main theorem, it is sufficient to show that

$$S_B(\lambda, A) = S_B(\lambda = 0, A'(\lambda, A)) \tag{2.17}$$

since the equality of determinants (part (iii)) then follows by (2.16). Equation (2.17) means that the bosonic action is a fixed point of the transformation (2.15) and follows if one can show that

$$R\, S_B(\lambda, A) = 0 \tag{2.18}$$

Writing out (2.18) explicitly, one gets two terms according to (2.8), namely

$$- \underbrace{\delta_\alpha S_B \cdot \Delta_\alpha}_{} = + \underbrace{\Delta_\alpha \; \delta_\alpha}_{} S_B =$$

$$= \frac{1}{2} \int dx\, p(A) \underbrace{\psi_\alpha \cdot \int dy\, \bar{\psi}_\alpha}_{} \,\Box A \; - \tag{2.19}$$

$$- \frac{1}{2} \int dx\, p(A) \underbrace{\psi_\alpha \cdot \int dy\, \bar{\psi}_\alpha}_{} \cdot \Big[m A^2 +$$

$$+ m\lambda \left(p(A) + A p'(A) \right) + \lambda^2 p(A) p'(A) \Big]$$

and

$$\frac{\partial S_B}{\partial \lambda} = \int dx \left(m A p(A) + \lambda p^2(A) \right) =$$

$$= \frac{1}{2} \delta_{\alpha\alpha} \int dx\, dy\, p(A(x))\, \delta(x-y) \left[m A(y) + \lambda p(A(y)) \right] =$$

$$= \frac{1}{2} \int dx\, p(A) \underbrace{\psi_\alpha \cdot \int dy\, \bar{\psi}_\beta}_{} \cdot \Big[- \overleftarrow{\not\partial} + m + \tag{2.20}$$

$$+ \lambda p'(A) \Big]_{\beta\alpha} \left(m A + \lambda p(A) \right)$$

where (2.11) has been used. Clearly, the terms without derivative in (2.20) cancel the second term in (2.19). To take care of the remainder, one integrates the derivative term in (2.20) by parts, using (2.10). This yields

$$\frac{1}{2}\int dx\, p(A)\,\gamma^{\mu}_{\alpha\beta}\underbrace{\psi_\beta \int dy\, \bar{\psi}_\alpha}\,\partial_\mu A\,(m+\lambda p'(A)) =$$

$$= \frac{1}{2}\int dx\,dy\; p(A(x))\,\gamma^{\mu}_{\alpha\beta}\,\partial_\mu A(y)\Big[\delta_{\alpha\beta}\,\delta(x-y) +$$

$$+ \underbrace{\psi_\beta(x)\,\partial^\nu \bar{\psi}_\gamma(y)}\,\gamma^{\nu}_{\gamma\alpha}\Big] = \tag{2.21}$$

$$= -\frac{1}{2}\int dx\, p(A)\,\underbrace{\psi_\alpha \cdot \int dy\, \bar{\psi}_\alpha}\;\Box A$$

Adding (2.21) and the uncancelled term in (2.19), we arrive at the desired result (2.18).

As an example, we take $p(A) = A^3$. Then, it is not difficult to show that

$$(RA)(x)\big|_{\lambda=0} = -m\int C(x-y)\,A^3(y)\,dy$$

$$(R^2A)(x)\big|_{\lambda=0} = 6m^2\int C(x-y)\,A^2(y)\,C(y-z)\,A^3(z)\,dy\,dz$$

$$+\,3\int \partial_\mu C(x-y)\,A^2(y)\,\partial_\mu C(y-z)\,A^3(z)\,dy\,dz \tag{2.22}$$

with the free field propagator

$$C(x) = \int \frac{e^{ikx}}{k^2 + m^2} \frac{d^2k}{(2\pi)^2} \qquad (2.23)$$

Substituting (2.22) into (2.15) and inverting, one obtains

$$(T_\lambda A)(x) = A'(x, \lambda, A) =$$
$$= A(x) + m\lambda \int C(x-y) A^3(y) dy -$$
$$- \tfrac{3}{2} \lambda^2 \int \partial_\mu C(x-y) A^2(y) \partial_\mu C(y-z) A^3(z) dy\, dz$$
$$+ O(\lambda^3) \qquad (2.24)$$

which coincides with the result given in [2], eq.(27).

Remarkably, the expansion (2.15) admits a graphical representation [19]. For instance, (2.24) corresponds to

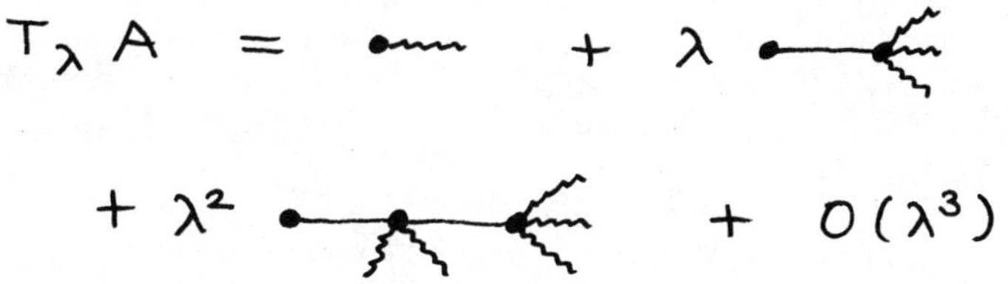

$$T_\lambda A = \cdots + \lambda \cdots$$
$$+ \lambda^2 \cdots + O(\lambda^3) \qquad (2.25)$$

with self-explanatory notation: wavy lines correspond to scalar fields and straight lines represent free spinor propagators which are contracted over their spinor indices. The action of R is schematically represented by two "insertions"

$$R(\cdots) = \cdots \qquad (2.26)$$

and

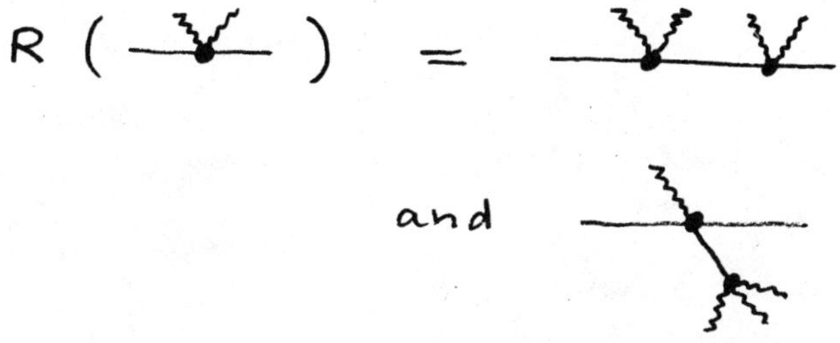

which are the graphical analogues of the two terms in the definition of R, see (2.9). A most important consequence of (2.26) is that only tree graphs occur in the expansion (2.15). Since the number of tree graphs with n vertices is bounded by $n!\, K^n$, where K is proportional to the number of wavy lines running into a vertex, it follows that the expansion (2.15) has a non-vanishing radius of convergence. This can be intuitively understood as the "combinatorial divergences" of ordinary perturbation theory (i.e. uncancelled factors of n!) are produced by functional integration, i.e. Wick's theorem, and the expansion (2.15) is done before integrating over the boson fields. Of course, the perturbation expansion (2.15) with subsequent integration over the boson fields as in (2.14) is entirely equivalent to the usual perturbation expansion. The problem of ultraviolet divergences appears in a somewhat different guise. In expressions such as (2.24), one has to deal with products of fields at the same space-time point. Since the latter are distributions (or possibly even more singular mathematical entities) such products are ill-defined unless a regularization is introduced, e.g. by smearing the fields with C^∞-functions of compact support. The existence of the renormalized theory is then equivalent to the existence of a suitably "renormalized" version of (2.15) as a nonlinear and non-local map of a distribution space into itself. Sometimes a finite volume cutoff is required which is effected by switching on the interaction only in a finite region $\Lambda \subset \mathbb{R}^d$. Since supersymmetry is then broken by surface terms all statements have to be modified. It is not difficult to show that the main theorem remains valid in this case up to terms which have support on the boundary $\partial\Lambda$ of the region Λ (for a discussion of these matters in the context of the Wess-Zumino model, see [23]).

Furthermore, the divergences which can appear can be read off from the expansion (2.25); performing the bosonic integration is equivalent to connecting wavy lines in all possible ways and summing over all these diagrams*. As an exercise the reader may verify that the Wess-Zumino model in four dimensions is only logarithmically divergent. Finally, "stochastic" models are naturally recovered in the present frame-work: for these, the expansion (2.14) turns out to be a geometric series whose inverse is local in the sense of e.g. eq.(1.14) [24]. Stochastic variables are nothing but local fixed points of (2.15).

*This approach to perturbation theory was first suggested in [7] and has been termed "infradiagrammar".

3. GLOBAL ASPECTS

The piecewise constancy of the vacuum functional (1.2) in a supersymmetric theory is evidently a consequence of the equivalence of the functional integration measure (1.7) to that of a free theory. However, this equivalence is valid locally in the space of bosonic field configurations, and, when integrating over this space, global properties will become significant. Substituting formula (2.16) into (1.2), one must take into account how many times the image space is covered when $A(x)$ assumes all possible values. Hence,

$$Z(\lambda) = \int d\mu(\lambda, A) = \int d\mu(\lambda = 0, A'(\lambda, A)) = \text{winding number of } T_\lambda \tag{3.1}$$

This shows that $Z(\lambda)$ is a topological invariant but of a different type from the one usually encountered in field theories since it characterizes mappings of infinite-dimensional spaces. It was first recognized by the authors of [7] , that (3.1) can be identified with the Witten index [25],

$$Z(\lambda) = n_B(E=0) - n_F(E=0) \tag{3.2}$$

Since winding numbers in infinite dimensional spaces are rather awkward to deal with, it is desirable to have a simplified prescription which allows one to compute (3.1) efficiently. It was argued in [26] that it is sufficient to determine the winding number in the ultralocal limit where kinetic terms are neglected*. This amounts to introducing a factor z in front of all kinetic terms and a factor of z^{-1} in front of the scalar field potential and then taking the limit $z \to 0$. E.g.,for the Wess-Zumino model in four dimensions [27], the transformation T_λ reads

$$\mathcal{A}'(x, \lambda, \mathcal{A}) = \mathcal{A}(x) + m\lambda \int C(x-y)\mathcal{A}^2(y)\,dy + O(\lambda^2) \tag{3.3}$$

*I.e., (3.1) is equal to an ordinary ("zero-dimensional") integral such as (3.10) below, or, equivalently, the winding number is continuous in the wave function renormalization parameter z at $z=0$. To be sure, a rigorous justification for this recipe remains to be given, and the reader should keep in mind the somewhat heuristic character of the arguments in this chapter.

with a complex field $\mathcal{A} = A + iB$ and the free (now Z -dependent) propagator

$$C(x) = \int \frac{e^{ikx}}{Z^2k^2+m} \frac{d^4k}{(2\pi)^4} \tag{3.4}$$

Making use of $\lim_{Z\to 0} C(x) = m^{-2}\delta(x)$ and the fact that all higher order terms in (3.3) contain at least one factor of Z and therefore disappear in the limit $Z \to 0$, one gets

$$\lim_{Z\to 0} \mathcal{A}'(x,\lambda,\mathcal{A}) = \mathcal{A}(x) + \frac{\lambda}{m}\mathcal{A}^2(x) \tag{3.5}$$

Since $\mathcal{A}$ is complex, the winding number of (3.5) is two. Assuming continuity of the winding number at $Z=0$, one arrives at the same conclusion for the full nonlocal transformation (3.3). Evidently, the mapping (3.5), although being locally invertible almost everywhere, is no longer globally invertible. This result is in agreement with the existence of two supersymmetric ground states in the Wess-Zumino model and the fact that the Witten index equals two in this case.

The above considerations are also relevant when supersymmetry is spontaneously broken. In this case, the vacuum energy no longer vanishes but (2.4) remains true, and therefore the construction (2.15) still makes sense. As an example, one may consider the O'Raifeartaigh model which exhibits spontaneous breakdown of supersymmetry [28]. By the previous arguments it is sufficient to study this model in the ultralocal limit. Moreover, I introduce an artificial parameter μ such that supersymmetry is restored for $\mu \neq 0$ whereas the original model is recovered by putting $\mu = 0$. The model contains three complex scalar fields $\mathcal{A}_1, \mathcal{A}_2, \mathcal{A}_3$; the relevant quantity is the superpotential

$$W(\mathcal{A}_1,\mathcal{A}_2,\mathcal{A}_3) = f\mathcal{A}_1 + m\mathcal{A}_2\mathcal{A}_3 + g\mathcal{A}_1\mathcal{A}_3^2 + \mu\mathcal{A}_1^2 \tag{3.6}$$

(I assume f, m and g to be non-zero), from which the ordinary potential can be derived in the usual fashion [1]. It is equal to the sum $\sum_{i=1,2,3} |\mathcal{A}_i'|^2$ where

$$\mathcal{A}_1' = \frac{\partial W}{\partial \mathcal{A}_1} = f + 2\mu \mathcal{A}_1 + g\mathcal{A}_3^2$$

$$\mathcal{A}_2' = \frac{\partial W}{\partial \mathcal{A}_2} = m\mathcal{A}_3 \tag{3.7}$$

$$\mathcal{A}_3' = \frac{\partial W}{\partial \mathcal{A}_3} = m\mathcal{A}_2 + 2g\mathcal{A}_1\mathcal{A}_3$$

Supersymmetry is spontaneously broken if there is no value of $\mathcal{A}_i^{(0)}$ for which $\mathcal{A}_i'(\mathcal{A}^{(0)}) = 0$. For $\mu \neq 0$, one sees that $\mathcal{A}_i' = 0$ if

$$\mathcal{A}_i^{(0)} = (-f/2\mu, 0, 0) \tag{3.8}$$

so the ground state is supersymmetric for any $\mu \neq 0$. For $\mu = 0$, however, the minimum (3.8) disappears and supersymmetry is spontaneously broken. To construct the transformation T, we first take $\mu \neq 0$, apply the procedure of the foregoing section and take the limit $\mu \to 0$ only afterwards. Again it can be shown that, in the ultralocal limit, higher order terms do not contribute and the transformation is reduced to the local mapping (3.7). It is quite instructive to exhibit the discontinuity at $\mu = 0$ explicitly by deriving the "zero-dimensional" vacuum functional from (3.7). In the ultralocal limit, the fermionic determinant is found to be (after a little algebra)

$$\det M(\mathcal{A})^{1/2} = \left| \det \frac{\partial \mathcal{A}_i'}{\partial \mathcal{A}_j} \right|^2 = 4m^4\mu^2 \tag{3.9}$$

The "zero-dimensional" vacuum functional is therefore

$$Z(f, \mu, m, g) = \int d\mathcal{A}\, d\bar{\mathcal{A}} \prod_{i=1}^{3} \delta(\mathcal{A}_i'(\mathcal{A})) \left| \det \frac{\partial \mathcal{A}_i'}{\partial \mathcal{A}_j} \right|^2 \tag{3.10}$$

(the δ-function comes from the factor Z^{-1} in front of the potential in the limit $Z \to 0$) and, by a well-known formula, one obtains

$$
\mathcal{Z}(f,\mu,m,g) = \int_{\mathbb{C}^3} d\mathcal{A}' d\bar{\mathcal{A}}' \sum_{\mathcal{A}'(\mathcal{A})=0} \prod_i \delta(\mathcal{A}'_i) = \tag{3.11}
$$

$$
= \begin{cases} 1 & \text{for} \quad \mu \neq 0 \\ 0 & \text{for} \quad \mu = 0 \end{cases}
$$

Note that (3.11) stays well defined for $\mu = 0$ although the determinant (3.9) vanishes there. Clearly, the derivative of $\mathcal{Z}$ with respect to any of its parameters vanishes, so $\mathcal{Z}$ is indeed piecewise constant, but discontinuous on the "critical surface" $\mu = 0$. (for f, m, g $\neq$ 0). The analysis of winding numbers leads to the same conclusion. For $\mu \neq 0$, the mapping (3.7) is one-to-one as one may easily verify, whereas, for $\mu = 0$, (3.7) maps $\mathbb{C}^3$ into a lower-dimensional submanifold of $\mathbb{C}^3$; since the image of $\mathbb{C}^3$ has measure zero in this case, the winding number equals zero for $\mu = 0$ in agreement with (3.11).

Quite generally, one might expect the cases where supersymmetry breaking occurs to be in one-to-one correspondence with the cases where the image space of the transformation (1.8) has lower dimension (although this "façon de parler" begs for some refinement if the relevant spaces are infinite dimensional), but this is not quite true. I have so far glossed over the fact that it is not the Jacobian but rather its modulus that appears in the textbook formula for the change of variables in an integral. In many cases of interest, one can actually prove that the Matthews Salam determinant is positive for arbitrary values of the scalar (or spin-1) fields*. If it is not , then another kind of global obstruction may arise in doing the integral (3.1). To have an example in this case as well, let us have a look at the ultralocal limit of supersymmetric quantum mechanics (1.10) [7,8]. The analogue of (3.10)

*E.g. (3.9) is manifestly positive as it is for all Wess-Zumino type models in the ultralocal limit.

now reads

$$Z = \int_{-\infty}^{+\infty} \delta(V(q))\, V'(q)\, dq \tag{3.12}$$

To evaluate (3.12), one must distinguish between two possibilities. If V runs from $-\infty$ to $+\infty$, the number of points q with $V(q)=0$ is odd, and, for simplicity, I assume that $V(q)$ is monotonous. (3.12) may thus be replaced by*

$$Z = \pm \int_{-\infty}^{+\infty} \delta(V(q))\, |V'(q)|\, dq = \pm 1 \tag{3.13}$$

In this case supersymmetry breaking is impossible. If, on the other hand, $V(q)$ covers a subset of $\mathbb{R}$ only, say the interval $[\xi_0, \infty]$, then

$$Z = \int_{-\infty}^{+\infty} \delta(V(q))\, \mathrm{sgn}\, V'(q)\, |V'(q)|\, dq =$$
$$= \int_{\xi_0}^{\infty} \delta(\xi)\, [1-1]\, d\xi = 0 \tag{3.14}$$

and supersymmetry breaking is possible. (3.13) and (3.14) correspond to convergence or non-convergence of the integral $\int_{-\infty}^{+\infty} \exp[\pm W(q)]\, dq$ where $W(q)$ is the superpotential; this is just the criterion given in [25]. As the foregoing discussion shows, Z may vanish although the image of (1.8) is <u>not</u> lower dimensional. This phenomenon is, of course, due to the change in sign of the determinant, but I do not know whether a similar effect can occur for more interesting models in higher dimensions.

4. SUPERSYMMETRIC GAUGE THEORIES

Supersymmetric gauge theories [29] are more interesting but also more difficult to deal with in the present framework than the models considered so far. The main source of difficulties is, of

---

*The occurrence of a negative winding number is evidently related to the orientation of the real axis and somewhat artificial. It is, however, not relevant for this argument.

course, the additional gauge symmetry which necessitates a gauge fixing procedure [30] that either explicitly violates supersymmetry [31] or, through additional ghost supermultiplets, renders the theory considerably more complicated [32]. Nonetheless, the vacuum energy still vanishes for supersymmetric ground states, and one therefore expects some version of the main theorem to remain true. Before stating the revised version of the theorem, however, some more notation must be introduced.

For the sake of clarity, only the pure N=1 Yang Mills theory in four dimensions will be considered in the following. For the same reason, I will adopt a non-supersymmetric gauge fixing procedure. The model is based on a supermultiplet containing one spin-1 field $A_\mu^a(x)$ and a Majorana spinor $\lambda^a(x)$, both in the adjoint representation of some gauge group. As before, auxiliary fields are assumed to have been eliminated since the question of off-shell versus on-shell representations is not relevant in the present context. In Euclidean signature, the invariant Lagrangian is

$$\mathcal{L}_0 = \frac{1}{4} F_{\mu\nu}^a F_{\mu\nu}^a + \frac{1}{2} \bar{\lambda}^a \gamma^\mu D_\mu \lambda^a \tag{4.1}$$

with

$$\begin{aligned} F_{\mu\nu}^a &\equiv \partial_\mu A_\nu^a - \partial_\nu A_\mu^a + g f^{abc} A_\mu^b A_\nu^c \\ D_\mu \lambda^a &\equiv \partial_\mu \lambda^a + g f^{abc} A_\mu^b \lambda^c \end{aligned} \tag{4.2}$$

(4.1) is invariant up to a total derivative under the supersymmetry transformations

$$\delta A_\mu^a = \varepsilon_\alpha \delta_\alpha A_\mu^a \quad , \quad \delta \lambda_\beta^a = \varepsilon_\alpha \delta_\alpha \lambda_\beta^a \tag{4.3}$$

where

$$\begin{aligned} \delta_\alpha A_\mu^a &= -(\bar{\lambda}^a \gamma_\mu)_\alpha \\ \delta_\alpha \lambda_\beta^a &= (\sigma^{\mu\nu})_{\beta\alpha} F_{\mu\nu}^a \end{aligned} \tag{4.4}$$

The Matthews Salam determinant $D(g, A_\mu^a)$ is defined from (4.1) as in (1.5)

$$\det M(g, A)^{1/2} = \det M(0, A)^{1/2} D(g, A) \tag{4.5}$$

where

$$\begin{aligned} M^{ac}_{\alpha\beta}(x, y; g, A) &= \\ &= \left[\delta^{ac}\gamma^\mu_{\alpha\beta}\partial_\mu + g f^{abc}\gamma^\mu_{\alpha\beta}A^b_\mu(x)\right]\delta(x-y) \end{aligned} \tag{4.6}$$

The superfluous gauge degrees of freedom in (4.1) are eliminated by adding a gauge fixing term $\frac{1}{2\alpha} G^a[A_\mu]^2$ to the Lagrangian (4.1). The gauge fixing function may be taken as, for example,

$$G^a[A_\mu] = \partial^\mu A^a_\mu \tag{4.7}$$

Other choices such as the axial gauge are possible, but in any case, I will impose the requirement

$$\frac{\partial}{\partial g} G^a[A_\mu] = 0 \tag{4.8}$$

The explicit breaking of gauge invariance must be compensated for by weighting the functional integration measure with the Fadeev-Popov determinant [30]

$$\Delta_{F.P.}(g, A_\mu) = \det \frac{\delta G^a[A_\mu]}{\delta A^b_\mu} D_\mu \delta(x-y) \tag{4.9}$$

(In the following I will not explicitly distinguish between $\Delta_{F.P.}(g, A)$ and $\Delta_{F.P.}(g, A) \cdot \Delta^{-1}_{F.P.}(0, A)$). Rewriting (4.9) by means of the Fadeev-Popov ghosts C^a, $\bar{C}^a$ in the usual way, the gauge fixing procedure results in the additional contribution to the Lagrangian (4.1)

$$\mathcal{L}_1 = \frac{1}{2\alpha} G^a[A_\mu]^2 + \bar{C}^a \frac{\delta G^a}{\delta A^b_\mu} D_\mu C^b \tag{4.10}$$

The combined Lagrangian $\mathcal{L}_0 + \mathcal{L}_1$ is then invariant under the BRS-transformations [33]

$$s A_\mu^a = D_\mu C^a \quad , \quad s\lambda^a = -g f^{abc} C^b \lambda^c$$

$$s C^a = -\frac{g}{2} f^{abc} C^b C^c \quad , \quad s\bar{C}^a = -\frac{1}{\alpha} G^a[A_\mu] \tag{4.11}$$

The full bosonic integration measure of the theory is obtained by integrating out the fermionic variables, i.e. the Majorana spinors and the ghosts; it is

$$d\mu(g, A_\mu) = \exp\left[-\int dx \left(\frac{1}{4}(F_{\mu\nu}^a)^2 + \frac{1}{2\alpha} G^a[A_\mu]^2\right)\right]$$

$$D(g, A_\mu)\, \Delta_{F.P.}(g, A)\, dA \tag{4.12}$$

The revised version of the theorem can then be stated as follows:

Theorem: Supersymmetric gauge theories are characterized by the existence of a transformation T_g of the Yang-Mills fields

$$T_g : A_\mu^a(x) \longrightarrow A_\mu'^a(x, g, A_\mu) \tag{4.13}$$

which is invertible at least in the sense of formal power series such that

(i) The Yang-Mills action without gauge-fixing term is mapped into a bosonic (i.e. abelian) action;

(ii) The Jacobi determinant of T_g is equal to the product of the Matthews-Salam and Fadeev-Popov determinants, i.e.

$$\det \frac{\delta A_\mu'^a(x, g, A)}{\delta A_\nu^b(y)} = \Delta_{F.P.}(g, A)\, D(g, A) \tag{4.14}$$

at least order by order in perturbation theory;

(iii) The gauge fixing function $G^a[A_\mu]$ is a fixed point of T_g.

For the proof, we again follow the prescription of ref. [19] *). One first notes that the supersymmetric Ward identity is no longer valid in the form (2.7) but rather, owing to the non-invariance of (4.10) under (4.4),

$$\langle \delta_\alpha Y \rangle = \langle \delta_\alpha S_1 \cdot Y \rangle \tag{4.15}$$

where $S_1 = \int \mathcal{L}_1 \, dx$ and the brackets stand for integration over all fields including fermions and ghosts. After some rearrangement, (4.15) can be written in the form

$$\langle \delta_\alpha Y \rangle = - \langle \int dx \, \bar{C}^a \, \delta_\alpha G^a[A_\mu] \, s(Y) \rangle \tag{4.16}$$

As in section 2, one is interested in constructing a suitable "R-operator". This can be derived by acting with the derivative on the expectation value of an arbitrary string of bosonic fields There are two terms from the exponential of the full action, namely

$$\frac{\partial}{\partial g} S_0 = \frac{1}{2} f^{abc} \int dx \left(F^a_{\mu\nu} A^b_\mu A^c_\nu + \bar{\lambda}^a \gamma^\mu A^b_\mu \lambda^c \right) =$$

$$= \delta_\alpha \Delta_\alpha - \frac{1}{4} f^{abc} \int dx \, \bar{\lambda}^a \gamma^\mu A^b_\mu \lambda^c \tag{4.17}$$

with the "operator insertion"

$$\Delta_\alpha \equiv - \frac{1}{4} f^{abc} \int dx \, (\sigma^{\mu\nu} \lambda^a)_\alpha A^b_\mu A^c_\nu \tag{4.18}$$

and

$$\frac{\partial}{\partial g} S_1 = \int dx \, \bar{C}^a \frac{\delta G^a}{\delta A^b_\mu} f^{bcd} A^c_\mu C^d \tag{4.19}$$

*I am indebted to O. Lechtenfeld for informing me about the results of [24], on which this derivation up to (4.20) is based, prior to publication.

where (4.8) has been taken into account. After a somewhat lengthy calculation, the correct form of the R-operator is found to be [13, 19, 24]

$$
\begin{aligned}
RX = {} & \frac{\partial X}{\partial g} + \underbrace{\delta_\alpha X \cdot \Delta_\alpha} + \\
& + \int dx \, \underbrace{\bar{C}^a \, \underbrace{\delta_\alpha G^a[A_\mu] \, \Delta_\alpha} \, s(X)}
\end{aligned}
\tag{4.20}
$$

In complete analogy with (2.9), this is again a purely bosonic expression. The fermion and ghost propagators appearing in (4.20) are given by, respectively,

$$
\begin{aligned}
\underbrace{(\not{D}\lambda)^a_\alpha(x) \, \bar{\lambda}^b_\beta}(y) &= \delta_{\alpha\beta} \, \delta^{ab} \, \delta(x-y) \\
\underbrace{(\bar{\lambda} \overleftarrow{\not{D}})^a_\alpha(x) \, \lambda^b_\beta}(y) &= \delta_{\alpha\beta} \, \delta^{ab} \, \delta(x-y)
\end{aligned}
\tag{4.21}
$$

and

$$
\begin{aligned}
\frac{\delta G^a}{\delta A^b_\mu} D_\mu \underbrace{C^b(x) \, \bar{C}^c}(y) &= \delta^{ac} \, \delta(x-y) \\
D_\mu \Big(\frac{\delta G^a}{\delta A^b_\mu} \underbrace{\bar{C}^b(x) \Big) \, C^c}(y) &= - \delta^{ac} \, \delta(x-y)
\end{aligned}
\tag{4.22}
$$

The crucial difference between (2.9) and (4.20) is the third term on the right side of (4.20) that disappears only for gauge invariant operators X which obey $s(X) = 0$.

The next step is to apply (4.20) to prove parts (i) and (iii) of the revised theorem; (ii) then follows as before. Let us begin with $X_0 = \frac{1}{4} \int F^a_{\mu\nu} F^a_{\mu\nu}$; the relevant expressions are

$$
\begin{aligned}
& \underbrace{\delta_\alpha X_0 \cdot \Delta_\alpha} = \\
& = \frac{1}{4} \int dx \, F^a_{\mu\nu} (D_\mu \underbrace{\bar{\lambda}^a \gamma_\nu)_\alpha \, f^{bcd} \int dy \, (\sigma^{\rho\sigma} \lambda^b}{)_\alpha} \, A^c_\rho A^d_\sigma
\end{aligned}
\tag{4.23}
$$

and

$$\frac{\partial}{\partial g} X_0 = \frac{1}{2} f^{abc} \int dx\, F^a_{\mu\nu} A^b_\mu A^c_\nu =$$

$$= \frac{1}{2} \delta^{ab} \delta^{\rho\sigma}_{\mu\nu} f^{bcd} \int dx\,dy\, F^a_{\mu\nu}(x)\, \delta(x-y)\, A^c_\rho(y)\, A^d_\sigma(y) \tag{4.24}$$

This can be reexpressed by writing (4.21) in the form

$$D_\tau \underbrace{\bar{\lambda}^a(x)\, \gamma^\tau \sigma^{\mu\nu} \sigma_{\rho\sigma} \lambda^b(y)} = -2 \delta^{ab} \delta^{\rho\sigma}_{\mu\nu}\, \delta(x-y) \tag{4.25}$$

and using the identity

$$\gamma^\tau \sigma^{\mu\nu} = \delta^{\tau[\mu} \gamma^{\nu]} + \gamma^{[\tau} \sigma^{\mu\nu]} \tag{4.26}$$

In this way, (4.24) becomes, after partial integration,

$$\frac{\partial}{\partial g} X_0 = -\frac{1}{4} \int dx\, F^a_{\mu\nu} (D_\mu \underbrace{\bar{\lambda}^a \gamma_\nu)_\alpha f^{bcd} \int dy\, (\sigma^{\rho\sigma} \lambda^b}\!)_\alpha A^c_\rho A^d_\sigma$$

$$+ \frac{1}{8} f^{bcd} \int dx\, D_{[\tau} F^a_{\mu\nu]}\, \underbrace{\bar{\lambda}^a \gamma^\tau \sigma^{\mu\nu} \int dy\, \sigma^{\rho\sigma} \lambda^b}\, A^c_\rho A^d_\sigma \tag{4.27}$$

The first term in (4.27) cancels (4.23) while the second term vanishes by the Bianchi identity. Finally, there is no contribution from the third term in (4.20) as X_0 is a gauge invariant operator. Altogether, it has been shown that

$$R X_0 = R\left(\frac{1}{4} \int F^a_{\mu\nu} F^a_{\mu\nu}\, dx \right) = 0 \tag{4.28}$$

which proves part (i) of the theorem. Part (iii) can be shown in a similar fashion; this time the first term in (4.20) does not contribute because of (4.8). The second term gives

$$\underbrace{\delta_\alpha G^a[A_\mu(x)]\, \Delta_\alpha} =$$

$$= \frac{1}{4} \frac{\delta G^a}{\delta A^b_\mu} (\underbrace{\bar{\lambda}^b(x) \gamma_\mu)_\alpha f^{cde} \int dy\, (\sigma^{\nu\rho} \lambda^c}\!)_\alpha A^d_\nu A^e_\rho \tag{4.29}$$

while the third one gives

$$\int dy\, \bar{C}^b\, \delta_\alpha G^b\, \Delta_\alpha\; s\left(G^a[A_\mu(x)]\right) =$$

$$= \frac{1}{4} \int dy\, \bar{C}^f\, \frac{\delta G^f}{\delta A^g_\nu} (\bar{\lambda}^g \gamma_\nu)_\alpha\, f^{cde} \,\cdot$$

$$\cdot \int dz\, (\sigma^{\rho\sigma} \lambda^c)_\alpha\, A^d_\rho A^e_\sigma \cdot \frac{\delta G^a}{\delta A^b_\mu} D_\mu C^b(x) \tag{4.30}$$

$$= -\frac{1}{4} \frac{\delta G^a}{\delta A^b_\mu} (\bar{\lambda}^b(x) \gamma_\mu)_\alpha\, f^{cde} \int dz\, (\sigma^{\rho\sigma} \lambda^c)_\alpha\, A^d_\rho A^e_\sigma$$

where, in the last line, the definition (4.22) of the ghost propagator has been used. Combining (4.29) and (4.30), we get

$$R\, G^a[A_\mu(x)] = 0 \tag{4.31}$$

which is part (iii) of the theorem. Note that (4.31) is a local statement whereas (4.28) expresses the fixed point property of an integrated quantity (in the second case, the integral was needed for the partial integration in (4.27)). The importance of (4.31) resides in the fact that it allows us to choose the gauge fixing parameter α in (4.10) arbitrarily and to take the limit $\alpha \to 0$; all results remain valid in the singular gauge where the functional integration measure acquires a factor $\prod \delta(G^a[A_\mu(x)])$. Thus, the gauge fixing function $G^a[A_\mu]$ may be considered as a stochastic variable for arbitrary gauges.

It is possible to obtain another stochastic variable by slightly modifying the prescription (4.20) [13]. To do so, one defines

$$R_\pm X \equiv \frac{\partial X}{\partial g} + 2\, \delta_\alpha X \cdot \Delta^\pm_\alpha + \\ + 2 \int dx\, \bar{C}^a\, \delta_\alpha G^a[A_\mu]\, \Delta^\pm_\alpha\, s(X) \tag{4.32}$$

with the chirally projected insertions

$$\Delta_\alpha^\pm \equiv \frac{1}{2}(1 \pm \gamma^5)_{\alpha\beta}\Delta_\beta \tag{4.33}$$

Let us apply (4.32) to $F_{\mu\nu}^{a(\pm)} \equiv \frac{1}{2}F_{\mu\nu}^a \pm \frac{1}{4}\varepsilon_{\mu\nu\rho\sigma}F_{\rho\sigma}^a$ and fix the sign for definiteness. As in (4.23-27), the first two terms in (4.32) give

$$2\,\underbrace{\delta_\alpha F_{\mu\nu}^{a+}\cdot\Delta_\alpha^-} + \frac{\partial}{\partial g}F_{\mu\nu}^{a+} =$$

$$= \frac{1}{4}\underbrace{(D_{[\mu}\bar{\lambda}^a\gamma_{\nu]} + \frac{1}{2}\varepsilon_{\mu\nu\alpha\beta}D_\alpha\bar{\lambda}^a_\beta)}f^{bcd}\underbrace{\int dy\,\sigma^{\rho\sigma}\lambda_-^b}A_\rho^c A_\sigma^d$$

$$-\frac{1}{8}D_\tau\underbrace{\bar{\lambda}^a(\gamma^{\tau\mu\nu} + \frac{1}{2}\gamma^{\tau\alpha\beta}\varepsilon_{\alpha\beta\mu\nu})f^{bcd}\int dy\,\sigma^{\rho\sigma}\lambda_-^b}A_\rho^c A_\sigma^d \tag{4.34}$$

where, of course, $\lambda_- \equiv \frac{1}{2}(1-\gamma^5)\lambda$. Now, in contrast to (4.27), the last term in (4.34) obviously cannot be eliminated by partial integration. Instead, one must use the identity

$$\gamma^{\tau\mu\nu} + \frac{1}{2}\gamma^{\tau\alpha\beta}\varepsilon_{\alpha\beta\mu\nu} = -2\left(\delta_{\mu\nu}^{\tau\lambda} + \frac{1}{2}\varepsilon_{\mu\nu\tau\lambda}\right)\gamma^\lambda\gamma^5 \tag{4.35}$$

together with $\gamma^5\lambda_- = -\lambda_-$, to see that (4.34) vanishes (note the factors of 2 in (4.32)). The third term in (4.32) produces a gauge rotation on $F_{\mu\nu}^{a+}$ which vanishes on the hypersurface $G[A_\mu] = 0$ [13]. Altogether, this implies that the (anti)self-dual field strengths

$$F_i^{a(\pm)} \equiv F_{0i}^a \pm \frac{1}{2}\varepsilon_{ijk}F_{jk}^a\,,\quad i,j,k = 1,2,3 \tag{4.36}$$

constitute fixed points of R_- and R_+,respectively.

It was already pointed out in [11] that the stochastic variables (4.36) may be employed to "take the square-root of the Yang-Mills action" because

$$\frac{1}{4}\int F_{\mu\nu}^a F_{\mu\nu}^a\,dx = \frac{1}{2}\int F_i^a F_i^a\,dx \pm Q(A) \tag{4.37}$$

with the topological charge Q(A). (The analogy of (4.37) with (1.10 - 15) is quite striking). It has furthermore been shown by exploiting "stochastic identities" that either $F_i^{a(+)}$ or $F_i^{a(-)}$ satisfy free field vacuum expectation values and therefore are indeed the Gaussian varibales we have been looking for [12]. In fact, the equality (4.14) may be demonstrated directly in Minkowski space by cleverly choosing a light-cone gauge [12,14] ; however, the relevant "Jacobian" there is complex and not obviously intepretable as an ordinary Jacobian. There is another interesting aspect of (4.37) and the present approach in general [11]. If one regards the bosonic action $S_B(\lambda, A)$ or (4.37) as a "Morse function" [34] on the manifold of bosonic field configurations it follows from the main theorem that, among all possible Morse transformations, there exists one and only one whose Jacobian equals a Matthews-Salam determinant (or (4.14)) if the bosonic action can be "supersymmetrized". Ordinarily, in Morse theory, such Morse functions are used to classify the topological propeties of certain (finite-dimensional) manifolds [34]. In contradistinction to the case of scalar field theories considered in section two, for which the relevant "manifolds" are topologically trivial, this aspect may deserve some study for gauge theories. There, the relevant manifold $\mathcal{O}\!\mathcal{L}/\mathcal{O}\!\mathcal{J}$, i.e. the space of gauge-equivalent gauge field configurations, does exhibit non-trivial topology, and (4.36) may be regarded as the Morse transformation which renders the "Morse-function (4.37) Gaussian.

To conclude this section, I give the explicit form of the transformation to second order in the gauge (4.7). It is [2]

$$\begin{aligned} A'^{a}_{\mu}(x,g,A) &= A^{a}_{\mu}(x) + g f^{abc}\int dy\, \partial_\nu C(x-y) A^{b}_{\mu}(y) A^{c}_{\nu} \\ &+ \frac{1}{2} g^2 f^{abc} f^{bde}\int dy\, dz\, \partial_\rho C(x-y) A^{c}_{\lambda}(y) \cdot \\ &\cdot \left[2\, \partial_{[\rho} C(y-z) A^{d}_{\mu]}(z) A^{e}_{\lambda}(z) + \right. \\ &\left. + \partial_\lambda C(y-z) A^{d}_{\rho}(z) A^{e}_{\mu}(z) \right] + O(g^3) \end{aligned} \tag{4.38}$$

from which the reader may verify the theorem directly up to second order. In particular, it is easily seen that

$$\partial^\mu A'^{a}_{\mu}(x) = \partial^\mu A^{a}_{\mu}(x) + O(g^3) \tag{4.39}$$

From the structure of the perturbative construction, of course nothing can be said about possible non-perturbative aspects, and it would be interesting to see what might happen in presence of anomalies of the type encountered in [35] or other anomalies. Nevertheless, the comments made at the end of section two are also relevant here.

ACKNOWLEDGEMENTS

I have benefitted from and enjoyed discussions with R. Flume, S. Fubini, L. Girardello, and O. Lechtenfeld.

REFERENCES

1 P. Fayet and S. Ferrara, Phys. Rep. C32 (1977) 249; J. Wess and J. Bagger, "Supersymmetry and Supergravity", (Princeton University Press, 1982).
2 H. Nicolai. Phys. Lett. 89B (1980) 341; Nucl. Phys. B176 (1980) 419.
3 B. Zumino, Nucl. Phys. B89 (1975) 535.
4 T.Matthews and A. Salam, Nuovo Cim. 12 (1954) 563; Nuovo Cim. 2 (1955) 120.
F.A. Berezin, "The Method of Second Quantization", (Academic Press, New York 1966);
E. Seiler, Comm. Math. Phys. 42 (1975) 163.
5 E. Witten, Nucl. Phys. B188 (1981) 513.
6 H. Nicolai, J. Phys. A9 (1976) 1497.
7 S. Cecotti and L. Girardello, Phys. Lett. 110B (1982) 39; Ann. Phys. 145 (1983) 81.
8 P. Salomonsen and J. W. van Holten, Nucl. Phys. B196 (1982)509.
9 G. Parisi and Wu Yangshi, Sci. Sinica (1981) 483.
10 G. Parisi and N. Sourlas, Phys. Rev. Lett. 43 (1979) 244; Nucl. Phys. B206 (1983) 321.
11 H. Nicolai, Phys. Lett. 117B (1982) 408.
12 V. de Alfaro, S. Fubini, G. Furlan, and G. Veneziano, Phys. Lett. 142B (1984) 399.
13 O. Lechtenfeld, Bonn University preprint HE-84-9 (1984).
14 V. de Alfaro, S. Fubini, G. Furlan, and G. Veneziano, preprint CERN-TH 4021 (1984).
15 H. Ezawa and J. R. Klauder, Bielefeld preprint ZIF No. 74 (1984).
16 S. Cecotti and L. Girardello, Nucl. Phys. B226 (1983) 417; N. Sakai and M. Sakamoto, Nucl. Phys. B229 (1983) 173.
17 H. Nicolai, Nucl. Phys. B140 (1978) 294.
18 K. Osterwalder and R. Schrader, Comm. Math. Phys. 31 (1973)83.
19 R. Flume and O. Lechtenfeld, Phys. Lett. 135B (1984) 91.
20 M. Goltermann, Phys. Lett. 124B (1983) 51.
21 H. Nicolai, Phys. Lett. 101B (1981) 396.
22 B. Zumino, in "Renormalization and Invariance in Quantum Field Theory", (Plenum Press, New York 1973);
S. Ferrara,Lett. Nuovo Cim. 13 (1975) 629;
S. Browne, Phys. Lett. 59B (1979) 253.
23 H. Nicolai, Nucl. Phys. B156 (1979) 157; 177.
24 K. Dietz and O. Lechtenfeld, in preparation.
25 E. Witten, Nucl. Phys. B202 (1982) 253.
26 L. Girardello, C. Imbimbo, and S. Mukhi, Phys. Lett. 132B (1983) 69.

27 J. Wess and B. Zumino, Nucl. Phys. B70 (1974) 39.
28 L. O'Raifeartaigh, Nucl. Phys. B96 (1975) 331.
29 J. Wess and B. Zumino, Nucl. Phys. B78 (1974) 353;
R. Delbourgo, A. Salam, and J. Strathdee, Phys. Lett 51B (1974) 475.
30 L. D. Fadeev and V. N. Popov, Phys. Lett. 25B (1967) 29;
G. 't Hooft, Nucl. Phys. B33 (1971) 173.
31 B. de Wit and D. Z. Freedman, Phys. Rev. D12 (1975) 2286.
32 B. de Wit, Phys. Rev. D12 (1975) 1268;
S. Ferrara and O. Piguet, Nucl. Phys. B93 (1975) 361;
F. Honerkamp, F. Krause, M. Scheunert, and M. Schlindwein, Nucl. Phys. B95 (1975) 397.
33 C. Becchi, A. Rouet, and R. Stora, Comm. Math. Phys. 42 (1975) 127; Ann. Phys. 98 (1976) 287.
34 J. Milnor, "Morse Theory" (Princeton University Press, Princeton, NJ. 1963).
35 E. Witten, Phys. Lett 117B (1982) 324.

SELECTED TOPICS FROM THE REPRESENTATION THEORY OF LIE SUPERALGEBRAS

Manfred Scheunert

Department of Physics
University of Freiburg
7800 Freiburg, W.Germany

1. INTRODUCTION

At present, the most profound results in the representation theory of Lie superalgebras [1,2] are due to Kac. He has shown [3,4] that a considerable part of the representation theory of semi-simple Lie algebras can be extended to the basic classical Lie superalgebras, provided one only considers the so-called typical irreducible representations (i.e., irreducible representations which are "in general position"). The characters, and in particular the dimensions of these representations can be calculated (generalization of Weyl's classical formulae), and any (finite-dimensional) representation, all of whose irreducible subquotients are typical, is completely reducible. Let us remark, however, that the tensor product of two typical representations may be not completely reducible. Regrettably, a satisfactory theory of the non-typical representations is still missing. This is particularly intriguing since the non-typical representations seem to be most important in the physical applications.

In the present work I am going to describe, under certain simplifying assumptions, some of Kac's results. After recalling a few basic items in section 2, I shall introduce in section 3 the enveloping algebra of a Lie superalgebra and use it to define the so-called induced representations. In section 4, the induction technique is applied to investigate the irreducible representations of consistently Z-graded Lie superalgebras of the type $L = L_{-1} \oplus L_0 \oplus L_1$. In the final section I shall show, on the example of the general linear Lie superalgebras, how one is led to the typical representations. Special attention is paid to the close relationship between the complete reducibility of a representation and the eigenvalues of the Casimir operators. Again the discussion is based on the induced representations. At the end of the present work I have added a bibliography which, hopefully, covers a major part of the more recent literature on the theory of Lie superalgebras and their representations.

I close this introduction by collecting some conventions as well as some basic definitions.

a) The base field will be any commutative field K of characteristic zero. Neither the algebras nor the vector spaces and modules are assumed to be finite-dimensional, unless otherwise stated. In fact, it is one of my aims to show that infinite-dimensional representations can be useful even if, at the end, one is only interested in the finite-dimensional ones.

b) Apart from some exceptions, I am going to use the notation of reference [2] (see also [5]). In the following, we shall take advantage of the fact that the Lie superalgebras under consideration admit a natural Z-gradation. Let us, therefore, recall some of the pertinent definitions.

c) Consider a vector space V. Suppose we are given a Z_2-gradation $(V_\alpha)_{\alpha \in Z_2}$ and a Z-gradation $(V_r)_{r \in Z}$ of V such that

$$V_{\bar{o}} = \bigoplus_{s \in Z} V_{2s} \, , \quad V_{\bar{1}} = \bigoplus_{s \in Z} V_{2s+1} \, .$$

Then we say that the Z-gradation $(V_r)_{r \in Z}$ is consistent with the Z_2-gradation $(V_\alpha)_{\alpha \in Z_2}$, and that the latter is induced from the given Z-gradation.

d) Let Γ be one of the abelian groups Z_2 or Z and let V and W be Γ-graded vector spaces. A linear mapping g of V into W is said to be homogeneous of degree $\gamma \in \Gamma$ if

$$g(V_\alpha) \subset W_{\alpha+\gamma} \quad \text{for all } \alpha \in \Gamma \, .$$

The vector space of all such mappings will be denoted by $Lgr(V,W)_\gamma$. It is a subspace of L(V,W), the vector space of all linear mappings of V into W. Obviously, the sum of these subspaces, denoted by Lgr(V,W), is direct:

$$Lgr(V,W) = \bigoplus_{\gamma \in \Gamma} Lgr(V,W)_\gamma \quad .$$

Thus Lgr(V,W) is a Γ-graded vector space.

If $\Gamma = Z_2$, the vector space Lgr(V,W) is equal to L(V,W), and the name Lgr(V,W) is only used to remind the reader of the fact that it is considered as a Z_2-graded vector space. For $\Gamma = Z$, however, Lgr(V,W) may be a proper subspace of L(V,W). Nevertheless, these spaces coincide if (for example) only finitely many of the homogeneous subspaces of V and W are different from $\{0\}$.

e) Let us next repeat the definition of a Lie superalgebra.

Definition
Consider a Z_2-graded algebra L, whose homogeneous subspaces are denoted by L_α ; $\alpha \in Z_2$, and whose multiplication is denoted by a pointed bracket $\langle\ ,\ \rangle$; thus

$$L = \bigoplus_{\alpha \in Z_2} L_\alpha$$

$$\langle L_\alpha , L_\beta \rangle \subset L_{\alpha+\beta} \quad \text{for all} \quad \alpha , \beta \in Z_2 \ .$$

L is called a Lie superalgebra if the following identities are satisfied:

$$\langle A,B \rangle = - (-1)^{\alpha\beta} \langle B,A \rangle$$

super-skewsymmetry

$$(-1)^{\gamma\alpha} \langle A, \langle B,C \rangle\rangle + (-1)^{\alpha\beta} \langle B, \langle C,A \rangle\rangle + (-1)^{\beta\gamma} \langle C, \langle A,B \rangle\rangle = 0$$

super-Jacobi-identity

for all $A \in L_\alpha$, $B \in L_\beta$, $C \in L_\gamma$; $\alpha , \beta , \gamma \in Z_2$.

Of course, this definition is well-known. It is included in order to point out that it makes sense if Z_2 is replaced by Z. A Z-graded algebra of this type is called a consistently Z-graded Lie superalgebra. This name is legitimate, for if L is endowed with the induced Z_2-gradation, then it is a Lie superalgebra in the sense of the foregoing definition.

Actually, most of the simple Lie superalgebras admit a natural consistent Z-gradation. Occasionally, therefore, I shall note explicitly that a certain definition, construction, result,.... has an obvious extension to the Z-graded case.

f) Let Γ be equal to Z_2 or Z and let V be a Γ-graded vector space. Then Lgr(V,V) is an associative Γ-graded algebra. On Lgr(V,V), regarded as a Γ-graded vector space, we introduce a new multiplication, the supercommutator $\langle\ ,\ \rangle$, by the requirement that

$$\langle A,B \rangle = A\,B - (-1)^{\alpha\beta}\, B\,A$$

for all $A \in \mathrm{Lgr}(V,V)_\alpha$, $B \in \mathrm{Lgr}(V,V)_\beta$; $\alpha , \beta \in \Gamma$.

If $\Gamma = Z_2$, we obtain a Lie superalgebra which is called the general linear Lie superalgebra of the Z_2-graded vector space V and which will be denoted by $gl(V_{\bar{0}}, V_{\bar{1}})$.
In case Γ is equal to Z, we obtain a consistently Z-graded Lie superalgebra which is called the general linear Lie superalgebra of the Z-graded vector space V and which will be denoted by $gl(V_r; r \in Z)$.
If $(V_{\bar{0}}, V_{\bar{1}})$ is the Z_2-gradation induced from $(V_r)_{r \in Z}$, the Lie super-

algebra $gl(V_r; r \in Z)$, endowed with its own induced Z_2-gradation, will, in general, only be a graded subalgebra of $gl(V_{\bar{0}}, V_{\bar{1}})$. However, if only finitely many of the V_r are different from $\{0\}$, then these two algebras coincide. In this case, therefore, the Z-gradation of V leads to a consistent Z-gradation of $gl(V_{\bar{0}}, V_{\bar{1}})$.

g) A particularly simple and useful example for the latter remark is obtained if we start from a Z_2-gradation $(V_\alpha)_{\alpha \in Z_2}$ of V and define a Z-gradation $(V_r)_{r \in Z}$ of V by

$$V_o = V_{\bar{0}}, \quad V_1 = V_{\bar{1}}, \quad V_r = \{0\} \quad \text{if} \quad r \neq 0,1 .$$

The corresponding Z-gradation of $gl(V_{\bar{0}}, V_{\bar{1}})$ is called canonical, it satisfies

$$gl(V_{\bar{0}}, V_{\bar{1}})_r = \{0\} \quad \text{if} \quad r \neq 0, \; \pm 1 .$$

All this can easily be visualized in a matrix notation. Suppose that V is finite-dimensional and that

$$\dim V_{\bar{0}} = n , \quad \dim V_{\bar{1}} = m .$$

Choose a basis $e_1, \ldots, e_n$ of $V_{\bar{0}}$ and a basis $e_{n+1}, \ldots, e_{n+m}$ of $V_{\bar{1}}$. Any linear mapping of V into itself is uniquely determined by its matrix with respect to the basis $e_1, \ldots, e_{n+m}$ of V. It is advantageous to write these matrices in a block form

$$\begin{pmatrix} A & B \\ C & D \end{pmatrix} ,$$

with A an n×n matrix, B an n×m matrix
C an m×n matrix, D an m×m matrix .

Then the "diagonal" block matrices $\begin{pmatrix} A & 0 \\ 0 & D \end{pmatrix}$ correspond to the elements of the Lie algebra $gl(V_{\bar{0}}, V_{\bar{1}})_o$, and the matrices of the form $\begin{pmatrix} 0 & B \\ 0 & 0 \end{pmatrix}$ (resp. $\begin{pmatrix} 0 & 0 \\ C & 0 \end{pmatrix}$) correspond to the elements of $gl(V_{\bar{0}}, V_{\bar{1}})_{-1}$ (resp. $gl(V_{\bar{0}}, V_{\bar{1}})_1$).

The Lie superalgebra of block matrices which corresponds to $gl(V_{\bar{0}}, V_{\bar{1}})$ is usually denoted by $gl(n,m)$. We remark that the representation of $gl(n,m)_o$ in $gl(n,m)_{\pm 1}$ is given by

$$\left\langle \begin{pmatrix} A & 0 \\ 0 & D \end{pmatrix} , \begin{pmatrix} 0 & B \\ C & 0 \end{pmatrix} \right\rangle = \begin{pmatrix} 0 & AB - BD \\ DC - CA & 0 \end{pmatrix}$$

and that the supercommutator of two odd elements is

$$\left\langle \begin{pmatrix} 0 & B \\ C & 0 \end{pmatrix} , \begin{pmatrix} 0 & B' \\ C' & 0 \end{pmatrix} \right\rangle = \begin{pmatrix} BC' + B'C & 0 \\ 0 & CB' + C'B \end{pmatrix} .$$

2. ELEMENTARY REPRESENTATION THEORY

Definition

A graded representation of a Lie superalgebra L in a Z_2-graded vector space V is a homomorphism of L into the general linear Lie superalgebra $gl(V_{\bar{0}},V_{\bar{1}})$. A Z_2-graded vector space which is endowed with a graded representation of L is called a graded L-module.

More explicitly, a graded representation of L in a Z_2-graded vector space V is a linear mapping $\varrho : L \rightarrow gl(V_{\bar{0}},V_{\bar{1}})$ which is homogeneous of degree zero, i.e.,

$$\varrho(L_\alpha)V_\beta \subset V_{\alpha+\beta} \quad \text{for all} \quad \alpha,\beta \in Z_2 ,$$

and such that

$$(*) \qquad \varrho(\langle A,B\rangle) = \varrho(A)\,\varrho(B) - (-1)^{\alpha\beta}\,\varrho(B)\,\varrho(A)$$
$$\text{for all} \quad A \in L_\alpha ,\ B \in L_\beta ;\ \alpha,\beta \in Z_2 .$$

Example

The assignment $A \rightarrow \text{ad } A = \langle A,\cdot\rangle$ defines a graded representation of L in the Z_2-graded vector space L. This representation is called the adjoint representation and L, endowed with this representation, is called the adjoint L-module.

For any graded L-module V, the representative of an element $A \in L$ will frequently be denoted by A_V. If x is an element of V, we shall even write $A\cdot x$ instead of $A_V x$.

Let V and W be two graded L-modules. A linear mapping $g : V \rightarrow W$ which is homogeneous of degree zero and satisfies

$$A_W \circ g = g \circ A_V \quad \text{for all} \quad A \in L$$

is called a homomorphism of the graded L-module V into the graded L-module W.

Concepts like graded submodules, graded quotient modules (with respect to graded submodules), direct sums of graded L-modules, graded irreducibility of a graded L-module, ... are defined in the obvious way.

Remarks

1) The reader will notice that we have simplified the language and speak of graded representations,...., graded irreducibility instead of Z_2-graded representations,...., Z_2-graded irreducibility. To avoid confusion it may occasionally be necessary to reintroduce the specification. This will be the case if the Z_2-graded notions are

to be confronted with the Z-graded ones, as defined in the subsequent remark.

2) The definition above and the discussion which follows can immediately be extended to the Z-graded case and then lead to the concepts of Z-graded representations of a consistently Z-graded Lie superalgebra L, of Z-graded L-modules,...., and of Z-graded irreducibility.

3) Let us next observe that the notion of a representation of a Lie superalgebra L in a vector space V can be defined even if V is not graded: This is a linear mapping ϱ which assigns to any element A of L a linear mapping $\varrho(A)$ of V into itself such that equation (*) is satisfied. These non-graded (i.e., not necessarily graded) representations and modules are of minor importance since several of the standard constructions cannot be carried out with them. Nevertheless, notions like submodules, direct sums of modules, and irreducibility still make sense.

One of the basic results of representation theory is Schur's lemma. In the Z_2-graded case, it takes the following form.

Schur's lemma
Suppose the base field K is algebraically closed. Let L be a Lie superalgebra and let V be a finite-dimensional graded-irreducible L-module. Denote by S the set of all elements of Lgr(V,V) which supercommute with all A_V; $A \in L$.

Then S is a graded subalgebra of Lgr(V,V) and we have

$$S_{\bar{0}} = K \cdot id$$

$$S_{\bar{1}} = \{0\} \quad \text{or} \quad S_{\bar{1}} = K \cdot j \quad \text{with} \quad j^2 = -\, id \, .$$

Proof
It is easy to see that S is a graded subalgebra of Lgr(V,V). Obviously, all non-zero homogeneous elements of S are bijective, since their kernel and image are graded submodules of V.
The equation $S_{\bar{0}} = K \cdot id$ now follows immediately: Let $g \in S_{\bar{0}}$ and let λ be an eigenvalue of g; then $g - \lambda\, id$ is a non-bijective element of $S_{\bar{0}}$ and hence equal to zero.
Suppose next that $S_{\bar{1}} \neq \{0\}$ and let j be a non-zero element of $S_{\bar{1}}$. Then j^2 is a non-zero scalar multiple of id, hence we may assume that $j^2 = -\, id$. If j' is any element of $S_{\bar{1}}$, the product $j \circ j'$ is a scalar multiple of id and hence j' is a scalar multiple of j.

Remarks

4) Let V be any graded-irreducible L-module. A mapping j as described in Schur's lemma exists if and only if the L-module V is reducible in the non-graded sense. (This holds even if K is not algebraically closed).

Proof
Define a linear mapping g of V into itself by

$$g(x) = (-1)^{\xi} x \quad \text{if} \quad x \in V_{\xi} \; ; \; \xi \in Z_2 .$$

Obviously, a subspace of V is graded if and only if it is invariant with respect to g, moreover, a linear mapping h of V into itself is homogeneous of degree $\eta \in Z_2$ if and only if $g \circ h = (-1)^{\eta} h \circ g$.

Now let W be any (non-graded) L-invariant subspace of V which is different from {0} and V. Then g(W) and hence W+g(W) and $W \cap g(W)$ are likewise. But the latter are invariant with respect to g and hence Z_2-graded. It follows that

$$W + g(W) = V \; , \quad W \cap g(W) = \{0\} \; ,$$

i.e., that V is the direct sum of W and g(W). Consequently, there is a unique linear mapping j of V into itself such that

$$j(w) = g(w) \; , \quad j(g(w)) = -w \quad \text{for all} \quad w \in W .$$

This mapping has the required properties: It is odd, its square is equal to -id, and it supercommutes with all A_V; $A \in L$.
Conversely, if a linear mapping j of V into itself has the properties in question, then $g \circ j$ is odd, its square is equal to id, and it commutes with all A_V; $A \in L$. It follows that

$$W = \{ w \in V \,|\, (g \circ j)(w) = w \}$$

is an L-invariant subspace of V which is different from {0} and V.

5) In the Z-graded case, Schur's lemma reads as follows:
Suppose the base field K is algebraically closed. Let L be consistently Z-graded Lie superalgebra and let V be a finite-dimensional Z-graded L-module which is irreducible in the Z-graded sense. Then any linear mapping of V into itself which supercommutes with all A_V; $A \in L$, is a scalar multiple of the identity.

Let V be a graded L-module. We recall that a bilinear form b on V is said to be homogeneous of degree β if b(x,y) vanishes whenever $x \in V_{\xi}$, $y \in V_{\eta}$; $\xi, \eta \in Z_2$, and $\xi + \eta + \beta \neq \bar{0}$, that b is said to be supersymmetric/super-skewsymmetric if

$$b(x,y) = \pm (-1)^{\xi\eta} \, b(y,x)$$

for all $x \in V_{\xi}$, $y \in V_{\eta}$; $\xi, \eta \in Z_2$, and that b is said to be L-invariant if

$$b(A \cdot x, y) + (-1)^{\alpha\xi} \, b(x, A \cdot y) = 0$$

for all $A \in L_\alpha$, $x \in V_\xi$, $y \in V$; $\alpha, \xi \in Z_2$.
Schur's lemma implies the following corollary.

Corollary
Suppose the base field K is algebraically closed. Let V be a finite-dimensional graded-irreducible module over a Lie superalgebra L.

a) Any two L-invariant bilinear forms on V are proportional.

b) The non-zero L-invariant bilinear forms on V are non-degenerate, homogeneous, and either supersymmetric or super-skewsymmetric.

Even for most of the simple Lie superalgebras, a finite-dimensional graded representation is not necessarily completely reducible. A weak substitute for complete reducibility is provided by the classical Jordan-Hölder theorem, as follows.
Let V be a graded L-module. A Jordan-Hölder sequence of V is a strictly increasing finite sequence $(V_r)_{o \leq r \leq n}$ of graded submodules of V such that $V_o = \{0\}$, $V_n = V$, and such that the quotients V_r/V_{r-1}; $1 \leq r \leq n$, are irreducible in the graded sense. Obviously, every finite-dimensional graded L-module has a Jordan-Hölder sequence.
The classical theorems by Schreier and Jordan-Hölder [6] now imply the following. Let $(V_r)_{o \leq r \leq n}$ be a Jordan-Hölder sequence of V and let $(V'_r)_{o \leq r \leq m}$ be any strictly increasing finite sequence of graded submodules of V. Then we have $m \leq n$ and equality holds if and only if $(V'_r)_{o \leq r \leq m}$ is a Jordan-Hölder sequence of V. In the latter case, there exists a permutation π of $\{1,2,\ldots,n\}$ such that the graded L-modules V'_r/V'_{r-1} and $V_{\pi(r)}/V_{\pi(r)-1}$ are isomorphic for $1 \leq r \leq n$.

Let us now describe some of the basic constructions which can be carried out with graded modules over a Lie superalgebra L. All these constructions are governed by the following rule of thumb. Consider first only homogeneous objects. Start from the classical equations, fix a standard ordering of the occurring objects, and add a factor of $(-1)^{\alpha\beta}$ whenever an object of degree α and an object of degree β have been inverted.

For example, if V and W are graded L-modules, then Lgr(V,W) admits a canonical structure of a graded L-module: The Z_2-gradation has already been defined and the action of L is fixed by the equation

$$(A \cdot g)\ (x) = A \cdot g(x) - (-1)^{\alpha\gamma}\ g(A \cdot x)$$

for all $A \in L_\alpha$, $g \in Lgr(V,W)_\gamma$, $x \in V$; $\alpha, \gamma \in Z_2$.

Let us next consider n graded L modules V_r; $1 \leq r \leq n$. We want to define a canonical graded L-module structure on the tensor product $V_1 \otimes \ldots . \otimes V_n$.
For any $\gamma \in Z_2$, we denote by $(V_1 \otimes \ldots . \otimes V_n)_\gamma$ the subspace of $V_1 \otimes \ldots . \otimes V_n$ which is generated by all tensors of the form

$x_1 \otimes \ldots\ldots \otimes x_n$, where $x_r \in V_r$ is homogeneous of degree ξ_r, $1 \leq r \leq n$, and $\xi_1 + \ldots\ldots + \xi_n = \gamma$. Then the subspaces $(V_1 \otimes \ldots\ldots \otimes V_n)_\gamma$; $\gamma \in Z_2$, form a Z_2-gradation of $V_1 \otimes \ldots\ldots \otimes V_n$, and the action of L on this space is fixed by the equation

$$A\cdot(x_1 \otimes \ldots \otimes x_n) = \sum_{r=1}^{n} (-1)^{\alpha(\xi_1 + \ldots\ldots + \xi_{r-1})} x_1 \otimes \ldots \otimes (A\cdot x_r) \otimes \ldots \otimes x_n$$

for all $A \in L_\alpha$, $x_r \in (V_r)_{\xi_r}$; $\alpha \in Z_2$, $\xi_r \in Z_2$.

Let π be a permutation of $\{1,2,\ldots\ldots,n\}$. Then there exists a unique linear mapping

$$S_\pi : V_1 \otimes \ldots\ldots \otimes V_n \longrightarrow V_{\pi^{-1}(-1)} \otimes \ldots\ldots \otimes V_{\pi^{-1}(n)}$$

such that

$$S_\pi (x_1 \otimes \ldots\ldots \otimes x_n) = \prod_{\substack{r<s \\ \pi(r)>\pi(s)}} (-1)^{\xi_r \xi_s} x_{\pi^{-1}(1)} \otimes \ldots\ldots \otimes x_{\pi^{-1}(n)}$$

for all elements $x_r \in (V_r)_{\xi_r}$; $\xi_r \in Z_2$, and S_π is an isomorphism of graded L-modules.

Suppose now that the V_r are all equal to a fixed graded L-module V. In this case, $\pi \longrightarrow S_\pi$ is a representation of the symmetric group $\mathcal{S}_n$ by automorphisms of the graded L-module $V^{\otimes n} = V \otimes \ldots\ldots \otimes V$ (n factors). Consequently, the standard representation theory of $\mathcal{S}_n$ can be applied. In particular, we can decompose $V^{\otimes n}$ into the direct sum of graded submodules corresponding to the various symmetry types. This procedure is of great practical importance, in spite of certain shortcomings which appear in the graded case. For more details, I refer the reader to the literature [7,8] .

Remark 6
The application of the Jordan-Hölder theorem and the subsequent constructions can immediately be extended to the Z-graded case.

3. THE ENVELOPING ALGEBRA

INDUCED REPRESENTATIONS

As in the Lie algebra case, the enveloping algebra of a Lie superalgebra is a particularly useful tool in representation theory. In the present section we are going to discuss the basic properties of this algebra and then use it to construct the so-called induced representations [1,2,4] . Henceforth, L denotes a finite-dimensional Lie superalgebra.

Qualitatively, the enveloping algebra U(L) of L consists of all (non-commutative) polynomials in the elements of any basis of L, subject to the sole condition that the (anti)commutation relations provided by L are satisfied. More precisely, U(L) can be characterized by the following properties a) - d) which fix it up to isomorphism.

a) U(L) is an associative Z_2-graded algebra and has a (non-zero) unit element.

b) The Lie superalgebra L is embedded in U(L) in such a manner that $L_\alpha = L \cap U(L)_\alpha$ for all $\alpha \in Z_2$, and

$$\langle A,B \rangle = AB - (-1)^{\alpha\beta} BA$$

for all $A \in L_\alpha$, $B \in L_\beta$; $\alpha, \beta \in Z_2$.
(More formally: L is a graded subalgebra of the Lie superalgebra associated with U(L).)

c) The algebra U(L) is generated by L and the unit element.

d) Let S be an associative algebra with unit element and let ϱ be a linear mapping of L into S such that

$$\varrho(\langle A,B\rangle) = \varrho(A)\,\varrho(B) - (-1)^{\alpha\beta}\,\varrho(B)\,\varrho(A)$$

for all $A \in L_\alpha$, $B \in L_\beta$; $\alpha, \beta \in Z_2$.
Then ϱ can be extended to a homomorphism $\bar{\varrho}$ of U(L) into S such that $\bar{\varrho}(1) = 1$.
Because of c), the extension $\bar{\varrho}$ is unique, since it follows that

$$\bar{\varrho}(A_1 \ldots A_r) = \varrho(A_1) \ldots \varrho(A_r)$$

for all elements $A_1, \ldots, A_r \in L$.
If S is a Z_2-graded algebra and if ϱ is homogeneous of degree zero, then so is $\bar{\varrho}$.

e) Let V be a Z_2-graded vector space. When applied to the algebra $S = \mathrm{Lgr}(V,V)$ the foregoing property implies:
Any graded representation of the Lie superalgebra L in a Z_2-graded vector space V can be extended to a graded representation of the associative algebra U(L) in V such that 1 is mapped onto id_V.

The converse is obvious: By restriction to L, any graded representation of U(L) yields a graded representation of L.
In view of these facts, the concepts of a "graded representation of L" and a "graded representation of U(L)", and, similarly, those of a "graded L-module" and a "left graded U(L)-module", are completely equivalent. It depends on the circumstances which language is to be preferred.

f) Let $E_1, \ldots, E_n$ be a homogeneous basis of L (i.e., a basis consisting of homogeneous elements). The products $E_{i_1} E_{i_2} \ldots E_{i_r}$, with $1 \leq i_1 \leq i_2 \leq \ldots \leq i_r \leq n$, but $i_s < i_{s+1}$ if E_{i_s} and $E_{i_{s+1}}$ are odd, form a

basis of the vector space U(L). (The case $r = 0$ is allowed and yields the unit element.)
This is the super-version of the famous Poincaré, Birkhoff, Witt theorem. Using b) and c) it is easy to see that the above-mentioned products generate the vector space U(L). The main content of the theorem is that these products are linearly independent.

g) Let L' be a graded subalgebra of L. Then the Z_2-graded algebra U(L') can be identified with that graded subalgebra of U(L) which is generated by L' and the unit element.
This is an easy consequence of the foregoing results.

Remark 1
We have already noted that the representation theory for L can be reduced to the representation theory for U(L). The algebra U(L) has the advantage of being associative, however, in general, it is not finite-dimensional.

The enveloping algebra will now be used to develop a technique by which any graded representation of a graded subalgebra of L can be enlarged to become a graded representation of L itself.
Thus let L' be a graded subalgebra of L. Because of g), the enveloping algebra U(L') of L' can be identified with a graded subalgebra of U(L). Let V be a graded L'-module. In view of e), V can also be regarded as a left graded U(L')-module. Let $\overline{V}$ denote the tensor product of the right graded U(L')-module U(L) with the left graded U(L')-module V,

$$\overline{V} = U(L) \underset{U(L')}{\otimes} V \,.$$

This tensor product can be defined to be the quotient of $U(L)\otimes V$ (the usual tensor product of vector spaces over K) modulo the subspace W which is generated by all tensors of the form

$$X\cdot X'\otimes y - X\otimes X'\cdot y \,,$$

with $X\in U(L)$, $X'\in U(L')$, and $y\in V$.

If $X\in U(L)$ and $y\in V$, the residue class of $X\otimes y$ modulo W will be denoted by $X\overline{\otimes} y$. This notation will be kept throughout the present section. Later on we shall follow the usual notation and drop the overbar.
It follows that

$$X\cdot X'\overline{\otimes}\, y = X\overline{\otimes} X'\cdot y$$

for all $X\in U(L)$, $X'\in U(L')$, and $y\in V$.
Since W is a Z_2-graded subspace of $U(L)\otimes V$, the quotient $\overline{V}$ admits a natural Z_2-gradation: If $X\in U(L)$ is homogeneous of degree ξ and $y\in V$

is homogeneous of degree η, then $X\bar{\otimes}y$ is homogeneous of degree $\xi+\eta$. It is easy to see that the Z_2-graded vector space $\bar{V}$ admits a natural structure of a left graded U(L)-module such that

$$X(Y\bar{\otimes}y) = X\cdot Y\bar{\otimes}y$$

for all $X,Y \in U(L)$ and $y \in V$.
The Z_2-graded vector space $\bar{V}$, endowed with the corresponding graded L-module structure, is called the graded L-module induced from the graded L'-module V.
From the Poincaré, Birkhoff, Witt theorem we deduce the following basic information on the structure of $\bar{V}$.
Let $E_1,\ldots,E_m$ be homogeneous elements of L whose residue classes modulo L' form a basis of the vector space L/L'.
Denote by H the set of all finite sequences $(j_1,\ldots,j_r)$ of positive integers such that $1\leq j_1\leq j_2\leq \ldots \leq j_r\leq m$, but $j_s< j_{s+1}$ if E_{j_s} and $E_{j_{s+1}}$ are odd. Set

$$E_J = E_{j_1} E_{j_2} \ldots E_{j_r}$$

if $J = (j_1,\ldots,j_r)$ is in H. (By definition, $E_\phi = 1$.)
Then for any $J \in H$, the mapping $y \longrightarrow E_J\bar{\otimes}y$ of V into $\bar{V}$ is injective, and $\bar{V}$ is the direct sum of its graded subspaces $E_J\bar{\otimes}V$; $J\in H$.
In particular, the mapping $y \longrightarrow 1\bar{\otimes}y$ of V into $\bar{V}$ is injective and we have

$$X'(1\bar{\otimes}y) = X'\bar{\otimes}y = 1\bar{\otimes}X'\cdot y$$

for all $X'\in U(L')$ and $y\in V$. Consequently, the graded L'-module V can be identified with the graded L'-submodule $1\bar{\otimes}V$ of $\bar{V}$. In this sense, the graded L-module $\bar{V}$ is an enlargement of the graded L'-module V.
The case where L' includes $L_{\bar{0}}$ is particularly interesting. In that case, the set H is finite and has 2^m elements. Thus if V is finite-dimensional, so is $\bar{V}$ and we have

$$\dim \bar{V} = 2^m \dim V.$$

We note that this observation applies to $L' = L_{\bar{0}}$ and then yields a proof of the Ado theorem in the super-case.

Remark 2
In closing we remark that all the results of this section can immediately be extended to the Z-graded case. In particular, if L is a consistently Z-graded Lie superalgebra, then its enveloping algebra U(L) has a unique Z-gradation such that $L_r = L\cap U(L)_r$ for all $r\in Z$, and this Z-gradation induces the original Z_2-gradation of U(L).

4. IRREDUCIBLE REPRESENTATIONS OF CONSISTENTLY Z-GRADED LIE SUPERALGEBRAS OF THE TYPE $L = L_{-1} \oplus L_o \oplus L_1$

In the present section we consider the finite-dimensional irreducible representations of a finite-dimensional consistently Z-graded Lie superalgebra L such that

$$L_r = \{0\} \quad \text{if} \quad r \neq 0, \pm 1.$$

Note that the Lie superalgebras gl(n,m), sl(n,m), sl(n,n)/center, osp(2,2r), gp(n) [9] , P(n-1) [1] are of this type. Let $V \neq \{0\}$ be a finite-dimensional L-module. We consider three cases simultaneously:

1) V is a Z-graded L-module and irreducible in the Z-graded sense.

2) V is a Z_2-graded L-module and irreducible in the Z_2-graded sense.

3) V is not graded and irreducible in the non-graded sense.

Our first result is that, in all three cases, V has a Z-gradation which converts it into a Z-graded L-module and that, again in all three cases, V is irreducible in the non-graded sense. More precisely, we can prove the following:

a) Define

$$V^o = \{x \in V \mid Ax = 0 \quad \text{for all} \quad A \in L_{-1}\} .$$

Then V^o is an irreducible L_o-submodule of V (in particular, $V^o \neq \{0\}$), and in the first and second cases, V^o is contained in a homogeneous subspace of V.

b) We set

$$V^r = \{0\} \quad \text{for all integers} \quad r < 0$$

and define by induction

$$V^{r+1} = L_1 V^r \quad \text{for all integers} \quad r \geqslant 0 .$$

(According to common usage, $L_1 V^r$ is the subspace of V generated by all elements of the form $A \cdot x$, with $A \in L_1$ and $x \in V^r$.)
Then $(V^r)_{r \in Z}$ is a Z-gradation of V, i.e.,

$$V = \bigoplus_{r \in Z} V^r .$$

Endowed with this Z-gradation, V is a Z-graded L-module, i.e.,

$$L_s V^r \subset V^{r+s} \quad \text{for all} \quad r,s \in Z .$$

c) If $(V'_r)_{r \in Z}$ is any Z-gradation of V which converts V into a Z-graded L-module, then there exists an integer m such that

$$V'_{r+m} = V^r \quad \text{for all} \quad r \in Z .$$

In the first case, this result applies to the original Z-gradation.

d) In the second case, one of the homogeneous subspaces $V_{\bar{0}}$, $V_{\bar{1}}$ is equal to $\bigoplus_{r\in Z} V^{2r}$ and the other is equal to $\bigoplus_{r\in Z} V^{2r+1}$.

e) In all three cases, V is irreducible in the non-graded sense.

The idea for the proof can be taken from reference [1] , the details are left to the reader.

In view of these results, we shall now restrict our attention to the finite-dimensional irreducible Z-graded L-modules V. Furthermore, we may assume that the Z - gradation of V is normalized in the sense that

$$V_r = \{0\} \quad \text{if} \quad r < 0$$

but

$$V_o \neq \{0\} \ .$$

This implies that

$$V^r = V_r \quad \text{for all} \quad r \in Z \ .$$

Proposition

Let W be a finite-dimensional irreducible L_o-module. Then there exists one and, up to isomorphism, only one finite-dimensional irreducible Z-graded L-moduleVwith a normalized Z-gradation such that the L_o-module V_o is isomorphic to W.

Proof

The method of proof is well-known.

Existence. Obviously,

$$L' = L_{-1} \oplus L_o$$

is a Z-graded subalgebra of L. We convert W into a Z-graded L'-module through the requirements that

$$W_o = W \ , \quad W_r = \{0\} \quad \text{if} \quad r \neq 0$$

and

$$L_{-1} W = \{0\} \ .$$

Let

$$\overline{W} = U(L) \otimes_{U(L')} W$$

denote the Z-graded L-module induced from the Z-graded L'-module W; it is finite-dimensional. Obviously, we have

$$\overline{W}_r = \{0\} \quad \text{if} \quad r < 0 \ .$$

On the other hand, we know that the L_o- module $\overline{W}_o = 1 \otimes W$ is isomorphic to W, hence irreducible; moreover, $\overline{W}_o$ generates the L-module $\overline{W}$. It follows that all proper Z-graded L-submodules of $\overline{W}$ must be contained in $\bigoplus_{r \geq 1} \overline{W}_r$. Consequently, there exists a proper Z-graded L-submodule $\overline{W}'$ of $\overline{W}$ which contains all the others.
The quotient

$$\hat{W} = \overline{W} / \overline{W}'$$

is an irreducible Z-graded L-module. Of course, we have

$$\hat{W}_r = \{0\} \quad \text{if} \quad r < 0 .$$

Moreover, the relation $\overline{W}_o \cap \overline{W}' = \{0\}$ implies that the L_o-module $\hat{W}_o$ is isomorphic to $\overline{W}_o$, hence isomorphic to W.

Uniqueness. Let V be any finite-dimensional irreducible Z-graded L-module with a normalized Z-gradation. Then there exists a homomorphism of Z-graded L-modules

$$f: U(L) \underset{U(L')}{\otimes} V_o \longrightarrow V$$

such that

$$f(X \otimes y) = X \cdot y$$

for all $X \in U(L)$ and $y \in V_o$.

The image of f is a Z-graded L-submodule of V and contains V_o, the kernel of f is a proper Z-graded L-submodule of $\overline{(V_o)}$ and hence contained in $\overline{(V_o)}'$. Since V is irreducible, it follows that f is surjective and that its kernel is equal to $\overline{(V_o)}'$. Consequently, f induces an isomorphism of the Z-graded L-module $(V_o)^\wedge$ onto the Z-graded L-module V.

Remark
Let us keep the notation of the existence part of the proof. It is worth while to investigate the Z-graded L-module $\overline{W}$ and, in particular, to describe its maximal submodule $\overline{W}'$ more explicitly.

For example, it is easy to see that the homogeneous component $\overline{W}'_r$; $r \geq 1$, is equal to the set of all elements $v \in \overline{W}_r$ such that $A_1 \ldots . A_r\, v = 0$ for all $A_1, \ldots, A_r \in L_{-1}$.

Let us also note that in case $\overline{W}' \neq \{0\}$ the L-module $\overline{W}$ cannot be completely reducible in the Z-graded sense. Apparently, some additional information is needed if this is to hold even in the Z_2-graded sense.

5. REPRESENTATION THEORY FOR gl(n,m)

In the present final section we are going to develop the representation theory for the general linear Lie superalgebra gl(n,m); $n,m \geq 0$. The reader will recall that the results of the foregoing section apply to this algebra, hence, in principle, all finite-dimensional irreducible graded representations of gl(n,m) are known. Nevertheless, it is necessary to resume the discussion from a more advanced point of view, for there are certain phenomena which are typical of the graded case and which we have not yet mentioned.

What we are going to do is to mimic the classical approach to the representation theory of semi-simple Lie algebras, to see how far we can get, and to point to some peculiarities of the graded case [1,3,4] . The example of the gl(n,m) algebras has been chosen in order to keep the technicalities at a minimum.

In the following, the indices i,j,k,ℓ run through the values $1,2,\dots,n+m$, unless otherwise stated. We set

$$\sigma_i = \begin{cases} 1 & \text{if} \quad 1 \leq i \leq n \\ -1 & \text{if} \quad n+1 \leq i \leq n+m \end{cases} .$$

A. The root system [2,10]

Let E_{ij} denote the $(n+m) \times (n+m)$ matrix defined by

$$(E_{ij})_{k\ell} = \delta_{ik}\,\delta_{j\ell} .$$

Obviously, the E_{ij} form a basis of gl(n,m). More precisely, the E_{ij} with $1 \leq i,j \leq n$ or $n+1 \leq i,j \leq n+m$ form a basis of $gl(n,m)_0 \simeq gl(n) \times gl(m)$ those with $1 \leq i \leq n$; $n+1 \leq j \leq n+m$ form a basis of $gl(n,m)_{-1}$, and those with $n+1 \leq i \leq n+m$; $1 \leq j \leq n$ form a basis of $gl(n,m)_1$.

Let h denote the subspace of gl(n,m) consisting of all diagonal matrices. It is well-known that h is a splitting Cartan subalgebra of $gl(n,m)_0$. Proceeding as in the Lie algebra case, let us construct the root space decomposition of gl(n,m) with respect to h.

This is easily achieved. Define a linear form ε_j on h by the requirement that $\varepsilon_j(H)$ is the j the diagonal element of H, for all $H \in h$, or, equivalently,

$$\varepsilon_j(E_{ii}) = \delta_{ij} .$$

Then we have, for all $H \in h$,

$$\langle H, E_{ij} \rangle = (\varepsilon_i - \varepsilon_j)(H) E_{ij} .$$

In analogy with the Lie algebra case, we say that $\varepsilon_i - \varepsilon_j$; $i \neq j$, is a root of gl(n,m) with respect to h and that E_{ij} is a corresponding root vector. A root is called even/odd depending on whether E_{ij} is even/odd, that is, if $\sigma_i \sigma_j = \pm 1$. The set Δ of all roots will be called the root system of gl(n,m) with respect to h,

$$\Delta = \{\varepsilon_i - \varepsilon_j \mid i \neq j\} .$$

Let us next define the linear form α_q on h by

$$\alpha_q = \varepsilon_q - \varepsilon_{q+1} \quad \text{for} \quad 1 \leq q < n+m .$$

The α_q form a basis of the root system Δ in the usual sense: Any root $\varepsilon_i - \varepsilon_j$; $i \neq j$, is a linear combination of the α_q with integral coefficients which are either all positive (if $i<j$) or all negative (if $i>j$). Correspondingly, the roots $\varepsilon_i - \varepsilon_j$ with $i<j$ are said to be positive, those with $i>j$ negative, and the sets of these roots will be denoted by Δ^+ and Δ^-, respectively.

Later on, we shall need the linear form ρ on h which is defined to be half the sum of the even positive roots minus half the sum of the odd positive ones. It is easy to see that

$$\rho = \sum_i \sigma_i r_i \varepsilon_i ,$$

with

$$2 r_i = \begin{cases} n-m+1-2i & \text{if} \quad 1 \leq i \leq n \\ -3n-m-1+2i & \text{if} \quad n+1 \leq i \leq n+m . \end{cases}$$

All this looks the same as in the case of the general linear Lie algebra. Of course, this is not a miracle since the adjoint action of h is the same in both cases. A crucial difference arises if we introduce an adequate "metric" on the dual h^* of h. To do so we recall that the bilinear form

$$(A,B) \longrightarrow \mathrm{Str}(AB)$$

on gl(n,m) is invariant, supersymmetric, and non-degenerate. Consequently, its restriction to h is symmetric and non-degenerate. The inverse of this latter form is the bilinear form $(\ |\ )$ on h^* we are looking for; it is given by

$$(\varepsilon_i \mid \varepsilon_j) = \sigma_i \delta_{ij} .$$

Note that the odd roots are isotropic (have "length" zero) with respect to this form.

B. Highest weight modules [11]

Our next task is to introduce the concept of a graded highest weight module over gl(n,m). By definition, this is a graded gl(n,m)-module W such that the following conditions are satisfied:
There exist a homogeneous non-zero element $w \in W$ and a linear form Λ on h such that

1) w generates W as a gl(n,m)-module.

2) w is annihilated by all E_{ij} with $i<j$ (that is, by the root vectors corresponding to the positive roots).

3) $H \cdot w = \Lambda(H)\, w$ for all $H \in h$.

It is easy to see that W is the direct sum of its (graded) weight subspaces

$$W^{\mu} = \{ x \in W | H \cdot x = \mu(H)\, x \text{ for all } H \in h \} \; ; \; \mu \in h^{*} ,$$

and that W^{μ} may be different from $\{0\}$ only if μ takes the form

$$\mu = \Lambda - \sum_{q=1}^{n+m-1} s_q \alpha_q$$

with some integers $s_q \geq 0$. In particular, we have

$$W^{\Lambda} = K \cdot w .$$

The foregoing remarks imply that, for a given highest weight module W, the linear form Λ is uniquely fixed and that w is determined up to a scalar factor. We call Λ the highest weight and w a generator of W.

Using the technique of induced representations, we can show that for any linear form $\Lambda \in h^{*}$ there exist graded highest weight modules with highest weight Λ .

In fact, let Λ be any element of h* and $\alpha \in Z_2$. The linear combinations of the E_{ij} with $i \leq j$ form a graded subalgebra L' of gl(n,m) containing h. We convert K into a graded L'-module $K(\Lambda, \alpha)$ by the requirements that all its elements be homogeneous of degree α and that

$$H \cdot 1 = \Lambda(H) \quad \text{for all} \quad H \in h$$

$$E_{ij} \cdot 1 = 0 \quad \text{if} \quad i < j .$$

Let U denote the enveloping algebra of gl(n,m). The induced gl(n,m)-module

$$M(\Lambda, \alpha) = U \otimes_{U(L')} K(\Lambda, \alpha)$$

is a graded highest weight module over gl(n,m) with highest weight Λ. The generators of $M(\Lambda, \alpha)$ are the non-zero scalar multiples of $1 \otimes 1$, they are homogeneous of degree α.

As in the proof of the proposition in section 4 we can now show:

a) Let W be any graded highest weight module over gl(n,m) with highest weight Λ, whose generators are homogeneous of degree α. If w is any generator of W, then there exists a unique homomorphism of the graded gl(n,m)-module $M(\Lambda, \alpha)$ into the graded gl(n,m)-module W which maps $1 \otimes 1$ onto w. This homomorphism is surjective, hence W is isomorphic to a quotient of $M(\Lambda, \alpha)$.

b) The graded gl(n,m)-module $M(\Lambda, \alpha)$ contains a proper graded submodule $N(\Lambda, \alpha)$ which includes all proper submodules of $M(\Lambda, \alpha)$. The quotient

$$I(\Lambda, \alpha) = M(\Lambda, \alpha)/N(\Lambda, \alpha)$$

is an irreducible graded gl(n,m)-module with highest weight Λ, whose generators are homogeneous of degree α. Any such gl(n,m)-module is isomorphic to $I(\Lambda, \alpha)$.

Remark
Of course, the two graded gl(n,m)-modules $M(\Lambda, \alpha)$; $\alpha \in Z_2$, are "essentially the same", they can be converted into each other by a simple "shift" of the Z_2-gradation. The same holds for $I(\Lambda, \alpha)$.

It is easy to see that a finite-dimensional irreducible graded gl(n,m)-module V is a graded highest weight module if and only if, for every central element C of the Lie algebra $gl(n,m)_o$, the eigenvalues of C_V are in the base field K.

We shall now have to answer the question as to when the gl(n,m)-modules $I(\Lambda, \alpha)$ are finite- dimensional.

Let us make the natural agreement that a root of $gl(n,m)_o \simeq gl(n) \times gl(m)$ with respect to h is positive if and only if it is an element of Δ^+. Then we have:
The gl(n,m)-module $I(\Lambda, \alpha)$ is finite-dimensional if and only if the irreducible highest weight module $J(\Lambda)$ over $gl(n,m)_o$ with highest weight Λ is finite-dimensional.

Proof
Obviously, $I(\Lambda, \alpha)$ contains a highest weight module over $gl(n,m)_o$ with highest weight Λ. Thus if $I(\Lambda, \alpha)$ is finite-dimensional, so is $J(\Lambda)$.

Conversely, if $J(\Lambda)$ is finite-dimensional, then the construction in section 4 yields a finite-dimensional (irreducible) graded highest weight module over gl(n,m) with highest weight Λ . Hence $I(\Lambda, \alpha)$ is finite-dimensional.

The criterion for $J(\Lambda)$ to be finite - dimensional is well-known: $J(\Lambda)$ is finite-dimensional if and only if $\Lambda(E_{i,i} - E_{i+1,i+1})$ is integral ≥ 0 for $1 \leq i < n$ and for $n+1 \leq i < n+m$. (Note that there is no condition on $\Lambda(E_{n,n} - E_{n+1,n+1})$.)

C. <u>Casimir elements</u> [11]

Beside the induced representations, the second important tool in our discussion of the finite-dimensional graded gl(n,m)-modules will be the Casimir elements.

Let U denote the enveloping algebra of gl(n,m). By definition, an element C of U is called a (generalized) Casimir element if it supercommutes with all elements of U or, equivalently, if it supercommutes with all elements of gl(n,m),

$$\langle C, A \rangle = 0 \quad \text{for all} \quad A \in gl(n,m).$$

Obviously, the set of all Casimir elements is a graded subalgebra of U which will be denoted by Z. According to our definition, it is conceivable that Z might contain odd elements. However, it can be shown that all Casimir elements of gl(n,m) are even and hence commute with all elements of U.

The well-known construction of Casimir elements for the general linear Lie algebras can immediately be generalized to the super-case: For any integer $r \geq 1$, the element

$$C_r = \sum_{i_1,\ldots,i_r} \sigma_{i_1}\sigma_{i_2}\cdots\sigma_{i_{r-1}} E_{i_r i_1} E_{i_1 i_2} \cdots E_{i_{r-1} i_r}$$

is a Casimir element of gl(n,m). (Note that the right hand side is to be calculated in U.)

Now let V be any graded gl(n,m)-module. As we know, the representation of L in V can uniquely be extended to a graded representation of U in V such that 1 is mapped onto the identity id_V. The representative of an element $X \in U$ will be denoted by X_V. It follows that if C is a Casimir element of gl(n,m), the Casimir operator C_V will commute with all A_V ; $A \in L$.

Suppose next that V is a graded highest weight module with highest weight $\Lambda \in h^*$. Then it is easy to see that, for any Casimir element C of gl(n,m), the representative C_V is a scalar multiple of the identity,

$$C_V = \chi_\Lambda(C)\, id_V ,$$

where the scalar factor depends only on Λ and C.

Obviously, for any fixed $\Lambda \in h^*$, the mapping $C \longrightarrow \chi_\Lambda(C)$ is an algebra homomorphism of Z onto K (i.e., it is linear and multiplicative). On the other hand, for any fixed Casimir element C, the mapping $\Lambda \longrightarrow \chi_\Lambda(C)$ is a polynomial function on h* which uniquely determines the Casimir element C.

Let us reformulate the latter statement. First of all, it will be convenient to consider $\chi_\Lambda(C)$ as a function of $\lambda = \Lambda + \rho$. Next we introduce coordinates in h* by setting

$$\ell_i = \sigma_i \lambda(E_{ii}).$$

Then we have

$$\chi_\Lambda(C) = \tilde{C}(\ell)$$

where $\tilde{C}$ is a polynomial in $\ell = (\ell_1, \ell_2, \ldots, \ell_{n+m})$, and C is uniquely determined by $\tilde{C}$.

It has been shown by Kac [3,4] that the polynomial $\tilde{C}$ has the following basic properties:

1) $\tilde{C}$ is symmetric in $\ell_1, \ldots, \ell_n$ and in $\ell_{n+1}, \ldots, \ell_{n+m}$ (invariance under the Weyl group).

2) Let $1 \leq i \leq n$ and $n+1 \leq j \leq n+m$. If we set $\ell_i = \ell_j$, then $\tilde{C}(\ell)$ is independent of this variable.

Conversely [12], any polynomial in ℓ which has these properties is equal to $\tilde{C}$ for exactly one Casimir element C.

Proof of 1) and 2)
The proof of these statements follows the classical lines [13]. We include it in order to demonstrate the usefulness of induced representations and the importance of the Poincaré, Birkhoff, Witt theorem.

1) Let $1 \leq q < n$ or $n+1 \leq q < n+m$; we are going to show that $\tilde{C}(\ell_1, \ldots, \ell_{n+m})$ remains unaltered if ℓ_q and ℓ_{q+1} are interchanged. Since $\tilde{C}$ is a polynomial, it is sufficient to prove this in the case where $\Lambda(E_{q,q} - E_{q+1,q+1}) = r$ is a positive integer.

Let us set

$$E_+ = E_{q,q+1} , \quad E_- = E_{q+1,q} , \quad H = E_{q,q} - E_{q+1,q+1} .$$

Obviously, these elements satisfy the commutation relations

$$\langle H, E_\pm \rangle = \pm 2 E_\pm , \quad \langle E_+, E_- \rangle = H .$$

Consequently, the following relation holds in U, for all integers $s \geq 0$,

$$\langle E_+, E_-^s \rangle = s E_-^{s-1}(H - s + 1) .$$

Now consider the induced module $M(\Lambda, \alpha)$ and let w denote a generator of it. We know that the element $E_-^{r+1}w$ is different from zero (Poincaré, Birkhoff, Witt theorem). On the other hand, we have

$$E_{ij}E_-^{r+1}w = 0 \quad \text{if} \quad i < j .$$

If $(i,j) \neq (q,q+1)$ this follows from the general form of the weights of $M(\Lambda, \alpha)$, whereas for $(i,j) = (q,q+1)$ the foregoing relation implies that

$$E_+ E_-^{r+1}w = \langle E_+, E_-^{r+1} \rangle w = (r+1)E_-^r(H - r)w = 0 .$$

Correspondingly, the submodule of $M(\Lambda, \alpha)$ generated by $E_-^{r+1}w$ is a graded highest weight module with highest weight $\Lambda - (r+1)\alpha_q$. It follows that

$$\chi_\Lambda(C) = \chi_{\Lambda-(r+1)\alpha_q}(C) \quad \text{for all} \quad C \in Z .$$

Transcribed to the coordinates ℓ_i this is just the relation we wanted to prove.

2) According to 1) it is sufficient to verify 2) for $i = n$; $j = n+1$. In this case, the proof is similar to that for 1).

Let $\Lambda \in h^*$ be such that $\ell_n = \ell_{n+1}$. Using the notation above, we now remark that $E_{n+1,n}w$ is a non-zero element of $M(\Lambda, \alpha)$ such that

$$E_{ij}E_{n+1,n}w = 0 \quad \text{for} \quad i < j .$$

Once again this is obvious if $(i,j) \neq (n,n+1)$, whereas

$$E_{n,n+1}E_{n+1,n}w = \Lambda(E_{n,n} + E_{n+1,n+1})\, w = 0$$

since

$$\Lambda(E_{n,n} + E_{n+1,n+1}) = \sigma_n(\ell_n - r_n) + \sigma_{n+1}(\ell_{n+1} - r_{n+1}) = 0 .$$

It follows that

$$\chi_\Lambda(C) = \chi_{\Lambda-\alpha_n}(C) \quad \text{for all} \quad C \in Z .$$

But if the n th and n+1 st coordinates of Λ coincide, the same holds for $\Lambda - t\alpha_n$, for all $t \in K$. Since $\chi_\Lambda(C)$ is a polynomial function of Λ, all this implies that $\chi_{\Lambda - t\alpha_n}(C)$ is independent of $t \in K$, for all $C \in Z$. Transcribed to the coordinates ℓ_i, this is just the statement 2).

The polynomials $\tilde{C}_r$ can be calculated by means of a generating function [10]. Here we only want to describe the general structure of the $\tilde{C}_r$.
For any integer $r \geq 1$, let P_r denote the polynomial in ℓ defined by

$$P_r(\ell) = \sum_i \sigma_i \ell_i^r .$$

Then $\tilde{C}_r - P_r$ has a total degree $\leq r-1$ and can be written as a polynomial in the P_s. Of course, this implies that $P_r - \tilde{C}_r$ can also be written as a polynomial in the $\tilde{C}_s$.

D. Reduction of finite-dimensional representations

We are now ready to discuss the complete reducibility of finite-dimensional graded gl(n,m)-modules. To motivate our procedure, let us assume for a moment that the base field is algebraically closed. As is well-known, even a finite-dimensional representation of a general linear Lie algebra is not necessarily completely reducible. This is simply due to the fact that the center may be represented by non-diagonalizable mappings.

Of course, something analogous is to be expected for the gl(n,m) algebras. Let us first consider a finite-dimensional irreducible graded gl(n,m)-module V. Since the base field is algebraically closed, V is a graded highest weight module. Consequently, the representatives H_V are diagonalizable for all $H \in h$. Obviously, this will also hold if V is only assumed to be completely reducible (in the graded sense).

For those elements $H \in h$ which lie in the semi-simple part of $gl(n,m)_o$ (isomorphic to $sl(n) \times sl(m)$) this condition is automatically fulfilled. On the other hand, by use of the induction technique, it is easy to construct finite-dimensional graded representations of gl(n,m) for which every non-zero element of the center of $gl(n,m)_o$ is represented by a non-diagonalizable mapping.

These remarks motivate the following

Convention

All gl(n,m)-modules that we are going to investigate in the sequel are assumed to be graded, finite-dimensional (unless otherwise stated), and to satisfy the following condition:

(*) H_V is diagonalizable for all central elements H of $gl(n,m)_o$.

The base field is allowed to be arbitrary (of characteristic zero).

Note that the condition (*) is stronger than might have been anticipated from the classical case: We require H_V to be diagonalizable for all elements H of the two-dimensional center of $gl(n,m)_o$, not only for those of the one-dimensional center of gl(n,m).

Let us also mention that the class of modules under consideration is closed under the formation of subquotients (i.e., quotients of submodules, or, equivalently, submodules of quotients), of finite direct sums, and of tensor products.

As before, the assumption (*) implies that H_V is diagonalizable for all $H \in h$. Consequently, we can construct the well-known weight-space-decomposition of V with respect to h. Define for any $\omega \in h^*$

$$V^{\omega} = \{ y \in V | H \cdot y = \omega(H) y \text{ for all } H \in h \} .$$

Then the V^{ω} ; $\omega \in h^*$, are graded subspaces of V and V is their direct sum.

We shall now use the Casimir elements to decompose the gl(n,m)-module V. Some care is needed since it is not known whether the Casimir operators C_V; $C \in Z$, are diagonalizable.

Let P_{++} denote the set of all $\Lambda \in h^*$ such that the corresponding irreducible graded highest weight modules $I(\Lambda, \alpha)$ are finite-dimensional, i.e., such that $\Lambda(E_{i,i} - E_{i+1,i+1})$ is an integer ≥ 0 for $1 \leq i < n$ and for $n+1 \leq i < n+m$.

Using a Jordan-Hölder sequence of the gl(n,m)-module V we notice that, for every Casimir element C, the eigenvalues of C_V are of the form $\chi_\Lambda(C)$ with $\Lambda \in P_{++}$ and hence lie in K. It is now easy to construct a decomposition of V into a direct sum of non-zero submodules V_q; $1 \leq q \leq p$, such that, for all $C \in Z$ and $q \in \{1, \ldots, p\}$, the Casimir operator C_{V_q} has only one eigenvalue. Of course, it is sufficient to investigate the submodules V_q. Stated differently, we may now assume that, for all $C \in Z$, the operator C_V has only one eigenvalue. If Ω is the highest weight of some irreducible subquotient of V, it follows that, for all $C \in Z$, the sole eigenvalue of C_V is equal to $\chi_\Omega(C)$.

Let us now distinguish two cases:

I) All irreducible subquotients of the gl(n,m)-module V are isomorphic, hence their highest weight is equal to Ω .

II) The gl(n,m)-module V has non-isomorphic irreducible subquotients, i.e., irreducible subquotients with different highest weights.

In the first case, V is the direct sum of irreducible submodules with highest weight Ω and we are done.

Proof

Consider a Jordan-Hölder sequence

$$\{0\} = V_o \subset V_1 \subset V_2 \subset \dots \subset V_r = V$$

of the gl(n,m)-module V. We show by induction on s; $1 \leq s \leq r$, that V_s is the direct sum of irreducible submodules with highest weight Ω. The case s = 1 is obvious. Assume that our claim has been proved for some s; $1 \leq s < r$. By assumption, V_{s+1}/V_s is irreducible with highest weight Ω. According to [14] we can choose an element $v \in V_{s+1}^{\Omega}$ whose canonical image in V_{s+1}/V_s is non-zero and hence generates the gl(n,m)-module V_{s+1}/V_s. (Note that at this point the condition (*) has been used.)

Let W_{s+1} denote the submodule of V_{s+1} generated by v. It is easy to see that W_{s+1} is a highest weight module with highest weight Ω and generator v. We know that

$$V_s + W_{s+1} = V_{s+1} \quad ;$$

let us show that

$$V_s \cap W_{s+1} = \{0\} \quad .$$

If not so, $V_s \cap W_{s+1}$ contains an irreducible submodule W whose highest weight must be equal to Ω. But every weight vector of W_{s+1} corresponding to the weight Ω is proportional to v. It follows that $v \in W \subset V_s$, a contradiction.

In case II, we are stuck: As the subsequent discussion will show, V is not necessarily completely reducible in this case. Note that in this case there exists a weight $\Lambda \in P_{++}$ such that $\Lambda \neq \Omega$ but $\chi_\Lambda(C) = \chi_\Omega(C)$ for all $C \in Z$. This motivates the following definition.

Let us say that two elements $\Lambda, \Lambda' \in h^*$ are separated by a Casimir element C if

$$\chi_\Lambda(C) \neq \chi_{\Lambda'}(C) \quad .$$

Consider now two elements $\Lambda, \Lambda' \in h^*$ which cannot be separated by any Casimir element, i.e., such that

$$\chi_\Lambda(C) = \chi_{\Lambda'}(C) \quad \text{for all} \quad C \in Z \ .$$

Using the coordinates

$$\ell_i = \sigma_i(\Lambda + \rho)(E_{ii}) \ , \quad \ell'_i = \sigma_i(\Lambda' + \rho)(E_{ii})$$

this is equivalent to the requirement that

$$\tilde{C}(\ell) = \tilde{C}(\ell') \quad \text{for all} \quad C \in Z \;,$$

which, in turn, is equivalent to the apparently weaker condition

$$\tilde{C}_i(\ell) = \tilde{C}_i(\ell') \quad \text{for} \quad 1 \leq i \leq n+m \;.$$

In fact, this latter condition is equivalent to

$$P_i(\ell) = P_i(\ell') \quad \text{for} \quad 1 \leq i \leq n+m \;,$$

and this condition means that the sequences $\ell_1, \ldots, \ell_n, \ell'_{n+1}, \ldots, \ell'_{n+m}$ and $\ell'_1, \ldots, \ell'_n, \ell_{n+1}, \ldots, \ell_{n+m}$ agree up to a permutation. But this implies that $\tilde{C}(\ell) = \tilde{C}(\ell')$ for all $C \in Z$.

Suppose now that Λ is such that

$$\ell_i \neq \ell_j \quad \text{if} \quad 1 \leq i \leq n;\; n+1 \leq j \leq n+m \;.$$

Then the foregoing discussion shows that there exist a permutation σ of $\{1, \ldots, n\}$ and a permutation τ of $\{n+1, \ldots, n+m\}$ such that

$$\ell'_i = \ell_{\sigma(i)} \quad \text{for} \quad 1 \leq i \leq n$$
$$\ell'_j = \ell_{\tau(j)} \quad \text{for} \quad n+1 \leq j \leq n+m \;.$$

If Λ and Λ' are elements of P_{++}, this is possible only if σ and τ are equal to the identity, i.e., if $\ell' = \ell$.

An element $\Lambda \in h^*$ such that

$$\ell_i \neq \ell_j \quad \text{if} \quad 1 \leq i \leq n; \quad n+1 \leq j \leq n+m$$

is said to be typical (in general position).
Thus we have shown:
Let $\Lambda, \Lambda' \in P_{++}$. If Λ is typical and different from Λ', then Λ and Λ' can be separated by a suitable Casimir element.

Returning to the discussion of our gl(n,m)-module V we see that if the highest weight Ω is typical, then we are necessarily in the case I and V is completely reducible.

It remains to investigate whether for non-typical Ω the module V may in fact be not completely reducible. This is actually the case. Indeed, let $\Lambda \in P_{++}$ and let $\bar{J}(\Lambda)$ be the gl(n,m)-module derived from the irreducible $gl(n,m)_0$-module $J(\Lambda)$ (with highest weight Λ) as described in section 4. Since $\bar{J}(\Lambda)$ is a highest weight module with highest weight Λ , any Casimir element C is represented by $\chi_\Lambda(C)$ id. According to Kac [4] , $\bar{J}(\Lambda)$ is irreducible if and only if Λ is typical, moreover, if $\bar{J}(\Lambda)$ is reducible, then it is not completely reducible.

Proof [4]
Let w denote a generator of the highest weight module $\bar{J}(\Lambda)$. By definition, the sole submodule of $\bar{J}(\Lambda)$ containing w is $\bar{J}(\Lambda)$ itself. This implies that $\bar{J}(\Lambda)$ is either irreducible or else not completely reducible, and we are in the former case if Λ is typical.

Now let T_+ and T_- denote the products of all E_{ij} with $1 \leq i \leq n$; $n+1 \leq j \leq n+m$ or with $n+1 \leq i \leq n+m$; $1 \leq j \leq n$, respectively, calculated in U in some fixed order (an inversion of two factors leads to a change of sign). Note that T_+ and T_- commute with all elements of $sl(n) \times sl(m)$, and that $T_+ T_-$ even commutes with all elements of $gl(n,m)_o$.

It is easy to see that any non-zero submodule of $\bar{J}(\Lambda)$ contains the non-zero(!) element $T_- w$. Consequently, $\bar{J}(\Lambda)$ is irreducible if and only if $T_+ T_- w$ is different from zero. Of course, $T_+ T_- w$ is proportional to w, hence all we have to do is to calculate the scalar factor.

In the following, Λ may be an arbitrary element of h*. Consider any (not necessarily finite-dimensional) highest weight module W with highest weight Λ and let w_Λ denote a generator of it. Once again, $T_+ T_- w_\Lambda$ is proportional to w_Λ , and the scalar factor depends only on Λ . More precisely, we have

$$T_+ T_- w_\Lambda = Q(\Lambda) w_\Lambda ,$$

where Q is a polynomial function on h* of total degree $\leq$ nm.

We express $Q(\Lambda)$ in terms of the coordinates

$$\ell_i = \sigma_i(\Lambda + \rho)(E_{ii})$$

and write

$$Q(\Lambda) = \hat{Q}(\ell) .$$

Recalling that $T_+ T_-$ commutes with all elements of $gl(n,m)_o$, we can now argue as in the proof of the statements 1) and 2) in subsection C and show that $\hat{Q}(\ell)$ is symmetric in $\ell_1, \ldots, \ell_n$ and in $\ell_{n+1}, \ldots, \ell_{n+m}$, and that $\hat{Q}(\ell)$ vanishes if $\ell_n = \ell_{n+1}$. It follows that

$$\hat{Q}(\ell) = c \prod_{\substack{1 \leq i \leq n \\ n+1 \leq j \leq n+m}} (\ell_i - \ell_j) ,$$

with some constant c which, according to our introductory remarks, is different from zero. Thus we have shown that $Q(\Lambda)$ vanishes if and only if Λ is non-typical.

E. Concluding remarks

Of course, the results of the present and the foregoing sections only form a starting point for a more detailed and explicit investigation of the gl(n,m)-modules. For example, the characters of gl(n,m)-modules should be calculated, the branching rules for the restriction of a representation to the most important graded subalgebras of gl(n,m) have to be determined, and the tensor products of gl(n,m)-modules must be decomposed. These problems are still under investigation, but substantial partial results have already been obtained. For more details, and for extensions to other Lie superalgebras, I refer the reader to the literature (see the subsequent bibliography).

REFERENCES

1. V. G. Kac, Lie superalgebras, Adv. Math. 26 (1977) 8
2. M. Scheunert, "The theory of Lie superalgebras", Lecture Notes in Mathematics 716; Springer; Berlin, Heidelberg, New York (1979)
3. V. G. Kac, Characters of typical representations of classical Lie superalgebras, Comm. Alg. 5 (1977) 889
4. V. G. Kac, Representations of classical Lie superalgebras, p. 597 in "Differential Geometrical Methods in Mathematical Physics II", Bonn 1977; K. Bleuler, H. R. Petry, and A. Reetz eds.; Lecture Notes in Mathematics 676; Springer; Berlin, Heidelberg, New York (1978)
5. M. Scheunert, Graded tensor calculus, J. Math. Phys. 24 (1983) 2658
6. N. Bourbaki, "Algèbre", chap. I, second edition; Hermann; Paris (1964)
7. P. H. Dondi and P. D. Jarvis, Diagram and superfield techniques in the classical superalgebras, J. Phys. A: Math. Gen. 14 (1981) 547
8. A. B. Balantekin and I. Bars, Dimension and character formulas for Lie supergroups, J. Math. Phys. 22 (1981) 1149
9. P. D. Jarvis and M. K. Murray, Casimir invariants, characteristic identities and tensor operators for "strange" superalgebras, J. Math. Phys. 24 (1983) 1705
10. M. Scheunert, Eigenvalues of Casimir operators for the general linear, the special linear, and the orthosymplectic Lie superalgebras, J. Math. Phys. 24 (1983) 2681
11. M. Scheunert, Casimir elements of ε Lie algebras, J. Math. Phys. 24 (1983) 2671

12. V. G. Kac, Laplace operators of infinite-dimensional Lie algebras and theta functions, Proc. Natl. Acad. Sci. USA 81 (1984) 645, Physics
13. J. Dixmier, "Algèbres enveloppantes"; Gauthier-Villars; Paris, Bruxelles, Montréal (1974)
14. N. Bourbaki, "Groupes et algèbres de Lie", chapitres 7,8; Hermann; Paris (1975)

BIBLIOGRAPHY

In the subsequent bibliography I have collected most of the more recent papers on Lie superalgebras and their representations which I became aware of. No attempt at completeness has been made, and I apologize to everybody whose papers are not appropriately included. Nevertheless, I hope that by also taking into account the references in the papers cited, a majority of the pertinent publications is covered. Note that all those papers are excluded which are already contained in the references of my monograph [2] or of the following paper by Kaplansky:
I. Kaplansky, Superalgebras, Pacific J. Math. 86 (1980) 93.

Agrawala, V. K., Wigner-Eckart theorem for graded generalized Lie algebras, Hadronic J. 2 (1979) 830

Agrawala, V. K., Invariants of generalized Lie algebras, Hadronic J. 4 (1981) 444

Alekseevskii, D. V., Leites, D. A., and Ščepočkina, I. M., Examples of simple infinite-dimensional Lie superalgebras of vector fields (Russian), C. R. Acad. Bulgare Sci. 33 (1980) 1187

Backhouse, N. B., On representations of the superalgebra osp(2,1); Group Theoretical Methods in Physics, Canterbury 1981; Phys. A. 114 (1982) 410

Balantekin, A. B., Character expansion for U(N) groups and U(N|M) supergroups, J. Math. Phys. 25 (1984) 2028

Balantekin, A. B., and Bars, I., Dimension and character formulas for Lie supergroups, J. Math. Phys. 22 (1981) 1149

Balantekin, A. B., and Bars, I., Representations of supergroups, J. Math. Phys. 22 (1981) 1810

Balantekin, A. B., and Bars, I., Branching rules for the supergroup SU(N|M) from those of SU(N+M), J. Math. Phys. 23 (1982) 1239

Bars, I., and Günaydin, M., Unitary representations of non-compact supergroups, Commun. Math. Phys. 91 (1983) 31

Bars, I., Morel, B., and Ruegg, H., Kac-Dynkin diagrams and supertableaux, J. Math. Phys. 24 (1983) 2253

Bednář, M., McKellar, B. H. J., and Šachl, V. F., Micu-type invariants for the Casimir operators of the Lie superalgebras osp(1,2n): application to the algebra osp(1,4), J. Phys. A: Math. Gen. 17 (1984) 1579

Berele, A., and Regev, A., Hook Young diagrams, combinatorics and representations of Lie superalgebras, Bull. Amer. Math. Soc. 8 (1983) 337

Bernstein, I. N., and Leites, D. A., Irreducible representations of finite-dimensional Lie superalgebras of series W and S, C. R. Acad. Bulgare Sci. 32 (1979) 277

Bernstein, I. N., and Leites, D. A., A formula for the characters of the irreducible finite-dimensional representations of Lie superalgebras of series gl and sl (Russian), C. R. Acad. Bulgare Sci. 33 (1980) 1049

Bernstein, I. N., and Leites, D. A., Invariant differential operators and irreducible representations of Lie superalgebras of vector fields, Selecta Math. Soviet. 1(2) (1981) 143

Bernstein, I. N., and Leites, D. A., The superalgebra Q(n), the odd trace and the odd determinant, C. R. Acad. Bulgare Sci. 35 (1982) 285

Bernstein, I. N., and Leites, D. A., Irreducible representations of finite-dimensional Lie superalgebras of type W, Selected translations Selecta Math. Soviet. 3(1) (1983/84) 63

Bincer, A. M., Eigenvalues of Casimir operators for the general linear and orthosymplectic Lie superalgebras, J. Math. Phys. 24 (1983) 2546

Blank, J., Havlicek, M., Exner, P., and Lassner, W., Boson-fermion representations of Lie superalgebras: The example of osp(1,2), J. Math. Phys. 23 (1982) 350

Chen, Jin-Quan, and Chen, Xuan-Gen, The Gel'fand basis and matrix elements of the graded unitary group U(m|n), J. Phys. A: Math. Gen. 16 (1983) 3435

Chen, Jin-Quan; Chen, Xuan-Gen; and Gao, Mei-Juan; The Clebsch Gordan coefficients and isoscalar factors of the graded unitary group SU(m|n), J. Phys. A: Math. Gen. 16 (1983) L47

Chen, Jin-Quan; Chen, Xuan-Gen; and Gao, Mei-Juan; The Casimir invariants and Gel'fand basis of the graded unitary group SU(m|n), J. Phys. A: Math. Gen. 16 (1983) 1361

Chen, Jin-Quan; Gao, Mei-Juan; and Chen, Xuan-Gen; The Clebsch-Gordan coefficient for SU(m|n) Gel'fand basis, J. Phys. A: Math. Gen. 17 (1984) 481

Chen, Jin-Quan; Gao, Mei-Juan; and Chen, Xuan-Gen; Coefficients of fractional parentage for $U(m+p|n+q) \supset U(m|n) \times U(p|q)$ and $U(m|n) \supset U(m) \times U(n)$, J. Phys. A: Math. Gen. 17 (1984) 1941

Delduc, F., and Gourdin, M., Mixed supertableaux of the superunitary groups I - SU(n|1), J. Math. Phys. 25 (1984) 1651

DeWitt, B. S., and van Nieuwenhuizen, P., Explicit construction of the exceptional superalgebras F(4) and G(3), J. Math. Phys. 23 (1982) 1953

Djoković, D. Ž., Superlinear algebra or two-graded algebraic structures, J. reine angew. Math. 305 (1979) 65

Dondi, P. H., and Jarvis, P. D., Diagram and superfield techniques in the classical superalgebras, J. Phys. A: Math. Gen. 14 (1981) 547

Farmer, R. J., and Jarvis, P. D., Representations of low-rank orthosymplectic superalgebras by superfield techniques, J. Phys. A: Math. Gen. 16 (1983) 473

Farmer, R. J., and Jarvis, P. D., Representations of orthosymplectic superalgebras II. Young diagrams and weight space techniques, J. Phys. A: Math. Gen. 17 (1984) 2365

Flato, M., and Fronsdal, C., Representations of conformal supersymmetry, Lett. Math. Phys. 8 (1984) 159

Green, H. S., and Jarvis, P. D., Casimir invariants, characteristic identities, and Young diagrams for color algebras and superalgebras, J. Math. Phys. 24 (1983) 1681

Gruber, B., Santhanam, T. S., and Wilson, R., A unified description of the representations of the graded Lie algebra Gsl(2), J. Math. Phys. 25 (1984) 1253

Han, Qi-Zhi, The finite-dimensional star and grade star irreducible representations of SU(n|1), Nuovo Cimento 64A (1981) 391

Han, Qui-Zhi; Song, Xing-Chang; Li, Gen-Dao; and Sun, Hong-Zhou; Irreducible representations of the graded Lie algebra SU(m|n), Phys. Energ. Fortis Phys. Nucl. 5 (1981) 546

Han, Qui-Zhi, and Sun, Hong-Zhou, Irreducible representations of superalgebras B(1,1) and B(0,2), Comm. Theoret. Phys. 2 (1983) 1137

Hlavatý, L., and Niederle, J., Casimir operators of the simplest supersymmetry superalgebras, Lett. Math. Phys. 4 (1980) 301

Hurni, J.-P., and Morel, B., Irreducible representations of the superalgebrae type II, J. Math. Phys. 23 (1982) 2236

Hurni, J.-P., and Morel, B., Irreducible representations of SU(M|N), J. Math. Phys. 24 (1983) 157, Erratum ibid. p. 2250

Inaba, I., Maekawa, T., and Yamamoto, T., Infinite dimensional representations of the graded Lie algebra (Sp(4):4). Representation of the para-Bose operators with real order of quantization, J. Math. Phys. 23 (1982) 954

Ivanov, E. A., and Sorin, A. S., Structure of representations of the conformal supergroup in the OSp(1,4) basis, Theor. Math. Phys. 45 (1981) 862 (Russ. orig. 45 (1980) 30)

Jarvis, P. D., and Green, H. S., Casimir invariants and characteristic identities for generators of the general linear, special linear and orthosymplectic graded Lie algebras, J. Math. Phys. 20 (1979) 2115

Jarvis, P. D., and Murray, M. K., Casimir invariants, characteristic identities and tensor operators for "strange" superalgebras, J. Math. Phys. 24 (1983) 1705

Kac, V. G., Representations of classical Lie superalgebras, p. 597 in "Differential Geometrical Methods in Mathematical Physics II", Bonn 1977; K. Bleuler, H. R. Petry, and A. Reetz eds.; Lecture Notes in Mathematics 676; Springer; Berlin, Heidelberg, New York (1978)

Kac, V. G., Contravariant form for infinite-dimensional Lie algebras and superalgebras, p. 441 in "Group Theoretical Methods in Physics", Austin 1978; W. Beiglböck, A. Böhm, and E. Takasugi eds.; Lecture Notes in Physics 94; Springer; Berlin, Heidelberg, New York (1979)

Kac, V. G., Laplace operators of infinite-dimensional Lie algebras and theta functions, Proc. Natl. Acad. Sci. USA 81 (1984) 645, Physics

Kaplansky, I., The Virasoro algebra, Commun. Math. Phys. 86 (1982) 49

King, R. C., Generalized Young tableaux for Lie algebras and superalgebras, p. 41 in "Group Theoretical Methods in Physics", Istanbul 1982; M. Serdaroğlu and E. İnönü eds.; Lecture Notes in Physics 180; Springer; Berlin, Heidelberg, New York, Tokyo (1983)

Kobayashi, Y., Nagamachi, S., and Miyamoto, H., Lie groups and Lie algebras with generalized Bose-Fermi symmetric parameters, Proc. Japan Acad. Ser. A Math. Sci. 59(5) (1983) 203

Leites, D. A., New Lie superalgebras and mechanics, Soviet Math. Dokl. 18 (1977) 1277 (Russ. orig. Dokl. Akad. Nauk SSSR 236(4) (1977))

Leites, D. A., Formulas for the characters of irreducible finite-dimensional representations of simple Lie superalgebras, Funkts. Anal. Prilozhen. 14(2) (1980) 35 (Engl. transl. p. 106)

Leites, D. A., A formula for the characters of the irreducible finite-dimensional representations of Lie superalgebras of series C (Russian), C. R. Acad. Bulgare Sci. 33 (1980) 1053

Leites, D. A., Irreducible representations of Lie superalgebras of vector fields and invariant differential operators, Funkts. Anal. Prilozhen. 16(1) (1982) 76

Leites, D. A., Representations of Lie superalgebras, Theor. Math. Phys. 52 (1983) 764 (Russ. orig. 52 (1982) 225)

Leites, D. A., and Serganova, V. V., Solutions of the classical Yang-Baxter equation for simple superalgebras, Theor. Math. Phys. 58 (1984) 16 (Russ. orig. 58 (1984) 26)

Lemire, F. W., and Patera, J., Congruence classes of finite representations of simple Lie superalgebras, J. Math. Phys. 23 (1982) 1409

Lyakhovskii, V. D., Stability of semisimple superalgebras, Theor. Math. Phys. 38 (1979) 78 (Russ. orig. 38 (1979) 115)

Marcu, M., The representations of spl(2,1) - an example of representations of basic superalgebras, J. Math. Phys. 21 (1980) 1277

Marcu, M., The tensor product of two irreducible representations of the spl(2,1) superalgebra, J. Math. Phys. 21 (1980) 1284

Mickelsson, J., Discrete series of Lie superalgebras, Rep. Math. Phys. 18 (1980) 197

Mitra, B., and Tripathy, K. C., The cohomology of generalized Lie algebras, J. Math. Phys. 25 (1984) 2550

Mosolova, M. V., Functions of noncommuting operators that generate a graded Lie algebra, Math. Notes Acad. Sci. USSR 29(1,2) (1981) 17 (Russ. orig. 29(1) (1981) 35)

Ne'eman, Y., and Sternberg, S., Internal supersymmetry and unification, Proc. Natl. Acad. Sci. USA 77 (1980) 3127, Physics

Palev, T. D., Fock space representations of the Lie superalgebra A(0,n), J. Math. Phys. 21 (1980) 1293

Palev, T. D., A trace formula for realizations of Lie superalgebras with creation and annihilation operators, C. R. Acad. Bulgare Sci. 33 (1980) 1195

Palev, T. D., Para-Bose and para-Fermi operators as generators of orthosymplectic Lie superalgebras, J. Math. Phys. 23 (1982) 1100

Palev, T. D., and Stoytchev, O. Ts., Finite-dimensional induced representations of the Lie superalgebra sl(1,n), C. R. Acad. Bulgare Sci. 35 (1982) 733

Parker, M., Classification of real simple Lie superalgebras of classical type, J. Math. Phys. 21 (1980) 689

Parker, M., Formes reelles des superalgebres de Lie simples classiques, Bull. Soc. Math. Belg. 33 (1981) 129

Rittenberg, V., A guide to Lie superalgebras, p. 1 in "Group Theoretical Methods in Physics", Tübingen 1977; P. Kramer and A. Rieckers eds.; Lecture Notes in Physics 79; Springer; Berlin, Heidelberg, New York (1978)

Rittenberg, V., and Scheunert, M., A remarkable connection between the representations of the Lie superalgebras osp(1,2n) and the Lie algebras o(2n+1), Commun. Math. Phys. 83 (1982) 1

Rittenberg, V., and Wyler, D., Generalized superalgebras, Nucl. Phys. B139 (1978) 189

Rittenberg, V., and Wyler, D., Sequences of $Z_2 \oplus Z_2$ graded Lie algebras and superalgebras, J. Math. Phys. 19 (1978) 2193

Ruegg, H., Labelling of irreducible representations of super Lie algebras, p. 265 in "Group Theoretical Methods in Physics", Istanbul 1982; M. Serdaroğlu and E. Ínönü eds., Lecture Notes in Physics 180; Springer; Berlin, Heidelberg, New York, Tokyo (1983)

Scheunert, M., Generalized Lie algebras, J. Math. Phys. 20 (1979) 712

Scheunert, M., Graded tensor calculus, J. Math. Phys. 24 (1983) 2658

Scheunert, M., Casimir elements of ε Lie algebras, J. Math. Phys. 24 (1983) 2671

Scheunert, M., Eigenvalues of Casimir operators for the general linear, the special linear, and the orthosymplectic Lie superalgebras, J. Math. Phys. 24 (1983) 2681

Scheunert, M., Casimir elements of Lie superalgebras, p. 115 in "Differential Geometric Methods in Mathematical Physics", Jerusalem 1982; S. Sternberg ed.; Mathematical Physics Studies 6; Reidel; Dordrecht, Boston, Lancaster (1984)

Serganova, V. V., Classification of real simple Lie superalgebras and symmetric superspaces, Funkts. Anal. Prilozhen. 17(3) (1983) 46 (Engl. transl. p. 200)

Sergeev, A. N., Invariant polynomial functions on Lie superalgebras (Russian), C. R. Acad. Bulgare Sci. 35 (1982) 573

Sergeev, A. N., The centre of enveloping algebra for Lie superalgebra $Q(n,\mathbb{C})$, Lett. Math. Phys. 7 (1983) 177

Shchepochkina, I. M., Exceptional simple infinite-dimensional Lie superalgebras (Russian), C. R. Acad. Bulgare Sci. 36 (1983) 313

Shmelev, G. S., Irreducible representations of infinite-dimensional Hamiltonian and Poisson Lie superalgebras, and invariant differential operators (Russian), Serdica 8 (1982) 408

Shmelev, G. S., Differential operators that are invariant with respect to the Lie superalgebra $H(2,2;\lambda)$ and its irreducible representations (Russian), C. R. Acad. Bulgare Sci. 35 (1982) 287

Shmelev, G. S., Classification of indecomposable representations of the Lie superalgebra W(0,2) (Russian), C. R. Acad. Bulgare Sci. 35 (1982) 1025

Shmelev, G. S., Contact Lie superalgebras are rigid (Russian), C. R. Acad. Bulgare Sci. 36 (1983) 569

Shmelev, G. S., Irreducible representations of Poisson-Lie superalgebras, and invariant differential operators, Funct. Anal. Appl. 17 (1983) 76 (Russ. orig. 17(1) (1983) 91)

Shmelev, G. S., Differential H(2n,m)-invariant operators and indecomposable osp(2,2n)-representations, Funkts. Anal. Prilozhen. 17(4) (1983) 94 (Engl. transl. p. 323)

Sudbery, A., Octonionic description of exceptional Lie superalgebras, J. Math. Phys. 24 (1983) 1986

Sun, Hong-Zhou, and Han, Qi-Zhi, Irreducible representations of graded Lie algebra $SU(n|1)$, Scientia Sinica 24 (1981) 914

Thierry-Mieg, J., Irreducible representations of the $SU(1|N)$ Lie superalgebras, Phys. Lett. 138B (1984) 393

Thierry-Mieg, J., Irreducible representations of the basic classical Lie superalgebras $SU(m|n)$; $SU(n|n)/U(1)$; $OSp(m|2n)$; $D(2|1;\alpha)$; G(3); F(4), p. 94 in "Group Theoretical Methods in Physics", Trieste 1983; G. Denardo, G. Ghirardi, and T. Weber eds.; Lecture Notes in Physics 201; Springer; Berlin, Heidelberg, New York, Tokyo (1984)

Wybourne, B. G., Characters, dimensions and branching rules for covariant irreps of $U(N|M)$, J. Phys. A: Math. Gen. 17 (1984) 1573

GRAVITY BEFORE SUPERGRAVITY

Philippe Spindel

Faculté des Sciences
Université de l'Etat
Avenue Maistriau, 15, 7000 Mons (Belgium)

I. INTRODUCTION

These lectures are intended for students not familiar with the methods used in many developments of general relativity, cosmology and supergravity. None of the matters discussed in this text are new, but they are so dispersed in the literature that it seems to me useful to collect them in the framework of an introductory course.

As much as possible these notes will be self-contained. The stress is mainly put on the geometrical significance of the concepts. However super-abstract notations have been avoided.

The first part of this course deals with geometry before gravity. Manifolds, tensors, spinors and their derivatives are defined. The rules of Cartan's exterior differential calculus are established. Basic formulas of Riemannian geometry are proved with the method of the moving frame (veilbein). Some aspects of the de Rham cohomology are touched lightly. By the way, these different notions are illustrated in the framework of Lie groups. The second part of the text begins with an analysis of the physical meaning of the curvature tensor which leads to the Einstein equations. Then the Weyl's and Palitini's variational principle are introduced and compared. The extension of first integrals for field equations on curved spaces is discussed. Finally a brief description of homogeneous cosmologies, in particular the anti-de Sitter space, completes these notes.

The choice of these topics just reflects the opinion of the author on what he considers as an essential cultural background to approach (super) gravity locally i.e. before to grapple with technical, topological and quantum difficulties. There exist innumberable text-books on geometry and gravity which cover the subject of these lectures. A few of them that I have used intensively are presented in the bibliography but, in order to lighten the text, they are not systematically quoted. Of course, there are many others as well recommandable.

II. THE RENUNCIATION OF MINKOWSKI SPACE TIME

In 1911 Einstein[1] showed that conservation of energy implies that the frequency ν of a photon travelling in a uniform gravitational field g must vary with the height z according to the relation

$$\nu(z) = \nu(z_o) \ [1 + c^{-2} g(z - z_o)] \quad . \tag{2.1}$$

This prediction has been confirmed experimentally up to 1 % accuracy by Pound, Rebka and Snider[2].

From this redshift experience Schild[3] concluded that the geometry of the universe could not be described by a flat Minkowski space-time. His argument is the following. Imagine two observers at rest located at different heights in the gravitational field of the Earth. Their world-lines are two parallel lines. Suppose that the bottom observer emits successive pulses of monochromatic electromagnetic waves. If flat geometry prevails, whatever the laws of propagation of light in a gravitational field are, the world lines of the pulses are parallel curves and as a consequence the interval of time for the emission of a given number of pulses must be equal to the interval of time for reception. But on the other hand this interval of time can be expressed as the ratio of the number of cycles of the pulses divided by their frequencies. This leads to a contradiction. Indeed although the number of cycles remains constant for each observes the frequencies are redschifted. This shows that the intervals of time for emission and reception must be different, a result incompatible with the assumed flat geometry of the space-time.

Since the beginning of the XVII th. century we know from Galilei experiences that bodies fall with an acceleration independent of their mass. According to Newton's second law this implies that the inertial mass of a body (which measures its amount of matter) is equal to its gravitational mass (its coupling to the gravitational field). This equality has been verified experimentally up to 1 part in 10^{11} for aluminium and gold[4] and even up to 1 part in 10^{12} for aluminimum and platinium[5] falling in the gravitational field of the Sun. These measurements are the most precise ever performed in physics!

From this equality of inertial and gravitational masses we can infer, in the framework of Newtonian particle mechanics, the equivalence of accelerated frames and uniform gravitational fields. Consider a system of particles interacting through forces $\vec{F}$ depending only on their relative positions in an external constant gravitational field $\vec{g}$. Their equations of motions, with respect to an inertial frame, are

$$m_i^I \ddot{\vec{x}}_i = \vec{F}_i(\vec{x}_j - \vec{x}_k) + m_i^G \vec{g} \tag{2.2}$$

With respect to an accelerated frame defined by the following coordinate change

$$\vec{x}' = \vec{x} - \frac{1}{2} \vec{g}\, t^2 \tag{2.3}$$

the contribution of the external gravitational field vanishes. An observer attached to this accelerated frame cannot detect through any mechanical experience the existence of the gravitational field $\vec{g}$.

The great achievement of Einstein was to postulate the *equivalence principle* that is no physical experiment enables one to distinguish an accelerated frame from a uniform gravitational field.

This statement however must be revisited for true gravitational fields, since they can be considered as constant fields only on small scales so that the equivalence principle only refers to local experiences. Indeed inhomogeneities are felt, and for instance particles falling freely towards the center of the Earth will converge to each other during their motion. So the equivalence principle should be interpreted in the same way as when we identify a curved surface with its tangent plane. It is valid up to second order effects.

III. CALCULUS ON MANIFOLDS

III.1. Manifolds and tensors

Einstein's general relativity describes the effects of gravitation by the equivalence principle which states that at every space-time point in an arbitrary gravitational field, a local frame exists such that the local laws of physics take their usual special-relativistic form. This principle constitutes the cornerstone of the theory of general relativity. It leads to a very deep relation between the geometry of the universe and the description of the gravitational field. Indeed, the equivalence principle asserts that on a small scale space-time looks like Minkowski space-time. The framework which describes the universe is constituted of small regions glued together but the resulting structure obtained after this pasting is in general non trivial.

Of course, this description is terribly vague. How small are these regions ? How are they linked up ? How to work on them ? Answers to these questions are given by the differential geometry. The basic structure needed to describe classically space-time is a 4-dimensional *differentiable manifold*. We recall that a n-dimensional manifold M_n is a topological space such that around any point P one can find a neighbourhood V_P in one-to-one continuous correspondence with an open subset of $\mathbb{R}^n$. These local isomorphisms ϕ_P define *local coordinates* by assigning to each point Q belonging to V_P a n-uple of numbers $x^\alpha = X^\alpha_P(Q)$ which are the components of the image-point $\phi_P(Q)$ in a fixed basis of $\mathbb{R}^n$. Moreover it is required that when two neighbourhoods overlap ($V_P \cap V_{P'}$ nonvoid), the mappings $\psi_{PP'} = \phi_P \circ \phi^{-1}_{P'}$ define one-to-one differentiable transformations between open subsets of $\mathbb{R}^n$. It is this last assumption which allows to define tensors and differentiable objects on a manifold.

A real-valued *function* f on M_n is called *differentiable* if and only if $f_P = f \circ \phi^{-1}_P$ is a differentiable mapping from $\mathbb{R}^n$ to $\mathbb{R}$ in the usual sense of calculus. A *curve* $C(t)$ defined by a mapping from $\mathbb{R}$ into M_n is *differentiable* if and only if $X^\alpha_P[C(t)]$ is a usual differentiable mapping from $\mathbb{R}$ to $\mathbb{R}^n$. As a general rule a geometrical object will be said differentiable if and only if its expression in local coordinates is differentiable in the usual sense. Once we have defined a differentiable curve we may consider its *tangent vectors*. The vector $\vec{v}_o$ tangent at the curve $C(t)$ at a point $P = C(t_o)$ is a first order differential operator that acts on functions defined on a neighborhood of P as :

$$\vec{v}_o(f) = \frac{d}{dt} f \circ C(t) \Big|_{t=t_o} \tag{3.1}$$

The number $\vec{v}_o(f)$ is called the *Pfaffian derivative* of f in the direction $\vec{v}_o$. In terms of local coordinates this formula reads :

$$\vec{v}_o(f) = v^\alpha_P(t_o) \frac{\partial f_P}{\partial x^\alpha} \Big|_{x^\alpha = X^\alpha_P[C(t_o)]} \tag{3.2a}$$

where

$$v^\alpha_P(t) = \frac{d}{dt} X^\alpha_P [C(t)] \tag{3.2b}$$

are the components of the vector $\vec{v}_o$ in the local chart (V_P, X^α_P).

All these constructions and definitions have a geometrical meaning because they are "gauge invariant" : that means that although we are working with a specific coordinate system, the results do not depend on our particular choice of V_P, ϕ_P or X^α_P (even if the values taken by v^α_P or if the analytical expression of f_P actually depend on them). In what follows we shall drop the label indicating which coordinate system is chosen and conform to use by denoting for instance f_P by $f(x)$ or $f(x^\alpha)$.

A differentiable manifold is characterized by a topological space M and a family of mappings X_P^α whose domain V_P cover M. This family of pairs (V_P, X_P^α) covering M and satisfying the consistency relation on the overlapping domains define an *atlas* on M. One can increase an atlas by adding a pair (V, X^α) which satisfies the consistency relation with all the previous members of the atlas. This addition of new local coordinate systems does not change any geometrical property of the manifold. In fact each atlas is included in a maximal atlas A which contains all possible compatible local coordinate systems. It is this pair (M,A) which defines precisely a differentiable manifold. It is not difficult to show that a topological space admits numerous distinct incompatible atlases. Nevertheless two "distinct" differentiable manifolds (M,A) and (M',A') are considered as equivalent — *diffeomorphic* — when there exists a one-to-one mapping ψ from M onto M' such that $(V', X^{\alpha'})$ belongs to A' if and only if $(\psi^{-1}(V'), X^{\alpha'} \circ \psi)$ belongs to A. Diffeomorphic differentiable manifolds represent a unique differentiable structure. Obviously if (M,A) and (M',A') are diffeomorphic M and M' are homeomorphic. However the converse is false. For instance on can show that all C^∞ differentiable structures on $\mathbb{R}^n$ are equivalent, except for $\mathbb{R}^4$ which admits at least two non-equivalent differentiable structures and for which the existence of a uncountable infinity of inequivalent structure is conjectured (*).

Consider now all the differentiable curves crossing at the point P. We see that their tangent vectors generate an n-dimensional vector space : the *tangent space* at P. Each coordinate system defines a natural basis of vectors $\frac{\partial}{\partial x^\alpha}$ tangent to the curves which are the inverse of coordinate lines under the map X_P^α. Of course, a canonical natural basis does not exist. The dual of the tangent space defines the *cotangent space* : it is the space of covariant vectors. Tensor products of tangent and cotangent spaces provide us with all kinds of "tangent" tensor-spaces.

III.2. Differentials forms and exterior derivative

In general, tensors at different points cannot be compared in a gauge invariant way without adding an extra structure on the manifold. However for completely antisymmetric covariant p-tensors fields — *differential p-forms* — it is possible to go further and

(*) Is there a connection between this property of $\mathbb{R}^4$ and the effective dimensionality of our space-time ? The question is open. Notice however that S_7 is the first sphere which admits more than one differentiable structure, actually 28.

to develop something analogous to the usual calculus : the *exterior differential calculus* à la Cartan[6].

On an n-dimensional manifold 0-forms are functions, 1-forms covariant vector fields and p-forms with degree $p > n$ are identically zero. In the following we shall denote by $\{\theta^a\}$ a set of n independent differentiable 1-forms (*). In general these forms are only locally defined, on the domain of a chart for instance. Some care is required when the full manifold is considered. Nevertheless these forms may also exist on domains which cannot be covered by a single coordinate patch; this is the case, e.g., on group manifolds (see further), on any 3-dimensional manifold, on the 7-dimensional sphere, etc... which are *parallelizable* which means that they admit a global field of frames.

From a basis $\{\theta^a\}$ of Λ^1, the space of 1-forms, we can construct a basis for Λ^p, the space of p-forms, by forming the $\binom{n}{p}$ independent completely antisymmetric tensorial product of θ^a. These products define the *exterior product* "$\wedge$" of the 1-forms considered above. In this way we obtain for instance :

$$\theta^a \wedge \theta^b = \theta^a \otimes \theta^b - \theta^b \otimes \theta^a$$

$$\theta^a \wedge \theta^b \wedge \theta^c = \theta^a \otimes \theta^b \otimes \theta^c + \text{five other terms weighted by } \pm 1$$

according to the evenness of the permutations of the indices a, b, c.

Linearity extends the definition of exterior products to any arbitrary forms.
If $\alpha = \frac{1}{p!}\, \alpha_{a_1 \ldots a_p}\, \theta^{a_1} \wedge \ldots \wedge \theta^{a_p}$ is a p-form of "components" $\alpha_{a_1 \ldots a_p}$ and β a q-form, the components of the (p+q)form $\omega = \alpha \wedge \beta$ are :

$$\omega_{a_1 \ldots a_{p+q}} = \frac{(p+q)!}{p!q!}\, \alpha_{[a_1 \ldots a_p}\, \beta_{a_{p+1} \ldots a_{p+q}]} \tag{3.3}$$

In this formula the bracket implies a complete antisymmetrization :

(*) From now on we shall not be worried by precise assumptions on the differentiability conditions. We assume that all the objects considered are smooth i.e. sufficiently differentiable in order for the calculations to be meaningful.

$$\varphi_{[a_1 \ldots a_p]} \equiv \frac{1}{p!} \sum_{\pi \in S_p} \operatorname{sgn} \pi \, \varphi_{a_{\pi(1)} \ldots a_{\pi(p)}}$$

where S_p is the group of permutations of p objects and sgn π is +1 or −1 according to wether the permutation π is even or odd.
It follows from eq. (3) that the exterior product is associative, bilinear but not commutative :

$$\alpha \wedge \beta = (-1)^{pq} \beta \wedge \alpha \quad . \tag{3.4}$$

Equation (1) may be interpreted in a dual way : to a function f we associate a 1-form df such that :

$$df(\vec{v}) \equiv \vec{v}(f) \quad . \tag{3.5}$$

This 1-form df is the exterior derivative of the 0-form f. In particular the exterior derivatives of the coordinates constitute the dual basis of the natural basis :

$$dx^\alpha\left(\frac{\partial}{\partial x^\beta}\right) = \frac{\partial x^\alpha}{\partial x^\beta} = \delta^\alpha_\beta \quad . \tag{3.6}$$

By comparing eqs (2) and (5,6) we get the expression for df in local coordinates :

$$df = \frac{\partial}{\partial x^\alpha} f(x) \, dx^\alpha \quad . \tag{3.7a}$$

This formula is written in natural coordinates. With respect to an arbitrary basis it becomes :

$$df = \vec{e}_a(f)\theta^a \equiv \partial_a f \, \theta^a \tag{3.7b}$$

where $\{\vec{e}_a\}$ and $\{\theta^a\}$ are dual basises of the tangent and cotangent spaces and ∂_a is just another notation for the vectors $\vec{e}_a$.

The definition of the *exterior derivative* is extended to arbitrary p-forms as an exterior product antiderivation with the following assumptions :

(1°) The operator d maps p-forms into (p+1)-forms.

(2°) It is linear with respect to the constants :

$$d(a\alpha + b\beta) = ad\alpha + bd\beta \qquad \text{(a,b being constants)} \tag{3.8a}$$

(3°) nilpotent (*) :

$$d^2 \equiv 0 \tag{3.8b}$$

(*) A p-form α such that $d\alpha = 0$ is called *closed*, if moreover $\alpha = d\beta$ then it is said to be *exact*. Obviously all exact forms are closed but the conserve is not necessarily true.

(4°) and obeys to the "anti-Leibniz rule" :

$$d(\alpha \wedge \beta) = d\alpha \wedge \beta + (-1)^p \alpha \wedge d\beta \quad , (\alpha \in \Lambda^p) \quad . \tag{3.8c}$$

Using these rules in local coordinates we obtain :

$$d\alpha = d(\frac{1}{p!} \alpha(x)_{\mu_1 \dots \mu_p} dx^{\mu_1} \wedge \dots \wedge dx^{\mu_p})$$

$$= \frac{1}{p!} d\alpha(x)_{\mu_1 \dots \mu_p} \wedge dx^{\mu_1} \wedge \dots \wedge dx^{\mu_p} + 0$$

$$= \frac{1}{p!} \frac{\partial}{\partial x^{\mu_o}} \alpha(x)_{\mu_1 \dots \mu_p} dx^{\mu_o} \wedge \dots \wedge dx^{\mu_p} \tag{3.9a}$$

i.e. $$(d\alpha)_{\mu_o \dots \mu_p}(x) = (p+1) \frac{\partial}{\partial x_{[\mu_o}} \alpha(x)_{\mu_1 \dots \mu_p]} \quad . \tag{3.9b}$$

It remains to be checked that this expression when obtained in a particular gauge satisfies the four postulates. This is left to the reader.

III.3. Pull back and push forward

Suppose we have a map ψ from M into $\widetilde{M}$. If $\widetilde{f}$ is a function defined on $\widetilde{M}$ we can build a new function f on M by the composition (fig. 1) : $f = \widetilde{f} \circ \psi$

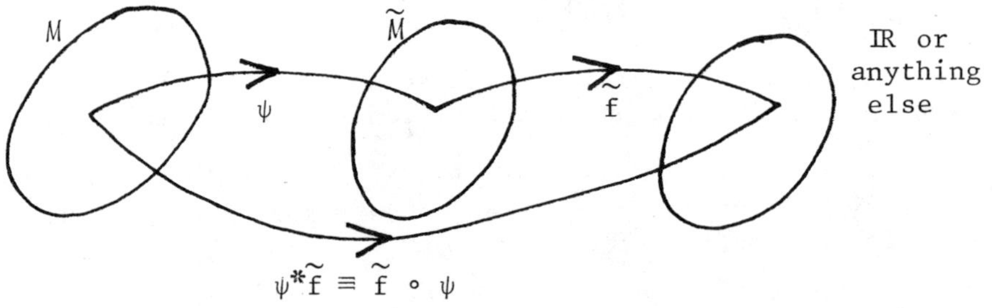

FIG. 1

This new function f defined on M is called the *pull back* of $\widetilde{f}$ and is denoted by $\psi^*\widetilde{f}$. From this definition we can construct the *push forward* of a vector from M to $\widetilde{M}$: if $\vec{v}$ is a vector belonging to the tangent space at the point P of M we obtain a new vector $\psi_*\vec{v}$ on $\widetilde{M}$ at $\psi(P)$ by defined its action on functions. If $\widetilde{f}$ is a function defined on a neighbourhood of $\psi(P)$ the pushed forward vector $\psi_*\vec{v}$ acts on it as follows :

$$\psi_*\vec{v}(\widetilde{f}) = \vec{v}(\psi^*f) \tag{3.10}$$

The geometrical significance of this definition is the following. Suppose that $\vec{v}$ is the tangent vector of a curve $C(t)$ at $t = t_o$. The map ψ transports this curve from M to $\tilde{M}$ as $\tilde{C}(t) = \psi_o C(t)$ (fig.2).

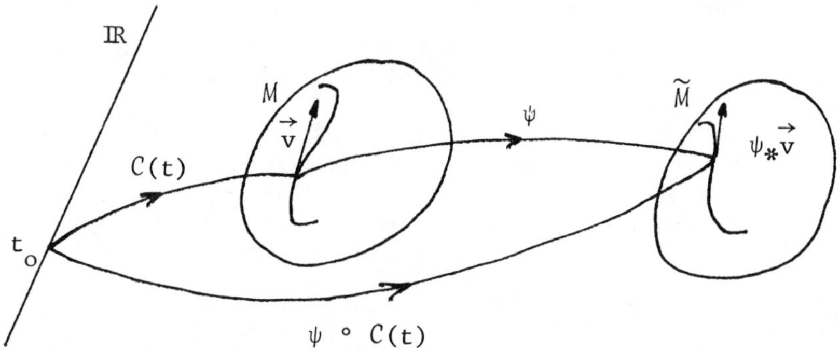

FIG. 2

The pushed forward vector $\psi_* \vec{v}$ is just the tangent vector of the curve $\tilde{C}(t)$ at $t = t_o$. Indeed we have

$$\frac{d}{dt} \tilde{f} \circ \tilde{C}(t) \Big|_{t=t_o} = \frac{d}{dt} (\tilde{f} \circ \psi) \circ C(t) \Big|_{t=t_o} = \frac{d}{dt} (\psi^* \tilde{f}) \circ C(t) \Big|_{t=t_o} .$$

More generally if $\tilde{\omega}$ is a p-covariant tensor field defined on $\tilde{M}$ its pull back $\psi^* \tilde{\omega}$ defined by specifying its action as a linear operator on vectors of M. More precisely if $\tilde{\omega}$ is defined at $\psi(P)$, $\psi^* \tilde{\omega}$ acts on vectors tangent to M at P as follows :

$$\psi^* \tilde{\omega}(\vec{v}_1, \ldots, \vec{v}_p) = \tilde{\omega}(\psi_* \vec{v}_1, \ldots, \psi_* \vec{v}_p) \quad . \tag{3.11}$$

Finally if T is a p-contravariant tensor defined at P its push forward at $\psi(P)$ is given by

$$\psi_* T(\tilde{\alpha}^1, \ldots, \tilde{\alpha}^p) = T(\psi^* \tilde{\alpha}^1, \ldots, \psi^* \tilde{\alpha}^p) \tag{3.12}$$

where $\tilde{\alpha}^1, \ldots, \tilde{\alpha}^p$ are covariant vectors on $\tilde{M}$ at $\psi(P)$.

Exercice I : Show that the pull back (push forward) of a tensorial product is equal to the tensorial product of the transported factors. Show also that these operations preserve the symmetries of the tensors.

We must stress the fact that while the pull back is well defined for covariant tensor fields, the push forward is not in general : if the points P_1 and P_2 of M are mapped on to the same point $\tilde{P} = \psi(P_1) = \psi(P_2)$ of $\tilde{M}$ there is nothing in general insuring that vectors defined at P_1 and P_2 will be pushed forward onto the same tangent vector at $\tilde{P}$. However when ψ^{-1} exists (and is also smooth i.e. ψ is a diffeomorphism) the push forward of contravariant tensor fields is well defined; moreover, by use of ψ^{-1}, one can define the push forward and pull back of any kind of tensor fields.

Exercice II : Derive the expressions for the pull back and push forward in local coordinate systems.

The exterior derivative has a remarkable behaviour with respect to the pull back : they commute. From exercice I it is obvious that this statement needs a proof only for 0-forms and exact 1-forms. If $\vec{v}$ is a vector on $\tilde{M}$ and f a 0-form on $\tilde{M}$, using sucessively the formulas (11), (5), (10) and again (11), we obtain

$$\psi^* d\tilde{f}(\vec{v}) = d\tilde{f}(\psi_* \vec{v}) = \psi_* \vec{v}(\tilde{f}) = \vec{v}(\psi^* \tilde{f}) = d\psi^* \tilde{f}(\vec{v})$$

i.e. since the vector $\vec{v}$ is arbitrary we deduce the property for 0-forms :

$$\psi^* d\tilde{f} = d\psi^* \tilde{f} \qquad . \tag{3.13}$$

On the other hand if $d\tilde{f}$ is an exact 1-form on $\tilde{M}$ we have

$$dd\tilde{f} = 0 \Rightarrow \psi^* dd\tilde{f} = 0$$

but from eq. (13) we also deduce that

$$d(\psi^* d\tilde{f}) = dd\psi^* \tilde{f} = 0 \quad \text{i.e.} \quad \psi^* dd\tilde{f} = d\psi^* d\tilde{f} \qquad . \tag{3.14}$$

In general an arbitrary p-form $\tilde{\alpha}$ is written in local coordinates $\tilde{x}$ as

$$\tilde{\alpha} = \frac{1}{p!} \alpha(x)_{\mu_1 \ldots \mu_p} dx^{\mu_1} \wedge \ldots \wedge dx^{\mu_p}$$

and from the previous results we deduce

$$d(\psi^* \tilde{\alpha}) = d(\frac{1}{p!} \psi^* \alpha(\tilde{x})_{\mu_1 \ldots \mu_p} \psi^* d\tilde{x}^{\mu_1} \wedge \ldots \wedge \psi^* d\tilde{x}^{\mu_p})$$

$$= \frac{1}{p!} (d\psi^* \alpha(x)_{\mu_1 \ldots \mu_p} \wedge \psi^* dx^{\mu_1} \wedge \ldots \wedge \psi^* dx^{\mu_p}) + 0$$

$$= \psi^* d\tilde{\alpha} \tag{3.15}$$

which completes the proof.

III.4. One parameter transformation groups

A standard problem in theoretical physics is to verify the invariance of a field under the action of a 1-parameter (continuous pseudo-) group of transformations ψ_t acting on a manifold M. These transformations are defined on a neighbourhood V of each point P of M for an interval I of values of the parameter t, depending on P and V. These transformations obey locally the composition rule $\psi_t \circ \psi_s = \psi_{t+s}$ for t and s small enough. On compact domains a non trivial interval of values of t always exists on which ψ_t is defined. When we want to define ψ_t on the whole manifold M a difficulty may arise. We must indeed restrict ourself to the common intersection of all the intervals associated to a covering of M. If the manifold is non compact the resulting interval of definition of ψ_t may collapse to the single point 0. However when this is not the case (for instance on compact manifolds where only a finite number of intersections have to be considered) ψ_t is defined on the whole real line by the relation : $\psi_{t+s} = \psi_t \circ \psi_s$.

For each point P of M the transformations ψ_t define curves $\psi_t(P)$ through P. These curves are called the *orbits of the group*. Their tangent vectors at the origin define a vector field $\vec{\xi}$ on M by

$$\frac{d}{dt}\psi_t(P)\Big|_{t=o} = \vec{\xi}(P) \quad . \qquad (3.16)(*)$$

Conversely if $\vec{\xi}$ is a C^1 vector field on M, eq. (16) defines locally the integral curves of $\vec{\xi}$ and a (pseudo-)(+) group of transformations called the *flow* of the vector field $\vec{\xi}$. The flow of $\vec{\xi}$ is locally one-to-one and as a consequence the pull back of any kind of tensor is well defined locally. This allows us to compare the values of a tensor field along the orbit of the flow. A tensor T will be said invariant with respect to the flow if it satisfies the relation

$$\psi_t^* T = T \quad . \qquad (3.17)$$

This formula may look confused at first sight. It is nevertheless unambiguous. The right hand side is just the value of the tensor field say at the point P. The left hand side represents also a tensor defined at the same point. This latter is obtained by pulling back from $\psi(P)$ to P the value of the tensor field at $\psi(P)$. In a more cumbersome notation eq. (17) can be written as :

$$\psi_t^* [\, T(\psi_t(P)\,] \equiv [\, \psi_t^* T\,](P) = T(P) \quad . \qquad (3.18)$$

(*) The meaning of this formula is given by eqs. (1) and (2).
(+) A group when M is compact or when $\vec{\xi}$ vanishes outside a compact set.

In practice, when we want to check the invariance of a tensor, it suffices to look at its behaviour under infinitesimal transformations or more precisely at its rate of transformation . This rate is just what we call the *Lie derivative* of T with respect to the vector field $\vec{\xi}$:

$$\pounds_{\vec{\xi}} T\Big|_P = \lim_{t\to o} \frac{1}{t} [\psi_t^* T - T]\Big|_P = \lim_{t\to o} \frac{1}{t}[T - \psi_{t*} T]\Big|_{\psi_t(P)} \qquad .(3.19)$$

In order to have a better understanding of this formula we shall compute the Lie derivative of some tensor fields.

If T is the function f then

$$\pounds_{\vec{\xi}} f\Big|_P = \lim_{t\to o} \frac{1}{t} [\psi_t^* f(P) - f(P)] = \lim_{t\to o} \frac{1}{t}[f \circ \psi_t(P) - f(P)] = \vec{\xi}(f)\Big|_P$$

i.e. in local coordinates

$$\pounds_{\vec{\xi}} f = \xi^\alpha \frac{\partial f(x)}{\partial x^\alpha} \qquad . \qquad (3.20)$$

If T represents the vector field $\vec{v}$ then obviously $\pounds_{\vec{\xi}} \vec{v}$ is also i.e. a differential operator acting or functions. Using the previous definitions we obtain directly

$$\begin{aligned}
\pounds_{\vec{\xi}} \vec{v}(f)\Big|_P &= \lim_{t\to o} \frac{1}{t}[[\psi_t^* \vec{v}](f) - \vec{v}(f)]\Big|_P \\
&= \lim_{t\to o} \frac{1}{t}[\vec{v}(f \circ \psi_{-t})\Big|_{\psi_t(P)} - \vec{v}(f)\Big|_P] \\
&= \lim_{t\to o} \frac{1}{t}[(\vec{v}(f\circ\psi_{-t}) - \vec{v}(f))\Big|_{\psi_t(P)} - \vec{v}(f)\Big|_P + \vec{v}(f)\Big|_{\psi_t(P)}] \\
&= -\vec{v}(\vec{\xi}(f)) + \vec{\xi}(\vec{v}(f)) \quad .
\end{aligned}$$

This last formula needs some comments. First it is meaningfull because if f is a function and $\vec{v}$ a vector field then v(f) is also a function on which $\vec{\xi}$ may act. Secondly $\vec{\xi}(\vec{v}(\ ))$ is in general not a vector, being not a first order differential operator. However, the commutator

$$[\vec{\xi}, \vec{v}](\ ) = \vec{\xi}(\vec{v}(\ )) - \vec{v}(\vec{\xi}(\ )) \qquad (3.21)$$

— which is also called the *Lie bracket* of the two vector fields $\vec{\xi}$ and $\vec{v}$ — satisfies the Liebniz rule

$$[\vec{\xi},\vec{v}](fg) = f[\vec{\xi},\vec{v}](g) + g[\vec{\xi},\vec{v}](f) \tag{3.22}$$

hence it is a vector field. Thirdly since the Lie derivative of $\vec{v}$ along the direction of $\vec{\xi}$ is expressed as a Lie bracket it is obviously skew-symmetric with respect to the exchange of the two vectors fields :

$$\pounds_{\vec{\xi}}\,\vec{v} = [\vec{\xi},\vec{v}] = -\pounds_{\vec{v}}\,\vec{\xi} \quad . \tag{3.23}$$

The vanishing of the Lie bracket of two vector fields $\vec{u}$ and $\vec{v}$ has an interesting geometrical interpretation. Suppose that U_λ and V_σ are the 1-parameter groups generated by $\vec{u}$ and $\vec{v}$. The invariance of the integral curves $V_\sigma(P)$ of $\vec{v}$ with respect to the transformations U_λ means that curvilinear parallelograms built on these flows close (fig. 3).

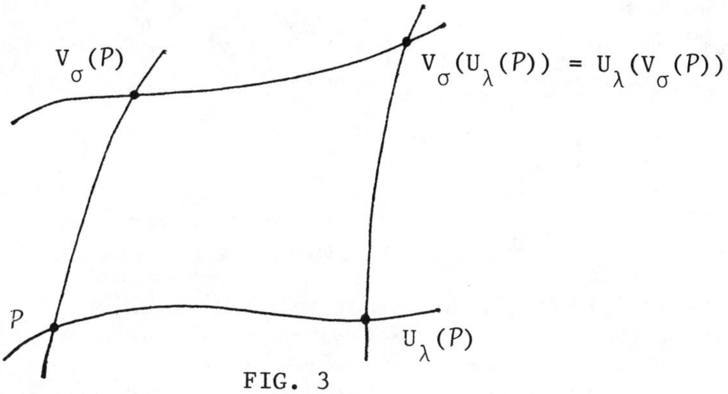

FIG. 3

In other words the transformations U_λ and V_σ commute. The parameters λ and σ can be used together as local coordinates. More generally if p vectors fields $\vec{u}_i$ commute : $[\vec{u}_i,\vec{u}_j] = 0$ for all (i,j) then local coordinate system exists such that $\vec{u}_i = \frac{\partial}{\partial x^i}$.

The expression for the Lie bracket in local coordinates can immediately be obtained from its definition :

$$[\vec{\xi},\vec{v}](f) = \xi^\alpha\,\partial_\alpha(v^\beta\,\partial_\beta f) - v^\alpha\,\partial_\alpha\,(\xi^\beta\,\partial_\beta f)$$

$$= (\xi^\alpha\,\partial_\alpha\,v^\beta - v^\alpha\,\partial_\alpha\,\xi^\beta)\,\partial_\beta f$$

i.e. $$[\vec{\xi},\vec{v}]^\beta = \xi^\alpha\,\partial_\alpha\,v^\beta - v^\alpha\,\partial_\alpha\,\xi^\beta \quad . \tag{3.24}$$

This result can be deduced from direct computation of the Lie derivative in a coordinate system. Suppose that the 1-parameter group generated by $\vec{\xi}$ is represented by the set of equations :

$$x^\alpha_t = X^\alpha(x^\beta, t)$$

i.e. $\frac{d}{dt} X^\alpha(x^\beta, t) = \xi^\alpha(x^\beta_t)$.

Then, for small values of t we have

$$x^\alpha_t = x^\alpha + t\, \xi^\alpha(x^\beta) + 0(t^2)$$

and the components of $\psi^*_t \vec{v}$ are given in the same approximation by

$$(\psi_t \vec{v})^\alpha = \frac{\partial x^\alpha}{\partial x^\beta_t} v^\beta(x^\gamma_t)$$

$$= v^\alpha(x^\beta) + t\xi^\beta \frac{\partial v^\alpha}{\partial x^\beta} - t\frac{\partial \xi^\alpha}{\partial x^\beta} v^\beta + 0(t^2) \quad .$$

This leads directly to the desired result :

$$(\pounds_{\vec{\xi}}\, \vec{v})^\alpha = \lim_{t\to o} \frac{1}{t}\left[v^\alpha(x^\beta) + t(\xi^\beta \frac{\partial v^\alpha}{\partial x^\beta} - v^\beta \frac{\partial \xi^\alpha}{\partial x^\beta}) - v^\alpha(x^\beta) + 0(t^2) \right]$$

$$= [\, \vec{\xi}\, , \vec{v}\,]^\alpha \quad .$$

Exercice III : Show that the Lie bracket satisfies the Jacobi identity : $[\vec{u}, [\vec{v},\vec{\omega}]] + [\vec{v}, [\vec{\omega},\vec{u}]] + [\vec{\omega}, [\vec{u},\vec{v}]] = 0$.

Finally we shall compute the Lie derivative of a 1-form ω. The answer will be a 1-form i.e. a linear operator acting on vectors. Suppose that $\vec{v}_o$ is a tangent vector at P. Define on a neighbourhood of P a vector field $\vec{v}$ such that $\vec{v}(P) = \vec{v}_o$. Now we can compute the Lie derivative of ω just as the Lie derivative of vector fields :

$$\pounds_{\vec{\xi}}\, \omega(\vec{v})\Big|_P = \lim_{t\to o} \frac{1}{t}\left[[\psi^*_t \omega](\vec{v}) - \omega(\vec{v}) \right]\Big|_P$$

$$= \lim_{t\to o} \frac{1}{t}\left[(\omega(\psi_{t*}\vec{v}) - \omega(\vec{v}))\Big|_{\psi_t(P)} + \omega(\vec{v})\Big|_{\psi_t(P)} - \omega(\vec{v})\Big|_P \right]$$

$$= - \omega\, (\pounds_{\vec{\xi}}\, v)\Big|_P + \vec{\xi}\, (\omega(\vec{v}))\Big|_P \qquad (3.25)$$

This formula is amazing. Its left hand side depends only on the value of $\vec{v}_o$, while the right hand side requires a continuation of $\vec{v}_o$ around P! However consistency is assured because if the latter is expressed in a coordinate system :

$$- \omega(\pounds_{\vec{\xi}}\, \vec{v}) + \vec{\xi}(\omega(\vec{v})) = (\xi^\beta \frac{\partial \omega_\alpha}{\partial x^\beta} + \omega_\beta \frac{\partial \xi^\beta}{\partial x^\alpha})\, v^\alpha_o \qquad (3.26)$$

we see that its value is in fact independent of way in which $\vec{v}_o$ is continued.

The knowledge of the Lie derivative of functions, vector and covector fields allows one to compute the Lie derivative of any tensor. The Lie derivative indeed obeys the Leibniz rule for tensorial products. For instance if $T = T^{\alpha}{}_{\beta}{}^{\gamma}\, \partial_\alpha \otimes dx^\beta \otimes \partial_\gamma$ then :

$$(\pounds_{\vec{\xi}} T)^{\alpha}{}_{\beta}{}^{\gamma} = \xi^\mu \frac{\partial T^{\alpha}{}_{\beta}{}^{\gamma}}{\partial x^\mu} - \frac{\partial \xi^\alpha}{\partial x^\mu} T^{\mu}{}_{\beta}{}^{\gamma} + \frac{\partial \xi^\mu}{\partial x^\beta} T^{\alpha}{}_{\beta}{}^{\gamma} - \frac{\partial \xi^\gamma}{\partial x^\mu} T^{\alpha}{}_{\beta}{}^{\gamma} \; . \tag{3.27}$$

<u>Exercice IV</u> : If $\vec{\xi}$ and $\vec{u}_i$ are vectors fields and ω a (p+1)-form the contraction of ω with ξ, denoted by $i(\vec{\xi})\omega$, is the p-form such that

$$i(\vec{\xi})\mu\ (\vec{u}_1, \ldots, \vec{u}_p) = \omega(\vec{\xi}, \vec{u}_1, \ldots, \vec{u}_p)$$

i.e. in terms of components

$$i(\vec{\xi})\omega_{\mu_1 \ldots \mu_p} = \omega_{\mu_o \mu_1 \ldots \mu_p} \xi^{\mu_o} \; .$$

Moreover if ω is a 0-form we assume that $i(\vec{\xi})\omega = 0$.

1°) Using eqs (3) show that the contraction of $\vec{\xi}$ with an exterior product can be expressed as :

$$i(\vec{\xi})(\alpha \wedge \beta) = (i(\vec{\xi})\alpha) \wedge \beta + (-1)^{\deg \alpha}\, \alpha \wedge (i(\vec{\xi})\beta) \; .$$

2°) Show that the Lie derivative of a differential form ω can be written as

$$\pounds_{\vec{\xi}}\, \omega = i(\vec{\xi})d\omega + d(i(\xi)\omega) \; .$$

Hint : For 0-form it is obvious. Assume that the formula is for p-forms. By use of 1° and the properties of the Lie derivative of tensorial products verify that the result remains true for (p+1)-forms.

3°) As the exterior derivative commutes with the pull back of differential forms it must also commute with the Lie derivative. Check this statement directly from the previous result.

4°) Show that if α is a 1-form then

$$d\alpha(\vec{u}_1, \vec{u}_2) = \pounds_{\vec{u}_1} [\alpha(\vec{u}_2)] - \pounds_{\vec{u}_2} [\alpha(\vec{u}_1)] - \alpha([\vec{u}_1, \vec{u}_2])$$

and more generally for a p-form ω :

$$d\omega(\vec{u}_1, \ldots, \vec{u}_{p+1}) = \sum_{1 \leq i \leq p+1} (-1)^{i-1}\, \vec{u}_i [\omega(\vec{u}_1, \ldots, \hat{\vec{u}}_i, \ldots, \vec{u}_{p+1})] + \sum_{1 \leq i < j \leq p+1} (-1)^{i+j}\, \omega([\vec{u}_i, \vec{u}_j], \vec{u}_1, \ldots, \hat{\vec{u}}_i, \ldots, \hat{\vec{u}}_j, \ldots, \vec{u}_{p+1})$$

where a hat over a vector indicates that it is omitted.

III.5. Invariant forms on Lie groups

In this section we shall apply all the previous machinery in the framework of Lie groups. This study is far from being purely academic. The geometry of Lie groups plays a key-rôle is cosmology. On the other hand their study can serve as an introduction to the theory of coset spaces the importance of which in modern Kaluza-Klein theory has still been emphasized by numerous works[7,8].

Let G be an n-dimensional Lie group and $H \subset G$ a one parameter subgroup. This subgroup defines on G two groups of transformations :

$$\textit{left translations} \quad L_{h(t)} : G \to G \text{ by } g \mapsto h(t)g$$

$$\textit{right translations} \quad R_{h(t)} : G \to G \text{ by } g \mapsto gh(t) \quad .$$

Obviously, since the group multiplication is associative, left and right translations commute :

$$L_h R_{h'} g = (h'g)h = h'(gh) = R_{h'} L_h g \quad . \tag{3.28}$$

This trivial remark leads to important consequences. Consider the vectors fields generated by a 1-parameter subgroup H :

$$\vec{\ell}(g) = \frac{d}{dt} gh(t)\Big|_{t=o} \tag{3.29a}$$

$$\vec{r}(g) = \frac{d}{dt} h(t)g\Big|_{t=o} \quad . \tag{3.29b}$$

The vector field generated by a one parameter subgroup of right (resp. left) translations is *left* (resp. *right*) *invariant*. The proof of this property just amounts to show that the action of $\vec{\ell}$(kg) on a function f at the point kg coincides with the action of the push forward of $\vec{\ell}(g)$, from g to kg, under the left translation defined by left multiplication by k :

$$L_{k^*}[\vec{\ell}\,](f)\Big|_{kg} = \vec{\ell}(L_k^* f)\Big|_g = \vec{\ell}(g)(f(kg))$$

$$= \frac{d}{dt} f(kg\, h(t))\Big|_{t=o} = \vec{\ell}(f)\Big|_{kg}$$

$$\text{i.e. } L_{k^*}\vec{\ell} = \vec{\ell} \quad . \tag{3.30}$$

In particular, if Id denotes the identity of G, we have

$$L_{g^*}[\vec{\ell}(Id)] = \vec{\ell}(g) \tag{3.31}$$

which shows that a left invariant vector field is completely determined by its value at the identity. This last property shows also that Lie group manifolds are parallelizable. In fact, if $\{\vec{e}_a\}$ is a basis of the tangent space at the identity, the n vectors fields :

$$\vec{\ell}_a(g) = L_{g*}[\vec{e}_a] \qquad (a = 1, \ldots, n) \tag{3.32}$$

define a global field of frames called a *left invariant (vector) basis*. In the same way, we may build a right invariant basis as

$$\vec{r}_a(g) = R_{g*}[\vec{e}_a] \qquad (a = 1, \ldots, n) \quad . \tag{3.33}$$

Since right and left translations commute, the Lie bracket of any two left and right invariant vector fields vanishes i.e.

$$[\vec{\ell}_a, \vec{r}_{a'}] = 0 \qquad \forall a,a' \quad . \tag{3.34}$$

On the other hand, it results immediately from the definition (21) that the Lie bracket of left (right) invariant vector fields is also left (right) invariant :

$$[L_{k*}[\vec{\ell}_a], L_{k*}[\vec{\ell}_{a'}]](f)\Big|_{kg} = \vec{\ell}_a(\vec{\ell}_{a'}(f)) - \vec{\ell}_{a'}(\vec{\ell}_a(f))\Big|_{kg}$$

$$= L_{k*}([\vec{\ell}_a, \vec{\ell}_{a'}])(f)\Big|_{kg} \quad . \tag{3.35}$$

As a consequence such a bracket can be written :

$$[\vec{\ell}_a, \vec{\ell}_b] = C^c_{ab}\, \vec{\ell}_c \tag{3.36}$$

where the coefficients C^c_{ab} are constant. These are the *structure constants* of the Lie group with respect to the basis $\vec{e}_a$ of the tangent space at the identity of the group.

Equations (31) and (33) show how to obtain left and right invariant basis. In the following we shall suppose that these invariant basises are built from a given common basis of the tangent space at the identity. At each point of the manifold we can express the vector of the right invariant basis in terms of the left invariant basis as :

$$\vec{r}_a(g) = \Lambda_a{}^b(g)\, \vec{\ell}_b(g) \quad \text{with} \quad \Lambda_a{}^b(Id) = \delta_a{}^b \quad . \tag{3.37}$$

The elements of the transformation matrices $\Lambda_a{}^b(g)$ obey interesting relations. From the commutation relation (34) we obtain

$$0 = [\vec{\ell}_a, \Lambda_a{}^c \vec{\ell}_c] = (\vec{\ell}_a(\Lambda_b{}^c) + \Lambda_b{}^d C^c_{ad})\vec{\ell}_c$$

$$\text{i.e. } \vec{\ell}_a(\Lambda_b{}^c) = \Lambda_b{}^d C^c_{da} \quad . \tag{3.38}$$

By introducing, this result in the computation of the Lie bracket of two vector fields belonging to the right invariant frame, we obtain

$$[\vec{r}_a,\vec{r}_b] = (\Lambda_a{}^c\Lambda_b{}^d C^f_{cd} + \Lambda_a{}^c \vec{\ell}_c(\Lambda_b{}^f) - \Lambda_b{}^c \vec{\ell}_c(\Lambda_a{}^f))\vec{\ell}_f$$

$$= -(\Lambda_b{}^d \Lambda_a{}^c C^f_{cd} \Lambda^{-1}{}_f{}^k)\vec{r}_k \quad .$$

But the Lie bracket of right invariant fields is right invariant; so, its components, with respect to a right invariant basis, are constants. Now, evaluating them at the identity we obtain the relation :

$$[\vec{r}_a,\vec{r}_b] = - C^c_{ab}\,\vec{r}_c \quad , \qquad (3.39)$$

which is actually valid everywhere on the manifold.

When we study the geometry of Lie groups, *invariant forms* play a more important rôle than vectors. Invariant (1-form) basis are obtained by considering the dual basis of the invariant (vector) basis. Their invariance is almost obvious. We obtain from the definition of the dual basis and by use of the relation (30) :

$$\theta^a[\vec{\ell}_b]\Big|_g = \delta^a_b = \theta^a[\vec{\ell}_b]\Big|_{kg} = \theta^a(kg)[L_{k*}(\ell_b(g))]$$

$$= L_k{}^*(\theta^a(kg))[\vec{\ell}_b(g)] = L_k{}^*(\theta^a)[\vec{\ell}_b]\Big|_{kg}$$

$$\text{i.e. } \theta^a = L_k{}^*\,\theta^a \quad . \qquad (3.40)$$

As the exterior derivative commutes with the pull back, the exterior derivative of an invariant 1-form is an invariant 2-form. Let us compute the components of the exterior derivative of the elements of a left invariant 1-form basis (θ^a). These components, in a left invariant 2-form basis, are easily obtained by use of the result of exercice (IV.4°) :

$$d\,\theta^a(\vec{\ell}_b,\vec{\ell}_c) = \vec{\ell}_b(\theta^a(\vec{\ell}_c)) - \vec{\ell}_c(\theta^a(\vec{\ell}_b)) - \theta^a([\vec{\ell}_b,\vec{\ell}_c]) \quad .$$

As $\theta^a(\ell_b) = \delta^a_b$ the first two terms of the right hand side of this equation vanish while the last one is given by eq. (36) :

$$d\,\theta^a(\vec{\ell}_b,\vec{\ell}_c) = - C^a_{bc} \quad .$$

We may reexpress this result as

$$d\,\theta^a + \frac{1}{2} C^a_{bc}\,\theta^b \wedge \theta^c = 0 \quad , \qquad (3.41)$$

which are called the *Maurer-Cartan structure equations*. If, instead of left invariant forms, we consider right invariant ones, the Maurer-Cartan equations become :

$$d\,\sigma^a - \frac{1}{2} C^a_{bc}\, \sigma^b \wedge \sigma^c = 0 \qquad (3.42)$$

with σ^a defined by $\sigma^a(r_b) = \delta^a_b$.

The Lie groups encountered in physics are often represented by a set of matrices A depending on n parameters x^a. These parameters define local coordinates on a neighbourhood of the identity which corresponds (by assumption) to $x^a = 0$. The Lie algebra of the group is then represented by the matrices

$$X_{(c)} = \left.\frac{\partial A}{\partial x^c}\right|_{x^a=0} \qquad (3.43)$$

corresponding to infinitesimal transformations. On the local chart x^a the invariant vector fields are obtained by solving the equations

$$\ell^c_{(a)}(x)\ \frac{\partial A}{\partial x^c} = A X_{(a)} \quad , \qquad (3.44a)$$

$$r^c_{(a)}(x)\ \frac{\partial A}{\partial x^c} = X_{(a)} A \quad . \qquad (3.44b)$$

The invariant 1-form basis are given by identifying the terms proportional to the infinitesimal generators $X_{(a)}$ on both sides of the equations :

$$X_{(a)}\ \theta^a = A^{-1}\ \frac{\partial A}{\partial x^b}\ dx^b \quad , \qquad (3.45a)$$

$$X_{(a)}\ \theta^a = -A\ \frac{\partial A^{-1}}{\partial x^b}\ dx^b \quad . \qquad (3.45b)$$

Exercice V : 1°) Prove eqs. (35) and (40) by computing Lie derivatives with respect to vector fields generating left translations. (Hint : Use the Jacobi identity and the invariance of δ^a_b).
2°) Interpret the integrability condition $d^2\,\theta^a \equiv 0$.
3°) Suppose that a Lie group is represented by matrices and that its Lie algebra structure constants are defined by $[X_{(a)}, X_{(b)}] = C^c_{ab}\, X_{(c)}$. Show that the generators (29) are given by eqs. (44). Check directly their invariance properties (35), (36) and (39). Prove then eqs. (45) and deduce the Maurer-Cartan structure equations.

III.6. Integration

Another property enjoyed by differential forms is that they can be integrated. Suppose that I^p is the unit cube in $\mathbb{R}^p$ and ψ a mapping from I^p into M. If ω is a p-form on I^p it can be written in terms of coordinates as

$$\omega = \omega(x)dx^1 \wedge \ldots \wedge dx^p \quad . \qquad (3.46)$$

The integral of ω on I^p is defined as a usual Riemann integral :

$$\int_{I^p} \omega = \int_{I^p} \omega(x)dx^1 \ldots dx^p \quad . \qquad (3.47)$$

The gauge invariance of this definition trivially results from the rule of interchanging integration variables. More generally if ω is a p-form on M defined on the image $\mathcal{D}$ of I^p by ψ, its integral on $\mathcal{D}$ is obtained via a pull back towards I^p :

$$\int_{\mathcal{D}} \omega = \int_{I^p} \psi^* \omega \quad . \qquad (3.48)$$

This defines the integral of forms on domains $\mathcal{D}$ which can be represented as the image of a unit cube under a single map. For a more general domain we need several maps in order to cover it. Then some care is needed to avoid multiple counting on the overlaps of different images. However in practice this is often obvious.

An n-dimensional manifold M is said to be orientable if we can build on it a nowhere vanishing n-form η :

$$\eta = \frac{1}{n!} \eta_{a_1 \ldots a_n} \underline{\theta}^{a_1} \wedge \ldots \wedge \underline{\theta}^{a_n} \quad . \qquad (3.49)$$

This n-form may be interpreted as an oriented volume element. It defines an orientation on M. Obviously on an orientable manifold there exists 2^κ distinct orientations where κ is the number of connected components of M. The existence of a volume element allows one to integrate a function f as the integral of the n-form $f\eta$ on M. Moreover such an n-form also defines an isomorphism between the space Λ^p of p-forms and the space $A^{(n-p)}$ of (n-p) completely antisymmetric contravariant tensor fields :

$$F = F^{[a_{p+1} \ldots a_n]} \vec{e}_{a_{p+1}} \otimes \ldots \otimes \vec{e}_{a_n} \rightarrow \langle F.\eta \rangle$$

$$= \frac{1}{p!} F^{a_{p+1} \ldots a_n} \eta_{a_{p+1} \ldots a_n a_1 \ldots a_p} \underline{\theta}^{a_1} \wedge \ldots \wedge \underline{\theta}^{a_p} \qquad (3.50)$$

This isomorphism is not a canonical one but it can be used to define the integral of (n-p) antisymmetric contravariant tensor fields on p dimensional submanifolds.
The integration of differential forms enjoys properties closely related to those of the exterior derivative.

1°) If ψ is an orientation preserving diffeomorphism from (M,η) to $(\widetilde{M},\widetilde{\eta})$ then

$$\int_{\mathcal{D}\subset M} \psi^* \widetilde{\eta} = \int_{\psi(\mathcal{D})} \widetilde{\eta} \qquad (3.51)$$

and more generally, if ψ is such that the pull back of any compact of $\widetilde{M}$ is a compact set of M, then there exists an *integer* deg(ψ) such that for all n-forms $\widetilde{\omega}$ on $\widetilde{M}$ with compact support we have :

$$\int_M \psi^*\widetilde{\omega} = \deg(f) \int_{\widetilde{M}} \widetilde{\omega} \quad . \tag{3.52}$$

2°) If ω is a p-form and $\mathcal{D}$ an orientable (p+1) dimensional submanifold then

$$\int_{\partial\mathcal{D}} \omega = \int_{\mathcal{D}} d\omega \qquad [\textit{Stokes theorem}] \tag{3.53}$$

if the orientation of the boundary $\partial\mathcal{D}$ of the domain $\mathcal{D}$ is the one which is canonically induced by the orientation of $\mathcal{D}$.

IV. CONNECTIONS, CURVATURE AND ALL THAT ...

IV.1. Parallel transport

At the beginning of the previous paragraph we noticed that in general there is no canonical way to compare tensors at different points on a manifold. We must specify from the outset how a tensor behaves when it undergoes a (parallel) transport along a given path.

The "modern" (1950) treatment of the problem rests on the theory of connections on principal fibre bundles and on their associated bundles[9]. This approach represents a great improvement compared to the classical one but it needs a too long digression through differential geometry in order to be discussed here. We shall instead restrict ourself to the method of the moving frame elaborated by Elie Cartan[10]. The main advantage of this method lies in his first prescription : choose in the most suitable way, according to the problem studied, a differentiable field of frames $\{\vec{e}_a\}$. This defines the *moving frame*.

The parallel transport rules stipulate how a frame is deformed with respect to the moving frame when it is transported along a given path. We emphasize that the deformation undergone by a frame depends on the path followed and not only on their end points.

Once the behaviour of a frame $\{\vec{e}_a\}$ under parallel transport is given, the most natural prescription for its dual frame $\{\theta^a\}$ is to assume that the coframe deforms into the dual of the transported vector frame. As any tensorial basis can be expressed as a product of elements of a frame and a coframe we define their parallel transport just as the product of their transported factors. Now we are in position to define the parallel transport of tensors along a curve by requiring that their components with respect to a transported basis remains constant. For scalars we just impose that their value doesn't change. We impose moreover to rules of parallel transport to satisfy natural consistency conditions. The deformation of the transported frame is smooth and depends only on the shape of the curve but not on its parametrization. If $T_{P_o \to P_1}[T(P_o)]$ is the tensor of P_1 obtained by transporting the tensor $T(P_o)$ from P_o to P_1 along a given path then :

$$T_{P_1 \to P_2}[T_{P_o \to P_1}[T(P_o)]] = T_{P_o \to P_2}[T(P_o)] \quad . \tag{4.1}$$

IV.2. Covariant derivative and connection

The rules of parallel transport allow us to generalize the concept of directionnal derivative used on Euclidean spaces. Indeed we have now prescriptions for comparing tensors defined at different points. The *covariant derivative* of a tensor field T in the direction of the vector $\vec{\xi} = \left.\frac{dC}{dt}\right|_{t=t_o}$ tangent to the curve C at $t=t_o$ is given by :

$$\nabla_{\vec{\xi}} T \equiv \frac{D}{\partial t} T = \lim_{t \to t_o} \frac{1}{t-t_o} \{T_{C(t)\to C(t_o)}[T(C(t))] - T(C(t_o))\} \ . \tag{4.2}$$

The different steps defining the parallel transport express themselves in the following properties of the covariant derivative :

1°) The covariant derivative is linear and preserves the tensorial nature of the object on which it acts :

$$\nabla_{\vec{\xi}}(aT + bU) = a\, \nabla_{\vec{\xi}} T + b\, \nabla_{\vec{\xi}} U \quad \text{(a,b being constants).} \tag{4.3a}$$

2°) It commutes with contractions.

3°) It obeys the Leibniz rule for tensor products

$$\nabla_{\vec{\xi}}(T \otimes U) = (\nabla_{\vec{\xi}} T) \otimes U + T \otimes (\nabla_{\vec{\xi}} U) \quad . \tag{4.3b}$$

4°) If f is a scalar field

$$\nabla_{\vec{\xi}} f = \vec{\xi}(f) \quad . \tag{4.3c}$$

5°) It is tensorial with respect to the direction of derivation :

$$\nabla_{(f\vec{\xi}_1 + g\vec{\xi}_2)} = f\, \nabla_{\vec{\xi}_1} + g\, \nabla_{\vec{\xi}_2} \quad \text{(f,g being functions).} \tag{4.3d}$$

These properties may be considered as defining the covariant derivative i.e. infinitesimal parallel transports.

A tensor field T is constant along a curve C if it satisfies the equation

$$\nabla_{\vec{\xi}} T = 0 \quad . \tag{4.4}$$

Conversely the parallel transport of a tensor along a curve is obtained by solving the first order equation :

$$\frac{D}{\partial t} T = 0 \tag{4.5}$$

with initial conditions given by the value of the tensor at the starting point. In particular if a curve C(t) is such that its tangent vector $\vec{\xi}$ is constant :

$$\nabla_{\vec{\xi}} \vec{\xi} = 0 \tag{4.6a}$$

the curve is called a geodesic and the parameter t which defines the tangent vector is called an affine parameter. More generally a curve is a geodesic if its tangent vector obeys the equation

$$\nabla_{\vec{\xi}} \vec{\xi} \div \vec{\xi} \quad . \tag{4.6b}$$

In such a case it is always possible to redefine the parameter describing the curve in such a way that the new tangent vector field satisfies eq. (6a).

Suppose we choose a moving grame $\{\vec{e}_a\}$. The covariant derivative will then be completely defined once we know the covariant derivative of the n vector fields $\vec{e}_a$. These objects are vector fields whose components depend linearly on the components of the vector $\vec{\xi}$ that specify the direction with respect to which the derivative is computed. In general we have :

$$\nabla_{\vec{\xi}} \vec{e}_a = \vec{e}_b \, \omega^b{}_{a\mu} \, \xi^\mu = \vec{e}_b \, \gamma^b{}_{ac} \, \xi^c \quad . \tag{4.7}$$

The n^2 numbers $\omega^b{}_{a\mu} \xi^\mu = \gamma^b{}_{ac} \xi^c$ give the rate at which the vectors $\vec{e}_a$ are modified when they are parallel transported in the direction $\vec{\xi}$. Note that in the first expression we refer to natural coordinates for the components of $\vec{\xi}$ whereas in the second one we use their components with respect to the moving frame. So if the moving frame is expressed as $\vec{e}_a = e_a{}^\mu \partial_\mu$ we obtain

$$\omega^b{}_{a\mu} \, e_c{}^\mu = \gamma^b{}_{ac} \quad . \tag{4.8}$$

These n^3 numbers : $\omega^b{}_{a\mu}$ (or $\gamma^b{}_{ac}$) define a *connection* on the manifold assigned to a given moving frame. Their transformations laws with respect to a change of frame will be established later.

As we just mentioned the connection coefficients measure deformation rate of a frame transported along a curve i.e. if

$$T_{C(t)\to C(t_o)} [\vec{e}_a(C(t))] = \vec{e}_a^{\,/\!/}(C(t_o)) = \vec{e}_b(C(t_o)) A^b{}_a(t-t_o)$$

then

$$\omega^{b}{}_{a\mu}\,\xi^{\mu} = \frac{d}{dt} A^{b}{}_{a}(t-t_o)\Big|_{t=t_o} \quad . \tag{4.9}$$

As by definition of the parallel transport of the dual frame

$$T_{C(t)\to C(t_o)}\,[\theta^{a}(C(t))] = \theta^{a\,/\!/}(C(t_o)) = A^{-1\,a}{}_{b}(t-t_o)\,\theta^{b}(C(t_o))$$

we get :

$$\nabla_{\vec{\xi}}\,\theta^{a} = -\,\omega^{a}{}_{b\mu}\,\theta^{b}\,\xi^{\mu} \quad . \tag{4.10}$$

Since the covariant derivative is tensorial with respect to the direction of derivation we may introduce a new operator $\underline{\nabla}$, called the *covariant differential*, which maps (p,q) tensor fields on (p,q) tensor valued one-form in such a way that

$$i(\vec{\xi})\underline{\nabla}\,T \equiv \nabla_{\vec{\xi}}\,T \quad . \tag{4.11}$$

When $\underline{\nabla}$ acts on functions it coincides with the exterior derivative

$$\underline{\nabla}\,f = df \tag{4.12}$$

and when acting on the basis vector fields of the moving frame it defines the *connection 1-forms* $\underline{\omega}^{a}{}_{b}$:

$$\underline{\nabla}\,\vec{e}_a = \vec{e}_b \otimes \underline{\omega}^{b}{}_{a} \tag{4.13}$$

where components in a naturel basis are

$$\underline{\omega}^{b}{}_{a} = \omega^{b}{}_{a\mu}\,d\,x^{\mu} \quad . \tag{4.14}$$

At this stage it is usefull to introduce a matrix notation. The n vectors defining a frame will be considered as given by a vector valued row vector

$$\vec{e} = (\vec{e}_1, \ldots, \vec{e}_n) \tag{4.15}$$

and the connection 1-forms as the components of a square 1-form valued matrix. With these conventions eq. (13) becomes

$$\underline{\nabla}\,\vec{e} = \vec{e} \otimes \underline{\omega} \quad . \tag{4.16}$$

The connection 1-forms $\omega^{a}{}_{b}$ refer to a moving frame $\{\vec{e}_a\}$. If we use another moving frame $\vec{e}\,'$ related to the previous one by :

$$\vec{e}\,' = \vec{e}\,\Lambda \qquad \text{i.e.} \quad \vec{e}\,'_a = \vec{e}_b\,\Lambda^{b}{}_{a} \tag{4.17}$$

than the corresponding connection 1-forms are given by

$$\underline{\nabla}\,\vec{e}\,' = \vec{e}\,' \otimes \underline{\omega}' = \vec{e} \otimes d\Lambda + \vec{e} \otimes \underline{\omega}\,\Lambda$$

i.e. $\underline{\omega}' = \Lambda^{-1}\,\underline{\omega}\,\Lambda + \Lambda^{-1}\,d\Lambda$

or $$\underline{\omega}'^{a}{}_{b} = (\Lambda^{-1})^{a}{}_{c}\,\underline{\omega}^{c}{}_{d}\,\Lambda^{d}{}_{b} + (\Lambda^{-1})^{a}{}_{c}\,d\Lambda^{c}{}_{b} \qquad . \qquad (4.18)$$

Clearly the transformation laws for the connection are not tensorial. However, the difference of two connections defines a tensor and, conversely, any connection can be expressed as a given connection plus a (1,1) tensor valued 1-form. This trick is often used in supergravity.

Exercice VI : 1°) The identity tensor is the (1-1) tensor $\vec{e}_a \otimes \theta^a$. His components with respect to $\{\vec{e}_a\}$ and $\{\theta^b\}$ are obviously δ^a_b. Show that the covariant differential of this tensor is zero. Translate this result in term of the components of the identity tensor expressed with respect to $\{\vec{e}_a\}$ and $\{dx^\mu\}$.
2°) Check the consistency of the transformation laws for $\underline{\omega}^a{}_b$. Determine the transformation rules for the components $\omega^a{}_{b\mu}$ and $\gamma^a{}_{bc}$ under a change of the moving frame and coordinates. When the moving frame is built from a natural basis the components of the connection 1-form are usually written as $\Gamma^\alpha_{\beta\gamma}$ i.e. $\underline{\nabla}\partial_\beta = \partial_\alpha \otimes \Gamma^\alpha_{\beta\gamma}\, dx^\gamma$. Compute the transformation rules for $\Gamma^\alpha_{\beta\gamma}$ under change of coordinates. Deduce the following expression for its Lie derivative :

$$L_{\vec{\xi}}\,\Gamma^\alpha_{\beta\gamma} = \partial_\beta\partial_\gamma\xi^\alpha + \Gamma^\alpha_{\beta\rho}\,\partial_\gamma\xi^\rho + \Gamma^\alpha_{\rho\gamma}\,\partial_\beta\xi^\rho - \Gamma^\rho_{\beta\gamma}\,\partial_\rho\xi^\alpha + \xi^\rho\partial_\rho\Gamma^\alpha_{\beta\gamma} \qquad .$$

IV.3. Torsion and curvature forms

Each moving frame $\{\vec{e}_a\}$ defines a moving dual frame $\{\theta^a\}$ which can be considered as a column 1-form valued matrix : $\underline{\theta}$. With respect to the transformation (17) of the moving frame it transforms as

$$\underline{\theta}' = \Lambda^{-1}\,\underline{\theta} \qquad . \qquad (4.19)$$

The exterior derivative of this relation leads to

$$d\underline{\theta}' = d\Lambda^{-1} \wedge \underline{\theta} + \Lambda^{-1} d\underline{\theta}$$

which, combined with eq. (18), gives us :

$$d\underline{\theta}' + \underline{\omega}' \wedge \underline{\theta}' = d\Lambda^{-1} \wedge \underline{\theta} + \Lambda^{-1} d\underline{\theta} + \Lambda^{-1}\underline{\omega} \wedge \underline{\theta} + \Lambda^{-1} d\Lambda \wedge \Lambda^{-1}\underline{\theta}$$
$$= \Lambda^{-1}(d\underline{\theta} + \underline{\omega} \wedge \underline{\theta}) \qquad . \qquad (4.20)$$

A similar computation also shows us that :

$$d\underline{\omega}' + \underline{\omega}' \wedge \underline{\omega}' = \Lambda^{-1}(d\underline{\omega} + \underline{\omega} \wedge \underline{\omega})\,\Lambda \qquad . \qquad (4.21)$$

As a consequence these expressions transforms as the components of tensors. They are called respectively

- the *torsion 2-form* :

$$\Theta^a \equiv \frac{1}{2} T^a{}_{bc}\, \underline{\theta}^b \wedge \underline{\theta}^c = d\underline{\theta}^a + \underline{\omega}^a{}_b \wedge \underline{\theta}^b \tag{4.22}$$

- the *curvature 2-form :*

$$\Omega^a{}_b \equiv \frac{1}{2} R^a{}_{bcd}\, \underline{\theta}^c \wedge \underline{\theta}^d = d\underline{\omega}^a{}_b + \underline{\omega}^a{}_c \wedge \underline{\omega}^c{}_b \tag{4.23}$$

and they define

the *torsion tensor* :

$$\mathbb{T} = \vec{e}_a \otimes \Theta^a \tag{4.24}$$

and the *curvature tensor :*

$$\mathbb{R} = \vec{e}_a \otimes \theta^b \otimes \Omega^a{}_b \tag{4.25}$$

The curvature tensor is related to the commutators of covariant derivatives. Let X, Y and Z be three vector fields. We obtain by use of exercice (IV.4°)

$$\begin{aligned}
\mathbb{R}(\vec{X},\vec{Y})\vec{Z} &= \vec{e}_a\ \theta^b(\vec{Z})\ \Omega^a{}_b(\vec{X},\vec{Y}) \\
&= \vec{e}_a\ R^a{}_{bcd}\ Z^b X^c Y^d \\
&= \vec{e}_a\ Z^b\ [\, d\underline{\omega}^a{}_b + \underline{\omega}^a{}_c \wedge \underline{\omega}^c{}_b\,]\,(\vec{X},\vec{Y}) \\
&= \vec{e}_a\ Z^b\ [\, \vec{X}(\underline{\omega}^a{}_b(\vec{Y})) - \vec{Y}(\underline{\omega}^a{}_b(\vec{X})) - \underline{\omega}^a{}_b([\vec{X},\vec{Y}]) \\
&\quad + \underline{\omega}^a{}_c(\vec{X})\underline{\omega}^c{}_b(\vec{Y}) - \underline{\omega}^a{}_c(\vec{Y})\underline{\omega}^c{}_b(\vec{X})\,]
\end{aligned}$$

which is equal to :

$$= \nabla_{\vec{X}} \nabla_{\vec{Y}} \vec{Z} - \nabla_{\vec{Y}} \nabla_{\vec{X}} \vec{Z} - \nabla_{[\vec{X},\vec{Y}]} \vec{Z} \tag{4.26}$$

as the reader may convince himself.

In particular if $\vec{X} = \partial_\mu$ and $\vec{Y} = \partial_\nu$ are vector fields generating coordinate lines but $\vec{Z} = \vec{e}_b$ is a vector field of the moving frame we obtain for the components of the curvature tensor :

$$R^a{}_{b\mu\nu} = \partial_\mu \omega^a{}_{b\nu} - \partial_\nu \omega^a{}_{b\mu} + \omega^a{}_{c\mu}\omega^c{}_{b\nu} - \omega^a{}_{c\nu}\omega^c{}_{b\mu} \quad . \tag{4.27a}$$

However if we express the component of $\mathbb{R}$ only with respect to moving (co-)frames we obtain with $\vec{X} = \vec{e}_c$, $\vec{Y} = \vec{e}_d$ and $\vec{Z} = \vec{e}_b$:

$$R^a{}_{bcd} = \vec{e}_c(\gamma^a{}_{bd}) - \vec{e}_d(\gamma^a{}_{bc}) + \gamma^a{}_{pc}\,\gamma^p{}_{bd} - \gamma^a{}_{pd}\gamma^p{}_{bc}$$
$$- [\vec{e}_c , \vec{e}_d]^p \, \gamma^a{}_{bp} \qquad . \qquad (4.27b)$$

The components of the torsion tensor measure the antisymmetry of the connection :

$$T(\vec{X},\vec{Y}) = \vec{e}_a \, \Theta^a \, (\vec{X},\vec{Y})$$
$$= \vec{e}_a \, [\, d\theta^a + \underline{\omega}^a{}_b \wedge \theta^b \,] (\vec{X},\vec{Y})$$
$$= \vec{e}_a \, [\vec{X}(Y^a) - \vec{Y}(X^a) - [\vec{X},\vec{Y}]^a + Y^b \, \underline{\omega}^a{}_b(\vec{X}) - X^b \, \underline{\omega}^a{}_b(\vec{Y})]$$
$$= \nabla_{\vec{X}} \vec{Y} - \nabla_{\vec{Y}} \vec{X} - [\vec{X},\vec{Y}] \qquad (4.28)$$

i.e. $$T^a{}_{\mu\nu} = \omega^a{}_{b\mu} \, \theta^b{}_\nu - \omega^a{}_{b\nu} \, \theta^b{}_\mu \qquad . \qquad (4.29)$$

Previously we have defined the covariant differential of a (p,q) tensor field as a (p,q+1) tensor field. This allows us to define, by induction, higher order covariant differentials. For instance the second covariant differential of a function f is the (0,2) tensor field, the components of which, $f_{;ab}$, are defined by :

$$\underline{\nabla}(\underline{\nabla} \, f) \equiv f_{;ab} \, \theta^a \otimes \theta^b = \underline{\nabla}(f_{,a} \, \theta^a)$$
$$= \vec{e}_b(\vec{e}_a(f))\theta^a \otimes \theta^b - \vec{e}_a(f)\theta^b \otimes \underline{\omega}^a{}_b \qquad .$$

These components satisfy the *Ricci identity :*

$$f_{;ab} - f_{;ba} = [\vec{e}_b,\vec{e}_a](f) + \vec{e}_c(f)T^c{}_{ab} \qquad . \qquad (4.30)$$

The components of the second differential of a vector field are

$$\underline{\nabla}(\underline{\nabla} \, \vec{X}) = X^a{}_{;bc} \, \vec{e}_a \otimes \theta^b \otimes \theta^c$$

and satisfy the Ricci identity

$$X^a{}_{;bc} - X^a{}_{;cb} = - X^d \, R^a{}_{dbc} + X^a{}_{;d} \, T^d{}_{bc} \qquad . \qquad (4.31)$$

This last result shows clearly that

$$\vec{e}_a \, X^a{}_{;bc} \neq \nabla_{\vec{e}_c} \, \nabla_{\vec{e}_b} \, \vec{X} \qquad .$$

Exercice VII : Prove eq. (31) and show that

$$X_{a;bc} - X_{a;cb} = X_d \, R^d{}_{abc} + X_{a;d} \, T^d{}_{bc} \qquad .$$

The torsion and curvature 2-forms verify other relations called the *Bianchi identities* which are just a consequence of $d^2 = 0$. The first identity is given by :

$$\begin{aligned} d\,\Theta^a &= d\,[\,d\theta^a + \underline{\omega}^a{}_b \wedge \theta^b\,] = d\underline{\omega}^a{}_b \wedge \theta^b - \underline{\omega}^a{}_b \wedge d\theta^b \\ &= \Omega^a{}_b \wedge \theta^b - \underline{\omega}^a{}_c \wedge \underline{\omega}^c{}_b \wedge \theta^b - \underline{\omega}^a{}_b \wedge \Theta^b + \underline{\omega}^a{}_b \wedge \underline{\omega}^b{}_c \wedge \theta^c \\ &= \Omega^a{}_b \wedge \theta^b - \underline{\omega}^a{}_b \wedge \Theta^b \end{aligned}$$

i.e. $$d\,\Theta^a + \underline{\omega}^a{}_b \wedge \Theta^b = \Omega^a{}_b \wedge \underline{\theta}^b \quad . \qquad (4.32)$$

The second Bianchi identity is obtained from :

$$\begin{aligned} d\Omega^a{}_b &= d\,[\,d\,\underline{\omega}^a{}_b + \underline{\omega}^a{}_c \wedge \underline{\omega}^c{}_b\,] = d\underline{\omega}^a{}_c \wedge \underline{\omega}^c{}_b - \underline{\omega}^a{}_c \wedge d\underline{\omega}^c{}_b \\ &= \Omega^a{}_c \wedge \underline{\omega}^c{}_b - \underline{\omega}^a{}_c \wedge \Omega^c{}_b \end{aligned}$$

i.e. $$d\Omega^a{}_b + \Omega^a{}_c \wedge \underline{\omega}^c{}_b - \underline{\omega}^a{}_c \wedge \Omega^c{}_b = 0 \quad . \qquad (4.33)$$

The "translation" of these identities in terms of components is not difficult but quite long. The first one leads to :

$$T^a{}_{[\,bc;d\,]} + T^a{}_{e\,[\,b}\; T^e{}_{cd\,]} = R^a{}_{[\,bcd\,]} \qquad (4.34)$$

which reduces to a symmetry property of the components of the Riemann tensor when the torsion vanishes. The second identity becomes :

$$R^a{}_{b\,[\,cd;e\,]} + R^a{}_{bf\,[\,c}\; T^f{}_{de\,]} = 0 \quad . \qquad (4.35)$$

Exercice VIII : 1°) Convince yourself of the correctness of these formulas, at least when then torsion vanishes.
2°) Show that

$$\begin{aligned} \pounds_{\vec{\xi}}\, \Gamma^\alpha{}_{\beta\gamma} &= \nabla_\beta \nabla_\gamma \xi^\alpha + T^\alpha{}_{\sigma\beta}\, \nabla_\gamma \xi^\sigma + T^\sigma{}_{\beta\gamma}\, \nabla_\sigma \xi^\alpha \\ &\quad + (\nabla_\gamma\, T^\alpha{}_{\sigma\beta} - R^\alpha{}_{\sigma\beta\gamma} - R^\alpha{}_{\beta\gamma\sigma})\,\xi^\sigma \quad . \end{aligned}$$

IV.4. Metric connections

Einstein's general relativity describes the gravitational field as a Lorentzian metric :

$$g = g_{ab}\, \theta^a \otimes \theta^b \qquad (4.36)$$

on a four-dimensional manifold. Moreover it is assumed that the lenght of vectors remains constant during parallel transport. As a consequence the scalar product of two vectors does not change when they are transported. Mathematically, this implies that the covariant differential of the metric tensor vanisches :

$$\underline{\nabla}\, g = (dg_{ab} - g_{cb}\, \underline{\omega}^{c}{}_{a} - g_{ac}\, \underline{\omega}^{c}{}_{b}) \otimes \theta^{a} \otimes \theta^{b} = 0 \quad . \tag{4.37}$$

As a consequence, the connection 1-forms are restricted by the equation :

$$dg_{ab} = \underline{\omega}_{ba} + \underline{\omega}_{ab} \tag{4.38}$$

where

$$\underline{\omega}_{ab} = g_{ac}\, \underline{\omega}^{c}{}_{b} \quad .$$

In terms of components eq. (38) reads as follows :

$$\partial_{c}\, g_{ab} = \gamma_{bac} + \gamma_{abc} \quad . \tag{4.39}$$

By combining eq. (39) with the equation for the torsion we can find the connection 1-form. The torsion equation indeed implies that :

$$\gamma^{a}{}_{[bc]} = \frac{1}{2}\,[\beta^{a}{}_{bc} - T^{a}{}_{bc}] \tag{4.40}$$

where $\beta^{a}{}_{bc}$ are the components of

$$d\,\theta^{a} = \frac{1}{2}\,\beta^{a}{}_{bc}\, \theta^{b} \wedge \theta^{c} \quad . \tag{4.41}$$

Lowering the upper index in eq. (40) and cyclically permuting a, b, c leads to

$$\begin{aligned} \gamma_{abc} &= \frac{1}{2}(\partial_{b}\, g_{ac} + \partial_{c}\, g_{ab} - \partial_{a}\, g_{bc}) \\ &+ \frac{1}{2}(\beta_{abc} + \beta_{bca} - \beta_{cab}) \\ &- \frac{1}{2}(T_{abc} + T_{bca} - T_{cab}) \quad . \end{aligned} \tag{4.42}$$

It follows that, in absence of torsion, the metric connection is completely determined by the metric tensor; it is then called the *Levi-Civita connection.* This is the connection which is used in classical general relativity. Its components are the usual *Christoffel symbols* when the moving frame is given by a field of natural frames.

Different advantages of the Cartan formalism will now come into light. If we choose as moving frame an orthonormal frame - a *vielbein* - i.e. a frame in which the components of the metric tensor are constants : $g_{ab} = \eta_{ab}$, then, all fields have their components expressed with respect to a physical frame. On the other hand the Levi-Civita connection enjoys then the antisymmetry property :

$$\underline{\omega}_{ab} = - \underline{\omega}_{ab} \tag{4.43}$$

and thanks to its uniqueness it is often more easy to guess its expression from eq. (22) with $\Theta^a = 0$ than to compute it from eq. (42). Moreover, the computation of the curvature tensor from eq. (23) is also greatly facilitated because the exterior product and the exterior derivative often reduce the lenght of the calculation by avoiding unnecessary operations.

The metric postulate (37) leads to new identities satisfied by the curvature 2-form. Let us define $\Omega_{ab} = g_{ac}\,\Omega^c{}_b$. We deduce from the definition of the curvature 2-form (23) :

$$\begin{aligned}
\Omega_{ab} &= g_{ac}\, d\underline{\omega}^c{}_b + \underline{\omega}_{ac} \wedge \underline{\omega}^c{}_b \\
&= d\underline{\omega}_{ab} - dg_{ac} \wedge \underline{\omega}^c{}_b + \underline{\omega}_{ac} \wedge \underline{\omega}^c{}_b \\
&= d\underline{\omega}_{ab} - \underline{\omega}_{cd} \wedge \underline{\omega}^c{}_b \\
&= -(d\underline{\omega}_{ba} - \underline{\omega}_{cb} \wedge \underline{\omega}^c{}_a)
\end{aligned}$$

i.e. $R_{abcd} = - R_{bacd}$. (4.44)

This last identity together with the obvious one $R_{abcd} = -R_{abdc}$ imply :

$$R_{abcd} - R_{cdab} = \frac{3}{2}(R_{a\,[bcd]} - R_{b\,[acd]} - R_{c\,[dab]} + R_{d\,[abc]}). \tag{4.45}$$

As the right-hand side of this relation only depends on the torsion by eq. (34), it vanishes for a torsion-free connection. In this case a new symmetry of the curvature tensor, which is not independent of the previous one, is obtained. By the way, notice that the *Ricci tensor*

$$R_{bd} = R^a{}_{bad} \tag{4.46}$$

is not in general a symmetric tensor, unless the torsion vanishes.

IV.5. Connections on Lie groups

We have seen previously that on a Lie group G we could define globally (up to a global linear transformation) two particular fields of frames which are right and respectively left invariant.

Each of these moving frames defines a global parallelism which is called the left and the right parallelism and it defines also parallel transport rules. The connection 1-forms associated to them vanish when their components are expressed with respect to the invariant frames. As a consequence, the curvature 2-forms also vanish but the torsion 2-forms do not. They are obtained directly from the structure equations; according to the parallelism considered we obtain :

$$\Theta^a_R = \frac{1}{2} C^a{}_{bc}\ \theta^b \wedge \theta^c \tag{4.47a}$$

for the right parallelism and

$$\Theta^a_L = -\frac{1}{2} C^a{}_{bc}\ \sigma^b \wedge \sigma^c \tag{4.47b}$$

for the left parallelism. Here, as in section (3.5), θ^a and σ^a denote the left and right invariant coframes.

<u>Exercice IX</u> : Show that $\mathbb{T}_R = -\mathbb{T}_L$.

The affine geodesic curves defined by these two parallelims are the same. They are the orbits of the one-parameters subgroup of transformations of G.

In order to fix the notations let us suppose we are considering the right parallelism. The so defined connection is metric with respect to any metric tensor with components g_{ab} constants in the left invariant frame. Moreover, these metric tensors are the only left invariant ones. They depend on $\frac{n(n+1)}{2}$ parameters(*).

If $\vec{\ell}_c$ is a vector field generating a one parameter subgroup of right translations, a left invariant metric

$$g = g_{ab}\ \theta^a \otimes \theta^b$$

will be invariant with respect to these transformations if and only if :

$$\mathcal{L}_{\vec{\ell}_c}\, g = 0 \qquad \text{i.e.} \quad g_{db}\, C^d{}_{ca} + g_{ad}\, C^d{}_{cb} = 0 \quad (a,b,d = 1,\dots,n). \tag{4.48}$$

As a consequence a metric tensor will be *bi-invariant* (simultaneously left and right invariant) if and only if the metric is invariant with respect to the adjoint group i.e. if and only if the metric tensor defines a Casimir operator in the Lie algebra of G. In particular if G is a simple group, it admits only one (up to a scale factor) bi-invariant metric, which is the Killing metric :

(*) dim G = n.

$$g_{ab} = C^d{}_{ac}\, C^a{}_{db} \quad . \tag{4.49}$$

If $\overset{R}{\nabla}$ (resp. $\overset{L}{\nabla}$) denotes the covariant differentiation associated to the right (resp. left) parallelism, we may introduce a third connection $\overset{\circ}{\nabla}$ given by :

$$\overset{\circ}{\nabla} = \frac{1}{2}(\overset{R}{\nabla} + \overset{L}{\nabla}) \quad . \tag{4.50}$$

The expression for the associated connection 1-form is easily computed. Using a left invariant moving frame we obtain :

$$\begin{aligned}
\overset{\circ}{\nabla}_{\vec{\ell}_a} \vec{\ell}_b &= \overset{\circ}{\gamma}{}^c{}_{ba}\, \vec{\ell}_c \\
&= \frac{1}{2} \overset{L}{\nabla}_{\vec{\ell}_a} \vec{\ell}_b && \text{(because } \overset{R}{\nabla} \vec{\ell}_a = 0) \\
&= \frac{1}{2} \overset{L}{\nabla}_{\vec{\ell}_a} (\Lambda_b^{-1\ c}\, \vec{r}_c) \\
&= \frac{1}{2} \vec{\ell}_a (\Lambda_b^{-1\ c}) \vec{r}_c && \text{(because } \overset{L}{\nabla} \vec{r}_a = 0) \\
&= -\frac{1}{2} C^d{}_{ba}\, \vec{\ell}_d && \text{(see eq. (3.38))}
\end{aligned}$$

i.e. $$\overset{\circ}{\underset{\sim}{\omega}}{}^a{}_b = \frac{1}{2} C^a{}_{cb}\, \theta^c \quad . \tag{4.51}$$

From the Maurer-Cartan structure equation we get that this connection is torsion-free and that the components of the curvature tensor are given by :

$$\overset{\circ}{R}{}^a{}_{bcd} = \frac{1}{4} C^a{}_{bk}\, C^k{}_{cd} \quad . \tag{4.52}$$

Moreover this connection is metric with respect to a bi-invariant metric. Indeed, in such a case, the condition $\overset{\circ}{\nabla} g = 0$ reduces to eq. (48) from which we conclude that

$$C_{abc} = g_{ad}\, C^d{}_{bc} = C_{[abc]} \quad \text{and} \quad C^a{}_{ab} = 0 \quad . \tag{4.53}$$

This last condition is necessary for the existence of a bi-invariant metric. In particular it is always satisfied for compact Lie groups because for such groups a bi-invariant metric can always be defined. Finally note that the Ricci tensor is given by

$$\overset{\circ}{R}_{bd} = \frac{1}{4} C^a{}_{bk}\, C^k{}_{ad} \tag{4.54}$$

i.e. semi-simple groups are Einstein manifolds when metrized with their Killing metric.

V. A BRIEF FLIGHT OVER DE RHAM COHOMOLOGY

V.1. Basic definitions

The subject discussed in this section is related to some problems that can appear in Kaluza-Klein theories. In fact, when performing dimensional reduction, we are often faced with harmonic analysis on a Riemannian compact space. For antisymmetric tensors subtleties related to the topology of the extra-dimensional space sometimes appear.

In the following of this section we shall suppose that we are working on an orientable compact Riemannian n-dimensional manifold without boundary M. All covariant derivatives will be defined using the Levi-Civita connection induced by the metric, and the volume element n-form used will also be the metric one which, expressed in local coordinates, in terms of the square root of the determinant of the metric, is

$$\eta = \sqrt{g}\; dx^1 \wedge \dots \wedge dx^n \quad . \tag{5.1}$$

The metric tensor defines an isomorphism between antisymmetric p-covariant and contravariant tensors by just lowering and raising indices. This allows us to introduce an inner product on the space of p-forms as follows :

$$(\alpha,\beta) = \int_M \alpha^{\mu_1\dots\mu_p}\beta_{\mu_1\dots\mu_p}\,\eta = (\beta,\alpha)\ ,\ \alpha \text{ and } \beta \in \Lambda^p. \tag{5.2}$$

We may now introduce the adjoint operator of the exterior differentiation d. It is called the *co-differentiation* and denoted by δ. This operator maps (p+1)-forms into p-forms :

$$\delta : \Lambda^{p+1} \to \Lambda^p \;\|\; (\delta\omega)_{\mu_1\dots\mu_p} = -\nabla_{\mu_o}\,\omega^{\mu_o}{}_{\mu_1\dots\mu_p} \quad . \tag{5.3a}$$

In natural coordinates we obtain the expression

$$(\delta\omega)_{\mu_1\dots\mu_p} = -\frac{1}{\sqrt{g}}\,\partial_{\mu_o}\sqrt{g}\,\omega^{\mu_o}{}_{\mu_1\dots\mu_p} \quad . \tag{5.3b}$$

If α is a p-form and β a (p-1)form we have :

$$(\alpha,d\beta) = (\delta\alpha,\beta) \tag{5.4}$$

and obviously

$$\delta^2 = 0 \quad . \tag{5.5}$$

The *Laplace-Beltrami operator* Δ is defined by :

$$\Delta = d\delta + \delta d \quad . \tag{5.6}$$

It maps p-forms on p-forms. In local coordinates it can be expressed in terms of the Dalembertian operator and the curvature tensor as[11] :

$$(\Delta\omega)_{\mu_1\cdots\mu_p} = -\nabla_\alpha \nabla^\alpha \omega_{\mu_1\cdots\mu_p} + \sum_{i=1}^{p} R^\alpha{}_{\mu_i}\, \omega_{\mu_1\cdots\mu_{i-1}\alpha\,\mu_{i+1}\cdots\mu_p} - \frac{1}{2}\sum_{i=1}^{p}\sum_{j=1}^{p} R^{\alpha\beta}{}_{\mu_i\mu_j}\, \omega_{\mu_1\cdots\mu_{i-1}\alpha\mu_{i+1}\cdots\mu_{j-1}\beta\mu_{j+1}\cdots\mu_p} . \tag{5.7}$$

A p-form ω is said to be harmonic if it is *closed* i.e.

$$d\omega = 0 \tag{5.8}$$

and *co-closed* :

$$\delta\omega = 0 . \tag{5.9}$$

Harmonic forms satisfy the Laplace-Betrami equation :

$$\Delta\omega = (\delta d + d\delta)\omega = 0 .$$

Conversely if ω satisfies this equation then :

$$0 = (\Delta\omega,\omega) = (\delta d\omega,\omega) + (d\delta\omega,\omega) = (d\omega,d\omega) + (\delta\omega,\delta\omega)$$

and as a consequence :

$$d\omega = 0 \qquad \text{and} \qquad \delta\omega = 0 .$$

V.2. Betti numbers

The space of closed differential forms of degree p, modulo the subspace of exact forms, is called the *p^{th} cohomology space* $H^p(M)$. Its dimension is the *p^{th} Betti number* $b^p(M)$. De Rham has proved that these numbers are topological concepts. Moreover Hodge has proved that, on an orientable compact Riemannian manifold, $b^p(M)$ is finite and is equal to the number of independent harmonic forms of degree p. These results are actually surprising as they show that what at first sight seems to be a metrical property or related to the differentiable structure of the manifold is in fact a topological property of the manifold.

Roughly speaking this results from the fact that the Betti numbers counts the number of independent p-dimensional surfaces which are not boundaries of some p+1 surface, that is the "number of p dimensional holes" of the manifold. These spaces of holes can be considered as the duals of the cohomology spaces, the product of elements belonging of each of these spaces being given by integration of p-forms on p-dimensional submanifolds.

Hodge has also shown that the equation

$$\Delta\alpha = \beta \tag{5.10}$$

admits solutions if and only if β is orthogonal to the space of harmonic forms. The necessity of this condition is very simple to establish. If γ is an harmonic form we have

$$(\beta,\gamma) = (\Delta\alpha,\gamma) = (\alpha,\Delta\gamma) = 0 \quad .$$

The proof of the sufficiency of the condition and of the finiteness of the space H^p is much more complicated[12]. However, if we admit this theorem we obtain immediately the *Hodge-de Rham decomposition theorem*. Let β be a p-form. As the space H^p is finite dimensional we can project β on the orthogonal of this space just by subtracting its harmonic part β_H. Then there exists a p-form α such that

$$d\delta\alpha + \delta d\alpha = \beta - \beta_H \quad .$$

As a consequence β admits a decomposition as a sum of an exact part $\beta_d = d(\delta\alpha)$, a coexact part $\beta_\delta = \delta(d\alpha)$ and an harmonic part β_H :

$$\beta = \beta_d + \beta_\delta + \beta_H \quad . \tag{5.11}$$

The uniqueness of this decomposition is proved as follows. Suppose that these exists a combination of exact, coexact and harmonic forms such that :

$$\alpha_d + \alpha_\delta + \alpha_H = 0 \quad .$$

It results that $d\alpha_\delta = 0$ and $\delta\alpha_d = 0$. But as by definition $\alpha_\delta = \delta\gamma$ and $\alpha_d = d\beta$ we obtain

$$0 = (d\alpha_\delta,\gamma) = (d\delta\gamma,\gamma) = (\delta\gamma,\delta\gamma) = (\alpha_\delta,\alpha_\delta)$$

i.e. $\alpha_\delta = 0$ and in the same way we can show that $\alpha_d = 0$. As a consequence $\alpha_H = 0$ which proves the uniqueness of the de Rham decomposition.

Hodge's theorem relates the pth Betti number of a manifold to solutions of a differential operator built from a metric defined on the manifold. The existence of a metric induces an isomorphism between the space Λ^p and Λ^{n-p}, the Hodge-star isomorphism defined as follows :

$$\Lambda^p \to \Lambda^{n-p} : \omega \to \star\omega \quad \|(\star\omega)_{\mu_{p+1}\cdots\mu_n} = \frac{1}{p!}\,\eta_{\mu_1\cdots\mu_p\mu_{p+1}\cdots\mu_n}\,\omega^{\mu_1\cdots\mu_p} \tag{5.12}$$

Exercice X : Show that $\star\star\omega = (-1)^{np+p+s}\omega$ where $p = \deg(\omega)$ and s is the number of negative eigenvalue of the metric tensor.

By using the $\star$ operator the formulas of the previous section can be written as follows :

$$(\alpha,\beta) = \int_M \alpha \wedge \star\beta \tag{5.13}$$

and

$$\delta = (-1)^{np+n+1} \star d \star \,. \tag{5.14}$$

As a consequence a p-form α will be harmonic if and only if $\star\alpha$ is an harmonic (n-p)-form, so that the Betti numbers of a compact orientable manifold are related by the duality property :

$$b_p(M) = b_{n-p}(M) \qquad [\,\textit{Poincaré}\,] \,. \tag{5.15}$$

The computation of the Betti numbers of an arbitrary manifold is in general difficult. However, having in mind applications to supergravity and Kaluza-Klein theories we shall restrict ourselves to special cases.

A fundamental formula which connects curvature and Betti numbers has been obtained by Yano[13]. Suppose that ω is an harmonic p-form. By using the expression (7) of the Laplace-Beltrami operator and integrating by part we obtain :

$$\begin{aligned} 0 = (\Delta\omega,\omega) = \int \Big(& \nabla_{\alpha_o}\omega_{\mu_1\cdots\mu_p}\nabla^{\alpha_o}\omega^{\mu_1\cdots\mu_p} \\ & + p\, R^{\alpha}{}_{\beta}\,\omega_{\alpha\mu_2\cdots\mu_p}\,\omega^{\beta\mu_2\cdots\mu_p} \\ & - \frac{1}{2}\,p(p-1)R^{\alpha\beta}{}_{\gamma\delta}\,\omega_{\alpha\beta\mu_3\cdots\mu_p}\,\omega^{\gamma\delta\mu_3\cdots\mu_p}\Big)\eta \,. \end{aligned} \tag{5.16}$$

As the metric is Riemannian there results that, if the quadratic form

$$F(\omega_p) \equiv R^{\alpha}{}_{\beta}\,\omega_{\alpha\mu_2\cdots\mu_p}\,\omega^{\beta\mu_2\cdots\mu_p} - \frac{(p-1)}{2}\,R^{\alpha\beta}{}_{\gamma\delta}\,\omega_{\alpha\beta\mu_3\cdots\mu_p}\,\omega^{\gamma\delta\mu_3\cdots\mu_p} \tag{5.17}$$

is positive definitive, the pth Betti number of the manifold (*)

(*) From eqs. (8,9) there results obviously that $b_o(M)$ is equal to the number of connected components of M.

is zero. Immediate consequences of this theorem are the following :
- The first Betti number of a compact orientable Riemannian manifold admitting a metric such that the Ricci tensor is positive definite is zero. This is the case, for instance, on semi-simple Lie groups.
- All the Betti numbers of the sphere S_n (excepted $b_o = b_n = 1$) are zero. Indeed the curvature tensor of S_n with respect to the metric canonically induced by the (n+1) dimensional Euclidean space is equal to

$$R^{\alpha\beta}{}_{\gamma\delta} = \frac{1}{R^2}\,(\delta^{\alpha}{}_{\gamma}\,\delta^{\beta}{}_{\delta} + \delta^{\alpha}{}_{\delta}\,\delta^{\beta}{}_{\gamma})$$

and as a consequence $\quad F(\omega_p) = \frac{1}{R^2}\,(n-p)\,\omega_{\mu_1\ldots\mu_p}\,\omega^{\mu_1\ldots\mu_p}$

is positive definite.
- Harmonic p-forms on the torus are covariantly constants :

$$\nabla_{\alpha}\,\omega_{\mu_1\ldots\mu_p} = 0 \qquad .$$

The maximum number of solutions of these equations is equal to $\frac{n!}{p!(n-p)!}$ and is actually the value of the pth Betti number of T_n .
The previous examples illustrate how the properties of the metric affect the existence of harmonic forms. In the following we shall thoroughly study metrical properties of harmonic forms and show how Betti numbers of homogeneous spaces can be computed just with algebraic techniques.

The infinitesimal generators of the isometry group (*) of a manifold are vector fields $\vec{\xi}$ such that :

$$\pounds_{\vec{\xi}}\, g = 0 \tag{5.18}$$

or equivalently

$$\nabla_{\alpha}\,\xi_{\beta} + \nabla_{\beta}\,\xi_{\alpha} = 0 \qquad . \tag{5.19}$$

They are called *Killing vector fields* and eqs. (19) are know as Killing equations. By applying the Ricci identity (4.31) to a Killing vector and using the Bianchi identity (4.34) we obtain the integrability condition

$$\nabla_{\alpha}\,\nabla_{\beta}\,\xi^{\gamma}{}_{\gamma} = R^{\rho}{}_{\alpha\beta\gamma}\,\xi_{\rho} \qquad . \tag{5.20}$$

By differentiating once again this last equation and using the Ricci identity we conclude that :

$$\pounds_{\xi}\, R^{\alpha}{}_{\beta\gamma\delta} = 0 \qquad . \tag{5.21}$$

(*) We limit ourselves to the identity connected component of the isometry group.

Actually this result is obvious. Indeed if we adapt the coordinate system to the Killing vector field eq. (18) means that the components of the metric tensor are independent of the coordinate associated to the orbit of ξ and eqs. (20) and (21) reflect the same property of the Levi-Civita connection coefficients and of the components of the curvature tensor. Finally as eq. (19) show us that the divergence of a Killing vector field is zero we obtain that the metrical volume element is invariant :

$$\mathcal{L}_{\vec{\xi}}\, \eta = 0 \tag{5.22}$$

i.e. the Lie derivative with respect to Killing vector field commutes with the Hodge star operator.

We are now in position to prove the first fundamental property of harmonic forms which is that they are invariant with respect to isometries : if ω is an harmonic p form :

$$\Delta\omega = (\star d\star d + d\star d\star)\omega = 0$$

and

$$\mathcal{L}_{\vec{\xi}}\, \Delta\omega = \Delta\mathcal{L}_{\vec{\xi}}\, \omega = 0$$

i.e. $\mathcal{L}_{\vec{\xi}}\,\omega$ is also harmonic. But from exercice IV and eq. (8) we obtain

$$\mathcal{L}_{\vec{\xi}}\, \omega = d\, i(\vec{\xi})\omega$$

which implies, thanks to Hodge-de Rham decomposition theorem, that

$$\mathcal{L}_{\vec{\xi}}\, \omega = 0$$

<u>Exercice XI</u> : Prove that if α is an exact invariant ($\mathcal{L}_{\vec{\xi}}\alpha = 0$) p-form there exists an invariant (p-1)-form β such that $\alpha = d\beta$. (Hint : Use Hodge-de Rham decomposition theorem which allows to write $\alpha = d\delta\gamma$).

In order to go ahead we consider now a Lie group manifold G and assume that the Laplace-Beltrami operator is built from a Riemannian bi-invariant metric. In such a case we can prove that bi-invariant p-form on G are harmonic. The idea of the proof is to show that if ω is a bi-invariant p-form then $d\omega$ and $\delta\omega$ are zero. We recall the notation of section (3.5). The generators of left and right translations on G are denoted respectively by $\vec{r}_a$ and $\vec{\ell}_a$. They define right and left invariant moving frames on the whole group. Their dual frames are denoted by σ^a and θ^a and are also right and left invariant. Consider now a left-invariant p-form ω. With respect to the coframe $\{\theta^a\}$ it can be written as follows :

$$\omega = \frac{1}{p!} \omega_{a_1 \ldots a_p} \theta^{a_1} \wedge \ldots \wedge \theta^{a_p}$$

where the components $\omega_{a_1 \ldots a_p}$ are constants. In order to be bi-invariant ω must also satisfy the equations

$$\pounds_{\vec{\ell}_a} \omega = 0$$

i.e. $$\sum_{r=1}^{p} \omega_{a_1 \ldots a_r \ldots a_p} \theta^{a_1} \wedge \ldots \wedge \pounds_{\vec{\ell}_a} \theta^{a_r} \wedge \ldots \wedge \theta^{a_p} = 0 \quad (5.23)$$

for all left-invariant vector field $\vec{\ell}_a$. Let us compute this Lie derivative for the left invariant from θ^a. We obtain from eq. (3.41)

$$\pounds_{\vec{\ell}_a} \theta^b = i(\ell_a) d\theta^b + di(\ell_a)\theta^b$$

$$= -C^b{}_{ac} \theta^c \quad .$$

As a consequence eq. (23) is equivalent to

$$\sum_{r=1}^{p} \omega_{a_1 \ldots a_r \ldots a_p} \theta^{a_1} \wedge \ldots \wedge C^{a_r}{}_{a\,b} \theta^b \wedge \ldots \wedge \theta^p = 0 \,. \quad (5.24)$$

On the other hand, the exterior derivative of ω is given by

$$d\omega = \frac{1}{p!} \sum_{r=1}^{p} (-1)^{r+1} \omega_{a_1 \ldots a_r \ldots a_p} \theta^{a_1} \wedge \ldots \wedge d\theta^{a_r} \wedge \ldots \wedge \theta^p$$

$$= -\frac{1}{p!} \sum_{r=1}^{p} \frac{1}{2} \theta^a \wedge \ldots \wedge \omega_{a_1 \ldots a_r \ldots a_p} \theta^1 \wedge \ldots \wedge C^{a_r}{}_{a\,b} \theta^b \wedge \ldots \wedge \theta^p$$

$$= 0$$

Notice that we have proved that the exterior derivative operator acting on left-invariant p-forms can be expressed as

$$d \equiv \frac{1}{2} \theta^a \wedge \pounds_{\vec{\ell}_a} \quad . \quad (5.25)$$

In order to complete the proof of harmonicity of b-invariant p-forms there remains to compute $\delta\omega$. As the Hodge star operator commutes with the Lie derivative with respect to Killing vectors, the dual of a left-invariant p-form is also invariant. Hence we obtain by using eq. (25) :

$$\star d \star \omega = \frac{1}{2} \star \theta^a \wedge \pounds_{\vec{\ell}_a} \star \omega = \frac{1}{2} \star \theta^a \wedge \star \pounds_{\vec{\ell}_a} \omega = 0$$

i.e. $\delta\omega = 0$ if ω is bi-invariant.

Therefore the determination of the Betti number of a compact Lie group reduces to the computation of the number of independent solutions of the equations

$$\sum_{r=1}^{p} \omega_{a_1 \ldots a_{r-1}\, b\, a_{r+1} \ldots a_p} C^b{}_{a_2 a} = 0 \quad . \qquad (5.26)$$

Notice that these equations are automatically antisymmetric in the indices $a_1 \ldots a_p$.

On semi-simple compact Lie groups it is well known that the first and second Betti numbers are zero. Indeed, if ω is a 1-form eq. (26) becomes

$$\omega_b C^b{}_{a_1 a} = 0 \quad .$$

By using the Killing metric (4.49) and the antisymmetry property (4.53) of the structure constants we obtain after contraction with $C^{a_1 a}{}_d$ that $\omega_b = 0$. This implies that $b_1(G) = 0$. If ω is a 2-form then eq. (26) becomes

$$\omega_{b a_2} C^b{}_{a_1 a} + \omega_{a_1 b} C^b{}_{a_2 a} = 0 \quad .$$

By adding to this equation the similar ones obtained by permutation of a, a_1 and a_2 we obtain

$$\omega_{b a_1} C^b{}_{a_2 a} = 0$$

i.e. $\omega_{b_1 a_2} = 0$ which confirms that $b_2(G) = 0$.

A non trivial solution of eq. (26) is given by the torsion tensor. Its covariant components (4.47) are completely antisymmetric (4.53) and solve eqs. (26) by virtue of the Jacobi identity. This shows that on semi-simple groups $b_3(G) \geqslant 1$. Actually it can be proved that if G is equal to the product of k simple groups then $b_3(G) = k$.

Finally let us briefly describe — without proof — how to compute the Betti numbers of a coset space G/H where G is compact and connected and H is a closed subgroup of G. On such a coset space we can always find a G invariant metric. As harmonic forms on homogeneous spaces are invariant the pth de Rham cohomology space $H^p(G/H)$ is isomorphic to the space of closed invariant p-forms on G/H modulo the space of exterior derivative of invariant (p-1)-forms. But invariant p-forms on G/H are in one-to-one correspondance with left-invariant forms on G, invariant under the adjoint transformations generated by H on G and which annihilates the tangent space to H at the identity of G. This allows to reduce the computation of Betti numbers of a coset space G/H to algebraic operations on the Lie algebra of G. The final result (which generalized our previous result for Lie groups) assures that $H^k(G/H)$ is isomorphic to

$$H^p(G/H) \approx \frac{\text{kernel } d : \Omega^p(\mathcal{G}/\mathcal{H})^H \to \Omega^{p+1}(\mathcal{G}/\mathcal{H})^H}{d(\Omega^{p-1}(\mathcal{G}/\mathcal{H})^H)} \tag{5.27}$$

where $\mathcal{G}$ is the Lie algebra of G,
$\mathcal{H}$ is the Lie algebra of H imbedded in G,
$\Omega^p(\mathcal{G}/\mathcal{H})^H$ is a sub-space of the space of p-forms defined on $\mathcal{G}$ such that

$$\omega \in \Omega^p(\mathcal{G}/\mathcal{H})^H \text{ if and only if}$$

1°) $\omega(X_1, \ldots, X_p) = 0$ if some $X_i \in \mathcal{H}$

2°) ω is invariant with respect to the adjoint transformations defined by the elements of H

and d is defined by

$$d\omega(X_1,\ldots,X_{p+1}) = \sum_{1\leqslant i\leqslant j\leqslant p+1} \omega(\, [X_i,X_j]\, ,\, X_1,\ldots,\hat{X}_i,\ldots,\hat{X}_j,\ldots,X_{p+1})$$

VI. SPINOR FIELDS

VI.1. Definition

In addition to tensor fields, a physically acceptable space-time must also admit globally defined spinor fields. This implies topological restrictions on the manifold which is supposed to describe the universe. The origin of these restrictions arises from the 1-2 correspondance between the isometry group O of the tangent spaces of the manifold and its covering group : the spin group Spin.

A spinor on the tangent space at P consists, at least, in a equivalence class of pairs $(\rho,\psi)_P$ where ρ is an orthonormal frame and ψ belongs to a vector space on which acts a representation of the spin group. The equivalence relation is defined by $(\rho,\psi) \sim (\rho',\psi')$ if and only if $\rho' = R\rho$, $\psi' = \mathcal{D}(\Lambda)\psi$, $R = H(\Lambda)$ with $R \in$ O, $\Lambda \in$ Spin and H denotes the 2-1 covering homomorphism of Spin onto O. However with this definition a sign ambiguity appears since $(\rho,\psi) \sim (\rho,-\psi)$. It can be resolved if we distinguish a frame from the one obtained by a 360° rotation. This is implemented by introducing an equivalence relation on the pairs (Λ,ρ) of elements of Spin and orthonormed frames :

$$(\Lambda,\rho) \sim (\Lambda',\rho') \text{ if and only if } \rho' : H(\Lambda'\Lambda^{-1})\rho \quad .$$

Each such equivalence class define a local *spinorial frame*.

A *spin structure* — when it exists — is then obtained by a

covering of the manifold with open subsets U_i on which is given a smooth field of pairs $(\rho,\Lambda)^i_P$ such that if $P \in U_i \cap U_j$ we have $(\rho,\Lambda)^i_P \sim (\rho,\Lambda)^j_P$. It is this last condition which may prevent the construction of a global spinor frame. Of course if the manifold considered possesses a global field of orthonormed frames ρ it admits a spin structure, for instance, by taking as spinor frame (ρ, Identity) on the whole manifold. In the frame work of general relativity Geroch [14] has proved that each noncompact 4-dimensional space-time on which spinors may be defined carries a global field of orthonormed tetrads. The interested reader will find enlightening discussion on these questions in references[15,16,17]. Now we are in position to present a definition of a *spinor field* on a manifold M : it is a field of equivalence classes of triplet $(\Lambda,\rho,\psi)_P$ such that :

$(\Lambda,\rho,\psi) \sim (\Lambda',\rho',\psi')$ if and only if $\rho' = R\rho$, $\psi' = \mathcal{D}(\Lambda'\Lambda^{-1})\psi$ and $R = H(\Lambda'\Lambda^{-1})$.

VI.2. Covariant derivative and other ones

Suppose we fixe (locally at least) a gauge by giving a moving frame on an open subset of M. In the previous definition the representation $\mathcal{D}$ defines the "tensorial" nature of the spinor field. In order to fixe the ideas and to simplify the notation we assume from now that ψ is an array of numbers defining a covariant spinor i.e. $\mathcal{D}(\Lambda) = \Lambda$. The covariant derivative of a spinor field is built from the isomorphism between the Lie algebras of O and Spin. Let us suppose that a metric connection has been defined on M. Then under parallel transport, an orthonormal frame remains orthonormed but, in general, rotates with respect to the moving frame. For an infinitesimal parallel transport the associated infinitesimal rotation of the frame is obtained from eq. (4.13) :

$$\nabla \vec{e}_a = \vec{e}_c \otimes \underline{\omega}^c{}_a \quad .$$

If we assume that during a parallel transport the components of a spinor remain constants with respect to the transported frame their components with respect to the original moving frame undergo the infinitesimal transformation

$$\delta_{/\!/}\psi = \frac{1}{8}\underline{\omega}^c{}_a \; [\gamma^a,\gamma_c]\,\psi \tag{6.1}$$

where γ^a denotes the usual constant Dirac matrices :

$$\gamma_a\gamma_b + \gamma_b\gamma_a = 2\eta_{ab} \qquad (*) \quad . \tag{6.2}$$

The *covariant derivative* of a spinor field is then defined as

(*) We use the convention that euclidean metric is positive definite.

follows :

$$\nabla\psi = d\psi - \frac{1}{8}\underline{\omega}^{c}{}_{a}\;[\gamma^{a},\gamma_{c}]\,\psi \equiv \psi_{;a}\,\theta^{a} \quad . \tag{6.3}$$

The elements of the dual of the spinor space are contravariant spinors $\overline{\chi}$. They transform contragrediently to the spinors i.e.

$$(\Lambda,\rho,\overline{\chi})\sim(\Lambda',\rho',\overline{\chi}) \text{ if and only if } (\Lambda,\rho)\sim(\Lambda',\rho') \text{ and } \overline{\chi}' = \tau\overline{\chi}\,\Lambda\Lambda'^{-1}$$

where τ represents the relative time orientation of the frame ρ with respect to ρ'. The covariant derivative of such a spinor field is obtained by a same reasoning as previously :

$$\nabla\overline{\chi} = d\overline{\chi} + \overline{\chi}\,\frac{1}{8}\underline{\omega}^{c}{}_{a}\;[\gamma^{a},\gamma_{c}] \quad . \tag{6.4}$$

From these definitions it results that :

$$\nabla\gamma^{a} = d\gamma^{a} + \underline{\omega}^{a}{}_{c}\gamma^{c} + \frac{1}{8}\underline{\omega}^{b}{}_{c}\;\{\gamma^{a},[\gamma^{c},\gamma_{b}]\} = 0 \tag{6.5}$$

in agreement with $\nabla g = 0$. Our definition of a spinor is indeed a metrical concept and the spin connection has been built from a metric connection.

Straightforward calculations lead to the following Ricci identities (compare with eqs. (4.26) and (4.31)) :

$$\nabla_{\vec{X}}\nabla_{\vec{Y}}\psi - \nabla_{\vec{Y}}\nabla_{\vec{X}}\psi - \nabla_{[\vec{X},\vec{Y}]}\psi = -\frac{1}{4}\Omega^{d}{}_{c}(\vec{X},\vec{Y})\gamma^{c}\gamma_{d}\psi$$

$$\psi_{;ab} - \psi_{;ba} = \frac{1}{4}R^{d}{}_{cab}\,\gamma^{c}\gamma_{d}\psi + T^{c}{}_{ab}\,\psi_{;c}$$

which show that the square of the Dirac operator is given by :

$$\begin{aligned}\gamma^{a}\nabla_{a}\gamma^{b}\nabla_{b}\psi &= \nabla^{a}\nabla_{a}\psi + \frac{1}{2}\gamma^{a}\gamma^{b}(\nabla_{a}\nabla_{b} - \nabla_{b}\nabla_{a})\psi \\ &= \Box\psi + \frac{1}{8}R_{cdab}\,\gamma^{a}\gamma^{b}\gamma^{c}\gamma^{d}\psi - \frac{1}{2}\gamma^{a}\gamma^{b}\,T^{c}{}_{ab}\nabla_{c}\psi \\ &= \Box\psi - \frac{1}{4}R\psi + \frac{1}{8}R_{[abcd]}\,\gamma^{a}\gamma^{b}\gamma^{c}\gamma^{d}\psi + \frac{1}{2}R_{[ab]}\gamma^{a}\gamma^{b}\psi - \frac{1}{2}\gamma^{a}\gamma^{b}\,T^{c}_{ab}\nabla_{c}\psi \; .\end{aligned} \tag{6.6}$$

Notice that in the particular case of vanishing torsion this expression reduces to :

$$\gamma^{a}\nabla_{a}\gamma^{b}\nabla_{b}\psi = \Box\psi - \frac{1}{4}R\psi \quad . \tag{6.7}$$

The reasoning leading to the definition of the covariant derivative of a spinor field can be generalized[18] to an arbitrary derivation which preserves the orthogonality of the vielbein.

Suppose that $\mathcal{D}_{\vec{\xi}}$ is such a derivation operator i.e. the matrix defined by the scalar products of vectors of a moving frame with their derivatives are antisymmetric

$$\lambda_{ab} = \mathcal{D}_{\vec{\xi}}\, \vec{e}_a \cdot \vec{e}_b = \lambda_{[ab]} \quad . \tag{6.8}$$

Then this operator is naturally defined on spinor field as follows :

$$\mathcal{D}_{\vec{\xi}}\, \psi = \xi^\alpha \partial_\alpha \psi - \frac{1}{8} \lambda_{ab} \, [\gamma^a, \gamma^b]\, \psi \quad . \tag{6.9}$$

On the other hand a derivation of tensor can always be expressed as a covariant derivative plus a contraction with a tensor. Derivation of spinor can be decompesed in the same way. For instance the Lie derivative[19] of vectors defining a vielbein with respect to a Killing vector $\vec{\xi}$ satisfies eq. (8) and can be written in term of torsion-free covariant derivative as follows :

$$\pounds_{\vec{\xi}}\, \vec{e}_a = \nabla_{\vec{\xi}}\, \vec{e}_a - \nabla_{\vec{e}_a} \vec{\xi} \tag{6.10}$$

hence the Lie derivative of a spinor field will be written :

$$\begin{aligned} \pounds_{\vec{\xi}}\, \psi &= \xi^\alpha \partial_\alpha \psi - \frac{1}{8} \pounds_{\vec{\xi}}\, \vec{e}_a \cdot \vec{e}_b \, [\gamma^a, \gamma^b]\, \psi \\ &= \nabla_{\vec{\xi}}\, \psi + \frac{1}{8} \xi^b{}_{;a} [\gamma^a, \gamma_b]\, \psi \quad . \end{aligned} \tag{6.11}$$

Even when eq. (8) is not satisfied we can use (9) as a definition for the derivative of a spinor field. Adopting this rule[19] for the Lie derivative with respect to an arbitrary vector field ξ, we obtain in general[20] :

$$[\pounds_{\vec{\xi}_1}, \pounds_{\vec{\xi}_2}]\, \psi = \pounds_{[\vec{\xi}_1, \vec{\xi}_2]}\, \psi - \frac{1}{2} g^{\mu\nu} \pounds_{\vec{\xi}_1} g_{\mu\alpha} \cdot \pounds_{\vec{\xi}_2} g_{\nu\beta} \frac{1}{8} [\gamma^\alpha, \gamma^\beta]\, \psi \tag{6.12}$$

while for tensor fields we have always :

$$[\pounds_{\vec{\xi}_1}, \pounds_{\vec{\xi}_2}] = \pounds_{[\vec{\xi}_1, \vec{\xi}_2]} \quad . \tag{6.13}$$

Notice that if the vector fields $\vec{\xi}_1$ and $\vec{\xi}_2$ are the generators of conformal transformations the last relation remains valid for spinor fields.

VI.3. Some properties of Dirac matrices in arbitrary space-time dimensions

As the properties of Dirac matrices have been discussed with great details during Prof. Freedman's lectures I will be satisfied with just reminding briefly some of them[21,22].

On even n-dimensional space-times all representations of Dirac matrices are equivalent. Moreover we can introduce a matrix

$$\Gamma^{n+1} \div \gamma^{1} \ldots \gamma^{n} \tag{6.14}$$

which anticommutes with all the Dirac matrices, and whose square is the identity matrix. This matrix allows us to define *Weyl spinors* by the condition : $\Gamma^{n+1} \psi_W = \pm \psi_W$ which is compatible only with the massless Dirac equation. On odd dimensional space-times a representation of the Dirac matrices is obtained from the previous one by adding to them $\pm \Gamma^{n+1}$ or $\pm i\Gamma^{n+1}$ according to the new dimension being timelike or spacelike. Each choice of sign in front of Γ^{n+1} defines a class of inequivalent representations of the Clifford algebra and no Weyl spinor exists.

On the other hand we may also define the *Dirac adjoint* of a spinor ψ by :

$$\overline{\psi} = \psi^{+} \beta_{\mu} \tag{6.15}$$

where the matrix β_{μ} is defined up to a real factor by the conditions :

$$\beta_{\mu} = \beta_{\mu}^{+} \quad \text{and} \quad \beta_{\mu} \gamma_{a} = \mu \gamma_{a}^{+} \beta_{\mu} \quad \text{with} \quad \mu = \pm 1 \ . \tag{6.16}$$

If ψ is a solution of the Dirac equation

$$(\gamma^{a} \nabla_{a} - m)\psi = 0 \tag{6.17a}$$

then the Dirac adjoint spinor satisfies the equation

$$\nabla_{a} \overline{\psi} \gamma^{a} - \mu m^{*} \overline{\psi} = 0 \tag{6.17b}$$

and the vector

$$J^{a} = \overline{\psi} \gamma^{a} \psi \tag{6.18}$$

is conserved when $m^{*} = -\mu m$.

Other matrices which play an important rôle are the complex conjugaison matrices C_{σ} defined by the relations

$$C_{\sigma} \gamma_{a} = \sigma \gamma_{a}^{*} C_{\sigma} \quad \text{with} \quad \sigma = \pm 1 \tag{6.19}$$

which imply that

$$C_{\sigma} C_{\sigma}^{*} = C_{\sigma}^{*} C_{\sigma} = \lambda \, \mathbb{1} \ . \tag{6.20}$$

The real factor λ which appears in this also equation can always be normalized to ± 1. However it is only when $\lambda = +1$ that Majorana spinors can be defined through the condition :

$$\psi_M = C_\sigma \, \psi^*_M \tag{6.21}$$

in which case we shall denotes the complex conjugaison matrices as C^M_σ . The existence of these matrices depends on the dimensionality n of the space-time and on the number t of timelike directions of the metric (t is the number of negative eigenvalues of the matrix of metric components). On even dimensional space-times the matrices β_μ and C_σ always exist. When the dimension n is odd β_μ exists only for $\mu = (-1)^t$ while C_σ exists only for $\sigma = (-1)^{(n-1)/2+t}$. Finally Majorana spinors can be defined with respect to C_+ if and only if $n - 2t = 0$, 1 or 2 (mod 8) and with respect to C_- if and only if $n - 2t = 0$, 6 or 7 (mod 8) whatever is the evenness of the space-time dimension. These results are summarized in table I:

TABLE I

t (mod 4) \ n (mod 8)	0	1	2	3	4	5	6	7
0	β_+,β_- C^M_+,C^M_-	β_+ C^M_+	β_+,β_- C^M_+,C_-	β_+ C_-	β_+,β_- C_+,C_-	β_+ C_+	β_+,β_- C_+,C^M_-	β_+ C^M_-
1	β_+,β_- C_+,C^M_-	β_- C^M_-	β_+,β_- C^M_+,C^M_-	β_- C^M_+	β_+,β_- C^M_+,C_-	β_- C_-	β_+,β_- C_+,C_-	β_- C_+
2	β_+,β_- C_+,C_-	β_+ C_+	β_+,β_- C_+,C^M_-	β_+ C^M_-	β_+,β_- C^M_+,C^M_-	β_+ C^M_+	β_+,β_- C^M_+,C_-	β_+ C_-
3	β_+,β_- C^M_+,C_-	β_- C_-	β_+,β_- C_+,C_-	β_- C_+	β_+,β_- C_+,C^M_-	β_- C^M_-	β_+,β_- C^M_+,C^M_-	β_- C^M_+

Summary of the existence conditions, with respect to the space and time dimensions of the space-time, for the matrices β_μ and $C^{(M)}_\sigma$ defined in the text.

Exercice XII : Show that on a Riemannian compact manifold whose scalar curvature is positive there is no solution for the massless Dirac equation $\gamma^a \nabla_a \psi = 0$ if the connexion is torsion-free[23]. (Hint : Integrate by part $\bar{\psi}\gamma^a \nabla_a \psi^b \nabla_b \psi = 0$).

VII. GRAVITATIONAL FIELD EQUATIONS

VII.1. Einstein equations

Einstein's general relativity describes the gravitational field by introducing a curved geometry on space-time. Since the equivalence principle asserts that, locally, one can always choose a Minkowskian frame : $g_{\mu\nu} = \eta_{\mu\nu}$ and $\Gamma_{\alpha\beta}{}^{\gamma} = 0$, the physical effects of the gravitational field will be reflected through second order deviation from flatness terms, that is through the curvature tensor. In order to exhibit the physical meaning of curvature we shall compare the equations of motion of a flow of non-interacting particles submitted to a gravitational field in both the framework of Newtonian mechanics and on a curved space-time.

Consider, on a Newtonian space-time, a flow of particles. In Cartesian coordinates the equations of their trajectories are :

$$x^{o} = \tau \qquad x^{i} = X^{i}(\tau,\xi^{j}) \qquad i,j = 1,\,2,\,3 \quad . \tag{7.1}$$

The first of these equations means that the time coordinate coincides with the proper time (after a suitable choice of the origin and units) for all the particles of the flow. This just reflects the absolute meaning of time in classical mechanics. The three parameters ξ^{j} label smoothly the different trajectories. They remain constant on each flow line (Lagrangian coordinates). The 4-velocity of these particles is given by :

$$V^{o} = 1 \qquad V^{i} \equiv \frac{\partial}{\partial\tau} X^{i}(\tau,\xi^{j}) \tag{7.2}$$

and their acceleration obeys Newton's second law :

$$A^{o} = 0 \qquad A^{i} \equiv \frac{\partial^{2}}{\partial\tau^{2}} X^{i}(\tau,\xi^{j}) = -\left.\frac{\partial\phi}{\partial x^{i}}\right|_{x^{i}=X^{i}(\tau,\xi^{j})} \tag{7.3}$$

where ϕ is the gravitational field.

When selecting one of the trajectories of the flow : $\xi^{j} = \xi^{j}_{o}$, the relative position of the neighbour-particles will be given by the vector Δ

$$\Delta^{o} = 0 \qquad \text{and} \qquad \Delta^{i} = \delta^{j} \left.\frac{\partial X^{j}}{\partial\xi^{j}}\right|_{\xi^{j}=\xi^{j}_{o}} \tag{7.4}$$

which connects points on neigbour-trajectories with the same value of τ. The relative velocity of these neighbour-particles is

$$\dot{\Delta}^{o} = 0 \qquad \text{and} \qquad \dot{\Delta}^{i} = \frac{\partial\Delta^{i}}{\partial\tau} = \delta^{j} \left.\frac{\partial^{2}}{\partial\tau\,\partial\xi^{j}}\right|_{\xi^{j}=\xi^{j}_{o}} \quad . \tag{7.5}$$

These equations are written in Lagrangian coordinates. We can express them in terms of Eulerian coordinates (x^o, x^i) by inverting eqs. (1). This leads to :

$$\tau = x^o \qquad \xi^i = \Xi^j(x^o, x^j) \tag{7.6a}$$

$$\Delta^o = 0 \qquad \Delta^i = \Delta^i(x^o, \Xi^j(x^o, x^k)) = D^i(x^o, x^j) \tag{7.6b}$$

$$V^o = 1 \qquad V^i = v^i(x^o, x^j) \tag{7.6c}$$

$$\dot{\Delta}^o = 0 \qquad \dot{\Delta}^i = \frac{d}{dx^o} D^i(x^o, x^j) = \Delta^j \frac{\partial v^i}{\partial x^j} \tag{7.6d}$$

$$\ddot{\Delta}^o = 0 \qquad \ddot{\Delta}^i = \Delta^j \frac{\partial}{\partial x^j} \frac{dv^i}{dx^o} = - \Delta^j \frac{\partial^2}{\partial x^j \, \partial x^i} \phi \tag{7.6e}$$

The key of these calculations relies upon the commutation of the partial derivatives with respect to τ and ξ^j. In other words we have used the commutativity of the velocity vector field $\frac{\partial}{\partial \tau}$ and the separation vector field $\delta^i \frac{\partial}{\partial \xi^i}$.

We shall now mimic this analysis in the framework of general relativity. A flow of noninteracting test particles is described by a familly of timelike geodesics. Their tangent vectors $\vec{u}$ represent the 4-velocities of the particles when normalized such that :

$$\nabla_{\vec{u}} \vec{u} = 0 \quad \text{and} \quad g(\vec{u}.\vec{u}) = \vec{u}.\vec{u} = -1 \quad . \tag{7.7}$$

The separation vector field $\vec{\Delta}$ of neighbour-curves is defined by :

$$[\vec{\Delta}, \vec{u}\,] = 0 \tag{7.8}$$

i.e., since the torsion is zero,

$$\nabla_{\vec{u}} \vec{\Delta} = \nabla_{\vec{\Delta}} \vec{u} \quad . \tag{7.9}$$

As a consequence, we obtain the *Jacobi equation for geodesic deviation* :

$$\begin{aligned} \nabla_{\vec{u}} \nabla_{\vec{u}} \vec{\Delta} &= \nabla_{\vec{u}} \nabla_{\vec{\Delta}} \vec{u} \\ &= R(\vec{u}, \vec{\Delta})\vec{u} + \nabla_{\vec{\Delta}} \nabla_{\vec{u}} \vec{u} + \nabla_{[\vec{u}, \vec{\Delta}\,]} \vec{u} \\ &= R(\vec{u}, \Delta)\vec{u} \end{aligned} \tag{7.10}$$

by deriving eq. (9) with respect to $\vec{u}$ and by making use of eq. (4.26) and eqs. (7,8) respectively.

Although this equation describes the relative separation of neigbour-geodesics, it cannot directly be compared with the Newtonian equation (6c). In the first scheme, the separation vector was characterized through $\Delta^o = 0$ with the spatial components expressed in an inertial frame. Now, in general relativity, such a global frame does not exist. Its analogue will be obtained by considering, along the reference trajectory, a field of non rotating orthonormal frames the timelike vector of which coincides everywhere with the 4-velocity vector tangent to the geodesic. Such a frame is built by parallel transporting along the geodesic three orthogonal spacelike vectors. They define a field of vierbein $\{\vec{e}_\alpha\}$ such that

$$\vec{e}_o = \vec{u} \quad , \quad \vec{u} \cdot \vec{e}_i = 0 \quad , \quad \vec{e}_i \cdot \vec{e}_j = \delta_{ij} \tag{7.11a}$$

and

$$\nabla_{\vec{u}} \vec{e}_i = 0 \quad . \tag{7.11b}$$

Moreover, the spatial separation vector defined by (4) lies in the 3-dimensional spacelike subspace. As a consequence we should not consider the vector $\vec{\Delta}$, used to obtain the Jacobi equation, but its projection $\vec{\Delta}_\perp$ onto the 3-space orthogonal to $\vec{u}$:

$$\vec{\Delta}_\perp = \vec{\Delta} + (\vec{u} \cdot \vec{\Delta})\vec{u} = \Delta^i_\perp \, \vec{e}_i \quad . \tag{7.12}$$

Its is quite easy to verify that the spatial relative velocity vector $\nabla_{\vec{u}} \vec{\Delta}_\perp$ obeys the relations :

$$\nabla_{\vec{u}} \vec{\Delta}_\perp = \dot{\Delta}^i_\perp \, \vec{e}_i = \nabla_{\vec{u}} \vec{\Delta} = \nabla_{\vec{\Delta}} \vec{u} = \nabla_{\vec{\Delta}_\perp} \vec{u} \tag{7.13}$$

and that the components of the spatial relative acceleration satisfy the equations :

$$\ddot{\Delta}^i_\perp = R^i{}_{ooj} \, \Delta^\delta_\perp \quad . \tag{7.14}$$

Here a dot indicates a derivation with respect to the proper time measured along the reference geodesic and in the last equation all indices refer to the vierbein (11). Comparing eq. (14) with its Newtonian version (6c) we see that, in the limit of weak fields,

$$R_{iooj} = \frac{\partial^2}{\partial x^i \, \partial x^j} \, \phi \quad . \tag{7.15}$$

The trace of this equation leads to

$$R_{oo} = \Delta\phi = 4\pi \, G\rho \quad , \tag{7.16a}$$

or in usual units ($x^o = ct$) :

$$R_{oo} = \frac{4\pi \, G}{c^2} \, \rho . \tag{7.16b}$$

where ρ is the mass density, source of the Newtonian gravitational field.

The *Einstein equations* that generalize this result are

$$E_{\mu\nu} = \kappa\, T_{\mu\nu} \quad . \tag{7.17}$$

The right-hand side of these equations is proportional to the energy-momentum tensor of all the matter fields present in the universe. Since this tensor is conserved : $\nabla_\mu T^{\mu\nu} = 0$, the geometrical tensor $E_{\mu\nu}$ must also be divergenceless. Cartan (again) has demonstrated that the most general second order tensor, built from the metric and the curvature tensors, and satisfying the divergence condition : $\nabla_\mu E^{\mu\nu} = 0$ is written as :

$$E_{\mu\nu} = R_{\mu\nu} - \frac{1}{2} g_{\mu\nu} R + \Lambda g_{\mu\nu} \tag{7.18}$$

up to an overall factor.

This tensor consists on two pieces :

$$G_{\mu\nu} = R_{\mu\nu} - \frac{1}{2} g_{\mu\nu} R \tag{7.19}$$

and a term proportional to the metric. The factor Λ is called the cosmological constant. Experimentally its absolute value is less then 10^{-56} cm^{-2}. Its physical meaning is quite far from being understood. It plays a crucial rôle in cosmological models and appears as a vacuum energy counterterm in some problems where quantized matter fields are coupled to gravity.

The coupling constant κ between matter and the gravitational field can be obtained by looking at the Newtonian limit of Einstein equations. Suppose that the gravitational field is generated by a low density perfect fluid. The energy-momentum tensor associed to this fluid is given by

$$T^\mu_\nu = (\rho c^2 + p) u^\mu u_\nu + p\, \delta^\mu_\nu \quad , \tag{7.20}$$

where u^μ denotes the 4-velocity of the fluid elements. By rewriting the Einstein equations (18) as

$$R_{\mu\nu} = \kappa\, [T_{\mu\nu} - \frac{1}{2} g_{\mu\nu} T^\alpha_\alpha] + \Lambda g_{\mu\nu} \tag{7.21}$$

we obtain, in the rest frame of the fluid,

$$\begin{aligned} R_{oo} &= \kappa\, [T_{oo} + \frac{1}{2} T^\alpha_\alpha] - \Lambda \\ &= \kappa\, [\rho c^2 - \frac{1}{2}(\rho c^2 - 3p)] - \Lambda \\ &\approx \frac{\kappa}{2} \rho c^2 \quad \text{(in the limit of a weak field)} \end{aligned} \tag{7.22}$$

from which we conclude that

$$\kappa = \frac{8\pi G}{c^4} = 2.076\ 10^{-48}\ \sec^2/\mathrm{g\ cm} \quad . \tag{7.23}$$

The first order corrections to Newton's gravitation predicted by Einstein equations have been experimentaly verified for a long time ago by the observations of the advance of the perihelions of Mercury, Venus, Icarus and the Earth[24,25,26] on the one hand and of the deflection of light rays by the Sun[27] on the other hand. More interresting are the observations of the binary pulsar PSR 1913+16 discovered in 1974[28]. This system presents an advance of the periapsis of 4.2 degree per year (*) and a slowing down of the orbital period[29] $(\frac{dp}{dt})_{obs} = (-2.3 \pm 0.2)10^{-12}$. This value is remarkably close to the value predicted by Einstein's gravitation theory : $(\frac{dp}{dt})_{Th} = -2.4\ 10^{-12}$. The great interest of this last experimental test of general relativity lies in the fact that the non linear structure of Einstein equations is to be taken into account, the slowing down of the orbital period can only be obtained theoretically[30] after an expansion which goes up with terms in c^{-5} .

VII.2. Variational principles

Einstein fields equations express the Ricci tensor, built from the Levi-Civita connection, in terms of the energy-momentum tensor. So, we are fronted with the problem of defining this last tensor. When the matter fields are described by a Lagrangian L_m which only depends on the fields their first covariant derivatives and the metric tensor, there exists a recipe which directly leads to a symmetric, divergenceless (modulo the field equations) energy-momentum tensor. It suffices to require that the action functional

$$S(\mathcal{D}) = "\int_{\mathcal{D}} \left[\frac{1}{2\kappa}(R - 2\Lambda) + L_m\right]\eta \tag{7.24}$$

remains stationary under arbitrary variations of the fields which vanish, with all their derivatives, at the boundary of $\mathcal{D}$. However this procedure is for from being the most general. First, there exists Lagrangians which requires intrinsically the use of vielbein instead of the metric tensor; secondly, nothing prevents us to examine what happens if the connection is considered as an independent field. (Actually, this last point of vue is almost always adopted in the framework of supergravity).

In order to obtain a better understanding of this problem

(*) For Mercury, the perihelic shift observed is of 5600 seconds per century, where 5557" are explained by the Newtonian theory and 43" is predicted by general relativity.

let us first compute the variation of the gravitational part of the action. Different choices of independent variables are conceivable. For instance, we may use the components $g_{\mu\nu}$ of the metric tensor and the connection coefficients $\Gamma^{\mu}_{\alpha\beta}$, or, the components of the veilbein e^{μ}_{a} and the spin connection coefficients $\omega^{a}{}_{b\mu}$. According to the one choice or the other, the gravitational action will be written as

$$S_G(g,\Gamma) = \frac{1}{2\kappa}\int_{\mathcal{D}} (g^{\alpha\beta}\, R^{\mu}{}_{\alpha\mu\beta}(\Gamma) - 2\Lambda)\ \eta \tag{7.25}$$

or

$$S_G(e,\omega) = \frac{1}{2\kappa}\int_{\mathcal{D}} (e^{\mu}_{a}\, e^{b\nu}\, R^{a}{}_{b\mu\nu}(\omega) - 2\Lambda)\ \eta \quad . \tag{7.26}$$

However, since the relation between $\omega^{a}{}_{b\mu}$ and $\Gamma^{\mu}{}_{\alpha\beta}$ is linear :

$$\omega^{a}{}_{b\mu} = e^{\nu}_{b}\ \Gamma^{\rho}{}_{\nu\mu}\ \theta^{a}_{\rho} + \theta^{a}_{\rho}\ \partial_{\mu}\ e^{\rho}_{b} \quad , \tag{7.27}$$

the Euler-Lagrange equations obtained by varying (25) or (26) with respect to the connection will be equivalent. On the other hand, since the variation of the fields considered are arbitrary, we can always restrict them to a domain $\mathcal{D}$ included in a single coordinate patch..

Let us compute the variational derivative δ_C of $S_G(g,\Gamma)$ with respect to $\Gamma^{\mu}_{\alpha\beta}$. The variation of the Ricci tensor is obtained directly from the expression of the curvature tensor :

$$R^{\mu}{}_{\alpha\nu\beta} = \partial_{\nu}\ \Gamma^{\mu}{}_{\alpha\beta} - \partial_{\beta}\ \Gamma^{\mu}{}_{\alpha\nu} + \Gamma^{\mu}{}_{\rho\nu}\ \Gamma^{\rho}{}_{\alpha\beta} - \Gamma^{\mu}{}_{\rho\beta}\ \Gamma^{\rho}{}_{\alpha\nu} \quad . \tag{7.28}$$

After a suitable regrouping of the various terms it can be written as

$$\delta_C\ R_{\alpha\beta} = \nabla_{\mu}\ \delta\ \Gamma^{\mu}{}_{\alpha\beta} - \nabla_{\beta}\ \delta\ \Gamma^{\mu}{}_{\alpha\mu} + T^{\mu}{}_{\nu\beta}\ \delta\ \Gamma^{\nu}{}_{\alpha\mu} \quad . \tag{7.29}$$

The variational derivative of the gravitational action is then obtained by putting (29) into (25) and integrating by part. The resulting integrant is :

$$g^{\alpha\beta}\ \delta_C\ R_{\alpha\beta} = \nabla_{\mu}(2g^{\alpha[\beta}\ \delta^{\mu]}_{\nu}\ \delta\ \Gamma^{\nu}{}_{\alpha\beta}) + (T_{\nu\sigma}\ g^{\sigma\alpha} - \nabla_{\mu}2g^{\alpha[\beta}\delta^{\mu]}_{\nu})\delta\ \Gamma^{\nu}{}_{\alpha\beta} \tag{7.30}$$

The first term in the right-hand side of this expression is the covariant divergence of a vector. However, when an arbitrary connection is used, we cannot just drop it by use of Stokes theorem. In such a general case we have for an arbitrary vector V^{μ} :

$$\begin{aligned} \nabla_{\mu}\ V^{\mu}.\sqrt{|g|} &= (\partial_{\mu}\ V^{\mu} + \Gamma^{\mu}{}_{\nu\mu}\ V^{\nu})\sqrt{|g|} \\ &= \partial_{\mu}(V^{\mu}\sqrt{|g|}) + (\Gamma^{\mu}{}_{\nu\mu}\ \sqrt{|g|} - \partial_{\nu}\ \sqrt{|g|}\ )\ V^{\nu} \end{aligned}$$

i.e. if V^μ vanishes at the boundary of $\mathcal{D}$ we obtain :

$$\int_{\mathcal{D}} \nabla_\mu V^\mu \sqrt{|g|}\, d^n x = \int_{\mathcal{D}} (\Gamma^\mu{}_{\nu\mu} \sqrt{|g|} - \partial_\nu \sqrt{|g|}\,)\, V^\nu\, d^n x \,. \tag{7.31}$$

The integrant in the last integral can be expressed in an explicit tensorial form by considering the covariant derivative of the metric tensor :

$$\nabla_\nu g_{\mu\sigma} = \partial_\nu g_{\mu\sigma} - \Gamma^\rho{}_{\mu\nu} g_{\rho\sigma} - \Gamma^\rho{}_{\sigma\nu} g_{\nu\rho} \qquad .$$

Contraction of this relation with $g^{\mu\sigma}$ gives us :

$$\frac{1}{2} g^{\mu\sigma} \nabla_\nu g_{\mu\sigma} = \partial_\nu \ln \sqrt{|g|} - \Gamma^\rho{}_{\rho\nu} \tag{7.32}$$

and as a consequence, we obtain finally :

$$\int_{\mathcal{D}} \nabla_\mu V^\nu \eta = \int_{\mathcal{D}} (T^\mu{}_{\mu\nu} - \frac{1}{2} g^{\mu\sigma} \nabla_{\mu\sigma}) V^\nu \eta \quad + \text{Boundary terms} \quad . \tag{7.33}$$

This is a "covariant version" of Stokes formula, valid on an arbitrary domain $\mathcal{D}$. Thanks to this formula we can factorize the variation of the connection in the variation of the gravitational action as

$$\delta_C S_G = \frac{1}{2\kappa} \int [\, (T^\beta{}_{\nu\sigma} g^{\sigma\alpha} + 2T^\rho{}_{\rho\mu} 2g^{\alpha\,[\,\beta} \delta^{\mu\,]}{}_\nu$$
$$-(2 \nabla_\mu g^{\alpha\,[\,\beta} \delta^{\mu\,]}{}_\nu + g^{\sigma\rho} g^{\alpha\,[\,\beta} \delta^{\mu\,]}{}_\nu \nabla_\mu g_{\sigma\rho})]\, \delta\, \Gamma^\nu{}_{\alpha\beta}\, \eta \,. \tag{7.34}$$

Following Lichnerowicz[31], we can simplify this expression by introducing the tensor (in n dimension) :

$$K^{\alpha\beta}{}_\nu = \nabla_\nu g^{\alpha\beta} + \frac{1}{2} g^{\alpha\beta} g^{\rho\sigma} \nabla_\nu g_{\rho\sigma} + g^{\alpha\sigma} T^\beta{}_{\sigma\nu} - g^{\alpha\beta} T^\rho{}_{\rho\nu} - \frac{1}{n-1} \delta^\beta{}_\nu g^{\sigma\beta} T^\rho{}_{\rho\sigma} \tag{7.35}$$

which leads to

$$\delta_C S_G = \frac{1}{2\kappa} \int_{\mathcal{D}} (\delta^\beta{}_\nu K^{\alpha\rho}{}_\rho - K^{\alpha\beta}{}_\nu)\, \delta\, \Gamma^\nu{}_{\alpha\beta}\, \eta \qquad . \tag{7.36}$$

As a consequence, the Euler-Lagrange equations resulting from the variation of the total action with respect to the connection are

$$K^{\alpha\beta}{}_\nu - \delta^\beta{}_\nu K^{\alpha\rho}{}_\rho = 2\kappa \frac{\delta L_m}{\delta \Gamma^\nu{}_{\alpha\beta}} \tag{7.37}$$

or in terms of spin coefficients and vielbein

$$K^{\alpha\beta}{}_{\nu} - \delta^{\beta}{}_{\nu} K^{\alpha\rho}{}_{\rho} = \kappa\, 2 \frac{\delta L_m}{\delta\omega^{a}{}_{b\beta}} \theta^{a}{}_{\nu} e_{b}^{\alpha} \quad . \tag{7.38}$$

In the case of pure gravity these equations reduce to $K^{\alpha\beta}{}_{\nu} = 0$. Without any extra assumption the general solution of this equation can be written as

$$\Gamma^{\alpha}{}_{\beta\gamma} = \frac{1}{2} g^{\alpha\sigma}(g_{\sigma\gamma,\beta} + g_{\sigma\beta,\gamma} - g_{\beta\gamma,\sigma}) + \pi^{\alpha}{}_{\beta\gamma} + \delta^{\alpha}{}_{\beta} V_{\gamma} \tag{7.39}$$

where the tensor $\pi^{\alpha}{}_{\beta\gamma}$ is compelled to satisfy the equations :

$$\pi_{\alpha\beta\gamma} + \pi_{\beta\gamma\alpha} = 0 \tag{7.40}$$

$$\pi^{\alpha}{}_{\beta\alpha} - \pi^{\alpha}{}_{\alpha\beta} = 0 \tag{7.41}$$

and V_{γ} is an arbitrary vector defining the trace of the torsion tensor :

$$V_{\gamma} = \frac{1}{n-1} T^{\alpha}{}_{\gamma\alpha} \quad . \tag{7.41}$$

However when we ask the connection to be torsion free (Palatini's assumption) or the connection to be metric (Weyl's assumption) then, the solution of the equation $K^{\alpha\beta}{}_{\nu} = 0$ reduce to the usual Levi-Civita connection ($\pi^{\alpha}{}_{\beta\gamma} = 0$ and $V_{\gamma} = 0$).

The Euler-Lagrange equations resulting from the variation with respect to the metric variables are not equivalent in all cases. If we adopt for the gravitational action the expression (25) where the volume element is :

$$\eta = \sqrt{|g|}\, dx^{1} \quad \ldots \quad dx^{n} \quad \text{with} \quad g = \det g_{\mu\nu} \, , \tag{7.43}$$

we obtain as field equations by varying with respect to $g^{\alpha\beta}$

$$R_{(\alpha\beta)}(\Gamma) - \frac{1}{2} g_{\alpha\beta} R(g,\Gamma) + \Lambda g_{\alpha\beta} = -2\kappa \frac{\delta L_m}{\delta g^{\alpha\beta}} \quad . \tag{7.44}$$

On the other hand the gravitational (26) where the volume element is expressed as a function of the determinant e of the vielbein θ^{a}_{μ} :

$$\eta = e\, dx^{1} \wedge \ldots \wedge dx^{n} \quad \text{with} \quad e = \det \theta^{a}_{\mu} \tag{7.45}$$

leads, after variation with respect to e^{μ}_{a}, to the field equations :

$$R^{a}{}_{\mu}(e,\omega) + \dot{K}^{a}{}_{\mu}(e,\omega) - e^{a}_{\mu} R(e,\omega) + 2e^{a}_{\mu} \Lambda = -2\kappa \frac{\delta L_m}{\delta e_{a}{}^{\mu}} \tag{7.46}$$

where

$$R^a{}_\mu(e,\omega) = \eta^{ab}\, e_c{}^\nu\, R^c{}_{b\nu\mu}(\omega) \text{ and } \dot{R}^a{}_\mu(e,\omega) = \eta^{cb}\, e_c{}^\nu\, R^a{}_{b\mu\nu}(e,\omega) \quad . \tag{7.47}$$

Notice that in general the tensor

$$G_{\mu\nu}(e,\omega) = \frac{1}{2}\, e_{a\mu}(R^a{}_\nu(e,\omega) + \dot{R}^a{}_\nu(e,\omega)) - \frac{1}{2}\, g_{\mu\nu}\, R(e,\omega) \tag{7.48}$$

is non-symmetric. It reduces to

$$G_{\alpha\beta}(g,\Gamma) = R_{\alpha\beta}(g,\Gamma) - \frac{1}{2}\, g_{\alpha\beta}\, R(g,\Gamma) \tag{7.49}$$

when $R_{ab\mu\nu} = - R_{ba\mu\nu}$, e.g. for metrical connections, and is symmetric only when the torsion vanishes.

We can also assume, from the begining, that the connection is metric and torsion-free. For such a choice the tensor $K^{\alpha\nu}{}_\nu$ identically vanishes so that the left-hand side of the fields equations will not change but the right-hand side now must include a term induced by the variation of the metric variables via the connection. This last term is easily obtained by use of the relations :

$$\delta\, \Gamma^\rho{}_{\nu\mu} = \frac{1}{2}\, g^{\rho\sigma}((\delta g_{\sigma\mu})_{;\nu} + (\delta g_{\nu\sigma})_{;\mu} - (\delta g_{\nu\mu})_{;\sigma}) \tag{7.50}$$

which combined with eq. (27) leads to

$$\delta\omega_{ab\mu} = (e_{[a}{}^\rho\, e_{b]}{}^\sigma\, \delta_\mu{}^\nu + e_{[a}{}^\sigma\, e_{b]}{}^\nu\, \delta_\mu{}^\rho + e_{[a}{}^\rho\, e_{b]}{}^\nu\, \delta_\mu{}^\sigma)(e^k_\sigma\, \delta e_{k\rho})_{;\nu} \tag{7.51}$$

These considerations lead us to define different matter fields tensors[32,33]. Suppose we have a matter lagrangian L_m depending on the vielbein $e^a{}_\mu$, the spin connection coefficients $\omega_{ab\nu} = \omega_{[ab]\nu}$ and matter fields ψ :

$$L_m = L(e,\omega,\psi) \quad .$$

The *canonical energy momentum tensor* is given by the variational derivative of L_m with respect to the vielbein leaving the connection fixed :

$$_cT^{\mu\nu} = \frac{\delta e\, L_m(e,\omega,\psi)}{\delta\, e^a_\nu}\; e^{a\mu} \quad . \tag{7.52}$$

The *spin tensor* is obtained by varying the matter lagrangian with respect to the spin connection :

$$\frac{1}{2}\, S^{[\mu\nu]\sigma} = \frac{\delta\, L_m(e,\omega,\psi)}{\delta\, \omega_{ab\sigma}}\; e^\mu_a\; e^\nu_b \tag{7.53}$$

while the *symmetric energy-momentum tensor* is defined through a full variation with respect to the vielbein :

$$ {}_sT^{\mu\nu} = \frac{1}{e}\frac{\delta\, e\, L_m(e,\omega(e),\psi)}{\delta\, e^a_\nu}\; e^{a\mu} $$

$$ = {}_cT^{\mu\nu} + \frac{1}{2}\nabla_\rho(S^{[\mu\nu]\rho} + S^{[\rho\nu]\mu} + S^{[\rho\mu]\nu}) \qquad (7.54) $$

Using standard arguments of calculus of variation, it can be shown that if L_m is Lorentz invariant with respect to the vielbein indices then ${}_sT^{\mu\nu m}$ is symmetric

$$ {}_sT^{\mu\nu} = {}_sT^{\mu\nu} \qquad (7.55) $$

modulo the field equations. If the matter lagrangian is a scalar with respect to the coordinate transformations then this tensor is also divergenceless

$$ \nabla_\nu\; {}_sT^{\mu\nu} = 0 \qquad (7.56) $$

modulo the field equations. These two properties result immediately from the invariance of the matter action

$$ \delta S_m = 0 = \int \left(\frac{\delta e\, L_m}{\delta\psi}\,\delta\psi + \frac{\delta e\, L_m}{\delta e^a_\mu}\,\delta e^a_\mu\right)\frac{1}{e}\,\eta \qquad (7.57) $$

with respect to arbitrary rotations of the vielbein :

$$ \delta e^a_\mu = \Omega^a{}_b\, e^b_\mu \qquad \text{with} \qquad \Omega_{ab} = -\Omega_{ba} \qquad (7.58) $$

and arbitrary local diffeomorphisms written as

$$ \delta e^a_\mu = \pounds_{\vec{\xi}}\, e^a_\mu \qquad (7.59) $$

where $\vec{\xi}$ is an arbitrary vector field.

From eq. (54) we see immediately that the canonical energy momentum tensor is in general not symmetric :

$$ {}_cT^{[\mu\nu]} + \frac{1}{2}\nabla_\rho S^{[\mu\nu]\rho} = 0 \qquad . \qquad (7.60) $$

This relation is just a local version of the total angular momentum conservation law, well known in flat space. The non conservation of this tensor results from eqs. (54,56) and the non commutativity of covariant derivatives :

$$ \nabla_\nu\; {}_cT^{\mu\nu} + \frac{1}{2} R^\mu{}_{\alpha\beta\gamma}\, S^{\alpha\gamma\beta} = 0 \qquad . \qquad (7.61) $$

We must emphasize that these results are only valid for matter lagrangians which contain at most first order derivative of the vielbien. For a more general analysis we refer the reader to the work of J. Géhéniau[34].

VII.3. An example : Weyl's theory[35]

Typically the use of vielbein is unavoidable for fermionic lagrangians. In this section we shall illustrate the previous analysis by studying briefly a model of Einstein-Dirac lagrangian in four dimensions.

The matter lagragian of spin $\frac{1}{2}$ field can be written as

$$L_{\frac{1}{2}} = \frac{1}{2}(\bar{\psi}\,\gamma^a\,\nabla_a\,\psi - \nabla_a\,\bar{\psi}\,\gamma^a\psi) - m\,\bar{\psi}\,\psi \tag{7.62}$$

or equivalently

$$L_{\frac{1}{2}} = \frac{1}{2}(\bar{\psi}\,\gamma^a\,e^\mu_a\,\partial_\mu\,\psi - e^\mu_a\,\partial_\mu\,\bar{\psi}.\gamma^a\psi - \frac{1}{8}\,e^\mu_a\,\omega^b{}_{c\mu}\,\bar{\psi}\{\gamma^a,[\gamma^c,\gamma_b]\}\psi) - m\,\bar{\psi}\,\psi \quad . \tag{7.63}$$

The canonical energy momentum tensor defined by this lagrangian is

$${}_cT_a{}^\nu = -\frac{1}{2}\,e_a^\rho\,e_b{}^\nu\,(\bar{\psi}\,\gamma^b\,\nabla_\rho\,\psi - \nabla_\rho\,\bar{\psi}\,\gamma^b\,\psi)$$

i.e.

$${}_cT^{\mu\nu} = -\frac{1}{2}\,(\bar{\psi}\,\gamma^\nu\,\nabla^\mu\,\psi - \nabla^\mu\,\bar{\psi}\,\gamma^\nu\,\psi) \tag{7.64}$$

which is obviously not symmetric (*). The spin tensor is, given by :

$$S^{[\mu\nu]\sigma} = \frac{1}{8}\,\bar{\psi}\,\{\gamma^a,[\gamma^b,\gamma^c]\}\psi\,e^\sigma_a\,e^\mu_b\,e^\nu_c \quad . \tag{7.65}$$

When the connection is supposed to be metric the equations (38) become algebraic. Their solution gives the torsion :

$$T^\alpha{}_{\beta\gamma} = \kappa\,S^\alpha{}_{[\gamma\beta]} \tag{7.66}$$

and as a consequence, thanks to eqs. (4-8, 4-42), we obtain the connection

$$\omega_{ab,\mu} = \overset{\circ}{\omega}_{ab,\mu} + \frac{\kappa}{4}\bar{\psi}\,\gamma_{[a}\,\gamma_b\,\gamma_{c]}\,\psi\,e^c_\mu \tag{7.67}$$

when $\overset{\circ}{\omega}_{ab,\mu}$ denotes the Levi-Civita connection coefficients.

If we introduce this expression for the connection in the Einstein equations we see that they induces quartic spinor field terms. In the same way this connection leads to a non-linear cubic term in the Dirac equations.

(*) Notice that we have used the matter field equations that imply $L_{1/2} = 0$.

As the equations of motion for the connection can be solved algebraically, we can substitute directly their solutions into the action and just consider as independent fields the vielbein and the spinor field. The resulting action becomes then :

$$S = \int_{\mathcal{D}} \{\frac{1}{2\kappa} [R(e,\overset{\circ}{\omega}(e))-2\Lambda] + [\frac{1}{2}(\overline{\psi}\, \gamma^a \overset{\circ}{\nabla}_a \psi - \overset{\circ}{\nabla}_a \overline{\psi}\, \gamma^a\psi)-m\, \overline{\psi}\, \psi] \}\, \eta$$

$$+ \int_{\mathcal{D}} \frac{\kappa}{32} \{\overline{\psi}\, \gamma_{[a}\gamma_b\gamma_{c]}\, \psi\, \psi\, \gamma^{[a}\gamma^b\gamma^{c]}\, \psi\}\eta \qquad (7.68)$$

where all covariant derivatives and connection coefficients refer to the Levi-Civita connection. This action differs from the usual Einstein-Dirac action through a quartic fermionic term the contribution of which can in general be shown negligible excepted in presence of a very huge matter density[36].

EXERCICE XIII: Show that if the Lagrangian $L(y^A, y^A_{,\mu}; z^B, z^B_{,\mu})$ depending on the fields y^A and z^B, in such that the Euler-Lagrange equations $\frac{\delta eL}{\delta z^B} = 0$ admit for their general solution

$$z^B = Z^B(y^A, y^A_{,\mu})$$

then the variational problem

$$\frac{\delta eL}{\delta z^B} = 0 \quad , \quad \frac{\delta eL}{\delta y^A} = 0$$

is equivalent to the problem $\frac{\delta e\overline{L}}{\delta y^A} = 0$ where the new lagrangian $\overline{L}(y, y^A_{,\mu})$ is obtained from L by substituting for the field variable z^B, its expression in terms of y^A and $y^A_{,\mu}$

VII.3. Constants for motion

The construction of first integrals of the equations of motion plays a crucial rôle as well in the classical theory as in the quantum theory. In this section we shall examine how to construct them on a curved space[37].

Assuming that the dynamics of a system can be described from a Lagrangian invariant under the action of some symmetry group, Noether's theorem associates a constant of motion to each generator of the group. As we are meanly interested in invariances with geometrical significance, we shall first suppose that the space-time, we are working on, admits a Killing vector $\vec{\xi}$. Once such a vector exists, we can built a first integral for the equation of motion of a charged particle :

$$\nabla_{\vec{u}}\, u_\alpha = \frac{e}{m} F_{\alpha\beta}\, u^\beta \quad , \quad u^\alpha u_\alpha = -1 \tag{7.69}$$

as $$\xi^\alpha (u_\alpha - \frac{e}{m} A_\alpha) \tag{7.70}$$

if the potential A is invariant with respect to the isometries generated by $\vec{\xi}$ i.e. :

$$\frac{e}{m} \pounds_{\vec{\xi}}\, A_\alpha = 0 \quad . \tag{7.71}$$

Since the exterior derivative commutes with the Lie derivative, this condition implies that the electromagnetic field also is invariant :

$$\frac{e}{m} \pounds_{\vec{\xi}}\, F_{\alpha\beta} = 0 \quad . \tag{7.72}$$

Conversely, if this condition is satisfied we can always locally find a gauge such that the potential obeys eq. (71). This result is the generalization to curved spaces of the well known conservation laws of energy, momentum and angular momentum formulated on flat space. In the framework of classical field theory, Killing vectors allow to construct a divergenceless current from the conserved symmetric energy-momentum tensor as :

$$J^\alpha = {}_S T^{\alpha\beta}\, \xi_\beta \quad , \quad \nabla_\alpha J^\alpha = 0 \quad . \tag{7.73}$$

When the fields vanish quickly enough at infinity, the flux of this current accross a spacelike surface Σ is conserved and defines a constant of motion :

$$K_\xi = \int_\Sigma {}_S T^{\alpha\beta}\, \xi_\beta\, d\sigma_\alpha \quad . \tag{7.74}$$

In first quantization theory, constants of motion are defined through differential operators which commute with the wave equation (representing the (super)-hamiltonian operator). It is not difficult to check that the most general first order operator $i\, \xi^\alpha \nabla_\alpha$ which commutes with the Klein-Gordon operator

$$(\hbar \nabla_\alpha + ie A_\alpha) g^{\alpha\beta} (\hbar \nabla_\beta + ie A_\beta) + m^2 \tag{7.75}$$

is just as previously defined by a Killing vector $\vec{\xi}$ such that eq. (71) is still satisfied.

In second quantization theory, standard arguments[18] show that space-time invariances are implemented on the space of states by an unitary operator, defined as the exponential of a self-adjoint generator Ξ . The commutation rules of this generator with the field operator $\hat{\phi}$ is expressed in terms of Lie derivative as :

$$[\, \Xi , \hat{\phi}\,] = -i\, \pounds_{\vec{\xi}}\, \hat{\phi} \quad . \tag{7.76}$$

Moreover if the fields are quantized via the canonical commutation rules of bosonic fields, anticommutation rules of fermionic fields, we can, at least formally, represent the generator Ξ by an operator $\hat{K}$ obtained by substituting a quantum field operator to the classical field in eq. (74). Without any normal product rule the resulting operator is ill-defined. For free fields it is defined up to an additive (infinite)constant having no influence on the commutation relation (76) which remains satisfied. As a consequence this operator $\hat{K}$ could be identified with Ξ if we assume the existence of an invariant vacuum state $|0>$ such that

$$\Xi\ |0> = \hat{K}\ |0> = 0 \quad . \tag{7.77}$$

Unfortunatelly the definition of such a state is far from being obvious and constitutes the main difficulty of a quantum field theory on curved background.

More general constants of motion can be defined. For instance we may consider homogeneous constants of the form :

$$S = \Sigma_{\alpha_1 \ldots \alpha_n} u^{\alpha_1} \ldots u^{\alpha_n} \quad . \tag{7.78}$$

By expressing the conservation of S along any integral curve of eqs. (69), we obtain the conditions

$$\Sigma_{(\alpha_1 \ldots \alpha_n ; \alpha_o)} = 0 \quad , \tag{7.79}$$

$$\frac{e}{m} \Sigma_{\alpha_o}{}^{(\alpha_1 \ldots \alpha_{n-1}} F^{\alpha_n)\alpha_o} = 0 \quad . \tag{7.80}$$

Tensors satisfying conditions (79) are called *Stackel tensors*. A well known example of second order Stackel tensor is given by the square of the angular momentum vector on spherically symmetric spaces. A less trivial example has been obtained by in the framework of the Kerr solution[38]. Restricting ourself to second order constants of motion, we can examine the more general non homogeneous quadratic form :

$$S = \Sigma_{\alpha\beta}\, u^{\alpha_1} u^{\alpha_2} + \Sigma_\alpha u^\alpha + \Sigma \quad .$$

This will define a constant of motion if and only if $\Sigma_{\alpha\beta}$ is a Stackel tensor and Σ_α a vector such that

$$2\frac{e}{m} \Sigma_\alpha{}^{(\beta} F^{\gamma)\alpha} = \Sigma^{(\beta;\gamma)} \quad , \tag{7.81}$$

$$\frac{e}{m} F_{\alpha\beta} \Sigma^\beta = \Sigma_{,\alpha} \quad . \tag{7.82}$$

The analogue of this constant of motion in quantum mechanics will be a second order differential operator

$$S = (\hbar \nabla_\alpha + ie A_\alpha)\Sigma^{\alpha\beta}(\hbar \nabla_\beta + ie A_\beta)$$

$$+ \frac{1}{2}[\Sigma^\alpha (\nabla_\alpha + ie A_\alpha) + (\nabla_\alpha + ie A_\alpha)\Sigma^\alpha]$$

$$+ \Sigma \qquad (7.83)$$

commuting with the Klein-Gordon operator. This implies the following necessary and sufficient conditions :

$$\Sigma_{(\alpha \beta ; \gamma)} \qquad (7.84)$$

$$2e \Sigma_\alpha{}^{(\beta} F^{\gamma)\alpha} = \Sigma^{(\beta;\gamma)} \qquad (7.85)$$

$$e \Sigma^\beta F_{\beta\alpha} + \frac{2}{3} \hbar^2 \{\Sigma_{[\alpha} R_{\beta]}{}^\gamma\}^{;\beta} = \Sigma_{,\alpha} \qquad (7.86)$$

$$\hbar^2 \{e \Sigma_\alpha{}^\beta F^{\alpha\gamma}{}_{;\gamma} - \frac{1}{2} \Sigma^\beta R\}_{;\beta} = 0 \quad . \qquad (7.87)$$

We must emphasize that even for an homogeneous second order differential operator ($\Sigma^\alpha = 0$, $\Sigma = 0$), in absence of electromagnetic interaction ($A = 0$), the commutation conditions (84,86) are still stronger than the classical ones. Second order tensors satisfying these stronger conditions are called *Killing tensors*.
We can also determine for the Dirac operator :

$$\gamma^\alpha(\nabla_\alpha + ie A_\alpha) \qquad (7.88)$$

the most general first order differential operator which commutes with it[39]. In four dimensions, the general solution of this problem is given by :

$$K = F^\alpha \nabla_\alpha + G \qquad (7.89)$$

where $F^\alpha = B^\alpha \mathbb{1} + C \gamma^\alpha + D^{\alpha\beta} \gamma_S \gamma_\beta + E^{\alpha\beta\gamma} \gamma_{[\beta} \gamma_{\gamma]}$ (7.90)

and $G = \Phi \mathbb{1} - \frac{3}{4} \star E^\alpha{}_{;\alpha} \gamma_S + \frac{1}{3} \star D^\alpha{}_{\beta;\alpha} \gamma^\beta - \frac{1}{4} B_{\beta;\gamma} \gamma^{[\beta} \gamma^{\gamma]}$. (7.91)

Each tensor appearing in these expressions is antisymmetric :

$$D^{\alpha\beta} = D^{[\alpha\beta]} \qquad (7.92)$$

$$E^{\alpha\beta\gamma} = E^{[\alpha\beta\gamma]} \quad . \qquad (7.93)$$

They must satisfy the necessary and sufficient conditions :

$$C_{,\alpha} = 0 \quad , \tag{7.94}$$

$$B_{(\alpha;\beta)} = 0 \quad , \tag{7.95}$$

$$D_{\alpha(\beta;\gamma)} = 0 \quad , \tag{7.96}$$

$$E_{\alpha\beta(\gamma;\delta)} = 0 \quad , \tag{7.97}$$

$$e\, F_{\alpha[\beta} D^{\alpha}{}_{\gamma]} = 0 \quad , \tag{7.98}$$

$$e\, F_{\alpha[\beta} E^{\alpha}{}_{\gamma\delta]} = 0 \quad , \tag{7.99}$$

$$e\, \pounds_{\vec{B}} \vec{A} = d\phi \quad . \tag{7.100}$$

Antisymmetric tensors $Y_{\alpha_1 \ldots \alpha_n}$ such that

$$Y_{\alpha_o \ldots (\alpha_p;\beta)} = 0 \tag{7.101}$$

have already been considered by Yano who has shown that they are the only ones such that the tensors

$$k_{\alpha_1 \ldots \alpha_p} = Y_{\alpha_o \ldots \alpha_p} u^{\alpha_o} \tag{7.102}$$

are geodesically parallel along any geodesic whose tangent vector $\vec{u}$ is defined by an affine parameter :

$$\nabla_{\vec{u}} \vec{u} = 0 \quad . \tag{7.103}$$

The original motivation for the study of constants of motion was the unexpected separability properties shared by the Hamilton-Jacobi[37], Klein-Gordon[40], Dirac[41,42,43], etc... equations on Kerr-Newman metric which are related to the existence of Killing and Yano tensors[44] on this space. Nevertheless, in the framework of an arbitrary spacetime, the connection between the existence of such tensors and the separability of the wave equations seems not yet to be completely elucidated.

Finally I would also mention that for the special case of massless particles and fields, conformal generalizations of the previous tensors permit to obtain extra constants of motion[45].

VIII. HOMOGENEOUS COSMOLOGIES

VIII.1. Canonical metrics

Numerous exact solutions of Einstein equations are known today[46]. Among them two classes have been extensively studied : the stationary axisymmetric solutions and the spatially homogeneous ones. The former are related to the theory of black-holes[37,38] while the latter are supposed to describe the large scale behaviour of the universe[47].

In the standard four-dimensional cosmological models, space-time is supposed to be not only spatially homogeneous but also isotropic. This assumption is well supported on a large scale by present day observations, but it is not clear that such a geometry stays adequate close to the initial singularity. This problem has motived a lot of work on models with few symmetries, and in particular on spatially homogeneous but anisotropic cosmological models[47,48].

A spatially homogeneous cosmology is a space-time model which admits, at least, a simply-transitive group of isometries G_3 the orbits of which are 3-dimensional spacelike surfaces. In the following we shall assume that we have chosen a coordinate system in which the equations of the transitivity surfaces simply reduces to $x^o = C^{te}$. The space-time metric will be given by :

$$ds^2 = -N^2\, dx^o \otimes dx^o + g_{k\ell}(N^k\, dx^o + dx^k) \otimes (N^\ell\, dx^o + dx^\ell) \quad (k,\ell = 1,\ 2,\ 3) \tag{8.1}$$

and, in these coordinates, the time components of each Killing vector $\vec{\xi}_{(i)}$ generating G_3 will vanish. The existence of these Killing vectors impose restrictions on the metric components. From the Killing equations (5.19) and the coordinate condition $\xi^o = 0$ we obtain three sets of equations

$$\pounds_{\vec{\xi}_{(i)}}\, g_{k\ell} \equiv \xi^j_{(i)}\, \partial_j\, g_{k\ell} + g_{j\ell}\, \partial_k\, \xi^j_{(i)} + g_{kj}\, \partial_\ell\, \xi^j_{(i)} = 0 \tag{8.2a}$$

$$\pounds_{\vec{\xi}_{(i)}}\, g_{ko} \equiv \xi^j_{(i)}\, \partial_j\, g_{ko} + g_{jo}\, \partial_k\, \xi^j_{(i)} + g_{kj}\, \partial_o\, \xi^j_{(i)} = 0 \tag{8.2b}$$

$$\pounds_{\vec{\xi}_{(i)}}\, g_{oo} \equiv \xi^j_{(i)}\, \partial_j\, g_{oo} + 2g_{oj}\, \partial_o\, \xi^j_{(i)} = 0 \quad . \tag{8.2c}$$

The first of these equation tell us that the induced metric on the transitivity surfaces is invariant. The second one can be written as

$${}^{(3)}\pounds_{\vec{\xi}_{(i)}}\, N_k + N_\ell\, \partial_o\, \xi^\ell_{(i)} = 0 \tag{8.3}$$

with $N_\ell = g_{o\ell} = g_{\ell k}\, N^k$ and, as a consequence, the third equation becomes :

$$\xi^j_{(i)}\, \partial_j\, N^2 = 0 \quad . \tag{8.4}$$

Since we have supposed these equations to be satisfied for three independent Killing vectors $\vec{\xi}_{(i)}$, there results that the quantity $-N^2 = g_{oo}$ only depends on the coordinate x^o. Moreover, the spatial coordinates x^k can be chosen such that the components of $\xi^\ell_{(i)}$ become x^o-independent. This is achieved by choosing a curve C which crosses each surface of homogeneity only once and by assigning the same spatial coordinates to all the points obtained by moving the curve by G_3 isometrics. With this choice of coordinates eq. (3) simplifies to

$$^{(3)}\pounds_{\vec{\xi}_{(i)}} N_k = 0 \tag{8.5}$$

i.e. N_k is given by arbitrary, x^o-dependent, linear combinations of components of G_3 invariant 1-forms θ^a.
Now the metric element can be rewritten as :

$$ds^2 = -N^2(x^o)dx^o \otimes dx^o + \gamma_{k\ell}(x^o)(\nu^k(x^o)dx^o + \theta^k)\otimes(\nu^\ell(x^o)dx^o + \theta^\ell). \tag{8.6}$$

This expression can further be simplified. Indeed, there still remains a large freedom in the choice of the curve C which defines the spatial coordinates. If C is choosen orthogonal to the surfaces of homogeneity, the non-diagonal terms N_k of the metric vanish along C and, as a consequence, they are zero everywhere. Finally the time coordinate x^o can be rescaled as $dt = N(x^o)dx^o$; we obtain, in this way, a canonical expression of the metric :

$$ds^2 = -dt \otimes dt + \gamma_{k\ell}(t)\underline{\theta}^k \otimes \underline{\theta}^\ell \quad . \tag{8.7}$$

However, for practical purpose, it is sometimes more easy to work with the expression (6) for the metric and to fixe the arbitrary functions N, ν^k so as to simplify (for instance) the integration of Einstein equations. On the other hand even if we assume, from the begining, that the homogeneity surfaces are spacelike, the dynamical equations may be such that during their evolution they become null on timelike. When such a phenomenon (which is quite common) appears the expression (7) of the metric will present a (coordinate) singularity which can be suppressed by using (6). This is quite analogous to the situation encountered with the Schwarzschild metric where the "spacelike coordinate" r becomes timelike when $r < 2m$. The singularity of the metric is just a coordinate singularity which can be eliminate for instance by the use of Eddington-Kruskal null coordinates.

Exercice XIV: Show by using the expression (6) of the metric that the surface of homogeneity are geodesically parallel when they are non-null. (Hint : This exercice is suggested just in order to give you practice with the geodesic equation. The answer is obvious from expression 7).

VIII.2. Bianchi classification

Spatially homogeneous cosmological models are caracterized by a transitive 3-dimensional group of isometries. This assumption insures the existence of three Killing vectors ξ^{α}_{A} such that the rank of the matrix (ξ^{α}_{A}) is equal to 3. These vectors obey the commutation relations

$$[\vec{\xi}_A, \vec{\xi}_B] = C^C_{AB}\,\vec{\xi}_C \quad , \qquad (8.8)$$

$$C^C_{AB} = -\,C^C_{BA} \qquad (8.9)$$

where the structure constants satisfy the Jacobi identity :

$$C^C_{[AB}\,C^E_{D]F} = 0 \quad . \qquad (8.10)$$

According to Lie's third theorem, any set of constants satisfying these last equations defines a unique simply-connected Lie group. As a consequence the study of spatially homogeneous cosmologics splits into three parts. First, we must enumerate all possible non isomorphic 3-dimensional Lie algebras. Secondly, we have to integrate eqs. (8) and to determine an invariant basis. Finally we must, eventually, integrate Einstein equations which in this case reduce to a set of coupled differential equations for one variable only instead of partial differential equations.

The first classification of all real 3-dimensional Lie algebras has been made by Bianchi[49] in 1897. Later Behr et al.[50] and Ellis and MacCallum[51] have reconsidered the problem and solved it, in an original way that we shall examine now. In three dimensions, the structure constants can be written as :

$$C^C_{AB} = \varepsilon_{ABD}\, n^{DC} + a_A\,\delta^C_B - a_B\,\delta^C_A \qquad (8.11)$$

where $a_A = \frac{1}{2} C^B_{AB}$, $n^{DC} = n^{(DC)} = \frac{1}{2}\varepsilon^{AB(D}C^{C)}_{AB}$

and $\varepsilon_{123} = \varepsilon^{123} = 1$.

Then the Jacobi identity (10) reduces to :

$$n^{AB}\, a_B = 0 \quad . \qquad (8.12)$$

The Killing vectors defining the isometry group are defined up to a $GL(3,\mathbb{R})$ transformation. Under a change of basis

$$\xi_A \rightarrow M_A^{\ B}\,\xi_B \quad , \qquad M_A^{\ B} \in GL(3,\mathbb{R}) \qquad (8.13)$$

the structure constants transform as

$$C^{C}_{AB} \to M_A{}^K M_B{}^L M^{-1\,C}{}_N C^N_{KL}$$

$$\text{i.e. } n^{DC} \to \det M \, M^{-1D}{}_K M^{-1C}{}_L n^{KL}$$

$$a_A \to M_A{}^K a_K \quad . \tag{8.14}$$

Using these transformations the matrix n^{AB} can be diagonalized

$$n^{AB} = \begin{pmatrix} n^1 & 0 & 0 \\ 0 & n^2 & 0 \\ 0 & 0 & n^3 \end{pmatrix} \tag{8.15}$$

and the vector a_A can be chosen pointing in the 1-direction :

$$a_A = (a,0,0) \quad . \tag{8.16}$$

This form of the structure constants is preserved under rescaling of the Killing vectors : $\xi_A \to \lambda_A \xi_A$ while

$$a \to \lambda_1 a$$
$$n^1 \to \frac{\lambda_2 \lambda_3}{\lambda_1} n^1 \text{ , etc...} \tag{8.17}$$

These transformations can be used to construct the canonical expressions for n^{AB} and a shown in table II :

Table II : Canonical expression and Bianchi type of all 3-dimensional *real* Lie algebras.

Type	a	n^1	n^2	n^3	
					Class A
I	0	0	0	0	Translations
II	0	1	0	0	Galilean
VII_o	0	1	1	0	Euclidean
VI_o	0	1	-1	0	Poincaré
IX	0	1	1	1	Rotation
VIII	0	1	1	-1	Lorentz
					Class B
V	1	0	0	0	
IV	1	0	1	0	
VII_a	$a > 0$	0	1	1	
$III_{a=1}$, $VI_{a\neq 1}$	$a > 0$	0	1	-1	

With respect to the canonical expression of the structure constants, the commutation relations of the Killing vectors become :

$$[\vec{\xi}_1,\vec{\xi}_2] = n^3\,\vec{\xi}_3 + a\,\vec{\xi}_2$$
$$[\vec{\xi}_2,\vec{\xi}_3] = n^1\,\vec{\xi}_1$$
$$[\vec{\xi}_3,\vec{\xi}_1] = n^2\,\vec{\xi}_2 - a\,\vec{\xi}_3 \quad . \tag{8.18}$$

We have to distinguish two main classes (A and B) according to $a = 0$ or $a \neq 0$. On the other hand all these Lie algebras are solvable excepted Type VIII and IX which are semi-simple. Type III appears also as a special case of type VI because its derived algebra is one dimensional instead of two.

Taub[52] has integrated eqs. (18) for each Bianchi type. Here we shall just work out one example which will also illustrate all our previous considerations about the geometry of Lie groups. Let us consider for instance the type III; so that the equations to be integrated are :

$$[\vec{\xi}_1,\vec{\xi}_2] = \vec{\xi}_2 - \vec{\xi}_3 \tag{8.19a}$$
$$[\vec{\xi}_2,\vec{\xi}_3] = 0 \tag{8.19b}$$
$$[\vec{\xi}_3,\vec{\xi}_1] = \vec{\xi}_2 - \vec{\xi}_3 \quad . \tag{8.19c}$$

Since $\vec{\xi}_2$ and $\vec{\xi}_3$ commute we can choose suited coordinates such that

$$\vec{\xi}_2 = \partial_2 \quad \text{and} \quad \vec{\xi}_3 = \partial_3 \quad .$$

Writting $\vec{\xi}_1$ as

$$\vec{\xi}_1 = a\,\partial_1 + b\,\partial_2 + c\,\partial_3$$

(with $a \neq 0$ in order to satisfy the transitivity assumption) we deduce from eqs. (19a) and (19c)

$$-a_{,2}\,\partial_1 - b_{,2}\,\partial_2 - c_{,2}\,\partial_3 = \partial_2 - \partial_3$$
$$a_{,3}\,\partial_1 + b_{,3}\,\partial_2 + c_{,3}\,\partial_3 = \partial_2 - \partial_3 \quad .$$

As a consequence we obtain

$$a = A(x^1) \quad , \quad b = B(x^1) - x^2 + x^3 \quad , \quad c = C(x^1) + x^2 - x^3$$

where A, B and C are arbitrary functions of the single variable x^1. The most general coordinate transformation which preserves the expression of $\vec{\xi}_2$ and $\vec{\xi}_3$ is

$$x^1 \to \tilde{x}^1 = f(x^1) \tag{8.20a}$$

$$x^2 \to \tilde{x}^2 = x^2 + g(x^1) \tag{8.20b}$$

$$x^3 \to \tilde{x}^3 = x^3 + h(x^1) \quad . \tag{8.20c}$$

Under this transformation the Killing vectors change into

$$\vec{\xi}_1 = A(f' \frac{\partial}{\partial \tilde{x}^1} + g' \frac{\partial}{\partial \tilde{x}^2} + h' \frac{\partial}{\partial \tilde{x}^3})$$

$$+ [B + g - h - \tilde{x}^2 + \tilde{x}^3] \frac{\partial}{\partial \tilde{x}^2} + [C - g + h + \tilde{x}^2 - \tilde{x}^3] \frac{\partial}{\partial \tilde{x}^3}$$

$$\vec{\xi}_2 = \frac{\partial}{\partial \tilde{x}^2} \quad \text{and} \quad \vec{\xi}_3 = \frac{\partial}{\partial \tilde{x}^3}$$

where a prime indicates a derivation with respect to x^1.
By choosing

$$f' = A^{-1}$$

$$g' = A^{-1} [h - g - B]$$

$$h' = A^{-1} [g - h - C]$$

we obtain an expression of $\vec{\xi}_1$ in which, finally,

$$a = 1 \quad \text{and} \quad b = -c = x^3 - x^2 \quad .$$

At this stage the coordinate system is not yet completely fixed. We can still perform the following coordinate transformations

$$x'^1 = x^1 + \alpha \tag{8.21a}$$

$$x'^2 = \beta e^{-2x^1} + \gamma \tag{8.21b}$$

$$x'^3 = -\beta e^{-2x^1} + \gamma \quad . \tag{8.21c}$$

These transformations constitute a 3 parameter group generated by the vectors fields

$$\vec{Y}_\alpha = \partial_1 \; , \; \vec{Y}_\beta = e^{-2x^1}(\partial_2 - \partial_3) \; \text{ and } \; \vec{Y}_\gamma = \partial_2 + \partial_3$$

which define an invariant frame

$$[\vec{Y}_A, \vec{\xi}_B] = 0 \qquad A = \alpha, \beta, \gamma \; ; \; B = 1, 2, 3 \quad .$$

In fact this frame is defined yp to a $GL(3,\mathbb{R})$ transformation.
If we impose the vectors of the invariant frame to coincide with

the Killing vectors, say at the point $x^i = 0$, the invariant frame gets completely fixed and is given by

$$\vec{X}_1 = Y_\alpha \quad , \tag{8.22a}$$

$$\vec{X}_2 = \frac{1}{2}(\vec{Y}_\beta + \vec{Y}_\gamma) \quad , \tag{8.22b}$$

$$\vec{X}_3 = \frac{1}{2}(\vec{Y}_\gamma + \vec{Y}_\beta) \tag{8.22c}$$

with the commutation relations

$$[\vec{X}_A, \vec{X}_B] = - C^C_{AB}\, \vec{X}_C \quad . \tag{8.23}$$

In other words, on the 3-dimensional Lie group defined as Bianchi type III we have computed the expressions for the left generators as $\vec{\xi}_A$ and for the right one as $\vec{X}_A$. Finally, from the previous expressions of these vector fields we directly obtain the dual invariant 1-forms :

$$\theta^1 = dx^1 \quad , \tag{8.24a}$$

$$\theta^2 = dx^2 + (x^2 - x^3)dx^1 \quad , \tag{8.24b}$$

$$\theta^3 = dx^3 - (x^3 - x^2)dx^2 \quad , \tag{8.24c}$$

and
$$\sigma^1 = dx^1 \quad , \tag{8.25a}$$

$$\sigma^2 = \frac{1}{2}\,[(1 + e^{2x^1})dx^2 + (1 - e^{2x^1})dx^3\,] \quad , \tag{8.25b}$$

$$\sigma^3 = \frac{1}{2}\,[(1 - e^{2x^1})dx^2 + (1 + e^{2x^1})dx^3\,] \quad . \tag{8.25c}$$

A part from the class of metrics described in the Bianchi classification, there exist other spatially homogeneous metrics characterized by a 4-parameter group of isometries without 3-parameter subgroup acting transitively. Since each G_4 contains an subgroup G_3 (Egorov's theorem)[53], this means that we are looking here for spaces on which G_3 acts (multiply transitively) on 2-dimensional surfaces. These surfaces correspond respectively to the Euclidean sphere S_2, the Euclidean plane and the sphere with negative curvature (*). Their metrics can be expressed as

$$ds^2_{(2)} = d\rho \otimes d\rho + \begin{Bmatrix} \sin^2\rho \\ \rho^2 \\ \mathrm{sh}^2\,\rho \end{Bmatrix} d\theta \otimes d\theta \tag{8.26}$$

(*) Spheres with negative curvature cannot be globally embedded in a 3-dimensional Euclidean space. However, in a 3-dimensional Minkowskian space with metric $ds^2_{(3)} = -dT^2 + dX^2 + dY^2$ they correspond to one connected component of the hyperboloïd : $-T^2 + X^2 + Y^2 = -R^2$.

and their G_3 groups are respectively of Bianchi type IX, VII_0 or VIII.

Kantowski[54,55] has proved that only those spaces for which the 2-dimensional orbits of G_3 are two-spheres are relevant. All the others admit a simply-transitive G_3. Moreover, from the Jacobi identity, the fourth generator should commute with all the generators of G_3. As a consequence, the only metrics to consider are :

$$ds^2 = -dt \otimes dt + a^2(t)dx \otimes dx + b^2(t)(d\theta \otimes d\theta + \sin^2\theta\, d\varphi \otimes d\varphi). \tag{8.27}$$

They have been studied by Kantowski and Sachs[55] and the corresponding homogeneous cosmologies are known as *Kantowski-Sachs models*. Notice that the Schwarzschild metric, for $r < 2m$, is a special case of (27).

VIII.3. The Anti-de Sitter Space

Among the numerous exact solutions describing spatially homogeneous cosmological models, the simplest ones are the *Friedman*[56]-*Lemaître*[57]-*Robertson*[58]-*Walker*[59] *spaces*. They are models for a universe which admits spacelike hypersurfaces homogeneous and isotropic. A motivation for the consideration of such spaces is that, on large scale, the universe looks isotropic around the Earth. Galaxies are more or less uniformely distribued in the sky (up to 30 %), [48] anisotropies in radio-sources distribution and in the cosmic X-ray background are less then a few per cent, and the cosmic microwave radiation is isotropic up to a fraction of a per cent. Adopting the democratic point of vue, saying that the Earth should not be in a special position in the universe, we conclude that the space is isotropic around each of his points. This implies that the Riemann tensor of the spatial 3-dimensional geometry must be of the form

$$R^{ij}{}_{k\ell} = K(\delta^i_k\, \delta^j_\ell - \delta^i_\ell\, \delta^j_k) \quad . \tag{8.28}$$

Moreover K must be a constant, since a non vanishing gradient of K will break the isotropy. Actually the constancy of K also directly results — in dimension greater then 2 — from the expression (28) of the Riemann tensor thanks to the contracted Bianchi identities :

$$\nabla_i\, G^{ij} = 0 \Leftrightarrow K_{,i} = 0 \quad . \tag{8.29}$$

As a consequence the geometry of the space is the same everywhere i.e. the space is homogeneous.

There exists only three possible geometries describing an homogeneous and isotropic space. They are the geometries of Euclidean spaces with constant curvature : the flat space and the spheres of positive or negative curvature. All these spaces are *maximally symmetric* i.e. they admit — at least locally — a maximal number of Killing vectors (6 in three dimensions, $\frac{n(n+1)}{2}$ in n dimensions). The metric of these spaces is given by a three-dimensional generalization of (26) :

$$d\sigma^2_{(k)} = d\rho \otimes d\rho + f^2(\rho)(d\theta \otimes d\theta + \sin^2\theta\, d\varphi \otimes d\varphi) \tag{8.30}$$

with $$f(\rho) = \begin{cases} \sin\rho & (k = +1) \\ \rho & (k = 0) \\ \mathrm{sh}\,\rho & (k = -1) \end{cases}$$

and the general form of a Friedman-Lemaître-Robertson-Walker space is

$$ds^2 = -dt \otimes dt + a^2(t)\, d\sigma^2_{(k)} \quad . \tag{8.31}$$

When $k = +1$, the spatial curvature is positive and if ρ runs from 0 to π the space is geodesically complete. The corresponding 3-space in a sphere S_3 and, as a consequence, it is compact (*closed universe*). When $k \leqslant 0$, the spacial curvature is null or negative, ρ goes from 0 to ∞, and the 3-space is diffeomorphic to $\mathbb{R}^3$ (*open universe*). However we can identify points on these manifolds in a way compatible with the metric and so build compact spaces of zero or negative curvature. This compactification is achieved by identifying points belonging to the orbits of an infinite discrete subgroup of the initial 6-parameter isometry group. Starting from a flat space we obtain, in this way, a torus. In this special case, the original isometry group of motions reduces to an abelian 3-dimensional group (Bianchi I). On spaces of negative curvature no continuous isometry survives. This is a consequence of the more general result which says that on a compact Riemannian space C with negative Ricci curvature :

$$R_{\alpha\beta}\, u^\alpha u^\beta \leqslant 0 \qquad \forall u^\alpha \tag{8.32}$$

globally defined Killing vector fields do not exist. The proof of this statement is analogous to the proofs presented in the section devoted to the Betti numbers. Suppose we have a solution ξ^α of the Killing equations (5.19). It satisfies the second order equation :

$$\Box\, \xi_\alpha + R_{\alpha\beta}\, \xi^\sigma = 0 \quad . \tag{8.33}$$

Contracting this equation with ξ^α and integrating by parts we obtain :

$$0 = \int_C \xi^\alpha(\Box\xi_\alpha + R_{\alpha\beta}\,\xi^\sigma)\eta = \int_C (-\nabla_\beta\,\xi_\alpha\,\nabla^{\beta\xi\alpha} + R_{\alpha\beta}\xi^\alpha\xi^\beta)\eta \quad .$$

But the last integrant is negative definite and null only if ξ^α vanishes identically. By the way, we see that, on a compact flat space, isometries are generated by covariantly constant vector fields :

$$\nabla_\alpha\, \xi_\beta = 0 \quad . \tag{8.34}$$

A special form of the metric (31) is :

$$ds^2 = -dt \otimes dt + (a \cos \frac{t}{a})^2 \, d\sigma^2_{(-1)} \qquad (8.35)$$

which solves the vacuum Einstein field equations with cosmological constant :

$$R_{\mu\nu} - \frac{1}{2} g_{\mu\nu} R + \Lambda g_{\mu\nu} = 0 \quad \text{with} \quad \Lambda = -\frac{3}{a^2} \quad . \qquad (8.36)$$

The space-time described by this metric is the *anti-de Sitter*[60] one. Though not very interesting in the framework of the classical cosmology, this space-time is often considered as describing "vacuum states" in supergravity. Its geometry is simple enough to be studied through elementary methods but at the same time complicated enough to offer unusual global properties to the reader used to work in the framework of Minkowski space.

The global structure of anti-de Sitter space becomes more transparent once we observe that it can be looked at as a "sphere" H_4 of radius a^2 embedded in a five dimensional pseudo-Minkowskian space whose metric is :

$$ds^2_{(5)} = -dU \otimes dU - dV \otimes dV + dR \otimes dR + R^2(d\theta \otimes d\theta + \sin^2 \theta \, d\varphi \otimes d\varphi). \qquad (8.37)$$

The equation for H_4 is

$$U^2 + V^2 - R^2 = a^2 \qquad (8.38)$$

and the expression (35) of the metric results from the local parametrization :

$$U = a \sin \frac{t}{a} \quad ,$$

$$V = a \operatorname{ch} \rho \cos \frac{t}{a} \quad ,$$

$$R = a \operatorname{sh} \rho \cos \frac{t}{a} \quad . \qquad (8.39)$$

We may visualize a 2-dimensional anti-de Sitter space H_2 as an hyperboloid embedded in a 3-dimensional space (fig. 4). In such a two-dimensional case the coordinate ρ is allowed to run on the whole real line, the sign of ρ being the analogous to an "angular coordinate".

The geodesics on anti-de Sitter space are depicted by the "circles" obtained by considering the intersections of (three) diametral (hyper-)planes with the "sphere" $H_{2(4)}$. In particular null geodesics correspond to linear generators of $H_{2(4)}$.

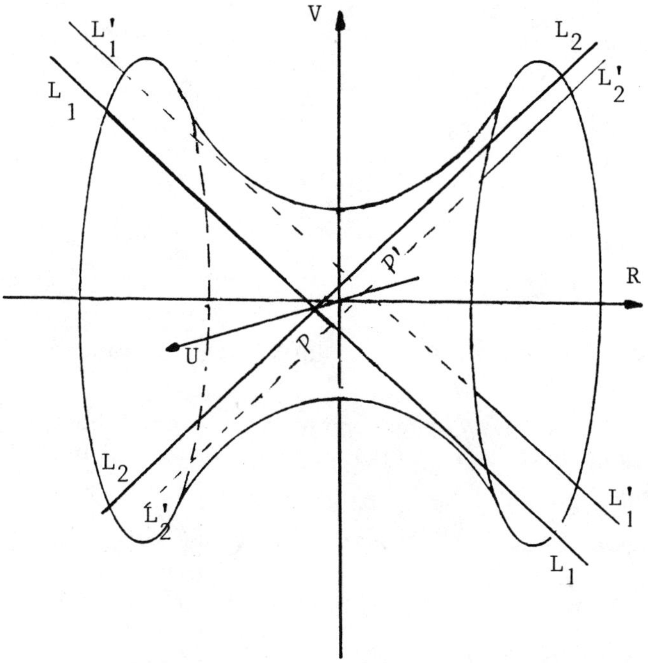

Fig. 4 : A 2-dimensional anti-de Sitter space represented as an hyperboloïd in a 3-dimensional Minkowski space.

Obviously the coordinates (t,ρ,θ,φ) cover only the regions inside the two light cones from P and P' bounded by the rays L_1, L_2 and L'_1 , L'_2 whose equations are

$$U = \pm a \quad ; \quad V = \pm R \quad .$$

The surfaces $t = t_o$ are represented by the intersection of H_4 with planes $U = a \sin t_o/a$ i.e. by drawing hyperboles inside the two light cones on H_2. The time-like curves orthogonal to these surfaces are closed geodesics. They are represented by the intersections of H_2 with planes :

$$1 > \frac{R}{V} = \text{constant} > -1 \quad .$$

The singularities $t = \pm \frac{a\pi}{2}$ correspond to null surfaces defined by the planes $U = \pm a$ which are the boundary of the domain covered by the coordinate system.

It is obvious, from the representation of the anti-de Sitter space as a sphere H_4, that its isometry group is SO(3,2). A global coordinate system (excepted for trivial spherical coordinates singularities) is given by the parametrization :

$$
\begin{aligned}
R &= \rho \quad , \\
V &= a\left(1 + \frac{\rho^2}{a^2}\right)^{1/2} \cos \frac{t'}{a} \quad , \\
U &= a\left(1 + \frac{\rho^2}{a^2}\right)^{1/2} \sin \frac{t'}{a} \quad ,
\end{aligned}
\tag{8.40}
$$

which leads to the metric

$$
ds^2 = -\left(1 - \frac{\Lambda}{3}\rho^2\right) dt' \otimes dt' + \left(1 - \frac{\Lambda}{3}\rho^2\right)^{-1} d\rho \otimes d\rho + \rho^2 (d\theta \otimes d\theta + \sin^2\theta \, d\varphi \otimes d\varphi). \tag{8.41}
$$

This metric is static and corresponds to the limit case of a Schwarzschild metric with a negative cosmological constant but a vanishing mass.

Performing the following coordinate transformation

$$
\tau = \frac{t'}{a} \quad \text{and} \quad \rho = a \,\mathrm{tg}\, \chi \quad (\chi \in [\,0\,, \tfrac{\pi}{2}\,[\,) \ , \tag{8.42}
$$

we obtain as an expression for the metric

$$
ds^2 = \frac{a^2}{\cos^2\chi} \left[-d\tau \otimes d\tau + d\chi \otimes d\chi + \sin^2\chi \, (d\theta \otimes d\theta + \sin^2\theta \, d\varphi \otimes d\varphi) \right] \tag{8.43}
$$

which shows explicitly that the anti-de Sitter space can conformally be embedded in an Einstein static universe. This expression of the metric has the advantage that is permits easily the description of the causal structure[61] of the anti-de Sitter space. Suppressing again the angular coordinates θ and φ, we can represent the anti-de Sitter space in a (τ,χ) plane in which each point represents a 2-dimensional sphere, excepted these on the line $\chi = 0$ (fig. 5).

On this (τ,χ) plane, null geodesics are depicted by straight lines : $\tau \pm \chi = \text{constant}$. The time-like geodesics orthogonal to the surfaces $t = \text{constant}$ all converge to the points P and P' and are closed curves. This last pathology can be avoided by considering the *universal covering* of the anti-de Sitter that is by assumming that the coordinate τ (or t') runs from $-\infty$ to $+\infty$.

The foccussing property of all the timelile geodesics from P and P' is general because the space-time is homogeneous (all points are equivalent). On the universal covering space, all the timelike geodesics from any point start diverging then reconverge to an image point, diverge again, reconverge to a new image point etc... So we see that there exist regions causaly connected to a point

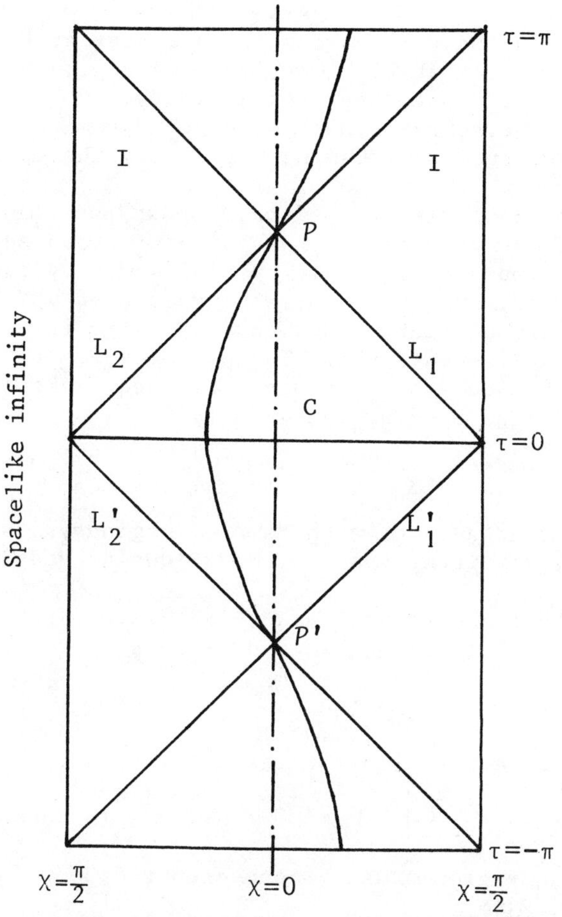

Fig. 5 : Causal description of anti-de Sitter space. All timelike geodesics from P' reconverge at P and are orthogonal to the surface $\tau = 0$. Region I is causaly connected to P' but can only be reached from P' be accelereated observed.

which cannot be reached by timelike geodesics from the point considered but only by accelerated curves. For instance the region outside the light cone from P is causaly connected to P' but can only be reached by accelerated observers starting from P'.

Another unusual property in that the region C, on which data given on the hypersurface $\tau = 0$ lead to a well defined Cauchy problem for causal equations, is a bounded domain. It is the domain included inside the past light-cone with vertex at P and the future light-cone with vertex at P'. Outside this region we have no control on informations coming from infinity. This property is not peculiar to the surface considered. The same phenomenon occurs for any spacelike surface crossing completely the space-time.

Breitenlohner and Freedman[62] have shown how this lack of unicity for the Cauchy problem of the Klein-Gordon equation can be circumvented by imposing suitable spatial boundary conditions. More precisely they have built Hilbert spaces on which the Cauchy problem has a unique solution and on which a positive definite energy functional is conserved. These spaces are represented by complete sets of mode functions spanning inequivalent representations of SO(3,2) and obeying the equation

$$\Box \varphi - m^2 \varphi = 0 \qquad \text{with} \qquad m^2 \geqslant \frac{3}{4}\Lambda = -\frac{9}{4a^2} \quad .$$

In other words fluctuations with "small imaginary masses" are allowed on anti-de Sitter space without loosing stability.

Exercice XV : Show that, on anti-de Sitter space, the stereographic coordinates (τ,ρ) obtained by solving the equations

$$(a,0,0) + \lambda(U-a,V,R) = (-a,\tau,\rho)$$

lead to the local expression for the metric

$$ds^2 = [\, 1 - \frac{\Lambda(\tau^2-\rho^2)}{12}]^{-2} \, [\, d\tau\otimes d\tau + d\rho\otimes d\rho + \rho^2 (d\theta\otimes d\theta + \sin^2\theta d\varphi\otimes d\varphi)\,]$$

valid inside the region causaly connected to the point P defined by U = a in eq. (38).

IX. ACKNOWLEDGMENTS

I am happy to thank M. Cousyns and C. Defise for their encouragements during the preparation of these notes, P. Castoldi, M. Henneaux, D. Polarski and especially C. Schomblond for numerous enlightening discussions and comments on the manuscript, and last but not least the organizators of the school for their warm hospitality. Thanks also go to L. Bouchez for typing my unreadable handwritten notes.

BIBLIOGRAPHY

B. Carter, Underlying Mathematical Structures of Classical Gravitation Theory, in Recent Developments in Gravitation, Cargèse 1978, eds. M. Lévy and S. Deser, Plenum Press, New York London, B 44, 41 (1979).

Y. Choquet-Bruhat, C. De Witt-Morette with M. Dillard-Bleick, Analysis, Manifolds and Physics, North-Holland,

Amsterdam New York Oxford (1982).
S.I. Goldberg, Curvature and Homology, Academic Press, New York (1962).
S.W. Hawking and G.F.R. Ellis, The Large Scale Structure of Space-Time, University Press, Cambridge (1973).
C.W. Misner, K.S. Thorne and J.A. Wheeler, Gravitation, W.H. Freeman and Cie, San Francisco (1973).
M. Spivak, A Comprehensive Introduction to Differential Geometry, Publish or Perish, Inc. Berkeley (1979).
S. Weinberg, Gravitation and Cosmology : Principles and Applications of the General Theory of Relativity, J. Wiley & Sons, Inc. New York London Sydney Toronto (1972).
K. Yano and S. Bochner, Curvature and Betti Numbers, Princeton Univ. Press, Princeton N.J. (1953).

REFERENCES

1. A. Einstein, Ann. Phys. (Germany) 35, 898 (1911) translated in The Principle of Relativity : A collection of Original Memoirs, Dover, New York.
2. R.V. Pound and G.A. Rebka, Phys. Rev. Lett. 4, 337 (1960), R.V. Pound and J.L. Snider, Phys. Rev. Lett. 13, 539 (1965).
3. A. Schild, Texas Quartely, 3, n° 3, 42 (1960), "Lectures on General Relativity Theory", in Relativity Theory and Astrophysics, ed. by J. Ehlers, Am. Math. Soc., Providence, R.I., 1 (1967).
4. P.G. Roll, R. Krotkov and R.H. Dicke, Ann. Phys. (U.S.A.), 26, 442 (1964).
5. V.B. Braginsky and V.I. Panov, Sov. Phys. J.E.T.P., 34, 464 (1971).
6. E. Cartan, Leçons sur les invariants intégraux, Hermann, Paris (1922); Les systèmes différentiel extérieurs et leurs applications géométriques, Hermann, Paris (1945).
7. A. Salam and J. Strathdee, Ann. Phys. 141, 316 (1982).
8. P. van Nieuwenhuizen, General Theory of Coset Manifolds and Antisymmetric Tensors Applied to Kaluza-Klein supergravity, preprint.
9. C. Ehresmann, Les connexions infinitésimales dans un espace fibré différentiable, Colloque de Topologie, Brussels, 29 (1950).
10. E. Cartan, Ann. Ecole Normale Supérieure 40, 325 (1923) and 41, 1 (1924); La Méthode du Repère Mobile, la Théorie des groupes continus et les Espaces Généralisés, Hermann, Paris (1935).
11. R. Weitzenböck, Invariant entheorie, Groningen, Noordhoff (1923).
12. P.B. Gilkey, The Index Theorem and the Heat Equation, Publish or Perish, Boston (1974).
13. K. Yano, Ann. of Math. 55, 328 (1952).
14. R. Geroch, J. Math. Phys. 9, 1739 (1968).

15. J.W. Milnor, L'Enseignement Math. 9, 198 (1963).
16. S.J. Avis and C.J. Isham, Quantum field theory and fibre bundles, in Recent Developments in Gravitation, Cargèse, 1978, eds. M. Lévy and S. Deser, Plenum Press, New York London, B 44, 347 (1979).
17. L. Castellani, L.J. Romans and N.P. Warner, A. Classification of Compactifying Solutions for d = 11 Supergravity, Preprint CALT-68-1055 (1984).
18. M. Henneaux, C. Schomblond and P. Spindel, Bull. Acad. Roy. Belg., 63, 179 (1977).
19. Y. Kosmann, Ann. di Math. pura ed. applicata, 91, 317 (1972).
20. M. Henneaux, Sur la Geométrodynamique avec Champs Spinoriels, Thesis U.L.B., Brussels (1980).
21. J. Scherk, Extended Supersymmetry and Extended Supergravity Theories, in Recent Developments in Gravitation, Cargèse 1978, eds. M. Lévy and S. Deser, Plenum Press, New York London, B 44, 479 (1979).
22. P. van Nieuwenhuizen, Simple supergravity and the Kaluza-Klein program, in Relativity, Groups and Topology II, Les Houches 1983, ed. by B.S. De Witt and R. Stora, North-Holland, Amsterdam, 823, (1984).
23. A. Lichnérowicz, Spineurs harmoniques, reprinted in Choix d'Oeuvres Mathématiques, Hermann, Paris (1982).
24. G.M. Clemence, Astron. Papers Am. Ephemeris, 11, part 1 (1943); Rev. Mod. Phys. 19, 361 (1947).
25. R.L. Duncombe, Astron. J. 61, 174 (1956).
26. I.I. Shapiro, W.B. Smith, M.E. Ash and S. Herrick, Astron. J. 76, 588 (1971).
27. H. von Klüber, The Determination of Einstein's Light-deflection in the Gravitational Field of the Sun, in Vistas in Astronomy, ed. by A. Beer, Pergamon Press, New York (1960).
28. R.A. Hulse and J.H. Taylor, Astrophys. J. 195, 51 (1975).
29. J.H. Taylor, L.A. Fowler and P.M. Mc Culloch, Nature, 277, 437 (1979).
30. T. Damour, Gravitational Radiation and Motion of Compact Bodies, in Gravitational Radiation ed. by N. Deruelle and T. Piran, Les Houches 1982, North-Holland, Amsterdam (1983).
31. A. Lichnérowicz, Théories Relativistes de la Gravitation et de l'Electromagnétisme, Masson, Paris (1955).
32. F. Belinfante, Physica 6, 887 (1939).
33. L. Rosenfeld, Mém. Acad. Roy. Belg., Cl. Sc. 18, n°6 (1940).
34. J. Géhéniau, Acad. Roy. Belg., Cl. Sc., 45, 447 (1959).
35. H. Weyl, Phys. Rev. 77, 699 (1950).
36. T.W.B. Kibble, J. Math. Phys., 2, 212 (1961).
37. B. Carter, Phys. Rev. D, 16, 3395 (1977).
38. B. Carter, Phys. Rev. 174, 1559 (1968); Analytic and Geometric Properties of the Kerr Solutions, in Black Holes, Les Houches 1972, ed. by C. De Witt and B.S. De Witt,

Gordon and Breach, New York, 61 (1973).
39. R.G. McLeneghan and P. Spindel, Phys. Rev. D, 20, 409 (1979); Bull. Soc. Math. Belg. 50, 65 (1979).
40. B. Carter, Commun. Math. Phys., 10, 280 (1968).
41. S. Chandrasekhar, Proc. Roy. Soc. London, A 349, 571 (1976).
42. D.N. Page, Phys. Rev. D, 14, 1509 (1976).
43. R. Güven, Proc. Roy. Soc. London, A 356, 465 (1977).
44. B. Carter and R.G. McLenaghan, Phys. Rev. D, 19, 1093 (1979).
45. N. Kamran and R.G. McLenaghan, Lett. Math. Phys. 7, 381 (1983).
46. D. Kramer, H. Stephani, E. Herlt and M. MacCallum, Exact Solutions of Einstein's Field Equations, ed. by. E. Schmutzer, Univ. Press, Cambridge (1980).
47. M.P. Ryan and L.C. Shepley, Homogeneous Relativistic Cosmologies, Princeton Univ. Press, Princeton New Jersey (1975).
48. M.A.H. MacCallum, Anisotropic and Inhomogeneous Relativistic Cosmologies, in General Relativity : an Einstein Centenary Survey, ed. by S.W. Hawking and W. Israel, Univ. Press, Cambridge, 533 (1979).
49. L. Bianchi, Soc. Ital. Sci. Mem. di Math. 11, 267 (1897).
50. C.G. Behr, F.B. Estabrook and H.D. Wahlquist, J. Math. Phys., 9, 497 (1968).
51. G.F.R. Ellis and M.A.H. MacCallum, Commun. Math. Phys., 12, 108 (1969).
52. A.H. Taub, Ann. Math., 53, 472 (1951).
53. Quoted by A.Z. Petrov, Einstein spaces, Pergamon Press (1964).
54. R. Kantowski, Some Relativistic Cosmological Models, Thesis Univ. of Texas (1966).
55. R. Kantowski and R.K. Sachs, J. Math. Phys. 7, 443 (1966).
56. A.A. Friedmann, Z. Phys. 21, 326 (1924).
57. G. Lemaître, Ann. Soc. Sci. Bruxelles, A 47, 2949 (1927); Mon. Not. Roy. Astron. Soc. 91, 483 (1931); Ann. Soc. Sci. Bruxelles, A 53, 51 (1933).
58. H.P. Robertson, Proc. Not. Acad. Sci., 15, 822 (1929); Rev. Mod. Phys. 5, 62 (1933); Astroph. J. 82, 284 (1935); Astroph. J. 83, 187 & 257 (1936).
59. A.G. Walker, Quart. J. Math. Oxford, 6, 81 (1935); Proc. Lond. Math. Soc. 42, 90 (1936).
60. W. de Sitter, Mon. Not. Roy. Astron. Soc. 78, 3 (1917).
61. R. Penrose, Conformal treatment of infinity, in Relativity, Groups and Topology, Les Houches 1963, ed. by C.M. de Witt and B.S. de Witt, Gordon and Breach, New York, 565, (1964).
62. P. Breitenlohner and D.Z. Freedman, Ann. of Phys., 144, 249 (1982).

SUPERSYMMETRIC YANG-MILLS THEORIES

K.S. Stelle*
Institute for Theoretical Physics
University of California
Santa Barbara, California 93106

ABSTRACT

The classical and quantum mechanical features of super Yang-Mills theories are reviewed with particular attention to the cancellations of ultraviolet infinities and to realization of the full set of internal symmetries of the theory in superspace. Chapter 1 summarizes the superspace formulation of $N = 1$ super Yang-Mills theory. Chapter 2 discusses in detail the ultraviolet cancellations in extended super Yang-Mills theories. Chapter 3 reviews the formalism of harmonic superspace and its application to super Yang-Mills theories.

1. INTRODUCTION

In order to describe a massless supersymmetric gauge theory,[1] we must start from the irreducible representation of supersymmetry containing a vector V_μ and its spin 1/2 partner λ_α, *i.e.*, the superspin 1/2 multiplet. The condition $\partial^\mu V_\mu = 0$, which is necessary for the irreducibility of this multiplet, indicates that in order to find an action for the multiplet in terms of unconstrained fields, it is necessary to add another supermultiplet containing the divergence of the vector field. In the final action, this extra multiplet must be eliminable by gauge transformations, so we can see that a supersymmetric gauge theory will have a whole multiplet's worth of gauge invariances.

The only other irreducible multiplet that can be used as the multiplet of gauge components is the superspin 0 chiral multiplet. The combination of the "physical" and the "gauge" multiplets may be described by a real general scalar superfield $V(x,\theta,\bar{\theta}) = V^*(x,\theta,\bar{\theta})$. Since the gauge variation of this superfield must itself be real, we conclude that for a chiral superfield of gauge transformation parameters Λ, in an Abelian theory we must have

$$\delta V = i(\Lambda - \Lambda^*), \qquad \bar{D}_{\dot{\alpha}}\Lambda = 0 \tag{1.1}$$

where the choice of the imaginary part of Λ is necessary for the vector V^μ to be a parity-even object. If the components of Λ are denoted $(A+iB, \varsigma, E-iF)$, then the components of the real multiplet (1.1) are $(B, \varsigma, E, F, \partial_\mu A, 0, 0)$, as can be worked out using the rules given in the last chapter. If the components of the superfield V are denoted $(C, \chi, H, K, V_\mu, \lambda, D)$, then the transformation (1.1) allows us to gauge away C, χ, H, K, and $\partial^\mu V_\mu$. The remaining gauge invariant quantities are described by a superfield whose lowest component is the Abelian gauge invariant spinor λ_α. This gauge invariant superfield is given by

$$W_\alpha = \bar{D}^2 D_\alpha V. \tag{1.2}$$

It is easy to check the gauge invariance of W_α:

$$\delta W_\alpha = i\bar{D}^2 D_\alpha(\Lambda - \Lambda^*) = i\bar{D}^2 D_\alpha \Lambda = -2i\partial_\mu \sigma^\mu_{\alpha\dot{\beta}} \bar{D}^{\dot{\beta}}\Lambda = 0. \tag{1.3}$$

Due to the derivative factor $\bar{D}^2$, W_α is a chiral spinor superfield,

$$\bar{D}_{\dot{\beta}} W_\alpha = \bar{D}_{\dot{\beta}} \bar{D}^2 D_\alpha V = 0, \tag{1.4}$$

and, in addition, the reality of V implies the reality of the scalar component of W_α:

$$D \equiv (D^\alpha W_\alpha)|_{\theta=\bar{\theta}=0} = (D^\alpha \bar{D}^2 D_\alpha V)\big|_{\theta=\bar{\theta}=0} = (\bar{D}_{\dot{\alpha}} D^2 \bar{D}^{\dot{\alpha}} V)\big|_{\theta=\bar{\theta}=0} = D^*. \tag{1.5}$$

Conditions (1.4) and (1.5) define an irreducible superfield, whose components are $(\lambda_\alpha, \partial_\mu V_\gamma - \partial_\nu V_\mu, D, \sigma^\mu_{\alpha\dot{\beta}} \lambda^{\dot{\beta}} \partial_\mu \lambda^{\dot{\beta}})$. This multiplet is known as the supersymmetric field strength multiplet, since it contains the field strength $F_{\mu\nu} = \partial_\mu V_\nu - \partial_\nu V_\mu$ of the vector field. It possesses in a local expression the same content as the superspin 1/2 multiplet $(\frac{1}{\Box}D, \frac{\not{\partial}}{\Box}\lambda, V_\mu{}^T)$.

Since the superfield W_α is chiral, we may form the gauge invariant action

$$I = \mathrm{Re} \int d^4x\, d^2\theta W^\alpha W_\alpha. \tag{1.6}$$

Expressing this in components, we have, after some rescalings, the action for the vector V_μ, spinor λ, and auxiliary field D:

$$I = \int d^4x \left(-\frac{1}{4} F_{\mu\nu} F^{\mu\nu} + \frac{i}{2} \bar{\lambda} \not{\partial} \lambda + \frac{1}{2} D^2 \right). \tag{1.7}$$

If we had taken the imaginary part of the integral in (1.6) instead of the real part, we would have obtained a total divergence whose spin 1 part is the Pontryagin index $\int d^4x\, F^*_{\mu\nu} F^{\mu\nu}$.

In order to describe non-Abelian supersymmetric gauge theories, we must generalize the above discussion. The clue that tells us how to do this is provided by the gauge transformation of some doublet of chiral matter superfields ϕ^i covariantly coupled to the gauge superfield V. The natural generalization of the ordinary gauge transformation is

$$\phi^i \rightarrow [e^{-i\Lambda T}]^{ij} \phi^j, \qquad T = \begin{bmatrix} 0 & 1 \\ -1 & 0 \end{bmatrix}. \tag{1.8}$$

The chirality of the gauge parameter superfield Λ is essential for the consistency of (1.8). If we let $\delta V = i(2g)^{-1}(V - V^*)$, we can write the gauge invariant kinetic term for the ϕ^i as

$$I_{kinetic} = \int d^4x\, d^4\theta \bar{\phi}^i [e^{2gVT}]^{ij} \phi^j. \tag{1.9}$$

This may now be expanded into components using the rules given in the last chapter. Unlike the kinetic term for the gauge multiplet (1.6), however, the gauge coupled matter kinetic term depends upon all of the components of the gauge multplet, not just the gauge invariant parts contributing to W_α. In order to simplify the component expression, it is convenient to make a non-supersymmetric gauge choice known as the Wess-Zumino gauge, setting the low dimension components of V to zero: $C = \chi = H = K = 0$. The only remaining unfixed gauge transformation is then the ordinary Maxwell gauge transformation for the vector field V_μ. In this gauge, the action (1.9) becomes

$$I_{kinetic} = \int d^4x \Big[-8(\nabla^\mu a^*)^i(\nabla_\mu a)^i + 4i\chi^{\alpha i}\sigma^\mu_{\alpha\dot\beta}(\nabla_\mu\bar\chi^{\dot\beta})^i + h^{*i}h^i + \mathrm{Re}(4g\lambda^\alpha\chi^i_\alpha T^{ij}a^{*j}) + 2ga^iT^{ij}a^{*j}D\Big] \quad (1.10)$$

where $\nabla^\mu = \partial^\mu - igV^\mu T$ is the usual covariant derivative.

The generalization to non-Abelian matter couplings is now straightforward, for matter fields ϕ belonging to some gauge group representation with generator matrices T_r must now transform as

$$\phi \to e^{-i\Lambda}\phi, \qquad \begin{array}{c}\Lambda = \Lambda^r \mathrm{T}_r \\ \bar D^{\dot\alpha}\Lambda = 0\end{array}. \quad (1.11)$$

which still preserves the chirality of ϕ, $\bar D_{\dot\alpha}\phi = 0$. In the matter kinetic term, we need to cancel a factor $e^{i\Lambda^*}$ coming from $\bar\phi$ and a factor $e^{-i\Lambda}$ coming from ϕ. Thus, if we again write the gauge coupled kinetic action as

$$I_{kinetic} = \int d^4x\, d^4\theta \bar\phi e^{2gV}\phi, \quad (1.12)$$

then the V^r must undergo nonlinear gauge transformations such that

$$e^{2gV} \to e^{-i\Lambda^*}e^{2gV}e^{i\Lambda}. \quad (1.13)$$

Taking the Abelian limit of this expression, we find that it agrees with (1.1).

In the non-Abelian case, there is a generalization of the field strength superfield (1.2):

$$W_\alpha = \frac{1}{2g}\bar D^2(e^{-2gV}D_\alpha e^{2gV}). \quad (1.14)$$

This is still a chiral superfield: $\bar D_{\dot\beta}W_\alpha = 0$, and it transforms covariantly in the adjoint representation of the gauge group:

$$W_\alpha \to \frac{1}{2g}\bar D^2\left[e^{-i\Lambda}e^{-2gV}e^{i\Lambda^*}D_\alpha(e^{-i\Lambda^*}e^{2gV}e^{i\Lambda})\right]$$

$$= \frac{1}{2g} e^{-i\Lambda} (\bar{D}^2 e^{-2gV} D_\alpha e^{2gV}) e^{i\Lambda} + \frac{1}{2g} e^{-i\Lambda} \bar{D}^2 D_\alpha e^{i\Lambda}$$
$$= e^{-i\Lambda} W_\alpha e^{i\Lambda}. \tag{1.15}$$

In order to expand the field strength multiplet into its components, we take repeated fermionic covariant derivatives as before. In this case, however, we want to preserve the gauge covariance as well as the supersymmetric covariance. Thus, we need a gauge covariantized fermionic derivative ∇_α. As one can see from the calculation (1.15), such a derivative is provided by

$$\nabla_\alpha \equiv e^{-2gV} D_\alpha e^{2gV} \equiv D_\alpha - igA^r_\alpha \mathrm{T}_r, \tag{1.16}$$

where the D_α is taken to act upon everything to the right. The superfield A^r_α is known as the superspace connection. Note that $\bar{D}_{\dot\alpha}$ is still covariant without a connection since Λ is chiral. Expanding W_α using ∇_α, we find the set of gauge-covariant components

$$W_\alpha \leftrightarrow (\lambda_\alpha, F_{\mu\nu} = \partial_\mu V_\nu - \partial_\nu V_\mu + ig[V_\mu, V_\nu], D, \sigma^\mu_{\alpha\dot\alpha} \nabla_\mu \bar\lambda^{\dot\alpha}), \tag{1.17}$$

where ∇_μ is the ordinary spacetime covariant derivative. Every term in (1.17) is actually a contraction of a set of gauge component fields into the Hermitian generators T_r; thus $D = D^r T_r$ is a Hermitian matrix. Note that in terms of the gauge covariantized ∇_α and $\nabla_{\dot\alpha} = \bar{D}_{\dot\alpha}$, the algebra of covariant derivatives becomes

$$\{\nabla_\alpha, \bar\nabla_{\dot\alpha}\} = -2i\sigma^\mu_{\alpha\dot\alpha}\nabla_\mu \tag{1.18}$$
$$\{\nabla_\alpha, \nabla_\beta\} = \{\bar\nabla_{\dot\alpha}, \bar\nabla_{\dot\beta}\} = 0 \tag{1.19}$$
$$[\nabla_\mu, \bar\nabla_{\dot\beta}] = i\sigma_\mu{}^\beta{}_{\dot\beta} W_\beta \tag{1.20}$$

where W_α appears as a superspace field strength.

The action for the supersymmetric Yang-Mills theory may be written down as before, since W_α is chiral as before:

$$I_{sym} = \mathrm{Re} \int d^4x\, d^2\theta\, \mathrm{tr}\, W^\alpha W_\alpha, \tag{1.21}$$

where the trace is over the gauge group indices. Expanding (1.21) into components, we find after some rescalings

$$I_{sym} = \mathrm{tr} \int d^4x \left(-\frac{1}{4} F_{\mu\nu} F^{\mu\nu} - \frac{i}{2} \bar{\lambda} \nabla\!\!\!\!/\, \lambda + \frac{1}{2} D^2 \right). \tag{1.22}$$

In writing down the form (1.22), it is important that we expanded W^α into components using ∇_α. If we had performed a non-gauge-covariant expansion with D_α, we would have obtained the form (1.22) only after going into a Wess-Zumino gauge. As in the Abelian case, if we had taken the imaginary part of the integral in (1.21), we would have obtained a total divergence, with spin-one part now the Pontryagin index for the Yang-Mills field.

The algebra of Yang-Mills gauge covariantized superspace derivatives (1.18)-(1.20) does not reveal the asymmetry between ∇_α and $\nabla_{\dot{\alpha}} = \bar{D}_{\dot{\alpha}}$. This suggests that there should be a more symmetrical way to include the Yang-Mills fields. Indeed, we may also start in direct analogy with ordinary non-supersymmetric Yang-Mills, and introduce connections for all the derivatives:

$$\begin{aligned} \nabla_\alpha &= D_\alpha - igA_\alpha^k T_k \\ \bar{\nabla}_{\dot{\alpha}} &= \bar{D}_{\dot{\alpha}} - ig\bar{A}_{\dot{\alpha}}{}^k T_k \\ \nabla_\mu &= \partial_\mu - igA_\mu^k T_k. \end{aligned} \tag{1.23}$$

Matter fields ϕ can be taken to transform with a general scalar superfield parameter

$$\phi \rightarrow e^{-iK^r T_r} \phi = e^{-iK} \phi \tag{1.24}$$

where K^r is a real general scalar. To preserve covariance, we require that the covariant derivatives ∇_A transform as

$$\nabla_A \rightarrow e^{-iK} \nabla_A e^{iK}. \tag{1.25}$$

Chiral matter fields transforming as in (1.24) must satisfy *covariant* constraints

$$\bar{\nabla}_{\dot{\alpha}} \phi = 0. \tag{1.26}$$

The covariant constraint (1.26) is an overdetermined equation. By applying another covariant derivative and symmetrizing, we derive the integrability condition

$$\{\bar{\nabla}_{\dot{\alpha}}, \bar{\nabla}_{\dot{\beta}}\} \phi = 0 \tag{1.27}$$

requiring

$$F_{\dot{\alpha}\dot{\beta}} = 0 \tag{1.28}$$

where $F_{\dot{\alpha}\dot{\beta}}$ is one of the field strength superfields

$$F_{AB} = D_A A_B \pm D_B A_A - ig\{A_A, A_B] - 2i\sigma_{AB}^C A_C \tag{1.29}$$

in which the plus sign occurs for fermi-fermi components and the only nonvanishing components of σ_{AB}^C are $\sigma^c_{\alpha\dot{\beta}}$, $\bar{\sigma}^c_{\dot{\alpha}\beta}$.

The constraint (1.28) and its conjugate

$$F_{\alpha\beta} = 0 \tag{1.30}$$

are thus required in order to allow consistent coupling of chiral superfields to the Yang-Mills multiplet. These constraints are also needed to give the correct field content to the Yang-Mills supermultiplet itself. Since they set fermi-fermi field strengths to zero, they can easily be solved:

$$\begin{aligned} A_\alpha &= \frac{i}{g} e^{-gS} D_\alpha e^{gS} \\ A_{\dot{\alpha}} &= \frac{i}{g} e^{g\bar{S}} \bar{D}_{\dot{\alpha}} e^{-g\bar{S}} \end{aligned} \tag{1.31}$$

where S is a complex general scalar prepotential. Under the K-gauge transformations (1.25), we require

$$\begin{aligned} e^{gS} &\to e^{gS} e^{iK} \\ e^{g\bar{S}} &\to e^{-iK} e^{g\bar{S}}. \end{aligned} \tag{1.32}$$

There is now an additional invariance, however, that only arises upon solving the constraints as in (1.31), since

$$e^{gS} \to e^{i\bar{\Lambda}} e^{gS}, \qquad D_\alpha \bar{\lambda} = 0 \tag{1.33}$$

leaves A_α invariant. The complete gauge transformation of e^{gS}, $e^{g\bar{S}}$ is thus

$$\begin{aligned} e^{gS} &\to e^{i\bar{\Lambda}} e^{gS} e^{iK}, \qquad D_\alpha \bar{\Lambda} = 0 \\ e^{g\bar{S}} &\to e^{-iK} e^{g\bar{S}} e^{-i\Lambda}, \quad \bar{D}_{\dot{\alpha}} \Lambda = 0 \end{aligned} \tag{1.34}$$

where K is a general scalar superfield, and Λ is chiral (both contracted with Hermitian group generators).

The K-gauge transformations can be used to gauge away the imaginary part of the complex scalar prepotential S. Alternatively, one may form the K-gauge invariant combination

$$e^{2gV} = e^{gS} e^{g\bar{S}}, \tag{1.35}$$

which transforms only under the Λ-transformations,

$$e^{2gV} \to e^{i\bar{\Lambda}} e^{2gV} e^{-i\Lambda}. \tag{1.36}$$

The complex object $e^{-g\bar{S}}$ can be used to tranform a K-gauge transforming object such as ϕ in (1.24) into an object transforming under the chiral Λ-transformation:

$$\phi_{(\Lambda)} = e^{-g\bar{S}} \phi_{(K)}. \tag{1.37}$$

Moreover, this field is chiral with an ordinary superspace derivative,

$$\bar{D}_{\dot{\alpha}} \phi_{(\Lambda)} = 0, \tag{1.38}$$

and, in general, when acting on objects like $\phi_{(\Lambda)}$, the covariant derivatives must be rewritten in the new Λ-basis,

$$\nabla_\alpha = e^{-2gV} D_\alpha e^{2gV} \tag{1.39}$$

$$\nabla_{\dot{\alpha}} = \bar{D}_{\dot{\alpha}}, \tag{1.40}$$

which is just what we had in (1.16).

There is one remaining constraint that can be imposed on the fermi-fermi field strengths. If we set

$$F_{\alpha\dot{\beta}} = 0, \tag{1.41}$$

then, as can be seen from (1.29), it is possible to solve for A_μ:

$$A_\mu = -\frac{i}{4}\bar{\sigma}_\mu^{\dot{\beta}\alpha}(D_\alpha \bar{A}_{\dot{\beta}} + \bar{D}_{\dot{\beta}} A_\alpha - ig\{A_\alpha, \bar{A}_{\dot{\beta}}\}). \tag{1.42}$$

This constraint is called the conventional constraint. It is analogous to the vanishing of the torsion in general relativity. From (1.28), (1.30), and (1.41) we recover the algebra (1.18)-(1.20).

Combining the superymmetric Yang-Mills theory with a general gauge coupled matter sector, the general supersymmetric matter plus Yang-Mills theory can be written

$$I = \int d^4x\, d^2\theta \bar{\phi} e^{2gV}\phi + \mathrm{Re}\int d^4x\, d^2\theta(\mathrm{tr}W^\alpha W_\alpha + f(\phi)), \tag{1.43}$$

where $f(\phi)$ is a group invariant superpotential. If the gauge group contains some Abelian $U(1)$ factors, then it is possible to add also the Fayet-Iliopoulos terms

$$I_{FI} = \int d^4x\, d^4\theta\, C^{\hat{k}} V_{\hat{k}} \tag{1.44}$$

where $C^{\hat{k}}$ are a set of coefficients, with $\hat{k}$ running over the number of Abelian $U(1)$ factors. These terms play an important role in the study of spontaneous gauge symmetry and supersymmetry breaking.

2. ULTRAVIOLET CANCELLATIONS IN SUPERSYMMETRIC GAUGE THEORIES

Apart from the group-theoretical aspects of unifying space-time and internal symmetries, the most striking characteristic of supersymmetric theories is their greatly improved ultraviolet behavior. Many cancellations take place between the infinities in diagrams with loops of bosons and those with loops of fermions. This is seen in the case of $N = 1$ supersymmetry in the absence of interaction or mass counterterms in the Wess-Zumino model, requiring that the corresponding Z_g and Z_m just cancel the wave function renormalizations in these terms.

A basic tool for the analysis of such situations is the background field method.[2] The essential features of the method can be illustrated by ordinary Yang-Mills theory. The first step is to split the gauge field A_μ into a background and a quantum part

$$A_\mu = A_\mu^B + a_\mu, \tag{2.1}$$

where the gauge fields are contracted with the Yang-Mills hermitian generators T_r

$$A_\mu = A_\mu^r T_r. \tag{2.2}$$

The standard gauge transformation

$$A_\mu \rightarrow \bar{e}^{igK} A_\mu e^{igK} + \frac{i}{g} \bar{e}^{igK} \partial_\mu e^{igK} \tag{2.3}$$

can be interpreted in two ways. It can be thought of as a background transformation under which the background field transforms as a connection and the quantum field transforms covariantly,

$$\begin{aligned} A_\mu^B &\rightarrow \bar{e}^{igK} A_\mu^B e^{igK} + \frac{i}{g} \bar{e}^{igK} \partial_\mu e^{igK} \\ a_\mu &\rightarrow \bar{e}^{igK} a_\mu e^{igK}. \end{aligned} \tag{2.4}$$

Alternatively, the gauge transformation can be interpreted as a "quantum" transformation in which the role of the connection is played by the quantum field. As long as the correct form for the *total* transformation is preserved, the assignment of the homogeneous part of the transformation to the background and quantum fields is arbitrary. One possibility is to leave the background unchanged under a quantum gauge transformation:

$$\begin{aligned} A_\mu^B &\rightarrow A_\mu^B \\ a_\mu &\rightarrow \bar{e}^{igK_{(q)}} a_\mu e^{igK_{(q)}} + \frac{i}{g} \bar{e}^{igK_{(q)}} \mathcal{D}_\mu^B e^{igK_{(q)}} \end{aligned} \tag{2.5}$$

where

$$\mathcal{D}_\mu^B e^{igK_{(q)}} = \partial_\mu e^{igK_{(q)}} - ig[A_\mu^B, e^{igK_{(q)}}]. \tag{2.6}$$

The theory is quantized by integrating over the quantum field only, and the background field plays the role of a source. In order to factor out the volume of the gauge group at each space-time point, we have to do gauge fixing for the quantum gauge transformations (2.5) and supply a Faddeev-Popov ghost action. An essential point of the background field method is to pick a quantum gauge fixing term that is *covariant* with respect to the background gauge transformations, *e.g.*,

$$I_{GF} = \int d^4x \left(-\frac{1}{2\alpha} tr\, F^2 \right) \tag{2.7}$$

where

$$F = \mathcal{D}^B_\mu a^\mu = \partial_\mu a^\mu - ig[A^B_\mu, a^\mu]. \tag{2.8}$$

In this case, the full quantum formalism remains manifestly covariant with respect to the background gauge transformations.

The basic object that one wishes to calculate in perturbation theory is the effective action Γ, which is calculated from the one-particle-irreducible (1PI) graphs. From the effective action may be derived the connected Green's functions by the standard procedure, writing a series of tree graphs with quantum-corrected vertices obtained by expanding Γ. In the background field method, the object to be calculated is $\Gamma(A^B_\mu)$, *i.e.*, from 1PI graphs with only external background legs. That this is sufficient for the renormalization program is *not* manifest, but follows from the Slavnov-Taylor identities for the Q-gauge invariances. Thus, in three-loop and higher order diagrams where there are divergent subdiagrams involving quantum fields a_μ, the renormalizations of the coupling constants for vertices involving quantum fields are correctly taken from lower order divergent diagrams with only external background fields. Due to the presence of the background field A^B_μ in the gauge-fixing term (2.7), the Green's functions derived in the background field method are not the same as those of ordinary field theory, but they give rise to the same results for gauge-invariant matrix elements[3] (such as the S-matrix, if one sets aside the usual problems of infrared divergences in Yang-Mills theories).

In order to calculate the effective action, we write down the classical action together with its gauge fixing and ghost action terms, and separate this into terms homogeneous in the quantum fields. From the terms quadratic in the quantum fields are derived the propagators and background-quantum vertices with two quantum legs; terms of higher order in the quantum fields occur as vertices in two- and higher-loop diagrams. Terms involving only the background field or linear in quantum fields are not used in the Feynman rules for calculating the quantum corrections to the effective action. Of course, the *tree-level* contribution to the effective action consists in just the terms in the action without quantum fields. In a renormalizable theory such as Yang-Mills theory, the counterterms must be of the same form as this tree-level contribution to $\Gamma(A^B_\mu)$. The possibility of non-renormalization theorems arises from the fact that the Feynman rules are derived from different terms in the expansion of the action from those which must serve as counterterms.

Since the background field transformations (2.4) must remain a manifest invariance of $\Gamma(A_\mu^B)$, the counterterm in Yang-Mills theory can only be some divergent coefficient times the original action. This can only be achieved by a multiplicative renormalization of fields and coupling constants such that the product gA_μ^B remains unrenormalized, so

$$Z_g = Z_A^{-1/2}. \tag{2.9}$$

Another consequence of the background field method in Yang-Mills theory is the non-renormalization of the Chern-Simons form,

$$K_\mu = \epsilon^{\mu\nu\rho\sigma} tr\,(A_\nu \partial_\rho A_\sigma - \frac{i}{3} g\, A_\nu [A_\rho, A_\sigma]), \tag{2.10}$$

the operator whose divergence is $tr\, F^*_{\mu\nu} F^{\mu\nu}$. Green's functions with insertions of this operator may be calculated from Feynman rules derived from an action including the term

$$I_{CS} = \int d^4x\, \xi^\mu K_\mu. \tag{2.11}$$

While this is clearly not locally gauge invariant, it is nonetheless renormalizable. One might expect it to mix with other non-gauge-invariant operators of dimension three. However, the term (2.11) does become locally gauge invariant when the source ξ^μ is taken to be a longitudinal vector, and this requires the two terms in (2.10) to be renormalized with the same coefficient. Consideration of this special case with longitudinal ξ^μ would not be sufficient if there were a purely transverse operator of dimension three, but there is no such operator that is a rigid gauge scalar.

If we insert the background-quantum splitup into (2.10), the relative factor of -1/3 is just what is needed to make the terms quadratic and cubic in quantum fields background gauge invariant:

$$I_{CS}^{(2)} = \frac{1}{2} \int d^4x \xi^\mu \epsilon_{\mu\nu\rho\sigma}\, tr\{a^\nu (\mathcal{D}_B^\rho a^\sigma - \mathcal{D}_B^\sigma a^\rho)\} \tag{2.12}$$

$$I_{CS}^{(3)} = -\frac{ig}{3} \int d^4x \xi^\mu \epsilon_{\mu\nu\rho\sigma}\, tr\{a^\nu [a^\rho, a^\sigma]\}. \tag{2.13}$$

Thus, the Feynman rules derived in the background field method will be fully background covariant, despite the lack of local invariance of the total term

(2.11). The lack of local gauge invariance of (2.11) does show up, however, in the parts of the action of zeroth and first order in the quantum fields. Since (2.11) is a renormalizable term, the counterterm for renormalizing $\Gamma(A^\mu_B, \xi^\nu)$ would be just the part of (2.11) dependent only on background fields. But since this part is not background gauge invariant, it cannot occur as a counterterm, so (2.11) must remain unrenormalized.[4]

As in the mass and interaction terms in the Wess-Zumino model, the non-renormalization of the term (2.11) happens because the renormalization of the source field ξ^μ just cancels the wave function renormalizations of the background gauge fields. Alternately, since the product gA^B_μ is unrenormalized, the product $g^2 K_\mu$ is finite. Thus, in a minimal subtraction scheme where renormalization constants are taken only from the poles, *e.g.*, in dimensional regularization, $g^2 K_\mu$ has no anomalous dimension. Writing the renormalization group equation for this operator, we have

$$\left(\mu\frac{\partial}{\partial\mu} + \beta(g)\frac{\partial}{\partial g}\right)(g^2 K_\mu) = 0 \tag{2.14}$$

where μ is the renormalization scale. Letting the anomalous dimension of K_μ be γ_K, we have

$$-\gamma_K g^2 K_\mu + 2\beta(g) g K_\mu = 0 \tag{2.15}$$

or

$$\gamma_K = \frac{2\beta(g)}{g}. \tag{2.16}$$

The result (2.16) for the anomalous dimension of the Chern-Simons term is the basis for the well-known Adler-Bardeen theorem[5] for the anomalous divergence of the $U(1)$ chiral current in a Yang-Mills theory with spinors. If one couples the Yang-Mills field to spinors, then the Chern-Simons form can mix under renormalization with the chiral current

$$J^5_\mu = \bar{\lambda}^b \gamma_5 \gamma_\mu \lambda_b \tag{2.17}$$

where b is a representation index for the Yang-Mills group. Due to the possibility of mixing, we now have a matrix of anomalous dimensions

$$\gamma \begin{pmatrix} J^5_\mu \\ K_\mu \end{pmatrix} = \begin{pmatrix} \gamma_J & 0 \\ \gamma_{KJ} & \gamma_K \end{pmatrix} \begin{pmatrix} J^5_\mu \\ K_\mu \end{pmatrix}, \tag{2.18}$$

where the zero in the upper right-hand corner is due to the lack of local gauge invariance of K_μ. J^5_μ may mix into K_μ, on the other hand, so there is, in general, an off-diagonal γ_{KJ}.

If there is a chiral anomaly, the divergence of J^5_μ is

$$\partial^\mu J^5_\mu = RF^*_{\mu\nu}F^{\mu\nu} = R\partial^\mu K_\mu. \tag{2.19}$$

Although not gauge invariant, the current $J^5_\mu - RK_\mu$ is nonetheless conserved, and thus must have vanishing anomalous dimension if it is to correctly generate transformations on fields. Thus,

$$\left(\mu\frac{\partial}{\partial\mu} + \beta(g)\frac{\partial}{\partial g}\right)(J^5_\mu - RK_\mu) = 0. \tag{2.20}$$

Substituting the renormalization group operators for J^5_μ andK_μ, we have

$$(-\gamma_J + R\gamma_{KJ})J^5_\mu + \left(-\beta(g)\frac{\partial R}{\partial g} + R\gamma_K\right)K_\mu = 0, \tag{2.21}$$

and since J^5_μ and K_μ are independent, we have

$$\gamma_J = R\gamma_{KJ} \tag{2.22}$$

$$\frac{\beta(g)}{R}\frac{\partial R}{\partial g} = \gamma_K. \tag{2.23}$$

Relation (2.22) provides a convenient way to calculate the anomalous dimension γ_J by calculating γ_{KJ} to one loop order less. In a supersymmetric theory, if J^5_μ is the chiral current component of the supercurrent, which contains other component currents that remain conserved, then it must have zero anomalous dimension and similarly for γ_{KJ}. Anomalous chiral currents that are not in the supercurrent may have a nonvanishing anomalous dimension.

The result (2.16) for the anomalous dimension γ_K still holds in the presence of coupling to spinors, for the Chern-Simons term must still remain unrenormalized. Inserting (2.16) into (2.23), we obtain the Adler-Bardeen theorem

$$\left(g\frac{\partial R}{\partial g} - 2R\right) = 0 \tag{2.24}$$

i.e.,

$$R = cg^2 \tag{2.25}$$

where

$$c = \frac{C_2(G)}{16\pi^2} \tag{2.26}$$

given by the well-known one-loop result.

A second example of a non-renormalization theorem in the background field method is the familiar non-renormalization of chiral superpotentials in $N = 1$ supersymmetry. Here one splits up the chiral superfield Φ into background and quantum parts

$$\Phi = \Phi_B + \phi. \tag{2.28}$$

All these superfields satisfy the chirality constraint

$$\bar{D}_{\dot{\alpha}}\Phi_B = \bar{D}_{\dot{\alpha}}\phi = 0. \tag{2.29}$$

In the case of chiral superfields, it is simple enough to work directly with the constrained superfield[6] ϕ by going to a chiral basis where it depends only on $z^\mu = x^\mu - i\theta^\alpha\sigma^\mu_{\alpha\dot{\alpha}}\bar{\theta}^{\dot{\alpha}}$ and θ^α.

Working directly with constrained superfields is not something that is generally feasible with the more complicated constraints that occur in extended supersymmetry. Therefore, we proceed to solve the constraint on ϕ[7,8] and work in terms of unconstrained superfields. The chirality constraint on ϕ is solved by

$$\phi = \bar{D}^2 X \tag{2.30}$$

where X is a dimension zero "prepotential" superfield. Since not all of the superspins in X contribute to ϕ, there is a gauge transformation

$$\delta X = \bar{D}^{\dot{\alpha}}\Lambda_{\dot{\alpha}} \tag{2.31}$$

that leaves ϕ unchanged.

To develop the Feynman rules, we insert the background-quantum split into the action and keep the terms quadratic or higher in the quantum fields. Taking the example of a single chiral superfield with mass and interaction terms, we have

$$I^B = \int d^4xd^4\theta\bar{\Phi}_B\Phi_B - \mathrm{Re}\int d^4xd^2\theta(m\Phi_B^2 + \frac{1}{3}h\Phi_B^3) \tag{2.32}$$

$$I^2_{(kinetic)} = \int d^4xd^4\theta(D^2\bar{X}\bar{D}^2X) \tag{2.33}$$

$$I^2_{(mass)}+I^3 = -\mathrm{Re}\int d^4xd^2\theta(m\bar{D}^2X\bar{D}^2X+\frac{h}{3}(3(\bar{D}^2X)^2\Phi_B+(\bar{D}^2X)^3). \tag{2.34}$$

The kinetic term (2.33) still possesses the gauge invariance (2.31), which needs to be fixed. A convenient gauge fixing function is

$$F_\alpha = D_\alpha X, \tag{2.35}$$

and the most convenient gauge fixing term is[7]

$$I_{GF} = \int d^4xd^4\theta\bar{F}^{\dot{\alpha}}(D_\alpha\bar{D}_{\dot{\alpha}} + \frac{3}{4}\bar{D}_{\dot{\alpha}}D_\alpha)F^\alpha \tag{2.36}$$

so that the resulting kinetic term is

$$I^2_{(kinetic)} + I_{GF} = \int d^4xd^4\theta\ \bar{X}\Box X. \tag{2.37}$$

There is no ghost action, since the transformation (2.31) is Abelian.

The mass and interaction terms (2.34) may be rewritten as a full superspace integral:

$$I^2_{(mass)} + I^3 = -\mathrm{Re}\int d^4xd^4\theta\left(mX\bar{D}^2X + \frac{h}{3}(3X(\bar{D}^2X)\Phi_B + X(\bar{D}^2X)^2)\right). \tag{2.38}$$

Notice that the background field enters as the constrained field Φ_B; it was never necessary to solve the constraint in the background.

The non-renormalization theorem is now apparent, for the Feynman rules derived from (2.37) and (2.38) give quantum corrections to the effective action that involve only full superspace integrals and couplings to the background in the constrained form Φ_B. Since it is not possible to rewrite the mass and interaction counterterms depending only on Φ_B (cf. 2.32) as full superspace integrals using only local operators, these counterterms are ruled out.

Several points need to be added to the above discussion for completeness, which we shall not dwell upon in detail here. Regularization of the quantum theory is a crucial issue that has received much debate. In practice, regularization by dimensional reduction[9] has been used, guaranteeing supersymmetry in the counterterms at least at low-loop orders. For the pure Wess-Zumino model, supersymmetric Pauli-Villars or higher derivative regularization can also be used. If dimensional reduction is used, care must be taken in massless theories to separate the ultraviolet from the infrared divergences, requiring the introduction of some form of infrared regularization.

When we come to treat supersymmetric gauge theories, we must make the background-quantum split in such a way as to preserve the maximal manifest background symmetry. Although $N = 1$ super Yang-Mills theory is ultimately expressed in terms of a dimension zero prepotential S, the best way to make the background-quantum split is at the level of the superspace gauge connection A_α[7]:

$$\nabla_C = D_C - igA_C, \quad C = \alpha, \dot{\alpha}; \mu \tag{2.39}$$

$$A_\alpha = A^B_\alpha + a_\alpha, \ etc. \tag{2.40}$$

In order to allow coupling to constrained matter superfields, we must still impose the representation-preserving constraints

$$F_{\alpha\beta} = F_{\dot{\alpha}\dot{\beta}} = 0 \tag{2.41}$$

and we shall continue to express the vector gauge connection A_μ in terms of $A_\alpha, A_{\dot{\alpha}}$ by solving the conventional constraint

$$F_{\alpha\dot{\beta}} = 0. \tag{2.42}$$

In order to maintain background gauge covariance, the background field must be constrained by

$$F^B_{\alpha\beta} = F^B_{\dot{\alpha}\dot{\beta}} = F^B_{\alpha\dot{\beta}} = 0 \tag{2.43}$$

where

$$F^B_{\alpha\beta} = \frac{i}{g}\{\mathcal{D}^B_\alpha, \mathcal{D}^B_\beta\} \tag{2.44}$$

$$\mathcal{D}^B_\alpha = D_\alpha - ig\ A^B_\alpha, \tag{2.45}$$

etc. The quantum fields a_α, $a_{\dot\alpha}$ must then be constrained by

$$\mathcal{D}^B_\alpha a_\beta + \mathcal{D}^B_\beta a_\alpha - ig\{a_\alpha, a_\beta\} = 0, \tag{2.46}$$

and similarly for $a_{\dot\alpha}$. As in the case of the chiral superfield, we solve only the constraints on the quantum fields in terms of a dimension zero complex prepotential U:

$$a_\alpha = \frac{i}{g}e^{-gU}\mathcal{D}^B_\alpha e^{gU} \tag{2.47}$$

$$a_{\dot\alpha} = \frac{i}{g}e^{g\bar{U}}\bar{\mathcal{D}}^B_{\dot\alpha} e^{-g\bar{U}}. \tag{2.48}$$

When the super Yang-Mills multiplet is coupled to matter, the constraints on the matter fields must be written with covariant derivatives. Since we wish to work with background covariant objects, we use the quantum gauge prepotential to define a background covariantly constrained chiral field:

$$\tilde{\Phi} = e^{-g\bar{U}}\Phi \tag{2.49}$$

where

$$\bar{\mathcal{D}}^B_{\dot\alpha}\tilde{\Phi} = 0. \tag{2.50}$$

The intermediate field $\tilde{\Phi}$ can now be separated into background and quantum parts, generalizing (2.28):

$$\tilde{\Phi} = \Phi_B + \bar{\mathcal{D}}^2_B X. \tag{2.51}$$

As before, there is a new gauge invariance for the matter system that arises from the solution (2.51),

$$\delta X = \bar{\mathcal{D}}^{\dot\alpha}_B \Lambda_{\dot\alpha}, \tag{2.52}$$

which leaves $\tilde{\Phi}$ invariant by virtue of the background constraint (2.43).

In order to derive the Feynman rules used to calculate the effective action $\Gamma(A^B_\alpha, \Phi_B)$, we must insert the background-quantum split into the classical action and keep terms quadratic or higher in quantum fields:

$$I^Q_{Y.M.} = \int d^4x d^4\theta \, \{\bar{a}^{\dot{\alpha}}(2\mathcal{D}^2_B \bar{a}_{\dot{\alpha}} + \mathcal{D}^B_\alpha \bar{\mathcal{D}}^B_{\dot{\alpha}} a^\alpha + i\mathcal{D}^B_{\alpha\dot{\alpha}} a^\alpha$$

$$-ig[a^\alpha, \bar{\mathcal{D}}^B_{\dot{\alpha}} a_\alpha] + \frac{g^2}{4}[a^\alpha, \{a_\alpha, \bar{a}_{\dot{\alpha}}\}]) + c.c.\Big\} \quad (2.53)$$

$$I^Q_M = \int d^4x d^4\theta \, \{(\bar{\Phi}^B + \mathcal{D}^2_B \bar{X})e^{2gV}(\Phi^B + \bar{\mathcal{D}}^2_B X)$$

$$- \bar{\Phi}^B(1 + 2gV)\Phi^B - \mathcal{D}^2_B \bar{X}\Phi_B - \bar{\Phi}^B \bar{\mathcal{D}}^2_B X$$

$$+ \left(mX\bar{\mathcal{D}}^2_B X + \frac{h}{3}(3(\bar{\mathcal{D}}^2_B X)\Phi_B + X(\bar{\mathcal{D}}^2_B X)^2)\right) + c.c.\Big\} \quad (2.54)$$

where

$$e^{2gV} = e^{gU} e^{g\bar{U}}. \quad (2.55)$$

The essential point to note in (2.53),(2.54) is that all integrals are over the full superspace, with the background fields appearing only through the constrained superfields A^B_α, Φ_B. The quantum corrections to the effective action must be written in the same way, and the ultraviolet counterterms must be gauge invariant local expressions written in this fashion.

Before stating the non-renormalization theorem, we must discuss a special feature of the background field method that occurs only in the one-loop graphs. In (2.53),(2.54), we have not included the gauge fixing terms for either the super Yang-Mills action or for the matter action. The Λ-gauge fixing function for the Yang-Mills action may be chosen to be

$$F = \bar{\mathcal{D}}^2_B V, \quad (2.56)$$

which remains covariant under background K-gauge transformations. Since the parameter Λ of a quantum gauge transformation is now background covariantly chiral, so are the ghost and antighost. In order to quantize in terms of unconstrained superfields, it is then necessary to solve the background covariant constraints on the ghost and antighost in the same way as for a chiral matter field. Note that since ghost fields never occur on the external legs of diagrams, we do not split them into background plus quantum parts. In addition to the Faddeev-Popov ghosts, the functional averaging of the gauge condition required to obtain the standard gauge fixing term introduces further chiral ghosts, the Nielsen-Kallosh ghosts,[10] which may be treated in the same way as the Faddeev-Popov ghosts.

For background covariantly chiral fields, whether matter fields or ghosts, we can fix the gauge as in (2.35),(2.36). However, we now have to include a ghost action, for the background covariant gauge fixing function

$$F_\alpha = \mathcal{D}_\alpha^B X \tag{2.57}$$

and the gauge transformation (2.52) cause the ghost $C_{\dot\alpha}^{(1)}$ to couple to the background field. Since this ghost occurs only in the combination $\bar{\mathcal{D}}_B^{\dot\alpha} C_{\dot\alpha}^{(1)}$, the ghost action has its own gauge invariance,

$$\delta C_{\dot\alpha}^{(1)} = \bar{\mathcal{D}}_B^{\dot\beta} \Lambda_{(\dot\alpha\dot\beta)}. \tag{2.58}$$

Fixing the gauge invariance (2.58) and providing a second level ghost action with $C_{\dot\alpha\dot\beta}^{(2)}$, we find another gauge invariance with parameter $\Lambda_{(\dot\alpha\dot\beta\dot\gamma)}$, and so forth ad infinitum. This problem is known as that of "ghosts for ghosts."[11]

In order to stop the above proliferation of ghosts, it is necessary to introduce a prepotential for the background gauge connnection. Thus, for this purpose only, we introduce the dimension zero background prepotential u^B:

$$A_\alpha^B = \frac{i}{g} e^{-gU^B} D_\alpha e^{gU^B} \tag{2.59}$$

$$A_{\dot\alpha}^B = \frac{i}{g} e^{g\bar U^B} \bar D_{\dot\alpha} e^{-g\bar U^B}. \tag{2.60}$$

Using $e^{g\bar U^B}$, the $C_{\dot\alpha}^{(1)}$ transformation can be written

$$\delta C_{\dot\alpha}^{(1)} = e^{g\bar U^B} \bar D^{\dot\beta} \tilde\Lambda_{(\dot\alpha\dot\beta)} e^{-g\bar U^B}. \tag{2.61}$$

The next ghost then appears only in the combination $\bar{D}^{\dot{\beta}}\tilde{C}^{(2)}_{\dot{\alpha}\dot{\beta}}$, and its gauge transformation involves only *flat* derivatives:

$$\delta\tilde{C}^{(2)}_{\dot{\alpha}\dot{\beta}} = \bar{D}^{\dot{\gamma}}\tilde{\Lambda}_{(\dot{\alpha}\dot{\beta}\dot{\gamma})}. \tag{2.62}$$

Then the gauge freedom (2.62) may be fixed using *flat* derivatives and all further ghosts decouple.

The price of decoupling the infinite series of ghosts has been the introduction of a background prepotential for the Yang-Mills multiplet. It was never necessary to introduce a prepotential for the background chiral superfields. Since it was introduced solely for the purpose of converting background covariant derivatives to flat ones in ghost actions, the background Yang-Mills prepotential can occur only in one-loop diagrams. The interactions with U^B occur only in ghost actions like that for $\tilde{C}^{(2)}_{\dot{\alpha}\dot{\beta}}$, derived using (2.61). Since the ghost $\tilde{C}^{(2)}_{\dot{\alpha}\dot{\beta}}$ couples *only* to background fields, it can occur only in one-loop diagrams.

The non-renormalization theorem for general $N = 1$ gauge and gauge-coupled matter theories can now be stated. The leading infinite parts of the effective action must be local in structure. Nonlocal infinities are removed by renormalizations at lower loop orders. At one loop, the divergent parts of Γ must be written as background gauge invariant integrals over a single $d^4\theta$ and the loop momentum of an expression involving the background super Yang-Mills prepotential U^B and the covariantly constrained matter fields Φ_B. At two loops and higher, the local divergent parts of Γ must be written as a single $d^4\theta$ integral and integrals over the loop momenta of a background gauge invariant expression involving the background connection A^B_α and the constrained matter fields Φ_B. Terms that cannot be written in accordance with these rules must remain unrenormalized, such as mass and interaction terms for chiral matter multiplets.

In extended supersymmetric theories, there are even more powerful non-renormalization theorems that can be derived along the lines detailed above. A basic problem that arises is the fact that, in many cases, the full supersymmetry of the field equations cannot be manifestly realized in the quantum formalism, at least with a finite number of component fields. Thus, one must first determine what is the maximal symmetry that can be manifestly realized. The chief example of a theory with this sort of problem is the maximally supersymmetric Yang-Mills theory,[12], which in four dimensions has $N = 4$ supersymmetry.

The classical dynamics of the $N = 4$ super Yang-Mills theory may be derived from the component action

$$I_{Y.M.4} = \frac{1}{g^2} \int d^4x \mathrm{tr} \left\{ -\frac{1}{4} F_{\mu\nu} F^{\mu\nu} + \frac{1}{2} \nabla^\mu \bar{\phi}^{ij} \nabla_\mu \phi_{ij} - i\lambda_i \sigma_\mu \nabla^\mu \bar{\lambda}^i \right.$$
$$- \frac{i}{2} \bar{\lambda}_{\dot{\beta}} [\bar{\lambda}^{\dot{\beta} j}, \phi_{ij}] - \frac{i}{2} \lambda_i^\alpha [\lambda_{\alpha j}, \bar{\phi}^{ij}]$$
$$\left. + \frac{1}{4} [\phi_{ij}, \phi_{k\ell}][\bar{\phi}^{ij}, \bar{\phi}^{k\ell}] \right\} \tag{2.63}$$

where $i, j, k, \ell : 1 \to 4$, and the four Weyl spinors λ^i_α and the six scalars $\phi_{ij} = \frac{1}{2}\epsilon_{ijk\ell}\bar{\phi}^{k\ell}$ are in the adjoint representation of the gauge group. The action (2.63) has a manifest $SU(4)$ invariance, but supersymmetry is not manifest and only forms a closed algebra on the fields present in (2.63) subject to the equations of motion.

In order to see the problem with deriving a full linearly realized supersymmetric formalism for the $N = 4$ theory,[13,8] suppose that there were an off-shell version involving a multiplet of fields $\mathbf{\Phi}$. The linearized action could then be written schematically in the form

$$I = \int d^4x \, \Phi \mathcal{O} \Phi \tag{2.64}$$

where $\mathcal{O}$ is some differential operator. With linearly realized supersymmetry, we can also form the higher derivative invariant

$$I' = \frac{1}{m^2} \int d^4x \Phi \mathcal{O} \Box \Phi \tag{2.65}$$

and analyze the spectrum of $I + I'$, which must fall into linear representations of $N = 4$ supersymmetry, with either physical or ghost signs for the residues of the propagators. Each massless physical fermion in the theory derived from I gives rise to two massless physical helicity states and four massive ghost states, while each auxiliary fermion pair in I contributes four massive physical states and four massive ghost states. It is assumed here that auxiliary fermions always occur in pairs. Since each massive linear $N = 4$ multiplet contains an integral multiple of 128 fermionic states we have for n auxiliary fermion pairs and the four original fermion fields,

$$\begin{aligned} 4n + 4.4 &= 128p \quad \text{massive physical states} \\ 4n &= 128q \quad \text{massive ghost states} \end{aligned} \tag{2.66}$$

where p and q are integers. Subtracting, one gets

$$p - q = \frac{1}{8} \tag{2.67}$$

which is not possible for finite integers p and q. It is possible to evade the above "no-go" theorem if one admits an infinite number of component fields, as in the harmonic superspace formalism,[14,37] which we shall discuss in the next chapter. To date, this formalism has not yet been developed to give a full quantization procedure.

In order to proceed with a finite number of component fields in a theory such as $N = 4$ super Yang-Mills, it is necessary to pick some part of the total symmetry of the theory that can be manifestly realized. One approach[8,15] that has been applied to the analysis of the $N = 4$ theory's infinities uses manifest Lorentz invariance and $N = 2$ supersymmetry. Another approach uses the light-cone gauge,[16] in which manifest Lorentz invariance is given up, but one can keep the $SU(4)$ invariance manifest. In both of these approaches, half of the total number of components of the supersymmetry transformations are linearly realized. Since the light-cone gauge precludes use of the background field method, it has not proven as useful in understanding ultraviolet cancellations where delicate issues involving gauge invariance arise.

When decomposed into representations of $N = 2$ supersymmetry, the propagating fields of $N = 4$ super Yang-Mills theory fall into two $N = 2$ massless supermultiplets, each carrying the adjoint representation of the gauge group. The two multiplets are the $N = 2$ Yang-Mills multiplet[17] ($A_\mu, \lambda_{\alpha i}(i = 1, 2), \phi$ (complex)) and the $N = 2$ matter representation, the hypermultiplet[18] ($\psi_{\alpha i}$, $L^{(ij)} = \epsilon^{ik}\epsilon^{jm}L_{km}(i, j, k, m = 1, 2)$, S (real)). In this formalism, there is a manifest $SU(2) \times U(1)$ symmetry. It is not necessary to discuss the $N = 4$ theory specifically in this context, for it turns out to be but one of a whole class of ultraviolet finite theories with $N = 2$ supersymmetry.

The superspace for $N = 2$ supersymmetry has fermionic coordinates θ^i_α, where i is an $SU(2)$ doublet index. In order to carry out background field quantization as outlined in the $N = 1$ case above, it is necessary to have a formulation of the theory in unconstrained $N = 2$ superfields. This formulation is considerably more complicated algebraically than the $N = 1$ formalism. For practical calculations, the $N = 1$ superfield formalism appears to be optimal. It has allowed a calculation by hand of the $N = 4$ theory's divergences to the three-loop order,[19] a result which was originally obtained by a component field calculation[20] starting from (2.63), but needing a computer. In contrast, in the $N = 2$ formalism, the non-renormalizaton theorems become transparent, although its complexity makes practical calculation difficult. Here, we shall only set out the salient features of the $N = 2$ quantization program. For more details, we refer the reader to Ref. 8.

The action for the $N = 2$ gauge multiplet can be written as a chiral integral over the square of the field strength superfield W^{21}:

$$I_{Y.M.2} = \mathrm{Re} \int d^4x d^4\theta \mathrm{tr} W^2. \tag{2.68}$$

The chiral integral for $N = 2$ supersymmetry is now over the four $d\theta_i^\alpha$; a full superspace integral would be over $d^8\theta$. The field strength superfield W satisfies the $N = 2$ covariant chirality condition

$$\bar{\nabla}_{\dot{\alpha}i} W = 0, \tag{2.69}$$

where

$$\bar{\nabla}_{\dot{\alpha}i} = \bar{D}_{\dot{\alpha}i} - igA_{\dot{\alpha}i} = -\frac{\partial}{\partial\bar{\theta}^{\dot{\alpha}i}} - i\theta_i^\alpha \sigma^\mu_{\alpha\dot{\alpha}} \partial_\mu - igA_{\dot{\alpha}i}. \tag{2.70}$$

The $\theta = \bar{\theta} = 0$ component of W is the complex scalar field of the multiplet. As in the $N = 1$ case, the gauge connection A^i_α must satisfy certain integrability conditions in order to allow gauge coupling to $N = 2$ matter multiplets (hypermultiplets). These representation-preserving constraints are

$$F^{(ij)}_{\alpha\beta} = F_{\dot{\alpha}(i\dot{\beta}j)} = 0 \tag{2.71}$$

$$F^i_{\alpha\dot{\beta}j} - \frac{1}{2}\delta^i_j F^k_{\alpha\dot{\beta}k} = 0. \tag{2.72}$$

In addition, there is a conventional constraint that determines the vectorial part of the connection, A_μ, in terms of $A_{\dot{\alpha}i}$, A^i_α:

$$F^k_{\alpha\dot{\beta}k} = 0. \tag{2.73}$$

The unconstrained part of $F^{ij}_{\alpha\beta}$ is the field strength superfield W:

$$F^{ij}_{\alpha\beta} = \epsilon^{\alpha\beta}\epsilon^{ij} W. \tag{2.74}$$

In the case of an Abelian gauge group, the constraints (2.71),(2.72) are solved by[22]

$$A^i_\alpha = D^i_\alpha P + D^{3i}_\alpha \bar{D}^{jk} V_{jk} - \frac{1}{2} D^i_\alpha \{D^j_k, \bar{D}^{\ell k}\} V_{j\ell} \tag{2.75}$$

where P is real and may be eliminated by a K-gauge transformation. The prepotential $V_{(jk)}$ is an $SU(2)$ triplet of dimension -2. The quadratic and cubic fermionic derivatives in (2.74) are defined by

$$D^i_\alpha = \epsilon_{\alpha\beta} D^{ij} + \epsilon^{ij} D_{\alpha\beta} \tag{2.76}$$

$$D^i_\alpha D^{jk} = \epsilon^{i(j} D^{3k)}_\alpha . \tag{2.77}$$

Extending the solution (2.75) for the constraints (2.71),(2.72) to the non-Abelian case is algebraically very complicated. However, since all we require in order to develop Feynman rules is an expansion of the solution in powers of the coupling constant g, we may employ the following trick.[23,8] Differentiating the constraints (2.71),(2.72) with respect to the coupling constant g, we obtain

$$\nabla^{(i}_\alpha B^{j)}_\beta + \nabla^{(j}_\beta B^{i)}_\alpha = 0 \tag{2.78}$$

$$\nabla^{(i}_\alpha B^{j)}_{\dot\beta} + \bar\nabla^{(j}_{\dot\beta} B^{i)}_\alpha = 0 \tag{2.79}$$

where

$$B^i_\alpha = \frac{\partial}{\partial g}(g A^i_\alpha) \tag{2.80}$$

$$B^i_{\dot\alpha} = \epsilon^{ij} B_{\dot\alpha j} . \tag{2.81}$$

Since (2.78),(2.79) have a similar form to the constraints in the Abelian gauge theory, they may be solved analogously:

$$\begin{aligned} B^i_\alpha = {} & \nabla^i_\alpha P + \nabla^{3i}_\alpha \bar\nabla^{jk} V_{jk} \\ & - \frac{1}{2} \nabla^i_\alpha \left(\{\nabla^j_k, \bar\nabla^{\ell k}\} V_{j\ell} + \frac{i}{2} [W, \nabla^{jk} V_{jk}] - \frac{i}{2} [\bar W, \bar\nabla^{jk} V_{jk}] \right) . \end{aligned} \tag{2.82}$$

The gauge connection A^i_α may then be reconstructed as a power series in g from the integral equation

$$A^i_\alpha = \frac{1}{g} \int_0^g dk \, B^i_\alpha(k) . \tag{2.83}$$

The split between background and quantum fields in the $N = 2$ case is done at the level of the connection A^i_α, in direct analogy with the $N = 1$ case. For everything but one-loop diagrams, only the quantum part of the connection needs to be expressed in terms of a prepotential. For details, and treatment of the quantum gauge invariances, their gauge fixing, and ghosts, see Ref. 8. The non-renormalization theorem derives from the fact that all superspace integrations in the Feynman rules are over the full superspace $d^8\theta$ and the background field enters only as the dimension 1/2 connection A^{Bi}_α, except in some one-loop ghost diagrams. Since the action (2.68) cannot be written this way as a local expression, it is ruled out as a counterterm at two loops and higher. What happens at the one-loop level must be settled by specific calculation in a given theory.

The hypermultiplet matter sector may also be given an off-mass-shell formulation in terms of $N = 2$ superfields[15] (without a central charge). In the free case, the action is

$$I_{H.M.} = \text{Re} \int d^4x d^8\theta (\rho^\alpha_i D_{\alpha j} L^{ij} + X_{ijk\ell} L^{ijk\ell}) \tag{2.84}$$

where ρ^α_i is a dimension -3/2 prepotential and $X_{(ijk\ell)}$ is a dimension -1 prepotential carrying the pseudo-real $\underset{\sim}{5}$ representation of $SU(2)$. The non-gauge physical and auxiliary fields are contained in the pseudo-real or real superfields $L^{(ij)}$, $L^{(ijk\ell)}$, and S, which satisfy the constraints

$$\begin{aligned} D^{(i}_\alpha L^{jk)} &= D_{\alpha\ell} L^{ijk\ell} \\ D^{(t}_\alpha L^{ijk\ell)} &= \bar{D}^{(t}_{\dot\alpha} L^{ijk\ell)} = 0 \\ D_{\alpha\beta} S &= [D^i_\alpha, \bar{D}_{\dot\alpha i}] S = 0. \end{aligned} \tag{2.85}$$

The solution to the constraints (2.85) is given in terms of the prepotentials ρ^α_i and $X_{ijk\ell}$:

$$\begin{aligned} L^{ij} &= \frac{3}{4} \bar{D}^{ij} D^{3k}_\alpha \rho^\alpha_k + \frac{1}{2} \bar{D}^{(i}_k D^{j)k} D^\ell_\alpha \rho^\alpha_\ell + \text{c.c.} \\ L^{ijk\ell} &= \frac{2}{5} D^{(ij} \bar{D}^{k\ell)} [D^m_\alpha \rho^\alpha_m + \bar{D}^{\dot\alpha}_m \bar{\rho}^m_{\dot\alpha}] \\ S &= D^{ij} \bar{D}^{k\ell} X_{ijk\ell}. \end{aligned} \tag{2.86}$$

The equations of motion following from (2.84) are

$$\begin{aligned} L_{ijk\ell} &= 0 \\ D_{\alpha j} L^{ij} &= D^i_\alpha S, \end{aligned} \tag{2.87}$$

which correctly describe the free hypermultiplet.

The $N = 2$ hypermultiplet has a non-renormalization theorem for its action (2.84). Making the background quantum split in L^{ij}, $L^{ijk\ell}$, and S analogously to the procedure followed in the chiral $N = 1$ multiplet, solving the constraints (2.85) only for the quantum fields, one finds that all quantum corrections to the effective action must be written as full superspace integrals over functionals of L^{ij}, $L^{ijk\ell}$, and S. Since the action (2.84) cannot be locally written in this way, it is ruled out as a counterterm and must remain unrenormalized. This is true at the one-loop and at higher loop orders, since, as with the chiral $N = 1$ multiplet, there is never any necessity of introducing a prepotential for the background hypermultiplet fields. The non-renormalization of the hypermultiplet action means that there is no wave function renormalization for the hypermultiplet fields. These conclusions are obtained equally in the case of coupling to the $N = 2$ gauge multiplet. For details, see Ref. 8.

The non-renormalization theorems for $N = 2$ supersymmetry are summarized by the statements that matter wave function renormalizations never occur, while the only possibility for a leading divergence of the form of the Yang-Mills action is at the one-loop order. As usual in the background field method, if such a divergence does occur, it requires coupling constant and wave function renormalizations such that the product (gA^i_α) remains unrenormalized. Once an infinity has occurred at the one-loop order, it also gives rise to non-leading local divergences in higher orders due to counterterm insertions to remove non-local divergences coming from divergent subgraphs. In regularization by dimensional reduction these non-leading divergences are of quadratic or higher order in $(d-4)^{-1}$, with coefficients that can be obtained from the renormalization group pole equations.

The condition for an $N = 2$ supersymmetric theory to be ultraviolet finite is that the one-loop gauge coupling constant renormalization vanish[24]

$$\sum_a m_a T(R_a) - C_2(G) = 0, \tag{2.88}$$

where there are m_a hypermultiplets in representations R_a of the gauge group, and for generators T^r

$$\begin{aligned} \mathrm{tr} T^r T^s &= C_2(G)\delta^{rs} \\ T^r T^r &= T(R)\mathbf{1}, \end{aligned} \tag{2.89}$$

where the conventional normalization has $T(R) = \frac{1}{2}$ for the fundamental representation. Equation (2.88) has many solutions beside the $N = 4$ theory, in which case R is the adjoint and $m = 1$. For example, for $SU(N)$ with hypermultiplets in the fundamental representation, $C_2(G) = N$, so we need $m = 2N$ hypermultiplets for a finite theory. In the formulation of the hypermultiplet that we have discussed, there must be an even number of hypermultiplets if they are to be given a complex representation of the gauge group, since the formulation with L^{ij}, $L^{ijk\ell}$, and S is pseudo-real (*i.e.*, satisfies a reality condition with indices raised and lowered with ϵ_{ij}). By explicit two-loop calculation[25], it has been found that finiteness is obtained even without this restriction, provided (2.88) is satisfied. A proof of finiteness in the general case may be afforded by the harmonic superspace formalism.[14,37]

In the case of the $N = 4$ theory, the cancellations may also be understood by an elegant heuristic argument[26] based upon the multiplet structure of the anomalies. Due to the pseudo-reality condition $\phi_{ij} = \frac{1}{2}\epsilon_{ijk\ell}\bar{\phi}^{k\ell}$ on the $SU(4)\,\underset{\sim}{6}$ of scalars, there is no way to nontrivially extend the $SU(4)$ to $U(4)$, unlike the case in the non-self-conjugate multiplets for $N = 2$. Thus, if there were a non-vanishing trace anomaly, supersymmetry would link this with a non-vanishing anomaly in some chiral current of the theory, which could only be one of the 9 chiral generators of $SU(4)$. Since the adjoint representation of $SU(4)$ which is carried by the spinors has been found to be anomaly free in all other contexts, this indicates that the trace anomaly should vanish as well. Applying a similar argument to the $N = 2$ supersymmetric finite theories requires use of the Adler-Bardeen theorem.[27] In this case, one also needs to address the question as to which chiral current is the one that satisfies the Adler-Bardeen theorem.

Starting from one of the massless theories that are finite by virtue of $N = 2$ supersymmetry and (2.88), it is possible to add soft supersymmetry-breaking mass and dimension three interaction terms that preserve the finiteness.[28,29] The soft breaking terms have coefficients carrying dimensions of mass and thus cannot upset the cancellations between the logarithmic divergences which are governed by the original supersymmetry. It must then be arranged for the soft breaking terms themselves not to generate divergences of their own forms or of other forms with dimension less than four. The most straightforward technique[30] for achieving these cancellations is to introduce a background "spurion" superfield in interactions that are protected by a non-renormalization theorem (*e.g.*, the $N = 1$ non-renormalization of chiral multiplet mass and interaction terms). Then one lets the spurion take a non-supersymmetric vacuum expectation value, thus introducing the soft symmetry breaking parameters. In general, soft breaking is characterized by a sum rule for the masses given to the various fields of spin J:

$$\sum_J (-1)^{2J+1}(2J+1)m_J^2 = 0. \tag{2.90}$$

There is an exception[29] to this sum rule in the case of a soft breaking term that gives mass terms of the form (scalar)^2 + (pseudoscalar)^2. In all of the soft breaking terms, there are also trilinear interactions with coefficients determined by the coefficients of the mass terms.

The finiteness of the $N = 2$ supersymmetric models relies upon an improvement in the ultraviolet power counting for extended supersymmetric theories. In order to test the correctness of this power counting, it is necessary to consider theories that are not renormalizable, so that eventually at some loop order the power counting allows infinities to occur. A remarkable set of calculations[31] of the divergences in higher dimensional extended super Yang-Mills theories enables one to test the predictions of power counting using the background field method against the actually occurring infinities.

In six dimensions, the maximally supersymmetric super Yang-Mills theory has $N = 2$ supersymmetry. $N = 1$ supersymmetric theories can be described using left-handed Weyl spinors carrying an additional $SU(2)$ doublet index and obeying an $SU(2)$ pseudo-reality condition

$$\theta_i^\alpha = \bar{\theta}^{\alpha j}\epsilon_{ji} \qquad \alpha = 1..4,\ i = 1,2. \tag{2.91}$$

For $N = 2$ supersymmetry, we need an additional right-handed spinor carrying a doublet index for another $SU(2)$

$$\theta_\alpha^{i'} = \epsilon^{i'j'}\bar{\theta}_{\alpha j'} \qquad \alpha = 1...4,\ i' = 1,2. \tag{2.92}$$

The rigid internal symmetry group for the six-dimensional $N = 2$ theory's field equations is then $SU(2) \times SU(2)$.

The six-dimensional gauge theories satisfy representation-preserving constraints just as in the four-dimensional case. These are[32]

$$\text{for } N = 1: \qquad F_{\alpha\beta}^{ij} = 0 \tag{2.93}$$

$$\begin{aligned} \text{for } N = 2: \qquad & F_{\alpha\beta}^{ij} = F_{i'j'}^{\alpha\beta} = 0 \\ & F_{\alpha j'}^{i\beta} = \delta_\alpha^\beta W_{j'}^i \\ & \bar{W}_i^{i'} = \epsilon^{i'j'} W_{j'}^j \epsilon_{ji}. \end{aligned} \tag{2.94}$$

In the $N = 1$ case, one may check via the Bianchi identities that (2.93) implies

$$F_{\alpha\mu}^i = (\Sigma_\mu)_{\alpha\beta} W^{\beta i}, \tag{2.95}$$

where $(\Sigma_\mu)_{\alpha\beta}$ is the $d = 6$ analog of $(\sigma_\mu)_{\alpha\dot{\beta}}$ in $d = 4$. The field strength multiplet is thus described in the $N = 1$ case by a spinor superfield, whose lowest component is the physical spinor $\lambda^{\beta i}$. The superfield $W^{\beta i}$ satisfies

$$\begin{aligned} \mathcal{D}^i_\alpha W^{\beta j} =& \delta^\beta_\alpha X^{(ij)} - \epsilon^{ij}(\Sigma^{MN})^\beta_\alpha F_{MN} \\ & \alpha, \beta : 1..4 \quad M, N : 1..6 \quad i, j : 1, 2 \end{aligned} \tag{2.96}$$

where X^{ij} is a pseudo-real triplet of auxiliary fields.

As in the case of the $N = 4$ theory in $d = 4$, the $N = 2$, $d = 6$ theory does not have an action formulation displaying its full supersymmetry with a finite number of auxiliary fields. One can describe the $N = 2$ theory as a coupling of the $N = 1$ theory to an $N = 1$ hypermultiplet matter sector in the adjoint representation of the gauge group. The $d = 6$ hypermultiplet is a straightforward generalization of the $d = 4$ hypermultiplet, with pseudo-real or real constrained superfields L^{ij}, $L^{ijk\ell}$, and S satisfying

$$\begin{aligned} \nabla^{(i}_\alpha L^{jk)} &= \nabla_{\alpha\ell} L^{ijk\ell} \\ \nabla^{(i}_\alpha L^{jk\ell m} &= 0 \\ \epsilon_{ij}\, \nabla^i_{(\alpha}\, \nabla^j_{\beta)} S &= 0. \end{aligned} \tag{2.97}$$

The naive degree of divergence of ℓ-loop graphs in d-dimensional Yang-Mills plus matter theories is

$$D = (d - 4)\ell + 4. \tag{2.98}$$

In calculating the power counting weight of a counterterm, derivatives, scalar and vector external field lines count as dimension 1; external spinors count as 3/2 and spinorial derivatives count as 1/2. The generic form of an ℓ-loop counterterm in $d = 6$, $N = 1$ super Yang-Mills is thus required by the non-renormalization theorem to be schematically

$$\Delta\Gamma^\ell = g^{2(\ell-1)} \int_{q=2\cdot\cdot 2\ell} d^6x d^8\theta (\nabla^{4\ell-2q}(gW^{\alpha i})^q). \tag{2.99}$$

Enumeration of the possibilities shows that there are no allowable essential counterterms at $\ell = 1$ or 2 loops for the pure $N = 1$ theory.[33] All the possible counterterms at 1 or 2 loops vanish subject to the classical field equations and the corresponding infinities may thus be eliminated by field redefinition renormalizations.

In a theory involving hypermultiplets, new counterterms are possible. For example, a generic theory could have counterterms of the form

$$\Delta\Gamma^{\ell} = g^{2(\ell-1)} \int d^6x d^8\theta (\nabla^{4\ell-2q}(gL)^q). \tag{2.100}$$

At one loop, all the candidates vanish subject to the classical field equations. At two loops, however, there could be a counterterm with $q = 4$. Thus, the generic coupling of $d = 6$, $N = 1$ super Yang-Mills to hypermultiplet matter is expected to have essential divergences in the matter sector from two loops onward.

In the case of the $d = 6$, $N = 2$ theory, counterterms must be off-mass-shell $N = 1$ supersymmetric but must acquire the full $N = 2$ supersymmetry when the background is required to satisfy the classical equations of motion. This links the matter terms that can occur to terms involving the gauge multiplet. Analysis of the only two-loop candidate that is $N = 2$ supersymmetric subject to the field equations but does not vanish subject to them shows that there must be a purely gauge multiplet part that, when written in $N = 1$ superfields, fails the test of the non-renormalization theorem. The first loop order at which an allowable essential divergence can occur is at three loops, where in the on-mass-shell full $N = 2$ superfield language it takes the form

$$\Delta\Gamma^{\ell=3}_{N=2} = g^8 \int d^6x d^6\theta^{[\alpha\beta]}_{(ij)} d^6\theta^{(i'j')}_{[\alpha\beta]} \operatorname{tr}[\bar{W}^k_{i'}, \bar{W}_{kj'}][W^i_{k'}, W^{jk'}]. \tag{2.101}$$

Testing for the occurrence of a divergence of the form (2.101) at the three-loop order would be a highly difficult calculation. On the other hand, analysis of the seven-dimensional theory shows that the counterterm (2.101) is just a dimensional reduction of an allowed $d = 7$ two-loop counterterm for the maximal $d = 7$ super Yang-Mills. This $d = 7$ calculation has been performed,[31] with the anticipated divergence being found. At the two-loop level in the six-dimensional theory, however, there are dramatic cancellations leaving no essential divergence, in accord with the non-renormalization theorem. These calculations constitute the most demanding test that has been met by our analysis of the mechanism of ultraviolet cancellations in supersymmetric theories and confirms its essential completeness.

3. HARMONIC SUPERSPACE

The principal technical problem in quantizing extended supersymmetric theories has been to keep control over the various types of internal symmetry present in these theories. While it turns out to be sufficient to consider only $N = 2$ supersymmetry in order to understand the ultraviolet cancellations in $N = 4$ super Yang-Mills theory, the remaining symmetry could still play an important role in governing the structure of the finite parts of the effective action. As we have seen, there is an obstacle to realizing this full symmetry in an action formulation with a finite number of component fields. In order to circumvent this obstacle, we now shall allow the number of fields to become infinite, but in a tightly controlled way using the properties of harmonic analysis on coset spaces.[34] The harmonic superspace[14] to be used has extra bosonic coordinates referring to a compact space of extra bosonic dimensions. In order to set the stage, we return to consideration of the role of constraints and integrability conditions in extended supersymmetric theories.

As we saw in the first chapter, when one tries to couple gauge representations of supersymmetry to constrained representations such as the chiral, the defining differential constraints must be covariantized, leading to integrability conditions on the gauge multiplet. For example, in $N = 1$ super Yang-Mills theory covariantly coupled to chiral matter, the constraint must be covariantized, $\bar{\mathcal{D}}_{\dot\alpha}\phi = (\bar{D}_{\dot\alpha} - ig\bar{A}_{\dot\alpha})\phi = 0$, where $\bar{A}_{\dot\alpha}$ is a superspace connection. Applying another covariant fermi derivative and symmetrizing gives $F_{\dot\alpha\dot\beta s}{}^{r}\phi^{s} = 0$, so the integrability condition is $F_{\alpha\beta} = F_{\dot\alpha\dot\beta} = 0$. This constraint on A_α cuts the component field content down to that required for the $N = 1$ super Yang-Mills action: A_μ λ, D (plus further gauge components).

In extended supersymmetry, one immediately starts running into problems in finding the appropriate constraints to be applied to find superspace actions. For example, the simplest $N = 2$ model is the hypermultiplet, with physical field components ϕ^i, λ, where ϕ^i is a complex $SU(2)$ doublet, and λ is a Dirac spinor. It would be natural to look for constraints in the standard $N = 2$ superspace x^μ, θ^i_α, $\bar\theta_{\dot\alpha i}, i = 1, 2$. Since the superfield $\phi^i(x, \theta^i, \bar\theta_i)$ contains too many fields, one needs constraints. The natural constraint would eliminate an unwanted complex spinor triplet, $D^{(i}_\alpha\phi^{j)} = \bar{D}^{(i}_{\dot\alpha}\phi^{j)} = 0$. However, this is already too strong[35], since use of the commutation relations between $D_{\alpha i}$ and $\bar{D}^i_{\dot\alpha}$ leads to $\not\partial\lambda = 0$, *i.e.* the field equations are already implied by the simplest constraint.

If one uses a pseudo-real formulation of the hypermultiplet with the scalars forming a triplet and a singlet under $SU(2)$, then the problem of a covariant formulation has been solved,[15] as we have seen in the last chapter. However, while staying within the context of a finite number of component fields without central charges and with the above particular complex form of the hypermultiplet, this problem has no solution.[36] To see this, note that the fermion fields carry an integral isospin representation, *i.e.*,they must have even numbers of symmetrized indices. Auxiliary fermions would have to carry integral isospin as well and would have to be paired so that one element of each pair could act as a Lagrange multiplier to prevent the other from propagating, *e.g.*,a pair with dimensions 3/2, 5/2. Taking into account the dimension (8) of a Dirac spinor and the dimension of an isospin I representation, the number of fermi components of such a pair would be $2 \cdot 8 \cdot (2I+1)$. Thus, the total number of fermi fields would be $16q+8$, wth the $16q$ coming from the auxiliary fields, and the 8 coming from the original propagating Dirac spinor.

On the other hand, irreducible representations of $N=2$ supersymmetry without central charges carrying integral isospin have $2 \cdot (2I'+1) \cdot 8 = 16p$ fermi components, where the first factor of 2 arises because the multiplet must be complex, and all $N=2$ multiplets without central charges contain multiples of 8 fermi components. Equality with the first count of fermi components would give $16p = 16q+8$, or $p-q=1/2$, which is not solvable for p and q finite integers.

In order to circumvent the above impasse, it is necessary to accept an infinite number of component fields. This is elegantly achieved in the harmonic superspace formalism.[14] Consider the following invariant subspace of $N=2$ superspace:

$$(z^\mu, (\theta_{\alpha 1} + i\theta_{\alpha 2})/\sqrt{2}\,, (\bar\theta^1_{\dot\alpha} + i\bar\theta^2_{\dot\alpha})/\sqrt{2})$$

where

$$z^\mu = x^\mu + (\frac{i}{2})\theta^1\sigma^\mu\bar\theta^2 + (\frac{i}{2})\theta^2\sigma^\mu\bar\theta^1 \tag{3.1}$$

Choice of this subspace breaks the $SU(2)$ covariance of the formalism, keeping only a residual $U(1)$.

In order to keep the full $SU(2)$ covariance manifest, one can generalize the above subspace, parametrizing the direction in $SU(2)/U(1)$ by a zweibein parameter $u^{\pm i} e\; SU(2)$. This parameter satisfies $u_i^- = \overline{u^{+i}}$, $u^{+i}u_i^- = 1$, where i is an $SU(2)$ index and + and - are $U(1)$ indices. If one identifies the $U(1)$ with the T^3 generator of $SU(2)$, then by a $U(1)$ gauge choice, $u^{\pm i}$ can be put into the form $u_1^- = \cos\theta$; $u_1^+ = i\sin\theta e^{-i\phi}$; $u_2^- = i\sin\theta e^{i\phi}$; $u_2^+ = \cos\theta$, so that there are effectively two extra bosonic coordinates θ, ϕ living on $SU(2)/U(1) = S^2$.

The essence of the harmonic superspace formalism is to take the $u^{\pm i}$ as extra bosonic coordinates and to use them to convert all $SU(2)$ indices to $U(1)$ indices, *e.g.*,$\theta^+_\alpha \equiv \theta^i_\alpha u_i^+$, $\theta^-_\alpha \equiv \theta^i_\alpha u_i^-$. Then the invariant "analytic" superspace is $(z^\mu_A = x^\mu - 2i\theta^i\sigma^\mu\bar\theta^j u_i^+ u_j^-,\ \theta_\alpha,^+ \bar\theta^+_{\dot\alpha},\ u^{\pm i})$.

Just as in the case of chiral superfields, we may now consider functions of the analytic superspace variables only, *i.e.*,that don't depend upon θ^-_α, $\theta^-_{\dot\alpha}$. These analytic superfields will be taken to be $SU(2)$ scalars, but may carry a

$U(1)$ charge q: $\phi^q(z^\mu_A,\ \theta^+,\ \bar{\theta}^+,\ u^\pm)$. Expanding such an object in powers of $\theta^+,\ \bar{\theta}^+$, we obtain coefficient functions that are functions of z^μ and $u_i^\pm$:

$$F^q(z, u_i^\pm) = \sum_{n=0}^{\infty} f(z)^{(i_1\ldots i_{n+q}, j_1\ldots j_n)} u^+_{i_1}\ldots u^+_{i_{n+q}} u^-_{j_1}\ldots u^-_{j_n}. \tag{3.2}$$

Thus, there are infinite numbers of component fields in an analytic superfield.

The new internal bosonic dimensions require a new type of covariant derivative along the internal manifold. For the $N = 2$ case with $SU(2)/U(1)$, there are the covariant derivatives

$$\begin{aligned} D^{++} = u^{-i}\partial/\partial u^{-i} \quad ; \quad D^{--} = u^{-i}\partial/\partial u^{+i} \\ D^3 = (\frac{1}{2})(u^{+i}\partial/\partial u^{+i} - u^{-i}\partial/\partial u^{-i}) \end{aligned} \tag{3.3}$$

satisfying the algebra

$$\begin{aligned} [D^{++}, D^{--}] = 2D^3 \\ [D^3, D^{++}] = D^{++} \quad ; \quad [D^3, D^{--}] = -D^{--}. \end{aligned} \tag{3.4}$$

The action of D^{++} on the internal coordinates is simply

$$D^{++}u^{+i} = 0 \quad ; \quad D^{++}u^{-i} = u^{+i} \tag{3.5}$$

Since the internal space $SU(2)/U(1)$ is compact, integration over this space may be performed, and the measure normalized so that $\int du = 1$. Orthogonality of the harmonics on $SU(2)/U(1)$ requires

$$\int du u^+_{(i_1\ldots} u^+_{i_n} u^-_{j_1\ldots} u^-_{j_m)} = 0, \tag{3.6}$$

i.e.,integration extracts the singlet part of the expansion in the $u_i^\pm$. Integration by parts follows from the rule $\int du D^{++} f^{--}(u) = 0$. The superspace measure for integration over the analytic superspace combines $\theta^+, \bar{\theta}^+$ and u integrations, and removes 4 units of $U(1)$ charge, so is denoted $d\mathfrak{z}_A^{(-4)}du = d^4z_A d^2\theta^+ d^2\bar{\theta}^+ du$. In order to define reality of representations, one uses a combined conjugation operation of complex conjugation coupled with an operation * that acts only upon $U(1)$ indices:

$$(u^{+i})^* = u^{-i} \quad , \quad (u_i^-)^* = -u_i^+ \tag{3.7}$$

Under the combined $\underset{-}{*}$ operation, the analytic superspace is real:

$$\overset{\ast}{\overline{z_A^\mu}} = z_A^\mu \quad , \quad \overset{\ast}{\overline{\theta_\alpha^+}} = \bar{\theta}_{\dot{\alpha}}^+ \quad , \quad \overset{\ast}{\overline{(\bar{\theta}_{\dot{\alpha}}^+)}} = -\theta_\alpha^+ \tag{3.8}$$

In order to describe the hypermultiplet,[14] one uses a complex analytic superfield $\phi^+(\mathfrak{z}_A, u)$ carrying one unit of $U(1)$ charge. The action is

$$I = \int d\mathfrak{z}_A^{(-4)} du \overset{\ast}{\overline{\phi}}{}^+ D^{++}\phi^+ \tag{3.9}$$

This is real, $I = \overset{\ast}{\overline{I}}$, and gives the field equation

$$D^{++}\phi^+ = 0. \tag{3.10}$$

In order to recognize this as the field equation for the hypermultiplet, one needs to expand ϕ^+ into ordinary $N = 2$ superfields, the general formula being

$$\phi^q(\mathfrak{z}_A, u) = \sum_{n=0}^{\infty} \phi^{(i_1\ldots i_{n+q} j_1 \ldots j_n)}_{(x,\theta_i,\bar{\theta}^i)} u^+_{i_1\ldots} u^+_{i_{n+q}} u^-_{j_1\ldots} u^-_{j_n}, \tag{3.11}$$

and the constraint of analyticity links the ordinary superfields by constraints:

$$D_\alpha^{(i}\phi^{j_1\ldots j_{2n+q})} = \frac{n+1}{2n+q+3} D_{\alpha\ell}\phi^{(\ell i j_1 \ldots j_{2n+q})} \tag{3.12}$$

and similarly for the $\bar{D}^i_{\dot\alpha}$ derivative. The hypermultiplet field equation sets $\phi^{(ijk)}$ and higher terms to zero, thus reproducing the field equations $D^{(i}_{\alpha}\phi^{j)} = \bar{D}^{(i}_{\dot\alpha}\phi^{j)} = 0$. Thus, the no-go theorem is circumvented by having an infinite number of fields in the off-shell multiplet. If one were to truncate the off-shell multiplet by setting all superfields after a certain point in he above expansion to zero (but keeping at least the first two), then two hypermultiplets would propagate.[36]

The harmonic superspace formalism has also been applied to the off-shell formulation of the maximal super Yang-Mills theory.[37] In the form that has been successfully solved using harmonic superspace, the theory has $N = 3$ supersymmetry. On-shell, there is a full $N = 4$ supersymmetry that appears. As in the $N = 2$ case, one parametrizes a breakdown of the internal symmetry $SU(3)$, in this case to $U(1) \times U(1)$, by a set of 3×3 unitary matrices with determinant 1. These can be given indices $u_i^{(a,b)}$, where $i = 1, 2, 3$ and $(a, b) = (1, 1), (-1, 1)$ or $(0, -2)(\bar{u}^{i(a,b)}$ has conjugated indices), with a and b corresponding to the eigenvalues of the two commuting diagonal generators of $SU(3)$, which can be identified with the $U(1) \times U(1)$.

As in the $N = 2$ case, there is an analytic superspace $(z_A^{\alpha\dot\alpha} = x^{\alpha\dot\alpha} + 2i(\theta^{(0,-2)\alpha}\bar\theta^{(0,-2)\dot\alpha} - \theta^{(-1,-1)\alpha}\bar\theta^{(1,1)\dot\alpha}),\ \theta^{(1,-1)\alpha},\ \theta^{(0,2)\alpha},\ \bar\theta^{(1,1)\dot\alpha},\ \bar\theta^{(-1,1)\dot\alpha},\ u)$ where $\theta^{(a,b)\alpha} = \bar{u}^{i(a,b)}\theta^{\alpha}_i$, $\bar\theta^{(a,b)\dot\alpha} = u_i^{(a,b)}\bar\theta^{\dot\alpha i}$. There is also a $\underset{\sim}{*}$ conjugation $u_i^{(1,1)} \to u^{i(0,2)}$; $u_i^{(0,2)} \to u^{i(-1,-1)}$; $u_i^{(-1,-1)} \to -u^{i(1,-1)}$ under which the analytic superspace is invariant $\theta_\alpha^{(1,-1)} \to -\bar\theta_{\dot\alpha}^{(-1,1)}$; $\theta_\alpha^{(0,2)} \to \bar\theta_{\dot\alpha}^{(1,1)}$; $z_A^{\alpha\dot\alpha} \to z_A^{\alpha\dot\alpha}$. There are also covariant derivatives

$$
\begin{aligned}
D^{(1,3)} &= -u_i^{(1,1)}\partial/\partial u_i^{(0,-1)} + \bar{u}^{i(0,2)}\partial/\partial\bar{u}^{i(1,-1)} \\
D^{(-1,3)} &= u_i^{(-1,1)}\partial/\partial u_i^{(0,-2)} - \bar{u}^{i(0,2)}\partial/\partial\bar{u}^{i(1,-1)} \\
D^{(2,0)} &= u_i^{(1,1)}\partial/\partial u_i^{(-1,1)} - \bar{u}^{i(1,-1)}\partial/\partial\bar{u}^{i(-1,-1)}
\end{aligned}
\tag{3.13}
$$

and their complex conjugates, which together with the generators of the $U(1) \times U(1)$ make up the $SU(3)$ Lie algebra.

The $N = 3$ super Yang-Mills multiplet is described by gauge connections $A_\alpha^{(a,b)} = A_\alpha^i u_i^{(a,b)}$ and their conjugates $\bar{A}_{\dot\alpha}^{(a,b)}$. In order to couple analytic superfields consistently to the Yang-Mills mulitplet, the covariant derivatives that enter into the analyticity constraint must be generalized to gauge covariant derivatives. As in the $N = 1$ super Yang-Mills case, this requires the integrability conditions

$$
\{\mathcal{D}_\alpha^{(1,1)}, \mathcal{D}_\beta^{(1,1)}\} = \{\bar{\mathcal{D}}_{\dot\alpha}^{(0,2)}, \bar{\mathcal{D}}_{\dot\beta}^{(0,2)}\} = \{\mathcal{D}_\alpha^{(1,1)}, \bar{\mathcal{D}}_{\dot\beta}^{(0,2)}\} = 0 \tag{3.14}
$$

Moreover, in the absence of gauge coupling, the analytic character of a superfield is perserved by operation with $D^{(\pm 1,3)}$ and $D^{(2,0)}$. These covariant derivatives in the internal space $SU(3)/U(1) \times U(1)$ can thus be used to construct analytic superfield Lagrangians. In order to preserve this property in the presence of gauge coupling, we require the further integrability conditions

$$
\begin{aligned}
[\mathcal{D}^{(\pm 1,3)}, \mathcal{D}_{\alpha}^{(1,1)}] &= [\mathcal{D}^{(\pm 1,3)}, \bar{\mathcal{D}}_{\dot{\alpha}}^{(0,2)}] = 0 \\
[\mathcal{D}^{(2,0)}, \mathcal{D}_{\alpha}^{(1,1)}] &= [\mathcal{D}^{(2,0)}, \bar{\mathcal{D}}_{\dot{\alpha}}^{(2,0)}] = 0
\end{aligned}
\tag{3.15}
$$

A final type of constraint on the gauge connections arises from the fact that the Yang-Mills gauge group may be taken to be local in $(x^{\mu}, \theta_{\alpha i}, \bar{\theta}^{i}_{\dot{\alpha}})$, but rigid with respect to the u variables, so that in the first instance, the $SU(3)/U(1) \times U(1)$ covariant derivatives will have no gauge connections, and will satisfy the "flat" algebra

$$
\begin{aligned}
[\mathcal{D}^{(\pm 1,3)}, \mathcal{D}^{(\mp 1,3)}] &= [\mathcal{D}^{(1,3)}, \mathcal{D}^{(2,0)}] = 0 \\
[\mathcal{D}^{(-1,3)}, \mathcal{D}^{(2,0)}] &= \mathcal{D}^{(1,3)}
\end{aligned}
\tag{3.16}
$$

The difficulty in achieving an invariant action formulation is revealed by the fact that requiring the natural conditions (3.14), (3.15) and (3.16) simultaneously implies the field equations. Of the three types of constraint, (3.15) is the most trivial, since it is essentially kinematical, being due only to the choice of local gauge group. Nonetheless, it is precisely this constraint which will be relaxed in the "off-shell" formalism and derived from an action.[37]

The formalism may be recast in a different form by noting that the constraints (3.14) may be solved in a fashion analogous to $N = 1$ super Yang-Mills:

$$
\begin{aligned}
(A_{\alpha}^{(1,1)})_K &= -ie^{-i\omega} D_{\alpha}^{(1,1)} e^{i\omega} \\
(\bar{A}_{\dot{\alpha}}^{(0,2)})_K &= -ie^{-i\omega} \bar{D}_{\dot{\alpha}}^{(0,2)} e^{i\omega}
\end{aligned}
\tag{3.17}
$$

where the "prepotential" ω is real, $\omega = \overline{\omega}$, so that $\bar{A}_{\dot{\alpha}}^{(0,2)} = \overline{A_{\alpha}^{(1,1)}}$. The object $e^{i\omega}$ has a new type of gauge invariance that arises from the solution to the constraints (3.14), which we denote Λ-gauge transformations, where the total set of gauge transformations is given by

$$
e^{i\omega'} = e^{i\Lambda} e^{i\omega} e^{-iK} \tag{3.18}
$$

and invariance of $A_\alpha^{(1,1)}$ and $\bar{A}_{\dot\alpha}^{(0,2)}$ follows if the parameter of the Λ transformation is analytic: $D_\alpha^{(1,1)}\Lambda = \bar{D}_{\dot\alpha}^{(0,2)}\Lambda = 0$. Under K-gauge transformations, $A_\alpha^{(1,1)}$ and $\bar{A}_{\dot\alpha}^{(0,2)}$ transform in the standard fashion as connections.

The object $e^{i\omega}$ can be used to convert K-covariant fields into Λ-covariant ones: $\phi_{(\Lambda)} = e^{i\omega}\phi_{(K)}$. If we wish to work with such Λ-covariant fields, the basis of covariant derivatives is changed:

$$
\begin{aligned}
(\mathcal{D}_\alpha^{(1,1)})_\Lambda &= e^{i\omega}(\mathcal{D}_\alpha^{(1,1)})_K e^{-i\omega} = D_\alpha^{(1,1)} \\
\bar{\mathcal{D}}_{\dot\alpha}^{(0,2)})_\Lambda &= e^{i\omega}(\bar{\mathcal{D}}_{\dot\alpha}^{(0,2)})_K e^{i\omega} = \bar{D}_{\dot\alpha}^{(0,2)}
\end{aligned}
\tag{3.19}
$$

where the change of basis has completely removed the connections from these derivatives. On the other hand, the change of basis gives rise to connections in $\mathcal{D}^{(\pm1,3)}$ and $\mathcal{D}^{(0,2)}$ which in the K-basis had none:

$$
\begin{aligned}
(\mathcal{D}^{(\pm1,3)})_\Lambda &= D^{(\pm1,3)} + iV^{(\pm1,3)} \quad , \quad V^{(\pm1,3)} = -ie^{i\omega}(D^{(\pm1,3)}e^{-i\omega}) \\
(\mathcal{D}^{(2,0)})_\Lambda &= D^{(2,0)} + iV^{(2,0)} \quad , \quad V^{(2,0)} = -ie^{i\omega}(D^{(2,0)}e^{-i\omega})
\end{aligned}
\tag{3.20}
$$

In the Λ-basis, the constraints (3.15) imply that the harmonic connections are analytic:

$$
D_\alpha^{(1,1)}V^{(a,b)} = \bar{D}_{\dot\alpha}^{(2,0)}V^{(a,b)} = 0 \tag{3.21}
$$

Since these constraints involve "flat" derivatives, they are trivially solvable and we may now take the harmonic connections $V^{(a,b)}$ as the fundamental objects of the theory.

While the constraints (3.15) become trivially solvable in the Λ-basis, the previously trivial constraints (3.16) now become more complicated:

$$
\begin{aligned}
D^{(1,3)}V^{(-1,3)} - D^{(-1,3)}V^{(1,3)} + [V^{(1,3)}, V^{(-1,3)}] &= 0 \\
D^{(1,3)}V^{(2,0)} - D^{(2,0)}V^{(1,3)} + i[V^{(1,3)}, V^{(2,0)}] &= 0 \\
D^{(-1,3)}V^{(2,0)} - D^{(2,0)}V^{(-1,3)} + i[V^{(-1,3)}, V^{(2,0)}] - V^{(1,3)} &= 0
\end{aligned}
\tag{3.22}
$$

The last of these can be solved algebraically for $V^{(1,3)}$ in terms of $V^{(-1,3)}$ and $V^{(2,0)}$. Considering these to be the fundamental objects of the theory, the first two equations in (3.22) are the dynamical equations of motion.

Surprisingly, the equations of motion (3.22) may be derived from an action:

$$I = \int d\mathfrak{Z}_A^{(-2,-6)} du\, tr\{V^{(2,0)}(D^{(1,3)}V^{(-1,3)} - D^{(-1,3)}V^{(1,3)})$$
$$- V^{(-1,3)}(D^{(1,3)}V^{(2,0)} - D^{(2,0)}V^{(1,3)}) + V^{(1,3)}(D^{(-1,3)}V^{(2,0)} - D^{(2,0)}V^{(-1,3)})$$
$$- (V^{(1,3)})^2 + 2i\, V^{(1,3)}[V^{(-1,3)}, V^{(2,0)}]\} \qquad (3.23)$$

This action is not written in terms of field strengths, but it is nonetheless invariant under Λ-gauge transformations. The Lagrangian transforms into a total harmonic derivative, which is annihilated by the *du* integration. Once more, the inclusion of an infinite number of ordinary component fields in the harmonic superspace formalism has allowed the construction of an action that otherwise could not exist.[13] The particular structure of this action is analogous to the Chern-Simons forms in Yang-Mills theory, which are not renormalized when taken as separate insertions in Green's functions.[4,38] This analogy should lead to a very direct proof of the ultraviolet finiteness of the $N = 3$ theory. The character of the field equations as integrability conditions is also very striking. Perhaps this will lead to further exact information about the quantum structure of the theory.

ACKNOWLEDGEMENT

This material is based in part upon research supported by the National Science Foundation under Grant No. PHY77-27084, supplemented by funds from the National Aeronautics and Space Administration.

References

* Permanent address: Blackett Laboratory, Imperial College, Prince Consort Road, London SW7 2AZ, England.

1. S. Ferrara and B. Zumino, Nucl. Phys. **B79** (1974) 413; A. Salam and J. Strathdee, Phys. Lett. **51**B (1974) 353; B. de Wit and D. Freedman, Phys. Rev. D**12** (1975) 2286.

2. G. 't Hooft, in *Proc. 12th Winter School of Theoretical Physics in Karpacz*, Acta Univ. Wratisl. No. 38 (1975); B.S. De Witt, in *Quantum Gravity 2*, eds. C.J. Isham, R. Penrose, and D.W. Sciama (Oxford University Press, 1981); D.G. Boulware, Phys. Rev. D**23** (1981) 389; L.F. Abbot, Nucl. Phys. B**185** (1981) 189.

3. L.F. Abbot, M.T. Grisaru, and R.K. Schaefer, Brandeis preprint BRX-TH35 (1983).

4. P. Breitenlohner, D. Maison, and K.S. Stelle, Phys. Lett. **134**B (1984) 63.

5. S. Adler, in Brandeis 1970 Lectures, eds. S. Deser, M.T. Grisaru, and H Pendleton (MIT Press, 1970).

6. M.T. Grisaru, W. Siegel, and M. Roček, Nucl. Phys. B**159** (1979) 429; M.T Grisaru and W. Siegel, Nucl. Phys. B**187** (1981) 149.

7. M.T. Grisaru and W. Siegel, Nucl. Phys. B**201** (1982) 292.

8. P.S. Howe, K.S. Stelle, and P.K. Townsend, Nucl. Phys. B**236** (1984) 125.

9. W. Siegel, Phys. Lett. **84**B (1979) 193; W. Siegel, P.K. Townsend, and P. va Nieuwenhuizen, in *Superspace and Supergravity*, eds. S.W. Hawking and M Roček (Cambridge University Press, 1981), p. 165.

10. N.K. Nielsen, Nucl. Phys. B**140** (1978) 499; R.E. Kallosh, Nucl. Phys. B**14** (1978) 141.

11. W. Siegel and S.J. Gates, Nucl. Phys. B**201** (1982) 292.

12. F. Gliozzi, J. Scherk, and D. Olive, Nucl. Phys. B**133** (1978) 253; L. Brink J. Schwarz, and J. Scherk, Nucl. Phys. B**121** (1977) 77.

13. M. Rocek and W. Siegel, Phys. Lett. **105**B(1981) 275; V. Rivelles and J.C Taylor, J. Phys. A: Math. Gen. **15** (1982) 163.

14. A. Galperin, E. Ivanov, S. Kalitzin, V. Ogievetsky, and E. Sokatchev, Clas Quantum Grav. **1** (1984) 469.

15. P.S. Howe, K.S. Stelle, and P.K. Townsend, Nucl. Phys. B**214** (1983) 51 K.S. Stelle, in *Proc. 21st Intl. Conf. on High Energy Physics*, eds. P. Petia and M. Porneuf, J. Phys. **12** (1982) 326.

16. S. Mandelstam, in *Proc. 21st Intl. Conf. on High Energy Physics*, eds. Petiau and M. Porneuf, J. Phys. **12** (1982) 331; L. Brink, O. Lindgren, ar B.E.W. Nielsen, Nucl. Phys. B**212** (1983) 401; Phys. Lett. **123**B (1983) 32

17. See the first paper of Ref. 1.

18. P. Fayet, Nucl. Phys. B**113** (1976) 135; M.F. Sohnius, Nucl. Phys. B1 (1978) 109; A. Galperin, E.A. Ivanov, and V.I. Ogievetsky, JETP Lett. (1981) 176.

19. M.T. Grisaru, M. Roček, and W. Siegel, Phys. Rev. Lett. **45** (1980) 106

W. Caswell and D. Zanon, Phys. Lett. **100**B (1980) 152.

20. L.V. Avdeev and O.V. Tarasov, Phys. Lett. **112**B (1982) 356.

21. R. Grimm, M. Sohnius, and J. Wess, Nucl. Phys. B**133** (1978) 275.

22. L. Mezincescu, JINR Report P2-12572 (1979).

23. P.S. Howe, K.S. Stelle, and P.K. Townsend, reported by K.S. Stelle in *Proc. Intl. Seminar on Group Theoretical Methods in Physics in Zvenigorod*, Nov. 1982; Imperial College preprint ICTP 82-83/13; J. Koller, Caltech preprint (1983).

24. P.S. Howe, K.S. Stelle, and P.C. West, Phys. Lett. **124**B (1983) 55.

25. P.S. Howe and P.C. West, King's College preprint (1984).

26. M.F. Sohnius and P.C. West, Phys. Lett. **100**B (1981) 245; S. Ferrara and B. Zumino, unpublished.

27. M.T. Grisaru and P.C. West, Brandeis preprint BRX-TH-141 (1983).

28. P.C. West, Lecture given at Shelter Island II Conf. June 1983 (King's College preprint, 1983); J.G. Taylor, Phys. Lett. **121**B (1983) 386; S. Rajpoot, J.G. Taylor, and M. Zaimi, Phys. Lett. **127**B (1983); King's College preprint (1983).

29. A. Parkes and P.C. West, King's College preprint (1983).

30. L. Girardello and M.T. Grisaru, Nucl. Phys. B**194** (1982) 65.

31. N. Marcus and A. Sagnotti, Phys. Lett. **135**B (1984) 85; Caltech preprint CALT-68-1129 (1984).

32. P.S. Howe, G. Sierra, and P.K. Townsend, Nucl. Phys. B**221** (1983) 331; J. Koller, Phys. Lett. **124**B (1983) 324; Nucl. Phys. B**222** (1983) 319.

33. P.S. Howe and K.S. Stelle, Phys. Lett. **137**B (1984) 175; A.E.M. van de Ven, Stony Brook preprint ITP-SB-84-26.

34. A. Salam and J. Strathdee, Ann. Phys. (N.Y.) **141** (1982) 316.

35. M.F. Sohnius, Nucl. Phys. B**138** (1978) 109.

36. P.S. Howe, K.S. Stelle, and P.C. West, in preparation.

37. A. Galperin, E. Ivanov, S. Kalitzin, V. Ogievetsky and E. Sokatchev, Dubna preprint E2-84-441 (Class Quantum Gravity, in press).

38. P.K. Townsend, in *Proc. 3rd Trieste School on Supersymmetry and Supergravity,* April 1984.

ON THE NON PERTURBATIVE BREAKDOWN OF GLOBAL SUPERSYMMETRY

Acharon Casher

Department of Physics and Astronomy
Tel Aviv University
Tel-Aviv, Israel

I INTRODUCTION

The problem of dynamical symmetry breaking in quantum field theory can be described succinctly as follows: given a classical Lagrangian characterized by a set of coupling constants and masses, do the mass spectrum and S-matrix preserve the global symmetries of the classical field equations?

In the case of asymptotically free gauge theories perturbation theory is reliable for short distance processes, and it follows that high momentum gauge invariant correlation functions obey the symmetry relations of the Lagrangian, including SUSY. (This is discussed in Sibold's lecture). The spectrum however is determined by the infra-red properties of the theory and a symmetry breaking may appear due to three different causes:

(a) Spontaneous breakdown due to the existence of a non symmetric vacuum solution of the classical field equations.
(b) Spontaneous breakdown due to the formation of non perturbative condensates.
(c) Explicit breakdown due to non perturbative quantum effects.

Possibility (a) is discussed in the lectures of S. Ferrara.

Possibility (b) is very difficult to investigate since its appearance is necessarily tied to the behaviour of a strongly interacting gauge theory. What may be done is to check its consistency with the general properties of the theory such as Ward identities, the SUSY algebra, the renormalization group, and the index theorems. The method of constructing effective Lagrangians compatible with the above properties has been used extensively and has led to some understanding of the conditions under which spontaneous SUSY breaking is allowed. We remark however that this method explicitly assumes that possibility (c) does not occur and in particular that the SUSY algebra is preserved by non perturbative effects. Both types of spontaneous SUSY breaking are accompanied by a massless Goldstone fermion. In what follows we shall investigate the third possibility and argue that it indeed occurs.

There are two distinct types of non perturbative phenomena in quantum field theory (and indeed in quantum mechanics). The first is due to effects which are non analytic in the coupling constant but are due to field configurations which may be reached from the classical vacuum by continuous deformation. In other words, we are concerned here with configurations whose perturbative measure is small (typically $\sim \exp(-1/g^2)$) but non zero. Clearly such effects cannot affect relations based on smooth variation of the Lagrangian; in particular the SUSY algebra cannot be violated by this type of effect. (In quantum mechanics tunnelling through a barrier is an example). The second and more interesting type of effect is specific to gauge theories and stems from the possible contribution of topologically non-trivial configuration to the path integral. Such field configurations cannot be reached from the classical vacuum by a smooth deformation, and their perturbative gaussian measure is strictly zero. The reason for the prohibition is gauge dependent - these configurations may involve a singular vector potential, or lead to infra-red divergences in the perturbative measure. The presence of such field configurations (instantons, fluxons, or monopoles) in the exact path integral means that the true phase space of a gauge theory is not the manifold of the canonical fields namely the Lie algebra valued vector potentials. Rather, at least in the presence of such fields the correct variables span the gauge group manifold. In particular, the action is rendered finite in a "singular" gauge by <u>defining</u> $F^2_{\mu\nu}$ as the limit of a small Wilson loop and not as the square of a (singular) curl of A_μ. We may thus anticipate that symmetry relations and operator algebras might change due to the change in the nature of the variables of the theory, and should be re-examined. A strong indication that SUSY may be broken by such effects is the lack of success in constructing lattice version of SUSY gauge theories. We conjecture that this is due to the necessity of using group manifold which do not have the required Kähler structure (see J. Bagger's lectures). The extent to which the "group" effects persist in the exact mass spectrum of the theory is precisely quantified by the so-called non perturbative effects. We thus may expect violations of the superalgebra and bose-fermi degeneracy proportional to the measure of the above configurations in the exact path integral. We remark here that SUSY violations due to topological effects of gauge fields do occur in quantum mechanics and follow precisely the pattern described above: the classical equations of motion are supersymmetric but the operator algebra is violated by singular gauge topologies[1].

A general examination of gauge theories along the previous line is very difficult, especially for massless (strongly interacting) models. We may however circumvent the problems by performing a direct calculation of physical quantities such as the vacuum energy and fermion-boson mass differences for SUSY-Higgs models. If the mass parameters which induce the spontaneous gauge symmetry breaking are sufficiently large and if all fields become massive, then perturbation theory is reliable (and supersymmetric!). The physical masses provide the necessary infra-red cutoff for non perturbative effects and the latter may be systematically expanded in powers of $\exp(-8\pi^2/g^2)$, where g is the on-shell small gauge coupling constant. When applied to the vacuum energy density (which is necessarily finite and calculable in a SUSY theory since it vanishes to all orders in perturbation theory) such an approach yields (as will be shown) an upper bound. We shall prove that this upper bound is negative thus rendering the SUSY algebra incompatible with the exact non perturbative definition of the path integral.

In sec. II we present the definition of the path integral using collective coordinates and a generalized Fadeev-Popov method, and prove the above mentioned "variational principle" for the vacuum energy.

In sec. III a SUSY-Higgs model which satisfies all the needed requirements is exhibited.

In sec. IV we evaluate instanton effects for this class of models and show that vacuum energy is indeed negative.

In sec. V we return to the general discussion.

II NON PERTURBATIVE EUCLIDEAN PATH INTEGRALS

The quantity we wish to evaluate is the partition function defined by:

$$Z = \exp(-V_{(4)}\epsilon_0) = \int d\Phi\, e^{-S_b[\Phi]} |\det D[\Phi]| \qquad (1)$$

$S_b[\Phi]$ is the action functional of all real bose fields Φ (scalar and vector) and $D[\Phi]$ is the full Dirac operator in the presence of these fields (the 1/2 is due to the Majorana nature of the fermion fields used in SUSY). All the necessary counter-terms, gauge fixing terms, and the log of the F-P determinant are lumped into $S_b[\Phi]$ for notational simplicity. Z is normalized by the requirement that it equals one for free fields - a normalization which is perserved to all orders in perturbation theory due to SUSY: the perturbative vacuum energy vanishes.

Our aim is to redefine the functional measure so that if Φ is a singular field with a finite action density the measure does not vanish. To this end, let us pick a set of field configurations $\phi(\zeta)$ characterized by a set of parameters $\{\zeta_j\}$ and shift the field:

$$\Phi = \phi(\zeta) + \beta_j\, \phi_{,j}(\zeta) + \alpha \quad ; \quad \phi_{,j} \equiv \frac{\partial \phi}{\partial \zeta_j} \tag{2}$$

$$(\phi_{,j} | \alpha) = 0 \tag{3}$$

The notation we introduced is:

$$(f | g) = \int dx\; f(x)\, g(x) \tag{4}$$

where summation over all indices implicit in (f,g) is implied. Introducing the matrix:

$$\Delta_{ij} = (\phi_{,i} | \phi_{,j}) \tag{5}$$

the functional measure is,

$$d\Phi = |\det \Delta|^{1/2}\, d\beta\, d\alpha \tag{6}$$

We now use the F-P procedure in order to change the variables from $\{\beta_j\}$ to $\{\zeta_j\}$.
Introduce the identity

$$1 = \int d\zeta\, \left|\det \frac{\partial f}{\partial \zeta}\right| \delta[f(\zeta)] \tag{7}$$

and choose for f the function:

$$f_j(\zeta) = (\phi_{,j} | \Phi - \phi) \tag{8}$$

We readily find:

$$\left.\frac{\partial f_i}{\partial \zeta_j}\right|_{f=0} = -\left[\Delta_{ij} - (\phi_{,ij} | \alpha)\right] \tag{9}$$

Noting the relation,

$$\delta[f] = |\det \Delta|^{-1}\, \delta[\beta] \tag{10}$$

we finally arrive at the desired representation $(|A| \equiv |\det A|)$:

$$Z = \int d\zeta \, |\Delta|^{1/2} |1 - \Delta^{-1}(\phi_{,\zeta\zeta}|\alpha)| \, |D[\phi+\alpha]|^{1/2} e^{-S_b(\phi+\alpha)} d\alpha$$

$$(\phi_{,\zeta}|\alpha) = 0 \qquad (11)$$

Equation (11) is completely general and exact. In particular the "classical" field ϕ, their collective coordinates ζ and the range of variation of the latter may be chosen at will. The only requirement is that all possible classes of configurations which cannot be continuously deformed into each other be included. Once such a class is identified, the α -integrations are conveniently referred to as quantum effects and their evaluation supplies a $\{\zeta\}$ -dependent effective action.

Note now that the integrand is positive definite. Therefore, once the α -integral has been performed there can be no cancellation between different classes of configurations or different regions of the ζ -integration. In other words, a reliable calculation or an expansion in a small parameter of the ζ -integrand supplies a lower bound on Z and an upper bound on $V_{(4)}\varepsilon_0$.

In order to arrive at such an expansion a judicious choice of $\phi(\zeta)$ has to be made, we allow a systematic expansion of the α -effects. We shall now perform this expansion formally. This is achieved by simply expanding every quantity in a power series and keeping in the action only the quadratic parts, treating the rest as perturbations.

Define,

$$S_b[\phi+\alpha] = S_b[\phi] + (\alpha|g) + \frac{1}{2}(\alpha|K|\alpha) + \ldots \qquad (12)$$

$$D[\phi+\alpha] = D[\phi] + \ldots \qquad (13)$$

The 0-th order terms in the expansion are thus,

$$Z_{1\text{-loop}} = \int d\zeta \, |\Delta|^{\frac{1}{2}} |D[\phi]|^{\frac{1}{2}} |K[\phi]|^{-\frac{1}{2}} \times \qquad (14)$$

$$\times \exp[-S_b[\phi] + \frac{1}{2}(g|K^{-1}|g)]$$

Note that K is defined only in the subspace of modes which are orthogonal to $\phi_{,\zeta}$ (eq. 11).

The expansion may be continued in the standard way, leading to a perturbation series where $K^{-1}(\phi)$ and $D^{-1}(\phi)$ play the role of propagators, the non linear terms in S_b and the higher order terms in $|\Delta|$ and $|D|$ supply the vertices, while g acts as an external current. The utility of this expansion and the reliability of eq. (14) as a leading approximation depend on three conditions:

(a) The coupling constants which appear in the non linear vertices should be small.
(b) The effects of g should be small compared to S_b .
(c) Ultra-violet and infra-red divergences should be absent.

Note that we need not require an a-priori manifest realization of any symmetry (including supersymmetry). If the conditions (a,b,c) are satisfied the expansion is reliable and automatically supplies a systematic approximation to Z.

III A SUSY-HIGGS MODEL

In order to make the considerations of the introduction more explicit we pause to exhibit an asymptotically free SUSY-gauge theory which defines a supersymmetric Higgs phase realized by massive supermultiplets. The simplest model of this type is an SU(2) gauge theory whose physical fields are

Gauge Multiplet: $(A_{\mu a}, \lambda_a)\,;\ a = 1,2,3$ (15)

where A_μ is the gauge field and λ_a the (Majorana) gaugino.

Matter Multiplets: $(\chi_{iA}, H_{iA}),\ (\chi_0, H_0)$ (16)

where i=1,2 is a flavour index and A=1,2 is the gauge group index. χ_{iA} is a pair of Majorana doublets while H_{iA} is the corresponding pair of complex scalar doublets. (χ_0, H_0) is a gauge singlet.

The superpotential is:

$$f(z) = y\, z_0 \left(\tfrac{1}{2} \epsilon_{ij} \epsilon_{AB} z_{iA} z_{jB} - v^2\right) \tag{17}$$

where y is a dimensionless Yukawa coupling constant and v is a mass parameter.

The scalar potential is thus,

$$V(H) = \frac{g^2}{2} \left| H^*_{iA}\, t_{aAB} H_{iB} \right|^2 + \left| \frac{\partial f}{\partial H} \right|^2 \tag{18}$$

where t_a are 1/2 the Paule matrices. V is minimized ($V=0$ at the minimum) by:

$$\langle H_0 \rangle = 0 \qquad \langle H_{iA} \rangle = v\, \delta_{iA} \tag{19}$$

The vacuum defined by eq. (19) generates a Higgs mechanism and an examination of the boson and fermion mass matrices leads to two Dirac supermultiplets:

massive gauge s.m.: $(A_{\mu a}, \Psi_a, H_a)$

$$\Psi_a = \frac{1+\gamma^5}{2}\lambda_a + \frac{1-\gamma^5}{2} t_{aAi} X_{iA} \; ; \; H_A = \sqrt{2}\, t_{aAi} H_{iA}$$

$$\text{mass} \equiv \mu = g v \tag{20}$$

massive matter s.m.: (X, H, H_0)

$$X = \frac{1+\gamma^5}{2} X_0 + \frac{1-\gamma^5}{2} \frac{X_{ii}}{\sqrt{2}} \quad ; \quad H = \frac{1}{\sqrt{2}} H_{ii}$$

$$\text{mass} \equiv M = y v \tag{21}$$

We also record here the 1-loop ß-function of the model:

$$\beta_1(g) = -\frac{g^3}{16\pi^2}\cdot 5 \tag{22}$$

Thus, the running coupling at momentum q is:

$$\frac{8\pi^2}{g^2(q)} = 5 \ln q/\Lambda \tag{23}$$

The perturbative analysis of the model in the Higgs phase is thus valid so long as the v.e.v. v is sufficiently large (and of course $y \ll 1$):

$$(yv, gv, v) \gg \Lambda \tag{24}$$

which means that the on-shell coupling determined by $q \sim v$ is small:

$$\frac{g^2(v)}{8\pi^2} = \frac{1}{5 \ln v/\Lambda} \ll 1 \tag{25}$$

Conversely, the generic non perturbative amplitude is:

$$\exp\left(-\frac{8\pi^2}{g^2(v)}\right) = \left(\frac{\Lambda}{v}\right)^5 \ll 1 \tag{26}$$

We note that for a background field composed of ($A_\mu = 0 = H_0$, $H_{iA} = \langle H_{iA} \rangle$) the determinants of the boson and fermions (including the effect of the F-P determinant) in (eq. 14) exactly cancel due to SUSY. Finally, note that all the Euclidean propagators decrease exponentially at large distances:

$$\Delta(x-y) \sim e^{-\kappa|x-y|} \qquad : \kappa = \mu, M \tag{27}$$

IV THE INSTANTON BACKGROUND

Before introducing our favorite candidates for $\phi(\zeta)$ we recall the form of the instanton in the singular gauge:

$$\begin{aligned} A^{inst.}_{\mu a} &= \frac{1}{g}\, a_{\mu a}(x-z, \rho) \\ a_{\mu a}(y,\rho) &= \frac{2\eta_{a\mu\nu}\, y_\nu\, \rho^2}{y^2(y^2+\rho^2)} \end{aligned} \tag{28}$$

$$\frac{1}{4}\left(F^{inst.}_{\mu\nu}\right)^2 = \frac{48\,\rho^4}{g(y^2+\rho^2)^4} \quad ; \quad \int \frac{F^2_{\mu\nu}}{4} = \frac{8\pi^2}{g^2} \tag{29}$$

ρ is a scale parameter and z_μ the center of the instanton. Our aim is to distribute the fields defined by eq. (28) in the Higgs vacuum and determine the modification (if any) in the vacuum energy density ϵ_0. To this end we define $\phi(\zeta)$ as follows:

$$\begin{aligned} |x-z_j| < r \; &: \; A^{cl}_\mu(x) = \frac{1}{g}\, a_\mu(x-z_j, \rho_j) \\ &\quad\; H^{cl}_{iA}(x) = H^{cl}_0(x) = 0 \end{aligned} \tag{30}$$

$$\begin{aligned} r < R < |x-z_j| \; &: \; A^{cl}_\mu(x) = 0 \\ &\quad\; H^{cl}_{iA}(x) = \delta_{iA} v, \; H^{cl}_0(x) = 0 \end{aligned} \tag{31}$$

$$\begin{aligned} r < |x-z_j| < R \; &: \; A^{cl}_\mu(x) = c(|x-z_j|)\,\frac{1}{g}\, a_\mu(x-z_j, \rho_j) \\ &\quad\; H^{cl}_{iA}(x) = (1-c)\,\delta_{iA} v, \; H^{cl}_0(x) = 0 \end{aligned} \tag{32}$$

The interpolating function c is arbitrary except for boundary conditions which insure the continuity of the fields and their derivatives:

$$x=r : c=1, \partial_\mu c=0 \; ; \; x=R : c=0, \partial_\mu c=0 \quad (33)$$

We also limit our considerations to instantons that are well separated and small:

$$|z_i - z_j| >> R \geq r >> \rho \quad (34)$$

The parameters (R,r) will not be treated as collective coordinates but as fixed parameter to be chosen so as to minimize the effects of the matching. Note that since we are only interested in an upper bound on ϵ_0 we may limit the ρ_i and z_i integrations by hand.

The expansion described in sec. II may now be carried out. We note first that the currents vanish everywhere except in the matching regions:

$$\frac{\delta S}{\delta H^{cl}} = \frac{\delta S}{\delta A_\mu^{cl}} = 0 : R < |x-z_j|, |x-z_j| < r \quad (35)$$

In the matching regions the function g_{i_A} and $g_{\mu a}$ are bounded by:

$$r < |x-z_j| < R : \quad |g_{i_A}| \leq \frac{v}{r^2} \quad , \; |g_0| = 0$$
$$|g_{\mu a}| \leq \frac{\rho^2}{g r^5} \quad (36)$$

where numerical constants and algebraic functions of r/R have been suppressed. Another property of the g's which is evident is that they are dominated by momenta $\sim r^{-1}$.

The 2nd order differential operators which define the bose propagators K^{-1} differ in the matching region from the Higgs-vacuum form by the replacement $v \to (1-c)v$ and by the presence of the small (see eq. 32) gauge field. Let us now choose for r (and R) the order of magnitude

$$R \geq r \sim \mu^{-1} \quad (37)$$

Under these conditions we easily estimate (for $|x-y| \sim r$):

$$|K_{\mu\nu}^{-1}(x,y)| \sim \mu^2 \quad (38)$$

$$|K^{-1}(x,y)| \sim \frac{\mu^4}{M^2} \quad (39)$$

Substituting the bounds (36) into $(g|K^{-1}|g)$ we find that the latter is $O(g^2)$ if

$$\rho \lesssim v^{-1} \sim g r \ll r \tag{40}$$

Note that the orthogonality conditions (eqs. 11) do not invalidate these estimates since the derivatives $\partial_\zeta A^{cl}, \partial_\zeta H^{cl}$ etc. are dominated in the matching region by the same momenta $\sim r^{-1} \sim \mu$.

Next, we note that for $|x-y| \gg R$ the propagators satisfy the vacuum equations defined by the mass matrices (20,21) and therefore decrease exponentially.

Finally, in the region $|x-y| < r$ the propagators are approximated by the single-instanton propagators. In all the regions the short distance behaviour is that of the free field propagators.

Before we continue we briefly discuss two subtleties due to the necessity of fixing the gauge. First, the fundamental shift of the fields (eq. 2) should be defined only up to a gauge transformation designed to preserve the gauge fixing conditions. This result in a covariantization of all the derivatives $\partial/\partial\zeta$:

$$\phi_{,\zeta} \longrightarrow \phi_{;\zeta} = \phi_{,\zeta} + \omega_{,\zeta} \cdot \phi \tag{41}$$

where $\omega(\zeta)$ is determined by requiring that the gauge condition be preserved by $\phi_{,\zeta}$.

Second, it is desirable to choose a gauge condition which leads to a massive F-P propagator with the same mass as the vector mesons. This is achieved by choosing:

$$(\nabla_\mu(A^{cl})A_\mu)_a + g\, t_{aAB} H^{cl}_{i_A} (H - H^{cl})_{i_B} = 0 \tag{42}$$

Note now that the matrix Δ_{ij} is non-zero only if i and j belong to the same instanton, since $\phi_{;i}$ has no overlap with $\phi_{;j}$ otherwise. This means that the determinant of factorizes:

$$|\Delta_{ij}|^{1/2} = \prod_i |\Delta(i)|^{1/2} = |\Delta|^{N/2} \tag{43}$$

where N is the total number of instantons, and $|\Delta|$ a constant determined by the parameters (μ, M, ρ, g). The next crucial observation is that since all the propagators (bose and fermi) decrease exponentially their determinants also factorize:

$$|K| = |K_0| \prod_j |K_j| \tag{44}$$

$$|D| = |D_0| \prod_j |D_j| \tag{45}$$

The property of SUSY means that $|K_0|^{-\frac{1}{2}}$ cancels with $|D_0|^{\frac{1}{2}}$, and thus after integrating over ρ and z we arrive at:

$$Z = \exp(-V_{(4)}\epsilon_0) \geqslant \exp V_{(4)} e^{-\frac{8\pi^2}{g^2}} v^4 \sigma \tag{46}$$

where σ is a dimensionless parameter which depends on y and g. Its precise dependence depends on the fermi and bose mass matrices and the matching. We observe that the previous estimates (36 - 40) and the smallness of g and y prove the validity of the conditions (a,b) of sec. II, while the renormalizability and the Higgs mechanism insure condition (c). We conclude that eq. (46) is indeed correct and provides an upper bound on ϵ_0 :

$$\epsilon_0 \leqslant - v^4 \sigma e^{-\frac{8\pi^2}{g^2(v)}} \tag{47}$$

The crucial issue is the value of σ. Had the background been an exact solution of the field equations, σ would be 0! The reason is that fermionic 0-modes (in particular) would have caused $|D|$ to vanish. The background we have chosen is not however an exact solution and due to the necessity of matching contains a finite density of regions in which the background fields do not solve the field equations. Thus, the would be normalizable 0-mode necessarily mix with continuum and become unstable. Moreover, even if for some value of the parameters 0-modes would exist, a variation of the scale ρ would change the external fields in the region $|x-z| < R$ and destroy the 0-modes. The reason 0-modes can exist for all values of ρ in the single instanton case is the scale invariance of the Yang-Mills and Dirac equations which is lost in the presence of masses. As explained above, if the contribution from any finite region of the parameter space is non-zero an upper bound on the vacuum energy results.

To summarize, we have shown that a diluted gas of instantons (and anti-instantons) lowers the vacuum energy density by an amount proportional to the instanton density. The main ingredients in the derivation are the absence of fermionic 0-modes, the factorization of instanton effects and the vanishing of the perturbative vacuum energy. The crucial property which insures all this is the existence of an infra-red scale determined by the mass spectrum. We conclude that the non perturbative definition of the path integral which includes topologically non trivial configurations is incompatible with the SUSY algebra.

V DISCUSSION

The preceding discussion concentrated on SUSY-Gauge-Higgs models which contain a built-in infra-red scale. Observe now that the mass parameter v may be varied without affecting the SUSY of perturbation theory. If v is decreased it can be shown from the r.g. equation (26) that the induced negative ϵ_0 will decrease (that is $|\epsilon_0|$ would increase!). When v approaches Λ, $g^2(v)$ necessarily approaches $O(1)$ and the perturbative treatment becomes unreliable. Note however that as $v \to 0$, Λ survives as the only candidate for a physical mass scale. We would therefore expect ϵ_0 to approach asymptotically a value of order $(-\Lambda^4)$. It cannot however be excluded a-priori that ϵ_0 would reverse its behaviour and vanish with v, or cross 0 and become positive. Even if either of these possibilities occur there is no obvious reason why SUSY should be restored at some critical value of v (it would of course be spontaneously broken if ϵ_0 becomes positive after restoration). As for all dynamical properties of strongly interacting gauge theories, a definite answer requires a solution of at least a qualitative understanding of the spectrum.

We have already discussed in the introduction some possible reasons for the breakdown. A detailed understanding of the interplay between the presence of classical fields and various "holes" in configuration space, and supersymmetry of the Lagrangian is clearly lacking and the resolution of the issue is a formidable challenge. We remark here that if a manifestly supersymmetric definition of the path integral existed the phenomenon we discovered could not happen. There have been some attempts to provide such a description, which necessarily raises the issue of "classical" fermi fields. Since these attempts have been made in the context of massless Yang-Mills theories it is difficult to make a comparison with our results for reasons which have already been discussed.

REFERENCES

1. Y. Aharonov, A. Casher, and S. Yankielowicz, SUSY breakdown due to topological effects in q.m., Phys. Rev. D30, 368 (1984).
2. The first suggestion that instantons may cause SUSY breaking is due to: L. Abbot, M. Grisaru, H. Schnitzer, Phys. Rev. D16 (1977)3002.
3. A detailed treatment of instanton determinants is given by G. 't Hooft, Phys. Rev. D14 (1976) 3432.
4. The effective Lagrangian approach was initiated by G. Veneziano, S. Yankielowicz, Nucl.Phys. B (1982).
5. An attempt to treat instantons in a manifestly supersymmetric way is given by A. I. Vainshtein, V. I. Zakharov, Nucl. Phys. B (1983).

SUPERGRAVITY COUPLINGS AND SUPERFORMS

Richard Grimm

L.A.P.P.
B.P. 909
74019 Annecy-le-Vieux Cedex, France

1. Introduction

The topics discussed in this seminar are based on work done in collaboration with G. Girardi, M. Müller and J. Wess[1,2,3].

In this collaboration a reducible multiplet[1] of N=1 supergravity consisting of 24 bosonic and 24 fermionic degrees of freedom was presented. It was shown that in superspace geometry[4,5,6] the reducible multiplet is characterised by a set of natural superspace constraints. This multiplet is reducible but not fully reducible. It contains as submultiplets the multiplets of minimal[7,8], non minimal[9,10] and new minimal[11,12] supergravities as well as a certain multiplet of 16+16 components[3]. The latter can be used to describe the truncation to N=1 supersymmetry of a variant[13] of N=4 supergravity[14] where one of the propagating spin zero degrees of freedom appears as an antisymmetric tensor gauge potential[15].

It was also shown[2,3] that the analogue of the Wess and Zumino construction of the chiral representation and the chiral density multiplet[16,6] of minimal supergravity exists for non minimal and new minimal supergravity and for our 16+16 multiplet and allows to construct explicitly the pure supergravity actions as well as matter and Yang Mills couplings.

In chapter 2 I will very briefly summarise the properties of N=1 superspace geometry that describes the reducible multiplet and indicate how the various submultiplets are obtained. The natural constraints will be stated and their consequences for the properties of torsion and curvature in superspace will be pointed out. The reductions will be achieved by imposing suitable additional constraints.

In chapters 3 and 4 one and two form gauge potentials will be described in curved superspace with natural constraints. The motivation is of course that such a formulation contains the couplings of the Yang-Mills and antisymmetric tensor gauge fields to the different supergravity multiplets as subcases. The field strengths, which are two and three forms, respectively, will be subject to the same constraints as in flat superspace[16,17]. The three form field strength in the present formulation, however, will be made to contain an extension of the Chern and Simon three form[18,19] to superspace. This will entail gauge transformations rotating the two form gauge potential into the Yang-Mills field strength[20,21] in order to render the (abelian) three form field strength invariant.

In chapter 5 a generic construction of the chiral representation and the chiral density multiplet will be explained.

In chapter 6 the methods developed in the preceding chapters will be used to construct an interaction of supersymmetric Yang-Mills theory with the 16+16 multiplet of supergravity on purely geometric grounds, using the properties of the superspace extension of the Chern and Simon three form.

Some observations on the possible connection of this theory with a truncation of N=4 extended supergravity with gauged[22] SU(2) × SU(2) will conclude this talk.

2. The reducible multiplet

In superspace geometry, constraints serve to eliminate superfluous degrees of freedom and to discriminate the various supergravity multiplets from each other. The desire of motivating and understanding superspace constraints has provoked a number of interesting investigations[23,24].

In ref. 1 it was shown how the reducible multiplet can be obtained from a set of natural constraints and a detailed and systematic analysis of superspace constraints (which I will not repeat here) was presented. One starts from a superspace with the Lorentz group as structure group (for notations and conventions see refs 5,6). Its non-vanishing curvature components

$$R_B{}^A \sim \left(R_b{}^a ,\ R_\beta{}^\alpha ,\ R^{\dot\beta}{}_{\dot\alpha} \right) \tag{1}$$

have the properties

$$R_{ba} = - R_{ab}$$

$$R_{\alpha}{}^{\alpha} = 0 \ , \ R^{\dot\alpha}{}_{\dot\alpha} = 0 \qquad (2)$$

$$R_{\beta\dot\beta\,\alpha\dot\alpha} = 2\epsilon_{\dot\beta\dot\alpha} R_{\beta\alpha} - 2\epsilon_{\beta\alpha} R_{\dot\beta\dot\alpha}$$

A consequence of (1) and (2) is that the curvature components can be expressed in terms of torsion components and their covariant derivatives[25]. The torsion components, in turn, are subject to the following constraints

$$T_{\gamma\beta}{}^{a} = 0 \ , \ T^{\dot\gamma\dot\beta\,a} = 0$$

$$T_{\gamma}{}^{\dot\beta\,a} = -2i(\sigma^{a}\epsilon)_{\gamma}{}^{\dot\beta} \qquad (3)$$

at canonical dimension 0,

$$T_{\gamma b}{}^{a} = 2n\,\delta_{b}^{a} T_{\gamma} \ , \ T^{\dot\gamma}{}_{b}{}^{a} = 2n\,\delta_{b}^{a} \bar{T}^{\dot\gamma}$$

$$T_{\gamma}{}^{\dot\beta}{}_{\dot\alpha} = (n-1)\,\delta_{\dot\alpha}^{\dot\beta} T_{\gamma} \ , \ T^{\dot\gamma}{}_{\beta}{}^{\alpha} = (n-1)\,\delta_{\beta}^{\alpha} \bar{T}^{\dot\gamma}$$

$$T_{\gamma\beta}{}^{\alpha} = (n+1)\left(\delta_{\beta}^{\alpha} T_{\gamma} + \delta_{\gamma}^{\alpha} T_{\beta}\right) \qquad (4)$$

$$T^{\dot\gamma\dot\beta}{}_{\dot\alpha} = (n+1)\left(\delta_{\dot\alpha}^{\dot\beta} \bar{T}^{\dot\gamma} + \delta_{\dot\alpha}^{\dot\gamma} \bar{T}^{\dot\beta}\right)$$

$n \in \mathbb{R}$, $3n+1 \neq 0$ (see also ref. 10), at (mass) dimension 1/2 and

$$T_{cb}{}^{a} = 0 \qquad (5)$$

at dimension 1.

The natural constraints (1) - (5) have consequences for the remaining torsion components as well. This can be most easily seen by using the superspace Bianchi identities. For instance, at dimension 1/2 one obtains

$$T_{\gamma\beta\,\dot\alpha} = 0 \ , \ T^{\dot\gamma\dot\beta\,\alpha} = 0 \qquad (6)$$

and the dimension 1 torsion components are restricted to be

$$T_{\gamma\,\beta\dot{\beta}\,\dot{\alpha}} = -2i\,\epsilon_{\gamma\beta}\,\epsilon_{\dot{\beta}\dot{\alpha}}\,R^{+}$$
$$T_{\dot{\gamma}\,\beta\dot{\beta}\,\alpha} = -2i\,\epsilon_{\dot{\gamma}\dot{\beta}}\,\epsilon_{\beta\alpha}\,R \tag{7}$$

$$T_{\gamma\,\beta\dot{\beta}\,\alpha} = i\,\epsilon_{\beta\alpha}\left(G_{\gamma\dot{\beta}} - Z_{\gamma\dot{\beta}}\right) - \frac{3i}{2}\,\epsilon_{\gamma\alpha}\,G_{\beta\dot{\beta}}$$
$$T_{\dot{\gamma}\,\beta\dot{\beta}\,\dot{\alpha}} = i\,\epsilon_{\dot{\beta}\dot{\alpha}}\left(G_{\beta\dot{\gamma}} + \bar{Z}_{\beta\dot{\gamma}}\right) - \frac{3i}{2}\,\epsilon_{\dot{\gamma}\dot{\alpha}}\,G_{\beta\dot{\beta}} \tag{8}$$

$$Z_{\alpha\dot{\alpha}} = -\left(n+\frac{1}{3}\right)\bar{\mathcal{D}}_{\dot{\alpha}}\,T_{\alpha} - \left(n-\frac{1}{3}\right)\mathcal{D}_{\alpha}\,\bar{T}_{\dot{\alpha}} + \frac{2}{3}(n-1)\,T_{\alpha}\,\bar{T}_{\dot{\alpha}}$$
$$\bar{Z}_{\alpha\dot{\alpha}} = -\left(n-\frac{1}{3}\right)\bar{\mathcal{D}}_{\dot{\alpha}}\,T_{\alpha} - \left(n+\frac{1}{3}\right)\mathcal{D}_{\alpha}\,\bar{T}_{\dot{\alpha}} - \frac{2}{3}(n-1)\,T_{\alpha}\,\bar{T}_{\dot{\alpha}} \tag{9}$$

It turns out that all the torsion (and thereby curvature) components can be expressed in terms of the superfields

$$T_{\alpha},\ \bar{T}^{\dot{\alpha}},\ R,\ R^{+},\ G_{\beta\dot{\beta}},\ W_{\underbrace{\gamma\beta\alpha}},\ W_{\underbrace{\dot{\gamma}\dot{\beta}\dot{\alpha}}} \tag{10}$$

The last two have dimension 3/2 and appear as part of the torsion components

$$T_{cb}{}^{\alpha},\ T_{cb\,\dot{\alpha}} \tag{11}$$

which, at $\theta = \bar{\theta} = 0$, contain the field strength of the Rarita-Schwinger (gravitino) field.

Furthermore, as a consequence of the superspace Bianchi identities, the superfields (10) are subject to certain constraint equations. For instance:

$$\mathcal{D}_{\beta}T_{\alpha} + \mathcal{D}_{\alpha}T_{\beta} = 0 \tag{12}$$

$$\mathcal{D}_{\alpha}R^{+} = -2(n+1)\,T_{\alpha}\,R^{+} \tag{13}$$

$$\bar{\mathcal{D}}_{\dot{\alpha}}\,W_{\underbrace{\gamma\beta\alpha}} = -(3n+1)\,\bar{T}_{\dot{\alpha}}\,W_{\underbrace{\gamma\beta\alpha}} \tag{14}$$

$$\bar{\mathcal{D}}_{\dot\alpha} R^{+} - \mathcal{D}^{\alpha} G_{\alpha\dot\alpha} = -\frac{1}{4}\left(n-\frac{1}{3}\right)\mathcal{D}^{\alpha}\bar{\mathcal{D}}_{\dot\alpha} T_{\alpha} - \frac{1}{4}\left(n+\frac{1}{3}\right)\bar{\mathcal{D}}^{\dot\alpha}\mathcal{D}_{\alpha}\bar{T}_{\dot\alpha} - \frac{i}{2}(n+1)\,\mathcal{D}_{\alpha\dot\alpha} T^{\alpha} + \text{some non linear terms} \tag{15}$$

The complete solution of the Bianchi identities is unfolded in full detail in ref. 1 and it is therefore not necessary to copy it here.

Also, the identification of the 24 bosonic and 24 fermionic components of the reducible multiplet together with their local supersymmetry transformation laws is minutely exposed in chapter III of the same reference, here I content myself with the reproduction of table 2 :

dim	superfield	lowest component	
$1/2$	$T_{\alpha}, \bar{T}^{\dot\alpha}$	$T_{\alpha}, \bar{T}^{\dot\alpha}$	
1	R, R^{+}	$-\frac{1}{6}M, -\frac{1}{6}M^{*}$	
	$G_{\alpha\dot\alpha}$	$-\frac{1}{3} b_{\alpha\dot\alpha}$	
	$S, \bar{S}$	$S, \bar{S}$	(16)
	$\bar{\mathcal{D}}_{\dot\alpha} T_{\alpha}, \mathcal{D}_{\alpha}\bar{T}_{\dot\alpha}$	$c_{\alpha\dot\alpha} + i d_{\alpha\dot\alpha}, -c_{\alpha\dot\alpha} + i d_{\alpha\dot\alpha}$	
$3/2$	$\mathcal{D}_{\alpha} R, \bar{\mathcal{D}}_{\dot\alpha} R^{+}$	$\chi_{\alpha}, \bar{\chi}_{\dot\alpha}$	
	$\mathcal{D}_{\alpha}\bar{S}, \bar{\mathcal{D}}_{\dot\alpha} S$	$\lambda_{\alpha}, \bar{\lambda}_{\dot\alpha}$	
2	$\mathcal{D}^{\alpha}\mathcal{D}_{\alpha}\bar{S}, \bar{\mathcal{D}}_{\dot\alpha}\bar{\mathcal{D}}^{\dot\alpha} S$	$J, \bar{J}$	

Together with the graviton and the gravitino fields, localised in lowest components of the supervielbein;

$$
\begin{aligned}
e_m{}^a(x) &= E_m{}^a(x,0,0) \\
\tfrac{1}{2}\psi_m{}^\alpha(x) &= E_m{}^\alpha(x,0,0) \\
\tfrac{1}{2}\bar{\psi}_{m\dot{\alpha}}(x) &= E_{m\dot{\alpha}}(x,0,0)
\end{aligned}
\tag{17}
$$

this constitutes the reducible multiplet. Note that the superfields S, $\bar{S}$ are defined as

$$
\begin{aligned}
S &= \mathcal{D}^\alpha T_\alpha - (n+1)T^\alpha T_\alpha \\
\bar{S} &= \bar{\mathcal{D}}_{\dot{\alpha}} \bar{T}^{\dot{\alpha}} - (n+1)\bar{T}_{\dot{\alpha}} \bar{T}^{\dot{\alpha}}
\end{aligned}
\tag{18}
$$

and therefore obey

$$
\begin{aligned}
\mathcal{D}_\alpha S &= 8 T_\alpha R^\dagger \\
\bar{\mathcal{D}}_{\dot{\alpha}} \bar{S} &= 8 \bar{T}_{\dot{\alpha}} R
\end{aligned}
\tag{19}
$$

The reductions to either of the submultiplets, the minimal, non-minimal, new minimal as well as our 16+16 multiplet can be obtained by imposing additional constraints, as visualized in the following diagram :

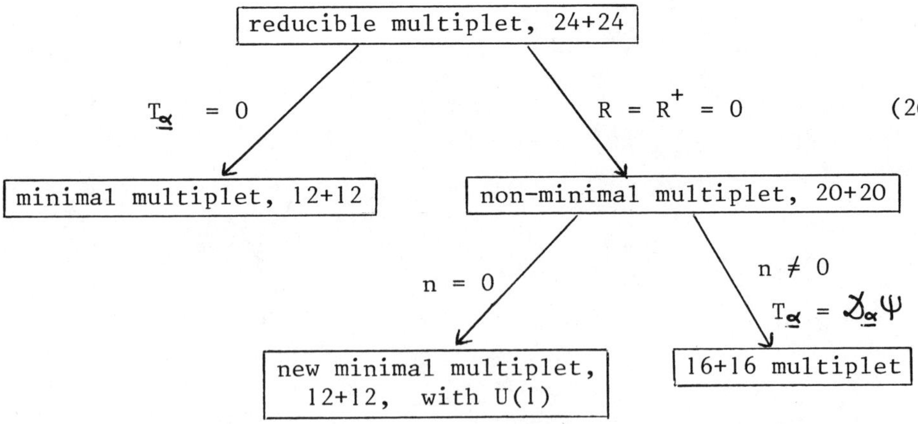

(20)

Detailed user's instructions for this diagram may be found in chapter IV of ref. 1 and in ref. 3.

One of the aims of this talk is to give yet another application of the reducible multiplet. To this end we need to describe supersymmetric Yang-Mills theory as well as the geometry of a two form gauge potential in a curved superspace with natural constraints. As already pointed out in the introduction this is a useful undertaking in itself and will be the subject of the next two chapters.

3. One form gauge potentials

The gauge potential A is a Lie algebra valued one form, subject to infinitesimal gauge variations :

$$\delta A = [A, \varepsilon] - d\varepsilon \tag{21}$$

Its field strength F,

$$F = dA + AA \tag{22}$$

is a two form in superspace,

$$F = \frac{1}{2} E^A E^B F_{BA} \tag{23}$$

and satisfies Bianchi identities

$$DF = 0 \tag{24}$$

involving the covariant exterior derivative D.

It is well known from flat superspace that the constraints

$$F_{\underline{\beta}\underline{\alpha}} = 0 \tag{25}$$

allow to describe supersymmetric Yang-Mills theory (an underlined spinor index can take dotted as well as undotted values). In particular the remaining components of F,

$$F_{\beta a}, \quad F^{\dot{\beta}}{}_{a}, \quad F_{ba} \tag{26}$$

can be expressed in terms of the superfields

$$W_\alpha, \quad \bar{W}^{\dot{\alpha}} \tag{27}$$

which in turn are subject to certain constraints.

In curved superspace with natural constraints this structure is realised by the following set of equations:

$$F_{\beta a} = i(\sigma_a \epsilon)_\beta{}^{\dot\beta} \bar{W}_{\dot\beta}$$
$$F^{\dot\beta}{}_a = i(\bar\sigma_a \epsilon)^{\dot\beta}{}_\beta W^\beta \tag{28}$$

$$F_{\beta\dot\beta\,\alpha\dot\alpha} = \frac{1}{2}\epsilon_{\dot\beta\dot\alpha} \sum_{\beta\alpha} \left(\mathcal{D}_\beta + (3n-1)T_\beta\right) W_\alpha - \frac{1}{2}\epsilon_{\beta\alpha} \sum_{\dot\beta\dot\alpha} \left(\bar{\mathcal{D}}_{\dot\beta} + (3n-1)\bar{T}_{\dot\beta}\right)\bar{W}_{\dot\alpha} \tag{29}$$

and:

$$\left(\mathcal{D}_\beta + (3n+1)T_\beta\right)\bar{W}_{\dot\alpha} = 0$$
$$\left(\bar{\mathcal{D}}_{\dot\beta} + (3n+1)\bar{T}_{\dot\beta}\right)W_\alpha = 0 \tag{30}$$

$$\left(\mathcal{D}^\alpha + (3n-1)T^\alpha\right)W_\alpha + \left(\bar{\mathcal{D}}_{\dot\alpha} + (3n-1)\bar{T}_{\dot\alpha}\right)\bar{W}^{\dot\alpha} = 0 \tag{31}$$

4. Two form gauge potentials

Two form gauge potentials in a curved superspace, constrained to describe new minimal supergravity, have been dealt with at full length in appendix C of ref. 1. Here I am going to generalise this in two respects. First of all the underlying curved superspace will be that of the reducible multiplet, subject to the natural constraints. On the other hand the field strength of the two form gauge potential will be defined to contain a (formal) generalisation of the Chern and Simon form to superspace, constructed out of A and F (as defined in the preceding chapter):

$$\omega_3 = k\left(AF - \frac{1}{3}AAA\right) \tag{32}$$

$$G = dB + \omega_3 \tag{33}$$

The two form gauge potential B is subject to the infinitesimal transformations

$$\delta B = d\Lambda + \mathrm{tr}(\varepsilon dA) \tag{34}$$

Λ describes a one form gauge parameter in superspace and the second term in (34) cancels the gauge variation of the Chern and Simon form.

Applying d to eq.(33) one obtains

$$dG = \mathrm{tr}(FF) \tag{35}$$

The three form field strength

$$G = \frac{1}{3!} E^A E^B E^C G_{CBA} \tag{36}$$

is subject to constraints (compatible with (25)):

$$G_{\underline{\gamma}\underline{\beta}\underline{\alpha}} = 0 \tag{37}$$

at dimension -1/2 and

$$G_{\gamma\beta a} = 0 \, , \quad G^{\dot{\gamma}\dot{\beta}}{}_a = 0 \tag{38}$$

$$G_\gamma{}^{\dot{\beta}}{}_a = -2i\,(\sigma_a\epsilon)_\gamma{}^{\dot{\beta}}\,\phi \tag{39}$$

at dimension zero.

It is well known, that in flat superspace and without ω_3 the constraints (37)-(39) describe a supermultiplet of 4+4 components, consisting of a real scalar, an antisymmetric tensor, a Majorana spinor and no auxiliary fields[26].

In the present case, as a consequence of (37)-(39) one obtains for the remaining field strengths

$$G_{\gamma\, ba} \, , \quad G^{\dot{\gamma}}{}_{ba} \, , \quad G_{cba} \tag{40}$$

at dimension 1/2 and 1, respectively:

$$G_{\gamma\beta\dot\beta\alpha\dot\alpha} = -2\epsilon_{\dot\beta\dot\alpha}\sum_{\beta\alpha}\epsilon_{\gamma\beta}\Phi_{;\alpha}$$
$$G_{\dot\gamma\beta\dot\beta\alpha\dot\alpha} = -2\epsilon_{\beta\alpha}\sum_{\dot\beta\dot\alpha}\epsilon_{\dot\gamma\dot\beta}\Phi_{;\dot\alpha} \tag{41}$$

with:

$$\Phi_{;\underline{\alpha}} = \left(\mathcal{D}_{\underline{\alpha}} + 4n\, T_{\underline{\alpha}}\right)\Phi \tag{42}$$

and:

$$G_{\gamma\dot\gamma\beta\dot\beta\alpha\dot\alpha} = -2i\,\epsilon_{\gamma\beta}\epsilon_{\dot\gamma\dot\alpha}W_{\alpha\dot\beta} + 2i\,\epsilon_{\gamma\alpha}\epsilon_{\dot\gamma\dot\beta}W_{\beta\dot\alpha},$$
$$W_{\alpha\dot\alpha} = \tfrac{1}{2}\left(\bar{\mathcal{D}}_{\dot\alpha} + (5n-1)\bar{T}_{\dot\alpha}\right)\Phi_{;\alpha} - \tfrac{1}{2}\left(\mathcal{D}_{\alpha} + (5n-1)T_{\alpha}\right)\Phi_{;\dot\alpha} \tag{43}$$
$$+ k\left(\mathcal{W}_\alpha\bar{\mathcal{W}}_{\dot\alpha}\right) + 2\Phi\left(G_{\alpha\dot\alpha} - \tfrac{1}{3}\mathcal{D}_\alpha\bar{T}_{\dot\alpha} + \tfrac{1}{3}\bar{\mathcal{D}}_{\dot\alpha}T_\alpha - \tfrac{2}{3}(n-1)T_\alpha\bar{T}_{\dot\alpha}\right)$$

These results are most easily obtained by using the Bianchi identities as defined in (35), in addition it turns out that the real superfield ϕ has to satisfy

$$\left(\mathcal{D}^\alpha + (5n+1)T^\alpha\right)\Phi_{;\alpha} = 8R^{+}\Phi + k\left(\bar{\mathcal{W}}_{\dot\alpha}\bar{\mathcal{W}}^{\dot\alpha}\right)$$
$$\left(\bar{\mathcal{D}}_{\dot\alpha} + (5n+1)\bar{T}_{\dot\alpha}\right)\Phi_{;}{}^{\dot\alpha} = 8R\Phi + k\left(\mathcal{W}^\alpha\mathcal{W}_\alpha\right) \tag{44}$$

As a matter of checking, one may convince oneself that, in the presence of constraints (37)-(39), eqs (41)-(44) constitute a complete solution of the Bianchi identities (35).

5. Chiral density multiplets

In minimal supergravity the concept of Wess and Zumino of the chiral representation and the chiral density multiplet[27,6] provides a very useful tool for constructing actions for pure supergravity and for interactions with matter and Yang-Mills theory. In ref. 2 the analogue of this procedure was carried through for non minimal supergravity and it was shown that the actions obtained this way can be specialised to describe new minimal and 16+16 supergravity as well. In this chapter, a generic construction of chiral representation and chiral density multiplet for the case of the natural constraints will be given, which can be used to reproduce any of the known examples and which, moreover, can accommodate other cases (see next chapter for an example).

As usual, one starts from a chiral superfield ϕ,

$$\bar{\mathcal{D}}^{\dot{\alpha}}\phi = 0 \tag{45}$$

and defines a chiral basis in terms of new Θ variables such that

$$\phi_1(x,\Theta) = A(x) + \Theta^\alpha \chi_\alpha(x) + \Theta^2 F(x) \tag{46}$$

The component fields being defined as

$$A(x) = \phi|_{\theta=\bar{\theta}=0}\ , \quad \chi_\alpha(x) = \mathcal{D}_\alpha\phi|_{\theta=\bar{\theta}=0}$$
$$F(x) = -\frac{1}{4}\mathcal{D}^\alpha\mathcal{D}_\alpha\phi|_{\theta=\bar{\theta}=0} \tag{47}$$

Next one introduces parameters

$$\eta^M(x,\Theta) = \eta^{(0)M}(x) + \Theta^\alpha\eta^{(1)}_\alpha{}^M(x) + \Theta^2\eta^{(2)M}(x) \tag{48}$$

such that the (known) local supersymmetry transformations of the component fields are reproduced by the transformation law

$$\delta\phi_1 = -\eta^M\partial_M\phi_1 \tag{49}$$

The (non-vanishing) parameters are determined to be

$$\eta^{(0)\mu} = \zeta^\mu(x) \tag{50}$$

the parameter of local component supersymmetry transformations and

$$\eta^{(1)}_\alpha{}^m = 2i(\bar{\zeta}\bar{\sigma}^m\epsilon)_\alpha$$

$$\eta^{(1)}_\alpha{}^\mu = -\frac{n+1}{2}\zeta^\beta(T_\beta\delta^\mu_\alpha + T_\alpha\delta^\mu_\beta)$$
$$-(n-1)\bar{\zeta}_{\dot{\beta}}\bar{T}^{\dot{\beta}}\delta^\mu_\alpha - i(\bar{\zeta}\bar{\sigma}^m\epsilon)_\alpha\psi_m{}^\mu$$

$$\eta^{(2)m} = -\frac{i}{2}(n+1)(\bar{\zeta}\bar{\sigma}^m\epsilon T) + (\bar{\zeta}\bar{\sigma}^n\sigma^m\epsilon\bar{\psi}_n) \tag{51}$$

$$
\begin{aligned}
\eta^{(2)\mu} = & - \zeta^{\mu}\left(2R^{+}| + \frac{n+1}{8} S + \frac{3}{16}(n+1)^{2} T^{2}\right) \\
& + \frac{1}{4}\left(2G^{\mu\dot{\beta}}| + (5n-\tfrac{1}{3})\bar{\mathcal{D}}^{\dot{\beta}} T^{\mu} + 4(n+\tfrac{1}{3})\mathcal{D}^{\mu}\bar{T}^{\dot{\beta}} - (n-1)(n-\tfrac{5}{3}) T^{\mu}\bar{T}^{\dot{\beta}}\right)\bar{\zeta}_{\dot{\beta}} \\
& - i\,(\bar{\zeta}\bar{\sigma}^{m})^{\beta}\psi_{m\beta}{}^{\mu} + \frac{i}{4}(n+1)(\bar{\zeta}\bar{\sigma}^{m})^{\mu}(\psi_{m}T) \\
& + \frac{i}{2}(n-1)(\bar{\zeta}\bar{\sigma}^{m})^{\mu}(\bar{\psi}_{m}\bar{T}) - \frac{1}{2}(\bar{\zeta}\bar{\sigma}^{m}\sigma^{n}\bar{\psi}_{m})\psi_{n}{}^{\mu}
\end{aligned}
$$

Again, a general chiral density

$$\Delta(x,\Theta) = a(x) + \Theta^{\alpha}\rho_{\alpha}(x) + \Theta^{2} f(x) \tag{52}$$

is defined by its transformation law

$$\delta\Delta = -(-1)^{m}\,\partial_{M}(\eta^{M}\Delta) \tag{53}$$

The general density will be particularised to the various supergravities by identifying

$$a(x) = e(x)\cdot e^{X}\Big|_{\theta=\bar{\theta}=0} \tag{54}$$

$e(x)$ is the determinant of the ordinary vierbein, $e(x) = \det e_{m}{}^{a}(x)$, and X is a generic superfield satisfying

$$\bar{\mathcal{D}}^{\dot{\alpha}} X = -2(3n+1)\bar{T}^{\dot{\alpha}} \tag{55}$$

It should be emphasised that X is not a new object. In any of the known examples it is constructed out of geometric quantities already present in the theory. In particular, in minimal supergravity, $X=0$.

Finally, the remaining components of the particular density $\mathcal{E}$ are determined to be

$$\rho_{\alpha} = e\cdot e^{X}\left|\left(\frac{13n-3}{2}T_{\alpha} + \mathcal{D}_{\alpha}X\right| + i(\sigma^{m}\bar{\epsilon}\bar{\psi}_{m})_{\alpha}\right) \tag{56}$$

and

$$f = e \cdot e^{X}| \Big[-M^{*} - \frac{3n-1}{2} S - \frac{1}{4} \mathcal{D}^{\alpha}\mathcal{D}_{\alpha}X| - \frac{1}{4}\mathcal{D}^{\alpha}X|\mathcal{D}_{\alpha}X|$$
$$- \frac{13n-3}{4} T^{\alpha}\mathcal{D}_{\alpha}X| - 4n(3n-1)T^{2} + i(3n-1)(\bar{\Psi}_{m}\bar{\sigma}^{m}T)$$
$$+ \frac{i}{2}(\bar{\Psi}_{m}\bar{\sigma}^{m}\mathcal{D}X|) - (\bar{\Psi}_{m}\bar{\sigma}^{mn}\epsilon\bar{\Psi}_{n}| \quad \Big] \tag{57}$$

Remark :

In many applications a slightly more compact notation turns out to be convenient. One splits $\mathcal{E}$ according to

$$\mathcal{E} = e^{X}\| \, \mathcal{E}_{0} \tag{58}$$

with the definitions

$$e^{X}\| = e^{X}| \Big[1 + \Theta^{\alpha}\mathcal{D}_{\alpha}X| - \frac{1}{4}\Theta^{2}\big(\mathcal{D}^{\alpha}\mathcal{D}_{\alpha}X| + \mathcal{D}^{\alpha}X|\mathcal{D}_{\alpha}X|\big) \Big] \tag{59}$$

and

$$a_{0} = e$$
$$\rho_{0\,\alpha} = e \Big[\frac{13n-3}{2} T_{\alpha} + i(\sigma^{m}\epsilon\bar{\Psi}_{m})_{\alpha} \Big] \tag{60}$$
$$f_{0} = e \Big[6R^{+}| - \frac{3n-1}{2} S - 4n(3n-1)T^{\alpha}T_{\alpha}$$
$$+ i(3n-1)(\bar{\Psi}_{m}\bar{\sigma}^{m}T) - (\bar{\Psi}_{m}\bar{\sigma}^{mn}\epsilon\bar{\Psi}_{n}) \Big]$$

The procedure of constructing supersymmetric densities consists now in multiplying $\mathcal{E}$ and a chiral superfield T and integrating over $d^{2}\Theta$:

$$\mathcal{L} = \int d^{2}\Theta \, \mathcal{E}\cdot T + h.c. \tag{61}$$

Examples :

In minimal supergravity, with $X=0$, one has $T=R$. In non-minimal supergravity, with

$$X = 2(3n+1) \frac{\bar{T}_{\dot{\alpha}} \bar{T}^{\dot{\alpha}}}{\bar{S}} \tag{62}$$

(see ref. 2), one has to take

$$T = \bar{S} \tag{63}$$

for the pure supergravity actions. Matter and Yang-Mills interactions are obtained by adding extra terms of the form (61), for examples see refs 2 and 27.

In the next chapter, as a further example, a theory will be outlined that describes 16+16 supergravity[3] in interaction with supersymmetric Yang-Mills theory and which will allow to obtain the full action from one and the same term (61), with T to be specified later.

6. Yang-Mills theory coupled to 16+16 supergravity

It has been observed in ref. 3 that the additional constraints

$$G_{\underline{\gamma}\, ba} = 0 \quad , \qquad n \neq 0 \quad , \tag{64}$$

on the field strength of the two form gauge potential causes the geometries of supergravity and of B and G to be patched together in the sense that some of the torsion components are expressed in terms of covariant superfields occuring in the geometry of two form potentials.

A similar mechanism works in the present case, when the (super) Chern and Simon three form is included. One imposes (64) on the geometric structures sketched in chapters 2, 3 and 4 and, as a consequence, obtains from (41), (42) and (44) that

$$T_{\underline{\alpha}} = \mathcal{D}_{\underline{\alpha}} \Psi \quad , \qquad \phi = e^{-4n\Psi} \tag{65}$$

as well as

$$\begin{aligned} R &= -\tfrac{1}{8} e^{4n\Psi}\, \mathrm{tr}(W^{\alpha} W_{\alpha}) \\ R^{+} &= -\tfrac{1}{8} e^{4n\Psi}\, \mathrm{tr}(\bar{W}_{\dot{\alpha}} \bar{W}^{\dot{\alpha}}) \end{aligned} \tag{66}$$

What are the consequences of these equations? First of all one checks that (65) and (66) are consistent and compatible with the superfield equations of natural superspace[1] and of chapter 3, like (13) and (30).

Furthermore the component fields (see (16))

$$d_{\alpha\dot{\alpha}}\,,\ \chi_\alpha\,,\ \bar{\chi}^{\dot{\alpha}}\,,\ \lambda_\alpha\,,\ \bar{\lambda}^{\dot{\alpha}}\,,\ J\,,\ \bar{J} \tag{67}$$

are no longer independent fields: they are expressed in terms of space time derivatives of other components and non linear terms. One thus arrives at a theory describing the coupling of supersymmetric Yang-Mills theory to 16+16 supergravity.

The Lagrangian is obtained by utilising the methods of chapter 5, taking X and T of eqs (54) and (61) to be

$$X = -2(3n+1)\Psi \tag{68}$$

$$T = -\frac{1}{8}\bar{\Delta}\, e^{2(3n+1)\Psi} \tag{69}$$

with the chiral projection operator

$$\bar{\Delta} = \bar{\mathcal{D}}_{\dot{\alpha}}\bar{\mathcal{D}}^{\dot{\alpha}} - 3(n+1)\bar{T}_{\dot{\alpha}}\bar{\mathcal{D}}^{\dot{\alpha}} - 2(n+1)\bar{S} - 8R \tag{70}$$

Taking into account (65) and (66) one obtains

$$T = e^{2(3n+1)\Psi}\left[\frac{1}{n}R - \frac{1}{2}\left(\bar{S} + 2(3n+1)\bar{T}_{\dot{\alpha}}\bar{T}^{\dot{\alpha}}\right)\right] \tag{71}$$

For computational purposes it is very convenient to split the chiral superfield T, similar to the splitting of the chiral density multiplet as defined in (58), (59):

$$T = e^{2(3n+1)\Psi} \,\|\, T_0 \tag{72}$$

The action is then given by

$$\mathcal{L} = \int d^2\Theta\, \mathcal{E}_0 T_0 + h.c. \tag{73}$$

and, with the information provided so far, it is a straightforward task to obtain its component form.

7. Conclusions

The reason for the rather dense presentation of the subject in this talk was to make as clear as possible the underlying (super) geometric ideas. This somewhat resulted in neglecting the explicit expressions in terms of component fields. It is however intended to make up for this in a separate and more detailed publication.

It remains to point out the connection of the theory described in chapter 6 with a truncation to N=1 supersymmetry of N=4 supergravity. Already in ref. 3 it was noted that for $n = -1/2$ the theory presented there was exactly the truncation to N=1 of the version of N=4 supergravity in which one of the scalars is replaced by an antisymmetric tensor gauge field.

In the theory of chapter 6 one might decide to take a Yang-Mills potential with values in the Lie algebra of SU(2). In this case the corresponding theory is expected to describe a truncation of a N=4 supergravity with an antisymmetric tensor gauge field and a gauged SU(2) × SU(2).

References

1. G. Girardi, R. Grimm, M. Müller and J. Wess, Karlsruhe preprint KA-THEP-84-5, LAPP-TH-102, to appear in Z. Phys. C.
2. R. Grimm, M. Müller and J. Wess, preprint TH.3881-CERN, LAPP-TH-107, to appear in Z. Phys. C.
3. G. Girardi, R. Grimm, M. Müller and J. Wess, preprint LAPP-TH-113, Karlsruhe KA-THEP-84-6, to appear in Phys. Lett. B.
4. V.P. Akulov, D. V. Volkov and V. A. Soroka, JETP Lett. 22:396 (1975) (English 187).
5. B. Zumino, Proceedings of the NATO Advanced Study Institute on Recent Developments in Gravitation, Cargèse, Corsica (Aug. 1978), NATO Advanced Study Institutes Series B (Plenum Press, New York, London), p. 405.
6. J. Wess and J. Bagger, Supersymmetry and Supergravity, Princeton Series in Physics, Princeton Univ. Press, Princeton, New Jersey, 1983.
7. K. S. Stelle and P. C. West, Phys. Lett. 74B:330 (1978).
8. S. Ferrara and P. van Nieuwenhuizen, Phys. Lett. 74B:333 (1978).
9. P. Breitenlohner, Nucl. Phys. B124:500 (1977).
10. W. Siegel and J. Gates, Nucl. Phys. B147:77 (1979).
11. V. Akulov, D. Volkov and V. Soroka, Theor. Math. Phys. 31:12 (1977).
12. M. F. Sohnius and P. C. West, Phys. Lett. 105B:353 (1981).
13. H. Nicolai and P. K. Townsend, Phys. Lett. 98B:257 (1981).
14. E. Cremmer, J. Scherk and S. Ferrara, Phys. Lett. 74B:61 (1978).
15. V. I. Ogievetsky and I.V. Polubarinov, Sov. J. Nucl. Phys. 4:156 (1966).

16. J. Wess, in Topics in Quantum Field Theory, J.A. de Azcarraga ed., Salamanca (1977); Lecture Notes in Physics 77, New York, Springer-Verlag (1978).
17. S. J. Gates Jr., Nucl. Phys. B184:381 (1981).
18. Shiing-shen Chern, Complex Manifolds Without Potential Theory, Second Edition 1979, Springer-Verlag New York, Heidelberg, Berlin.
19. B. Zumino, Les Houches lectures 1983, to be published by North-Holland, R. Stora and B. De Witt eds.
20. G. F. Chapline and N. S. Manton, Phys. Lett. 120B:105 (1983).
21. L. Baulieu, Phys. Lett. 126B:455 (1983).
22. D. Z. Freedman and J. H. Schwarz, Nucl. Phys. B137:333 (1978).
23. S. J. Gates Jr. and W. Siegel, Nucl. Phys. B163:519 (1980).
24. S. J. Gates Jr., K. S. Stelle and P. C. West, Nucl. Phys. B169: 347 (1980).
25. N. Dragon, Z. Phys. C2:29 (1979).
26. W. Siegel, Phys. Lett. 85B:333 (1979).
27. J. Wess and B. Zumino, Phys. Lett. 79B:394 (1978).

SUPERSYMMETRY IN NUCLEAR PHYSICS

F. Iachello

A.W. Wright Nuclear Structure Laboratory
Yale University
New Haven, Connecticut 06511

1. INTRODUCTION

About a decade ago, a new concept (supersymmetry) was introduced in physics. Originally developed for applications in elementary particle physics[1-3], this concept has been recently applied to a variety of problems, both in relativistic and non-relativistic physics. Up to now, the only field in which experimental examples of supersymmetries have been found is nuclear physics. Supersymmetry here has been used to classify simultaneously spectra of even-even and odd-even nuclei. In this article, I will briefly review applications of supersymmetry considerations to problems in nuclear physics.

Supersymmetry is used in nuclear physics in a way which is different form that used in elementary particle physics. It is a natural extension of the way in which "normal" symmetries had been previously used in the study of nuclear spectra. Suppose that, on the basis of some physics imput, one finds that states of a nucleus can be assigned to a representation $\mathcal{R}$ of some group G and that the Hamiltonian H describing that nucleus can be written in terms of the generators of G. The group G then serves two purposes: (i) it provides a basis in which H can be diagonalized and (ii), most importantly, if it happens that H contains only Casimir invariants of a complete chain of subgroups $G \supset G' \supset G''$ ... ,

$$H=\alpha\mathcal{C}(G) +\beta\, \mathcal{C}(G') +\gamma\, \mathcal{C}(G'') + \dots , \qquad (1)$$

it provides closed solutions to the corresponding eigenvalue problem (mass or energy formulas) since then

$$E = \alpha\langle C(G)\rangle + \beta\langle C(G')\rangle + \gamma\langle C(G'')\rangle + \dots , \tag{2}$$

where $\langle C(G)\rangle$ denotes the expectation value of $C(G)$ in the representation R . These mass formulas are very useful in comparing with experiment. An example of application of this type of symmetry (which I will call dynamic symmetry) to problems other than nuclear physics, is given by Gell-Mann- Ne'eman SU(3)[4]. Here

$$\begin{array}{ccccc} G & \supset & G' & \supset & G'' \\ \downarrow & & \downarrow & & \downarrow \\ SU(3) & & SU(2)\otimes U(1) & & SO(2)\otimes U(1) \end{array} \tag{3}$$

leading to the Gell-Mann- Okubo[5] mass formula

$$E(I,I_3,Y) = a + bY + c\left[\frac{Y^2}{4} - I(I+1)\right] . \tag{4}$$

In generalizing this concept to supersymmetry, one assumes that states in several nuclei (both even-even and even-odd) can be assigned to the representation R^* of some supergroup G^*. The Hamiltonian H (and other operators of interest) are constructed now from generators of G^*. Once more, G^* serves the purpose of (i) providing a basis in which H can be diagonalized, and (ii) if it happens that H contains only Casimir invariants of a complete chain of subgroups $G^*\supset G'^*\supset G''^*$ ...,

$$H = \alpha C(G^*) + \beta C(G'^*) + \gamma C(G''^*) + \dots , \tag{5}$$

it provides mass or energy formulas

$$E = \alpha\langle C(G^*)\rangle + \beta\langle C(G'^*)\rangle + \gamma\langle C(G''^*)\rangle + \dots . \tag{6}$$

These can be used now to classify simultaneously states in even-even and odd-even nuclei.

In applications of supersymmetries to nuclear physics, there are therefore two steps that must be performed: (i) the identification of the supergroup G^* appropriate to the problem; (ii) the study of the properties of its representations; in particular, one must know the branching rules $G^*\supset G'^*$, the algotitm for determining the eigenvalues of Casimir invariants and techniques for evaluating matrix elements of tensor operators between states.

Supersymmetries in nuclear physics are based on a model of nuclei, called interacting boson model. In this article, I will first briefly review this model, discussing its symmetries. I will then generalize it to include fermions and discuss the supersymmetries of the corresponding model. Finally, I will show some experimental examples of these supersymmetries and comment on future extensions and analogies.

2. DYNAMIC SYMMETRIES IN NUCLEI : BROKEN U(6)

The basic framework for discussing properties of nuclei is provided by the spherical shell model. Here protons and neutrons move in the average field due to all others. The single particle levels in this field bunch into clusters (shells), Fig. 1. The major closed shells are at nucleon number 2,8,20,28,50,82, and 126. The low-lying states of nuclei correspond to excitations within the last, unfilled shell (valence shell). Quite often, these display collective features. It has been suggested that these collective features arise from the fact that the effective nucleon-nucleon interaction

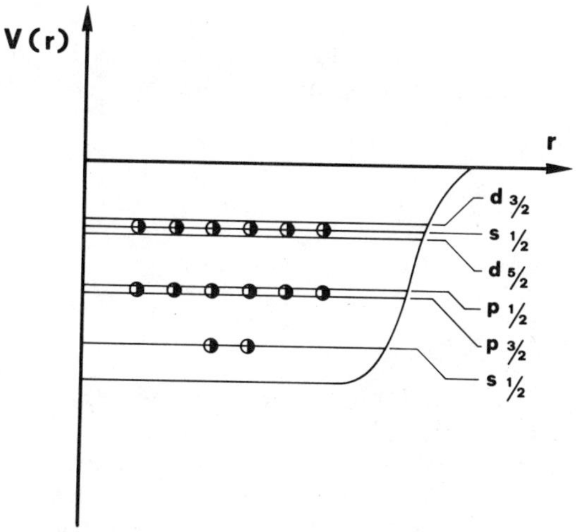

Fig. 1. Schematic representation of the average shell model potential in nuclei. The single particle levels are labelled by their spectroscopic notation.

inside nuclei tends to correlate nucleons in pairs. Two nucleons gain ≈1.5 MeV when coupled to angular momentum J=0 and ≈0.5 MeV when coupled to angular momentum J=2, while gaining no energy if coupled to larger values of J, Fig.2. It has thus been suggested that states in even-even nuclei be described by an assembly of bosons with angular momenta J=0 (s boson) and J=2 (d boson)[6,7]. The correlated pairs in nuclei (bosons) are similar to Cooper pairs in the electron gas[8]. The 1+5 components of the s and d boson span a six-dimensional space. It is customary to introduce boson creation and annihilation operators $s^\dagger, d^\dagger_\mu$ ($\mu=\pm2,\pm1,0$) , (s, d_μ) , altogether denoted by $b^\dagger_\alpha$($\alpha=1,\ldots,6$) , b_α . States are constructed by repeated application of creation operators $b^\dagger_\alpha$ on a vacuum

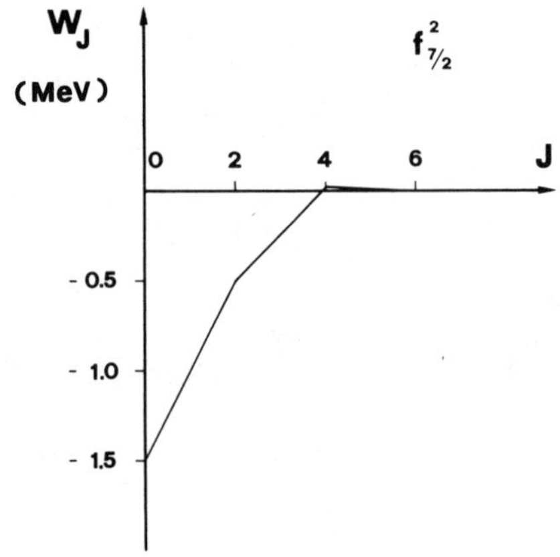

Fig.2. Matrix elements of the two-body residual interaction for two identical nucleons in the $f_{7/2}$ single particle level.

$$\mathcal{B} : \quad b_\alpha^\dagger \ldots |0\rangle \,. \tag{7}$$

Since the bosons span a six-dimensional space, the states in Eq.(7) form totally symmetric irreducible representations of the group U(6). They are characterized by the number of bosons, N. This is taken as the number of valence pairs. In this article, I will not distinguish between proton and neutron pairs, although this aspect has been extensively discussed in the literature[7]. The states (7) are characterized by a Young tableau

$$[N] \equiv \overbrace{\Box\,\Box}^{N\text{-times}} \ldots \Box \,. \tag{8}$$

The Hamiltonian describing a nucleus is written in terms of the generators $G_{\alpha\beta} = b_\alpha^\dagger b_\beta$ of U(6) as

$$H = E_{0B} + \sum_{\alpha\beta} \varepsilon_{\alpha\beta} G_{\alpha\beta} + \frac{1}{2} \sum_{\substack{\alpha\beta\\ \gamma\delta}} u_{\alpha\beta\gamma\delta}\, G_{\alpha\beta}\, G_{\gamma\delta} \,, \tag{9}$$

where only up to two-body terms have been included. This Hamiltonian can be separated into a U(6) invariant part plus the rest

$$H = E_0 + H' \,. \tag{10}$$

In the absence of H',all states would be degenerate. H' splits the degeneracy and from the point of view of a detailed understanding of nuclear spectra it is the most interesting part, leading to a classification scheme in terms of a broken U(6) symmetry.

The pattern of the U(6) symmetry breaking depends on the values of the coefficients $\varepsilon_{\alpha\beta}, u_{\alpha\beta\gamma\delta}$ appearing in Eq.(9). These in turn depend on the number of valence pairs, N (an U(6) unvariant). As a result, different nuclei display different breaking patterns. A study of the group structure of the problem shows that there are three and only three breaking patterns characterized by the group chains[9],

$$U(6) \begin{cases} \nearrow U(5) \supset O(5) \supset O(3) \supset O(2) \quad , & (I) \\ \longrightarrow SU(3) \supset O(3) \supset O(2) \quad , & (II) \\ \searrow O(6) \supset O(5) \supset O(3) \supset O(2) \quad . & (III) \end{cases} \qquad (11)$$

The fact that only three patterns are possible is a consequence of the condition that the rotation group, O(3), be contained in the chain. Using the arguments discussed in the introduction, one can associate to each chain a dynamic symmetry, by writing the Hamiltonian in terms only of Casimir invariants of the chain. For example, for chain III, one can write

$$H = E_0^{(III)} + AC_2(O6) + BC_2(O5) + CC_2(O3) \ , \qquad (12)$$

where $C_2(O6), \ldots$ denotes the quadratic Casimir operator of O(6), ... The Casimir invariant of O(2) (rotations around the z-axis) does not appear unless the nucleus is placed in an external field. Similarly, higher order invariants have not been included since H is chosen to be at most quadratic in the generators, Eq.(9).

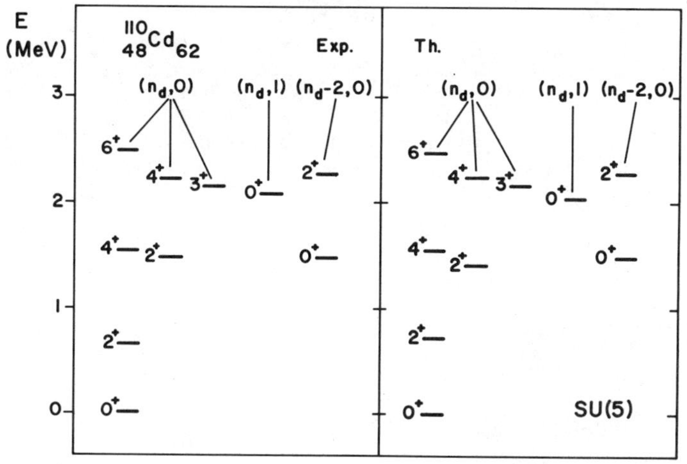

Fig.3. An example of U(5) Dynamic symmetry (chain I) in nuclei: the nucleus $^{110}_{48}Cd_{62}$ .On the left the experimental spectrum. On the right, the spectrum predicted by the mass formula (15) I.

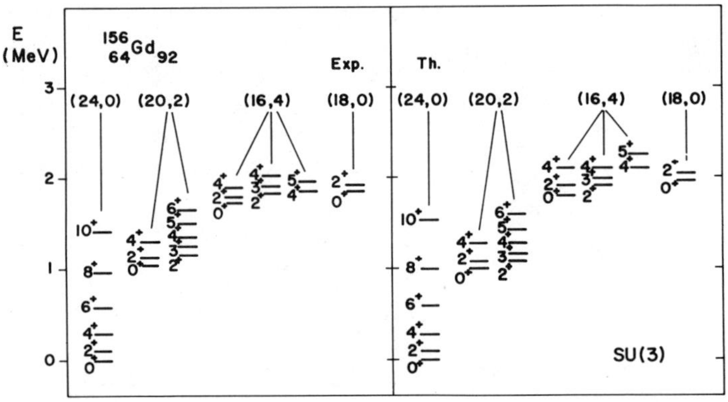

Fig.4. An example of SU(3) dynamic symmetry (chain II) in nuclei: the nucleus $^{156}_{64}Gd_{92}$. On the left the experimental spectrum. On the right, the spectrum predicted by the mass formula (15) II.

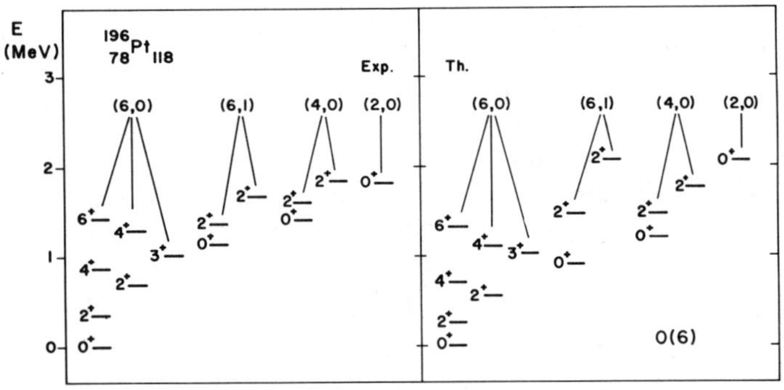

Fig.5. An example of O(6) dymanic symmetry (chain III) in nuclei: the nucleus $^{196}_{78}Pt_{118}$. On the left the experimental spectrum. On the right, the spectrum predicted by the mass formula (15) III.

The eigenvalues of H are

$$E(N,\sigma,\tau,\nu_\Delta,L,M_L) = E_0^{(III)} + A\sigma(\sigma+4) + B\tau(\tau+3) + CL(L+1), \quad (13)$$

where $N,\sigma,\tau,\nu_\Delta,L,M_L$, label the representations of the groups appearing in the chain III

$$\left| \begin{array}{ccccc} U(6) & O(6) & O(5) & O(3) & O(2) \\ \downarrow & \downarrow & \downarrow & \downarrow & \downarrow \\ N & \sigma & \tau(\nu_\Delta) & L & M_L \end{array} \right\rangle . \quad (14)$$

In a similar way, one can find mass formulas corresponding to the other chains. The complete results are

$$E^{(I)}(N,n_d,v,n_\Delta,L,M_L) = E_0^{(I)} + \varepsilon n_d + \alpha n_d(n_d+4) + \beta v(v+3) + \gamma L(L+1), \quad (I)$$

$$E^{(II)}(N,\lambda,\mu,K,L,M_L) = E_0^{(II)} + \kappa(\lambda^2+\mu^2+\lambda\mu+3\lambda+3\mu) + \kappa' L(L+1), \quad (II) \qquad (15)$$

$$E^{(III)}(N,\sigma,\tau,\nu_\Delta,L,M_L) = E_0^{(III)} + A\sigma(\sigma+4) + B\tau(\tau+3) + CL(L+1). \quad (III)$$

Several examples of all three dynamic symmetries have been found in the spectra of medium mass and heavy nuclei, three of which are shown in Figs.3,4 and 5.

3. GEOMETRY AND COSETS

As discussed above, states of even-even nuclei are described in terms of boson degrees of freedom interpreted as correlated pairs of particles (Cooper pairs). It is possible to give to the bosons a dual,geometric interpretation. This is done by introducing 5 classical variables α_μ ($\mu=\pm2,\pm1, 0$) through an intrinsic (or coherent state[10]

$$|N,\alpha_\mu\rangle = (s^\dagger + \sum_\mu \alpha_\mu d_\mu^\dagger)^N |0\rangle \quad . \qquad (16)$$

The variables α_μ are in the coset space $U(6)/U(5)\otimes U(1)$. In order to visualize the corresponding geometry, one can plot the variables α_μ on the surface of an ellipsoid of radius

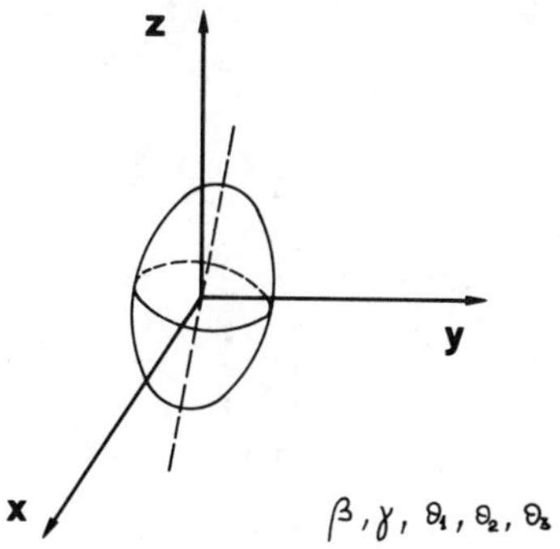

Fig.6. Shape description of nuclei. The variables $\beta,\gamma,\theta_1,\theta_2,\theta_3$, are called Bohr variables.

$$R = R_0(1+\sum_{\mu} \alpha_{\mu} \; Y_{2\mu}(\theta,\phi)) \quad , \tag{17}$$

as shown in Fig.6. The five variables α_μ can be transformed then into two variables, β and γ, describing the deformation of the ellipsoid and the Euler angles, $\theta_1, \theta_2, \theta_3$, describing the orientation of the intrinsic frame. This leads to a description of nuclear spectra in terms of shapes[11]. The breaking patterns, Eq.11, then provide a classification of shapes of nuclei in terms of symmetry groups, similar to the classification of the shape of cristals in terms of point groups.

4. DYNAMIC SUPERSYMMETRY IN NUCLEI: BROKEN U(6 /Ω)

In addition to nuclei with an even number of protons and neutrons in which all particles are paired together to J=0 and J=2, there are also nuclei with an odd number of neutrons (or protons) and states in even-even nuclei with two (or more) unpaired particles. The description of these nuclei and states requires the introduction of both bosonic degrees of freedom (pairs) and fermionic degrees of freedom (unpaired particles). The corresponding model is called interacting boson-fermion model[12]. The degeneracy of the fermionic space depends on the allowed values of the fermion angular momenta, j_i, and thus on the shell under consideration. It is given by $\Omega=\sum_i(2j_i+1)$. Again, it is convenient to introduce fermion creation (and annihilation) operators $a_i^\dagger$ $(i=1,\ldots,\Omega)$, (a_i). States can then be constructed as

$$\mathcal{BF} : \quad a_i^\dagger \ldots b_\alpha^\dagger \ldots |0\rangle \quad . \tag{18}$$

These states form representations of the group $U^{(B)}(6) \otimes U^{(F)}(\Omega)$, where an index B,F has been added in order to distinguish bosons from fermions. The states in Eq.(18) are characterized by the number of bosons N and fermions M. For example, in the low-lying states of even-odd nuclei, N is the number of valence pairs and M=1 (one unpaired particle). In general, the states (18) are characterized by Young tableaux which are completely symmetric for $U^{(B)}(6)$ and completely antisymmetric for $U^{(F)}(\Omega)$,

$$[N] \otimes \{M\} \equiv \overbrace{\square\;\square\ldots\square}^{N\text{-times}} \otimes \left.\begin{matrix}\square\\ \square\\ \vdots \\ \square\end{matrix}\right\} M\text{-times} \tag{19}$$

The Hamiltonian describing a nucleus with unpaired fermions is written in terms of the generators $G_{\alpha\beta}^{(B)}=b_\alpha^\dagger b_\beta$ and $G_{ij}^{(F)}=a_i^\dagger a_j$ as

$$H = H^{(B)} + H^{(F)} + V^{(BF)} , \qquad (20)$$

where

$$H^{(B)} = E_{OB} + \sum_{\alpha\beta} \varepsilon_{\alpha\beta} G^{(B)}_{\alpha\beta} + \frac{1}{2} \sum_{\alpha\beta\gamma\delta} u_{\alpha\beta\gamma\delta} \; G^{(B)}_{\alpha\beta} \; G^{(B)}_{\gamma\delta} ,$$

$$H^{(F)} = E_{OF} + \sum_{ij} \eta_{ij} G^{(F)}_{ij} + \frac{1}{2} \sum_{ijkl} v_{ijkl} \; G^{(F)}_{ij} \; G^{(F)}_{kl} ,$$

$$V^{(BF)} = \sum_{\alpha\beta ij} w_{\alpha\beta ij} \; G^{(B)}_{\alpha\beta} \; G^{(F)}_{ij} . \qquad (21)$$

The classification in terms of $U^{(B)}(6) \otimes U^{(F)}(\Omega)$ treats each nucleus separately. One may attempt to see whether or not several nuclei could be placed within a larger multiplet structure. Since even-even nuclei are bosonic and even-odd nuclei are fermionic, the larger structure must involve representations of supergroups. As a result, it has been suggested[13,14] that large regions of nuclei be assigned to specific representations of the supergroup $U(6,\Omega)$. In general, supergroups $U(n/m)$ are characterized by $(n+m)^2$ generators, which can be written, in the present context, as

$$\begin{array}{lll} G^{(B)}_{\alpha\beta} = b^{\dagger}_{\alpha} b_{\beta} & & n^2 \\ G^{(F)}_{ij} = a^{\dagger}_{i} a_{j} & & m^2 \\ F^{\dagger}_{i\alpha} = a^{\dagger}_{i} b_{\alpha} & & mn \\ F_{\alpha i} = b^{\dagger}_{\alpha} a_{i} & & mn \\ \hline & & (m+n)^2 \end{array} \qquad . \qquad (22)$$

Thus, imbedding $U^{(B)}(6) \otimes U^{(F)}(\Omega)$ into $U(6/\Omega)$, involves the introduction of the $2 \times 6 \times \Omega$ fermionic generators $F^{\dagger}_{i\alpha}, F_{\alpha i}$. The theory of supergroups and superalgebras can now be used to analyze the breaking patterns of $U(6/\Omega)$. The ingredients that one needs in this analysis are:

(i) a decomposition of the representations of $U(6/\Omega)$ into representations of subgroups, specifically $U(6) \otimes U(\Omega)$. The relevant representations here are the totally supersymmetric representations, denoted by

$$[\mathcal{N}\} \equiv \overbrace{\boxslash \; \boxslash \ldots \boxslash}^{\mathcal{N}\text{-times}} , \qquad (23)$$

characterized by the total number of bosons plus fermions, $\mathcal{N}=N+M$. Their decomposition is

$$[\mathcal{N}\} = (\overbrace{\square\,\square \ldots \square}^{\mathcal{N}} \otimes 1) \oplus (\overbrace{\square\,\square \ldots \square}^{\mathcal{N}-1} \otimes \square) \oplus$$

$$\otimes \cdots \otimes \left(\overbrace{\square\;\square \cdots \square}^{N-k} \otimes \left.\begin{matrix}\square\\ \square\\ \vdots\\ \square\end{matrix}\right\} k \right)$$

$$k \leq \Omega \,, \quad N - k \geq 0 \;; \qquad (24)$$

(ii) the evaluation of the expectation value of the Casimir operators and of the matrix elements of the generators for a specific group chain.

Supersymmetries are applied to the analysis of nuclear properties in the same way as symmetries are. The Hamiltonian,Eq. (20), is written as a $U(6/\Omega)$ invariant part plus the rest

$$H = E_0 + H' \quad . \qquad (25)$$

In the absence of H' all states of a given irreducible representation of $U(6/\Omega)$ are degenerate. H' splits the degeneracy. Whenever H' is written in terms only of Casimir operators of a complete chain of subgroups of $U(6/\Omega)$, the splitting can be calculated in closed form, and one obtains mass formulas which can be used to describe simultaneously both even-even and even-odd nuclei.

The number of breaking patterns that one can obtain starting from $U(6/\Omega)$ is rather large,since there are three breaking patterns for the boson part of the Hamiltonian, $H^{(B)}$, and several patterns for the fermion part, $H^{(F)}$. The breaking patterns appropriate to $H^{(F)}$ depend on the values of the fermion angular momenta,j_i. Up to now, only a limited number of cases has been investigated. The first case studied was that of a single value of j ($j=3/2,\Omega=4$). This case is relevant to odd-proton nuclei in the Os-Ir region[14]. It has been suggested that the breaking pattern appropriate to these nuclei is that described by the group chain

$$U(6/4) \supset U^{(B)}(6) \supset U^{(F)}(4) \supset O^{(B)}(6) \supset SU^{(F)}(4) \supset$$
$$\supset Spin(6) \supset Spin(5) \supset Spin(3) \supset Spin(2) \;, \qquad (26)$$

where Spin(n) denotes a spinor group[15]. By writing H as in Eq.(5) and calculating the expectation values of the appropriate Casimir operators, one obtains then the mass formula

$$E(N,N,M,\Sigma,\sigma_1,\sigma_2,\sigma_3,\tau_1,\tau_2,\nu_\Delta,J,M_J) = E_0 + E_1 N + E_2 N^2 +$$
$$+ A\Sigma(\Sigma+4) + A'\left[\sigma_1(\sigma_1+4)+\sigma_2(\sigma_2+2)+\sigma_3^2\right]+$$
$$+ B\left[\tau_1(\tau_1+3)+\tau_2(\tau_2+1)\right] + CJ(J+1) \;. \qquad (27)$$

The quantum numbers appearing in this formula label the irreduci-

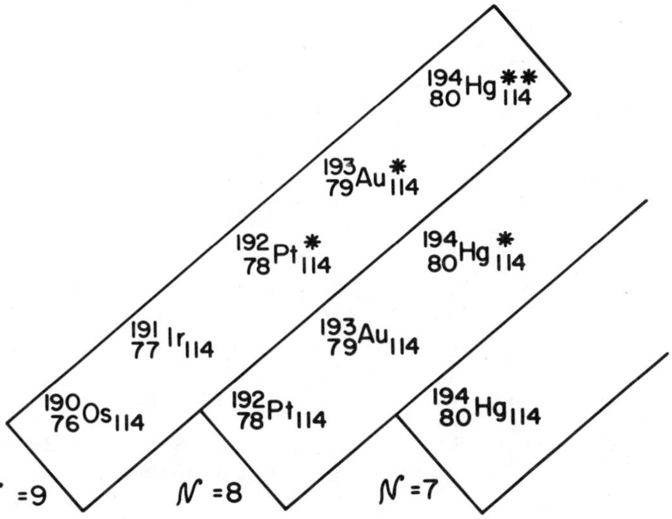

Fig.7. U(6/4) supermultiplets in the Os-Ir region.

ble representations of the groups appearing in Eq.(26),

$$\left| \begin{array}{ccccc} U(6/4) & U^{(B)}(6) & U^{(F)}(4) & O^{(B)}(6) & SU^{(F)}(4) \\ \downarrow & \downarrow & \downarrow & \downarrow & \\ \mathcal{N} & N & M & \Sigma & \\ & Spin(6) & Spin(5) & Spin(3) & Spin(2) \\ & \downarrow & \downarrow & \downarrow & \downarrow \\ & (\sigma_1,\sigma_2,\sigma_3) & (\tau_1,\tau_2(\nu_\Delta)) & J & M_J \end{array} \right\rangle . \qquad (28)$$

The mass formula (27) describes several nuclei (a supermultiplet), characterized by a given number of bosons plus fermions, $\mathcal{N}$. To each supermultiplet there belong states in five nuclei with $N=\mathcal{N}, M=0$; $N=\mathcal{N}-1, M=1$; $N=\mathcal{N}-2, M=2$; $N=\mathcal{N}-3, M=3$; $N=\mathcal{N}-4, M=4$. The quantum number M does not appear explicitly in Eq.(27), since it can be eliminated using the condition $N+M=\mathcal{N}$ and thus absorbed into the U(6/4) invariants E_0, E_1 and E_2. Typical supermultiplets in the Os-Ir region are shown in Fig.7. The predicted excitation spectra of a pair of nuclei (even-even and even-odd) are shown in Fig.8. They can be compared with the experimental spectra shown in Fig.9. The extent to which the two agree is evidence for the occurrence of dynamic supersymmetry in nuclei. The calculated excitation spectra in Fif.8 involve only two free parameters, B and C. Supersymmetry here provides a description of 22 levels (7 in the even-even and 15 in the odd-even nucleus) within $\approx$20% in terms of these two parameters.

Another case which has been discussed in detail in the literature is the case $j=1/2, 3/2, 5/2$ ($\Omega=12$). This case is relevant to

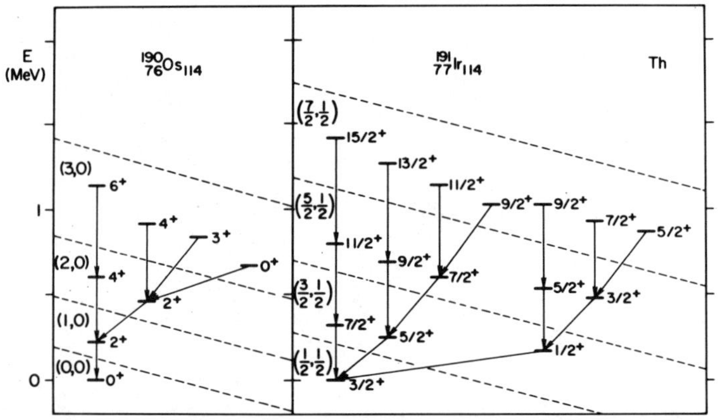

Fig.8. An example of U(6/4) supersymmetry in nuclei:excitation spectra predicted by the mass formula (27).

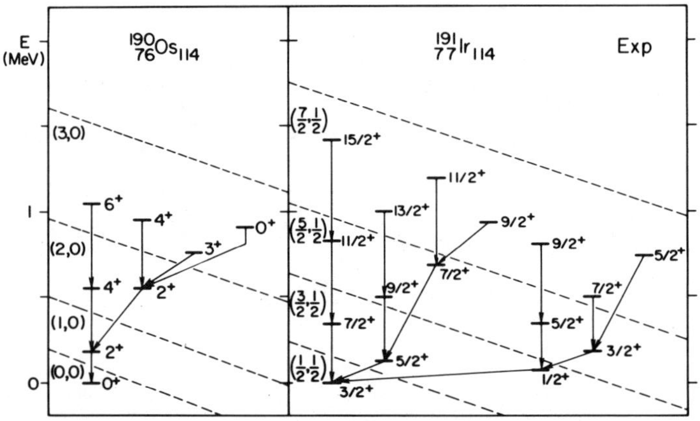

Fig.9. An example of U(6/4) supersymmetry in nuclei: experimental excitation spectra of the pair of nuclei ^{190}Os-^{191}Ir.

the description of odd-neutron nuclei in the Pt region[16,17]. Two breaking patterns have been suggested for these nuclei. One of these is

$$\begin{aligned} U(6/12) &\supset U^{(B)}(6) \otimes U^{(F)}(12) \supset U^{(B)}(6) \otimes U^{(F)}(6) \otimes SU_s(2) \supset \\ &\supset O^{(B)}(6) \otimes O^{(F)}(6) \otimes SU_s(2) \supset O^{(B+F)}(6) \otimes SU_s(2) \supset \\ &\supset O^{(B+F)}(5) \otimes SU_s(2) \supset O^{(B+F)}(3) \otimes SU_s(2) \supset \\ &\supset Spin(3) \supset Spin(2) \; . \end{aligned} \tag{29}$$

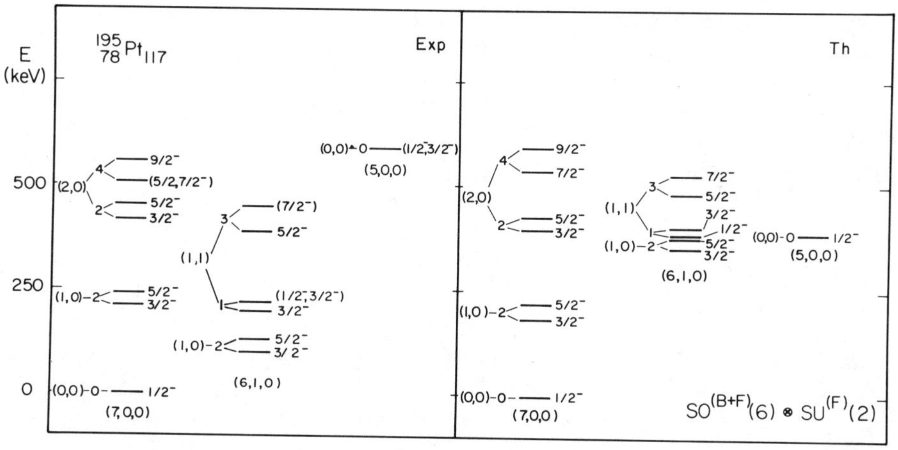

Fig.10. An example of U(6/12) supersymmetry in nuclei. Only the odd-even member of the supermultiplet, ${}^{195}_{78}Pt_{117}$, is shown. On the left, the experimental spectrum; on the right, the spectrum predicted by the corresponding mass formula.

Fig.10 shows how the breaking pattern (29) compares with experiment, although the comparison is limited to the odd-even member of the supermultiplet.

In addition to energy levels, supersymmetry provides a way to calculate in closed form other properties. Since all physical operators are constructed out of generators of the supergroup $U(6/\Omega)$ (or its powers) it is sufficient to consider these operators. It is convenient for this purpose to rewrite the generators (22) in matrix form

$$G = \begin{pmatrix} b^\dagger_\alpha b_\beta & b^\dagger_\alpha a_i \\ a^\dagger_i b_\alpha & a^\dagger_i a_j \end{pmatrix} , \qquad (30)$$

which distinguishes clearly between the Bose sector $b^\dagger_\alpha b_\beta, a^\dagger_i a_j$ and and the Fermi sector $b^\dagger_\alpha a_i, a^\dagger_i b_\alpha$ of the algebra. Matrix elements of the Bose sector of the algebra can be tested experimentally by considering electromagnetic decays. The operators inducing electromagnetic transitions can be written schematically as

$$T = \sum_{\alpha\beta} t_{\alpha\beta}\, G^{(B)}_{\alpha\beta} + \sum_{ij} t_{ij}\, G^{(F)}_{ij} . \qquad (31)$$

They connect states in the same nucleus. Several tests of supersymmetry selection rules and predictions have been performed, showing the same degree of accuracy ($\approx$20%) observed for energies.

It is interesting to note that also matrix elements of the Fermi sector of the algebra can be tested experimentally by studying transfer reactions, for example (^3He,^2H). The appropriate operator can be written as

$$P_+ = \sum_{i\alpha} p_{i\alpha} F^\dagger_{i\alpha} ,$$
$$P_- = \sum_{\alpha i} p_{\alpha i} F_{\alpha i} . \tag{32}$$

These operators connect states in different nuclei. Supersymmetry appears to describe also here the experimental situation reasonably well, although considerable discrepancies have been found in some cases.

5. SUPERGEOMETRY AND SUPERCOSETS

The geometric interpretation discussed in Sect.3 can be extended to include fermions by introducing super-coherent states. One needs here, in addition to commuting classical variables, also Grassman variables. For example, one can introduce coherent states of the form

$$|N;\alpha_\mu, g_m\rangle = (s^\dagger + \sum_\mu \alpha_\mu d^\dagger_\mu + \sum_m g_m a^\dagger_m)^N |0\rangle , \tag{33}$$

where the g_m's are Grassman variables. The α_μ and g_m's are in a supercoset space. Not much has been done up to now in the exploitation of the supergeometry associated with the supercoset space.

6. CONCLUSIONS

The use of supersymmetry considerations in nuclear physics has open the way for a detailed and unified description of even-even and even-odd nuclei. Several broken supersymmetries of the type $U(6/\Omega)$ have been identified experimentally, indicating that supersymmetry is a useful concept in physics. Supersymmetries used in nuclear physics are different from those mostly sought in particle physics. The basic idea behind supersymmetry applications in nuclear physics is that fermions and their associated Cooper pairs form a supermultiplet

$$\psi = \begin{pmatrix} b_\alpha \\ a_i \end{pmatrix} . \tag{34}$$

One may attempt to use this idea in other branches of physics. Nambu[18], for example, has considered applications to electrons moving in type II superconductors, where the supersymmetric partners are electrons and their associated Cooper pairs. The extent to which this idea can also be applied to particle physics is not clear at present ,although some suggestions[19] have been made to constuct supersymmetries by placing a quark and an antidiquark in the same supermultiplet

$$\psi = \begin{pmatrix} Q \\ \bar{D} \end{pmatrix} . \qquad (35)$$

ACKNOWLEDGEMENTS

I wish to thank I. Bars,F. Gürsey and Y. Nambu for discussions on the quark-diquark model of hadrons and type II superconductors. This work was supported in part under the Department of Energy Contract No. DE-AC 02-76 ER 03074.

REFERENCES

1. P.Ramond,Phys. Rev. D3 (1971) 2415;A.Neveu and J.H.Schwarz, Nucl. Phys. B31 (1971) 86.
2. D.V.Volkov and V.P.Akulov,Phys. Lett. 46B (1973) 109.
3. J.Wess and B.Zumino,Nucl. Phys. B70 (1974) 39.
4. M.Gell-Mann,Phys. Rev. 125 (1962) 1067; Y. Ne'eman,Nucl. Phys. 26 (1961) 222.
5. S. Okubo, Progr. Theor. Phys. 27 (1962) 949.
6. A. Arima and F.Iachello,Phys. Rev. Lett. 35 (1975) 1069.
7. A. Arima ,T.Otsuka,F.Iachello and I.Talmi,Phys. Lett. 66B (1977) 205; T.Otsuka,A.Arima,F.Iachello and I.Talmi,Phys. Lett. 76B (1978) 139.
8. L.N.Cooper,Phys. Rev. 104 (1956) 1189.
9. A.Arima and F.Iachello,Ann.Phys.(N.Y.) 99 (1976) 253;A.Arima and F.Iachello,Ann.Phys.(N.Y.) 111 (1978) 201; A.Arima and F.Iachello,Ann.Phys.(N.Y.) 123 (1979) 468.
10. A.Bohr and B.R.Mottelson,Physica Scripta 22 (1980) 468; A.E.L. Dieperink,O.Scholten and F.Iachello,Phys.Rev.Lett.44 (1980) 1747; J.N.Ginocchio and M.W.Kirson, Phys.Rev.Lett. 44 (1980) 1744; D.H.Feng,R.Gilmore and S.R.Deans,Phys. Rev. C23 (1980) 1254.
11. See,for example, A.Bohr and B.R.Mottelson,"Nuclear Structure", Vol.2,(Benjamin,New York,1975).
12. F.Iachello and O.Scholten,Phys.Rev.Lett. 43 (1979) 679.
13. F.Iachello,Phys.Rev.Lett. 44 (1980) 772.
14. A.B.Balantekin,I.Bars and F.Iachello,Phys.Rev.Lett. 47 (1981) 19; A.B.Balantekin,I.Bars and F.Iachello,Nucl.Phys. A370 (1981) 284.
15. R.Gilmore,"Lie Groups,Lie Algebras and Some of Their Applications", (Wiley, New York,1974),p.111.

16. A.B.Balantekin,I.Bars,R.Bijker and F.Iachello,Phys. Rev. C27 (1983) 1761.
17. D.D.Warner,R.Casten,M.L.Stelts,H.G.Borner and G.Barreau,Phys. Rev. C26 (1982) 1921; Sun Hong Zhou,A.Frank and P.van Isacker, Phys.Lett. 124B (1983) 275; Sun Hong Zhou,A.Frank and P.van Isacker,Phys.Rev. C27 (1983) 2430.
18. Y.Nambu, private communication.
19. S.Catto and F.Gürsey,Lett.Nuovo Cimento 35 (1982) 241; I.Bars and H.C.Tze,private communication.

MINIMAL SUPERSYMMETRIC MODELS : THE TOP QUARK MASS AND THE MASS OF THE LIGHTEST HIGGS*

Hans Peter Nilles**

Département de Physique Théorique
Université de Genève
1211 Genève 4, Switzerland

ABSTRACT

The predictions of the minimal model based on spontaneously broken supergravity are shown to depend strongly on the actual value of the top quark mass. This dependence is discussed and its consequences for the light particle spectrum are pointed out. In particular we will see that this model predicts an upper limit on the mass of the lightest neutral Higgs boson. This bound depends on m_{top}, but is always smaller than the mass of the toponium bound state.

In this seminar I shall report on recent results in the investigation of supersymmetric models obtained by M. Nusbaumer and myself[1]. They are concerned with predictions of the supersymmetric extension of the standard $SU(3) \times SU(2) \times U(1)$ model for the light particle spectrum.

The motivation to consider a supersymmetric version of the standard model comes from a theoretically unsatisfactory property of the Higgs sector of the standard model[2]. In fact one can regard the supersymmetric extension as the replacement of this Higgs sector by a more sophisticated one.

---

* Talk given at the 1984 Advanced Study Institute, University of Bonn, August 1984.

** Partially supported by the Swiss National Science Foundation.

The Higgs sector in the standard model consists of an SU(2)-doublet complex scalar h with potential

$$V(h) = -m^2|h|^2 + \lambda|h|^4 \tag{1}$$

where m^2 is chosen positive to trigger a spontaneous breakdown of $SU(2)\times U(1)$. The vacuum expectation value (vev) of the Higgs field is given by $(m^2/\lambda)^{1/2}$ and the parameters have to be chosen in such a way that the weak bosons $W^{\pm}$ and Z° aquire the correct masses. This then fixes one combination of m^2 and λ. The mass of the remaining physical scalar field is then given by

$$m_h = \sqrt{2}m \tag{2}$$

an undetermined quantity. The perturbative consistency of the model just tells us that m_h should be somewhere in the region between 7GeV and 1000GeV. Please remember this "prediction" of the standard model to judge the predictions of the supersymmetric model.

What is unsatisfactory about this whole sector is the absence of an explanation of the magnitude of the weak interaction scale set by the vev of the Higgs field. Theoretically there is no reason for this scale to be 100GeV. It could as well be 10^{10} or 10^{20}GeV. The scale is just a free input parameter. This property becomes even more pronounced when we consider one loop radiative corrections to m^2 in perturbation theory. These contributions due to the scalar self interactions are quadratically divergent

$$\delta m^2 \sim \lambda\Lambda^2 \tag{3}$$

where Λ is a cutoff. There is nothing wrong with these quadratic divergencies : we regularize these contributions and write the theory in terms of renormalized quantities and these are free input parameters as before.

Suppose now that in such a theory we would like to understand the scale m^2 in terms of more fundamental quantities. Suppose even that we would have found a reason why m is in the 100GeV region. If the theory stays as it is (m^2 being quadratically divergent) then our understanding would be spoiled and $m_{renormalized}$ would still be a completely free input parameter. With the quadratic divergencies new mass scales enter the theory that have nothing to do with the scales already present in the model. We can freely choose them but do not understand them.

To be able to understand the weak mass scale even in principle these quadratic divergencies have to be absent, which implies that the arbitrary cutoff in (3) has to be replaced by a physical cutoff Λ_P. This can only happen if the theory has additional structure in the form of new degrees of freedom at a scale Λ_P. To understand why the weak scale is of order of 100GeV this physical cutoff should then not exceed let us say 10TeV. One might now imagine several ways to add to the theory such a physical cutoff. One way out would be the absence of fundamental scalar fields in the theory. This is a rather radical approach in which, however, it turned out to be very difficult (if not impossible) to produce a working model. Facing these problems one was then led to a more conservative approach which allows the existence of fundamental scalar particles but eleminates quadratic divergencies with the help of a symmetry. The only symmetry known to do this is supersymmetry. The physical cutoff Λ_P is given by the supersymmetric partners of the particles in the standard model. In addition of the contribution discussed before to m^2 there is a contribution involving the fermionic partners and leading to

$$\delta m^2 \sim \lambda(m_B^2 - m_F^2) \ . \qquad (4)$$

In the supersymmetric limit $m_B = m_F$ this cancellation is exact, but we know already that supersymmetry has to be broken since none of these partners has been observed yet. These partners serve as a cutoff : i.e. in the present circumstances we would expect them to be in an energy range between 100GeV and 10TeV. More about these masses can only be said after we have turned this philosophy into a working model as we will see later. The first task is to show that there exists a model at all which fulfills these requirements. But we want to be more ambitious. First of all we want to construct a model with the absolute minimum of new structure and also we want to construct a model where we can really understand the weak scale in terms of the supersymmetry breakdown scale. This then would imply that in the limit of exact supersymmetry also $SU(2) \times U(1)$ would remain unbroken. The advantage of a minimal model is the fact that it makes the most precise predictions. This is true here as it is true for grand unified models where only the minimal model makes predictions that can be ruled out. Our second requirement on the relation of the weak scale and the supersymmetry breakdown scale is purely theoretical, but since the whole motivation for supersymmetry is theoretical we will stick to it.

The particle content of the minimal model is given by the quarks and leptons and their scalar partners to fill chiral supermultiplets. Every gauge boson receives a supersymmetric partner (gaugino) in a massless vector supermultiplet and we have to add two Higgs chiral superfields. The action is given by the supersymmetrized $SU(3) \times SU(2) \times U(1)$ gauge interactions and a superpotential

$$g = \mu H\bar{H} + g_E^{ij} H L_i \bar{E}_j + g_D^{ij} H Q_i \bar{D}_j + g_U^{ij} \bar{H} Q_i \bar{U}_j \qquad (5)$$

where H and $\bar{H}$ are the Higgs superfields, $L(\bar{E})$, Q, $(\bar{D},\bar{U})$ denote the lepton doublet (singlet), quark doublet (singlets) respectively and $i,j = 1,\ldots,3$ are family indices. At this stage neither supersymmetry nor $SU(2)\times U(1)$ are broken. Observe that of the two Higgses one (H) couples to down quark and leptons whereas the second $(\bar{H})$ couples to the up quarks. To have masses for all quarks and leptons it will therefore be necessary that both the scalars in H and $\bar{H}$ receive a vev.

The next step involves a discussion of supersymmetry breakdown which can in principle be done in various ways. The most economic way is the introduction of a hidden sector that breaks supergravity (the local form of supersymmetry) spontaneously. It does not lead to any new observable particles in the 100GeV region but introduces soft explicit supersymmetry breaking terms in the low energy theory given above. The size of the breakings is essentially given by the magnitude of the gravitino mass. With this approach the scalar potential of the minimal model is given by

$$V = \left|\frac{\partial g}{\partial \phi_a}\right|^2 + m_a^2 \phi_a \phi_a^* + \frac{1}{2} D^2 + m_{3/2}(A_E^{ij} g_E^{ij} h \ell_i \bar{e}_j \ldots) +$$

$$+ B\mu m_{3/2} h\bar{h} + h.c. \quad . \qquad (6)$$

The first and third term represent the supersymmetric potential and the other are supersymmetry breaking terms. The origin of the breaking parameters is given by the breakdown of supergravity at a large scale like the Planck mass. At this large scale they are given by

$$m_a^2 = m_{3/2}^2 \quad \text{for quark and lepton partners}$$

$$m_a^2 = m_{3/2}^2 + \mu^2 \quad \text{for Higgs scalars} \qquad (7)$$

$$A_E = A_D = A_U = B + 1 = A \; .$$

The values of these parameters in the 100GeV region have to be determined by renormalization group improved perturbation theory just as one has to compute the evolution of the $SU(3)\times SU(2)\times U(1)$ running coupling constants.

The model as discussed up to now thus has three additional input parameters $m_{3/2}$, A and μ. At a large scale one can also introduce a common Majorana mass for the gauge fermions which we call m_0. In addition to the Yukawa and gauge couplings the minimal model has thus four input parameters compared to the two (μ^2,λ) in the standard model. These are not very many if one takes into account that the model contains necessarily two Higgses.

In the next step we have to discuss the breakdown of $SU(2)\times U(1)$. The Higgs potential is given by

$$V_{Higgs} = m_1^2|h|^2 + m_2^2|\bar{h}|^2 - m_3^2(h\bar{h} + \bar{h}^* h^*) +$$
$$+ ((g_1^2+g_2^2)/8)(|h|^2 - |\bar{h}|^2)^2 \qquad (8)$$

where the last term cames from the D^2 contribution in (6) and g_2, g_1 denote the $SU(2)\times U(1)$ coupling constants. We have to impose the condition

$$2|m_3|^2 \leq m_1^2 + m_2^2 \qquad (9)$$

for the potential to be bounded from below. On the other hand we need

$$|m_3|^4 > m_1^2 \cdot m_2^2 \qquad (10)$$

for the absolute minimum to break $SU(2)\times U(1)$. With the parameters (7) given at the large scale we have thus no $SU(2)\times U(1)$ breakdown. The breakdown of the weak interactions has to come from radiative corrections. Let us look e.g. at the renormalization group equation for m_2^2:

$$\tilde{\mu}\frac{\partial}{\partial\tilde{\mu}}m_2^2 = \frac{3}{8\pi^2}(g_U^{33})^2[m_2^2 + m_{\bar{U}}^2 + m_Q^2 + m_{3/2}^2|A_U|^2] -$$
$$- \frac{1}{2\pi^2}[\frac{3}{4}|\tilde{m}_2|^2 g_2^2 + \frac{1}{4}|\tilde{m}_1|^2 g_1^2] - \frac{1}{8\pi^2}[3g_2^2 + g_1^2]\mu^2 . \qquad (11)$$

The first term has a positive sign which implies a decreasing m_2^2 with decreasing $\tilde{\mu}$. The crucial coefficient is the top quark Yukawa coupling g_U^{33} and we could imagine that with large enough g_U^{33} m_2^2 could become small enough to induce a breakdown of $SU(2)\times U(1)$. Contrary to the standard model where g_U^{33} was just a parameter to parametrize the top quark mass m_{top} it is here a crucial dynamical parameter. The predictions of the supersymmetric model will be different for different values of m_{top}. To extract these predictions we have to integrate some 25 renormalization group equations to determine the parameters at low energies which has to be done numerically. To compute the evolution we start with the parameters given in (7) at a scale $\tilde{\mu} = M_X \sim 3\times 10^{16}$GeV where the $SU(3)\times SU(2)\times U(1)$ coupling constants meet :

$\alpha = g^2/4\pi \sim 1/24$. The crucial parameters for the induction of $SU(2) \times U(1)$ breakdown are $m_{3/2}$, A, μ, m_o and g_U^{33} or equivalently m_{top}. The induction of this breakdown at the correct scale will then require one relation between these five parameters just in the same way as it fixed the ratio of m^2 and λ in the standard model. For this to happen one requires the parameters for $\tilde{\mu} \sim 100\text{GeV}$ to obey the relation

$$\frac{m_1^2 - \omega^2 m_2^2}{\omega^2 - 1} = M_Z^2/2 \tag{12}$$

where M_Z is the mass of the weak neutral boson $Z°$ and $\omega = \bar{v}/v$ denotes the ratio of the $\bar{h}$ and h vevs. It is related to the masses by

$$\frac{\omega}{\omega^2+1} = \frac{m_3^2}{m_1^2+m_2^2} \; . \tag{13}$$

We can now discuss the influence of the top quark mass on the light particle spectrum in the model. For the model to be consistent we have to have $m_{top} \leq 200\text{GeV}$. It comes from the fact that g_U^{33} at the large scale has to be infinite to give such a large m_{top} at low energies.

The qualitative effects of g_U^{33} on the model can already be guessed from the formulas given so far. If the top quark mass is large (i.e. $100\text{GeV} \leq m_{top} \leq 200\text{GeV}$) m_2^2 will decrease fast when we lower $\tilde{\mu}$. The condition for the breakdown of $SU(2) \times U(1)$ is given in (10) and in order not to overshoot we have to have m_3 small which in turn implies in general small μ. (By small we mean small compared to $m_{3/2}$.) From (12) and (13) on the other hand we then conclude that ω has to be very large : i.e. $\bar{v} >> v$, the vevs have to be asymmetric. Thus large m_{top} implies small μ and large ω. Small μ might be phenomenologically interesting since in the limit $\mu = 0$ there are three massless particles in the model. Charged and neutral Dirac fermions as a combination of Higgs and gauge fermions as well as a pseudoscalar axion. A more quantitative discussion in this mass region also reveals that m_o has to be moderately small. Since the mass of the photino is given by $m_o/2$ this also is of phenomenological interest. In the region of large m_{top} the model thus predicts a variety of states which have to be light compared to $m_{3/2}$. We do not really know what is the exact value of $m_{3/2}$ but even if it is as large as 1TeV these states might be within experimental reach during the next few years.

These predictions change when we lower the mass of the top quark. In an intermediate range $50\text{GeV} \leq m_{top} \leq 100\text{GeV}$ a quantitative discussion has to be performed to see what happens. I will not repeat this discussion here since it is given in detail in my Moriond lectures[3]. It turns out that in this intermediate range we expect m_o to be very large : typically $m_o \gtrsim 2m_{3/2}$. This does not really mean that it is impossible to have light supersymmetric particles in this case but one has certainly to "squeeze" some parameters to predict new light particles.

For even lower top quark masses $m_{top} \leq 50\text{GeV}$ at first glance it seems that the situation becomes even worse. But it pays off to look at things more closely. If m_{top} is small m_2^2 will not change very much compared to m_1^2 when we go from $\tilde{\mu} \sim 10^{16}\text{GeV}$ to $\tilde{\mu} \sim 100\text{GeV}$. From (10) we then need m_3 to be comparable to m_1 and m_2 which in turn by (13) implies ω to be close to unity. Thus for small m_{top} we need $\mu \sim m_{32}$ and $v \approx \bar{v}$. That this case is interesting can be seen from the potential (8) directly. If $m_1^2 = m_2^2 = m_3^2$ the potential has a flat direction and consequently there is a massless particle. Let us therefore look at the Higgs spectrum more closely[4]. We started with two complex doublets i.e. eight degrees of freedom. Three of them disappear because of the Higgs mechanism. There remain two charged Higgses with mass

$$m^2_{h^\pm} = m_1^2 + m_2^2 + M_W^2 \tag{14}$$

bounded from below by the mass of the weak W-bosons. There is one neutral pseudoscalar with mass

$$m_p^2 = m_1^2 + m_2^2 . \tag{15}$$

The mass of the two remaining scalar Higgses are given by

$$m^2_{S_{1,2}} = \frac{1}{2}\left[m_p^2 + M_z^2 \pm \left((m_p^2 + M_z^2)^2 - 4m_p^2 M_z^2 \frac{(\bar{v}^2 - v^2)^2}{(\bar{v}^2 + v^2)^2}\right)^{1/2}\right] \tag{16}$$

where again v and $\bar{v}$ denote the vevs of h and $\bar{h}$ respectively. For $\bar{v} = v$ we then have a massless scalar particle

$$m^2_{S_1} = 0$$
$$m^2_{S_2} = m_p^2 + M_z^2 . \tag{17}$$

This is the candidate for the light particle in the region $m_{top} \leq 50\text{GeV}$. We will come back to this point in a moment.

Observe first that (16) gives an absolute prediction of the

minimal model. The mass of the lightest neutral Higgs is bounded[4] from above by M_Z. This bound is reached if the vevs are as different as possible i.e. $v = 0$ or $\bar{v} = 0$ in which case we have

$$m^2_{S_1} = M^2_Z \qquad (18)$$
$$m^2_{S_2} = m^2_p$$

and this bound is independent of the values of the parameters $m_{3/2}$, m_0, A, μ and m_{top}. The upper bound is reached in a situation which corresponds to the region of large top quark masses where we had $\bar{v} >> v$. On the other hand for small m_{top} we have seen that in general v and $\bar{v}$ tend to become comparable in magnitude and from (16) we thus expect a stronger bound on the mass of the lightest neutral Higgs. Having made this observation we have studied this question in more detail. For this we varied for fixed m_{top} the parameters $m_{3/2}$, m_0, A and μ keeping the constraint (12). For $m_{top} \geq 55$GeV we then found that the general bound $m_H \lesssim M_Z$ can still be approximately reached. For smaller m_{top}, however, we obtain stronger bounds. The dependence of the bound on m_{top} is given in the figure and the bound is quite strict for small m_{top}. For a top quark mass of 30 (40) GeV for example the upper bound on the mass of the Higgs is given by 11 (24) GeV respectively. This is of particular interest since, according to recent results at the $p\bar{p}$-collider, the region $m_{top} \lesssim 50$GeV seems to be experimentally favored. We thus find that the mass of the Higgs is always sufficiently light so that it can be produced and detected in processes like toponium decaying in $H\gamma$ and or Z° decay.

Two comments are in order now. First, the bound given in the figure is the absolute bound in the minimal model. The Higgs could still be lighter and this bound is only reached for very special configurations of the parameters. Secondly, none of the two Higgs candidates seen in experiments, the $\xi(2.2)$ and $\xi(8.3)$, fits particularly well with a supersymmetric interpretation in the minimal model. The production rate for these states is two large. In a non-supersymmetric two-Higgs model this rate could be enhanced if the vevs v and $\bar{v}$ are sufficiently different. The bound on m_h for small m_{top}, however, comes from the fact that v and $\bar{v}$ have to be approximately equal. But I think it is still premature to draw any definite conclusions about the nature of the ξ or ζ states.

Nonetheless this bound on the Higgs masses is the only really precise prediction of the minimal supersymmetric model. Predictions

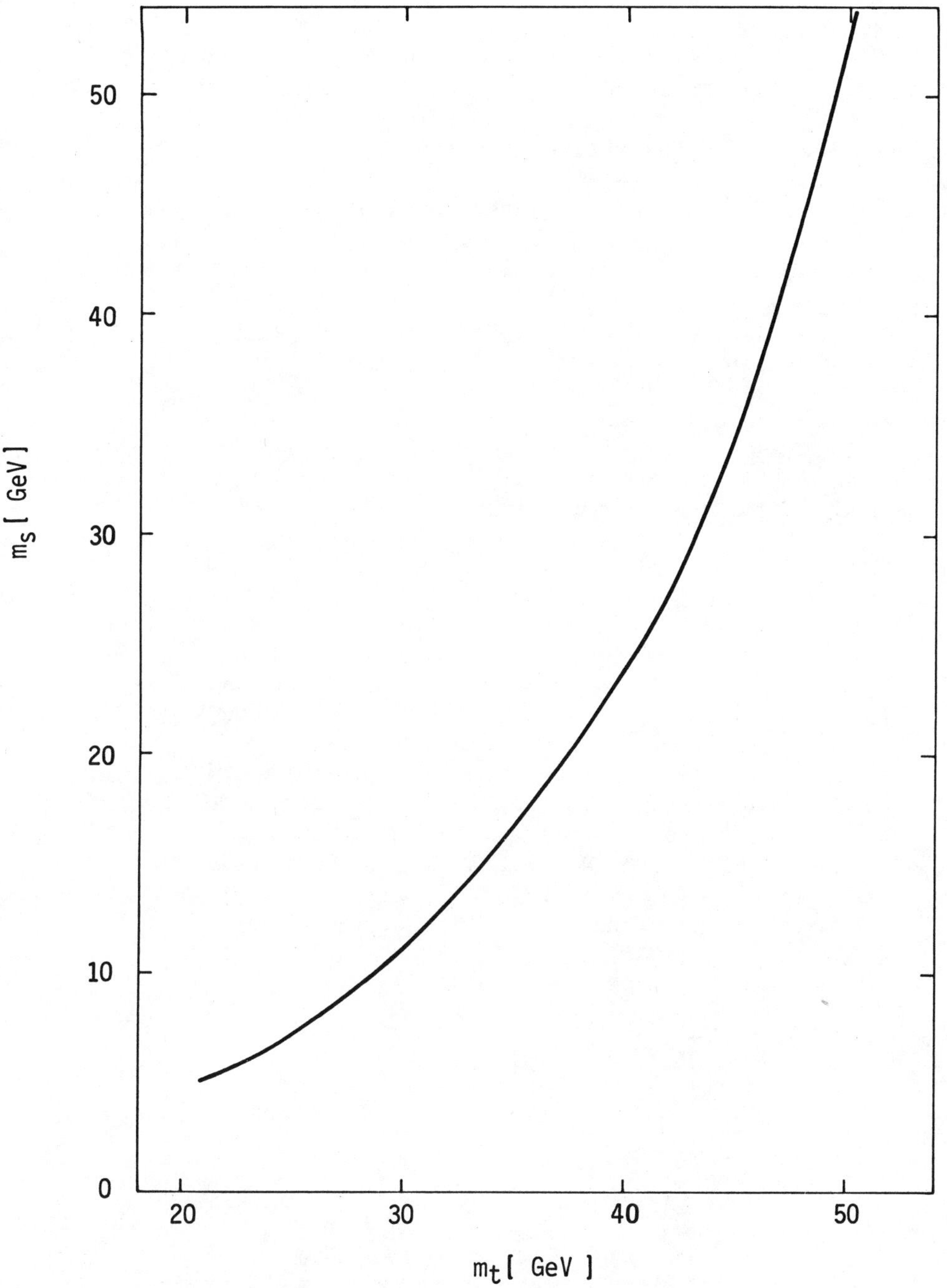

Fig. 1 : Bounds on the mass of the neutral Higgs as a function of the top quark mass m_t .

for other light states crucially depend on parameters like the gravitino mass which we do not know exactly. But given the bound on m_h the model is more predictive than the standard model. It is predictive enough to be ruled out experimentally in the near future. If, however, a neutral Higgs is found in the predicted range this will not really tell us whether supersymmetry is present or not. But if it fits one can search with more confidence for photinos, gluinos or scalar partner of quarks and leptons.

Note added : While preparing this written version of my talk I became aware of a paper by P. Majumdar and P. Roy[5] who, although in a slightly different context, derived bounds on the mass of the lightest Higgs which are consistent with ours.

ACKNOWLEDGEMENT

I would like to thank the organizers of this meeting for their efficient and successful efforts.

REFERENCES

1. H.P. Nilles and M. Nusbaumer, University of Geneva preprint UGVA-DPT 1984/05-431, Phys. Lett. 145B:73 (1984).
2. For a review and a comprehensive list of references see : H.P. Nilles, University of Geneva preprint UGVA-DPT 1983/12-412, to appear in Physics Reports C (1984).
3. H.P. Nilles, University of Geneva preprint UGVA-DPT 1984/03-419, to be published in the Proceedings of the XIX Rencontre de Moriond 1984, editor Tran Thanh Van.
4. K. Inoue, A. Kakuto, H. Komatsu and S. Takeshita, Progr. of Theor. Phys. 67:1889 (1983).
5. P. Majumdar and P. Roy, Saha Inst. of Nucl. Phys. preprint SINP/TNP/84/8 (1984).

SUPERGRAVITY GRAND UNIFICATION*

Burt A. Ovrut†,††

Department of Physics
University of Pennsylvania
Philadelphia, Pa. 19104

ABSTRACT

N=1 supergravity Grand Unified Theories, in which the gauge group and supersymmetry are spontaneously broken in the same sector, are discussed. We present two theories; one in which supersymmetry is broken using the O'Raifeartaigh mechanism, and one in which it is broken using a new method of symmetry breaking. Such theories all have a light $SU(3)_C$ octet, $SU(2)_L$ triplet, and $U(1)_Y$ singlet, in addition to the usual fields of the supersymmetric Weinberg-Salam model. The radiative breaking of $SU(2)_L \times U(1)_Y$ to $U(1)_{EM}$ is discussed.

In this paper we discuss Grand Unified Theories (GUTS) with gauge group G, coupled to N=1 supergravity. Such theories can be divided into two types:
A) those in which G and supersymmetry (SUSY) are spontaneously broken in the same sector, and
B) those in which G is spontaneously broken in one sector and SUSY is broken in another (so-called "hidden") sector. Clearly, theories of type A and B differ in their high energy behavior. These differences can have effects on such phenomena as the nucleon decay rate and early universe cosmology. At energies below the Grand Unification Mass (GUM) theories of both types can be described by effective Lagrangians.

---

*Invited talk at the NATO Advanced Study Institute, Bonn, West Germany (August 20-31, 1984).

†Work supported in part by the Department of Energy under Contract Number EY-76-C-02-3071.

††On leave of absence from the Rockefeller University, New York, N. Y., 10021.

We will demonstrate that the low energy effective Lagrangians of theories of type A and B differ in a fundamental and model independent way. Specifically, in contrast to hidden sector models, theories of type A always contain an $SU(3)_C$ octet, $SU(2)_L$ triplet, and $U(1)_Y$ singlet in addition to the usual fields of the supersymmetric Weinberg-Salam model. In this paper we consider type A models only.

Theories of type A can also be divided into two classes:
1) those in which SUSY is spontaneously broken using the O'Raifeartaigh mechanism [1], and
2) those in which SUSY is spontaneously broken using a new method of symmetry breaking [2].
The O'Raifeartaigh mechanism was introduced in globally supersymmetric theories, and extends to locally supersymmetric models as well. The new method of SUSY breaking is unique to locally supersymmetric theories. We will present specific models of class 1 and 2 below. First, however, we want to discuss a number of properties that all theories of class 1 and 2 have in common.
a) The input mass scales are the Planck mass, M_P, and the gravitino mass, $m_{3/2}$ (or equivalently, the SUSY breaking scale, $\mu = \sqrt{m_{3/2} M_P}$). The GUM, M_G, and the electroweak scale, m_{EW}, are induced by the theory.
b) The scale of G breaking is set by the only relevant mass at high energies, M_P. Therefore, M_G is of $O(M_P)$. It follows that the spontaneous breaking of G is a purely gravitational effect.
c) The scale of $SU(2)_L \times U(1)_Y$ breaking to $U(1)_{EM}$ is set by the only relevant mass scale at low energies, $m_{3/2}$. Therefore, m_{EW} is of $O(m_{3/2})$. It follows that the spontaneous breaking of $SU(2)_L \times U(1)_Y$ is a purely gravitational effect. The value of m_{EW} relative to $m_{3/2}$ depends on the top quark mass, the $SU(3)_C \times SU(2)_L \times U(1)_Y$ gaugino masses, and the ratio of the "down" to "up" Higgs doublet vacuum expectation values (VEV's).
d) The mass splitting of the Higgs $SU(3)_C$ triplets and $SU(2)_L$ doublets can be achieved either unnaturally or naturally in the group theoretical sense. The scale of the color triplet mass is then set by μ or M_P respectively. We emphasize that natural theories predict a long nucleon lifetime consistent with present experimental data.
e) The low energy spectrum always contains, in addition to the usual fields of the supersymmetric Weinberg-Salam model, an $SU(3)_C$ $\underline{8}$, $SU(2)_L$ $\underline{3}$, and $U(1)_Y$ $\underline{1}$ superfield, each with mass of $O(m_{3/2})$. Experimental discovery of these fields would be an unambiguous new signal of grand unification, independent of the nucleon lifetime.

I) Locally Supersymmetric Geometrical Hierarchy Model [3]

This theory uses the O'Raifeartaigh mechanism to spontaneously break SUSY. The gauge group is G=SU(5). The chiral superfields are 1) A, Σ, X. A and Σ transform as $\underline{24}$'s and X as a $\underline{1}$ under SU(5). These fields form the O'Raifeartaigh sector of the model. Both SUSY and

SU(5) are spontaneously broken, SU(5) being broken to $SU(3)_C \times SU(2)_L \times U(1)_Y$.

2) H, H'. H and H' transform as $\underline{5}$ and $\underline{\bar{5}}$ under SU(5). The color triplet fermions associated with $\bar{H}$ and $\bar{H}'$ mediate nucleon decay. The $SU(2)_L$ doublet scalars develop VEV's that spontaneously break $SU(2)_L \times U(1)_Y$ to $U(1)_{EM}$.

3) M'_J, M_J. M'_J and M_J transform as $\underline{\bar{5}}$'s and $\underline{10}$'s under SU(5). These are the usual lepto-quark superfields. We assume that J=1,3.
The superpotential for the theory is given by

$$W = \beta\mu^2 M + \lambda_1 X(\mathrm{Tr}\, A^2 - \mu^2) + \lambda_2\, \mathrm{Tr}\, \Sigma A^2 + \lambda_3 H'(A + m1)H + \lambda^U_{IJ} H M_I M_J + \lambda^D_{IJ} H' M_I M'_J \tag{1}$$

Parameters λ_i, λ^U, λ^D, and β are dimensionless and μ, m and M have dimension 1. Parameter β is fine-tuned to give zero cosmological constant, and $M=M_p/\sqrt{8\pi} \simeq 2.4 \times 10^{18}$GeV. The scale of SUSY breaking is set by μ, which is chosen to be of $O(10^{10-11}$GeV). Mass parameter m is fine-tuned to give zero Higgs doublet mass in the $M\to\infty$ limit. It follows that the color triplet Higgino mass is of $O(\mu)$. Hence, the mass splitting of the Higgs $SU(3)_C$ triplet and $SU(2)_L$ doublets is achieved unnaturally in this version of the model.

The potential energy of the theory is given by [4]

$$V = e^{\sum_i |\phi_i|^2/M^2} \left(\sum_i |D_{\phi_i} W|^2 - \frac{3}{M^2}|W|^2 \right) + \frac{1}{2}\,\mathrm{Tr}\, D^2 \tag{2}$$

where

$$D_{\phi_i} W = \frac{\partial W}{\partial \phi_i} + \phi_i^\dagger \frac{W}{M^2} \tag{3}$$

and D^a are the auxiliary fields associated with the SU(5) vector superfields. The absolute minimum of V is unique and given, to leading order, by

$$\langle X \rangle = \frac{\lambda_2}{\sqrt{\lambda_2^2 + 30\,\lambda_1^2}}\,(\sqrt{3}-1)M$$

$$\langle \Sigma \rangle = \frac{\lambda_1}{\sqrt{\lambda_2^2 + 30\,\lambda_1^2}}\,(\sqrt{3}-1)M \begin{pmatrix} 2 & & & & \\ & 2 & & & \\ & & 2 & & \\ & & & -3 & \\ & & & & -3 \end{pmatrix}$$

$$\langle A \rangle = \frac{\lambda_1}{\sqrt{\lambda_2^2 + 30\,\lambda_1^2}}\,\mu \begin{pmatrix} 2 & & & & \\ & 2 & & & \\ & & 2 & & \\ & & & -3 & \\ & & & & -3 \end{pmatrix} \tag{4}$$

All other VEV's are undetermined to this order. This minimum has zero cosmological constant if

$$\beta = - \frac{\lambda_1 \lambda_2}{\sqrt{\lambda_2^2 + 30 \lambda_1^2}} (2 - \sqrt{3}) \tag{5}$$

In the absence of supergravity, the tree level potential energy is constant in the X, Σ plane. It follows from (4) that the tree level potential develops a unique minimum when coupled to supergravity, with the scale of $\langle X\rangle$ and $\langle\Sigma\rangle$ set by M. VEV's $\langle\Sigma\rangle$ and $\langle A\rangle$ break SU(5) to $SU(3)_C \times SU(2)_L \times U(1)_Y$.

SUSY is spontaneously broken in the vacuum state if $\langle D_{\phi_i} W\rangle \neq 0$ for at least one field ϕ_i. To leading order, we find

$$\langle D_X W\rangle = - \frac{\sqrt{3}\ \lambda_1 \lambda_2}{\lambda_2^2 + 30 \lambda_1^2} \mu^2$$

$$\langle D_\Sigma W\rangle = - \frac{\sqrt{3}\ \lambda_1^2 \lambda_2}{\lambda_2^2 + 30 \lambda_1^2} \mu^2 \begin{pmatrix} 2 & & & & \\ & 2 & & & \\ & & 2 & & \\ & & & -3 & \\ & & & & -3 \end{pmatrix} \tag{6}$$

All other $\langle D_{\phi_i} W\rangle$ vanish to this order. It follows that SUSY is spontaneously broken at a scale of $O(\mu)$. Note that $\langle D_\Sigma W\rangle$ also breaks SU(5) to $SU(3)_C \times SU(2)_L \times U(1)_Y$. VEV's (6) are identical to those obtained in global SUSY. Hence, the O'Raifeartaigh mechanism is uneffected by supergravity.

Given vacuum state (4), and SUSY breaking VEV's (6), we can determine the effective potential at energies below the unification mass. We find that

$$V_{LE} = V_{LE}^{(1)}(y_a) + V_{LE}^{(2)}(\Sigma_3, \Sigma_8) + \tfrac{1}{2} D_\alpha (y, \Sigma)^2 \quad . \tag{7}$$

The first term on the right hand side of (7) is a function of the usual fields, y_a, of the supersymmetric Weinberg-Salam model, and is given by

$$V_{LE}^{(1)}(y_a) = \left| \frac{\partial \tilde{g}}{\partial y_a} \right|^2 + m_{3/2} A(\tilde{g} + h.c.) + m_{3/2}^2 |y_a|^2 \tag{8}$$

where

$$\tilde{g} = e^{2-\sqrt{3}}(\lambda^U_{IJ}Q_{LI}H_2\bar{u}_{RJ} + \lambda^D_{IJ}Q_{LI}H'_2\bar{d}_{RJ} + \lambda^D_{IJ}L_{LJ}H'_2\bar{e}_{RI})$$

$$m_{3/2} = e^{2-\sqrt{3}} \frac{\lambda_1 \lambda_2}{\sqrt{\lambda_2^2 + 30 \lambda_1^2}} \mu \left(\frac{\mu}{M}\right)$$

$$A = 3 - \sqrt{3} \tag{9}$$

This part of V_{LE} is identical in form to those derived in "hidden sector" models [5]. The second term on the right side of (7) is a function of Σ_3 and Σ_8, the $SU(3)_C \times SU(2)_L \times U(1)_y$ (1,3,0) and (8,1,0) components of Σ respectively. It is given by

$$V^{(2)}_{LE}(\Sigma_3,\Sigma_8) = \mathrm{Tr}\left|\frac{\partial\tilde{f}}{\partial\Sigma_3}\right|^2 + \mathrm{tr}\left|\frac{\partial\tilde{f}}{\partial\Sigma_8}\right|^2 + m_{3/2}A'(\tilde{f} + h.c.)$$

$$+ m^2_{3/2}(\mathrm{Tr}|\Sigma_3|^2 + \mathrm{Tr}|\Sigma_8|^2) \tag{10}$$

where

$$\tilde{f} = \frac{m_{3/2}}{\sqrt{3}-1}\left(\frac{9}{2}\mathrm{Tr}\,\Sigma_3^2 - \frac{4}{3}\mathrm{Tr}\,\Sigma_8^2\right)$$

$$A' = -\frac{5}{2}\left(\frac{3}{5}\sqrt{3} - 1\right) \tag{11}$$

The $SU(3)_C \times SU(2)_L \times U(1)_Y$ (1,1,0) component, Σ_1, of Σ is also light. However, to leading order it has no interactions, and hence, we do not include it in (7). This part of V_{LE} is unique to GUTs in which G and SUSY are broken in the same sector of the theory. The third term in (7) contains the usual $SU(3)_C$, $SU(2)_L$, and $U(1)_Y$ D-term functions of y_a, Σ_3 and Σ_8. Note that V_{LE} can be written as

$$V_{LE} = V_{SUSY} + V_{GRAVITY} \tag{12}$$

where

$$V_{SUSY} = \left|\frac{\partial\tilde{g}}{\partial y_a}\right|^2 + \sum_{3,8}\left|\frac{\partial\tilde{f}}{\partial\Sigma_A}\right|^2 + \frac{1}{2}D_\alpha(y,\Sigma)^2 \tag{13}$$

and $V_{GRAVITY}$ is everything else. V_{SUSY} is supersymmetric, whereas $V_{GRAVITY}$ is induced uniquely by supergravity and explicitly breaks SUSY. It can be shown [6] that V_{LE} (as well as the entire low energy Lagrangian) is renormalizable.

We calculate the values of M_G, $\sin^2\theta_w$ at 10^2GeV, and $\alpha_G(M_G)$ to the one-loop level. Taking $m_{3/2} = 10^3$ GeV, $\mu=10^{11}$GeV, $\alpha_3(10^2\mathrm{GeV}) = 1/10$, and $\alpha_{EM}(10^2\mathrm{GeV})=1/127$, we find that

$$M_G = 3.3 \times 10^{21} \text{ GeV}$$

$$\sin^2\theta_w \, (10^2 \text{GeV}) = .244$$

$$\alpha_G(M_G) = .22 \tag{14}$$

All of these values are too large. M_G is greater than M_p and, hence, corrections due to quantum gravitation (incalculable) cannot be ignored. The value of $\sin^2\theta_w$ exceeds the experimental upper bound of (approximately) .230. Finally $\alpha_G(M_G)$ is so large that perturbation theory cannot be trusted. We conclude that superpotential (1) does not properly account for M_G and $\sin^2\theta_w$. Fortunately, there is a simple solution to this problem. Any superfield which transforms as a $\underline{10}$ under SU(5) has the $SU(3)_C \times SU(2)_L \times U(1)_Y$ decomposition

$$10 = (\bar{3},1,-\tfrac{2}{3}) + (3,2,\tfrac{1}{6}) + (1,1,1) \tag{15}$$

If one can give the $(\bar{3},1,-2/3)$ and $(3,2,1/6)$ components mass of $O(M)$, and the $(1,1,1)$ component a mass of $O(\mu)$, then the addition of such a superfield to the theory will reduce the values of all quantities in (14). Henceforth, we add n families of $\underline{10}$, $\overline{10}$ representations. For each $\underline{10}$ we must introduce $\overline{10}$ to make the theory anomaly free. We denote these superfields by 10_i, $10'_i$, where i=1,n. To superpotential (1) we add the interaction

$$W = \ldots + \lambda_{4i} 10'_i (\Sigma + \tilde{m}1)10_i \tag{16}$$

where λ_{4i} are dimensionless and $\tilde{m}$ has dimension 1. Parameters λ_{4i} and $\tilde{m}$ can be fine-tuned so that all components of 10_i, $10'_i$ get mass of $O(M)$, except for components $(1,1,1)$, $(1,1,-1)$ which get mass of $O(\mu)$. Furthermore, $\langle 10_i \rangle = \langle 10'_i \rangle = 0$ for this choice of parameters. Taking n=2, we find, for the above values of $m_{3/2}$, μ, α_3, and α_{EM}, that

$$M_G = 1.8 \times 10^{17} \text{ GeV}$$

$$\sin^2\theta_w (10^2 \text{ GeV}) = .219$$

$$\alpha_G(M_G) = .13 \tag{17}$$

which are acceptable values.

We will postpone discussion of electroweak symmetry breaking until the end of the paper.

The second theory we would like to discuss is the locally supersymmetric generalization of the Dimopoulos-Georgi-Sakai (DGS) model [7].

II) Locally Supersymmetric DGS Model [8]

This theory spontaneously breaks SUSY by a new method (not O'Raifeartaigh or Polonyi). The gauge group is G=SU(5). The chiral superfields are

1) Σ. Σ is a $\underline{24}$ under SU(5). The scalar components of Σ develop a VEV which breaks $SU(5) \to SU(3)_C \times SU(2)_L \times U(1)_Y$. The Kahler derivatives associated with the auxiliary F fields of Σ develop a VEV that spontaneously breaks SUSY.

2) H, H'. H and H' transform as a $\underline{5}$ and $\underline{\overline{5}}$ respectively under SU(5). The $SU(3)_C$ triplet components mediate nucleon decay, whereas the $SU(2)_L$ doublets act as the usual Weinberg-Salam Higgs superfields. The scalar components associated with these doublets develop VEV's that break $SU(2)_L \times U(1)_Y \to U(\underline{1})_{EM}$.

3) M'_J, M_J. M'_J and M_J are $\underline{5}$'s and $\underline{10}$'s under SU(5). They are the usual lepto-quark families (J=1,3).

4) θ, θ'. θ and θ' transform as $\underline{50}$ and $\underline{\overline{50}}$ under SU(5). These superfields lead to a natural $SU(3)_C$ triplet/$SU(2)_L$ doublet mass splitting in H and H'.

5) 10, 10'. 10 and 10' transform as $\underline{10}$'s and $\underline{\overline{10}}$'s under SU(5). The $SU(3)_C \times SU(2)_L$ singlet components of these superfields reduce M_G, $\alpha_G(M_G)$, and $\sin^2\theta_w(10^2\text{GeV})$ to acceptable values.

6) ψ, ψ'. ψ and ψ' transform as $\underline{40}$ and $\underline{\overline{40}}$ under SU(5). These superfields lead to a natural $SU(3)_C$ triplet/singlet mass splitting in 10 and 10'.

The superpotential for the theory is given by

$$W = W_\Sigma + W_\theta + W_Y + W_\psi \qquad (18)$$

where

$$W_\Sigma = \beta m M^2 + \frac{m}{2}\,\mathrm{Tr}\Sigma^2 + \frac{\lambda}{3}1\mathrm{Tr}\Sigma^3 + \frac{\lambda_4}{4M}\,\mathrm{Tr}\Sigma^4 \qquad (19)$$

$$W_\theta = M_\theta \theta'\theta + \frac{\lambda_2}{M}\,\theta'\Sigma^2 H + \frac{\lambda_3}{M}\,\theta\Sigma^2 H' \qquad (20)$$

$$W_y = f_{IJ}\, H\, M_I M_J + g_{IJ}\, H' M_I M'_J \qquad (21)$$

$$W_\psi = M_\psi \psi'\psi + \frac{\lambda_5}{M}\,\psi'\Sigma^2 10 + \frac{\lambda_6}{M}\psi\Sigma^2 10' \qquad (22)$$

Parameters β, λ_i, f_{IJ}, and g_{IJ} are dimensionless, and M, M_θ, and M_ψ have dimension 1. The potential energy, V, can be calculated from (2) and (3). First, consider the potential energy, V_Σ, calculated from W_Σ alone. Taking

$$\beta = 0 \qquad (23)$$

we find that V_Σ has two minima. The first occurs at $\langle\Sigma\rangle = 0$, at which point $V_\Sigma(0) = 0$. The physically relevant second minimum occurs at

$$\langle\Sigma\rangle = \frac{1}{\sqrt{10}} M \begin{pmatrix} 2 & & & & \\ & 2 & & & \\ & & 2 & & \\ & & & -3 & \\ & & & & -3 \end{pmatrix} \tag{24}$$

We find that $V_\Sigma(\langle\Sigma\rangle) = 0$ if we take

$$\lambda_1 = \frac{5}{3}\sqrt{10}\left(\frac{m}{M}\right)$$

$$\lambda_4 = \frac{20}{21}\left(\frac{m}{M}\right) \tag{25}$$

Vacuum state (24) spontaneously breaks $SU(5) \to SU(3)_C \times SU(2)_L \times U(1)_Y$. Furthermore this state has vanishing cosmological constant. Of the two degenerate minima, we will assume that (24) is the true vacuum. The Kahler derivative associated with Σ, evaluated at (24), is given by

$$\langle D_\Sigma W\rangle = \frac{1}{3\sqrt{10}} mM \begin{pmatrix} 2 & & & & \\ & 2 & & & \\ & & 2 & & \\ & & & -3 & \\ & & & & -3 \end{pmatrix} \tag{26}$$

It follows that SUSY is spontaneously broken, and that the scale of the breaking is $\sqrt{mM}$. We emphasize that SUSY breaking occurs even though W_Σ is not an O'Raifeartaigh superpotential. Note that the "non-renormalizable" $M^{-1}\mathrm{Tr}\Sigma^4$ term in (19) is required in order to have vanishing cosmological constant when $\beta=0$. Now include the contribution of W_θ to the potential energy. Expanding Σ around (24) the quadratic part of V involving θ, θ', H, and H' only is given, to leading order, by

$$M^2(\hat{\theta}^\dagger\hat{\theta} + \hat{\theta}'^\dagger\hat{\theta}') + (\theta_3^\dagger, H_3^\dagger)\mathcal{M}^2\begin{pmatrix}\theta_3 \\ H_3\end{pmatrix} + (\theta_3'^\dagger, H_3'^\dagger)\mathcal{M}'^2\begin{pmatrix}\theta_3' \\ H_3'\end{pmatrix} \tag{27}$$

where

$$\mathcal{M}^2 = \begin{pmatrix} 8M_\theta^2 + \frac{25}{2}\lambda_3^2 M^2 & 10\sqrt{8}\,\lambda_2 M_\theta M \\ 10\sqrt{8}\,\lambda_2 M_\theta M & 100\,\lambda_2^2 M^2 \end{pmatrix} \tag{28}$$

and $\mathcal{M}'^2$ is the same as (28) with λ_2 and λ_3 exchanged. H_3, H'_3 are the $SU(3)_C$ triplet components of H and H', θ_3, θ_3' are the $SU(3)_C \times SU(2)_L \times U(1)_Y$ (3,1,-2), (3,1,2) components of θ and θ', and $\hat{\theta}$, $\hat{\theta}'$ are all other components of these fields.

The $SU(2)_L$ doublets in H and H' are massless to this order since there are no $SU(3)_C \times SU(2)_L \times U(1)_Y$ components of θ and θ' to which they can couple quadratically. This is called the "missing multiplet" method [9]. Therefore, the $SU(2)_L$ doublet components of H and H' are naturally light in this model. To reduce the number of mass scales in the theory, we take

$$M_\theta = M \tag{29}$$

It follows that the diagonalized masses are proportional to M. If we choose $\lambda_2 = \lambda_3 = 1$, then the diagonalized mass matrices are

$$\mathcal{M}_D = \mathcal{M}_D' = \begin{pmatrix} 3.39\ M & 0 \\ 0 & 10.44\ M \end{pmatrix} \tag{30}$$

Therefore, the eigenstates of $\mathcal{M}_D$ and $\mathcal{M}'_D$ are naturally heavy with masses of O(M). Note, however, that some of the masses of $\mathcal{M}_D$ and $\mathcal{M}'_D$ can be made arbitrarily small by choosing λ_2 and/or λ_3 to be $\ll 1$. The eigenstates of $\mathcal{M}_D$ and $\mathcal{M}'_D$ mediate nucleon decay, the dominant channel being $p(n) \to K^+(K^0)\bar{\nu}_\mu$.

We will assume that $m_{3/2} > m_{\tilde{W}}$ ($\tilde{\phi}$ denotes the superpartner of ϕ). Let m_c and m_s be the charm and strange quark masses respectively, θ_c the Cabibbo angle, and v and v' the VEV's of the doublet components of H and H'. Taking $m_c = 1.5$ GeV, $m_s = .15$ GeV, $\sin^2\theta_c = .2$, $v = v' = 250/\sqrt{2}$ GeV, $m_W = 83$GeV, $m_{\tilde{W}} = 80$ GeV, and $m_{3/2} = 250$ GeV, we find that

$$\tau_n \simeq 1.6 \times 10^{-4}\ m^2_{\tilde{H}_3} \tag{31}$$

Since SUSY is unbroken at scale M, $m_{\tilde{H}_3} = m_{H_3}$. Assuming that m_{H_3} is the smallest eiganvalue in (30), it follows from (31) that

$$\tau_n \simeq 1.06 \times 10^{34}\ y \tag{32}$$

Therefore, nucleon lifetimes are naturally long in our theory. Note, however, that τ_n can be made arbitrarily small by choosing λ_2 and/or λ_3 to be $\ll 1$.

Now include the contribution W_Y to the potential energy, and calculate the low energy, effective potential. We find that

$$V_{LE} = V^{(1)}_{LE}(y_a) + V^{(2)}_{LE}(\Sigma_3, \Sigma_8) + \frac{1}{2} D_\alpha(y, \Sigma)^2 \tag{33}$$

The first term on the right side of (33) is a function of the usual fields, y_a, of the supersymmetric Weinberg-Salam model and is given by

$$V^{(1)}_{LE}(y_a) = \left|\frac{\partial \tilde{g}}{\partial y_a}\right|^2 + m_{3/2}\, A(\tilde{g} + h.c.) + m^2_{3/2}\, |y_a|^2 \tag{34}$$

where

$$\tilde{g} = e^{3/2}(\lambda^U_{IU}\, Q_{LI}H_2\bar{u}_{RJ} + \lambda^D_{IJ}Q_{LI}H'_2\bar{d}_{RJ} + \lambda^D_{IJ}L_{LJ}H'_2\bar{e}_{RI})$$

$$m_{3/2} = e^{3/2}\ \frac{m}{3}$$

$$A = 3 \tag{35}$$

This part of V_{LE} is identical in form to those derived in "hidden sector" models. The second term on the right side of (33) is a function of Σ_3 and Σ_8, the $SU(3)_C \times SU(2)_L \times U(1)_Y$ (1,3,0) and (8,1,0) components of $\tilde{\Sigma}$ respectively. It is given by

$$V^{(2)}_{LE}(\Sigma_3,\Sigma_8) = Tr\left|\frac{\partial f_3}{\partial \Sigma_3}\right|^2 + Tr\left|\frac{\partial f_8}{\partial \Sigma_8}\right|^2 + m_{3/2}A_3(\tilde{f}_3 + h.c.)$$

$$+ m_{3/2}\, A_8(\tilde{f}_8 + h.c.) + m^2_{3/2}(Tr|\Sigma_3|^2 + Tr|\Sigma_8|^2) \tag{36}$$

where

$$\tilde{f}_3 = -\, m_{3/2}\left(\frac{135}{14}\right) Tr\ \Sigma^2_3$$

$$\tilde{f}_8 = m_{3/2}\left(\frac{185}{14}\right) Tr\ \Sigma^2_8$$

$$A_3 = \frac{124}{45}$$

$$A_8 = \frac{558}{185} \tag{37}$$

The $SU(3)_C \times SU(2)_L \times U(1)_Y$ (1,1,0) component, Σ_1, of Σ is also light. However, to leading order it has no interactions and, hence, we do not include it in (37). This part of V_{LE} is unique to GUTs in which G and SUSY are broken in the same sector of the theory. The third term in (33) contains the usual $SU(3)_C$, $SU(2)_L$, and $U(1)_Y$ D-term functions of y_a, Σ_3 and Σ_8. Note, once again that V_{LE} can be decomposed as in (12) and (13). V_{SUSY} is supersymmetric, whereas $V_{GRAVITY}$ is induced uniquely by supergravity and explicitly breaks SUSY. The effective Lagrangian is renormalizable.

We calculate the values of M_G, $\sin^2\theta_W$ at 10^2GeV, and $\alpha_G(M_G)$ to the one-loop level. Without the 10, 10' superfields, all of these quantities are too large.

Now include the contribution of W_ψ. To reduce the number of mass scales we take

$$M_\psi = M \tag{38}$$

The $SU(3)_C \times SU(2)_L \times U(1)_Y$ $(\bar{3},1,-2/3)$, $(3,2,1/6)$ components of 10 find partners in ψ' and get mass of $\theta(M)$. However, there is no component of ψ' to which the (1,1,1) component of 10 can couple quadratically. Hence, (1,1,1) only gets a gravity induced mass

$$m_{(1,1,1)} = m_{3/2} \tag{39}$$

and is, therefore, naturally light. The same is true for the (1,1,-1) component of 10'. This is another example of the "missing multiplet" method. Taking $m_{3/2}$=250GeV, $\alpha_3(10^2\text{GeV}) = 1/10$, and $\alpha_{EM}(10^2\text{GeV}) = 1/127$, we find that

$$\begin{aligned} M_G &= 3.43 \times 10^{18}\ \text{GeV} \\ \sin^2\theta_w\ (10^2\text{GeV}) &= .226 \\ \alpha_G(M_G) &= .082 \end{aligned} \tag{40}$$

which are acceptable values.

We now discuss the breaking of $SU(2)_L \times U(1)_Y$ to $U(1)_{EM}$. The mechanism for this breaking in theories I and II is identical. In both theories it can be shown that, at tree level, the absolute minimum of V_{LE} occurs at the origin of scalar field space (in theory II there is another minimum degenerate with the origin which we ignore). It would appear that $SU(2)_L \times U(1)_Y$ is unbroken, but this is not the case. A closer examination reveals that the tree level effective potential is only valid for field amplitudes less than, but of O(M). The low energy behavior of the potential can be determined by calculating the one-loop radiative corrections to V_{LE}, and using the renormalization group to scale the field amplitudes down from O(M) to $O(m_{EW})$. In this paper we assume that all $SU(3)_C \times SU(2)_L \times U(1)_Y$ gaugino masses vanish at tree level, and take $|\langle H'_2\rangle| < |\langle H_2\rangle|$. It can then be shown [10] that

$$m^2_{H_2}(t) \simeq \left(-\frac{1}{2} + \frac{3}{2}\, e^{-\frac{3}{4\pi^2}\int_t^o \lambda_t^2(t')dt'}\right) m^2_{3/2}$$

$$m^2_{Q_3}(t) = \frac{1}{3}\, m^2_{H_2}(t) + \frac{2}{3}\, m^2_{3/2}$$

$$m_{\bar{u}_3}^2(t) = \frac{2}{3} m_{H_2}^2 (t) + \frac{1}{3} m_{3/2}^2 \qquad (41)$$

where H_2, Q_3 and u_3 are the "up" Higgs doublet, and left and right chiral top squarks respectively, λ_t is the top quark Yukawa coupling, and $t = \ln(|y_a|/M)$. At high energy ($t=0$), $m^2_{H_2}=m^2_{Q_3}=m^2_{\bar{u}_3}=m^2_{3/2}$, the correct boundary condition. At low energy ($t\to-\infty$), $m^2_{H_2}\to -1/2\, m^2_{3/2}$, signaling the spontaneous breakdown of $SU(2)_L \times U(1)_Y$ to $U(1)_{EM}$. Note, however, that in this limit $m^2_{Q_3}\to 1/2 m^2_{3/2}$ and $m^2_{\bar{u}_3}\to 0$, so that baryon number and charge remain unbroken. It is clear from (41) that the magnitude of the electroweak symmetry breaking is a function of λ_t, and therefore of the top quark mass. Taking $m_{3/2} = 250$ GeV, we find that $SU(2)_L \times U(1)_Y$ is broken to $U(1)_{EM}$, inducing a Z-boson mass of 92.5 GeV, as long as

$$m_t(A = 3 - \sqrt{3}) = 165 \text{ GeV}$$

$$m_t(A = 3) = 107 \text{ GeV} \qquad (42)$$

If we introduce tree level gaugino masses, and (or) allow $|\langle H'_2\rangle|$ to be of $O(|\langle H_2\rangle|)$, then the predicted value of the top quark mass decreases. A top quark mass of O(40 GeV) can easily be accommodated in this class of theories.

Finally, we would like to show that light $SU(3)_C$ $\underline{8}$, $SU(2)_L$ $\underline{3}$, and $U(1)_Y$ singlet chiral superfields occur naturally in any theory in which SU(5) and SUSY are spontaneously broken in the same sector. Assuming that Σ is the only chiral superfield, the potential energy can be written as

$$V = M^6 e^G (G_{\Sigma\alpha} G^{\Sigma\alpha} - \frac{3}{M^2}) \qquad (43)$$

where

$$G = \frac{1}{M^2} \mathrm{Tr}|\Sigma|^2 + \ln\left(\frac{|W|^2}{M^6}\right) \qquad (44)$$

$$G_{\Sigma\alpha} = \frac{\partial}{\partial\Sigma_\alpha} G, \quad G^{\Sigma\alpha} = \frac{\partial}{\partial\Sigma^*_\alpha} G, \text{ and } \alpha = 1,\ldots, 24.$$

In this notation the gravitino mass is given by

$$m_{3/2} = M e^{\langle G\rangle/2}$$

Note that

$$G_{\Sigma\alpha}{}^{\Sigma\beta} = \frac{\delta_\alpha{}^\beta}{M^2} \quad , \quad G_{\Sigma_\alpha \Sigma_\beta}{}^{\Sigma_\gamma} = 0 \quad . \qquad (46)$$

Denote the scalar components of the $\underline{8}$, $\underline{3}$, and $\underline{1}$ components of Σ by O, T, S and the fermionic components by $\tilde{O}$, $\tilde{T}$, and $\tilde{S}$. Then S develops a VEV, $\langle\Sigma\rangle_S$, with the properties that

$$\langle V\rangle = 0 \tag{47}$$

$$\langle V_{\Sigma_\alpha}\rangle = 0 \tag{48}$$

The Σ (mass)^2 matrix is proportional to $\langle V_{\Sigma\beta}{}^{\Sigma\gamma}\rangle$ and $\langle V_{\Sigma_\beta\Sigma_\gamma}\rangle$. Using Equations (45) - (48) we find

$$\langle V_{\Sigma\beta}{}^{\Sigma\gamma}\rangle = m_{3/2}^2\,(M^4\langle G_{\Sigma_\alpha\Sigma_\beta}\rangle\langle G^{\Sigma_\alpha\Sigma_\gamma}\rangle - \delta_\beta{}^\gamma) \tag{49}$$

and

$$\langle V_{\Sigma_\beta\Sigma_\gamma}\rangle = m_{3/2}^2(M^4\langle G_{S\,\Sigma_\beta\Sigma_\gamma}\rangle\,\langle G^S\rangle + 2M^2\langle G_{\Sigma_\beta\Sigma_\gamma}\rangle) \quad . \tag{50}$$

It follows that the masses of O, T, and S are determined by $\langle G_{\Sigma_\alpha\Sigma_\beta}\rangle$ and $\langle G_{S\,\Sigma_\beta\Sigma_\gamma}\rangle$. The most general form for $\langle G_{\Sigma_\alpha\Sigma_\beta}\rangle$ is

$$\langle G_{\Sigma_\alpha\Sigma_\beta}\rangle = \frac{1}{M^2}\ \mathrm{Tr}\ \Big(\sum_{n=0}^{\infty} \lambda_\alpha\lambda_\beta Y^n\left(\frac{\langle\Sigma\rangle_S}{M}\right)^n a_n\Big) \quad , \tag{51}$$

where λ_α are the SU(5) generators, Y is the hypercharge matrix, and a_n are arbitrary coefficients. Assuming $\langle\Sigma\rangle_S$ is of $O(M)$, it follows that $\langle G_{OO}\rangle$, $\langle G_{TT}\rangle$, and $\langle G_{SS}\rangle$ are of the same order of magnitude unless there exists fine-tuning between large values of a_n. That is, $\langle G_{OO}\rangle$, $\langle G_{TT}\rangle$, and $\langle G_{SS}\rangle$ are "naturally" of the same order of magnitude. VEV $\langle G_{SS}\rangle$ can be calculated from Eqns. (47) and (48). Eqn. (47) implies that $\langle G^S\rangle$ is non-zero and of $O(1/M)$. Eqn. (48) gives

$$\langle G_{SS}\rangle = -\frac{1}{M^2}\,\frac{\langle G_S\rangle}{\langle G^S\rangle} \tag{52}$$

Hence, $\langle G_{SS}\rangle$, as well as $\langle G_{OO}\rangle$ and $\langle G_{TT}\rangle$, are naturally of $O(\frac{1}{M^2})$. Similarly, one can show that $\langle G_{SOO}\rangle$, $\langle G_{STT}\rangle$, and $\langle G_{SSS}\rangle$ are naturally all of $O(\frac{1}{M^3})$. Therefore, it follows from Eqns. (49) and (50) that the masses of O, T, and S (and, by a similar proof, their fermionic superpartners $\tilde{O}$, $\tilde{T}$, and $\tilde{S}$) are all naturally of $O(m_{3/2})$ as claimed. It is straightforward to extend this proof to the case where Σ interacts with other chiral superfields.

REFERENCES

[1] L. O'Raifeartaigh, Nucl. Phys. B96, 331 (1975).

[2] B. Ovrut and S. Raby, Phys. Lett. 134B, 51 (1984); B. Ovrut and P. Steinhardt, Phys. Lett. 133B, 161 (1983).

[3] B. Ovrut and S. Raby, Phys. Lett. 125B, 270 (1983); Phys. Lett. 130B, 277 (1983).

[4] E. Cremmer, S. Ferrara, L. Giradello, and T. van Proeyen, Nucl. Phys. B212, 413 (1983); J. Bagger and E. Witten, Phys. Lett. 118B, 103 (1982).

[5] H. Nilles, Nucl. Phys. 217B, 366 (1983); R. Arnowitt, A. Chamseddine, and P. Nath, Phys. Rev. Lett. 49, 970 (1982); L. Ibanez, Phys. Lett. 118B, 73 (1982); R. Barbieri, S. Ferrara, and C. Savoy, Phys. Lett. 119B, 343 (1982); L. Hall, J. Lykken, and S. Weinberg, Univ. of Texas Preprint (1983).

[6] B. Ovrut and J. Wess, Phys. Lett. 112B, 347 (1982).

[7] S. Dimopoulos and H. Georgi, Nucl. Phys. B193, 150 (1981); N. Sakai, Z. F. Phys. C11, 153 (1981).

[8] B. Ovrut and S. Raby, Phys. Lett. B (1984) to appear.

[9] S. Dimopoulos and F. Wilczek, Erice Summer Lectures, Plenum Press (1981).

[10] L. Alvarez-Gaume, M. Claudson, and M. Wise, Nucl. Phys. B207, 16 (1982); L. Ibanez, Madred Preprint FIUAM/ 82-8 (1983); J. Ellis, D. Nanopoulos, and K. Tamvakis, CERN Preprint TH. 3418 (1982); J. Polchinski, and M. Wise, Nucl. Phys. B221, 495 (1983); H. P. Nilles, Phys. Lett. 115B, 193 (1982); K. Inoue, A. Kakuto, H. Komatsu, and S. Takeshita, Prog. Theor. Phys. 68, 927 (1982); B. Ovrut and S. Raby, Phys. Lett. 130B, 277 (1983).

TOWARDS A THEORY OF FERMION MASSES: A TWO-LOOP FINITE N = 1 GLOBAL SUSY GUT

D. R. Timothy Jones

University of Colorado
Department of Physics
Boulder, CO 80309

Stuart Raby*

Los Alamos National Lab
Theoretical Division
T-8, MS B285
Los Alamos, NM 87545

One of the most outstanding problems of high energy physics is the family problem. Why are there (at least) three families of quarks and leptons with identical quantum numbers under $SU_3 \times SU_2 \times U_1$, and with mass ranging from .5 MeV to greater than ~30 GeV? Many attempts have been made to understand this puzzle, including global and local family groups or compositeness ideas. In the following, we discuss the constraints on fermion masses resulting from the requirement of finiteness in an N = 1 supersymmetric theory. Although SUSY may provide an explanation for the hierarchy $M_{p\ell}/M_W$, it has to date said little or nothing new about the family problem. This is a step in that direction.

This talk is organized as follows:

I. CONDITIONS FOR FINITENESS.

II. CONSTRUCTING A TWO-LOOP FINITE SU_5 GUT.

III. FERMION MASSES.

IV. TOP QUARK MASS.

V. A GENERALIZED TWO-LOOP FINITE SU_5 GUT.

* Lecture given at the 1984 NATO ADVANCED STUDY INSTITUTE on "SUPERSYMMETRY", BONN, Federal Republic of Germany, 20-31 August 1984.

I. FINITENESS CONDITIONS

N = 4 global SUSY theories have been shown to be finite to all orders in perturbation theory.[1] N = 2 global SUSY theories are known, in general, to have one-loop divergences. However, for particular choices of representations they can be made one-loop finite. In these cases they are also finite to all orders in perturbation theory.[2]

What is known about N = 1 global SUSY theories? We consider an N = 1 SUSY Yang-Mills theory with gauge group G and N matter multiplets ϕ_i, i = 1, ..., N, (chiral superfields) which transform in a representation R (possibly reducible) of G. We also define a superspace potential

$$W = \frac{1}{6} C^{ijk}\phi_i\phi_j\phi_k \quad , \tag{I.1}$$

where C^{ijk} are Yukawa couplings and W is, of course, invariant under G. The conditions for one-loop finiteness are, as follows:[3]

$$\beta_g^{(1)} = \frac{g^3}{16\pi^2}[T(R) - 3C_2(G)] \equiv 0 \quad , \tag{I.2}$$

$$\gamma^{i\,(1)}_{\;j} = \frac{1}{2}[F^i{}_j - 4g^2C_2(R_\beta)(E_\beta)^i{}_j] \equiv 0 \quad . \tag{I.3}$$

$\gamma^i{}_j$ are the anomalous dimensions (using a SUSY regularization prescription), g is the gauge coupling, $F^i{}_j \equiv C^{imn}C^*_{jmn}$, $T(R)\delta_{ab} \equiv T_r(T_aT_b)$, and $C_2(R) \equiv \sum_a T_aT_a$, where T_a is the generator in the representation R of G, and $C_2(G) = \sum_a T_aT_a$ with T_a in the adjoint representation. $(E_\beta)^i{}_j$ is a projection onto the irreducible representation R_β. Recently it has been shown that the conditions for two-loop finiteness are:[3] $\beta_g^{(2)} = \gamma^{i(2)}_{\;j} = 0$ if $\beta_g^{(1)} = \gamma^{i(1)}_{\;j} = 0$, i.e., if the theory is one-loop finite it is then necessarily two-loop finite. It is not known, however, whether these theories are finite beyond two-loops.

II. CONSTRUCTING A TWO-LOOP FINITE SU_5 GUT[4]

The gauge group $G = SU_5$ with $C_2(SU_5) = 5$. We consider the representations listed in Table I. In order to have a realistic theory we must include the following representations:

1) need 24 or 75 to break SU_5 to $SU_3 \times SU_2 \times U_1$;
2) need at least three families of quarks and leptons, i.e., $3(10 + \bar{5})$;
3) need Higgs fields in $\bar{5}$, 5 or $\overline{45}$, 45 to break $SU_2 \times U_1$ to U_1 and give quarks and leptons mass.

Consider the first condition for finiteness Eq. (I.2); we must satisfy $\sum_\beta T(R_\beta) = 3C_2(G) \equiv 15$. With one 24 and $3(10 + \bar{5})$ we have $\sum_\beta T(R_\beta) = 11$. We are then allowed $4(5 + \bar{5})$ of Higgs. It is clear from Table I that 45, 50, or 75 dimensional representations are much too large to obtain a finite theory. This is the minimal model. It will become clear later that a four family model cannot be made realistic.

In order to verify the second finiteness condition Eq. (I.3), we need to construct the superspace potential. We consider

$$W = q\mathrm{Tr}(24^3) + p\bar{K}24k + g_{ijk}10_i 10_j H_k + \bar{g}_{ijk}10_i \bar{5}_j \bar{H}_k$$

$$+ M\mathrm{Tr}(24^2) + M'\bar{K}K + \bar{K}(m_i H_i) + (\bar{m}_i \bar{H}_i)K + \bar{H}_i M_{ij} H_j \quad , \quad \text{(II.1)}$$

where $i,j = 1,2,3$ and $\bar{K},\bar{H}_i$ (K,H_i) are the four $\bar{5}(5)$ of Higgs. The parameters $q,p,g_{ijk},\bar{g}_{ijk}$ must satisfy the finiteness condition (I.3). M, M', M_{ij} are mass parameters of order M_{GUT} (to be determined) and $m_i,\bar{m}_i$ will be an intermediate scale. This is not the most general superspace potential. It defines a particular model. In section V we shall discuss the most general superspace potential

Table I. List of the Representations of SU(5) Which May Be Used to Build a Realistic Model. $C_2(R)$ and $T(R)$ Are Defined in the Text.

Representation	T(R)	$C_2(R)$
24	5	5
5	$\frac{1}{2}$	$\frac{12}{5}$
$\bar{5}$	$\frac{1}{2}$	$\frac{12}{5}$
10	$\frac{3}{2}$	$\frac{18}{5}$
$\overline{10}$	$\frac{3}{2}$	$\frac{18}{5}$
45	12	$\frac{32}{5}$
50	$\frac{35}{2}$	$\frac{42}{5}$
75	25	8

$d_R C_2(R) \equiv d_{Adj} T(R)$

with d_R the dimension of the representation R.

and the corrections on results obtained in sections II-IV. Now we shall discuss the mechanism for:

1) GUT breaking,
2) Higgs doublet, triplet splitting,
3) SUSY and $SU_2 \times U_1$ breaking.

1) GUT breaking: The terms in W, $qTr(24^3) + MTr(24^2)$ are necessary to affect the breaking of $SU_5 \to SU_3 \times SU_2 \times U_1$. In fact there is a SUSY minimum of the scalar potential for which

$$\langle 24 \rangle = M_G \begin{pmatrix} 2 & & & & \\ & 2 & & & \\ & & 2 & & \\ & & & -3 & \\ & & & & -3 \end{pmatrix}, \tag{II.2}$$

where M_G is a function of M and q. There are also other SUSY minima which break $SU_5 \to SU_4 \times U_1$, or even preserve SU_5. Once SUSY is broken one of these minima will become lower than the others. For now we shall assume that the $SU_3 \times SU_2 \times U_1$ minimum is a global minimum.

2) Higgs doublet, triplet splitting: Consider the following terms in W,

$$\bar{K}[p24 + M']K + (\bar{m}_i\bar{H}_i)K + \bar{K}(m_iH_i) + \bar{H}_iH_jM_{ij} \quad . \tag{II.3}$$

In the SU_3 triplet sector we have the 4 x 4 mass matrix:

$$\begin{array}{c c} & \begin{array}{cc} K & H_j \end{array} \\ \begin{array}{c} \bar{K} \\ \bar{H}_i \end{array} & \left(\begin{array}{c|c} 0 & m_j \\ \hline \bar{m}_i & M_{ij} \end{array}\right) \end{array} \tag{II.4}$$

where we have fine-tuned $M' = -2pM_G$. We choose M_{ij} to have two non-zero eigenvalues of order M_G and one zero eigenvalue. As a result we obtain two massive triplets at M_G and two triplets with mass of order m. We define

$$H_i = \alpha_iH+\cdots \quad ,$$

$$\bar{H}_i = \beta_iH+\cdots \quad , \tag{II.5}$$

such that H and $\bar{H}$ are the zero eigenstates of M_{ij}. Then the off-diagonal mass terms are

$$(\bar{m}_i\beta_i)\bar{H}K + (m_i\alpha_i)\bar{K}H \quad . \tag{II.6}$$

What is the value of the parameters $m_i, \bar{m}_i$? Consider the SU_2 doublet mass matrix:

$$\begin{array}{c} \\ \bar{K} \\ \bar{H}_i \end{array} \begin{array}{c} \begin{array}{cc} K & H_j \end{array} \\ \left(\begin{array}{c|c} \bar{M} & m_j \\ \hline \bar{m}_i & M_{ij} \end{array}\right) \end{array} \tag{II.7}$$

where $\bar{M} \equiv -3pM_G + M' = 5/2\ M'$. We have three eigenvalues of order M_G and one eigenvalue of order m^2/M_G (if $m << M_G$). In particular, in the 2 x 2 sector (including $K, \bar{K}$ and the massless eigenstates of M_{ij}) we have

$$\begin{array}{c} \\ \bar{K} \\ H \end{array} \begin{array}{c} \begin{array}{cc} K & H \end{array} \\ \left(\begin{array}{cc} \bar{M} & m_i\alpha_i \\ \bar{m}_i\beta_i & 0 \end{array}\right) \end{array} \tag{II.8}$$

with the approximate diagonalized mass terms

$$\bar{M}\bar{K}K + \frac{(m\cdot\alpha)(\bar{m}\cdot\beta)}{\bar{M}}\bar{H}H \tag{II.9}$$

We then choose

$$\frac{(m\cdot\alpha)(\bar{m}\cdot\beta)}{\bar{M}} \sim M_W \tag{II.10}$$

in order to have one pair of light Higgs doublets. Thus $m \equiv (m\cdot\alpha) \sim (\bar{m}\cdot\beta)$ is of order 10^{10} GeV.

With m of order 10^{10} GeV, the color triplet Higgs scalars $(H,\bar{H})$ will contribute to nucleon decay with the dominant decay modes

$$p \to k^0\mu^+ \quad , \quad k^+\bar{\nu} \quad , \quad n \to k^0\bar{\nu} \quad . \tag{II.11}$$

The color triplet Higgs fermions will <u>not</u> contribute to proton decay at the tree level. This is because the mass terms Eq. (II.6) are off-diagonal and only H and $\bar{H}$ couple to quarks and leptons at the tree level. Once SUSY is broken, radiative corrections will induce the necessary dimension five operators contributing to proton decay. These are, however, expected to be suppressed.

3) <u>SUSY and $SU_2 \times U_1$ breaking:</u> We intend to break SUSY via "soft" SUSY breaking terms of the form[5]

$$M_W^{-2}\delta\mathscr{L} \sim \frac{1}{3}\sum_i \phi_i^*\phi_i + \chi\chi - (\tfrac{1}{6}C^{ijk}\phi_i\phi_j\phi_k + \text{h.c.}) \ , \tag{II.12}$$

which are known to preserve the one-loop finiteness of the theory. Once SUSY is broken, the weak breaking can proceed in the standard way by utilizing the renormalization group running of the parameters from M_G to M_W. This part of our program is now in progress.[6]

III. FERMION MASSES

We must now satisfy the second finiteness condition (I.3). Consider first the states $\bar{K}$,K,24 and the parameters p,q. The finiteness condition for K,$\bar{K}$ (see fig. 1) gives

$$p^2 = g^2 \tag{III.1}$$

For 24 we obtain (see fig. 2)

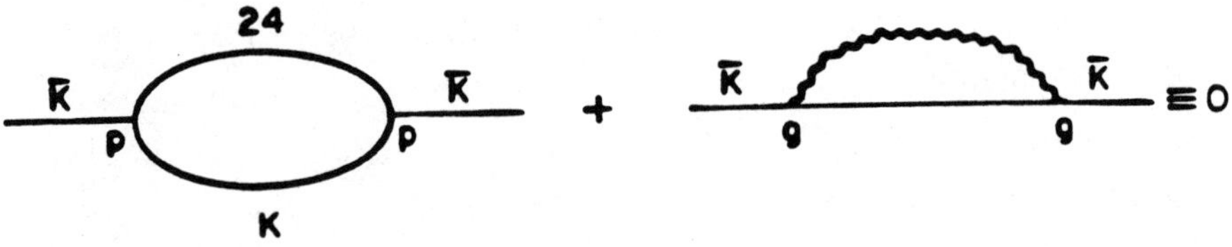

Fig. 1. One-loop wave function renormalization graphs for $\bar{K}$.

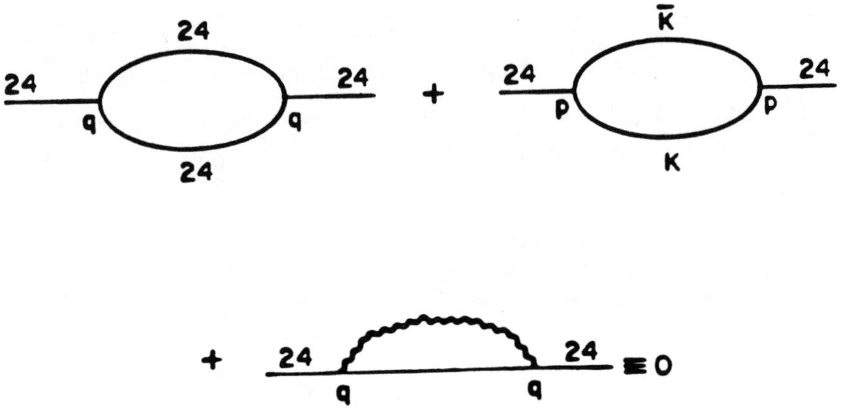

Fig. 2. One-loop wave function renormalization graphs for 24.

$$q^2 = \frac{5}{21} g^2 \quad . \qquad \text{(III.2)}$$

Consider now the Higgs $H_i, \bar{H}_i$ and the three families of quarks and leptons. The relevant graphs are given in fig. 3a-d. We obtain the finiteness conditions for;

$$\bar{5}_i \; : \quad 4\bar{g}^*_{\ell im}\bar{g}_{\ell jm} = 2\delta_{ij}\frac{12}{5} g^2 \qquad \text{(III.3)}$$

$$\bar{H}_i \; : \quad 4\bar{g}^*_{\ell mi}\bar{g}_{\ell mj} = 2\delta_{ij}\frac{12}{5} g^2 \qquad \text{(III.4)}$$

$$10_i \; : \quad 2\bar{g}^*_{i\ell m}\bar{g}_{j\ell m} + 3g^*_{i\ell m}g_{j\ell m} = 2\delta_{ij}\frac{18}{5} g^2 \qquad \text{(III.5)}$$

$$H_i \;:\; 3g^*_{\ell mi}g_{\ell mj} = 2\delta_{ij}\,\frac{12}{5}\,g^2 \quad . \tag{III.6}$$

Note that g_{ijk} and $\bar{g}_{ijk}$ <u>cannot</u> be set to zero!! The finiteness conditions have incorporated the Yukawa couplings into the general scheme. They are no longer completely independent or arbitrary parameters.

There are many ways of satisfying the conditions (III.3-6). We don't know if we have exhausted all possibilities. Consider, as an example, the 3 x 3 matrices (where the elements indicate the non-zero terms in g_{ijk} and $\bar{g}_{ijk}$, respectively):

<u>up</u>

$$\begin{array}{c} \\ 10_1 \\ 10_2 \\ 10_3 \end{array} \begin{array}{c} \begin{array}{ccc} 10_1 & 10_2 & 10_3 \end{array} \\ \begin{pmatrix} H_1 & 0 & 0 \\ 0 & H_2 & H_3 \\ 0 & H_3 & H_2 \end{pmatrix} \end{array} \tag{III.7}$$

<u>down</u>

$$\begin{array}{c} \\ \bar{5}_1 \\ \bar{5}_2 \\ \bar{5}_3 \end{array} \begin{array}{c} \begin{array}{ccc} 10_1 & 10_2 & 10_3 \end{array} \\ \begin{pmatrix} \bar{H}_1 & \bar{H}_2 & 0 \\ \bar{H}_2 & \bar{H}_1 & \bar{H}_3 \\ 0 & \bar{H}_3 & \bar{H}_2 \end{pmatrix} \end{array} \tag{III.8}$$

The above anzatz automatically satisfies the partial condition that the left-hand side of III.3-6 are proportional to δ_{ij}. In order to satisfy this partial condition, it is sufficient to use the following rule:

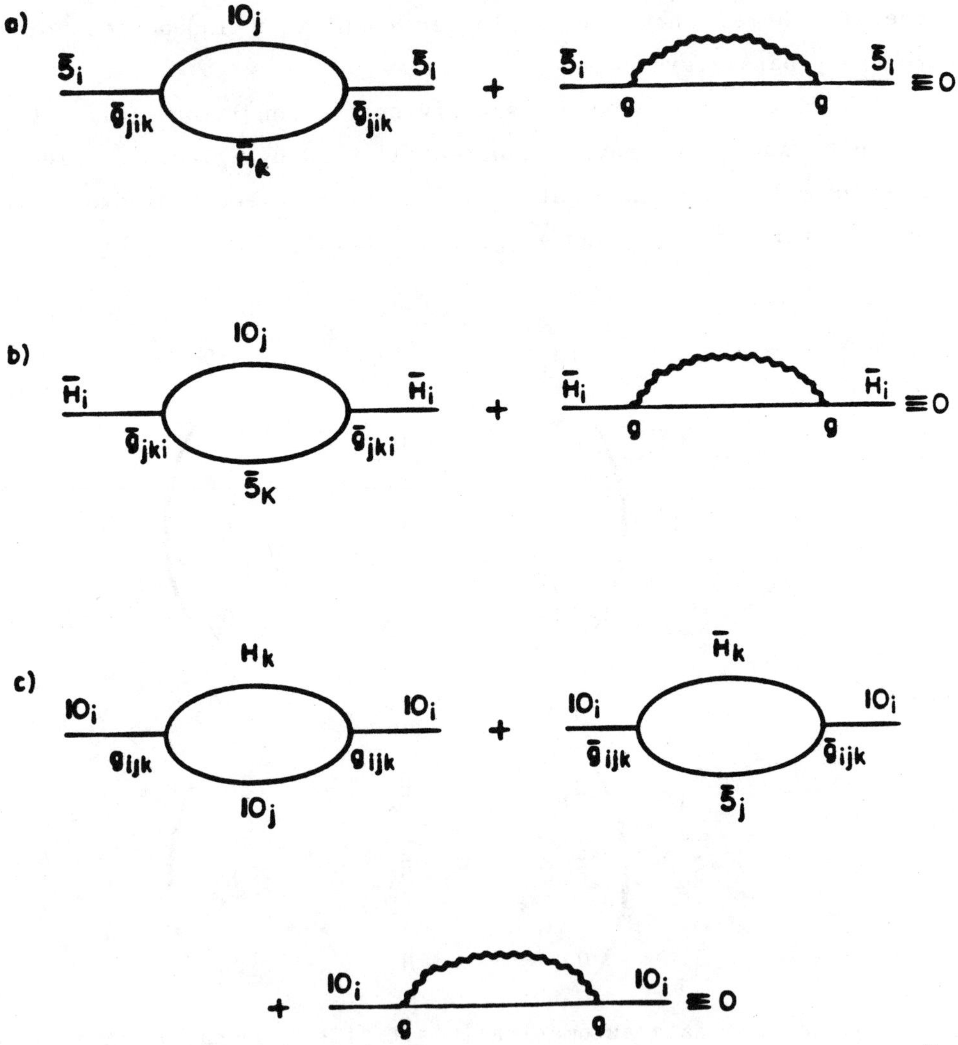

Fig. 3. One-loop wave function renormalization graphs for (a) $\bar{5}_i$, (b) $\bar{H}_i$, (c) 10_i and (d) H_i.

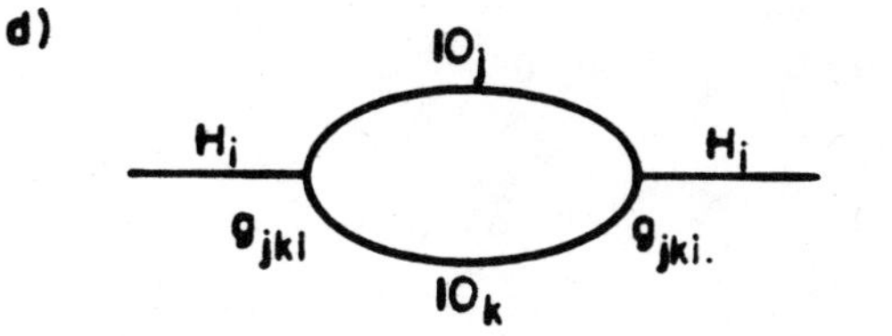

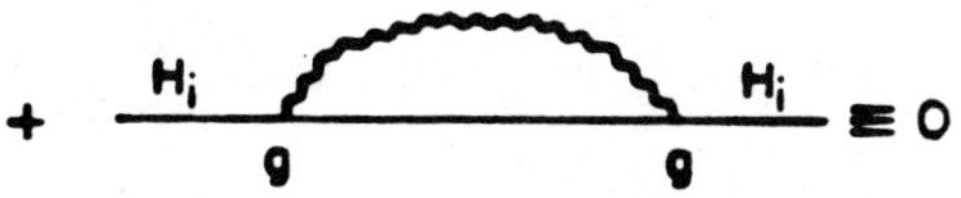

1) only one Higgs per element in the 3 x 3 matrix, and

2) any Higgs may appear only once in any row or column.

With the above rule, the set of non-linear conditions III.3-6 become a set of 12 linear equations for the non-vanishing couplings (in an obvious notation) of III.7,8, respectively.

up

$$\begin{pmatrix} g_{111} & 0 & 0 \\ 0 & g_{222} & g_{233} \\ 0 & g_{233} & g_{332} \end{pmatrix} \qquad \text{(III.9)}$$

down

$$\begin{pmatrix} \bar{g}_{111} & \bar{g}_{212} & 0 \\ \bar{g}_{122} & \bar{g}_{221} & \bar{g}_{323} \\ 0 & \bar{g}_{233} & \bar{g}_{332} \end{pmatrix} \qquad \text{(III.10)}$$

The solution, for the up and down mass matrices, is

up

$$\frac{8}{5}\,g\begin{pmatrix} \alpha_1 & 0 & 0 \\ 0 & \cos\Theta\,\alpha_2 & \frac{\alpha_3}{\sqrt{2}} \\ 0 & \frac{\alpha_3}{\sqrt{2}} & \sin\Theta\,\alpha_2 \end{pmatrix} \tag{III.11}$$

down

$$\frac{6}{5}\,g\begin{pmatrix} \frac{1+\cos^2\Theta}{2}^{1/2}\beta_1 & \frac{\sin\Theta}{\sqrt{2}}\,\beta_2 & 0 \\ \frac{\sin\Theta}{\sqrt{2}}\,\beta_2 & \frac{\sin\Theta}{\sqrt{2}}\,\beta_1 & \cos\Theta\,\beta_3 \\ 0 & \sin\Theta\,\beta_3 & \cos\Theta\,\beta_2 \end{pmatrix} \tag{III.12}$$

where Θ is an arbitrary parameter and α_i, β_i are defined in II.5.

There are many such solutions with up to nine free parameters (including α_i, β_i). We have sufficient freedom to fit the up and down quark masses. Note, we have the standard SU_5 result, that the lepton mass matrix is identical to the down quark matrix at M_{GUT}. This is clearly a problem. Moreover, clearly the finiteness conditions are, unfortunately, not powerful enough to explain the family hierarchy.

IV. TOP QUARK MASS

We now want to present the one loop results for α_G, M_G, $\sin^2\Theta$, m_b/m_τ and m_t. The model has three relevant scales M_G, m, and M_W. Between M_G and m we have two pair of color triplet Higgs and one pair of Higgs doublets; below m we have only one Higgs

leptons and the $SU_3 \times SU_2 \times U_1$ gauge multiplets. We find

$$M_G = 1.9 \times 10^{17} \text{ GeV} \quad ,$$
$$\alpha_G = .047 \quad ,$$
$$\sin^2 \Theta = .21 \quad . \qquad \text{(IV.1)}$$

using $\alpha_3 = .1$, $\alpha^{-1} = 128$ at M_W.

In general, if we obtain a family hierarchy, i.e., $m_t \gg m_c \gg m_u$ and $m_b \gg m_s \gg m_d$, then up to corrections of order $(m_c/m_t)^2$ or $(m_s/m_b)^2$, we have

$$\lambda_t^2 = \frac{8}{5} g^2 \quad ,$$

$$\lambda_b^2 = \lambda_\tau^2 = \frac{6}{5} g^2 \quad , \qquad \text{(IV.2)}$$

at M_G. $\lambda_t, \lambda_b, \lambda_\tau$ are Yukawa couplings for top, bottom, and tau, respectively. In order to obtain $\lambda_t, \lambda_b, \lambda_\tau$ at M_W, we use the one loop renormalization group equations given in ref 4. Between M_G and m we must use the low energy superspace potential, including the color triplet Higgs fields $H_3, \bar{H}_3$:

$$W \supset \lambda_t H Q \bar{T} + \lambda_b \bar{H} Q \bar{B} + \lambda_\tau \bar{H} L \bar{\tau}$$

$$+ \lambda_1 H_3 \bar{T} \bar{\tau} + \frac{\lambda_2}{2} H_3 Q_\alpha Q_\beta \varepsilon^{\alpha\beta} + \lambda_3 \bar{H}_3 \bar{T} \bar{B} + \lambda_4 \bar{H}_3 Q_\alpha L^\alpha \quad . \qquad \text{(IV.3)}$$

The boundary conditions at M_G are

$$\lambda_t = \lambda_1 = \lambda_2 = \sqrt{\frac{8}{5}}\, g \quad ,$$

$$\lambda_b = \lambda_\tau = \lambda_3 = \lambda_4 = \sqrt{\frac{6}{5}}\, g \quad . \qquad \text{(IV.4)}$$

$$g = \sqrt{4\pi\alpha_G} \quad .$$

We then obtain $\lambda_b, \lambda_\tau, \lambda_t$, at M_W.

The masses are given by the expressions:

$$m_b = \lambda_b \langle \bar{H}_0 \rangle \quad ,$$

$$m_\tau = \lambda_\tau \langle \bar{H}_0 \rangle \quad ,$$

$$m_t = \lambda_t \langle H_0 \rangle \quad . \tag{IV.5}$$

The ratio $(m_b/m_\tau)|_{M_W} = (\lambda_b/\lambda_\tau)|_{M_W}$ is completely determined. However in order to determine m_b or m_t we must know $\langle \bar{H}_0 \rangle$ and $\langle H_0 \rangle$. These are, in principle, determined by running the soft SUSY breaking scalar masses Eq. (II.12) from M_G to M_W. However, since we have not yet done this analysis, we use as input

$$M_W^2 = \frac{g_2^2}{2}(\langle H_0 \rangle^2 + \langle \bar{H}_0 \rangle^2) \quad ,$$

$$\frac{m_\tau}{\lambda_\tau} = \langle \bar{H}_0 \rangle \quad . \tag{IV.6}$$

We find

$$\left.\frac{m_b}{m_\tau}\right|_{M_W} = 1.8 \quad , \tag{IV.7}$$

and

$$m_t = 155 \pm 10\% \text{ GeV} \quad . \tag{IV.8}$$

Using $m_b(m_b) = 4.25$ GeV[7] and renormalizing to $m_b(M_W)$, we find the experimental result

$$\left.\frac{m_b}{m_\tau}\right|_{M_W,\ exp} = 1.8 \tag{IV.9}$$

in excellent agreement with (IV.7).

Finally the 10% error in (IV.8) takes into account thresholds we have ignored at M_G and M_W. We note that the renormalization group equations have an infra-red fixed point at $m_t \sim 180$ GeV. The result (IV.8) is strongly affected by the running between M_G and m.

V. GENERALIZED TWO-LOOP FINITE SU_5 GUT

In this section we shall discuss the question, how general are the results obtained in section IV? We consider the more general notation

$$\bar{5}_x \quad , \quad x = 1, \ldots, 7$$

$$5_\alpha \quad , \quad \alpha = 1, \ldots, 4$$

$$10_i \quad , \quad i = 1,2,3 \quad . \tag{V.1}$$

The most general superspace potential with these states (respecting SU_5) is given by

$$W = \lambda_{x\alpha}\bar{5}_x 24 5_\alpha + \bar{g}_{ixy}10_i\bar{5}_x\bar{5}_y + g_{ij\alpha}10_i 10_j 5_\alpha$$
$$+ M_{x\alpha}\bar{5}_x 5_\alpha + q\mathrm{Tr}(24^3) + M\mathrm{Tr}(24^2) \quad . \tag{V.2}$$

The last two terms in (V.2), once again, have a minimum which breaks SU_5 to $SU_3 \times SU_2 \times U_1$. This gives us the $\bar{5}5$ mass matrix

$$\begin{array}{c} \\ \bar{5}_1 \\ \vdots \\ \bar{5}_4 \\ \vdots \\ \bar{5}_7 \end{array} \begin{array}{c} 5_1 \; . \; . \; . \; 5_4 \\ \left(\begin{array}{c} \\ \langle 24\rangle \lambda_{x\alpha} + M_{x\alpha} \\ \\ \end{array} \right) \end{array} . \tag{V.3}$$

We can then always rotate the states $\bar{5}_x$ to $\tilde{\bar{5}}_x$ (which includes $\bar{5}_\alpha, \alpha=1,\ldots,4$ and $\bar{5}_i, i=1,\ldots,3$) defined by the matrix

$$\begin{array}{c} \\ \bar{5}_\alpha \\ \\ \bar{5}_i \end{array} \begin{array}{c} 5_\beta \\ \left(\begin{array}{c} \langle 24\rangle \tilde{\lambda}_{\alpha\beta} + \tilde{M}_{\alpha\beta} \\ \hline 0 \end{array} \right) \end{array} . \tag{V.4}$$

We then obtain

$$\begin{aligned} W = {} & \bar{5}_\alpha(\tilde{\lambda}_{\alpha\beta}24 + \tilde{M}_{\alpha\beta})5_\beta \\ & + \bar{5}_i(\tilde{\lambda}_{i\beta}24 + \tilde{M}_{i\beta})5_\beta \\ & + g_{i\alpha\beta}10_i\bar{5}_\alpha\bar{5}_\beta \\ & + \bar{g}_{ijk}10_i\bar{5}_j\bar{5}_k \\ & + \tilde{\bar{g}}_{ij\alpha}10_i\bar{5}_j\bar{5}_\alpha \\ & + g_{ij\alpha}10_i10_j5_\alpha \, , \end{aligned}$$

$$\text{where } \tilde{\bar{g}}_{ij\alpha} \equiv 2\bar{g}_{ij\alpha} \; . \tag{V.5}$$

Several terms must be deleted since they are dimension four baryon violating operators. We therefore set

$$\tilde{\lambda}_{i\beta},\ \tilde{M}_{i\beta},\ \bar{g}_{ijk},\ \bar{g}_{i\alpha\beta} \tag{V.6}$$

to zero. We now have

$$W = \bar{5}_\alpha(\tilde{\lambda}_{\alpha\beta}24 + \tilde{M}_{\alpha\beta})5_\beta + \tilde{\bar{g}}_{ij\alpha}10_i\bar{5}_j\bar{5}_\alpha + g_{ij\alpha}10_i10_j5_\alpha\ , \tag{V.7}$$

where $\bar{5}_\alpha, 5_\alpha$ are Higgs fields and $10_i, \bar{5}_i$ are the three families of quarks and leptons.

The finiteness conditions (Eq. I.3) are now given by

$$\begin{aligned}
\bar{5}_\alpha &: \quad \frac{24}{5}\tilde{\lambda}^*_{\alpha\gamma}\tilde{\lambda}_{\beta\gamma} + 4\tilde{\bar{g}}^*_{ij\alpha}\tilde{\bar{g}}_{ij\beta} = 2\delta_{\alpha\beta}\frac{12}{5}g^2\ , \\
\bar{5}_i &: \quad 4\tilde{\bar{g}}^*_{ki\alpha}\tilde{\bar{g}}_{kj\alpha} = 2\delta_{ij}\frac{12}{5}g^2\ , \\
5_\alpha &: \quad \frac{24}{5}\tilde{\lambda}^*_{\gamma\alpha}\tilde{\lambda}_{\gamma\beta} + 3g^*_{ij\alpha}g_{ij\beta} = \frac{24}{5}\delta_{\alpha\beta}g^2\ , \\
10_i &: \quad 2\tilde{\bar{g}}^*_{ik\alpha}\tilde{\bar{g}}_{jk\alpha} + 3g^*_{ik\alpha}g_{jk\alpha} = 2\delta_{ij}\frac{18}{5}g^2\ .
\end{aligned} \tag{V.8}$$

By taking the appropriate traces we find

$$\begin{aligned}
\mathrm{Tr}\ \bar{5}_i &: \quad \sum_{ki\alpha}|\tilde{\bar{g}}_{ki\alpha}|^2 = \frac{18}{5}g^2\ , \\
\mathrm{Tr}\ \bar{5}_\alpha &: \quad \sum_{\alpha\beta}|\tilde{\lambda}_{\alpha\beta}|^2 = g^2\ , \\
\mathrm{Tr}\ 5_\alpha &: \quad \sum_{ij\alpha}|g_{ij\alpha}|^2 = \frac{24}{5}g^2\ , \\
\mathrm{Tr}\ 10_i &: \quad \text{consistent}\ .
\end{aligned} \tag{V.9}$$

Thus, the four expressions in (V.8) are self-consistent. Note a theory with two pairs of Higgs fields and four families of quarks and leptons (satisfying I.2) cannot be made realistic. In this case the same equations (V.8) apply with now $\alpha,\beta,\gamma = 1,2$ and $i,j,k = 1,\ldots,4$. The trace conditions (V.9) are however, inconsistent in this case and the theory cannot be made finite.

Let us now assume that we have only one light pair of Higgs doublets in the low energy theory. Let $\alpha = 1$ be that light Higgs. We then obtain the analogous equations to (IV.2) for the quark and lepton Yukawa couplings to the light Higgs. We find

$$|\tilde{\bar{g}}_{ij1}|^2 = \frac{6}{5}\, g^2 \left(1 - \frac{|\tilde{\lambda}_{1\gamma}|^2}{g^2}\right) \sim \lambda_b^2 \quad ,$$

$$|g_{ij1}|^2 = \frac{8}{5}\, g^2 \left(1 - \frac{|\tilde{\lambda}_{\gamma 1}|^2}{g^2}\right) \sim \lambda_t^2 \quad . \qquad \text{(V.10)}$$

where $|\tilde{\lambda}_{1\lambda}|^2/g^2 \lesssim 1$ and $|\tilde{\lambda}_{\gamma 1}|^2/g^2 \leq 1$ [using (V.9)]. Thus λ_τ will in general be smaller than $\sqrt{8/5}$ g. Moreover, λ will in general start smaller than $\sqrt{6/5}$ g; thus $\langle\bar{H}_0\rangle$ will in general be larger and $\langle H_0\rangle$ smaller. Both effects will thus tend to decrease the value of m_t. As a result, the value m_t = 155 GeV is probably an "upper bound". Upper bound is in quotes since in general there may not be any color triplet Higgs at $m \sim 10^{10}$ GeV. These had a big effect on the running; tending to lower λ_t.

Finally if there are no intermediate mass Higgs triplets, then the GUT scale will in general be smaller. Thus $M_{GUT} = 1.9 \times 10^{17}$ GeV may be an "upper bound".

In conclusion, we have presented a realistic two-loop finite SUSY GUT. Finiteness conditions are seen to incorporate Higgs Yukawa couplings into the overall scheme; i.e., they are no longer completely arbitrary. In a particular model the top quark mass is

155 GeV ± 10%. However, a general analysis finds that there is still an enormous amount of freedom in the Higgs Yukawa couplings and 155 GeV may only be an upper bound for m_t. Finally, the constraints of finiteness give no clue to the origin of a family hierarchy.

REFERENCES

1. S. Mandelstam, Proc. of the XX1 International Conference on High Energy Physics, Paris (1982); A. Parkes and P. West, Phys. Lett. B122:365 (1983); Nucl. Phys. B222:269 (1983); A. Namazie, A. Salam and J. Strathdee, Phys. Rev. D28:1481 (1983); J. G. Taylor, Phys. Lett. 121B:386 (1983); J. J. van der Bij and Y. P Yao, Phys. Lett. 125B:171 (1983); S. Rajpoot, J. G. Taylor and M. Zaimi, Phys. Lett. 127B:347 (1983).
2. M. T. Grisaru and W. Siegel, Nucl. Phys. B210:29 (1982); L. Girardello and M. T. Grisaru, Nucl. Phys. B194:65 (1982); P. S. Howe, K. S. Stelle and P. K. Townsend, Nucl. Phys. B214:513 (1983); A. Parkes and P. West, Phys. Lett. 127B:35 (1983), J. M. Frere, L. Mezincescu and Y. P. Yao, Phys. Rev. D29:1196 (1984).
3. A. Parkes and P. West, Phys. Lett. 138B:99 (1984); P. West, Phys. Lett. 137B:371 (1984); D. R. T. Jones and L. Mezincescu, Phys. Lett. 136B:242 (1984); 138B:293 (1984).
4. D. R. T. Jones and S. Raby, to be published in Phys. Lett. B; S. Hamidi and J. H. Schwarz, CALT-68-1159, CAL-TECH (1984).
5. D. R. T. Jones, L. Mezincescu and Y. P. Yao, Univ. of Michigan preprint (1984).
6. J. Bjorkman, D. R. T. Jones and S. Raby, in preparation.
7. J. Gasser and H. Leutwyler, Phys. Rep. 87 (1982).

RECENT RESULTS IN THE RENORMALIZATION OF SUPERSYMMETRIC GAUGE THEORIES*

Olivier Piguet**

Département de Physique Théorique
Université de Genève (Switzerland)

Klaus Sibold***

Max-Planck-Institut für Physik und Astrophysik
Werner Heisenberg Institut für Physik
Munich (Fed.Rep.Germany)

INTRODUCTION

The linear realization of N=1 supersymmetric theories[1,2] has always been the most promising candidate for the extension of the symmetry to higher orders in perburbation theory. In fact in its superfield form e.g. in chiral models [2,3,4,5] it provides the most concise and elegant formalism for implementing manifest symmetry and exploiting its consequences. For instance the celebrated non-renormalization theorems[6,4,7] find in this formulation their most natural explanation. Despite of this beauty the linear version of supersymmetry leads in gauge models to non-trivial problems which have been solved only recently and not yet for all cases of interest.

The basic reason for these difficulties is the simple fact that gauge superfields[FN1]

$$\phi = C + \theta\chi + \bar\theta\bar\chi + \frac{1}{2}\theta^2 h + \frac{1}{2}\bar\theta^2\bar h + \theta\sigma^\mu\bar\theta v_\mu + \frac{1}{2}\bar\theta^2\theta\lambda + \frac{1}{2}\theta^2\bar\theta\bar\lambda + \frac{1}{4}\theta^2\bar\theta^2 D \qquad (1.1)$$

*Seminar given by K. Sibold
**Supported in part by the Swiss National Science Foundation
***Heisenberg Fellow
FN1 For the notation s. Refs. 14,20,26.

have canonical dimension zero[8,9,10]

Any function

$$\mathcal{F}(\phi) = \phi + c_2 \phi^2 + c_3 \phi^3 + \cdots \tag{1.2}$$

therefore has the same right to serve as fundamental variable, but different choices of $\mathcal{F}$ might a priori not be equivalent, already at the classical level[11]. And even if they were classically it is not automatic that they would be so after renormalization since the $\mathcal{F}$'s are composite operators. The task is to formulate consistently a theory having infinitely many parameters and to explain their role. Since in the quantized theory to these parameters correspond ultraviolet (UV) divergences we shall call this the UV-problem.

Of course we introduced this nomenclature only because there is also an infrared (IR) problem associated with the vanishing canonical dimension of ϕ : the propagator of the dimension zero field C has in a general gauge the form

$$\langle CC \rangle \sim \frac{\alpha - 1}{(k^2)^2} \tag{1.3}$$

(α denotes the gauge parameter)[10,11].

Feynman diagrams will therefore in general diverge at the infrared end of the internal momenta due to this non-integrable singularity. These divergences depend obviously on the choice of gauge and one therefore will guess that gauge independent quantities will not suffer from this divergence. But this guess has to be proved. In the present seminar we sketch the solution of these problems for a general massless N=1 supersymmetric non-abelian gauge theory.

Before proceeding let us digress and discuss shortly an alternative formulation: the so-called Wess-Zumino gauge. In this approach all fields with infracanonical dimensions ($C, \chi, \bar{\chi}$, M, $\bar{M}$) are eliminated from the very beginning and a supersymmetric gauge theory looks then just like an ordinary one (with fields of spin 1, 1/2, 0 and of canonical dimensions 1, 3/2, 1 resp.), the only peculiarity being the specific choice of representations and equality of some couplings[8,12]. Unfortunately the supersymmetry transformations become non-linear and therefore expressing the symmetry in higher orders is difficult and not yet achieved. Likewise the supersymmetry covariance of operators (e.g. currents) is an unsolved problem and it is therefore not clear whether supersymmetry does or does not restrict any further potential anomalies in current conservation equations or in the Slavnov identity.

Section 2 Exposition of the technique

Let us first of all convince ourselves that the symmetric theory indeed depends on infinitely many parameters. We are given a simple, compact group G whose fundamental representation is generated by τ^i, whereas an arbitrary unitary representation is generated by matrices T^i. We collect the gauge superfields ϕ^i in a matrix:

$$\phi = \phi^i \tau^i \qquad \phi^i = \mathrm{Tr}\, \tau^i \phi \qquad [\tau^i, \tau^j] = i f^{ijk} \tau^k \tag{2.1}$$

and let fields transform under rigid gauge transformations by demanding

$$\delta_\omega \varphi = i [\varphi, \omega] \qquad \omega \equiv \omega^i \tau^i \qquad \omega^i = \mathrm{const}. \tag{2.2}$$

for fields φ transforming under the adjoint representation,

$$\delta_\omega A_a = i T^i_{ab} \omega^i A_b \tag{2.3}$$

for the chiral matter fields.

It is well known[9,10] that the action

$$\begin{aligned} \Gamma_{cl} = & -\frac{1}{g^2} \mathrm{Tr} \int dS\, F^\alpha F_\alpha + \int dV\, \bar{A}_a (e^{\phi^i T^i})_{ab} A_b \\ & + \int dS\, \lambda\, d_{abc} A_a A_b A_c + \int d\bar{S}\, \bar{\lambda}\, d_{abc} \bar{A}_a \bar{A}_b \bar{A}_c \\ & F^\alpha \equiv \bar{D}\bar{D} (e^{-\phi} D^\alpha e^{\phi}) \end{aligned} \tag{2.4}$$

is invariant under the rigid transformations (2.2), (2.3) and under the local gauge transformations

$$\begin{aligned} \delta_g \phi & = i(\Lambda - \bar{\Lambda}) + \frac{i}{2}[\phi, \Lambda + \bar{\Lambda}] + \frac{i}{12}[\phi,[\phi, \Lambda - \bar{\Lambda}]] + \cdots \\ & \equiv i\, Q^s(\phi, \Lambda, \bar{\Lambda}) \end{aligned} \tag{2.5}$$

where the coefficients of the multiple commutators in Q^s (s for special) are derived from inverting the power series

$$\delta_g e^{\phi} = i(e^{\phi}\Lambda - \bar{\Lambda} e^{\phi}) . \tag{2.6}$$

It is now very useful to observe that the law of composing two transformations (2.5) to a third one is efficiently derived from the nilpotency of the Becchi-Rouet-Stora (BRS) transformations associated with (2.5). Let in $Q^s(\phi, \Lambda, \bar{\Lambda})$ the variables Λ, $\bar{\Lambda}$ go into the variables c_+, $\bar{c}_+$ which are chiral superfields and anti-commute and thus define

$$s^s \phi = c_+ - \bar{c}_+ + \frac{1}{2}[\phi, c_+ + \bar{c}_+] + \frac{1}{12}[\phi,[\phi, c_+ - \bar{c}_+]] + \cdots \tag{2.7}$$

Then the transformation

$$b^s = b^s_0 + b^s_1 + b^s_2 + \cdots \tag{2.8}$$

$$b^s_0 \phi = c_+ - \bar{c}_+ \qquad\qquad b_0 \tilde{c}_+ = 0$$

$$b^s_1 \phi = \frac{1}{2}[\phi, c_+ + \bar{c}_+] \qquad\qquad b_1 \tilde{c}_+ = -\tilde{c}_+\tilde{c}_+$$

$$b^s_2 \phi = \frac{1}{12}[\phi,[\phi, c_+ - \bar{c}_+]] \qquad\qquad b_2 \tilde{c}_+ = 0$$

$$- \tag{2.9}$$

$$b^s_k \phi = b_k \underbrace{[\phi,[\phi \cdots [\phi, c_+ - (-1)^k \bar{c}_+] \cdots]}_{k} \qquad\qquad b_k \tilde{c}_+ = 0$$

(b_k given from (2.5); $\tilde{c}_+$ denotes either c_+ or $\bar{c}_+$) is indeed nilpoten

$$(b^s)^2 \phi = 0 \qquad\qquad (b^s)^2 \tilde{c}_+ = 0 \tag{2.10}$$

and (2.10) embodies the composition law for transformations (2.6). It is now easy to make manifest the infinitely many parameters entering the theory. At each order k in the number of fields we are free to change the transformation law (2.9):

$$b^s_k \phi \quad \rightarrow \quad b_k \phi = b^s_k \phi + a_k b^s_0 (\phi)^k \tag{2.11}$$

And, indeed, since b^s was nilpotent one can check that b is also nilpotent i.e. two gauge transformations combine to a third one. The values of a_k affect in a well-determined way all higher terms of b but remain themselves arbitrary. (For a detailed exposition s. ref. 13).

Although the actual construction of invariants with respect to the transformations (2.11) is not trivial the result can be formulated easily: one just replaces in (2.4) ϕ by $\mathfrak{F}(\phi)$ of (1.2) where the coefficients c_k are uniquely determined by the given a_k[14]. So the arbitrariness in the transformation law b applied to ϕ is closely connected with the freedom of choosing functions $\mathfrak{F}(\phi)$ as fundamental variable instead of ϕ.

The task is now to control the dependence of the theory on the parameters a_k. To begin with let us explain our technique in the simpler case of the ordinary gauge parameter α. We add to the classical action (2.4) gauge-fixing and Faddeev-Popov terms

$$\Gamma_{\phi\pi} = \mathrm{Tr}\Big\{\int dS\, B\, \bar{D}\bar{D}\, DD\phi + \int d\bar{S}\, \bar{B}\, DD\bar{D}\bar{D}\phi$$
$$+ \int dV\big(\alpha B\bar{B} - (DDc_- + \bar{D}\bar{D}\bar{c}_-)\, Q^S(\phi, c_+, \bar{c}_+)\big)\Big\} \qquad (2.12)$$

where a Lagrange multiplier field B and the anti-ghost fields $c_-(\bar{c}_-)$ have been introduced. $\Gamma_{\phi\pi}$ is invariant if we postulate

$$s c_- = B \qquad s\bar{c}_- = \bar{B}$$
$$s B = 0 \qquad s\bar{B} = 0 \qquad (2.13)$$

Now we differentiate $\Gamma_{cl} + \Gamma_{\phi\pi}$ with respect to α since this derivative is the infinitesimal variation of α :

$$\partial_\alpha(\Gamma_{cl} + \Gamma_{\phi\pi}) = \mathrm{Tr}\int dV B\bar{B} = s\,\tfrac{1}{2}\int dV\, \mathrm{Tr}(c_-\bar{B} + \bar{c}_- B)\,. \qquad (2.14)$$

We note that the variation of α is a BRS variation which implies that it vanishes when tested with physical fields i.e. due to (2.14) the α -variation of a physical quantity vanishes. In order to establish equations of type (2.14) to all orders and also in more complicated cases we re-interprete (2.14). We understand it as expressing the BRS-invariance of

$$\Gamma = \Gamma_{cl} + \Gamma_{\phi\pi} + \tfrac{1}{2}\chi\, \mathrm{Tr}\int dV(c_-\bar{B} + \bar{c}_- B) \qquad (2.15)$$

under enlarged BRS-transformations, where α also transforms, namely in an anti-commuting variable χ .

$$s\alpha = \chi \qquad (2.16)$$

$$s\Gamma \equiv \int dV\, Q\frac{\delta\Gamma}{\delta\phi} - \int d\bar{S}\, \bar{c}_+\bar{c}_+\frac{\delta\Gamma}{\delta\bar{c}_+} + \chi\,\partial_\alpha\Gamma = 0 \qquad (2.17)$$

Differentiating (2.17) with respect to χ at $\chi = 0$ one finds exactly (2.14). Hence establishing (2.17) means establishing ordinary BRS-invariance ($\chi = 0$) and at the same time controlling the α -dependence.

Let us now discuss the dependence of the theory on the parameters a_k. The fact that they enter via BRS variation terms lets one hope that they are unphysical, more specifically, that they are gauge type parameters like the parameter α . To check this idea we introduce anti-commuting parameters x_k as BRS variations of the a_k

$$\mathcal{s}\, a_k = x_k \tag{2.18}$$

and postulate invariance under this generalized BRS transformation

$$\mathcal{s}\Gamma + \chi \partial_\alpha \Gamma + \sum_k x_k \partial_{a_k} \Gamma = 0 \tag{2.19}$$

If one can find such a Γ one has shown exactly like for α that the a_k are gauge parameters. By solving (2.19) one can indeed show on the classical level that there exists an action Γ satisfying (2.19)[14]. So much to the UV-problem, as preparation.

We draw our attention now to the IR-problem which, interestingly enough, permits an analogous treatment. First of all we note that the gauge $\alpha = 1$ in which the offending contribution to the ϕ-propagator vanishes is not stable in the non-abelian case, hence the problem cannot be circumvented but must be solved. Unfortunately the function $1/(k^2)^2$ does not lead uniquely to a well-defined distribution, so we have already at the classical level to make up our mind of what we mean with it. Choosing anyone of its possible distributional forms[15] one might treat other parts of a Feynman diagram as test function. This approach was not successful[16]. One might also proceed formally and try to compensate the IR-divergences arising for any α by non-local and non-linear field redefinitions (via corresponding counter-terms)[17]. Sample calculations of this type exist[18]. But this is a dangerous procedure since one may loose control over the subdivision into gauge and physical sector. One should remind oneself in this context of the Nicolai mapping[19] which is a non-local, non-linear transformation of an (hopefully!) interacting theory into a free one. Now in the context of a gauge theory such a mapping might indeed be an equivalence transformation since it might just be a gauge changing transformation. We shall therefore adopt another method which permits explicit control of the gauge sector and does not alter the definition of "physical" as dictated by the Slavnov-identity[20]. We observe that a transformation

$$\phi_i \to \phi_i' = c_1 \phi_i = (\hat{c}_1 + \tfrac{1}{2}\mu^2 \theta^2 \bar{\theta}^2)\, \phi_i \tag{2.20}$$

is of the type (1.2) i.e. compatible with BRS-invariance. Performed in all but the gauge fixing terms it regularizes $\langle T\phi\phi\rangle$! E.g.

$$\langle CC \rangle \sim \frac{\alpha - 1}{(k^2-\mu^2)^2} \tag{2.21}$$

Furthermore, since μ^2 is just a c-parameter, we know also that

$$\frac{\partial \Gamma}{\partial \mu^2} = \mathcal{s}(\cdots) \tag{2.22}$$

i.e. the variation of μ^2 will leave unaffected physical quantities. Of course we have broken explicitly supersymmetry, but only softly and since there is a general machinery available for the control of softly broken symmetries this does not seem to be dangerous. We present the main idea of this procedure. Softly broken supersymmetry of the classical theory is expressed by the Ward-identity

$$W_\alpha^{hom}\Gamma = \mu^2 Q_\alpha \tag{2.23}$$

where W_α^{hom} denotes the ordinary homogeneous supersymmetry transformations of all fields of the theory and Q_α a breaking term of dimension 5/2 (since μ^2 has dimension mass²). The usual trick (R. Stora in ref. 21) to keep track of the right covariance of such a breaking is to couple an external field u with suitable covariance to it and to homogenize the Ward-identity

$$W_\alpha \Gamma \equiv W_\alpha^{hom}\Gamma - i\int dV\, \theta_\alpha \bar\theta^2 \mu^2 \frac{\delta\Gamma}{\delta u} = 0 , \tag{2.24}$$

where

$$\frac{\delta\Gamma}{\delta u^\alpha_\lambda} = -i\, Q_\alpha . \tag{2.25}$$

(2.24) is the Ward-identity of spontaneously broken supersymmetry for a formally symmetric action depending on the shifted field

$$u' = u + \frac{1}{2}\mu^2 \theta^2 \bar\theta^2 . \tag{2.26}$$

In accordance with (2.22) we can achieve that $\delta\Gamma/\delta u$ is also a BRS-variation - which will imply that the inhomogeneous term in (2.24) is non-physical - by introducing for u' too a BRS-doublet partner v'

$$s u' = v' = v + \frac{1}{2}\nu^2 \theta^2 \bar\theta^2 \tag{2.27}$$

and demanding BRS-invariance taking (2.27) into account. ν^2 is a Grassmann variable, the BRS-transform of μ^2

$$s\mu^2 = \nu^2 \tag{2.28}$$

and u, v are gauge singlet vector superfields.

To summarize the above discussion we may say that it suggests proving the following Ward-identities:

$$W_\alpha \Gamma = 0 \tag{2.29}$$

with W_α having the form as in (2.24) and including the external fields u,v.

The Slavnov-identity

$$\Delta(\Gamma) \equiv \sum_{\alpha} \int \frac{\delta\Gamma}{\delta \rho_{\alpha}} \frac{\delta\Gamma}{\delta \varphi_{\alpha}} + z \frac{\partial\Gamma}{\partial p} = 0 . \qquad (2.30)$$

Here p stands for all gauge parameters needed and z for their BRS-variations. The Γ-non-linear first terms originated from the naive variation form (2.17) by coupling external fields ρ_{α} to all non-linear BRS-variations of fields. This is the only possibility to describe the renormalization of those in higher orders. We furthermore impose conformal R-invariance to ensure masslessness

$$W_{R} \Gamma = 0 \qquad (2.31)$$

and rigid invariance since we wish to describe such a theory

$$W_{\omega} \Gamma = 0 . \qquad (2.32)$$

In the classical approximation (2.29) means writing Γ in terms of shifted superfields, (2.31) and (2.32) are similarly trivial to satisfy. It is only (2.30) which requires work. Together with normalization conditions which we do not reproduce here a unique Γ can indeed be found in the tree approximation[20]. What about higher orders? Again, we expose the line of argument in short terms.

The method which we use is best illustrated for linear symmetry transformations but is also applicable for non-linear ones like, in our case, for the BRS-variations. So let us take as an example any linear symmetry transformation $W_a \equiv -i \int dx\, \delta_a \varphi / \delta \varphi$ acting on Γ (the generating functional for vertex functions). All decent subtraction schemes permit to express this action as insertion

$$W_a \Gamma = \Delta_{\alpha} \cdot \Gamma \qquad (2.33)$$

i.e. the transformed functional $W_a \Gamma$ is nothing but a new functional, namely the one of all vertex functions with the specified insertion (vertex) Δ_a as a vertex. (The "action principle"[21,22]). Δ_a is local (an integrated field monomial) and dimensionwise restricted by

$$\text{UV-dim}\,(\Delta_a) = 4 - d(\varphi) + d(\delta_a \varphi) \equiv \delta(\Delta_a)$$

$$\text{IR-dim}\,(\Delta_a) = 4 - r(\varphi) + r(\delta_a \varphi) \equiv \rho(\Delta_{\alpha}) \qquad (2.34)$$

where $d(\varphi)$, $r(\varphi)$ are the UV-dimension resp. IR-dimension of φ. Looking at the corresponding diagrams it is clear that

$$W_a \Gamma = \Delta_a \cdot \Gamma = \Delta_a + o(\hbar \Delta_a) \tag{2.35}$$

where $o(\hbar \Delta_a)$ stands for all genuine loop diagrams. Suppose that the W_a's satisfy an algebra

$$[W_a, W_b] = i f_{abc} W_c \tag{2.36}$$

Then (2.35) and (2.36) imply a classical consistency condition

$$W_a \Delta_b - W_b \Delta_a = i f_{abc} \Delta_c \tag{2.37}$$

i.e. the Δ_a's which were up to now only restricted by (2.34) and locality become much more constrained: they have to be solutions of (2.37). An algebraically trivial solution of (2.37) is

$$\Delta_a = W_a \hat{\Delta} \tag{2.38}$$

($\hat{\Delta}$ local). If the <u>general</u> solution of (2.37) is of the form (2.38) and furthermore

$$\delta(\hat{\Delta}) \leq 4 \qquad \rho(\hat{\Delta}) \geq 4 \tag{2.39}$$

then $-\hbar\hat{\Delta}$ can be added to the action and the functional $\Gamma - \hbar \hat{\Delta}$ satisfies obviously

$$W_a(\Gamma - \hbar \hat{\Delta}) = o(\hbar \Delta_a) \tag{2.40}$$

By induction one can then prove the Ward-identity

$$W_a \Gamma = 0 \tag{2.41}$$

to all orders.

Section 3 Results and discussion

Using the method of the last section one can easily prove the Ward-identities (2.31)(2.32)[21,23,5]. Furthermore there are no hard anomalies ($\delta(\hat{\Delta}) = 4$) in the supersymmetry Ward-identity[24] and also none in the Slavnov-identity if one restricts the matter representations T such that

$$d^{ijk} \, \mathrm{Tr}\, T^i T^j T^k = 0 \tag{3.1}$$

(here d^{ijk} is the completely symmetric tensor in the group G)[25]. In fact we are left with Ward identities of the form

$$W_\alpha \Gamma = W_\alpha \hat{\Delta} + o(\hbar \hat{\Delta}) \tag{3.2}$$

$$\mathcal{S}(\Gamma) = b\,\hat{\Delta} + o(\hbar \hat{\Delta}) \tag{3.3}$$

with $\rho(\hat{\Delta}) \leq 4$, i.e. there are at most IR-anomalies. Examples for elements in $\hat{\Delta}$ are: $\mathrm{Tr}\int dx\, v^\mu v_\mu$, $\mathrm{Tr}\int dx\, \chi\lambda$, $\int dx\, A_0 \bar{A}_0$. Here the last term stands symbolically for a rigidly invariant combination of the $\theta = 0$ components of the matter multiplets.

In a first step one can eliminate all of them from (3.2) by rewriting them as components of superfields, e.g.

$$\int dx\, A_0 \bar{A}_0 = \frac{1}{16\mu^2} \int dV\, u'\, A \bar{A} + \text{hard terms} \tag{3.4}$$

whereby only hard terms have to be absorbed. In a second step their coefficients in the r.h.s. of (3.3) can be shown to vanish by a very close analysis of the small momentum behaviour as dictated by the anomalous Slavnov-identity (3.3) and the renormalization group[20,26]. Hence, the result is: the Ward-identities (2.29)(2.30) (2.31)(2.32) can be proved to all orders.

The validity of the Slavnov-identity (2.30) implies that all physical quantities are indeed independent of all "gauge"-parameters (α, a_k, μ^2), this solves the UV- and IR-problem. The validity of the supersymmetry Ward-identity (2.29) means that all physical quantities are susy-covariant. The masses of the matter fields are susy-degenerate, namely zero. It is to be noted that we do not perform the limit $\mu^2 \rightarrow 0$ since it does not exist in general. We rather pick out the μ^2-independent quantities as the physical ones. In the present strictly massless model those are Greens functions of gauge invariant operators. (The existence of the latter has to be checked case by case.) The susy-covariance of off-shell Greens functions is violated by the inhomogeneous terms in the Ward-identity (2.29).

That this situation is the best one can hope for, shows a glance into the Wess-Zumino gauge. There the gauge fixing term is non-supersymmetric. Using the multiplier field B it is a BRS-variation and therefore, if Ward-identities can be proved at all - which has not yet been done - they are of the same nature as those above: deviation from supersymmetry by BRS-variations.

With the above results all N=1 massless supersymmetric models can be formulated and exist as such. In particular the N=2 "finite" models and the N=4 model fall into this class and therefore do exist. What has to be proved for those in addition are the Ward-identities for the higher N supersymmetries. These symmetry variations are non-linear and therefore that proof is difficult.

REFERENCES

1. J. Wess, B. Zumino, Nucl.Phys. B70 (1974) 39.
2. J. Wess, B. Zumino, Phys.Lett. 49B (1974) 52.
3. A. Salam, J. Strathdee, Nucl.Phys. B76 (1974) 477.
4. K. Fujikawa, W. Lang, Nucl.Phys. B88 (1975) 61.
5. T.E. Clark, O. Piguet, K. Sibold, Nucl.Phys. B143 (1978) 445.
6. J. Iliopoulos, B. Zumino, Nucl.Phys. B76 (1974) 310.
7. M.T. Grisaru, M. Rocek, W. Siegel, Nucl.Phys. B159 (1979) 429.
8. J. Wess, B. Zumino, Nucl.Phys. B78 (1974) 1.
9. A Salam, J. Strathdee, Phys.Lett. 51B (1974) 353.
10. S. Ferrara, O. Piguet, Nucl.Phys. B93 (1975) 261.
11. O. Piguet, K. Sibold, Nucl.Phys. B197 (1982) 257.
12. B. d. Wit, D.Z. Freedman, Phys.Rev. D12 (1975) 2286.
13. K. Sibold, Renormalized Rigid N=1 Supersymmetry; Geneva lecture notes 1983.
14. O. Piguet, K. Sibold, Gauge Independence in N=1 Supersymmetric Yang-Mills Theories. Geneva University UGVA-DPT 1984/02-416; to be published in Nucl.Phys. B.
15. E. d'Emilio, M. Mintche , Perturbative Confinement in 4 dimensions, Pisa SNS 9/1979.
16. O. Piguet, K. Sibold, unpublished.
17. J.W. Juer, D. Storey, Nucl.Phys. B216 (1983) 185.
18. L.F. Abbott, M.T. Grisaru, D. Zanon, Infrared divergences and a non-local gauge for superspace Yang-Mills theory, Brandeis BRX-TH-160.
19. H. Nicolai, These Proceedings.
20. O. Piguet, K. Sibold, The off-shell infrared problem in N=1 SYM theories; I. Pure SYM, Geneva, UGVA-DPT 1984/04-423, to be published in Nucl.Phys. B.
21. S. lectures by J.H. Lowenstein, C. Becchi, R.Stora, in "Renormalization Theory", ed. A.S. Wightman, G. Velo; Reidel Publ. 1976.
22. O. Piguet, A. Rouet, Phys.Rep. 76 No. 1 (1981) 1-77.
23. T.E. Clark, O. Piguet, K. Sibold, Nucl.Phys. B119 (1977) 292.
24. O. Piguet, M. Schweda, K. Sibold, Nucl.Phys. B174 (1980) 183.
25. O. Piguet, K. Sibold, Nucl.Phys. B247 (1984) 484.
26. O. Piguet, K. Sibold, The off-shell infrared problem in N=1 SYM theories; II. All massless models. MPI-PAE/PTh 57/84; to be published in Nucl.Phys. B.

EVIDENCE FOR A NARROW MASSIVE STATE IN THE RADIATIVE DECAYS OF THE UPSILON

Hans-Jochen Trost

Deutsches Elektronen-Synchrotron
Notkestr. 85, D-2000 Hamburg 52
Fed. Rep. of Germany

(representing the Crystal Ball Collaboration *)

ABSTRACT

Evidence is presented for a state, which we call ζ, with a mass $M = (8322 \pm 8 \pm 24)$ MeV and a line width $\Gamma < 80$ MeV (90% confidence level) obtained using the Crystal Ball NaI(Tl) detector at DORIS II. Radiative transitions to this state are observed from about 100'000 $\Upsilon(1S)$ decays in two independent sets of data: One in which $\zeta \to$ multiple hadrons, and one which is strongly biased towards $\zeta \to 2$ low multiplicity jets. The branching ratio to this state from the $\Upsilon(1S)$ is of order 0.5%.

1. Introduction

It has been realized for some time that precision measurements of the radiative decay of the various quarkonium states provide a powerful tool with which to search for hypothetical particles, such as gluonic mesons(1), Higgs bosons(2), or supersymmetric particles(3). We report here such an investigation using $\Upsilon(1S)$ and $\Upsilon(2S)$ data that were obtained using the Crystal Ball NaI(Tl) detector(4) installed in the DORIS II storage ring at DESY. The data samples consist of about 100K produced $\Upsilon(1S)$ ($\int \mathcal{L}\, dt$ = 10.7 pb^{-1}) and of about 200K produced $\Upsilon(2S)$ (64.5 pb^{-1}).

The ability of the Crystal Ball detector to resolve and measure monochromatic γ's in the DORIS II-environment has been demonstrated[4,5]. The detector has been shown to have the resolution and absolute energy measurement capabilities projected from previous SPEAR performance. In these earlier studies a key role was played by the ability of the detector to reduce background caused by π^0- decay γ's and from other particles faking single γ's. Our ability to identify photons and to eliminate background is energy dependent. The results reported here were obtained using algorithms and subtraction techniques optimized for the region of E_γ from $\sim$ 700 to $\sim$ 2000 MeV. Other energy regions are still under investigation.

Below we describe two analyses of our data searching for particles X in the reaction $\Upsilon \rightarrow \gamma X$. The first involves isolation of events in which X decays into many particles (including photons). Additional cuts based on an $X \rightarrow c\bar{c}$ decay model are subsequently applied. The second analysis isolates events in which X typically decays into a few particles (including photons); it uses cuts developed from an $X \rightarrow \tau\bar{\tau}$ decay model.

2. High multiplicity analysis

In the first analysis, from all the recorded triggers at the $\Upsilon(1S)$ energy multihadron events are selected by efficiently removing beam gas, cosmic rays, e^+e^-X and QED events (including radiative $\gamma\tau\bar{\tau}$ events). The efficiency for selecting multihadron events is found to be ε_h = (0.90 ± 0.05). The resulting sample of events contains contributions from $\Upsilon(1S)$ and continuum decays approximately in the ratio of 2.5 to 1.

The next set of cuts applied to the multihadron events was designed to obtain a good acceptance, a minimal background, and to optimize the energy resolution for photons from about 700 to 2000 MeV. These cuts remove charged particles, photons with showers contaminated by energy depositions from nearby particles, and photons resulting from π^0 decay. The π^0's were identified as either a pair of clearly separated photons or as a single cluster formed by the two merged photon showers. To illustrate our methods, a single photon and a π^0 , generated by Monte Carlo with 1.0 GeV particle energy and tracked through the detector using the EGS shower program, are displayed in fig. 1. For each crystal the measured deposited energy is indicated. To reconstruct a particle we first set a threshold (10 MeV is chosen) for energies in single crystals and draw a contour line around all crystals showing an energy above this threshold. Each connected contour line thus defines a "connected region"; in fig. 1 the module energies inside these regions are shown with larger type. Next we look for local maxima of the energy distribution inside the connected regions; the corresponding crystals are called "bump modules" (underlined energies in fig. 1). Their orientation already gives a crude measurement of the direction of the particle that caused the observed energy deposition. When locating bump modules one has to take care not to accept fluctuations in the tail of a shower as bumps, i.e. genuine particles. Thus the crystal with 131 MeV in fig. 1b is not called a bump module by our software. As we know that from the π^0 decay there is a second photon we find the problem of overlapping showers nicely illustrated. The quantities used to detect and distinguish showers are:

- the energy in the bump module "E1", underlined in fig. 1;
- the sum over the energies of the bump module and the three crystals sharing a whole face with the bump module, "E4", enclosed by dashed lines in fig. 1;
- the sum over the energies of the bump module and the (usually) twelve crystals sharing a face or an edge with the bump module, "E13", enclosed by solid lines;
- the energy sum over all crystals belonging to the connected regions, "ECR", indicated by larger size of the energy numbers;
- the invariant mass "$m_{\gamma\gamma}$" resulting from an attempt to fit two

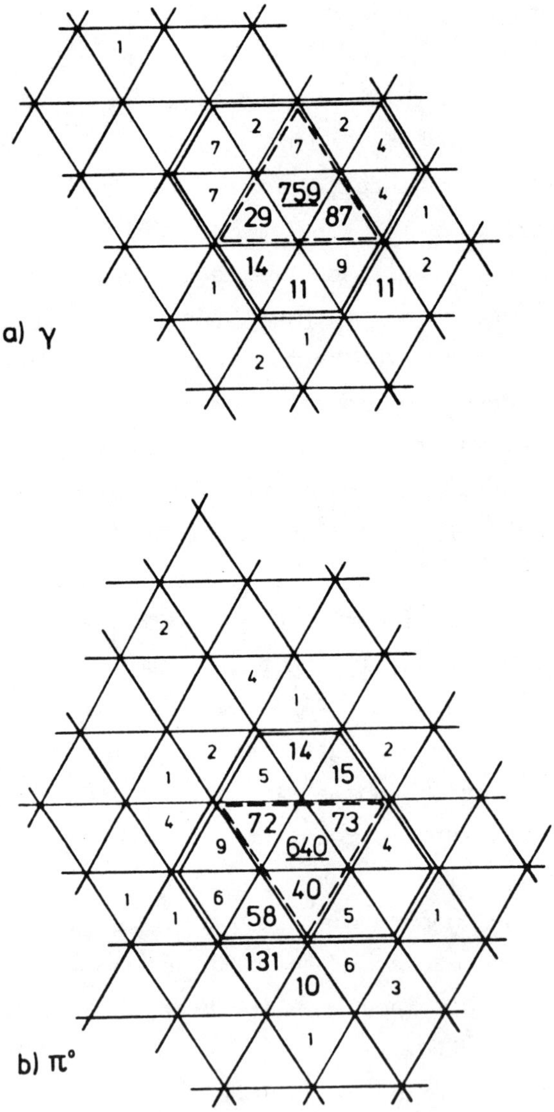

Fig. 1: Energy distributions of showers for a) a photon and b) a π^0, each generated with 1 GeV energy by Monte Carlo

shower distributions to the energy distribution inside the connected regions;

- the difference of the logarithms of likelihoods for the fit with two shower functions just mentioned and a fit with only one shower function, "DLL" = $\ln(L_{2\gamma}) - \ln(L_{1\gamma})$. This quantity is near 0.0 if the hypotheses cannot be distinguished.

The method and algorithm used for the last two quantities will be referred to as "PIFIT" below (see ref. 6 for applications in charmonium decays to e.g. $\gamma\pi^0$ and $\gamma\pi^0\pi^0$). The high lateral segmentation of our detector is an essential pre-requisite for this type of analysis. The separation power between γ's and π^0's reaches up to around 2 GeV particle energy.

The quantities actually tested in the photon selection are given in table 1 for the showers of fig. 1.

Table 1: Quantities for photon selection for the examples of fig. 1 ($E_\gamma = E_{\pi^0}$ = 1.0 GeV generated)

	γ	π^0
E1 / E4	0.86	0.78
E4 / E13	0.94	0.88
E13 / ECR	1.03	0.89
$m_{\gamma\gamma}$	55 MeV	138 MeV
DLL	0.13	8.6

The π^0 will be rejected only on the basis of the E13 / ECR and the PIFIT values, i.e. $m_{\gamma\gamma}$ and DLL. All of these tests are refinements since the analysis of the photon spectrum from the $\Upsilon(2S)$ for the bottonium states[4,5]. A useful side-effect of the PIFIT analysis is that showers contaminated from other closely passing particles that have been missed by our particle selection (the bump search in particular) are detected here, the whole cluster can then be removed. The whole procedure leaves us with clean, isolated photon showers. The final measured photon energy is derived from E13 by correcting for the average lateral leakage out of the 13 crystals over which the energy was originally summed.

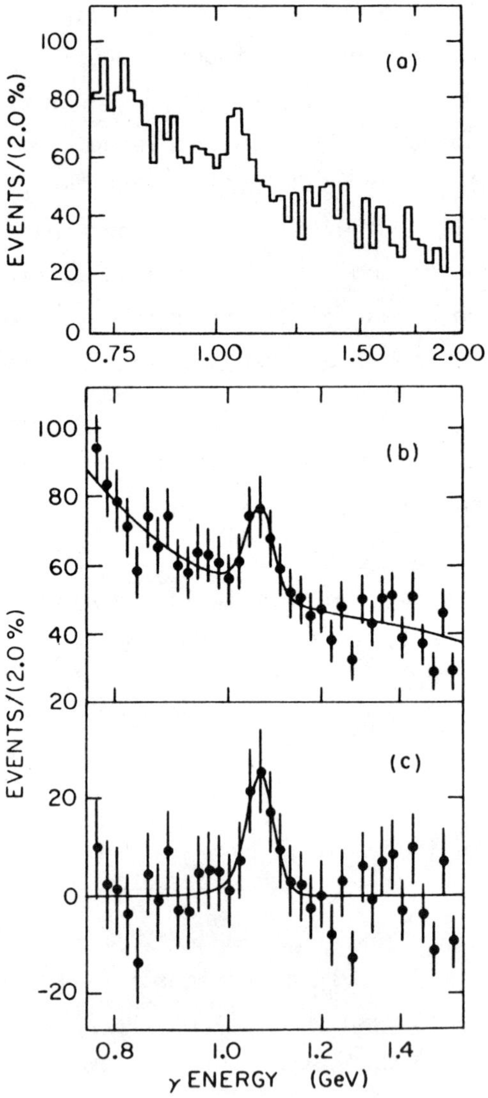

Fig. 2: a) Inclusive spectrum for $\Upsilon(1S) \rightarrow \gamma$ + multiple hadrons before physics oriented cuts

b) The ζ region of a) with fit (see text)

c) Same as b) after subtraction of the fitted background

Fig. 2a shows the resulting inclusive photon spectrum from the $\Upsilon(1S)$. The spectrum of fig. 2a was fitted in the region of E_γ between 750 and 1604 MeV using a line shape of the Crystal Ball measured at 1.5 GeV[4]. This line shape had variable amplitude and mean, a fixed $\sigma_E/E = 0.027/E^{1/4}$(GeV) (our expected resolution for photons in a multihadron environment), and was superposed on a background polynomial of order 3. The fit yielded a 4.0 standard deviation signal of (89.5 ± 22.5) counts at E_γ = (1074 ± 9) MeV, where only the statistical error is given (an overall scale error of ±2% on the energy is yet to be supplied). When allowed to vary in the fit, $\sigma_E/E = 0.028^{+0.013}_{-0.009}/E^{1/4}$(GeV) which is consistent with our expected resolution. No other potential line in the spectrum of fig. 2a can be fitted, consistent with our resolution, with a statistical significance of more than 2.2 standard deviations.

Additional cuts designed to enhance multihadronic decays of the ζ were then applied to the data to investigate the signal further. These cuts were developed by the use of Monte Carlo[7] calculations which simulated the process $\Upsilon(1S) \to \gamma\zeta$, $\zeta \to 2$ hadron jets. While charm quark jets were used as a model, jets due to lighter quarks or gluons lead to very similar results. This led to the following cuts (in the following particles are defined as energy clusters with energy deposition in the NaI(Tl) greater than 50 MeV): total multiplicity between 9 and 20; charged multiplicity ≧ 2; neutral multiplicity ≦ 12; total energy deposited in the Na(Tl) ≦ 8000 MeV; sphericity of the event ≧ 0.16.

Fig. 3a shows the inclusive photon spectrum for the $\Upsilon(1S)$ after the application of the above cuts. A fit to the spectrum of the same type as described above (see fig. 3b) now yields a significance of 4.2 standard deviations for the signal. The signal parameters become

$$\begin{aligned} E_\gamma &= (1072 \pm 8 \pm 21)\ \text{MeV} \\ M &= (8319 \pm 10 \pm 24)\ \text{MeV} \\ \text{Counts} &= 87.1 \pm 20.5 \\ \chi^2 &= 24.8 \text{ for 32 degrees of freedom,} \end{aligned} \qquad (1)$$

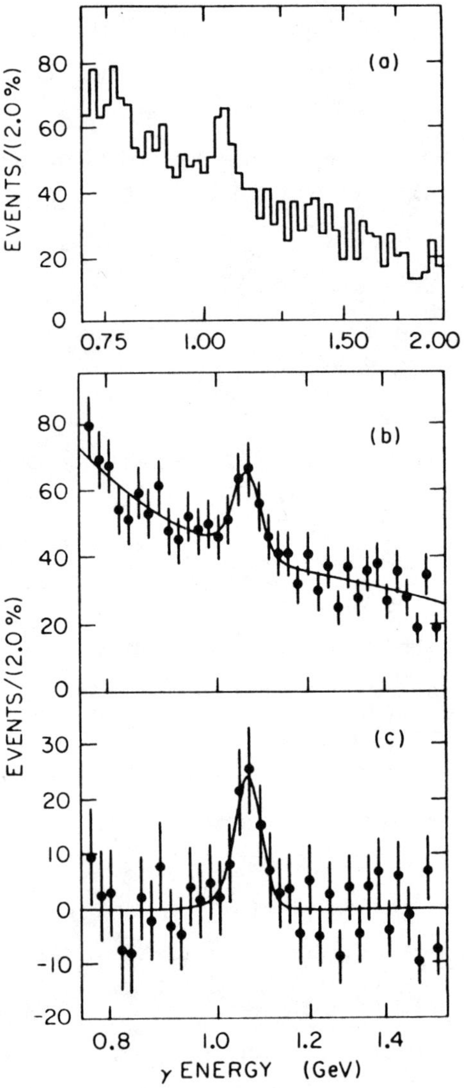

Fig. 3: For Υ(1S) as fig. 2 but after physics oriented cuts

where the first error in E_γ or M is statistical and the second is systematic.

The efficiency for this selection was investigated in various ways. First we used a γ-jet-jet Monte Carlo simulation for various fixed photon energies and jet-jet models ($u\bar{u}$, $c\bar{c}$, gg). Second we superimposed Monte Carlo generated photons onto real hadronic events at the c.m. energy of interest ($\Upsilon(1S)$ or $\Upsilon(2S)$). The methods show a systematic difference (fig. 4; the results for $\Upsilon(1S)$ are nearly identical to those displayed in fig. 4 for $\Upsilon(2S)$), causing a large contribution to the systematic error of the efficiency. Therefore we estimate a photon detection efficiency near 1 GeV of (18 ± 10)% leading to a branching ratio

$$B[\Upsilon(1S) \to \gamma\zeta] \cdot B[\zeta \to \text{hadrons}] = (0.47 \pm 0.11 \pm 0.26)\% \quad (2)$$

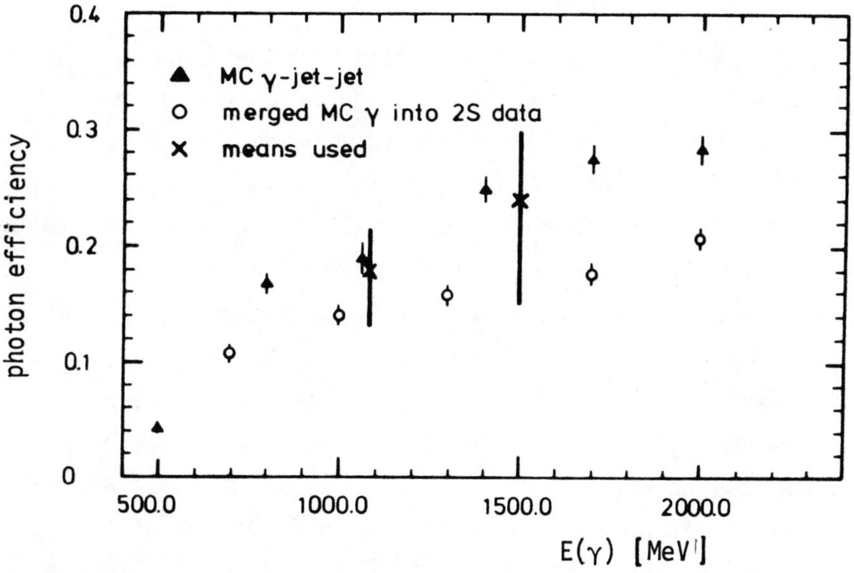

Fig. 4: Photon detection efficiency on $\Upsilon(2S)$ for the high multiplicity analysis

It is obvious from fig. 4 that we seriously loose efficiency below $E_\gamma \approx 700$ MeV. The upper end of the usefulness of our analysis is given by the loss in π^0 rejection power in the PIFIT algorithm above $E_\gamma \approx 2$ GeV.

A number of checks were made to ensure that the signal was not instrumental or an artifact of the analysis. First, all the cuts used to obtain the spectrum of fig. 2 were removed one at a time; this procedure indicated that none of the cuts used had anomalous effects. Second, by dividing the data appropriately, no preference for a particular period or geometrical region could be detected. Correlations between γ-energy and the triggers generated by the events containing the candidate γ's were found to be essentially constant moving from below to beyond the region of E_γ = 1 GeV. Off-resonance data, Monte Carlo 3g and $q\bar{q}$ events, and random beam cross events were subjected to the same analysis procedure and showed no significant fluctuations near 1 GeV. Finally, J/Ψ data taken at SPEAR and analyzed by the same program showed no narrow line at about 1 GeV.

The ϒ(2S) data set, analyzed similarily to the ϒ(1S) sample, does not show any narrow line (fig. 5). This is strong evidence that the signal from the 1S is not artificially induced. However, a signal for the ζ from the cascade $\Upsilon(2S) \to \pi\pi\Upsilon(1S)$, or $\gamma\gamma\Upsilon(1S)$, is expected. A fit using a fixed width $\sigma_E/E = 0.033$ (taking account of the Doppler broadening) leads to an upper limit of 70 events (90% c.l.) at E_γ = 1072 MeV. This is consistent with an expectation of 53 ± 13 events based on the observed signal on the ϒ(1S). No peak is observed either for the direct process $\Upsilon(2S) \to \gamma + \zeta'(8.32\ \text{GeV})$, i.e. at $E_\gamma = 1556$ MeV. While the detection efficiency (fig. 4) has rather large systematic uncertainties as mentioned before, the ratio of efficiencies ε(1079 MeV)/ε'(1560 MeV) is uncertain to about 10% of this ratio. Thus we find an upper limit $\frac{B[\Upsilon(2S) \to \gamma + \zeta]}{B[\Upsilon(1S) \to \gamma + \zeta]} < 0.22$ (90% c.l.).

3. Low multiplicity analysis

The second analysis consisted of looking for low multiplicity decays, motivated by a possible Higgs interpretation of the signal described above for which the decay into $\tau^+\tau^-$ (2) might be substantial. The data selection used previously tends to anti-select the $\gamma\tau^+\tau^-$ sample, in particular because of the bias towards multihadron states of high multiplicity. To obtain a data set biased in favor of $\gamma\tau^+\tau^-$ decay, a different set of cuts was employed. Disregarding the motivational bias, this data set can be viewed as an orthogonal set of events of low multiplicity.

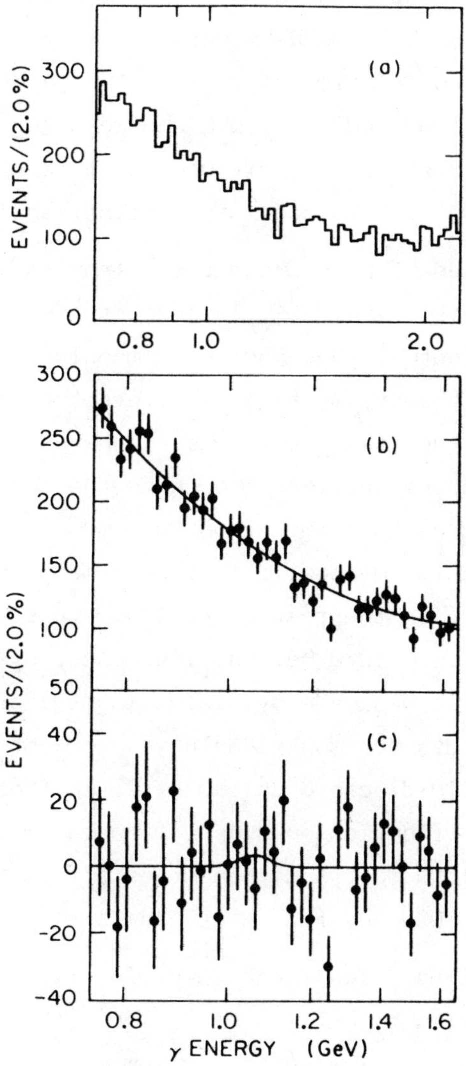

Fig. 5: Inclusive spectrum for Υ(2S), as fig. 2; the fit region is chosen for the Doppler-broadened 1 GeV line from Υ(1S)

A new pre-selection was performed on all the recorded $\Upsilon(1S)$-region triggers by requiring a total energy of at least 1200 MeV and at least 2 particles in the detector. As in the first analysis, an initial set of cuts was applied to arrive at an inclusive photon spectrum in the E_γ-region of 700 to 2000 MeV. Although similar to the cuts leading to fig. 2, the details of these cuts are different. Care was taken not to exclude low multiplicity events, using a Monte Carlo calculation(8) as a guide. QED-background was substantially reduced by exploiting the correlation of the γ with the beam direction (strong in the case of radiative QED and weak for a possible ζ related $\gamma\tau^+\tau^-$ final state). In addition cuts were applied to eliminate unwanted e^+e^-X events, beam gas interactions, and cosmic ray events.

The remaining series of cuts was derived from the Monte Carlo simulation of $\Upsilon(1S) \rightarrow \gamma\zeta \rightarrow \gamma\tau^+\tau^-$(8). In essence these cuts were just boundary tunings (both in one and two dimensional distributions) of such variables as thrust, multiplicity, event track-alignment, transverse momentum to the beam, etc. Some of these cuts were not new; they just strengthened earlier ones up to the more restrictive boundaries now allowed by the limitation to a $\gamma\tau^+\tau^-$ type configuration. In particular, a total multiplicity requirement of less than 9 guaranties no overlap with the results of (1). A check that no important bias towards $E_\gamma \sim 1070$ MeV was introduced by the cuts derived from the Monte Carlo was made by evaluating the efficiency of the sum of all these cuts for Monte Carlo $\gamma\tau^+\tau^-$ events with E_γ between 700 and 2000 MeV (10 discrete values of E_γ in the range were taken). The efficiency distribution obtained is approximately constant (at 24%) from 700 to 1500 MeV, and then drops off to $\sim$ 18% at 2000 MeV; no peaking in the 1000 MeV region is seen.

Fig. 6 shows the final signal obtained. The fit of 6b (similar to that in fig. 2), with σ_E/E fixed at $2.7\%/E^{1/4}$(GeV), yields a 3.3 standard deviation signal with the following parameters:

$$E_\gamma = (1062 \pm 12 \pm 21)\ \text{MeV} \qquad (3)$$
$$M = (8330 \pm 14 \pm 24)\ \text{MeV}$$

$$\text{Counts} = 23.8^{+7.9}_{-7.2}$$

$$\chi^2 = 29.9 \text{ for 41 degrees of freedom,}$$

in excellent agreement with the values recorded in (1). Fitting 6a with a variable width yields a $\sigma_E = 0.034^{+0.027}_{-0.012}/E^{1/4}$(GeV), consistent with the expected resolution.

These results are statistically independent of those shown in (1) due to the requirement of disjoint multiplicity distributions. The combined significance of both peaks is thus greater than 5 standard deviations. They also provide evidence that these signals do not derive from (either real $\Upsilon(1S)$ or background) events of a particular kinematic configuration. Of course this fact would not exclude a common experimental systematic error, producing a sharp anomaly at $E_\gamma \sim 1070$ MeV independent of the final state kinematics. However, such an error is unlikely because in the first analysis no such effect is seen in the closely-related $\Upsilon(2S)$-sample.

4. Further investigations

The two independent signals have been used to estmate an upper limit to the intrinsic width of the ζ. The observed peaks are consistent with the known Crystal Ball resolution function at $E_\gamma \sim 1$ GeV, which is an asymmetric Gaussian of FWHM (64 ± 5) MeV[4]. Unfolding this resolution from the combined observed FWHM (82 ± 23) MeV yields a 90% c.l. upper limit on the intrinsic ζ width of 80 MeV. This result is dominated by statistical precision in the observed width, not by the systematic error in the resolution function; if the resolution error is increased by a factor 3, the upper limit increases by only 10 MeV.

To obtain a value for $B[\Upsilon(1S) \rightarrow \gamma\zeta]$ which includes final states contributing to the second signal and not to the first it is necessary to use a model. We thus assume that the ζ has two kinds of decay, represented by the $c\bar{c}$ and $\tau\bar{\tau}$ Monte Carlo models. The data is found to be consistent with these models and indicates that inclusion of low

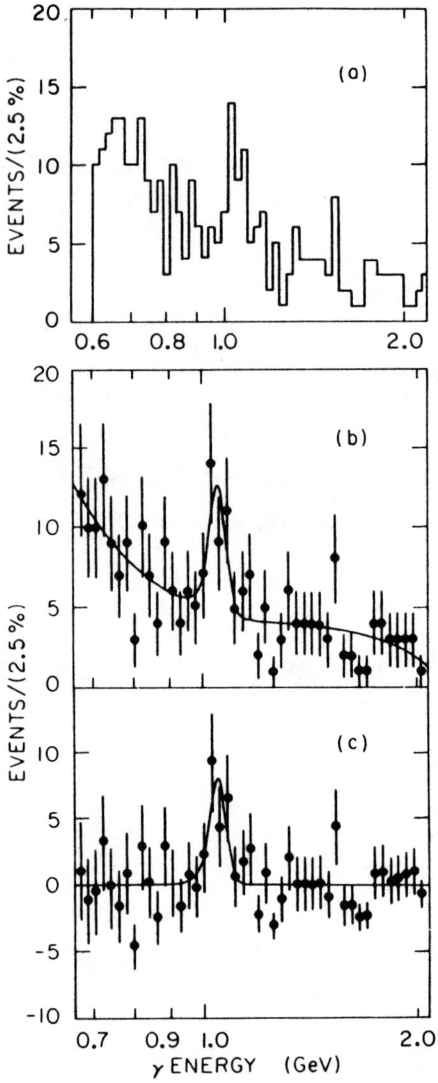

Fig. 6: Inclusive spectrum for $\Upsilon(1S) \rightarrow \gamma$ + low multiplicity, as fig. 2.

multiplicity $\tau\bar{\tau}$ like final states will increase the branching ratio (2) by about 20%. We have tried to model both signals by using the $\gamma c\bar{c}$ Monte Carlo alone, and this results in a poor fit to the data (2-3 standard deviation disagreement). However, this may be due to an inadequate $c\bar{c}$ Monte Carlo. It must be emphasized that we do not prove that the ζ decays into $c\bar{c}$ and $\tau\bar{\tau}$, we only show consistency with the model used as an aid in extracting the signal of (3).

We have also looked for the possible contributions from $\Upsilon(1S) \to \gamma\zeta \to \gamma\tau^+\tau^-$ followed by the decays $\tau^\pm \to e^\pm\nu\bar{\nu}$, $\tau^\pm \to \mu^\pm\nu\bar{\nu}$, showing no event at the γ mass (fig. 7). An upper limit of 0.2% (90% c.l.) for $B[\Upsilon(1S) \to \gamma\zeta]B[\zeta \to \tau^+\tau^-]$ has been found, compatible with the signal from the second analysis - even if we assume that this signal is entirely caused by a $\tau\bar{\tau}$-decay of the ζ. Additionally, an upper limit of 3×10^{-4} (90% c.l.) for the branching ratio $B[\Upsilon(1S) \to \gamma\zeta]B[\zeta \to e^+e^-]$ has been determined (fig. 8).

5. Conclusions and outlook

We have observed two statistically independent signals at the same mass; one of 4.2 and the other 3.3 standard deviations. The fact that both peaks appear at the same position with a compatible width supports the hypothesis that we are seeing the same state in two different data samples. We can thus combine the significance of both peaks; this yields a greater than 5 standard deviation effect. Both our signals have widths consistent with the detector energy resolution. Taking the weighted average of the radiated photon's fitted peak value and width in the two cases gives the best estimate of the mass and width of this new state, herein named ζ,

$$\begin{aligned} E_\gamma &= (1069 \pm 7 \pm 21)\ \text{MeV} \\ M &= (8322 \pm 8 \pm 24)\ \text{MeV} \\ \Gamma &< 80\ \text{MeV}\ (90\%\ \text{c.l.}) \\ B[\Upsilon(1S) &\to \gamma\zeta] \sim 0.5\%. \end{aligned} \qquad (4)$$

The interpretation of this new state as the neutral Higgs boson expected in the standard model gives a disagreement of approximately two orders of magnitude between this observed branching ratio and that predicted. This branching ratio can be accomodated in some extensions of the standard model, e.g. two Higgs-doublet models. A

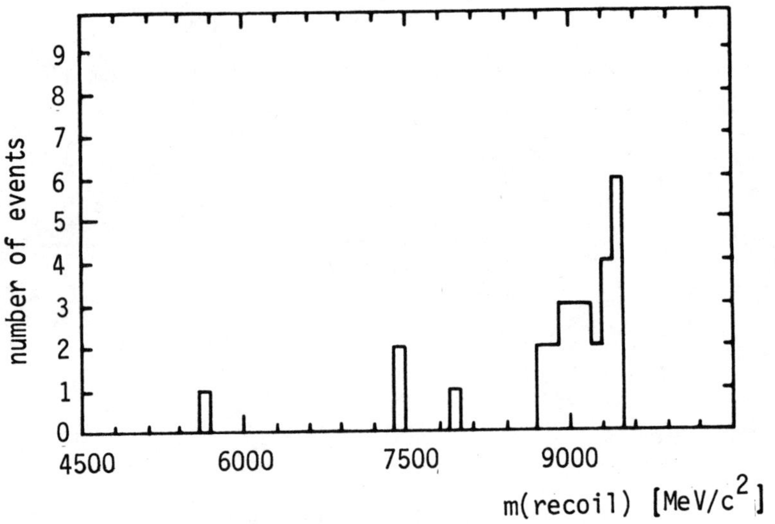

Fig. 7: Mass recoiling against the photon in the detected final state $\gamma e\mu$ of the process $e^+e^- \to \gamma\tau^+\tau^-$

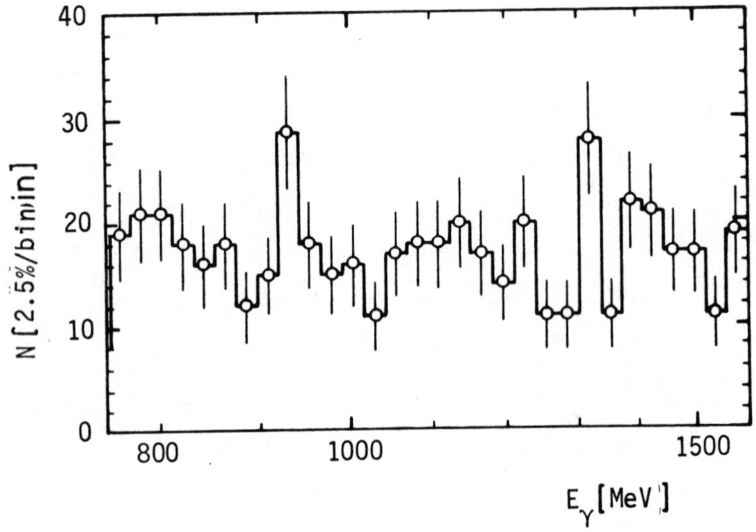

Fig. 8: Photon spectrum for the process $e^+e^- \to \gamma e^+e^-$

less model-dependent quantity is the ratio $\frac{B[\Upsilon(2S) \rightarrow \gamma\zeta]}{B[\Upsilon(1S) \rightarrow \gamma\zeta]}$, in which the strength of the Higgs' coupling to b-quarks cancels out; in either model this ratio is predicted to be ~ 1.0[2], while our upper limit is 0.22, in apparent disagreement. Further, given the limited statistics of the present experiment, it cannot be proven that the mode $\zeta \rightarrow \tau\bar{\tau}$ exists, although our analysis is consistent with it.

Some tasks are left for the future. Simplified analyses with fewer cuts have been started and shown the signal again, yet they need to be as thoroughly checked as the one described herein. More data on the $\Upsilon(1S)$ are being taken right now both at DORIS II and CESR.

(*) C. Peck et al., internal reports DESY 84-064, SLAC-PUB 3380(1984)

References

(1) S. Brodsky et al., Phys. Lett. B73, 203 (1978);
K. Koller, T. Walsh, Nucl. Phys. B140, 449 (1978);
J.D. Bjorken, 1980, Proc. Summer Inst. on Particle Physics, SLAC-224, Stanford, SLAC.

(2) F.Wilczek, Phys. Rev. Lett. 39, 1304 (1977);
J. Ellis et al., Phys. Lett. 83B, 339 (1979);
S. Weinberg, Phys. Rev. Lett. 36, 294 (1976);
H.E. Haber, G.L. Kane, Phys. Lett. 135B, 196 (1984).

(3) See e.g. H.E. Haber and G.L. Kane, Phys. Rep., to appear.

(4) J. Gaiser et al., SLAC-PUB-3232; J. Gaiser, Ph.D.thesis, SLAC-255 (1983).

(5) J. Irion et al., SLAC-PUB-3325.

(6) C.D. Edwards, Ph.D.thesis, CALT-68-1165 (1984).

(7) The events $\Upsilon(1S) \rightarrow \gamma\zeta(8.3$ GeV$)$, $\zeta \rightarrow c\bar{c}$ were generated. The ζ decayed isotropically into $c\bar{c}$ in its center-of-mass-system, and the $c\bar{c}$ pair was then hadronized using the symmetric Lund model (Monte Carlo version 5.2).
Ref: B. Andersson et al., Phys. Rep. 97, 33 (1983).

(8) The events $\Upsilon(1S) \rightarrow \gamma\zeta(8.3$ GeV$)$, $\zeta \rightarrow \tau^+\tau^-$ were generated. The ζ decayed isotropically into $\tau^+\tau^-$ in its center-of-mass-system, and the τ decays were modeled as described in C.A. Blocker et al., Phys. Rev. Lett. 49, 1369 (1982).

DISORDER, DIMENSIONAL REDUCTION AND SUPERSYMMETRY IN STATISTICAL MECHANICS

Franz Wegner

Institut für Theoretische Physik
Ruprecht-Karls-Universität
D-6900 Heidelberg, F.R. Germany

INTRODUCTION

During the last five years supersymmetric theories have become of interest for the explanation of dimensional reduction in disordered systems. In at least two cases the disordered system is closely related to a pure system in two fewer dimensions. In both cases this dimensional reduction can be explained as the consequence of an underlying hidden supersymmetry[1,2] which adds a pair of anticommuting coordinates to the d conventional real space ones. In both cases the Lagrangian is invariant under rotations in superspace and thus the expectation values are reduced to those of the (d-2)-dimensional system. These two systems are
(i) Ferromagnet in a random magnetic field. Although such a system can hardly be realized experimentally, there are equivalent systems in the form of dilute uniaxially anisotropic antiferromagnets[3]. Without magnetic field the magnetic moments of the two sublattices align in opposite directions. In a dilute antiferromagnet the sublattices will be occupied at random by magnetic ions. Depending on which sublattice in a given region is more strongly occupied a uniform magnetic field tends to order the system in one or the other direction and thus has the effect of a staggered magnetic field of random strength.
(ii) Electron in a strong magnetic field and random potential in the lowest Landau level. The study of such systems is motivated by the observation of the (integer) quantized Hall effect[4] according to which a two-dimensional electron gas (for example in the inversion layer of a MOSFET) in a strong magnetic field exhibits at low temperatures plateaus of the Hall conductivity $\sigma_{xy} = ne^2/h$, n being integer. The calculation of averaged Green's functions is considerably simplified if one restricts oneself to the lowest Landau level.

In this case the calculation of the density of states is reduced from a two-dimensional to a zero-dimensional problem.

In the following these two systems and their supersymmetric field theories will be reviewed.

FERROMAGNET IN A RANDOM MAGNETIC FIELD

Aharony, Imry and Ma[5] found in an ε expansion that the critical exponents of a ferromagnet in a random magnetic field in $d = 6 - \varepsilon$ dimensions agree with those of the system without magnetic field in $d = 4 - \varepsilon$ dimensions. Young[6] proved that the most infrared divergent diagrams for the random field ferromagnet in d dimensions yield the same values as the diagrams for a ferromagnet without field in d-2 dimensions.

Parisi and Sourlas[1] showed that this dimensional reduction can be understood in terms of a hidden supersymmetry of the corresponding field theory. The magnetic system with the local magnetization density $\phi(x)$ be governed by the Ginzburg-Landau Lagrangian

$$L\{\phi\} = \int d^dx(-\frac{1}{2}\phi(x)\Delta\phi(x) + V(\phi(x)) - h(x)\phi(x)) \tag{1}$$

where the magnetic field be stochastic and Gaussian distributed

$$\overline{h(x)} = 0, \quad \overline{h(x)h(x')} = W\delta(x - x'). \tag{2}$$

The ensemble averaged (indicated by a bar) thermal (indicated by brackets) expectation value of $A\{\phi\}$ is given by

$$\overline{< A\{\phi\} >} \sim \int Dh < A\{\phi\} >_h \exp(-\frac{1}{2W}\int d^dx\, h^2(x)). \tag{3}$$

The $\sim$ reminds one that the right-hand-side has to be normalized so that $\overline{1} = 1$.

Tree approximation

From Young's proof one knows that the most infrared divergent diagrams yield the complete reduction to two fewer dimensions. These are the diagrams with the maximum number of $\overline{h^2}$ insertions, since each insertion yields a "propagator" $G_o(q)\overline{h^2}G_o(q) \sim q^{-4}$. If only these diagrams are considered, the problem is reduced to the tree approximation, the solution of which is given by the saddle point equation

$$\delta L/\delta\phi(x) = 0, \tag{4}$$

thus

$$h(x) = -\Delta\phi(x) + V'(\phi(x)) \tag{5}$$

which is a stochastic differential equation, h being the stochastic function.

In order to eliminate the field h in (3) one introduces

$$\begin{aligned} Dh &= D\phi \cdot \det(\Delta - V''(\phi)) \\ &= D\phi \cdot \int D\bar{\psi}D\psi \exp(\psi(\Delta - V'')\bar{\psi}) \end{aligned} \tag{6}$$

where $\psi(x)$, $\bar{\psi}(x)$ are Grassmannian valued fields. Moreover,

$$\begin{aligned} &\exp(-\frac{1}{2W}\int h^2(x)d^dx) \\ &= \int D\omega \exp(\int d^dx(-\frac{W}{2}\omega^2 + \omega(-\Delta\phi(x) + V'(\phi)))) \end{aligned} \tag{7}$$

where the real field $\omega(x)$ has been introduced and use has been made of eq. (5). Adding all contributions one obtains in tree approximation

$$\overline{\langle A \rangle} \sim \int D\phi D\bar{\psi}D\psi D\omega\, A\{\phi\} \exp(-L_{ss}) \tag{8}$$

with

$$\begin{aligned} L_{ss} = \int d^dx[&\frac{W}{2}\omega^2(x) + \omega(x)(\Delta\phi(x) - V'(\phi(x))) \\ &+ \psi(x)(\Delta - V''(\phi(x)))\bar{\psi}(x)]. \end{aligned} \tag{9}$$

Supersymmetric theory

Parisi and Sourlas made the important observation that the Lagrangian L_{ss} can be brought in an explicit supersymmetric form by introducing the superfield

$$\Phi(x,\bar{\theta},\theta) = \phi(x) + \bar{\theta}\psi(x) + \bar{\psi}(x)\theta + \bar{\theta}\theta\,\omega(x) \tag{10}$$

which is a field of d real space components and of the pair of anticommuting components θ and $\bar{\theta}$ of a superspace. Indeed,

$$\int d\bar{\theta}d\theta\, V(\Phi) = -\omega V'(\phi) - \psi V''(\phi)\bar{\psi}, \tag{11}$$

$$\int d\bar{\theta}d\theta\, \Phi(\Delta + W\partial_\theta\partial_{\bar{\theta}})\Phi = -2\omega\Delta\phi - 2\psi\Delta\bar{\psi} - W\omega^2. \tag{12}$$

Thus

$$L_{ss} = \int d^dx d\bar{\theta} d\theta \left(-\frac{1}{2}\Phi\Delta_{ss}\Phi + V(\Phi)\right) \tag{13}$$

with the Superlaplacian

$$\Delta_{ss} = \Delta + W\partial_\theta \partial_{\bar{\theta}}. \tag{14}$$

The Lagrangian L_{ss} is invariant under rotations in superspace with metric $x^2 + 4\theta\bar{\theta}/W$. It has the same form as the original Lagrangian L, eq. (1), but the field ϕ has been replaced by the superfield Φ, space by superspace and the magnetic field has disappeared. Since this superspace is equivalent to a (d-2)-dimensional ordinary space (see, e.g. ref. 7), the expectation values of functions of ϕ in a (d-2)-dimensional plane of the d-dimensional system are given by the expectation value of the (d-2)-dimensional pure system.

It should be noted, however, that this result holds for the most divergent contributions only and that it is restricted to perturbation theory. The question of whether the dimensional reduction scheme holds down to the lower critical dimensionality is still controversial, in particular the question of whether the lower critical dimensionality for the random field Ising model is two or three[3].

It should also be noted that there may be more than one solution to the saddle point equation (5) for given h(x). Then all solutions will contribute to L_{ss}. Since in eq. (6) the Jacobian has been used and not its modulus, some contribute with positive, some with negative weight.

ELECTRON IN STRONG MAGNETIC FIELD AND RANDOM POTENTIAL

Free electron in a magnetic field

A free electron in two dimensions and in a perpendicular magnetic field B is governed by the Hamiltonian

$$H_o = \frac{1}{2m}\left(\underset{\sim}{p} - \frac{e}{c}\underset{\sim}{A}\right)^2, \quad \underset{\sim}{A} = \frac{1}{2}B(-y,x). \tag{15}$$

The eigenvalues of H_o are quantized and given by

$$E_n = \hbar\omega\left(n + \frac{1}{2}\right), \quad n = 0,1,2,... \tag{16}$$

where $\omega = eB/mc$ is the cyclotron frequency. With cyclotron length $\ell = (\hbar c/eB)^{1/2}$ and units $\hbar\omega = 1$, $\ell = \sqrt{1/2}$ the Hamiltonian reads

$$H_o = -(\partial_z - \frac{1}{2} z^*)(\partial_{z^*} + \frac{1}{2} z) + \frac{1}{2}, \quad (17)$$

where z is the complex coordinate

$$z = x + iy. \quad (18)$$

The eigenstates of the lowest Landau level can be written as

$$\phi_m = (\pi m!)^{-1/2} z^m \exp(-\frac{1}{2} zz^*), \quad m = 0,1,2,... \quad (19)$$

In general the eigenfunctions of the lowest Landau level are given by

$$\phi = u(z) \cdot \exp(-\frac{1}{2} zz^*), \quad \partial_{z^*}u = 0, \quad (20)$$

u being holomorphic. Restricting to the lowest Landau level one obtains the Green's function

$$G_o(r,r',E + i0) = < r|P \frac{1}{E-H_o+i0} P|r' > = \frac{1}{\varepsilon} C(z,z') \quad (21)$$

where P projects on the lowest Landau level, and

$$\varepsilon = E + i0 - \frac{1}{2} \hbar\omega \quad (22)$$

$$C(z,z') = \sum_m \phi_{o,m}(z)\phi^*_{o,m}(z') = \pi^{-1}\exp(-\frac{1}{2} zz^* - \frac{1}{2} z'z'^* + zz'^*). \quad (23)$$

Random potential V

We now add a random potential V to the kinetic energy H_o

$$H = H_o + V. \quad (24)$$

The effect of this random potential will be a broadening of the Landau levels. If this broadening is small in comparison to the spacing between the Landau levels $\hbar\omega$, then it is a good approximation to restrict oneself to the Hilbert space of one Landau level to calculate its contribution to the density of states. Thus for the lowest Landau level we replace (24) by

$$H = H_o + PVP. \quad (25)$$

If the potential consists of point-like scatterers, that is, if it is uncorrelated between different points, then the calculation of the one-particle Green's function is reduced to a calculation for a zero-dimensional system. (For a d-dimensional system it is reduced to a (d-2)-dimensional one.) This was shown for a white-noise potential

$$\bar{V}(r) = 0, \qquad \overline{V(r)V(r')} = W\delta(r - r') \tag{26}$$

by means of a diagrammatic expansion[8].

Brézin, Gross and Itzykson[2] have generalized the solution to systems of arbitrary point scatterers characterized by a function $g(\alpha)$

$$\overline{\exp(-i \int d^2r\, \alpha(r)V(r))} = \exp(\int d^2r\, g(\alpha(r))). \tag{27}$$

Moreover, they described the system in terms of a supersymmetric field theory and obtained in this way the dimensional reduction.

Supersymmetric formulation

They use the path-integral representation

$$< r| \frac{1}{E-H+i0} |r' > = -i \int D\phi D\phi^* D\psi D\bar{\psi}\; \phi(r)\phi^*(r') \exp(i\mathcal{H}), \tag{28}$$

$$\mathcal{H} = \int d^2r[\phi^*(E - H_o - V)\phi + \bar{\psi}(E - H_o - V)\psi] \tag{29}$$

where $\phi(r)$ is a complex field. The Grassmannian field $\psi(r)$ is introduced to provide the correct normalization[7,9]

$$\int D\phi D\phi^* D\psi D\bar{\psi}\; e^{i\mathcal{H}} = 1 \tag{30}$$

and to avoid the replica trick. $\mathcal{H}$ is invariant under unitary graded transformations between ϕ and ψ. As a consequence $\phi(r)\phi^*(r')$ in eq. (28) can be replaced by $\psi(r)\bar{\psi}(r')$. For systems which obey time-reversal invariance one obtains $\mathcal{H}$ invariant under unitary ortho-symplectic transformations[10].

Restriction to the lowest Landau level is made by the requirement (20) and

$$\psi = v(z) \cdot \exp(-\frac{1}{2} zz^*), \tag{31}$$

v being a Grassmann field holomorphic in z. Using (20), (27) and (31) the averaged Green's function reads

$$\begin{aligned} G(r,r',E + i0) &= < r|\, P \frac{1}{E-H+i0} P\, |r' > = \\ &= -ie^{-1/2zz^*-1/2z'z'^*} \int Du\ldots D\bar{v}\; u(z)u^*(z'^*) \exp(-L_o-L') \end{aligned} \tag{32}$$

where $u(z)u^*(z'^*)$ may be replaced by $v(z)\bar{v}(z'^*)$,

$$L_o = -i\varepsilon \int dzdz^*(u^*u + \bar{v}v)e^{-zz^*} \tag{33}$$

$$L' = - \int dzdz^* g(e^{-zz^*}(u^*u + \bar{v}v)). \tag{34}$$

Brézin et al. make the important observation that the holomorphic superfield

$$\Phi(z,\theta) = u(z) + \theta v(z) \tag{35}$$

$$\bar{\Phi}(z^*,\bar{\theta}) = u^*(z^*) + \bar{v}(z^*)\bar{\theta} \tag{36}$$

allows the representation

$$L_o = -i\varepsilon \int dzdz^* d\theta d\bar{\theta}\ e^{-zz^* - \theta\bar{\theta}} \bar{\Phi}\Phi \tag{37}$$

$$L' = - \int dzdz^* d\theta d\bar{\theta}\ h(e^{-zz^* - \theta\bar{\theta}} \bar{\Phi}\Phi) \tag{38}$$

with

$$h(\alpha) = \int_0^\alpha \frac{d\beta}{\beta} g(\beta), \tag{39}$$

where $L_{ss} = L_o + L'$ is invariant under rotation in superspace with metric $zz^* + \theta\bar{\theta}$.

Invariance under rotation in superspace

Infinitesimal rotations

$$\begin{aligned} \delta z &= \bar{\omega}\theta, \quad \delta\theta = \omega z \\ \delta z^* &= \bar{\theta}\omega, \quad \delta\bar{\theta} = z^*\bar{\omega} \end{aligned} \tag{40}$$

mix the components z and θ as well as z^* and $\bar{\theta}$. As a consequence of the invariance of L the correlation

$$\begin{aligned} &< \Phi(z,\theta)\bar{\Phi}(z'^*,\bar{\theta}') > = < u(z)u^*(z'^*) > + \theta < v(z)\bar{v}(z'^*) > \bar{\theta}' \\ &= f(z,z'^*)\cdot(1 + \theta\bar{\theta}') \end{aligned} \tag{41}$$

is invariant under rotation

$$\delta < \Phi\bar{\Phi} > = \frac{\partial f}{\partial z} \bar{\omega}\theta + \frac{\partial f}{\partial z'} \bar{\theta}'\omega + (\omega z\bar{\theta}' + \theta z'^*\bar{\omega})f = 0 \tag{42}$$

which implies

$$\partial f/\partial z - z'^{*} f = 0$$
$$\partial f/\partial z' - z f = 0 \tag{43}$$

and thus

$$f = k \cdot e^{zz'^{*}} . \tag{44}$$

Thus the averaged Green's function (32) is given by

$$G(r,r',E + i0) = -i\pi k \; C(z,z'). \tag{45}$$

The constant k is defined by

$$k = \langle \Phi(0,0)\bar{\Phi}(0,0) \rangle_{L_{ss}} = \frac{\langle \Phi\bar{\Phi} \; e^{-L'} \rangle_{L_o}}{\langle e^{-L'} \rangle_{L_o}} \tag{46}$$

which can be evaluated by means of superdiagrams. They contain propagators

$$\langle \Phi(z,\theta)\bar{\Phi}(z'^{*},\bar{\theta}') \rangle = \frac{i}{\pi\varepsilon} \; e^{zz'^{*} + \theta\bar{\theta}'} \tag{47}$$

and vertices

$$\sim \frac{\partial^{\ell} h}{\partial \alpha^{\ell}} \; e^{-\ell(zz^{*} + \theta\bar{\theta})} . \tag{48}$$

Since the superdiagrams in 2 real dimensions (x,y) and 2 Grassmann dimensions $(\theta,\bar{\theta})$ with n vertices yield[7]

$$\prod_{i=1}^{n} \int dz_i \ldots d\bar{\theta}_i \; f\{z,\theta\} = \pi^{n} f(0) \tag{49}$$

the original two-dimensional problem is reduced to the zero-dimensional one

$$k = \frac{\int d\phi^{*} d\phi \; \phi\phi^{*} \; e^{i\pi\varepsilon\phi^{*}\phi + \pi h(\phi^{*}\phi)}}{\int d\phi^{*} d\phi \; e^{i\pi\varepsilon\phi^{*}\phi + \pi h(\phi^{*}\phi)}} , \tag{50}$$

and the density of states is obtained from the imaginary part of the Green's function

$$\rho(E) = \mathrm{Re}\ k(E - \frac{1}{2}\hbar\omega + i0). \qquad (51)$$

I am not aware of a similar dimensional reduction for higher Landau levels or for two-particle Green's functions.

REFERENCES

1. G. Parisi, N. Sourlas, Phys. Rev. Lett. 43 (1979) 744
2. E. Brézin, D.J. Gross, C. Itzykson, Nucl. Phys. B235 [FS11] (1984) 24
3. S. Fishman, A. Aharony, J. Physics C12 (1979) L729; R.J. Birgeneau, R.A. Cowley, G. Shirano, H. Yoshizawa, J. Stat. Phys. 34 (1984) 817; Y. Imry, J. Stat. Phys. 34 (1984) 849
4. K. v. Klitzing, G. Dorda, M. Pepper, Phys. Rev. Lett. 45 (1980) 494; BMS, Physics Today 34, No. 6 (1981) 17.
5. Y. Imry, S. Ma, Phys. Rev. Lett. 35 (1975) 1399 A. Aharony, Y. Imry, S. Ma, Phys. Rev. Lett. 37 (1976) 1367
6. A.P. Young, J. Physics C10 (1977) L257
7. A.J. McKane, Phys. Lett. 76A (1980) 22
8. F. Wegner, Z. Physik B51 (1983) 279
9. K.B. Efetov, Zh. Eksp. Teor. Fiz. 82 (1982) 872; JETP 55 (1982) 514
10. F. Wegner, Z. Physik B49 (1983) 297

INDEX